INTRODUCTORY
TECHNICAL MATHEMATICS

FIFTH EDITION

INTRODUCTORY TECHNICAL MATHEMATICS

Robert D. Smith
John C. Peterson

THOMSON
™
DELMAR LEARNING

Australia Canada Mexico Singapore Spain United Kingdom United States

THOMSON

DELMAR LEARNING

Introductory Technical Mathematics, 5ᵗʰ Edition
Robert D. Smith and John C. Peterson

Vice President, Technology and Trades Business Unit:
David Garza

Editorial Director:
Sandy Clark

Executive Editor:
Stephen Helba

Development:
Mary Clyne

Marketing Director:
Deborah Yarnell

Channel Manager:
Dennis Williams

Marketing Coordinator:
Stacey Wiktorek

Production Director:
Mary Ellen Black

Production Manager:
Larry Main

Art & Design Coordinator:
Francis Hogan

Technology Project Manager:
Kevin Smith

Technology Project Specialist:
Linda Verde

Senior Editorial Assistant:
Dawn Daugherty

Library of Congress Cataloging-in-Publication Data:
Card Number:

ISBN: 1-4180-1543-1 Soft cover
ISBN: 1-4180-1545-8 Hard cover

NOTICE TO THE READER

Contents

Preface

Introductory Technical Mathematics is written to provide practical vocational and technical applications of mathematical concepts. Presentation of concepts is followed by applied examples and problems that have been drawn from diverse occupational fields.

Both content and method have been used by the authors in teaching related technical mathematics on both the secondary and postsecondary levels. Each unit is developed as a learning experience based on preceding units. The applied examples and problems progress from simple to those whose solutions are relatively complex. Many problems require the student to work with illustrations such as are found in trade and technical manuals, handbooks, and drawings.

The book was written from material developed for classroom use and it is designed for classroom purposes. However, the text is also very appropriate for self-instruction use. Great care has been taken in presenting explanations clearly and in giving easy-to-follow procedural steps in solving examples. One or more examples are given for each mathematical concept presented. Throughout the book, practical application examples from various occupations are shown to illustrate the actual on-the-job uses of the mathematical concept. Students often ask, "Why do we have to learn this material and of what practical value is it?" This question was constantly kept in mind in writing the book and every effort was made to continuously provide an answer.

An understanding of mathematical concepts is emphasized in all topics. Much effort was made to avoid the mechanical *plug-in* approach often found in mathematics textbooks. A practical rather than an academic approach to mathematics is taken. Derivations and formal proofs are not presented; instead, understanding of concepts followed by the application of concepts in real situations is stressed.

Student exercises and applied problems immediately follow the presentation of concept and examples. Exercises and occupationally related problems are included at the end of each unit. The book contains a sufficient number of exercises and problems to permit the instructor to selectively plan assignments.

Illustrations, examples, exercises, and practical problems expressed in metric units of measure are a basic part of the content of the entire text. Emphasis is placed on the ability of the student to think and to work with equal ease with both the customary and the metric systems.

An analytical approach to problem solving is emphasized in the geometry and trigonometry sections. The approach is that which is used in actual on-the-job trade and technical occupation applications. Integration of algebraic and geometric principles with trigonometry by careful sequencing and treatment of material also helps the student in solving occupationally-related problems.

The majority of instructors state that their students are required to perform basic arithmetic operations on whole numbers, fractions, and decimals prior to calculator usage. Thereafter, the students use the calculator almost exclusively in problem-solving computations. The structuring of calculator instructions and examples in this text reflect the instructors' preferences. The scientific calculator is introduced at the end of this Preface. Extensive calculator instruction and examples are given directly following each of the units on whole numbers, fractions and mixed numbers, and decimals. Further calculator instruction and examples are given throughout the text wherever calculator applications are appropriate to the material presented. Often there are differences in the methods of computation among various makes and models of calculators. Where there are two basic ways of performing calculations, both ways are shown.

An extensive survey of instructors using the fourth edition was conducted. Based on instructor comments and suggestions, significant changes were made. The result is an updated and improved fifth edition, which includes the following revisions:

- Throughout the book content has been reviewed and revised to clarify and update wherever relevant.
- Section VI, Basic Statistics, is a new section. This includes a new unit on statistics and a unit that consolidates all of the statistical graphing techniques of bar, line, and circle graphs.
- The metric and the customary systems of measure have been placed in separate units.
- New material on conversion between the metric and the customary systems of measure has been added to the unit on the metric system and to Appendix A.
- The use of spreadsheets for graphing has been included. Most students learn the basics of working with spreadsheets outside of the mathematics classroom. This material builds on that experience.

The following supplementary materials are available to instructors:

- Instructor's Guide consisting of solutions and answers to all problems.
- Student Solutions Manual for solutions to all odd-numbered exercises and problems.
- An e.resource containing:

    Computerized Test Bank

    PowerPoint Presentation Slides

    Image Library

About the Authors

Robert D. Smith was Associate Professor Emeritus of Industrial Technology at Central Connecticut State University, New Britain, Connecticut. Mr. Smith has had experience in the manufacturing industry as tool designer, quality control engineer, and chief manufacturing engineer. He has also been active in teaching applied mathematics, physics, and industrial materials and processes on the secondary school level and in apprenticeship programs. He is the author of Thomson Delmar Learning's *Mathematics for Machine Technology*.

John C. Peterson is a retired Professor of Mathematics at Chattanooga State Technical Community College, Chattanooga, Tennessee. Before he began teaching he worked on several assembly lines in industry. He has taught at the middle school, high school, two-year college, and university levels. Dr. Peterson is the author or coauthor of three other Thomson Delmar Learning books: *Technical Mathematics*, *Technical Mathematics with Calculus*, and *Math for the Automotive Trade*. In addition, he has had over 80 papers published in various journals, has given over 200 presentations, and has served as a vice president of The American Mathematical Association of Two-Year Colleges.

If you have any comments or corrections you may contact the author at SmithIntroTechMath @comcast.net.

Acknowledgments

The author and publisher wish to thank the following individuals for their contribution to the review process:

Andrew Bachman
Pottstown School District
Pottstown, PA

Susan Berry
Elizabethtown Community and Technical College
Elizabethtown, KY

John Black
Salina Area Technical School
Salina, KS

Vicky Ohlson
Trenholm Technical College
Montgomery, AL

Stephanie Craig
Newcastle School of Trades
Pulaski, PA

Steve Ottmann
Southeast Community College
Lincoln, NE

Dennis Early
Wisconsin Indianhead Technical College
New Richmond, WI

Dr. Julia Probst
Trenholm Technical College
Montgomery, AL

Debbie Elder
Triangle Tech
Pittsburgh, PA

Tony Signoriello
Newcastle School of Trades
Pulaski, PA

Steve Hlista
Triangle Tech
Pittsburgh, PA

John Shirey
Triangle Tech
Pittsburgh, PA

Todd Hoff
Wisconsin Indianhead Technical College
New Richmond, WI

William Strauss
New Hampshire Community Technical
  College
Berlin, NH

Mary Karol McGee
Metropolitan Community College
Omaha, NE

In addition, the following instructors reviewed the text and solutions for technical accuracy:

Chuckie Hairston
Halifax Community College
Weldon, NC

Todd Hoff
Wisconsin Indianhead Technical College
New Richmond, WI

The author and publisher also wish to extend their appreciation to the following companies for the use of credited information, graphics, and charts:

L. S. Starrett Company
Athol, MA 01331

Chicago Dial Indicator
Des Plaines, IL 60016

Texas Instruments, Inc.
P.O. Box 655474
Dallas, TX 75265

S-T Industries
St. James, MN 56081

The publisher wishes to acknowledge the following contributors to the supplements package:

Linda Willey and Stephen Ottmann: Technical review of the Student Solutions Manual
Susan Berry: PowerPoint presentations
Anthony Signoriello: Computerized Test Bank

Introduction to the Scientific Calculator

A scientific calculator is to be used in conjunction with the material presented in this textbook. Complex mathematical calculations can be made quickly, accurately, and easily with a scientific calculator.

Although most functions are performed in the same way, there are some differences among different makes and models of scientific calculators. In this book, generally, where there are two basic ways of performing a function, both ways are shown. However, not all of the differences among the various makes and models of calculators can be shown. It is very important that you become familiar with the operation of your scientific calculator. An owner's manual or reference guide is included with the purchase of a scientific calculator, explains the essential features and keys of the specific calculator, as well as providing detailed information on the proper use. *It is essential that the owner's manual be studied and referred to whenever there is a question regarding calculator usage.*

For use with this textbook, the most important feature of the scientific calculator is the Algebraic Operating System (AOS™). This system, which uses algebraic logic, permits you to enter numbers and combined operations into the calculator in the same order as the expressions are written. The calculator performs combined operations according to the rules of algebraic logic, which assigns priorities to the various mathematical operations. *It is essential that you know if your calculator uses algebraic logic.*

Most scientific calculators, in addition to the basic arithmetic functions, have algebraic, statistical, conversion, and program or memory functions. Some of the keys with their functions are shown in the above table. Most scientific calculators have functions in addition to those shown in the table.

SOME TYPICAL KEY SYMBOLS AND FUNCTIONS FOR A SCIENTIFIC CALCULATOR	
KEY(s)	**FUNCTION(s)**
$+$, $-$, $\times$, $\div$, $=$, or EXE, or ENTER	Basic Arithmetic
$+/-$ or $(-)$	Change Sign
π	Pi
$($, $)$	Parentheses
EE or EXP	Scientific Notation
Eng	Engineering Notation
STO, RCL, EXC	Memory or Memories
x^2, $\sqrt{x}$	Square and Square Root
$\sqrt[x]{y}$, $\sqrt[y]{\ }$	Root
y^x or x^y	Power
$1/x$ or x^{-1}	Reciprocal
$\%$	Percent
$a^{b/c}$ or $A^{b/c}$	Fractions and Mixed Numbers
DRG	Degrees, Radians, and Graduations
DMS or $°\,'\,''$	Degrees, Minutes, and Seconds
sin, cos, tan	Trigonometric Functions

General Information About the Scientific Calculator

Since there is some variation among different makes and models of scientific calculators, your calculator function keys may be different from the descriptions that follow. *To repeat, it is very important that you refer to the owner's manual whenever there is a question regarding calculator usage.*

- Solutions to combined operations shown in this text are performed on a calculator with algebraic logic (AOS™).

Turning the Calculator On and Off

- The method of turning the calculator on with battery-powered calculators depends on the calculator make and model. When a calculator is turned on, 0 and/or other indicators are displayed. Basically, a calculator is turned on and off by one of the following ways.

- With calculators with an on/clear, $\boxed{\text{ON/C}}$, key, press $\boxed{\text{ON/C}}$ to turn on. Press the $\boxed{\text{OFF}}$ key to turn off.

- With calculators with an all clear power on/power off, $\boxed{\text{AC}}$, key, press $\boxed{\text{AC}}$ to turn on. Generally, the $\boxed{\text{AC}}$ key is also pressed to turn off.

- With calculators that have an on-off switch, move the switch either on or off. The switch is usually located on the left side of the calculator.

- NOTE: In order to conserve power, most calculators have an automatic power off feature that automatically switches off the power after approximately five minutes of nonuse.

Clearing the Calculator Display and all Pending Operations

- To clear or erase *all* entries of previous calculations, depending on the calculator, either of the following procedures is used.

- With calculators with an on/clear, $\boxed{\text{ON/C}}$, key, press $\boxed{\text{ON/C}}$ *twice.*

- With calculators with the all clear, $\boxed{\text{AC}}$, key, press $\boxed{\text{AC}}$.

Erasing (Deleting) the Last Calculator Entry

- A last entry error can be removed and corrected without erasing previously entered data and calculations. Depending on the calculator, either of the following procedures is used.

- With calculators with the on/clear, $\boxed{\text{ON/C}}$, key, press $\boxed{\text{ON/C}}$.

- With calculators with a delete, $\boxed{\text{DEL}}$, key, press $\boxed{\text{DEL}}$. If your calculator has a backarrow, $\boxed{\blacktriangleleft}$, key, use it to move the cursor to the part you want to delete.

- With calculators with a clear, $\boxed{\text{CLEAR}}$, key, press $\boxed{\text{CLEAR}}$.

Alternate–Function Keys

- Most scientific calculator keys can perform more than one function. Depending on the calculator, the $\boxed{\text{2nd}}$ and $\boxed{\text{3rd}}$ keys or $\boxed{\text{SHIFT}}$ key enable you to use alternate functions. The alternate functions are marked above the key and/or on the upper half of the key. Alternate functions are shown and explained in the book where their applications are appropriate to specific content.

Decisions Regarding Calculator Use

The exercises and problems presented throughout the text are well suited for solutions by calculator. However, it is felt decisions regarding calculator usage should be left to the discretion of the course classroom or shop instructor. The instructor best knows the unique learning environment and objectives to be achieved by the students in a course. Judgments should be made by the instructor as to the degree of emphasis to be placed on calculator applications, when and where a calculator is to be used, and the selection of specific problems for solution by calculator. Therefore, exercises and problems in this text are *not* specifically identified as calculator applications.

Calculator instruction and examples of the basic operations of addition, subtraction, multiplication, and division of whole numbers, fractions, and decimals are presented at the ends of each of Units 1, 2, and 3. Further calculator instruction and examples of mathematics operations and functions are given throughout the text wherever calculator applications are appropriate to the material presented.

SECTION 1 ⠿

Fundamentals of General Mathematics

UNIT 1 ⣿ Whole Numbers

OBJECTIVES

After studying this unit you should be able to

- express the digit place values of whole numbers.
- write whole numbers in expanded form.
- estimate answers.
- arrange, add, subtract, multiply, and divide whole numbers.
- solve practical problems using addition, subtraction, multiplication, and division of whole numbers.
- solve problems by combining addition, subtraction, multiplication, and division.
- solve arithmetic expressions by applying the proper order of operations.
- solve problems with formulas by applying the proper order of operations.

All occupations, from the least to the most highly skilled, require the use of mathematics. The basic operations of mathematics are addition, subtraction, multiplication, and division. These operations are based on the decimal system. Therefore, it is important that you understand the structure of the decimal system before doing the basic operations.

The development of the decimal system can be traced back many centuries. In ancient times, small numbers were counted by comparing the number of objects with the number of fingers. To count larger numbers pebbles might be used. One pebble represented one counted object. Counting could be done more quickly when the pebbles were placed in groups, generally ten pebbles in each group. Our present number system, the decimal system, is based on this ancient practice of grouping by ten.

1–1 Place Value

In the decimal system, 10 number symbols or digits are used. The digits 0, 1, 2, 3, 4, 5, 6, 7, 8, and 9 can be arranged to represent any number. The value expressed by each digit depends on its position in the written number. This value is called the *place value.* The chart shows the place value for each digit in the number 2,452,678,932. The digit on the far right is in the units (ones) place. The digit second from the right is in the tens place. The digit third from the right is in the hundreds place. The value of each place is ten times the value of the place directly to its right.

Billions		Hundred Millions	Ten Millions	Millions		Hundred Thousands	Ten Thousands	Thousands		Hundreds	Tens	Units
2	,	4	5	2	,	6	7	8	,	9	3	2

EXAMPLES

Write the place value of the underlined digit in each number.

1. 23,164 . . . . Hundreds *Ans*

2. 523 . . . . Units *Ans*

3. 143,892 . . . . Hundred Thousands *Ans*

4. 89,874,726 . . . . Millions *Ans*

5. 7,623 . . . . Tens *Ans*

1–2 Expanding Whole Numbers

The number 64 is a simplified and convenient way of writing 6 tens plus 4 ones. In its expanded form, 64 is shown as $(6 \times 10) + (4 \times 1)$.

EXAMPLE

Write the number 382 in expanded form in two different ways.

382 = 3 hundreds plus 8 tens plus 2 ones

382 = 3 hundreds + 8 tens + 2 ones *Ans*

382 = $(3 \times 100) + (8 \times 10) + (2 \times 1)$ *Ans*

EXAMPLES

Write each number in expanded form.

1. 7,028 . . . . $(7 \times 1,000) + (0 \times 100) + (2 \times 10) + (8 \times 1)$ *Ans*

2. 52 . . . . 5 tens + 2 ones *Ans*

3. 734 . . . . 7 hundreds + 3 tens + 4 ones *Ans*

4. 86,279 . . . . $(8 \times 10,000) + (6 \times 1,000) + (2 \times 100) + (7 \times 10) + (9 \times 1)$ *Ans*

5. 345 . . . . $(3 \times 100) + (4 \times 10) + (5 \times 1)$ *Ans*

EXERCISE 1–2

Write the place value for the specified digit of each number given in the tables.

	Digit	Number	Place Value	
1.	7	6,732	Hundreds	*Ans*
2.	3	139		
3.	6	16,137		
4.	4	3,924		
5.	3	136,805		
6.	2	427		
7.	9	9,732,500		
8.	5	4,578,190		

	Digit	Number	Place Value
9.	1	10,070	
10.	0	15,018	
11.	9	98	
12.	7	782,944	
13.	5	153,400	
14.	9	98,600,057	
15.	2	378,072	
16.	4	43,728	

Write each whole number in expanded form.

17. 857 = $(8 \times 100) + (5 \times 10) + (7 \times 1)$ *Ans*

18. 32   **20.** 1,372   **22.** 5,047   **24.** 23,813

19. 942   **21.** 10   **23.** 379   **25.** 504

26. 6,376	**29.** 600	**32.** 7,500,000	**35.** 234,123
27. 333	**30.** 685,412	**33.** 97,560	**36.** 17,643,000
28. 59	**31.** 90,507	**34.** 70,001	**37.** 428,000,975

1–3   Estimating (Approximating)

For many on-the-job applications, there are times when an exact mathematical answer is not required. Often a rough mental calculation is all that is needed. Making a rough calculation is called estimating or approximating. Estimating is widely used in practical applications. A painter estimates the number of gallons of paint needed to paint the exterior of a building; it would not be practical to compute the paint requirement to a fraction of a gallon. In ordering plywood for a job, a carpenter makes a rough calculation of the number of pieces required. An electrician approximates the number of feet of electrical cable needed for a wiring job; there is no need to calculate the exact length of cable.

When computing an exact answer, it is also essential to estimate the answer before the actual arithmetic computations are made. Mistakes often can be avoided if approximate values of answers are checked against their computed values. For example, if digits are incorrectly aligned when doing an arithmetic operation, errors of magnitude are made. Answers that are $\frac{1}{10}$ or 10 times the value of what the answer should be are sometimes carelessly made. First estimating the answer and checking it against the computed answer will tell you if an error of this type has been made.

Examples of estimating answers are given in this unit. When solving exercises and problems in this unit, estimate answers and check the computed answers against the estimated answers. Continue to estimate answers for exercises and problems throughout the book.

It is important also to estimate answers when using a calculator. You can press the wrong digit or the wrong operation sign; you can forget to enter a number. If you have approximated an answer and check it against the calculated answer, you will know if you have made a serious mistake.

When estimating an answer, exact values are rounded. Rounded values are approximate values. Rounding numbers enables you to mentally perform arithmetic operations. When rounding whole numbers, determine the place value to which the number is to be rounded. Increase the digit at the place value by 1 if the digit that follows is 5 or more. Do not change the digit at the place value if the digit that follows is less than 5. Replace all the digits to the right of the digit at the place value with zeros.

EXAMPLES •——————————————————————————————

1. Round 612 to the nearest hundred.
   Since 1 is less than 5, 6 remains unchanged. 600 *Ans*

2. Round 873 to the nearest hundred.
   Since 7 is greater than 5, change 8 to 9. 900 *Ans*

3. Round 4,216 to the nearest thousand.
   Since 2 is less than 5, 4 remains unchanged. 4,000 *Ans*

4. Round 175,890 to the nearest ten thousand.
   Since 5 follows 7, change 7 to 8. 180,000 *Ans*

——•

EXERCISE I–3A

Round the following numbers as indicated.

1. 63 to the nearest ten   **4.** 2,587 to the nearest thousand

2. 540 to the nearest hundred   **5.** 8,480 to the nearest thousand

3. 766 to the nearest hundred   **6.** 32,403 to the nearest ten thousand

7. 46,820 to the nearest thousand

8. 53,738 to the nearest ten thousand

9. 466,973 to the nearest ten thousand

10. 949,500 to the nearest hundred thousand

Rounding to the Even

Many technical trades use a process called *rounding to the even.* Rounding to the even can be used to help reduce bias when several numbers are added. When using rounding to the even, determine the place value to which the number is to be rounded. (This is the same as in the previous method.) The only change is when the digit that follows is a 5 followed by all zeros. Then increase the digits at the place value by 1 if it is an odd number (1, 3, 5, 7, or 9). Do not change it if it is an even number (0, 2, 4, 6, or 8). In both cases, replace the 5 with a 0.

EXAMPLES •───────────────────────────────────

1. Round 4,250 to the nearest hundred.
Since 2 is an even number, it remains the same. 4,200 *Ans*

2. Round 673,500 to the nearest thousand.
Since 3 is an odd number, change the 3 to a 4. 674,000 *Ans*

EXERCISE 1–3B

Using rounding to the even to round the following numbers as indicated.

1. 785 to the nearest ten

2. 675 to the nearest ten

3. 1,350 to the nearest hundred

4. 5,450 to the nearest hundred

5. 31,500 to the nearest thousand

6. 24,520 to the nearest thousand

7. 26,455 to the nearest hundred

8. 26,455 to the nearest ten

1–4 Addition of Whole Numbers

A contractor determines the cost of materials in a building. A salesperson charges a customer for the total cost of a number of purchases. An air-conditioning and refrigeration technician finds lengths of duct needed. These people are using addition. Practically every occupation requires daily use of addition.

Definitions and Properties of Addition

The result of adding numbers (the answer) is called the *sum.* The *plus sign* (+) indicates addition.

Numbers can be added in any order. The same sum is obtained regardless of the order in which the numbers are added. This is called the *commutative property of addition.* For example, 2 + 4 + 3 may be added in either of the following ways:

$$2 + 4 + 3 = 9 \quad \text{or} \quad 3 + 4 + 2 = 9$$

The numbers can also be grouped in any way and the sum is the same. This is called the *associative property of addition.*

$$(2 + 4) + 3 \qquad \text{or} \qquad 2 + (4 + 3)$$
$$6 + 3 = 9 \qquad\qquad\qquad 2 + 7 = 9$$

Procedure for Adding Whole Numbers

Writing the numbers in expanded form shows why the numbers are lined up in the short form as described below.

EXAMPLE •——————————————————————————————————————

Add. 345 + 613.

	Expanded Form			Short Form

3 hundreds + 4 tens + 5 ones   345
6 hundreds + 1 ten   + 3 ones   +613
9 hundreds + 5 tens + 8 ones   958
      ↑   ↑   ↑

add hundreds   add tens   add ones

———•

In the short form, write the numbers to be added under each other. Place the units digits under the units digit, the tens digits under the tens digit, etc. Add each column of numbers starting from the column on the right (units column). If the sum of any column is ten or more, write the last digit of the sum in the answer. Mentally add the rest of the number to the next column. Continue the same procedure until all columns are added.

EXAMPLES •————————————————————————————————————

1. Add. 763 + 619

Estimate the answer. Round each number to the nearest hundred.

   800 + 600 = 1,400

Compute the answer.

   Write the numbers under each other, placing digits in proper place positions.

   Add the numbers in the units column: 3 + 9 = 12. Write 2 in the answer.

   Add the 1 to the numbers in the tens column: (1) + 6 + 1 = 8.

   Add the numbers in the hundreds column: 7 + 6 = 13. Write 13 in the answer.

```
    7 6 3
  + 6 1 9
  1 , 3 8 2 Ans
```

Check. The exact answer 1,382 is approximately the same as the estimate 1,400.

2. Add. 63,679 + 227 + 8,125 + 96

Estimate the answer. Round each number to the nearest thousand.

   64,000 + 0 + 8,000 + 0 = 72,000

Compute the answer.

   Check. The exact answer is 72,127. It is approximately the same as the estimate of 72,000.

```
    6 3 , 6 7 9
          2 2 7
        8 , 1 2 5
  +          9 6
  7 2 , 1 2 7 Ans
```

———•

EXERCISE I–4
———

Add the following numbers.

1. 33 + 88

2. 953 + 38

3. 53 + 12 + 951

4. 896 + 675 + 33

5. 73 + 1370 + 542

6. 3,653 + 8,063 + 47

7. 6,737 + 3,519 + 8,180

8. 9,734 + 10,505 + 91,613

9. 15,973 + 829 + 7,515

10. 17,392 + 2,085 + 1,670 + 13

11. 38 + 55,404 + 132,997 + 8

12. 18,768 + 3,023 + 7,787,030 + 544

1–5   Subtraction of Whole Numbers

A plumber uses subtraction to compute material requirements of a job. A machinist determines locations of holes to be drilled. A retail clerk inventories merchandise. An electrician estimates the profit of a wiring installation. Subtraction has many on-the-job applications.

Definitions

Subtraction is the operation which determines the difference between two quantities. It is the *inverse* or opposite of addition. The quantity subtracted is called the *subtrahend.* The quantity from which the subtrahend is subtracted is called the *minuend.* The result of the subtraction operation is called the *difference.* The *minus sign* ($-$) indicates subtraction.

Procedure for Subtracting Whole Numbers

Write the number to be subtracted (subtrahend) under the number from which it is subtracted (minuend). Place the units digit under the units digit, the tens digit under the tens digit, etc. Subtract each column of numbers starting from the right (units column). Writing the numbers in expanded form shows why the numbers are lined up when they are subtracted.

EXAMPLE •————————————————————

Subtract. $847 - 315$

		Expanded Form			**Short Form**
8 hundreds	+	4 tens	+	7 ones	847
3 hundreds	+	1 ten	+	5 ones	-315
5 hundreds	+	3 tens	+	2 ones	532 *Ans*
↑		↑		↑	
subtract hundreds		subtract tens		subtract ones	

If the digit in the subtrahend represents a value greater than the value of the corresponding digit in the minuend, it is necessary to regroup. Regroup the number in the minuend by taking or borrowing 1 from the number in the next higher place and adding 10 to the number in the place directly to the right. The value of the minuend remains unchanged. The value represented by each digit of a number is 10 times the value represented by the digit directly to its right. For example, 85 is a convenient way of writing 8 tens plus 5 ones, $(8 \times 10) + (5 \times 1)$. The 5 in 85 can be increased to 15 by taking 1 from the 8 without changing the value of 85. This process is called *regrouping,* or *borrowing,* since 8 tens and 5 units $(80 + 5)$ is regrouped as 7 tens and 15 units $(70 + 15)$.

The difference can be checked by adding the difference to the subtrahend. The sum should equal the minuend. If the sum does not equal the minuend, go over the operation to find the error.

EXAMPLES •————————————————————

1. Subtract. $917 - 523$

Estimate the answer. Round each number to the nearest hundred.

$900 - 500 = 400$

Compute the answer.

Write the subtrahend 523 under the minuend 917. Place the digits in the proper place positions.

$$\begin{array}{r} 9\ 1\ 7 \\ -\ 5\ 2\ 3 \\ \hline 3\ 9\ 4\ \textit{Ans} \end{array}$$

Subtract the units: $7 - 3 = 4$. Write 4 in the answer.

Subtract the tens. Since 2 is larger than 1, regroup the minuend 917.

$(8 \times 100) + (11 \times 10) + (7 \times 1)$

$11 - 2 = 9$. Write 9 in the answer.

Subtract the hundreds: $8 - 5 = 3$. Write 3 in the answer.

Check the answer to the estimate.

394 is approximately the same as 400.

Check by adding. Adding the answer 394 and the subtrahend 523 equals the minuend 917.

$$\begin{array}{r} 3\ 9\ 4 \\ +\ 5\ 2\ 3 \\ \hline 9\ 1\ 7\ \text{Ck} \end{array}$$

2. Subtract. $87,126 - 3,874$

Estimate the answer. Round each number to the nearest thousand.

$87,000 - 4,000 = 83,000$

Compute the answer.

Check the answer to the estimate. 83,252 is approximately the same as 83,000.

Check by adding. Adding the answer 83,252 and the subtrahend 3874 equals the minuend 87,126.

$$\begin{array}{r} 8\ 7\ ,1\ 2\ 6 \\ -\ \ 3\ ,8\ 7\ 4 \\ \hline 8\ 3\ ,2\ 5\ 2\ \textit{Ans} \end{array}$$

$$\begin{array}{r} 8\ 3\ ,2\ 5\ 2 \\ +\ \ 3\ ,8\ 7\ 4 \\ \hline 8\ 7\ ,1\ 2\ 6\ \text{Ck} \end{array}$$

EXERCISE I–5

Subtract the following numbers.

1. $35 - 18$
2. $98 - 29$
3. $76 - 67$
4. $312 - 97$
5. $673 - 558$

6. $1,570 - 988$
7. $7,803 - 5,905$
8. $49,406 - 5,498$
9. $19,135 - 11,236$
10. $707,353 - 533,974$

Use the chart shown in Figure 1–1 for problems 11–16. Find how much greater value A is than value B.

	A	B
11.	517 inches	298 inches
12.	779 meters	488 meters
13.	2,732 pounds	976 pounds
14.	8,700 days	5,555 days
15.	12 807 liters	9 858 liters
16.	4,464 acres	1,937 acres

Figure 1–1

Solve and check. The addition in parentheses is done first.

17. $87 - (35 + 19) = 33\ \textit{Ans}$
18. $908 - (312 + 6 + 88)$
19. $3,987 - (616 + 17 + 1,306)$
20. $(32 + 63 + 9) - 22$
21. $(503 + 7,877 + 6) - 2,033$

I–6   Problem Solving—Word Problem Practical Applications

It is important that you have the ability to solve problems that are given as statements, commonly called word problems. In actual practice, situations or problems have to be "figured out."

Often the relevant facts of the problem are written down and analyzed. The procedure for solving a problem is determined before arithmetic computations are made.

Whether the word problem is simple or complex, a definite logical procedure should be followed to analyze the problem. Some or all of the following steps may be required depending on the nature and complexity of the particular problem.

- Read the entire problem, several times if necessary, until you understand what it states and what it asks.
- Understand each part of the given information. Determine how the given information is related to what is to be found.
- If the problem is complex, break the problem down into simpler parts.
- It is sometimes helpful to make a simple sketch to help visualize the various parts of a problem.
- Estimate the answer.
- Calculate or compute the answer. Write your computations carefully and neatly. Check your work to make sure you have not made a computational error.
- Check the answer step-by-step against the statement of the original problem. Did you answer the question asked?
- Always ask yourself, "Does my answer sound sensible?" If not, recheck your work.

1–7   Adding and Subtracting Whole Numbers in Practical Applications

EXAMPLE •

The production schedule for a manufacturing plant calls for a total of 2,370 parts to be completed in 5 weeks. The number of parts manufactured during the first 4 weeks are 382, 417, 485, and 508, respectively. How many parts must be produced in the fifth week to fill the order?

Determine the procedure.

The number of parts produced in the fifth week equals the total parts to be completed minus the sum of 4 weeks' production

Estimate the answer.

$2,400 - (400 + 400 + 500 + 500)$

$2,400 - 1,800 = 600$

Compute the answer.

$2,370 - (382 + 417 + 485 + 508)$

$2,370 - 1792 = 578$, 578 parts *Ans*

Check the answer to the estimate.

578 is approximately the same as 600.

EXERCISE 1–7

1. A heavy equipment operator contracts to excavate 850 cubic yards of earth for a house foundation. How much remains to be excavated after 585 cubic yards are removed?

2. A sheet metal contractor has 124 feet of band iron in stock. An additional 460 feet are purchased. On June 2, 225 feet are used. On June 4, 197 feet are used. How many feet of band iron are left after June 4?

3. An automobile mechanic determines the total bill for both labor and materials for an engine overhaul at $463. The customer pays $375 by check and pays the balance by cash. What amount does the customer pay by cash?

4. Five stamping machines in a manufacturing plant produce the same product. Each machine has a counter that records the number of parts produced. The table in Figure 1–2 shows the counter readings for the beginning and end of one week's production.

	Machine 1	Machine 2	Machine 3	Machine 4	Machine 5
Counter Reading Beginning of Week	17,855	13,935	7,536	38,935	676
Counter Reading End of Week	48,951	42,007	37,881	72,302	29,275

Figure 1–2

a. How many parts are produced during the week by each machine?

b. What is the total weekly production?

5. A painter and decorator purchase 18 gallons of paint and 68 rolls of wallpaper for a house redecorating contract. The table in Figure 1–3 lists the amount of materials that are used in each room.

	Kitchen	Living Room	Dining Room	Master Bedroom	Second Bedroom	Third Bedroom
Paint	2 gallons	4 gallons	2 gallons	3 gallons	3 gallons	2 gallons
Wallpaper	8 rolls	14 rolls	10 rolls	12 rolls	10 rolls	9 rolls

Figure 1–3

a. Find the amount of paint remaining at the end of the job.

b. Find the amount of wallpaper remaining at the end of the job.

6. An electrical contractor has 5000 meters of BX cable in stock at the beginning of a wiring job. At different times during the job, electricians remove the following lengths from stock: 325 meters, 580 meters, 260 meters, and 65 meters. When the job is completed, 135 meters are left over and are returned to stock. How many meters of cable are now in stock?

7. A printer bills a customer $1,575 for an order. In printing the order, expenses are $432 for bond paper, $287 for cover stock, $177 for envelopes, and $26 for miscellaneous materials. The customer pays the bill within 30 days and is allowed a $32 discount. How much profit does the printer make?

8. The table in Figure 1–4 lists various kinds of flour ordered and received by a commercial baker.

	Bread Flour	Cake Flour	Rye Flour	Rice Flour	Potato Flour	Soybean Flour
Ordered	3,875 lb	2,000 lb	825 lb	180 lb	210 lb	85 lb
Received	3,650 lb	2,670 lb	910 lb	75 lb	165 lb	85 lb

Figure 1–4

a. Is the total amount of flour received greater or less than the total amount ordered?

b. How many pounds greater or less?

9. In order to make the jig shown in Figure 1–5, a machinist determines dimensions A, B, C, and D. All dimensions are in millimeters. Find A, B, C, and D in millimeters.

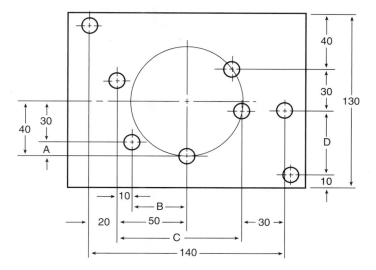

Figure 1–5

10. A small business complex is shown in the diagram in Figure 1–6. To provide parking space, a paving contractor is hired to pave the area not occupied by buildings or covered by landscaped areas. The entire parcel of land contains 41,680 square feet. How many square feet of land are paved?

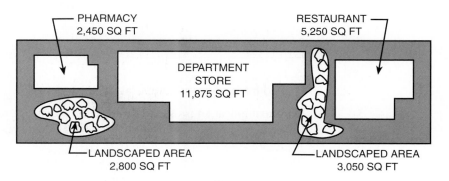

Figure 1–6

1–8 Multiplication of Whole Numbers

A mason estimates the number of bricks required for a chimney. A clerk in a hardware store computes the cost of a customer's order. A secretary determines the weekly payroll of a firm. A cabinetmaker calculates the amount of plywood needed to install a store counter. A garment manufacturing supervisor determines the amounts of various materials required for a production run. These are a few of the many occupational uses of multiplication.

Definitions and Properties of Multiplication

Multiplication is a short method of adding equal amounts. For example, 4 times 5 (4 × 5) means 4 fives or 5 + 5 + 5 + 5.

The number to be multiplied is called the *multiplicand.* The number by which it is multiplied is called the *multiplier. Factors* are the numbers used in multiplying. The multiplicand and the multiplier are both factors. The result or answer of the multiplication is called the *product.* The *times sign* (×) indicates multiplication.

Numbers can be multiplied in any order. The same product is obtained regardless of the order in which the numbers (factors) are multiplied. This is called the *commutative property of multiplication.* For example, 2 × 4 × 3 may be multiplied in either of the following ways:

$$2 \times 4 \times 3 = 24 \quad \text{or} \quad 3 \times 4 \times 2 = 24$$

The numbers can also be grouped in any way and the product is the same. This is called the *associative property of multiplication.*

$$(2 \times 4) \times 3 \quad \text{or} \quad 2 \times (4 \times 3)$$

$$8 \times 3 = 24 \quad\quad\quad 2 \times 12 = 24$$

Expanded Form of Multiplication

The expanded form for multiplication shows why the products are aligned as described later.

EXAMPLE •────────────────────────────────

Multiply. 386 × 7

	Expanded Form			Shorter Form
	3 hundreds +	8 tens +	6 ones	386
×	_____	_____	7	× 7
	21 hundreds +	56 tens +	42 ones	42
	2100 +	560 +	42	560
	2702			2100
				2702

•

Procedure for Short Multiplication

Short multiplication is used to compute the product of two numbers when the multiplier contains only one digit. A problem such as 7 × 386 requires short multiplication.

EXAMPLE •────────────────────────────────

Multiply. 7 × 386

Estimate the answer. Round 386 to 400.

$\quad$ 7 × 400 = 2,800

Compute the answer.

$\quad$ Write the multiplier under the units digit of the multiplicand.

$\quad$ Multiply the 7 by the units of the multiplicand.

$\quad$ 7 × 6 = 42

$\quad$ Write 2 in the units position of the answer.

$\quad$ Multiply the 7 by the tens of the multiplicand.

$\quad$ 7 × 8 = 56

$\quad$ Add the 4 tens from the product of the units.

$\quad$ 56 + 4 = 60

$$\begin{array}{r} 3\,8\,6 \\ \times \quad 7 \\ \hline 2\,,7\,0\,2 \ \textit{Ans} \end{array}$$

Write the 0 in the tens position of the answer.

Multiply the 7 by the hundreds of the multiplicand.

$7 \times 3 = 21$

Add the 6 hundreds from the product of the tens.

$21 + 6 = 27$

Write the 7 in the hundreds position and the 2 in the thousands position.

Check the answer to the estimate.

2,702 is approximately the same as 2,800.

Procedure for Long Multiplication

Long multiplication is used to compute the product of two numbers when the multiplier contains two or more digits. A problem such as $436 \times 7{,}812$ requires long multiplication.

When multiplying by a number that is not in the ones place, both the digits and the place values get multiplied. For example, 4 tens $\times$ 6 tens = 24 tens $\times$ tens = 24 hundreds.

EXAMPLE

Multiply. 243×60

	Expanded Form			**Shorter Form**
	2 hundreds +	4 tens +	3 ones	243
$\times$		6 tens +	0	$\times$ 60
	0 hundreds +	0 tens +	0 ones	0
12 thousand +	24 hundreds +	18 tens +	0 ones	180
12,000 +	2,400 +	180 +	0	2,400
14,580				12,000
				14,580

EXAMPLE

Multiply. $436 \times 7{,}812$

Estimate the answer. Round 436 to 400 and 7,812 to 8,000.

$400 \times 8{,}000 = 3{,}200{,}000$

Compute the answer.

Write the multiplier under the multiplicand, placing digits in proper place positions.

Multiply the multiplicand by the units of the multiplier, using the procedure for short multiplication. This answer is called the first partial product.

$6 \times 7{,}812 = 46{,}872$

Write the first partial product, starting at the units position and going from right to left.

Multiply the multiplicand by the tens of the multiplier to get the second partial product.

$3 \times 7{,}812 = 23{,}436$

Write this partial product under the first partial product, starting at the tens position and going from right to left.

$$\begin{array}{r} 7\,8\,1\,2 \\ \times\ \ 4\,3\,6 \\ \hline 4\,6\ \ 8\,7\,2 \\ 2\,3\,4\ \ 3\,6 \\ 3\ \ 1\,2\,4\ \ 8 \\ \hline 3\,,4\,0\,6\,,0\,3\,2\ \textit{Ans} \end{array}$$

Multiply the multiplicand by the hundreds of the multiplier for the third, and last, partial product.

$4 \times 7,812 = 31,248$

Write the last partial product under the second partial product, starting at the hundreds position and going from right to left.

Add the three partial products to get the product.

Check the answer to the estimate.

3,406,032 is approximately the same as 3,200,000.

Multiplication with a Zero in the Multiplier

The product of any number and zero is zero. This is called the *multiplicative property of zero.* For example, $0 \times 0 = 0, 0 \times 6 = 0, 0 \times 8,956 = 0$. When the multiplier contains zeros, the zeros must be written in the product to maintain proper place value.

EXAMPLES •

1.
```
    674
×   200
134,800
```

2.
```
   364
×  203
 1 092
72 80
73,892
```

Multiplying Three or More Factors

As previously stated by the associative property, the multiplication of three or more numbers, two at a time, may be done in any order or in any grouping. The factors are multiplied in separate steps. Two groupings are shown in the following example.

EXAMPLE •

$7 \times 5 \times 2 \times 3$ $7 \times 5 \times 2 \times 3$

$35 \times 2 \times 3$ $7 \times 5 \times 6$

70×3 7×30

210 *Ans*   210 *Ans*

EXERCISE I–8

Multiply and check.

1.
```
 75
× 8
```

2.
```
775
× 5
```

3.
```
1,877
×    9
```

4.
```
54,157
×     8
```

5. 6×523

6. $3 \times 1,804$

7. $5 \times 12,199$

8. $4 \times 456,900$

9. $8 \times 318,234$

10. $9 \times 2,132,512$

11.
```
 57
×81
```

12.
```
914
× 67
```

13.
```
12,737
×    79
```

14.
```
7,816
×  513
```

15.
```
15,553
×   999
```

16.
```
23,418
× 1,147
```

17.
```
327,800
×    274
```

18.
```
405,607
×    112
```

19. $419 \times 7{,}635$

20. $423 \times 63{,}940$

21. $2{,}561 \times 17{,}738$

22. $1{,}176 \times 62{,}347$

23. $4{,}214 \times 18{,}919$

24. 943
   $\times\ \ 70$

25. $1{,}798$
   $\times\ \ 507$

26. $7{,}100$
   $\times\ \ 590$

27. $8{,}009$
   $\times\ \ 400$

28. $6 \times 8 \times 15$

29. $12 \times 16 \times 7$

30. $63 \times 150 \times 15 \times 8$

1–9   Division of Whole Numbers

Division is used in all occupations. The electrician must know the number of rolls of cable to order for a job. A baker determines the number of finished units made from a batch of dough. A landscaper needs to know the number of bags of lawn food required for a given area of grass. A printer determines the number of reams of paper needed for a production run of circulars.

Division is the process of finding how many times one number is contained in another. It is a short method of subtracting. Dividing 24 by 6 is a way to find the number of times 6 is contained in 24.

$$24 - 6 = 18$$
$$18 - 6 = 12$$
$$12 - 6 = 6$$
$$6 - 6 = 0$$

Six is subtracted 4 times from 24; therefore, 4 sixes are contained in 24.

$$24 \div 6 = 4$$

Definitions

Division is the opposite, or *inverse,* of multiplication. In division, the number to be divided is called the *dividend.* The number by which the dividend is divided is called the *divisor.* The result of division is called the *quotient.* A difference left is called the *remainder.* The symbol for division is $\div$. The expression $21 \div 7$ can be written in fractional form as $\frac{21}{7}$. When written as a fraction, the dividend, 21, is called the numerator, and the divisor, 7, is called the denominator. The long division symbol, $\overline{)\ \ }$, is used when computing a division problem.

$$21 \div 7 \text{ is written as } 7\overline{)21}$$

Zero as a Dividend

Zero divided by a number equals zero. For example, $0 \div 5 = 0$. The fact that zero divided by a number equals zero can be shown by multiplication. The expression $0 \div 5 = 0$ means $0 \times 5 = 0$. Since 0×5 does equal 0, it is true that $0 \div 5 = 0$.

Zero as a Divisor

Dividing by zero is impossible. Students sometimes confuse division of a number by zero with division of zero by a number. It can be shown by multiplication that a number divided by zero is impossible; it is undefined. The expression $5 \div 0 = ?$ means that $? \times 0 = 5$. Since there is no real number that can be multiplied by 0 to equal 5, the division, $5 \div 0$, is not possible. In the case of $0 \div 0 = ?$, there is not a unique solution, but there are infinite solutions. The expression $0 \div 0 = ?$ means $? \times 0 = 0$. Since any number times 0 equals 0, the division $0 \div 0$ has no unique solution and is also not possible.

Procedure for Dividing Whole Numbers

Write the numbers of the division problem with the divisor outside the long division symbol and the dividend within the symbol. In any division problem, the answer multiplied by the divisor plus the remainder equals the dividend.

EXAMPLE •————————————————————————————————

Divide. 4,505 ÷ 6

Estimate the answer. Round 4,505 to 4,500.

4,500 ÷ 6. The answer will be between 700 and 800.

```
          7 5 0    R 5 Ans
       6)4 5 0 5
         4 2
         ———
          3 0
          3 0
          ———
            5
            0
          ———
            5
```

Compute the answer.

Write the problem with the divisor outside the long division symbol and the dividend within the symbol.

The divisor, 6, is not contained in 4, the number of thousands. The 6 will divide the 45, which is the number of hundreds. Write the 7 in the answer above the hundreds place. Multiply 7 × 6 = 42. Subtract 42 hundreds from 45 hundreds. Write the 3 hundreds remainder in the hundreds column, and add 0 tens from the dividend.

Divide 30 tens by 6. Write the 5 in the answer above the tens place. Subtract 30 tens from 30 tens. Write 5, from the dividend, in the units column.

Divide 5 by 6. Since 6 is not contained in 5, write 0 in the answer above the units place. Subtract 0 from the 5. The remainder is 5.

The answer is 750 R 5.

Check the answer to the estimate.

750 R 5 is between 700 and 800.

Check by multiplying the answer by the divisor and adding the remainder.

```
    7 5 0        4 , 5 0 0
  ×     6      +         5
  ———————      —————————————
  4 , 5 0 0      4 , 5 0 5 Ck
```

Selecting Trial Quotients

In solving long division problems, often the trial quotient selected is either too large or too small. When this occurs, another trial quotient must be selected.

EXAMPLE •————————————————————————————————

Divide 68,973 by 76.

Estimate the answer. Round 76 to 80 and 68,973 to 70,000.

70,000 ÷ 80 = 7,000 ÷ 8. The answer is approximately 900.

Compute the answer.

Write the divisor and dividend in the proper positions.

The divisor 76 is not contained in 6 or 68.

Divide 689 hundreds by 76. The partial quotient is estimated as 8.

```
            8          INCORRECT
  7 6)6 8 , 9 7 3      Trial quotient
      6 0  8           must be
      8    1           increased to 9.
```

Multiply: 8 × 76 = 608

Subtract: 689 − 608 = 81

The remainder 81 is greater than the divisor 76. The partial quotient is too small and must be increased to 9.

NOTE: The remainder 81 is greater than the divisor 76.

The problem is now correctly solved.

Divide: 689 ÷ 76

Write 9 in the partial quotient.

Multiply: 9 × 76 = 684

Subtract: 689 − 684 = 5

Bring down the 7.

Divide: 57 ÷ 76 = 0

Write 0 in the partial quotient.

Bring down the 3.

Divide: 573 ÷ 76

Estimate 8 as the trial divisor.

Multiply: 8 × 76 = 608

Since 608 cannot be subtracted from 573, the trial quotient, 8, is too large and must be decreased to 7.

Multiply: 7 × 76 = 532

Subtract: 573 − 532 = 41

The answer is 907 with a remainder of 41.

Check the answer to the estimate.

907 R 41 is approximately the same as 900.

Check by multiplying the answer by the divisor and adding the remainder.

$$\begin{array}{r} 907 \text{ R } 41 \text{ Ans} \\ 76\overline{)68,973} \\ 68\ 4 \\ \hline 573 \\ 532 \\ \hline 41 \end{array}$$

$$\begin{array}{r} 907 \\ \times\ 76 \\ \hline 5\ 442 \\ 63\ 49 \\ \hline 68,932 \end{array} \qquad \begin{array}{r} 68,932 \\ +\quad 41 \\ \hline 68,973 \text{ Ck} \end{array}$$

Maintain proper place value in division. Zeros must be shown in the quotient over their respective digits in the dividend.

EXAMPLE

Divide. 24,315,006 ÷ 4,863

$$\begin{array}{r} 5,000 \text{ R } 6 \text{ Ans} \\ 4,863\overline{)24,315,006} \\ 24\ 315 \\ \hline 006 \end{array}$$

Check:
$$\begin{array}{r} 4,863 \\ \times 5,000 \\ \hline 24,315,000 \end{array} \qquad \begin{array}{r} 24,315,000 \\ +\qquad 6 \\ \hline 24,315,006 \text{ Ck} \end{array}$$

EXERCISE 1–9

Divide and check.

1. 3)261
2. 9)405
3. 6)408
4. 9)1,962
5. 8)20,376
6. 4)26,356
7. 479,997 ÷ 7
8. 3,811 ÷ 2
9. 53,043 ÷ 5

10. $\dfrac{98,951}{8}$
11. $\dfrac{413,807}{3}$
12. $\dfrac{700,514}{9}$
13. 27)486
14. 43)559
15. 32)7,712

16. 46)9,522
17. 36,650 ÷ 68
18. 95,631 ÷ 122
19. 30,007 ÷ 604
20. 323)69,768
21. 618)78,486
22. 461,079 ÷ 924
23. 65,000 ÷ 800

24. $\dfrac{799,981}{542}$
25. $\dfrac{194,072}{2,624}$
26. $\dfrac{461,312}{2,176}$
27. $\dfrac{7,808,510}{3,776}$
28. $\dfrac{6,700,405}{4,062}$

I–I0   Multiplying and Dividing Whole Numbers in Practical Applications

EXAMPLE •————————————————————————————————

The total cost of fixtures and luminaries for an office lighting installation is found by an electrician. The following fixtures and luminaries are specified: 12 incandescent fixtures at $18 each, 22 semidirect fluorescent luminaries at $37 each, and 33 direct fluorescent luminaries at $28 each. Find the total cost.

Determine the procedure.

Multiply the required number of each luminary or fixture by the cost of each. The total cost is the sum of the products.

Estimate the answer.

Total cost = (10 × $20) + (20 × $40) + (30 × $30)

Total cost = $200 + $800 + $900 = $1,900

Compute the answer.

Total cost = (12 × $18) + (22 × $37) + (33 × $28)

Total cost = $216 + $814 + $924 = $1,954 *Ans*

Check the answer to the estimate.

$1,954 is approximately the same as $1,900.

EXERCISE I–I0

1. An offset press feeds at the rate of 2,050 impressions per hour. How many impressions can a press operator print in 14 hours?

2. A chef estimates that an average of 150 pounds of ground beef are prepared daily. How many pounds of ground beef should be ordered for a 4-week supply? The restaurant is closed only on Mondays.

3. A welder fabricates 22 steel water tanks for a price of $20,570. Find the cost of each tank.

4. A tractor-trailer operator totals diesel fuel bills for 185 gallons of fuel used in a week. The truck travels 1,665 miles during the week. How many miles per gallon does the truck average?

5. Two sets of holes are drilled by a machinist in a piece of aluminum flat stock as shown in Figure 1–7. All dimensions are in millimeters.

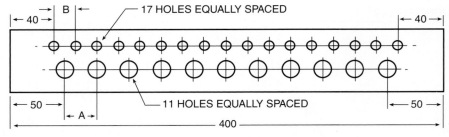

Figure I–7

a. Find, in millimeters, dimension A.

b. Find, in millimeters, dimension B.

NOTE: Whenever holes are arranged in a straight line, the number of spaces is one less than the number of holes.

6. In a commercial bakery, roll dividing machines produce 16,000 dozen rolls in 8 hours. Determine the number of single rolls produced per minute.

7. An architectural engineering assistant determines the total weight of I beams required for a proposed building. The table in Figure 1–8 lists the data used in finding the weight. Find the total weight of all I beams for the building.

	20″ × 7″ I Beams Weight: 80 lb/ft	18″ × 6″ I Beams Weight: 55 lb/ft	12″ × 5″ I Beams Weight: 32 lb/ft
Number of 10-foot lengths	15	0	24
Number of 16-foot lengths	12	18	7
Number of 20-foot lengths	8	32	25
Number of 24-foot lengths	17	8	0

Figure 1–8

NOTE: The table shows the cross-section dimensions of each type of I beam. The weights given are for 1 foot of length for each type of I beam.

8. An apartment complex is being built; it will have 318 apartments. Each workday heating and air-conditioning systems can be installed in 6 apartments. How many workdays are required to complete installations for the complete complex?

9. A gasoline dealer estimated that during the month of July (25 business days) an average of 6,500 gallons of gasoline would be sold each day. During July, a total of 175,700 gallons are actually sold. How many more gallons are sold than were estimated for the month?

10. A cosmetologist determines that an average of 3 ounces of liquid shampoo are required for each shampooing application. The beauty salon has 9 quarts of shampoo in stock. How many shampooing applications are made with the shampoo in stock?

NOTE: One quart contains 32 ounces.

11. An ornamental iron fabricator finds the material requirements for the railing shown in Figure 1–9. How many vertical pieces of 1-inch square wrought iron are needed for this job?

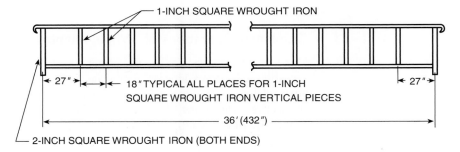

Figure 1–9

12. In estimating the time required to complete a proposed job, an electrical contractor determines that a total of 735 hours are needed. Three electricians each work 5 days per week for 7 hours per day. How many weeks are required to complete the job?

13. The size of air-conditioning equipment needed in a building depends on the number of windows and the location of the windows. The table in Figure 1–10 lists the number and the amount of square feet of four different sizes of windows. The heat gain through glass in Btu/h for each square foot of glass area is shown on the table. A Btu (British thermal unit) is a unit of heat. Find the total heat gain (Btu/h) for the building.

Window Size (Number of square feet)	NUMBER OF WINDOWS AT EACH SIDE OF BUILDING			
	North Side 25 Btu/h/sq ft	South Side 76 Btu/h/sq ft	East Side 90 Btu/h/sq ft	West Side 99 Btu/h/sq ft
15 sq ft	8	10	4	6
24 sq ft	9	6	2	7
32 sq ft	0	4	0	3
36 sq ft	2	2	1	1

Figure 1–10

14. An excavating contractor finds that a piece of land must be drained of water before work on a job can begin. Two pumps are used to drain the water. One pump operates at the rate of 70 liters per minute for 30 minutes. The second pump operates at a rate of 90 liters per minute for 45 minutes. How many liters of water are pumped by both pumps?

15. A chef plans the menu for a particular reception for 161 guests. The appetizer is a 6-ounce serving of tomato juice for each guest. How many 46-ounce cans of tomato juice are ordered for this reception?

16. A bookcase shown in Figure 1–11 is produced in quantities of 1,500 by a furniture manufacturer. All pieces, except the top and back, are made from 12-inch-wide lumber. One foot of stock is allowed for cutting and waste for each bookcase. Find the total number of feet of 12-inch stock needed to manufacture the 1,500 units.

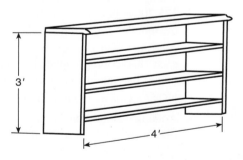

3'

4'

Figure 1–11

1–11   Combined Operations of Whole Numbers

Many occupations require the use of combined operations in solving arithmetic expressions. The arithmetic expressions are often given as formulas in occupational textbooks, manuals, and other related occupational reference materials. A *formula* uses symbols to show the relationship between quantities. The formula used in the electrical industry to find the number of kilowatts (kW) of power (P) in terms of voltage (E) and current (I) is

$$P = \frac{E \times I}{1,000}$$

where voltage is expressed in volts, and current is expressed in amperes.

Order of Operations

Following the proper order of operations is a basic requirement in solving problems involving the use of formulas. A given arithmetic expression must have a unique solution. The expression $3 + 5 \times 4$ must have only one answer. The correct answer, 23, is found by using the following order of operations rules.

Order of Operations

- First, do all operations within grouping symbols. Grouping symbols are parentheses ( ), brackets [ ], and braces { }. The fraction bar is also a symbol used as a grouping symbol. The numerator and denominator are each considered enclosed in parentheses.
- Raise to a power. This is sometimes called exponentiation and includes finding roots. It will be discussed later.
- Next, do multiplication and division operations in order from left to right.
- Last, do addition and subtraction operations in order from left to right.

Some people use the memory aid "**P**lease **E**xcuse **M**y **D**ear **A**unt **S**ally" to help them remember the order of operations. The **P** in "Please" stands for parentheses, the **E** for exponents or raising to a power, **M** and **D** for multiplication and division, and the **A** and **S** for addition and subtraction.

EXAMPLES •————————————————————————————

1. Find the value of $(15 + 6) \times 3 - 28 \div 7$.

Do the work in parentheses.	$(15 + 6) \times 3 - 28 \div 7$
Multiply and divide.	$21 \times 3 - 28 \div 7$
Subtract.	$63 - 4$
	$59 \; Ans$

2. Evaluate $(27 + 9) \times (12 - 7)$.

Do the work in parentheses.	$(27 + 9) \times (12 - 7) = 36 \times (12 - 7)$
	$= 36 \times 5$
Multiply.	$= 180 \; Ans$

3. Determine the value of $36 - (29 - (42 - (8 + 16)))$.

   Do the work in the innermost parentheses.

   $$36 - (29 - (42 - (8 + 16))) = 36 - (29 - (42 - 24))$$

Now work in the innermost of the remaining parentheses.	$= 36 - (29 - 18)$
Do the work in the final parentheses.	$= 36 - 11$
Subtract.	$= 25 \; Ans$

4. Find the value of $\dfrac{120 - 25 \times 3}{12 + 24 \div 8} + 10$.

   The fraction bar is the grouping symbol. Do all work above and below the bar first.

   $$\frac{120 - 25 \times 3}{12 + 24 \div 8} + 10$$

   $$\frac{120 - 75}{12 + 3} + 10$$

   Divide.

   $$\frac{45}{15} + 10$$

   Add.

   $$3 + 10$$

   $$13 \; Ans$$

EXERCISE I–II

Perform the indicated operations.

1. $7 + 8 - 6$

2. $26 - 9 + 3$

3. $57 + 18 - 14$

4. $94 - 87 + 32 - 27$

5. $26 + 16 \div 4$

6. $(28 + 16) \div 4$

7. $\dfrac{72}{8} + 40$

8. $\dfrac{72 + 40}{8}$

9. $(85 + 51) \div 4$

10. $\dfrac{25 - 13 + 4}{2}$

11. $25 - 13 + \dfrac{4}{2}$

12. $11 \times (8 - 5) + 9 \times (2 + 7)$

13. $11 \times 8 - 5 + 9 \times 2 + 7$

14. $\dfrac{3 \times (29 - 6)}{23}$

15. $\dfrac{324}{9} + \dfrac{288}{48}$

16. $\dfrac{253 - 17 \times 3}{85 + 52 - 36}$

17. $8 \times 12 + 60 \div (9 + 3)$

18. $8 \times (12 + 60) \div (9 + 3)$

19. $(8 \times 12 + 60) \div (9 + 3)$

20. $\left(\dfrac{81}{9} + 14\right) \times 5$

21. $\dfrac{81}{9} + (14 \times 5)$

22. $41 + 3 \times 7 - 6$

23. $41 + 3 \times (7 - 6)$

24. $(15 \times 6) \div (3 \times 5)$

25. $(142 - 37) \div (7 \times 5)$

26. $\dfrac{14 + 10 \times 7}{4 + 2}$

27. $(276 - 84) \times 8 \div 12$

28. $30 \times (10 \times 4 - 40) \div (18 - 7)$

29. $\dfrac{157 - 21 \times 3}{5 - 18 \div 6} + 17$

30. $86 + \dfrac{27 + 8 - 5}{6} - 31$

31. $(8 \times 6 - 20) \div (7 + 21)$

32. $(8 \times 6 - 20) \div 7 + 21$

33. $\dfrac{576 - 16 \times 10 \times 3}{44 - 18 \times 2} - (12 - 9)$

34. $\dfrac{16 + 6 \times 21}{157 - 12 \times 13} - \dfrac{10 + 3 - 7}{46 - 4 \times 10}$

I–I2   Combined Operations of Whole Numbers in Practical Applications

EXAMPLE •

An engineering technician is required to determine the size circle (diameter) needed to make the part shown in Figure 1–12. The part (a segment) must be contained within a circle. The length, dimension *c*, must be 20 inches and the height, dimension *h*, must be 5 inches. The technician looks up the formula in a handbook.

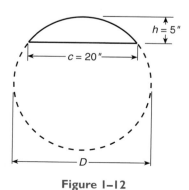

Figure I–I2

$$D = \frac{c \times c + 4 \times h \times h}{4 \times h}$$

where D = diameter
c = length of chord
h = height of segment

Substitute the numerical values for the variables.

$$\frac{20 \times 20 + 4 \times 5 \times 5}{4 \times 5}$$

The fraction bar is the grouping symbol. Do the work above and below the bar.

$$\frac{400 + 100}{20}$$

Divide.

$$\frac{500}{20}$$

25" Ans

EXERCISE 1–12

1. Electrical power (P) in kilowatts equals voltage (E) in volts times current (I) in amperes divided by 1,000.

$$P = \frac{E \times I}{1,000}$$

Find the number of kilowatts of power using the values given in the table in Figure 1–13.

	Voltage (E) (in volts)	Current (I) (in amperes)	Power (P) (in) kilowatts
a.	110	100	
b.	115	200	
c.	220	150	
d.	230	100	
e.	220	50	

Figure 1–13

2. A retailer borrows $4,700 from a bank for 30 months. Using an installment loan table, the monthly payment on $4,000 is $158. The monthly payment on $700 is $27.65. How much total interest must the retailer pay the bank?

Total interest = (monthly payment on $4,000 + monthly payment on $700)
× 30 − amount borrowed

3. An electrical circuit in which 3 cells are connected in series is shown in Figure 1–14.

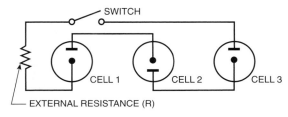

Figure 1–14

Compute the number of amperes of current in circuits a, b, and c using the values given in the table in Figure 1–15.

$$I = \frac{E \times ns}{r \times ns + R}$$

where E = voltage of one cell in volts
ns = number of cells in circuit
r = internal resistance of one cell in ohms
R = external resistance of circuit in ohms
I = current in amperes

	E	ns	r	R	I
a.	2 volts	3 cells	1 ohm	3 ohms	
b.	5 volts	3 cells	1 ohm	2 ohms	
c.	6 volts	3 cells	1 ohm	3 ohms	

Figure 1–15

4. Carpenters find amounts of lumber needed in board feet (bd ft). A board foot is the equivalent of a piece of lumber 1 foot wide, 1 foot long, and 1 inch thick.

$$\text{Bd ft} = \frac{T \times W \times L}{12}$$

where T = thickness in inches
W = width in inches
L = length in feet

Find the number of board feet in each piece of lumber shown in Figure 1–16.

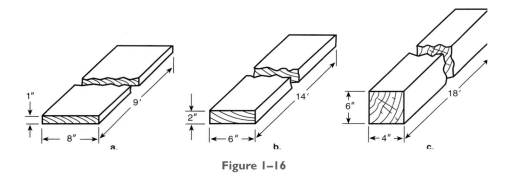

Figure 1–16

5. A comparison between Fahrenheit and Celsius scales is shown in Figure 1–17. Express the temperatures given as equivalent degrees Fahrenheit or degrees Celsius readings.

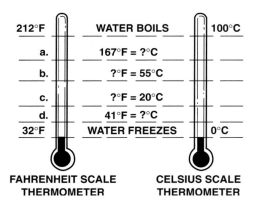

Figure 1–17

To express degrees Fahrenheit in degrees Celsius, use

$$°C = \frac{5 \times (°F - 32)}{9}$$

To express degrees Celsius in degrees Fahrenheit, use

$$°F = \frac{9 \times °C}{5} + 32$$

6. An electronics technician finds the total resistance (R_T) in ohms for the circuit shown in Figure 1–18.

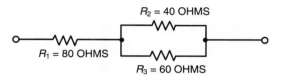

Figure 1–18

The individual resistances are represented by the symbols R_1, R_2, and R_3. Find the total resistance of the circuit.

$$R_T = R_1 + \frac{R_2 \times R_3}{R_2 + R_3}$$

⠿ UNIT EXERCISE AND PROBLEM REVIEW

PLACE VALUE

Write the place value for the specified digit of each number given in the table.

	Digit	Number	Place Value
1.	3	6,938	
2.	5	519	
3.	7	27,043	
4.	8	5,810,612	
5.	0	60,443	
6.	2	2,706	

EXPANDING WHOLE NUMBERS

Write each whole number in expanded form.

7. 48   **9.** 13,692   **11.** 5,103
8. 319   **10.** 863   **12.** 6,600,000

ADDITION OF WHOLE NUMBERS

Add and check.

13. 43 +54   **16.** 586 +787   **19.** 87,495 +96,986   **22.** 6,057 + 443 + 697
14. 67 +84   **17.** 8,463 + 388   **20.** 186,693 +557,935   **23.** 152,077 + 2,073 + 16,478
15. 123 + 96   **18.** 30,736 + 9,405   **21.** 707 + 932 + 13   **24.** 84 + 30,309 + 129,427

SUBTRACTION OF WHOLE NUMBERS

Subtract and check.

25. 47
    -14

26. 84
    -31

27. 90
    -37

28. 86
    -49

29. 787
    -612

30. 212
    -109

31. 700
    -387

32. 1,707
    $-\ 983$

33. $1,955 - 1,947$

34. $4,304 - 3,770$

35. $12,621 - 9,097$

36. $47,435 - 8,707$

37. $874,906 - 51,109$

38. $885,172 - 79,453$

MULTIPLICATION OF WHOLE NUMBERS

Multiply and check.

39.   412
    $\times\ \ 9$

40. 56
    $\times 19$

41. 8,055
    $\times\ \ 903$

42. 14,932
    $\times\ \ 8,206$

43. $7,778 \times 9,380$

44. $3,305 \times 5,617$

45. $70,000 \times 80,000$

46. $7 \times 10 \times 5 \times 2$

47. $3 \times 22 \times 20$

48. $8 \times 19 \times 78$

49. $55 \times 66 \times 77$

50. $8 \times 913 \times 72$

51. $61 \times 200 \times 816$

DIVISION OF WHOLE NUMBERS

Divide and check.

52. $8\overline{)624}$

53. $6\overline{)6,012}$

54. $67,393 \div 9$

55. $\dfrac{470,362}{9}$

56. $52\overline{)832}$

57. $16\overline{)4,848}$

58. $89\overline{)356,712}$

59. $814\overline{)13,838}$

60. $38,141 \div 177$

61. $59,492 \div 111$

62. $\dfrac{371,844}{2,817}$

63. $\dfrac{312,906}{3,981}$

COMBINED OPERATIONS OF WHOLE NUMBERS

Perform the indicated operations.

64. $9 + 15 - 14$

65. $30 + 21 \div 3$

66. $(46 + 26) \div 12$

67. $\dfrac{104 + 32}{8}$

68. $67 - 42 + \dfrac{15}{3}$

69. $31 + 7 \times (14 - 3)$

70. $31 + 7 \times 14 - 3$

71. $\dfrac{46 + 18 \times 5}{19 - 11}$

72. $(273 - 194) \times 16 \div 4$

73. $12 \times (10 - 3) + 7 \times (3 + 6)$

74. $\dfrac{6 \times (21 - 7)}{21}$

75. $\dfrac{140 - 21 \times 5}{122 - 119 + 4}$

76. $(9 \times 8 + 34) \div (42 + 11)$

77. $\dfrac{125}{5} + 12 \times 7$

78. $\dfrac{128 - 16 \times 2}{9 - 21 \div 3}$

79. $47 - \dfrac{83 + 16 - 51}{8} + 13$

80. $(19 \times 5 - 20) \div 5 - 3$

81. $40 \times (15 \times 4 - 42) \div 90$

82. $\dfrac{282 - 14 \times 3 \times 6}{6} - (21 - 16)$

83. $\dfrac{36 + 7 \times 21}{61} + \dfrac{34 - 18 - 6}{31 \times 8 - 243}$

WHOLE NUMBER PRACTICAL APPLICATION PROBLEMS

Solve the following problems.

84. During the first week of April, a print shop used the following paper stock: 5,570 sheets on Monday, 7,855 sheets on Tuesday, 7,236 sheets on Wednesday, 6,867 sheets on Thursday, and 6,643 sheets on Friday. During the following week, 4,050 more sheets are used than during the first week. Find the total sheets used during the first 2 weeks of April.

85. A 5-floor apartment building has 8 electrical circuits per apartment. There are 6 apartments per floor. How many electrical circuits are there in the building?

86. The invoice shown in Figure 1–19 is mailed to the Center Sports Shop by a billing clerk of the M & N Sports Equipment Manufacturing Company. (An invoice is a bill sent to a retailer by a manufacturer or wholesaler for merchandise purchased by the retailer.) The extension shown in the last column of the invoice is the product of the number (quantity) of units multiplied by the price of one unit. Find the extension amount for each item on the invoice and add the extensions to determine total cost.

	Quantity	Unit	Unit Price	Description	Extension
a.	15	dozen	$ 2	Floats	$30
b.	24	each	$11	Fishing rods	
c.	1	box	$18	Spools of line	
d.	18	each	$ 7	Reels	
e.	36	each	$ 5	Baseball bats	
f.	5	box	$36	Baseballs	
g.	24	package	$ 6	Golf balls	
h.	15	each	$14	Putters	
i.				Total Cost _____	

Figure 1–19

87. The drill jig shown in Figure 1–20 is laid out by a machine drafter. All dimensions are in millimeters.

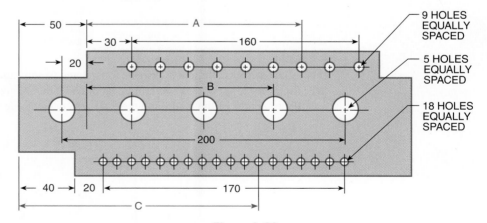

Figure 1–20

a. Find, in millimeters, dimension A.

b. Find, in millimeters, dimension B.

c. Find, in millimeters, dimension C.

88. Figure 1–21 shows the front view of a wooden counter that is to be built for a clothing store. All pieces of the counter except the top and back are to be made of the same thickness and width of lumber. How many total feet (1 foot = 12 inches) of lumber should be ordered for this job? Do not include the top or back. Allow 6 feet for waste.

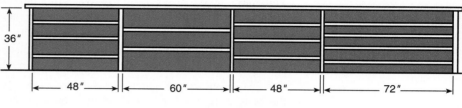

Figure 1–21

89. An 8-pound cut of roast beef is to be medium roasted at 350°F. Total roasting time is determined by allowing 15 minutes roasting time for each pound of beef. If the roast is placed in a preheated oven at 2:00 P.M., at what time should it be removed?

90. The accountant for a small manufacturing firm computes the annual depreciation of each piece of tooling, equipment, and machinery in the company. From a detailed itemized list, the accountant groups all items together that have the same life expectancy (number of years of usefulness) as shown in Figure 1–22. Find the annual depreciation for each group and the total annual depreciation of all tooling, equipment, and machinery, using the straight-line formula.

Annual depreciation = (cost − final value) ÷ number of years of usefulness

	Group	Cost	Final Value	Number of Years of Usefulness	Annual Depreciation
a.	Tooling	$14,500	$1,200	5 years	
b.	Equipment	$28,350	$3,750	6 years	
c.	Equipment	$17,900	$2,040	10 years	
d.	Machinery	$67,700	$7,940	8 years	
e.	Machinery	$80,300	$10,600	10 years	
f.	Total Annual Depreciation _____				

Figure 1–22

91. A landscaper contracts to provide topsoil and to seed and lime the parcel of land shown in Figure 1–23. In order to determine labor and material costs, the landscaper must first know the total area of the land. Find the total area in square feet.

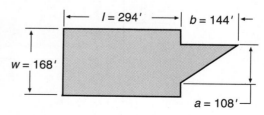

Figure 1–23

Total area (square feet) = $l \times w + (a \times b) \div 2$

92. The formula called Young's Rule is used in the health field to determine a child's dose of medicine.

Child's dose = (age of child) ÷ (age of child + 12) × average adult dose

What dose (number of milligrams) of morphine sulfate should be given to a 3-year-old child if the adult dose is 10 milligrams?

1–13   Computing with a Calculator: Whole Numbers

Basic Arithmetic Functions

The digit keys are used to enter any number into the display in a left-to-right order.

The operations of addition, subtraction, multiplication, and division are performed with the four arithmetic keys and the equals key. The equals key completes all operations entered and readies the calculator for additional calculations. Certain makes and models of calculators have the execute key |EXE| or the enter key |ENTER| instead of the equal key |=|. If your calculator has the execute key, substitute |EXE| for |=| in the examples that follow. If your calculator has the enter key, substitute |ENTER| for |=|.

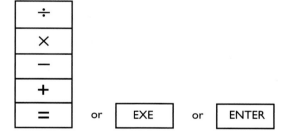

Examples of each of the four arithmetic operations of addition, subtraction, multiplication, and division are presented. Following the individual operation problems, combined operations expressions are given with calculator solutions. Make it a practice to estimate answers before doing calculator computations. Compare the estimate to the calculator answer. Also, an answer to a problem should be checked by doing the problem a second time to ensure that improper data was not entered. Remember to clear or erase previously recorded data and calculations before doing a problem. Depending on the make and model of the calculator, press |AC| or |CLEAR| *once* or |ON/C| *twice*.

Individual Arithmetic Operations

NOTE: Because each of the following examples is of a single type of arithmetic operation, combined operations are not involved. Therefore, it is not required that a calculator with algebraic logic be used to solve this set of problems.

EXAMPLES •——————————————————————————————

1. Add. 37 + 85

Solution. 37 |+| 85 |=| 122 *Ans*

2. Add. 95 + 17 + 102 + 44

Solution. 95 |+| 17 |+| 102 |+| 44 |=| 258 *Ans*

Some calculators have both a |−| and a |(−)| key. The |−| is used for subtraction. We will see later how to use the |(−)| key.

3. Subtract. 95 − 37

Solution. 95 |−| 37 |=| 58 *Ans*

4. Subtract. $126 - 84 - 15$

   Solution. $126 \boxed{-} 84 \boxed{-} 15 \boxed{=} 27$ *Ans*

5. Multiply. 216×13

   Solution. $216 \boxed{\times} 13 \boxed{=} 2808$ *Ans*

6. Multiply. $49 \times 7 \times 84 \times 12$

   Solution. $49 \boxed{\times} 7 \boxed{\times} 84 \boxed{\times} 12 \boxed{=} 345{,}744$ *Ans*

7. Divide. $378 \div 27$

   Solution. $378 \boxed{\div} 27 \boxed{=} 14$ *Ans*

Combined Arithmetic Operations

NOTE: Because the following problems are combined operations expressions, your calculator must have algebraic logic to solve the problems as shown. The expressions are solved by entering numbers and operations into the calculator in the same order as the expressions are written.

EXAMPLES •————————————————————————————

1. Evaluate. $28 + 16 \div 4$

   Solution. $28 \boxed{+} 16 \boxed{+} 4 \boxed{=} 32$ *Ans*

   Because the calculator has algebraic logic, the division operation $(16 \div 4)$ was performed before the addition operation (adding 28) was performed.

   NOTE: If the calculator does not have algebraic logic, the answer to the expression if solved in the order as shown is incorrect: $28 \boxed{+} 16 \boxed{\div} 4 \boxed{=} 11$ *INCORRECT ANSWER*. The calculator merely performed the operations in the order entered, without assigning priorities to various operations, and gave an incorrect answer. If your calculator does not have algebraic logic, then you will have to remember to put in the parentheses.

2. Evaluate. $11 \times 8 - 5 + 9 \times 2 + 7$

   Solution. $11 \boxed{\times} 8 \boxed{-} 5 \boxed{+} 9 \boxed{\times} 2 \boxed{+} 7 \boxed{=} 108$ *Ans*

3. Evaluate. $35 - \dfrac{21}{3} + 18 \times 12$

   Solution. $35 \boxed{-} 21 \boxed{\div} 3 \boxed{+} 18 \boxed{\times} 12 \boxed{=} 244$ *Ans*

4. Evaluate. $8 \times (20 - 7) - 6$

   As previously discussed in the section on order of operations, operations enclosed within parentheses are done first. A calculator with algebraic logic performs the operations within parentheses before performing other operations in a combined operations expression. If an expression contains parentheses, enter the expression in the calculator in the order in which it is written. The parentheses keys must be used.

   Solution. $8 \boxed{\times} \boxed{(} 20 \boxed{-} 7 \boxed{)} \boxed{-} 6 \boxed{=} 98$ *Ans*

   Parentheses keys

5. Evaluate. $46 + (5 + 7) \times (57 - 38)$

   Solution. $46 \boxed{+} \boxed{(} 5 \boxed{+} 7 \boxed{)} \boxed{\times} \boxed{(} 57 \boxed{-} 38 \boxed{)} \boxed{=} 274$ *Ans*

6. Evaluate. $\dfrac{14 + 10 \times 7}{4 + 2}$

   Recall that for a problem expressed in fractional form, the fraction bar is also used as a grouping symbol. The numerator and denominator are each considered as being enclosed in parentheses.

   $$\frac{14 + 10 \times 7}{4 + 2} = (14 + 10 \times 7) \div (4 + 2)$$

   Solution. $\boxed{(} 14 \boxed{+} 10 \boxed{\times} 7 \boxed{)} \boxed{\div} \boxed{(} 4 \boxed{+} 2 \boxed{)} \boxed{=} 14$ *Ans*

The expression may also be evaluated by using the [=] key to simplify the numerator without having to enclose the entire numerator in parentheses. However, parentheses must be used to enclose the denominator.

Solution. 14 [+] 10 [×] 7 [=] [÷] [(] 4 [+] 2 [)] [=] 14 *Ans*

7. Evaluate. $\dfrac{157 - 21 \times 3}{5 - 18 \div 6} + 17$

$$\frac{157 - 21 \times 3}{5 - 18 \div 6} + 17 = (157 - 21 \times 3) \div (5 - 18 \div 6) + 17$$

Solution. [(] 157 [−] 21 [×] 3 [)] [÷] [(] 5 [−] 18 [÷] 6 [)] [+] 17 [=] 64 *Ans*

Using the [=] key to simplify the numerator:

Solution. 157 [−] 21 [×] 3 [=] [÷] [(] 5 [−] 18 [÷] 6 [)] [+] 17 [=] 64 *Ans*

8. Evaluate. $\dfrac{100 - (17 + 13)}{112 - 77}$

$$\frac{100 - (17 + 13)}{112 - 77} = (100 - (17 + 13)) \div (112 - 77)$$

Observe these parentheses.

To be sure that the complete numerator is evaluated before dividing by the denominator, the complete numerator must be enclosed within parentheses. This is an example of an expression containing parentheses within parentheses.

Solution. [(] 100 [−] [(] 17 [+] 13 [)] [)] [÷] [(] 112 [−] 77 [)] [=] 2 *Ans*

Using the [=] key to simplify the numerator:

Solution. 100 [−] [(] 17 [+] 13 [)] [=] [÷] [(] 112 [−] 77 [)] [=] 2 *Ans*

9. Evaluate. $\dfrac{280 \div (32 - 27)}{(47 - 26) \div 3} \times 16$

$$\frac{280 \div (32 - 27)}{(47 - 26) \div 3} \times 16 = (280 \div (32 - 27)) \div ((47 - 26) \div 3) \times 16$$

Solution. [(] 280 [÷] [(] 32 [−] 27 [)] [)] [÷] [(] [(] 47 [−] 26 [)] [÷] 3 [)] [×] 16 [=] 128 *Ans*

Complete numerator   Complete denominator
enclosed in parentheses   enclosed in parentheses

Using the [=] key to simplify the numerator:

Solution. 280 [÷] [(] 32 [−] 27 [)] [=] [÷] [(] [(] 47 [−] 26 [)] [÷] 3 [)] [×] 16 [=] 128 *Ans*

Practice Exercise

Evaluate the following expressions. The expressions begin with basic single arithmetic operations and progress to combined operations, including problems requiring grouping with parentheses. Remember to check your answers by estimating the answer and doing each problem twice. The solutions to the problems directly follow the Practice Exercise. Compare your answers to the given solutions.

Evaluate the following expressions.

A. Individual Operations

  1. 58 + 109

  2. 73 + 18 + 315

  3. 22 + 7 + 219 + 55

  4. 314 − 249

5. 96×412

6. $112 \times 6 \times 8$

7. $33 \times 17 \times 5 \times 21$

8. $486 \div 27$

B. Combined Operations

1. $135 - 36 \div 9$

2. $9 \times 7 + 18 \times 3 - 16$

3. $100 - 44 + \dfrac{120}{6} \times 7$

4. $4 \times (16 + 23 - 9) + 12$

5. $183 + (27 - 14) \times (65 - 57)$

6. $\dfrac{348 - 18 \times 6}{67 + 13}$

7. $\dfrac{776 - 16 \times 5}{26 - 120 \div 6} - 83$

8. $\dfrac{432 \div (57 - 33)}{(52 + 38) \div 15} \times 49$

Solutions to Practice Exercise

A. Individual Operations

1. $58 \boxed{+} 109 \boxed{=} \ 167 \ Ans$

2. $73 \boxed{+} 18 \boxed{+} 315 \boxed{=} \ 406 \ Ans$

3. $22 \boxed{+} 7 \boxed{+} 219 \boxed{+} 55 \boxed{=} \ 303 \ Ans$

4. $314 \boxed{-} 249 \boxed{=} \ 65 \ Ans$

5. $96 \boxed{\times} 412 \boxed{=} \ 39 \ 552 \ Ans$

6. $112 \boxed{\times} 6 \boxed{\times} 18 \boxed{=} \ 12 \ 096 \ Ans$

7. $33 \boxed{\times} 17 \boxed{\times} 5 \boxed{\times} 21 \boxed{=} \ 58 \ 905 \ Ans$

8. $486 \boxed{\div} 27 \boxed{=} \ 18 \ Ans$

B. Combined Operations

1. $135 \boxed{-} 36 \boxed{\div} 9 \boxed{=} \ 131 \ Ans$

2. $9 \boxed{\times} 7 \boxed{+} 18 \boxed{\times} 3 \boxed{-} 16 \boxed{=} \ 101 \ Ans$

3. $100 \boxed{-} 44 \boxed{+} 120 \boxed{\div} 6 \boxed{\times} 7 \boxed{=} \ 196 \ Ans$

4. $4 \boxed{\times} \boxed{(} 16 \boxed{+} 23 \boxed{-} 9 \boxed{)} \boxed{+} 12 \boxed{=} \ 132 \ Ans$

5. $183 \boxed{+} \boxed{(} 27 \boxed{-} 14 \boxed{)} \boxed{\times} \boxed{(} 65 \boxed{-} 57 \boxed{)} \boxed{=} \ 287 \ Ans$

6. $\boxed{(} 348 \boxed{-} 18 \boxed{\times} 6 \boxed{)} \boxed{\div} \boxed{(} 67 \boxed{+} 13 \boxed{)} \boxed{=} \ 3 \ Ans$

    or

    $348 \boxed{-} 18 \boxed{\times} 6 \boxed{=} \boxed{\div} \boxed{(} 67 \boxed{+} 13 \boxed{)} \boxed{=} \ 3 \ Ans$

7. $\boxed{(} 776 \boxed{-} 16 \boxed{\times} 5 \boxed{)} \boxed{\div} \boxed{(} 26 \boxed{-} 120 \boxed{\div} 6 \boxed{)} \boxed{-} 83 \boxed{=} \ 33 \ Ans$

    or

    $776 \boxed{-} 16 \boxed{\times} 5 \boxed{=} \boxed{\div} \boxed{(} 26 \boxed{-} 120 \boxed{\div} 6 \boxed{)} \boxed{-} 83 \boxed{=} \ 33 \ Ans$

8. $\boxed{(} 432 \boxed{\div} \boxed{(} 57 \boxed{-} 33 \boxed{)} \boxed{)} \boxed{\div} \boxed{(} \boxed{(} 52 \boxed{+} 38 \boxed{)} \boxed{\div} 15 \boxed{\times} 49 \boxed{=} \ 147 \ Ans$

    or

    $432 \boxed{\div} \boxed{(} 57 \boxed{-} 33 \boxed{)} \boxed{=} \boxed{\div} \boxed{(} \boxed{(} 52 \boxed{+} 38 \boxed{)} \boxed{\div} 15 \boxed{)} \boxed{)} \boxed{\times} 49 \boxed{=}$
    $147 \ Ans$

UNIT 2 ⠿ Common Fractions

OBJECTIVES

After studying this unit you should be able to

- express fractions as equivalent fractions.

- express fractions in lowest terms.

- express mixed numbers as improper fractions.

- determine lowest common denominators.

- add, subtract, multiply, and divide fractions.

- add and subtract combinations of fractions, mixed numbers, and whole numbers.

- multiply and divide combinations of fractions, mixed numbers, and whole numbers.

- solve practical problems by combining addition, subtraction, multiplication, and division.

- simplify arithmetic expressions with combined operations by applying the proper order of operations.

- solve practical combined operations problems involving formulas by applying the proper order of operations.

Most measurements and calculations made on the job are not limited to whole numbers. Manufacturing and construction occupations require arithmetic operations using values from fractions of an inch to fractions of a mile. Food service employees prepare menus using fractions of ounces and pounds. Stock is ordered, costs are computed, and discounts are determined using fractions. Medical technicians and nurses deal with fractions when computing in the apothecaries' system. Fractional arithmetic operations are necessary in the agriculture and horticulture fields in computing liquid and dry measures.

2–1 Definitions

A *fraction* is a value that shows the number of equal parts taken of a whole quantity or unit. The symbol used to indicate a fraction is the slash (/) or the bar (———).

The *denominator* of the fraction is the number that shows how many equal parts are in the whole quantity. The *numerator* of the fraction is the number that shows how many equal parts of the whole are taken. The numerator and denominator are called the *terms* of the fraction.

$$\frac{5}{8} \begin{array}{l} \leftarrow \text{NUMERATOR} \\ \leftarrow \text{DENOMINATOR} \end{array}$$

A *proper fraction* is a number less than 1; the numerator is less than the denominator. Some examples of proper fractions written with a slash are ¾, ⅝, and ⁹⁹⁄₁₀₀. These same fractions written with a bar are $\frac{3}{4}$, $\frac{5}{8}$, and $\frac{99}{100}$. An *improper fraction* is a number greater than 1; the numerator is greater than the denominator. Some examples are $\frac{4}{3}$, $\frac{8}{5}$, and $\frac{100}{99}$.

33

Writing fractions with a slash can cause people to misread a number. For example, some people might think that 1¼ means $^{11}/_4 = \frac{11}{4}$ rather than $1\frac{1}{4}$. For this reason, the slash notation for fractions will not be used in this book.

2–2   Fractional Parts

A line segment shown in Figure 2–1 is divided into four equal parts.

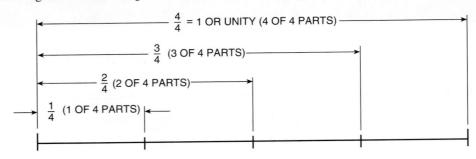

Figure 2–1

$$1 \text{ part} = \frac{1 \text{ part}}{4 \text{ parts}} = \frac{1}{4} \text{ of the length of the line segment}$$

$$2 \text{ parts} = \frac{2 \text{ parts}}{4 \text{ parts}} = \frac{2}{4} \text{ of the length of the line segment}$$

$$3 \text{ parts} = \frac{3 \text{ parts}}{4 \text{ parts}} = \frac{3}{4} \text{ of the length of the line segment}$$

$$4 \text{ parts} = \frac{4 \text{ parts}}{4 \text{ parts}} = \frac{4}{4} \text{ or } 1$$

NOTE: Four parts make up the whole $\left(\dfrac{4}{4} = 1\right)$.

EXERCISE 2–2

1. The total length of the line segment shown in Figure 2–2 is divided into equal parts. Write the fractional part of the total length that each length, A through F, represents.

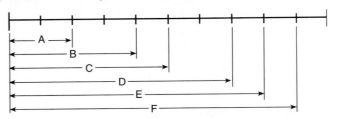

Figure 2–2

2. A riveted sheet metal plate is shown in Figure 2–3. Write the fractional part of the total number of rivets that each of the following represents.

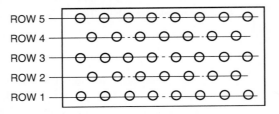

Figure 2–3

a. Row 1

b. Row 2

c. Row 2 plus row 3

d. The sum of rows 3, 4, and 5

2–3   A Fraction as an Indicated Division

A fraction indicates division.

$$\frac{3}{4} \text{ means 3 is divided by 4 or } 3 \div 4$$

When performing arithmetic operations, it is sometimes helpful to write a whole number as a fraction by placing the whole number over 1. To divide by 1 does not change the value of the number.

$$5 = \frac{5}{1}$$

2–4   Equivalent Fractions

Equivalent fractions and equivalent units of measure use the principle of multiplying by 1. Multiplication by 1 does not change the value. The 1 used is in the form of a fraction that has an equal numerator and denominator. The value of a fraction is not changed by multiplying both the numerator and denominator by the same number.

EXAMPLES

1. Express $\frac{3}{8}$ as thirty-seconds.

   Determine what number the original denominator is multiplied by to get the desired denominator. ($32 \div 8 = 4$)

   Multiply the numerator and denominator by 4.

   $$\frac{3}{8} = \frac{?}{32}$$

   $$\frac{3}{8} \times \frac{4}{4} = \frac{12}{32} \, Ans$$

2. $\dfrac{3}{4} = \dfrac{?}{64}$

   $64 \div 4 = 16$

   $\dfrac{3}{4} \times \dfrac{16}{16} = \dfrac{48}{64} \, Ans$

3. $\dfrac{2}{5} = \dfrac{?}{45}$

   $45 \div 5 = 9$

   $\dfrac{2}{5} \times \dfrac{9}{9} = \dfrac{18}{45} \, Ans$

EXERCISE 2–4

Express each fraction as an equivalent fraction as indicated.

1. $\dfrac{1}{2} = \dfrac{?}{16}$

2. $\dfrac{5}{8} = \dfrac{?}{16}$

3. $\dfrac{3}{4} = \dfrac{?}{32}$

4. $\dfrac{9}{16} = \dfrac{?}{32}$

5. $\dfrac{7}{8} = \dfrac{?}{32}$

6. $\dfrac{3}{8} = \dfrac{?}{64}$

7. $\dfrac{11}{16} = \dfrac{?}{64}$

8. $\dfrac{2}{3} = \dfrac{?}{18}$

9. $\dfrac{1}{4} = \dfrac{?}{20}$

10. $\dfrac{9}{10} = \dfrac{?}{60}$

11. $\dfrac{19}{12} = \dfrac{?}{72}$

12. $\dfrac{17}{18} = \dfrac{?}{270}$

2–5   Expressing Fractions in Lowest Terms

Multiplication and division are inverse operations. The numerator and denominator of a fraction can be divided by the same number without changing the value. A fraction is in its lowest terms when the numerator and denominator do not contain a common factor.

Arithmetic computations are usually simplified by using fractions in their lowest terms. Also, it is customary in occupations to write and speak of fractions in their lowest terms; it is part of the language of occupations. For example, a carpenter calls $\frac{6}{12}$ foot, $\frac{1}{2}$ foot; a machinist calls $\frac{12}{16}$ inch, $\frac{3}{4}$ inch; a chef calls $\frac{4}{6}$ cup, $\frac{2}{3}$ cup.

EXAMPLES •————————————————————————————————————

1. Express $\dfrac{12}{32}$ in lowest terms.

Determine a common factor in the numerator and denominator. The numerator and the denominator can be evenly divided by 2.

$$\frac{12 \div 2}{32 \div 2} = \frac{6}{16}$$

If the fraction is not in lowest terms, find another common factor in the numerator and the denominator. Continue until the numerator and denominator have no common factor.

$$\frac{6 \div 2}{16 \div 2} = \frac{3}{8} \; Ans$$

2. Express $\dfrac{16}{64}$ in lowest terms.   **3.** Express $\dfrac{18}{24}$ in lowest terms.

$$\frac{16 \div 16}{64 \div 16} = \frac{1}{4} \; Ans \qquad\qquad \frac{18 \div 6}{24 \div 6} = \frac{3}{4} \; Ans$$

EXERCISE 2–5

Express each fraction in lowest terms.

1. $\dfrac{2}{8}$ **5.** $\dfrac{90}{80}$ **9.** $\dfrac{48}{64}$

2. $\dfrac{9}{12}$ **6.** $\dfrac{15}{75}$ **10.** $\dfrac{36}{45}$

3. $\dfrac{6}{16}$ **7.** $\dfrac{7}{28}$ **11.** $\dfrac{81}{36}$

4. $\dfrac{28}{32}$ **8.** $\dfrac{9}{2}$ **12.** $\dfrac{128}{24}$

2–6   Expressing Mixed Numbers as Improper Fractions

A *mixed number* is a whole number plus a proper fraction. In problem solving, it is often necessary to express a mixed number as an improper fraction. To express the mixed number as an improper fraction, find the number of fractional parts contained in the whole number, then add the proper fractional part.

EXAMPLES •————————————————————————————————————

1. Express $5\dfrac{1}{4}$ as an improper fraction.

Find the number of fractional parts contained in the whole number.

$$\frac{5}{1} \times \frac{4}{4} = \frac{20}{4}$$

Add this fraction to the fractional part of the mixed number. Add the numerators, 20 + 1, and write their sum over the denominator, 4.

$$\frac{20}{4} + \frac{1}{4} = \frac{21}{4} \; Ans$$

2. Express $4\dfrac{5}{8}$ as an improper fraction.   **3.** Express $3\dfrac{3}{16}$ as an improper fraction.

$$\dfrac{4}{1} \times \dfrac{8}{8} = \dfrac{32}{8}$$ $$\dfrac{3}{1} \times \dfrac{16}{16} = \dfrac{48}{16}$$

$$\dfrac{32}{8} + \dfrac{5}{8} = \dfrac{37}{8} \; Ans$$ $$\dfrac{48}{16} + \dfrac{3}{16} = \dfrac{51}{16} \; Ans$$

To express a mixed number as an improper fraction with an indicated denominator, the mixed number is first expressed as an equivalent fraction. The equivalent fraction is then expressed as a fraction with the indicated denominator.

EXAMPLE

Express the mixed number as an equivalent improper fraction as indicated.

$$5\dfrac{1}{4} = \dfrac{?}{12}$$

Express $5\dfrac{1}{4}$ as an equivalent fraction. $$\dfrac{5}{1} \times \dfrac{4}{4} = \dfrac{20}{4}; \; \dfrac{20}{4} + \dfrac{1}{4} = \dfrac{21}{4}$$

Express the equivalent fraction as a fraction with the indicated denominator. $$\dfrac{21}{4} \times \dfrac{3}{3} = \dfrac{63}{12} \; Ans$$

2–7 Expressing Improper Fractions as Mixed Numbers

In problem solving, an answer may be obtained that should be expressed as a mixed number. For example, a drafter obtains an answer of $\frac{63}{32}$ inches. In order to make the measurement, $\frac{63}{32}$ inches should be expressed as the mixed number $1\frac{31}{32}$ inches.

EXAMPLES

1. Express $\dfrac{60}{32}$ as a mixed number.

To find how many whole units are contained, divide. Place the remainder over the denominator. $$\dfrac{60}{32} = 1\dfrac{28}{32}$$

Express the fractional part in lowest terms. $$1\dfrac{28}{32} = 1\dfrac{7}{8} \; Ans$$

2. Express $\dfrac{30}{8}$ as a mixed number.   **3.** Express $\dfrac{74}{10}$ as a mixed number.

$$\dfrac{30}{8} = 3\dfrac{6}{8}$$ $$\dfrac{74}{10} = 7\dfrac{4}{10}$$

$$3\dfrac{6}{8} = 3\dfrac{3}{4} \; Ans$$ $$7\dfrac{4}{10} = 7\dfrac{2}{5} \; Ans$$

EXERCISE 2–7

Express each mixed number as an improper fraction.

1. $2\dfrac{1}{2}$ **3.** $1\dfrac{5}{8}$ **5.** $1\dfrac{3}{16}$

2. $5\dfrac{3}{4}$ **4.** $9\dfrac{2}{5}$ **6.** $16\dfrac{3}{4}$

7. $4\dfrac{9}{32}$ **9.** $5\dfrac{31}{32}$ **11.** $43\dfrac{4}{5}$

8. $10\dfrac{3}{8}$ **10.** $59\dfrac{11}{16}$ **12.** $218\dfrac{7}{8}$

Express each improper fraction as a mixed number.

13. $\dfrac{14}{5}$ **17.** $\dfrac{79}{16}$ **21.** $\dfrac{318}{32}$

14. $\dfrac{10}{3}$ **18.** $\dfrac{96}{5}$ **22.** $\dfrac{217}{8}$

15. $\dfrac{63}{4}$ **19.** $\dfrac{87}{8}$ **23.** $\dfrac{451}{64}$

16. $\dfrac{47}{32}$ **20.** $\dfrac{133}{64}$ **24.** $\dfrac{412}{25}$

Express each mixed number as an equivalent improper fraction as indicated.

25. $1\dfrac{1}{2} = \dfrac{?}{8}$ **27.** $7\dfrac{5}{8} = \dfrac{?}{32}$ **29.** $81\dfrac{3}{8} = \dfrac{?}{32}$

26. $5\dfrac{3}{4} = \dfrac{?}{16}$ **28.** $9\dfrac{3}{4} = \dfrac{?}{64}$ **30.** $46\dfrac{9}{10} = \dfrac{?}{50}$

2–8   Division of Whole Numbers; Quotients as Mixed Numbers

In unit 1–9 the answer to a problem like $49 \div 6$ was written as 8 R1 because

$$\begin{array}{r} 8 \\ 6\overline{)49} \\ 48 \\ \hline 1 \end{array}$$

The answer to this type of problem is often written as a mixed number. Thus, $49 \div 6$ would be written as $8\dfrac{1}{6}$.

EXERCISE 2–8

Divide. Write the quotient as a mixed number.

1. $43 \div 7$ **5.** $258 \div 15$

2. $67 \div 9$ **6.** $543 \div 27$

3. $147 \div 11$ **7.** $1294 \div 24$

4. $196 \div 13$ **8.** $2465 \div 25$

2–9   Use of Common Fractions in Practical Applications

EXAMPLE

An automotive technician measured the tread depth of a tire as $\dfrac{6}{64}''$. What is the depth is lowest terms?

Solution

$$\dfrac{6 \div 2}{64 \div 2}'' = \dfrac{3}{32}'' \; Ans$$

EXERCISE 2–9

1. A patio is constructed of patio blocks all of which are the same size as shown in Figure 2–4. Four days are required to lay the patio. Sections 1, 2, 3, and 4 are laid on the first, second, third, and fourth days, respectively.

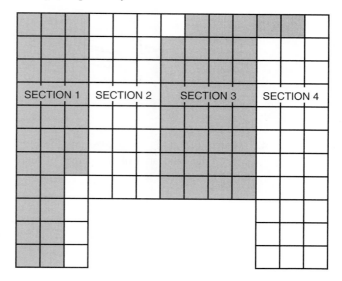

Figure 2–4

What fractional part of the finished patio does each of the following represent?

   a. The number of tiles laid the first day.

   b. The number of tiles laid the second day.

   c. The number of tiles laid the third day.

   d. The number of tiles laid the fourth day.

   e. The number of tiles laid the first and second days.

   f. The number of tiles laid the second, third, and fourth days.

2. The machined part shown in Figure 2–5 is redimensioned. Express dimensions A through N as mixed numbers or fractions in lowest terms. All dimensions are in inches.

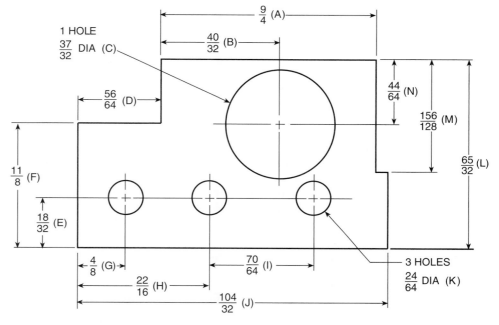

Figure 2–5

3. A welded support base shown in Figure 2–6 is cut in 4 pieces. What fractional part of the total length does each of the 4 pieces represent? All dimensions are in inches. Express answers in lowest terms.

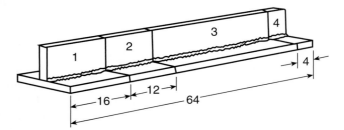

Figure 2–6

4. A parcel of land is subdivided into 5 building lots as shown in Figure 2–7. What fractional part of the total area of the parcel of land is represented by each of the 5 lots? Express answers in lowest terms.

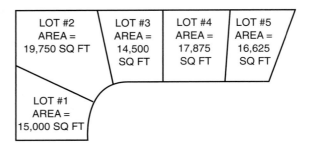

Figure 2–7

5. A retail television and appliance firm has a 3-day sale on television sets. The table shown in Figure 2–8 lists daily sales for each of 3 types of sets. What fractional part of the total 3-day sales does each day's sales represent? Express answers in lowest terms.

Type of Television Set	First Day Sales	Second Day Sales	Third Day Sales
Black and White Portable	$7,440	$8,680	$11,160
Color Portable	$17,360	$16,120	$18,600
Color Console	$13,640	$12,400	$14,880

Figure 2–8

6. An auto technician checked the tread depth of a tire. The depth was $\frac{12}{64}$ in. What is this depth in lowest terms?

2–10   Addition of Common Fractions

A welder determines material requirements for a certain job by adding steel plate lengths of $8\frac{1}{2}$ inches, $3\frac{1}{4}$ inches, and $10\frac{3}{16}$ inches. In finding the costs for a garment, a garment designer adds $\frac{3}{4}$ yard and $1\frac{7}{8}$ yards of woolen cloth. To straighten a particular automobile frame, an auto body specialist totals measurements of $42\frac{3}{16}$ inches, $2\frac{5}{8}$ inches, and $\frac{3}{32}$ inch. These computations require the ability to add fractions and mixed numbers.

Lowest Common Denominators

Fractions cannot be added unless they have a common denominator. Common denominator means that the denominators of each of the fractions are the same, as $\frac{3}{16}$, $\frac{7}{16}$, and $\frac{15}{16}$. In order to add fractions such as $\frac{1}{4}$ and $\frac{7}{8}$, it is necessary to determine a common denominator. The lowest common denominator is the smallest denominator that is evenly divisible by each of the denominators of the fractions being added. The lowest common denominator for the fractions $\frac{1}{4}$ and $\frac{7}{8}$ is 8, since 8 is the smallest number evenly divisible by both 4 and 8.

Procedure for Determining Lowest Common Denominators

When it is difficult to determine the lowest common denominator, a procedure using prime factors is used. A *factor* is a number being multiplied. A *prime number* is a whole number other than 0 and 1 that is divisible only by itself and 1. Some examples of prime numbers are 2, 3, 5, 7, 11, 13, 17, 19, and 23. A *prime factor* is a factor that is a prime number.

To determine the lowest common denominator, first factor each of the denominators into prime factors. List each prime factor as many times as it appears in any one denominator. Multiply all of the prime factors listed.

EXAMPLES

1. Find the lowest common denominator of $\frac{5}{6}$, $\frac{1}{5}$, and $\frac{3}{16}$.

Factor each of the denominators into prime factors.

The prime factors 3 and 5 are each used as factors only once. List these factors. The prime factor 2 is used once for denominator 6 and 4 times for denominator 16. List 2 as a factor 4 times. If the same prime factor is used in 2 or more denominators, it is listed only for the denominator for which it is used the greatest number of times. Multiply all prime factors listed to obtain the lowest common denominator.

$6 = 2 \times 3$
$5 = 5 \text{ (prime)}$
$16 = 2 \times 2 \times 2 \times 2$
$3 \times 5 \times 2 \times 2 \times 2 \times 2 = 240 \text{ } Ans$

2. Find the lowest common denominator of $\frac{9}{10}$, $\frac{3}{8}$, $\frac{4}{9}$, and $\frac{7}{15}$.

Factor each of the denominators into prime factors.

Multiply all prime factors listed to obtain the lowest common denominator.

$10 = 2 \times 5$
$8 = 2 \times 2 \times 2$
$9 = 3 \times 3$
$15 = 3 \times 5$
$2 \times 2 \times 2 \times 3 \times 3 \times 5 = 360 \text{ } Ans$

Comparing Values of Fractions

To compare the values of fractions with like denominators, compare the numerators. For fractions with like denominators, the larger the numerator, the larger the value of the fraction. For example, $\frac{9}{16}$ is greater than $\frac{7}{16}$ since $\frac{9}{16}$ contains 9 of 16 parts and $\frac{7}{16}$ contains only 7 of 16 parts.

To compare the values of fractions with unlike denominators, express the fractions as equivalent fractions with a common denominator and compare numerators.

EXAMPLE

List the following fractions in ascending order (increasing values with smallest value first, greatest value last).

$$\frac{3''}{4}, \frac{19''}{32}, \frac{7''}{8}, \frac{41''}{64}, \frac{13''}{16}$$

Express each fraction as an equivalent fraction with a denominator of 64.

$$\frac{3''}{4} = \frac{48''}{64}$$

$$\frac{19''}{32} = \frac{38''}{64}$$

$$\frac{7''}{8} = \frac{56''}{64}$$

$$\frac{41''}{64} = \frac{41''}{64}$$

$$\frac{13''}{16} = \frac{52''}{64}$$

Compare numerators and list in ascending order.

$$\frac{19''}{32}, \frac{41''}{64}, \frac{3''}{4}, \frac{13''}{16}, \frac{7''}{8} \; Ans$$

Observe the locations of these fractions on the enlarged inch scale shown in Figure 2–9.

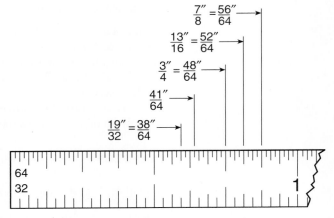

Figure 2–9

EXERCISE 2–10A

Determine the lowest common denominator for each of the following sets of fractions.

1. $\dfrac{1}{4}, \dfrac{1}{2}, \dfrac{3}{4}$

2. $\dfrac{3}{8}, \dfrac{1}{4}, \dfrac{3}{16}$

3. $\dfrac{1}{2}, \dfrac{4}{5}, \dfrac{7}{10}$

4. $\dfrac{1}{3}, \dfrac{14}{15}, \dfrac{2}{5}$

5. $\dfrac{2}{3}, \dfrac{3}{4}, \dfrac{1}{6}$

6. $\dfrac{5}{8}, \dfrac{2}{3}, \dfrac{7}{12}$

7. $\dfrac{1}{2}, \dfrac{7}{12}, \dfrac{1}{4}, \dfrac{7}{16}$

8. $\dfrac{3}{10}, \dfrac{1}{4}, \dfrac{4}{5}, \dfrac{5}{8}$

9. $\dfrac{3}{8}, \dfrac{1}{4}, \dfrac{5}{12}, \dfrac{1}{6}$

Using prime factors, determine the lowest common denominator for each of the following sets of fractions.

10. $\dfrac{7}{10}, \dfrac{1}{6}, \dfrac{8}{9}$

11. $\dfrac{5}{8}, \dfrac{9}{10}, \dfrac{7}{12}$

12. $\dfrac{2}{7}, \dfrac{1}{8}, \dfrac{1}{6}$

13. $\dfrac{7}{10}, \dfrac{2}{3}, \dfrac{3}{8}$

14. $\dfrac{13}{14}, \dfrac{3}{7}, \dfrac{5}{8}$

15. $\dfrac{9}{10}, \dfrac{1}{2}, \dfrac{5}{7}, \dfrac{2}{3}$

16. $\dfrac{4}{25}, \dfrac{7}{10}, \dfrac{1}{4}, \dfrac{5}{6}$

17. $\dfrac{3}{14}, \dfrac{7}{8}, \dfrac{10}{21}, \dfrac{3}{4}$

18. $\dfrac{8}{9}, \dfrac{1}{3}, \dfrac{7}{12}, \dfrac{3}{10}$

Arrange each set of fractions in ascending order (increasing values with smallest value first, greatest value last).

19. $\dfrac{1}{2}, \dfrac{3}{8}, \dfrac{9}{16}$

20. $\dfrac{7}{64}, \dfrac{9}{32}, \dfrac{11}{16}$

21. $\dfrac{7}{8}, \dfrac{5}{12}, \dfrac{3}{4}$

22. $\dfrac{7}{10}, \dfrac{2}{3}, \dfrac{4}{15}, \dfrac{1}{4}$

23. $\dfrac{7}{8}, \dfrac{5}{16}, \dfrac{1}{2}, \dfrac{9}{12}$

24. $\dfrac{4}{90}, \dfrac{19}{50}, \dfrac{43}{45}, \dfrac{3}{10}$

Adding Fractions

To add fractions, express the fractions as equivalent fractions having the lowest common denominator. Add the numerators and write their sum over the lowest common denominator. Express the fraction in lowest terms.

EXAMPLE

Add. $\dfrac{5}{6} + \dfrac{1}{3} + \dfrac{7}{10} + \dfrac{4}{15}$

Express the fractions as equivalent fractions with 30 as the denominator.	$\dfrac{5}{6} = \dfrac{25}{30}$
	$\dfrac{1}{3} = \dfrac{10}{30}$
Add the numerators and write their sum over the lowest common denominator, 30.	$\dfrac{7}{10} = \dfrac{21}{30}$
	$+\dfrac{4}{15} = \dfrac{8}{30}$
Express the fraction in lowest terms.	$\dfrac{64}{30} = 2\dfrac{2}{15}$ Ans

EXERCISE 2–10B

Add the following fractions. Express all answers in lowest terms.

1. $\dfrac{1}{8} + \dfrac{3}{8}$

2. $\dfrac{7}{32} + \dfrac{25}{32}$

3. $\dfrac{3}{64} + \dfrac{23}{64} + \dfrac{59}{64}$

4. $\dfrac{7}{8} + \dfrac{1}{2} + \dfrac{1}{4}$

5. $\dfrac{1}{2} + \dfrac{7}{12} + \dfrac{2}{3}$

6. $\dfrac{7}{15} + \dfrac{9}{10} + \dfrac{2}{3}$

7. $\dfrac{3}{4} + \dfrac{7}{8} + \dfrac{5}{16}$

8. $\dfrac{3}{16} + \dfrac{1}{8} + \dfrac{1}{4}$

9. $\dfrac{9}{32} + \dfrac{5}{16} + \dfrac{3}{8}$

10. $\dfrac{3}{5} + \dfrac{5}{7} + \dfrac{1}{35} + \dfrac{3}{7}$

11. $\dfrac{14}{15} + \dfrac{1}{3} + \dfrac{2}{3} + \dfrac{9}{10}$

12. $\dfrac{17}{20} + \dfrac{1}{5} + \dfrac{1}{4} + \dfrac{5}{12}$

13. $\dfrac{7}{12} + \dfrac{15}{24} + \dfrac{1}{18} + \dfrac{1}{4}$

14. $\dfrac{6}{7} + \dfrac{13}{14} + \dfrac{2}{3} + \dfrac{1}{2}$

Adding Fractions, Mixed Numbers, and Whole Numbers

To add fractions, mixed numbers, and whole numbers, express the fractional parts of the number using a common denominator. Add the whole numbers. Add the fractions. Combine the whole number and the fraction and express in lowest terms.

EXAMPLES

1. Add. $\dfrac{2}{3} + 2\dfrac{13}{24} + \dfrac{7}{12} + 15$

Express the fractional parts as equivalent fractions with 24 as the denominator.

$$\dfrac{2}{3} = \dfrac{16}{24}$$

$$2\dfrac{13}{24} = 2\dfrac{13}{24}$$

$$\dfrac{7}{12} = \dfrac{14}{24}$$

$$+\,15 = +\,15$$

Add the whole numbers.

$$= \dfrac{16}{24}$$

$$= 2\dfrac{13}{24}$$

$$= \dfrac{14}{24}$$

$$= +\,15$$

$$17$$

Add the fractions.

$$= \dfrac{16}{24}$$

$$= 2\dfrac{13}{24}$$

$$= \dfrac{14}{24}$$

$$= +\,15$$

$$17\dfrac{43}{24}$$

Combine the whole number and the fraction.
Express the answer in lowest terms.

$$18\dfrac{19}{24}\ Ans$$

2. Add $3\dfrac{3}{16} + 9 + 14\dfrac{27}{64}$ and express the answer in lowest terms.

$$3\dfrac{3}{16} = 3\dfrac{12}{64}$$

$$9 = 9$$

$$+\,14\dfrac{27}{64} = 14\dfrac{27}{64}$$

$$26\dfrac{39}{64}\ Ans$$

EXERCISE 2–10C

Add the following values. Where necessary, express answers in lowest terms.

1. $2 + \dfrac{3}{4}$

2. $5\dfrac{1}{4} + \dfrac{1}{4}$

3. $7\dfrac{3}{4} + \dfrac{5}{8}$

4. $3\dfrac{1}{16} + 15\dfrac{3}{8}$

5. $17\dfrac{1}{10} + 14\dfrac{3}{5}$

6. $8\dfrac{1}{8} + 9\dfrac{5}{32}$

7. $14\frac{15}{64} + 9\frac{9}{16} + 13$

8. $9\frac{3}{8} + 1\frac{1}{2} + 15 + \frac{3}{16}$

9. $\frac{4}{15} + 3\frac{1}{3} + 7\frac{2}{3} + 18\frac{9}{10}$

10. $13\frac{1}{4} + \frac{7}{8} + 12 + 15\frac{5}{12}$

11. $39 + 1\frac{7}{12} + \frac{2}{3} + 21\frac{1}{2}$

12. $14\frac{4}{5} + 107 + 5\frac{19}{35} + \frac{2}{7}$

13. $7\frac{17}{24} + 17\frac{5}{12} + 3\frac{5}{8} + 55\frac{1}{2}$

14. $18\frac{9}{10} + \frac{19}{25} + 72 + 14\frac{1}{5}$

2–11 Subtraction of Common Fractions

While making a part from a blueprint, a machinist often finds it necessary to express blueprint dimensions as working dimensions. Subtraction of fractions and mixed numbers is used to properly position a part on a machine, to establish hole locations, and to determine depths of cuts. Subtraction of fractions and mixed numbers is used in most occupations in determining material requirements, costs, and stock sizes.

Subtracting Fractions from Fractions

As in addition, fractions must have a common denominator in order to be subtracted. To subtract a fraction from a fraction, express the fractions as equivalent fractions with a common denominator. Subtract the numerators. Write their difference over the common denominator.

EXAMPLES

1. Subtract $\frac{1}{2}$ from $\frac{5}{8}$.

   Express the fractions as equivalent fractions with 8 as the denominator.

   Subtract the numerators.

   Write their difference 1 over the common denominator 8.

   $$\frac{5}{8} = \frac{5}{8}$$
   $$-\frac{1}{2} = \frac{4}{8}$$
   $$\frac{1}{8} \; Ans$$

2. Subtract $\frac{13}{15} - \frac{9}{20}$ and express the answer in lowest terms.

   $$\frac{13}{15} = \frac{52}{60}$$
   $$-\frac{9}{20} = \frac{27}{60}$$
   $$\frac{25}{60} = \frac{5}{12} \; Ans$$

EXERCISE 2–11A

Subtract the following fractions as indicated. Express the answers in lowest terms.

1. $\frac{4}{5} - \frac{1}{5}$

2. $\frac{7}{8} - \frac{3}{8}$

3. $\frac{97}{100} - \frac{89}{100}$

4. $\frac{15}{16} - \frac{7}{16}$

5. $\frac{19}{32} - \frac{1}{4}$

6. $\frac{17}{20} - \frac{3}{5}$

7. $\frac{4}{5} - \frac{13}{20}$

8. $\frac{15}{16} - \frac{1}{3}$

9. $\frac{7}{8} - \frac{3}{4}$

10. $\dfrac{3}{8} - \dfrac{1}{6}$ **12.** $\dfrac{5}{6} - \dfrac{3}{8}$ **14.** $\dfrac{5}{8} - \dfrac{3}{10}$

11. $\dfrac{3}{4} - \dfrac{7}{10}$ **13.** $\dfrac{15}{16} - \dfrac{31}{64}$ **15.** $\dfrac{19}{16} - \dfrac{5}{12}$

Subtracting Fractions and Mixed Numbers from Whole Numbers

To subtract a fraction or a mixed number from a whole number, express the whole number as an equivalent mixed number. The fraction of the mixed number has the same denominator as the denominator of the fraction that is subtracted. Subtract the numerators and whole numbers. Combine the whole number and fraction. Express the answer in lowest terms.

EXAMPLE •——————————————————————————

Subtract $\dfrac{13}{16}$ from 8.

Express the whole number as an equivalent mixed number.

$$8 = 7\dfrac{16}{16}$$

Subtract.

$$-\dfrac{13}{16} = -\dfrac{13}{16}$$

$$7\dfrac{3}{16} \ Ans$$

EXERCISE 2–11B

Subtract the following values as indicated. Express the answers in lowest terms.

1. $7 - \dfrac{1}{2}$ **5.** $6 - 3\dfrac{7}{8}$ **9.** $312 - 310\dfrac{11}{32}$

2. $8 - 2\dfrac{15}{32}$ **6.** $47 - 46\dfrac{7}{8}$ **10.** $119 - 107\dfrac{13}{32}$

3. $18 - \dfrac{7}{16}$ **7.** $75 - 68\dfrac{3}{5}$ **11.** $126 - 125\dfrac{61}{64}$

4. $23 - \dfrac{3}{4}$ **8.** $257 - \dfrac{29}{64}$ **12.** $138 - 2\dfrac{15}{16}$

Subtracting Fractions and Mixed Numbers from Mixed Numbers

To subtract a fraction or a mixed number from a mixed number, the fractional part of each number must have the same denominator. Express fractions as equivalent fractions having a common denominator. When the fraction subtracted is larger than the fraction from which it is subtracted, one unit of the whole number is expressed as a fraction with the common denominator. Combine the whole number and fractions. Subtract. Express the answer in lowest terms.

EXAMPLES •——————————————————————————

1. Subtract $\dfrac{5}{8}$ from $8\dfrac{3}{16}$.

Express the fractions as equivalent fractions with a common denominator of 16.

Since 10 is larger than 3, express one unit of the mixed number as a fraction. Combine the whole number and fractions. Subtract.

$$8\frac{3}{16} = 8\frac{3}{16} = \left(7 + \frac{16}{16} + \frac{3}{16}\right) = 7\frac{19}{16}$$
$$-\frac{5}{8} = \frac{10}{16} = \qquad\qquad\qquad \frac{10}{16}$$
$$7\frac{9}{16}\ Ans$$

2. Subtract $33\frac{7}{10} - 17\frac{3}{4}$ and express the answer in lowest terms.

$$33\frac{7}{10} = 33\frac{14}{20} = \left(32 + \frac{20}{20} + \frac{14}{20}\right) = 32\frac{34}{20}$$
$$-17\frac{3}{4} = 17\frac{15}{20} = \qquad\qquad\qquad 17\frac{15}{20}$$
$$15\frac{19}{20}\ Ans$$

EXERCISE 2–11C

Subtract the following values as indicated. Express the answers in lowest terms.

1. $7\frac{3}{8} - \frac{1}{8}$

2. $12\frac{15}{16} - \frac{27}{32}$

3. $7\frac{2}{3} - \frac{7}{8}$

4. $12\frac{3}{10} - \frac{3}{4}$

5. $2\frac{1}{3} - 1\frac{2}{5}$

6. $7\frac{1}{4} - 4\frac{5}{6}$

7. $9\frac{3}{5} - 4\frac{1}{5}$

8. $6\frac{5}{8} - 1\frac{1}{2}$

9. $10\frac{13}{16} - \frac{3}{16}$

10. $15\frac{31}{32} - \frac{3}{4}$

11. $23\frac{1}{2} - 21\frac{1}{4}$

12. $39\frac{1}{32} - \frac{9}{16}$

13. $63\frac{7}{10} - 37\frac{29}{50}$

14. $21\frac{13}{16} - 20\frac{13}{16}$

15. $79\frac{6}{7} - 8\frac{7}{8}$

16. $299\frac{3}{32} - 298\frac{3}{4}$

Combining Addition and Subtraction of Fractions and Mixed Numbers

Often on-the-job computations require the combination of two or more different arithmetic operations using fractions and mixed numbers. When solving a problem that requires both addition and subtraction operations, follow the procedures learned for each operation.

EXERCISE 2–11D

Refer to the chart shown in Figure 2–10. For each of the problems 1–3, find how much greater value A is than value B.

	A	B
1.	$3\frac{1}{2}$ inches + $\frac{3}{4}$ inch	$1\frac{5}{16}$ inches + $1\frac{1}{8}$ inches
2.	$5\frac{3}{10}$ miles + $\frac{1}{3}$ mile	$\frac{5}{6}$ mile + $\frac{4}{5}$ mile
3.	500 gallons + $60\frac{1}{4}$ gallons	$455\frac{1}{8}$ gallons + $\frac{1}{3}$ gallon

Figure 2–10

2–12   Adding and Subtracting Common Fractions in Practical Applications

EXAMPLE

A sheet metal worker scribes (marks) hole locations on an aluminum sheet shown in Figure 2–11. The locations are scribed in the order shown: locations 1, 2, 3, and 4. Find, in inches, the distance from location 3 to location 4. All dimensions are in inches.

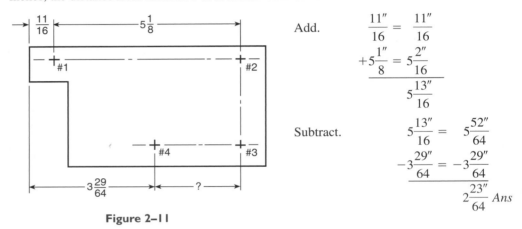

Add.

$$\frac{11''}{16} = \frac{11''}{16}$$
$$+5\frac{1''}{8} = 5\frac{2''}{16}$$
$$\overline{\qquad 5\frac{13''}{16}}$$

Subtract.

$$5\frac{13''}{16} = 5\frac{52''}{64}$$
$$-3\frac{29''}{64} = -3\frac{29''}{64}$$
$$\overline{\qquad\quad 2\frac{23''}{64}\ Ans}$$

Figure 2–11

EXERCISE 2–12

Solve the following problems.

1. An order for business forms that require 7 ruled columns is received by a printing shop. A printer lays out columns of the following widths: $\frac{3}{4}$ inches, $1\frac{5}{16}$ inches, $3\frac{1}{8}$ inches, $2\frac{17}{32}$ inches, $1\frac{1}{2}$ inches, $1\frac{5}{16}$ inches, and $1\frac{7}{8}$ inches. What width sheets are required for this job?

2. To find material requirements for a production run of oak chairs, an estimator finds the length of stock required for the chair leg shown in Figure 2–12. Find, in inches, the length.

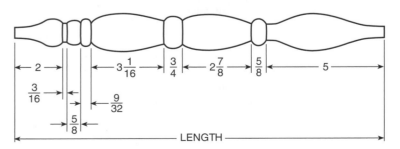

Figure 2–12

3. Determine, in inches, dimensions A, B, C, D, E, F, and G of the steel base plate shown in Figure 2–13.

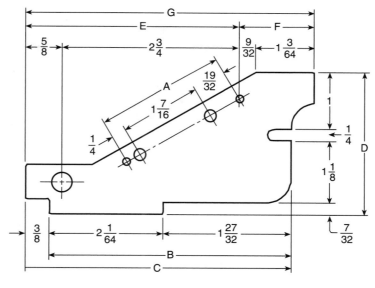

Figure 2–13

4. In finishing the interior trim of a building, a carpenter measures and saws a length of molding in 3 pieces as shown in Figure 2–14. Find, in inches, the length of the third piece.

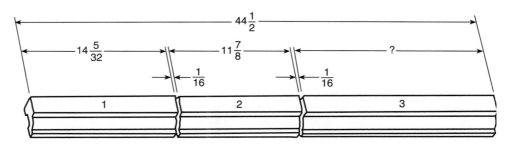

Figure 2–14

NOTE: $\frac{1}{16}$ inch is allowed for each saw cut.

5. A baker mixes a batch of dough that weighs 190 pounds. The dough consists of $115\frac{1}{4}$ pounds of flour, 2 pounds of salt, $3\frac{3}{4}$ pounds of sugar, $1\frac{1}{3}$ pounds of malt, $3\frac{1}{3}$ pounds of shortening, and water. How many pounds of water are contained in the batch?

6. A bolt of woolen cloth is shrunk before a garment maker uses it. Before shrinking, the bolt measures $28\frac{3}{8}$ yards long and $\frac{3}{4}$ yard wide. The material shrinks $\frac{2}{3}$ yard in length and $\frac{1}{36}$ yard in width.

   a. Find the length of the material after shrinking.

   b. Find the width of the material after shrinking.

7. Find, in inches, dimensions A, B, C, D, E, F, G, H, and I of the mounting plate shown in Figure 2–15 on page 50.

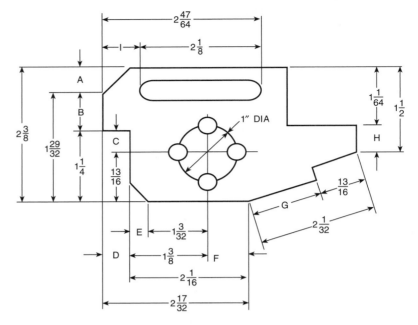

Figure 2–15

8. A plumber's piping plan shown in Figure 2–16 consists of 6 copper pipes and 7 fittings. Both ends of each pipe are threaded $\frac{1}{2}$ inch into the fittings. Find, in inches, the total length of the 6 pipes needed for this plan.

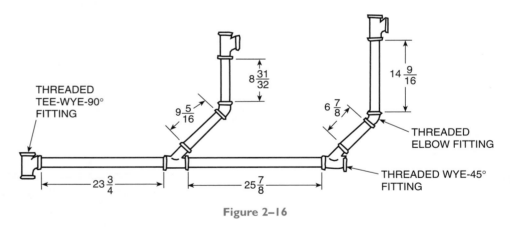

Figure 2–16

9. A sheet metal technician shears 5 pieces from a 54-inch length of sheet steel. The lengths of the sheared pieces are $5\frac{3}{4}$ inches, $11\frac{1}{8}$ inches, $9\frac{5}{32}$ inches, $10\frac{7}{16}$ inches, and $4\frac{3}{8}$ inches. What is the length of sheet steel left after all pieces are sheared?

10. Three views of a machined part are drawn by a drafter as shown in Figure 2–17 on page 51. A $2\frac{1}{2}$-inch margin from each of the 4 edges of the sheet of paper is allowed.

    a. Find, in inches, the distance from the left edge of the sheet to the right edge of the right side view (distance X).

    b. Find, in inches, the distance from the bottom of the sheet to the top edge of the top view (distance Y).

11. In estimating labor costs for a job, a bricklayer figures a total of 48 hours. The job takes longer to complete than estimated. The hours worked each day are as follows: $7\frac{3}{4}$, $7\frac{1}{6}$, 8, $9\frac{5}{6}$, $10\frac{1}{2}$, and $11\frac{2}{3}$. By how many hours is the job underestimated?

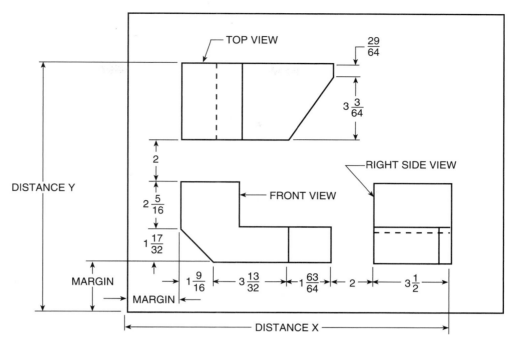

Figure 2-17

12. A tool and die maker bores 3 holes in a checking gauge. The left edge and bottom edge of the gauge are the reference edges from which the hole locations are measured, as shown in Figure 2–18. Sketch and dimension the hole locations from the reference edges according to the following directions.

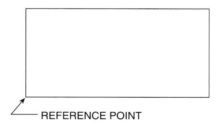

REFERENCE POINT

Figure 2-18

Hole #1 is $1\frac{3}{32}$ inches to the right, and $1\frac{5}{8}$ inches up.

Hole #2 is $2\frac{1}{64}$ inches to the right, and $2\frac{3}{16}$ inches up.

Hole #3 is $3\frac{1}{4}$ inches to the right, and $3\frac{1}{2}$ inches up.

a. Find, in inches, the horizontal distance between hole #1 and hole #2.

b. Find, in inches, the horizontal distance between hole #2 and hole #3.

c. Find, in inches, the horizontal distance between hole #1 and hole #3.

d. Find, in inches, the vertical distance between hole #1 and hole #2.

e. Find, in inches, the vertical distance between hole #2 and hole #3.

f. Find, in inches, the vertical distance between hole #1 and hole #3.

13. An offset duplicator operator prints forms and other material required by a company. In planning for 3 duplicating orders, the operator estimates the paper required. Four different types of bond paper are needed to print the 3 orders as shown in the table in Figure 2–19. Also shown is the amount of paper on hand.

a. How many total reams of each type of paper are required to complete the 3 orders?
b. How many additional reams of each type must be ordered?

	TYPE OF BOND PAPER			
	17 × 22 Substance 16	17 × 22 Substance 20	17 × 22 Substance 24	17 × 22 Substance 28
Paper required for Order 1	$3\frac{3}{4}$ reams	6 reams	$\frac{1}{2}$ ream	0
Paper required for Order 2	0	$8\frac{1}{3}$ reams	$3\frac{2}{3}$ reams	$5\frac{1}{4}$ reams
Paper required for Order 3	$7\frac{1}{8}$ reams	0	$\frac{7}{8}$ ream	$5\frac{3}{4}$ reams
Paper on hand	$6\frac{1}{2}$ reams	9 reams	$\frac{3}{4}$ ream	$10\frac{3}{8}$ reams

Figure 2–19

NOTE: Paper thickness is designated by the weight of 500 sheets (1 ream). For example, the 17 × 22 substance 16 bond paper listed in the table means that five hundred 17-inch-wide by 22-inch-long sheets weigh 16 pounds.

2–13   Multiplication of Common Fractions

A printer finds the width of a type page consisting of six $1\frac{7}{8}$-inch-wide columns. A dry goods clerk finds the purchase price of $5\frac{2}{3}$ yards of fabric at \$4 a yard. A welder determines the material needed for a job that requires 25 pieces of $\frac{13}{16}$-inch-long angle iron. Multiplication of fractions and mixed numbers is used for these computations.

Meaning of Multiplication of Fractions

Just as with whole numbers, multiplication of fractions is a short method of adding equal amounts. It is important to understand the meaning of multiplication of fractions. For example, $6 \times \frac{1}{4}$ means $\frac{1}{4}$ is added 6 times.

$$\frac{1}{4} + \frac{1}{4} + \frac{1}{4} + \frac{1}{4} + \frac{1}{4} + \frac{1}{4}$$

Adding the fractions gives the sum $\frac{6}{4}$ or $1\frac{1}{2}$:

$$6 \times \frac{1}{4} = 1\frac{1}{2}$$

The enlarged 1-inch scale in Figure 2–20 shows six $\frac{1}{4}$-inch parts.

$$6 \times \frac{1''}{4} = 1\frac{1''}{2}$$

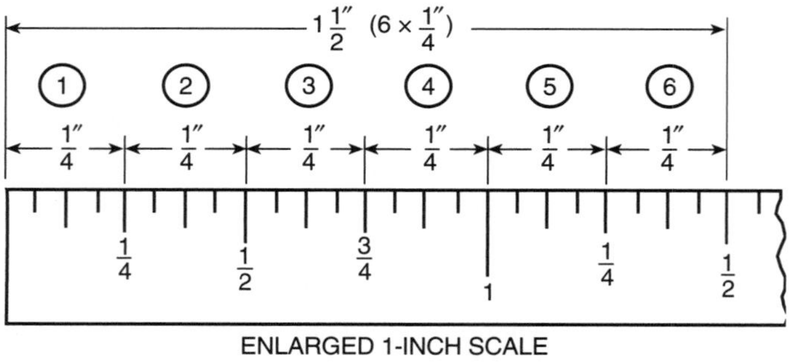

ENLARGED 1-INCH SCALE

Figure 2–20

Recall that multiplication of whole numbers can be done in any order. Multiplication of fractions can also be done in any order. The expression $6 \times \frac{1}{4}$ is the same as $\frac{1}{4} \times 6$. Six divided into 4 equal units is written as $\frac{1}{4}$ of 6 or $\frac{1}{4} \times 6$. The 6-inch scale in Figure 2–21 shows 6 inches divided into 4 equal parts and 1 of the 4 parts is taken.

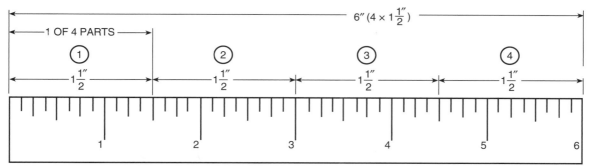

6-INCH SCALE

Figure 2–21

$$\frac{1}{4} \times 6'' = 1\frac{1''}{2}$$

2 *of the 4 parts equal* $1\frac{1''}{2} + 1\frac{1''}{2} = 3''$ *or* $\frac{2}{4} \times 6'' = 3''$

3 *of the 4 parts equal* $1\frac{1''}{2} + 1\frac{1''}{2} + 1\frac{1''}{2} = 4\frac{1''}{2}$ *or* $\frac{3}{4} \times 6'' = 4\frac{1''}{2}$

4 *of the 4 parts equal* $1\frac{1''}{2} + 1\frac{1''}{2} + 1\frac{1''}{2} + 1\frac{1''}{2} = 6''$ *or* $\frac{4}{4} \times 6'' = 6''$

The meaning of fractions multiplied by fractions and mixed numbers multiplied by fractions is the same as that which was described with a fraction multiplied by a whole number.

EXAMPLES •————————————————————————————

1. $\frac{3}{4} \times \frac{7}{8}$ means when $\frac{7}{8}$ is divided in 4 equal parts, 3 of the 4 parts are taken.

2. $\frac{15}{16} \times 2\frac{3}{32}$ means when $2\frac{3}{32}$ is divided in 16 equal parts, 15 of the 16 parts are taken.

————————————————————————————————————•

EXERCISE 2–13A ————————————————————————————————

For each of the following statements, insert the proper numerical values for a and b.

1. $\frac{2}{5} \times \frac{3}{4}$ means when $\frac{3}{4}$ is divided in *a* equal parts, *b* of the *a* parts are taken.

2. $\frac{17}{32} \times \frac{7}{8}$ means when $\frac{7}{8}$ is divided in *a* equal parts, *b* of the *a* parts are taken.

3. $\frac{5}{3} \times \frac{1}{2}$ means when $\frac{1}{2}$ is divided in *a* equal parts, *b* of the *a* parts are taken.

4. $\dfrac{9}{16} \times 3\dfrac{15}{32}$ means when $3\dfrac{15}{32}$ is divided in a equal parts, b of the a parts are taken.

5. $\dfrac{127}{64} \times 10\dfrac{9}{10}$ means when $10\dfrac{9}{10}$ is divided in a equal parts, b of a parts are taken.

Multiplying Fractions

To multiply two or more fractions, multiply the numerators. Multiply the denominators. Write as a fraction with the product of the numerators over the product of the denominators. Express the answer in lowest terms.

EXAMPLES •————————————————————————————

1. Multiply. $\dfrac{2}{3} \times \dfrac{4}{5}$

Multiply the numerators.

Multiply the denominators. $\qquad \dfrac{2}{3} \times \dfrac{4}{5} = \dfrac{8}{15}$ *Ans*

Write as a fraction.

2. Multiply. $\dfrac{1}{2} \times \dfrac{3}{4} \times \dfrac{5}{6}$

$$\dfrac{1}{2} \times \dfrac{3}{4} \times \dfrac{5}{6} = \dfrac{15}{48} = \dfrac{5}{16} \; Ans$$

3. A hole is to be drilled in a block shown in Figure 2–22 to a depth of $\frac{3}{4}$ of the thickness of the block. Find the depth of the hole.

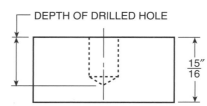

$$\dfrac{3}{4} \times \dfrac{15''}{16} = \dfrac{45''}{64} \; Ans$$

Figure 2–22

————————————————————————————————————•

EXERCISE 2–13B

Multiply the following fractions as indicated. Express the answers in lowest terms.

1. $\dfrac{1}{4} \times \dfrac{1}{2}$ **5.** $\dfrac{5}{8} \times \dfrac{5}{8}$ **9.** $\dfrac{3}{4} \times \dfrac{2}{3} \times \dfrac{9}{10}$

2. $\dfrac{1}{4} \times \dfrac{3}{4}$ **6.** $\dfrac{1}{6} \times \dfrac{2}{5}$ **10.** $\dfrac{5}{6} \times \dfrac{17}{20} \times \dfrac{1}{15}$

3. $\dfrac{1}{3} \times \dfrac{2}{3}$ **7.** $\dfrac{19}{20} \times \dfrac{3}{5}$ **11.** $\dfrac{11}{12} \times \dfrac{5}{6} \times \dfrac{3}{20}$

4. $\dfrac{4}{5} \times \dfrac{7}{8}$ **8.** $\dfrac{3}{8} \times \dfrac{15}{16}$ **12.** $\dfrac{8}{9} \times \dfrac{18}{21} \times \dfrac{3}{8} \times \dfrac{1}{9}$

Multiplying Any Combination of Fractions, Mixed Numbers, and Whole Numbers

To multiply any combination of fractions, mixed numbers, and whole numbers, write the mixed numbers as fractions. Write whole numbers over the denominator 1. Multiply numerators. Multiply denominators. Express the answer in lowest terms.

EXAMPLES •———

1. Multiply as indicated. $4 \times \dfrac{7}{8}$

Write the whole number, 4, over 1.

Multiply numerators.
Multiply denominators.

Express the answer in lowest terms.

$$\dfrac{4}{1} \times \dfrac{7}{8} = \dfrac{28}{8} = 3\dfrac{1}{2} \ Ans$$

2. Calculate. $6\dfrac{2}{3} \times 5\dfrac{1}{2}$

Write the mixed number $6\dfrac{2}{3}$ as the fraction $\dfrac{20}{3}$.

Write the mixed number $5\dfrac{1}{2}$ as the fraction $\dfrac{11}{2}$.

Multiply numerators.
Multiply denominators.

$$\dfrac{20}{3} \times \dfrac{11}{2} = \dfrac{20 \times 11}{3 \times 2} = \dfrac{220}{6}$$

Express as a mixed number in lowest terms

$$36\dfrac{2}{3} \ Ans$$

3. Multiply. $5\dfrac{1}{2} \times 3 \times \dfrac{7}{16}$

Write the mixed number, $5\dfrac{1}{2}$, as the fraction $\dfrac{11}{2}$.

Write the whole number, 3, over 1.

Multiply numerators.
Multiply denominators.

Express in lowest terms.

$$\dfrac{11}{2} \times \dfrac{3}{1} \times \dfrac{7}{16} = \dfrac{231}{32} = 7\dfrac{7}{32} \ Ans$$

Dividing by Common Factors (Cancellation)

Problems involving multiplication of fractions are generally solved more quickly and easily if a numerator and a denominator are divided by any common factors before the fractions are multiplied. This process is often called *cancellation.*

EXAMPLES •———

1. Multiply. $\dfrac{3}{4} \times \dfrac{8}{9}$

The factor 3 is common to both the numerator 3 and the denominator 9. Divide 3 and 9 by 3.

$$\dfrac{\overset{1}{\cancel{3}}}{4} \times \dfrac{8}{\underset{3}{\cancel{9}}}$$

The factor 4 is common to both the denominator 4 and the numerator 8. Divide 4 and 8 by 4.

$$\dfrac{1}{\cancel{4}} \times \dfrac{\overset{2}{\cancel{8}}}{3}$$

Multiply numerators.
Multiply denominators.

$$\dfrac{1 \times 2}{1 \times 3} = \dfrac{2}{3} \ Ans$$

2. Multiply. $2\frac{2}{5} \times 6\frac{7}{8}$

Express the mixed numbers as fractions.

Divide 5 and 55 by 5.
Divide 12 and 8 by 4.

Multiply numerators.
Multiply denominators.

Express the answer in lowest terms.

$$\overset{3}{\underset{1}{\cancel{12}}} \times \frac{\overset{11}{\cancel{55}}}{\underset{2}{\cancel{8}}} = \frac{33}{2} = 16\frac{1}{2} \; Ans$$

3. Multiply $\frac{11}{15} \times \frac{3}{4} \times \frac{9}{22} \times \frac{1}{2}$ and express the answer in lowest terms.

$$\frac{\overset{1}{\cancel{11}}}{\underset{5}{\cancel{15}}} \times \frac{3}{4} \times \frac{\overset{3}{\cancel{9}}}{\underset{2}{\cancel{22}}} \times \frac{1}{2} = \frac{9}{80} \; Ans$$

EXERCISE 2–13C

Multiply the following values as indicated. Express the answers in lowest terms.

1. $\frac{1}{2} \times 5$ **5.** $10 \times \frac{31}{32}$ **9.** $\frac{7}{12} \times \frac{8}{21}$ **13.** $6\frac{7}{8} \times 4\frac{1}{4}$

2. $\frac{3}{4} \times 12$ **6.** $\frac{7}{8} \times \frac{5}{4}$ **10.** $\frac{7}{8} \times 3\frac{1}{5}$ **14.** $83\frac{3}{4} \times 2\frac{2}{15}$

3. $\frac{7}{8} \times 11$ **7.** $\frac{2}{5} \times \frac{3}{7}$ **11.** $\frac{1}{2} \times 10\frac{3}{8}$ **15.** $\frac{3}{4} \times 12 \times 1\frac{5}{8}$

4. $15 \times \frac{3}{5}$ **8.** $\frac{3}{4} \times \frac{8}{9}$ **12.** $2\frac{2}{5} \times 1\frac{1}{3}$ **16.** $4\frac{4}{5} \times 25 \times 2\frac{1}{16}$

2–14   Multiplying Common Fractions in Practical Applications

When solving a problem that requires two or more different operations, think the problem through to determine the steps used in its solution. Then follow the procedures for each operation.

EXAMPLE •

Thirty welded pipe supports are fabricated to the dimensions shown in Figure 2–23. A cutoff and waste allowance of $\frac{3}{4}''$ is made for each piece. A total of 24'-3$\frac{3}{8}''$ $\left(291\frac{3}{8}''\right)$ of channel iron of the required size is in stock. How many feet of channel iron are ordered for the complete job?

Find the length of channel iron required for one support.

$$5 \times 2\frac{7}{16} = \frac{5}{1} \times \frac{39''}{16} = \frac{195''}{16} = 12\frac{3}{16}$$

$$1\frac{1''}{2} = 1\frac{8''}{16}$$

$$12\frac{3''}{16} = 12\frac{3''}{16}$$

$$+ \quad \frac{7''}{8} = \frac{14''}{16}$$

$$\overline{\qquad\qquad} $$

$$13\frac{25''}{16} = 14\frac{9''}{16}$$

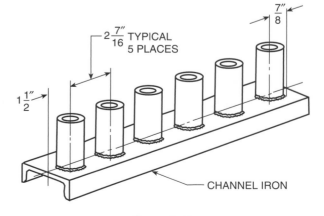

$2\frac{7''}{16}$ TYPICAL
5 PLACES

$\frac{7''}{8}$

$1\frac{1''}{2}$

CHANNEL IRON

Figure 2–23

Find the length of one support including the cutoff and waste allowance.

$$14\frac{9''}{16} = 14\frac{9''}{16}$$

$$+ \quad \frac{3''}{4} = \frac{12''}{16}$$

$$14\frac{21''}{16} = 15\frac{5''}{16}$$

Find the length of 30 supports.

$$30 \times 15\frac{5''}{16} = \frac{30}{1} \times \frac{245''}{16} = \frac{7,350''}{16} = 459\frac{3''}{8}$$

Find the amount of channel iron ordered.

$$459\frac{3''}{8}$$

$$-291\frac{3''}{8}$$

$$168''$$

Express the answer in feet.

$$168 \div 12 = 14 \qquad 14' \; Ans$$

EXERCISE 2–14

Solve the following problems.

1. The unified thread shown in Figures 2–24 and 2–25 may have either a flat or rounded crest or root. If the sides of the unified thread are extended, a sharp V-thread is formed. *H* is the height of a sharp V-thread. The pitch, *P*, is the distance between two adjacent threads.

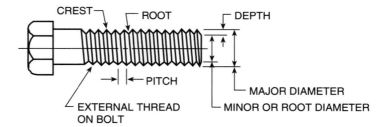

Figure 2–24

Refer to Figure 2–25.

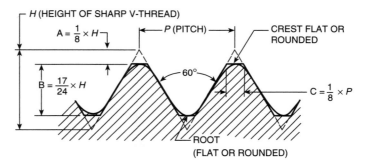

Figure 2–25

Find dimensions A, B, and C as indicated.

a. $H = \dfrac{5''}{16}$, A = ?, B = ?

b. $H = \dfrac{3''}{8}$, A = ?, B = ?

c. $H = \dfrac{15''}{64}$, A = ?, B = ?   g. $P = \dfrac{1''}{6}$, C = ?

d. $H = \dfrac{1''}{2}$, A = ?, B = ?   h. $P = \dfrac{1''}{20}$, C = ?

e. $H = \dfrac{3''}{4}$, A = ?, B = ?   i. $P = \dfrac{1''}{28}$, C = ?

f. $P = \dfrac{1''}{4}$, C = ?   j. $P = \dfrac{1''}{32}$, C = ?

2. The recipe for chicken salad shown in Figure 2–26 makes 4 servings. A chef finds the amount of each ingredient for a serving of 25 on one day and a serving of 42 on the next. Find the amount of each ingredient needed for 25 servings and 42 servings.

   Hint: For the 25 servings, multiply each ingredient by $\frac{25}{4}$ or $6\frac{1}{4}$.

		AMOUNT OF EACH INGREDIENT REQUIRED		
		4 Servings	25 Servings	42 Servings
a.	Diced cooked chicken	$1\frac{3}{4}$ cups	? cups	? cups
b.	Sliced celery	$1\frac{1}{4}$ cups	? cups	? cups
c.	Lemon juice	1 teaspoon	? teaspoons	? teaspoons
d.	Chopped green onions	2	?	?
e.	Salt	$\frac{1}{2}$ teaspoon	? teaspoons	? teaspoons
f.	Paprika	$\frac{1}{8}$ teaspoon	? teaspoons	? teaspoons
g.	Medium size avocado	1	?	?
h.	Cashew nuts	$\frac{1}{2}$ cup	? cups	? cups
i.	Mayonnaise	$\frac{1}{3}$ cup	? cups	? cups

Figure 2–26

3. An interior decorator applies the following steps in computing the width of nondraw sheer window draperies.
   - Measure the window width.
   - Triple the window width measurement.
   - Allow 4 inches $\left(\frac{1}{3}\text{ foot}\right)$ for each side hem (2 required).

   a. Find, in feet, the width of drapery fabric needed for three $6\frac{1}{4}$-foot-wide windows.

   b. Find, in feet, the width of drapery fabric needed for two $3\frac{1}{2}$-foot-wide windows.

   c. Find, in feet, the total width needed for all the windows.

4. At the end of summer, a home improvement center has an end-of-season clearance sale on outdoor materials. The amount of price markdown depends on the particular piece of merchandise. A customer purchases the items shown in the table in Figure 2–27. What is the total bill?

   NOTE: One-fourth off the list price means $\frac{1}{4}$ of the listed price is deducted, or the markdown price is $\frac{3}{4}$ of the list price.

Item	List Price	Markdown from List Price	Number Purchased
#1	$6	$\frac{1}{3}$ off	3
#2	$10	$\frac{1}{5}$ off	2
#3	$17	$\frac{1}{4}$ off	4
#4	$25	$\frac{1}{2}$ off	2

Figure 2–27

5. A sheet metal technician is required to cut twenty-five $3\frac{9}{16}$-inch lengths of band iron, allowing $\frac{3}{32}$-inch waste for each cut. The pieces are cut from a strip of band iron that is $121\frac{3}{4}$ inches long. How much stock is left after all pieces are cut?

6. A printer selects type for a book that averages $12\frac{1}{2}$ words per line and 42 lines per page. The book has 5 chapters. The number of pages in each chapter are as follows: $30\frac{1}{2}$, 37, $28\frac{2}{3}$, $40\frac{1}{4}$, and $43\frac{1}{3}$. How many estimated words are in the book?

7. A nurse practitioner gives a patient $\frac{1}{2}$ tablet of morphine for pain. If one morphine tablet contains $\frac{1}{4}$ grain (gr) of morphine, how much morphine does the patient receive?

8. A nurse practitioner gave a patient $2\frac{1}{2}$ tablets of ibuprofin. Each tablet contained 250 mg. How many mg of ibuprofin did the patient receive?

9. A certain model car requires $27\frac{3}{8}$ inches of $\frac{1}{2}$-inch air-conditioning hose. How many inches will be needed for seven cars?

10. A race car averages $2\frac{9}{16}$ miles per minute. How far will this car travel in $1\frac{1}{4}$ hours?

2–15 Division of Common Fractions

Division of fractions and mixed numbers is used by a chef to determine the number of servings that can be prepared from a given quantity of food. A painter and decorator find the number of gallons of paint needed for a job. A cosmetologist finds the number of applications that can be obtained from a certain amount of hair rinse solution. A cabinetmaker determines shelving spacing for a counter installation.

Meaning of Division of Fractions

As with division of whole numbers, division of fractions is a short method of subtracting. Dividing $\frac{1}{2}$ by $\frac{1}{8}$ means finding the number of times $\frac{1}{8}$ is contained in $\frac{1}{2}$.

$$\frac{4}{8} - \frac{1}{8} = \frac{3}{8}$$

$$\frac{3}{8} - \frac{1}{8} = \frac{2}{8}$$

$$\frac{2}{8} - \frac{1}{8} = \frac{1}{8}$$

$$\frac{1}{8} - \frac{1}{8} = 0$$

Since $\frac{1}{8}$ is subtracted 4 times from $\frac{1}{2}$, 4 one-eighths are contained in $\frac{1}{2}$.

$$\frac{1}{2} \div \frac{1}{8} = 4$$

The enlarged 1-inch scale in Figure 2–28 shows four $\frac{1}{8}$-inch parts in $\frac{1}{2}$ inch.

$$\frac{1''}{2} \div \frac{1''}{8} = 4$$

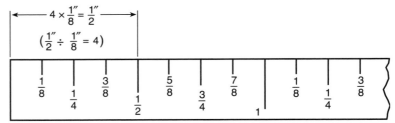

ENLARGED 1-INCH SCALE

Figure 2–28

The meaning of division of a fraction by a whole number, a fraction by a mixed number, and a mixed number by a mixed number is the same as division of fractions.

Division of Fractions as the Inverse of Multiplication of Fractions

Division is the inverse of multiplication. Dividing by 2 is the same as multiplying by $\frac{1}{2}$.

$$5 \div 2 = 2\frac{1}{2}$$

$$5 \times \frac{1}{2} = 2\frac{1}{2}$$

$$5 \div 2 = 5 \times \frac{1}{2}$$

Two is the multiplicative inverse of $\frac{1}{2}$, and $\frac{1}{2}$ is the multiplicative inverse of 2. The multiplicative inverse of a fraction is a fraction that has the numerator and denominator interchanged. The multiplicative inverse of $\frac{7}{8}$ is $\frac{8}{7}$.

Dividing Fractions

To divide fractions, invert the divisor, change to the inverse operation, and multiply.

EXAMPLES •────────────────────────────────────

1. Divide. $\frac{5}{8} \div \frac{3}{4}$

Invert the divisor and multiply

$$\frac{5}{8} \div \frac{3}{4} = \frac{5}{8} \times \frac{4}{3} = \frac{5}{6} \ Ans$$

2. Calculate. $3\frac{1}{2} \div \frac{5}{4}$

Write the mixed number $3\frac{1}{2}$ as the fraction $\frac{7}{2}$.

$$3\frac{1}{2} \div \frac{5}{4} = \frac{7}{2} \div \frac{5}{4}$$

Invert the divisor and multiply.

$$3\frac{1}{2} \div \frac{5}{4} = \frac{7}{2} \times \frac{4}{5}$$

$$= \frac{14}{5} = 2\frac{4}{5} \ Ans$$

3. Divide. $9\dfrac{3}{4} \div 2\dfrac{1}{6}$

Change each mixed number to a fraction.

$$9\dfrac{3}{4} = \dfrac{39}{4}$$

$$2\dfrac{1}{6} = \dfrac{13}{6}$$

$$9\dfrac{3}{4} \div 2\dfrac{1}{6} = \dfrac{39}{4} \div \dfrac{13}{6}$$

Invert the divisor and multiply.

$$9\dfrac{3}{4} \div 2\dfrac{1}{6} = \dfrac{39}{4} \times \dfrac{6}{13}$$

$$= \dfrac{9}{2} = 4\dfrac{1}{2} \ Ans$$

EXERCISE 2–15A

Divide the following fractions as indicated. Express the answers in lowest terms.

1. $\dfrac{1}{2} \div \dfrac{1}{4}$ **5.** $\dfrac{3}{16} \div \dfrac{3}{8}$ **9.** $\dfrac{4}{9} \div \dfrac{1}{6}$

2. $\dfrac{1}{4} \div \dfrac{1}{2}$ **6.** $\dfrac{4}{5} \div \dfrac{9}{20}$ **10.** $\dfrac{11}{15} \div \dfrac{33}{40}$

3. $\dfrac{5}{8} \div \dfrac{1}{8}$ **7.** $\dfrac{11}{12} \div \dfrac{5}{6}$ **11.** $\dfrac{10}{11} \div \dfrac{4}{9}$

4. $\dfrac{4}{5} \div \dfrac{7}{8}$ **8.** $\dfrac{5}{6} \div \dfrac{3}{10}$ **12.** $\dfrac{15}{64} \div \dfrac{75}{128}$

Dividing Any Combination of Fractions, Mixed Numbers, and Whole Numbers

To divide any combinations of fractions, mixed numbers, and whole numbers, write mixed numbers as fractions. Write whole numbers over the denominator 1. Invert the divisor. Change to the inverse operation and multiply.

EXAMPLES •————————————————————————

1. Divide. $\dfrac{19}{25} \div 3\dfrac{3}{10}$

Write $3\dfrac{3}{10}$ as a fraction

$$\dfrac{19}{25} \div \dfrac{33}{10} =$$

Invert the divisor, $\dfrac{33}{10}$.

Change to the inverse operation and multiply.

$$\dfrac{19}{\underset{5}{25}} \times \dfrac{\overset{2}{10}}{33} = \dfrac{38}{165} \ Ans$$

2. Divide. $3\dfrac{3}{64} \div 40$

Write $3\dfrac{3}{64}$ as a fraction.

$$\dfrac{195}{64} \div \dfrac{40}{1} =$$

Write 40 over 1.

Invert the divisor, $\dfrac{40}{1}$.

Change to the inverse operation and multiply.

$$\dfrac{\overset{39}{195}}{64} \times \dfrac{1}{\underset{8}{40}} = \dfrac{39}{512} \ Ans$$

3. Divide. $56\dfrac{2}{9} \div 8\dfrac{10}{21}$

Write $56\dfrac{2}{9}$ and $8\dfrac{10}{21}$ as fractions. $\dfrac{506}{9} \div \dfrac{178}{21} =$

Invert the divisor, $\dfrac{178}{21}$.

Change to the inverse operation and multiply. $\dfrac{\overset{253}{\cancel{506}}}{\underset{3}{\cancel{9}}} \times \dfrac{\overset{7}{\cancel{21}}}{\underset{89}{\cancel{178}}} = \dfrac{1771}{267} = 6\dfrac{169}{267}\ Ans$

EXERCISE 2–15B

Divide the following values as indicated. Express the answers in lowest terms.

1. $12 \div \dfrac{1}{2}$ **5.** $15 \div \dfrac{10}{3}$ **9.** $3\dfrac{9}{32} \div \dfrac{11}{64}$ **13.** $4 \div 6\dfrac{1}{2}$

2. $15 \div \dfrac{5}{8}$ **6.** $21 \div \dfrac{28}{31}$ **10.** $\dfrac{8}{15} \div 7\dfrac{27}{30}$ **14.** $50 \div 15\dfrac{9}{25}$

3. $\dfrac{1}{9} \div 3$ **7.** $\dfrac{7}{16} \div 2\dfrac{1}{4}$ **11.** $8\dfrac{5}{16} \div 2\dfrac{5}{8}$ **15.** $7\dfrac{1}{2} \div 3\dfrac{5}{6}$

4. $\dfrac{6}{5} \div 9$ **8.** $5\dfrac{3}{8} \div \dfrac{3}{16}$ **12.** $9 \div 2\dfrac{5}{8}$ **16.** $2\dfrac{31}{32} \div 102\dfrac{3}{4}$

2–16   Dividing Common Fractions in Practical Applications

EXAMPLE

The machine bolt shown in Figure 2–29 has a thread pitch of $\frac{1}{16}$ inch. The pitch is the distance between 2 adjacent threads or the thickness of 1 thread. Find the number of threads in $\frac{7}{8}$ inch.

The number of threads is found by dividing the total threaded length $\left(\frac{7''}{8}\right)$ by the pitch or thickness of 1 thread $\left(\frac{1''}{16}\right)$.

$$\dfrac{7''}{8} \div \dfrac{1''}{16} = \dfrac{7}{\underset{1}{\cancel{8}}} \times \dfrac{\overset{2}{\cancel{16}}}{1} = 14\ Ans$$

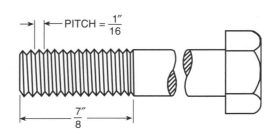

Figure 2–29

EXERCISE 2–16

Solve the following problems.

1. In order to drill the holes in the part shown in Figure 2–30, a machinist finds the horizontal distances between the centers of the holes. There are five sets of holes, A, B, C, D, and E. The holes within each set are equally spaced.

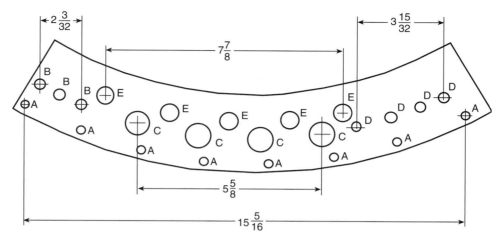

Figure 2–30

Find, in inches, the horizontal distance between the centers of the two consecutive holes listed.

a. A holes   **c.** C holes   **e.** E holes

b. B holes   **d.** D holes

2. How many full lengths of wire, each $1\frac{3}{4}$ feet long, can be cut from a 115-foot length of wire?

3. A plumber makes 4 pipe assemblies of different lengths as shown in Figure 2–31. For any one assembly, the distances between 2 consecutive fittings, distance B, are equal as shown. Determine distance B for each pipe assembly. Refer to the table in Figure 2–32.

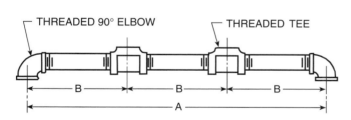

Figure 2–31

Pipe Assembly	A	B
#1	$18\frac{3}{4}''$	
#2	$21\frac{9}{16}''$	
#3	$32\frac{5}{8}''$	
#4	$40\frac{11}{16}''$	

Figure 2–32

4. A fast-service restaurant chain sells the well-known "quarter-pounder." The quarter-pounder is a hamburger containing $\frac{1}{4}$ pound of ground beef. Five area restaurants use a total of $3\frac{1}{2}$ tons of ground beef in a certain week to make the quarter-pounders. The restaurants are open seven days a week. On the average, how many quarter-pounders are sold per day by all five area restaurants?

NOTE: One ton equals 2,000 pounds.

5. The gasoline consumption of an automobile is measured in two categories, city driving and highway driving.

a. Using the data listed in the table in Figure 2–33 on page 64, find the gasoline consumption in miles per gallon (mi/gal or mpg) for each of the three types of automobiles.

b. How many more miles of highway driving can the compact 4-cylinder car travel than the V-8 engine full-size car when each car consumes $8\frac{3}{4}$ gallons of gasoline?

Type of Automobile	Distance Traveled City	Gasoline Consumed City	Miles per Gallon City	Distance Traveled Highway	Gasoline Consumed Highway	Miles per Gallon Highway
Full-Size Sedan V-8	266 mi	$15\frac{1}{5}$ gal		406	$19\frac{1}{3}$ gal	
Intermediate Size 6 Cyl.	262 mi	$13\frac{1}{10}$ gal		450	$18\frac{3}{4}$ gal	
Compact Size 4 Cyl.	300 mi	$11\frac{1}{4}$ gal		481	$14\frac{4}{5}$ gal	

Figure 2–33

6. An illustrator lays out a lettering job shown in Figure 2–34. Margins are measured, and spacing for lettering is computed. Each row of lettering is $\frac{7}{8}$ inch high with a $\frac{5}{16}$-inch space between each row. Determine the number of rows of lettering on the sheet.

   Hint: Notice that there is one less space between lettering rows than the number of rows.

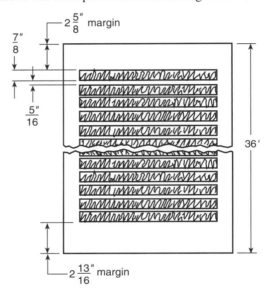

Figure 2–34

7. A radiologist experiments with smaller amounts of radiation to see how the quality of an X-ray will be affected. A normal amount is 96 mAs (milliampere-seconds). How many mAs will the patient receive with an exposure that is

   a. $\dfrac{3}{4}$

   b. $\dfrac{1}{2}$, and

   c. $\dfrac{1}{3}$

   of the normal radiation amount?

8. A nurse practitioner gave a patient $1\frac{1}{2}$ tablets of ibuprofin. Each tablet contained 250 mg. How many mg of ibuprofin did the patient receive?

9. A piece of electrical wire is $24\frac{1}{3}$ feet long. How many $2\frac{3}{4}$ foot long pieces can be cut from this wire?

10. A certain circuit is properly fused for a $7\frac{1}{2}$-horsepower motor. The motor is to be replaced by several $\frac{5}{8}$-horsepower motor. How many motors with the fuses be able to carry?

2–17   Combined Operations with Common Fractions

Combined operation problems given as arithmetic expressions require the use of the proper order of operations. Practical application problems based on formulas are found in occupational textbooks, handbooks, and manuals.

Order of Operations

- Do all the work in parentheses first. Parentheses are used to group numbers. In a problem expressed in fractional form, the numerator and the denominator are each considered as being enclosed in parentheses.

$$\frac{2\frac{5}{8} - \frac{3}{4}}{15 + 7\frac{9}{16}} = \left(2\frac{5}{8} - \frac{3}{4}\right) \div \left(15 + 7\frac{9}{16}\right)$$

- If an expression contains parentheses within brackets, do the work within the innermost parentheses first.
- Do multiplication and division next in order from left to right.
- Last, do addition and subtraction in order from left to right.

EXAMPLES •

1. What is the total area (number of square feet) of the parcel of land shown in Figure 2–35?

$$\text{Area} = 120\frac{3'}{4} \times 80' + \frac{1}{2} \times 120\frac{3'}{4} \times 42\frac{1'}{3}$$

$$\text{Area} = 9{,}660 \text{ sq ft} + 2{,}555\frac{7}{8} \text{ sq ft}$$

$$\text{Area} = 12{,}215\frac{7}{8} \text{ sq ft } Ans$$

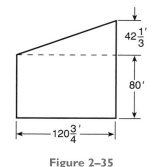

Figure 2–35

2. Find the value of $\frac{3}{8} + 5\frac{1}{2} \times 8 - 12\frac{1}{4} \div 2$.

$$\frac{3}{8} + 44 - 12\frac{1}{4} \div 2$$

$$\frac{3}{8} + 44 - 6\frac{1}{8}$$

$$38\frac{1}{4} \; Ans$$

3. Find the value of $\left(\frac{3}{8} + 5\frac{1}{2}\right) \times 8 - 12\frac{1}{4} \div 2$.

$$5\frac{7}{8} \times 8 - 12\frac{1}{4} \div 2$$

$$47 - 6\frac{1}{8} = 40\frac{7}{8} \; Ans$$

4. Find the value of $\dfrac{84\frac{3}{5} - 18\frac{7}{10} \times 3}{1\frac{8}{9} + 20\frac{2}{3} \div 2\frac{2}{5}} - 1\frac{4}{7}$.

$$\left(84\frac{3}{5} - 18\frac{7}{10} \times 3\right) \div \left(1\frac{8}{9} + 20\frac{2}{3} \div 2\frac{2}{5}\right) - 1\frac{4}{7}$$

$$28\frac{1}{2} \div 10\frac{1}{2} - 1\frac{4}{7}$$

$$2\frac{5}{7} - 1\frac{4}{7} = 1\frac{1}{7} \; Ans$$

EXERCISE 2–17

Perform the indicated operations.

1. $\dfrac{3}{4} + \dfrac{1}{2} - \dfrac{1}{8}$

2. $3 - \dfrac{7}{8} + 2\dfrac{3}{8}$

3. $\dfrac{7}{10} + \dfrac{3}{5} - \dfrac{2}{15}$

4. $2\dfrac{2}{3} + 12 - \dfrac{5}{6} + \dfrac{7}{12}$

5. $\left(\dfrac{5}{8} + \dfrac{3}{4}\right) \div \dfrac{1}{2}$

6. $\dfrac{7}{16} + \dfrac{1}{8} \div \dfrac{3}{4}$

7. $16 \div \dfrac{3}{4} - 2\dfrac{7}{16}$

8. $\dfrac{16 - 2\frac{7}{16}}{\frac{3}{4}}$

9. $\dfrac{\frac{3}{5} + \frac{9}{10} - \frac{4}{25}}{4}$

10. $\dfrac{5\frac{5}{8}}{2\frac{1}{4} - 1\frac{1}{8}}$

11. $25\dfrac{2}{3} - 18\dfrac{5}{6} + \dfrac{6\frac{1}{2}}{1\frac{1}{2}}$

12. $10\dfrac{31}{64} + \dfrac{9}{16} \times 4 - 1\dfrac{1}{2}$

13. $10\dfrac{31}{64} + \dfrac{9}{16} \times \left(4 - 1\dfrac{1}{2}\right)$

14. $\left(\dfrac{1}{25} \times 10\right) \div \left(\dfrac{3}{10} \times 20\right)$

15. $\left(\dfrac{1}{25} \times 10\right) \div \dfrac{3}{10} + 20$

16. $\dfrac{7\frac{3}{4} + 3 \times \frac{1}{2}}{10 - 5\frac{1}{4}}$

17. $\dfrac{\left(7\frac{3}{4} + 3\right) \times \frac{1}{2}}{10 - 5\frac{1}{4}}$

18. $\left(\dfrac{3}{16} - \dfrac{3}{32}\right) \times \dfrac{1}{8} \div \dfrac{1}{16}$

19. $\left(10\dfrac{2}{3} + 5\dfrac{1}{3} \times \dfrac{5}{6}\right) \div 8\dfrac{1}{6}$

20. $\dfrac{7}{15} \times \left(4 - \dfrac{7}{10}\right) + 5 \times 12\dfrac{2}{5}$

21. $\left(\dfrac{7}{15} \times 4 - \dfrac{7}{10} + 5\right) \times 12\dfrac{2}{5}$

22. $\dfrac{9\frac{1}{2} \times \left(3\frac{7}{8} - 2\frac{3}{8}\right)}{3\frac{1}{4}}$

23. $\dfrac{31\frac{1}{5}}{5\frac{1}{5}} + \dfrac{30\frac{5}{6}}{5}$

24. $\dfrac{120 - 20\frac{1}{2} \times 4}{40\frac{2}{3} + 12 - 6\frac{1}{2}}$

26. $8\frac{3}{4} \times 10 + 50\frac{7}{10} \div \left(12 + \frac{2}{3}\right)$

25. $\dfrac{\left(120 - 20\frac{1}{2}\right) \times 4}{40\frac{2}{3} + 12 - 6\frac{1}{2}}$

2–18   Combined Operations of Common Fractions in Practical Applications

EXAMPLE •────────────────────────────────

A semicircular-sided section shown in Figure 2–36 is fabricated in a sheet metal shop. A semi-circle is a half-circle that is formed on sheet metal parts by rolling. The sheet metal technician makes a stretchout of the section as shown in Figure 2–37. A stretchout is a flat layout that, when formed, makes the required part.

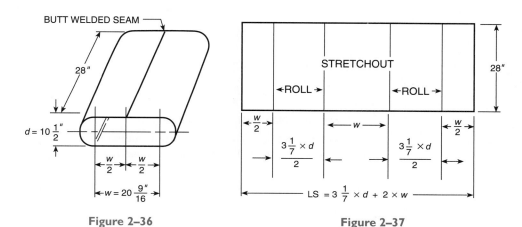

Figure 2–36   Figure 2–37

The length of the stretchout, which is called Length Size, is calculated from the following formula:

Length Size (LS) = $3\frac{1}{7} \times d + 2 \times w$ where d = diameter of roll (semicircle)

w = distance between centers of semicircles

Refer to Figure 2–36 for the values of d and w, substitute the values in the formula, and solve.

$$LS = 3\frac{1}{7} \times d + 2 \times w$$

$$LS = 3\frac{1}{7} \times 10\frac{1}{2}'' + 2 \times 20\frac{9}{16}''$$

$$LS = 33'' + 41\frac{1}{8}''$$

$$LS = 74\frac{1}{8} \; Ans$$

EXERCISE 2–18

	Torque (lb-ft)	r/min	hp
a.	$3\frac{1}{2}$	2,250	
b.	$4\frac{3}{8}$	2,400	
c.	$5\frac{1}{4}$	2,750	
d.	$4\frac{2}{3}$	3,750	

Figure 2–38

Solve the following problems.

1. The horsepower (hp) of a motor is found from the following formula:

$$hp = \frac{2 \times \pi \times T \times \text{r/min}}{33,000}$$

where T = torque

$$\pi = 3\frac{1}{7}$$

Torque is a turning effect. It is the tendency of the armature to rotate. Torque is expressed in pound-feet (lb-ft). Using the torque and revolutions per minute (r/min or rpm) of four electric motors, a through d in Figure 2–38, find the horsepower of each.

2. A simple parallel electrical circuit is shown in Figure 2–39. An environmental systems technician finds the total resistance of the circuit.

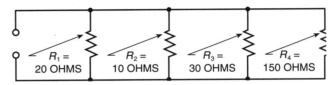

| $R_1 =$ 20 OHMS | $R_2 =$ 10 OHMS | $R_3 =$ 30 OHMS | $R_4 =$ 150 OHMS |

Figure 2–39

$$R_T = \cfrac{1}{\cfrac{1}{R_1} + \cfrac{1}{R_2} + \cfrac{1}{R_3} + \cfrac{1}{R_4}}$$

where R_1, R_2, R_3, and R_4 are the individual resistors and R_T is the total resistance.

Find the total resistance of the circuit.

3. Pulleys and belts are widely used in automotive, commercial, and industrial equipment. An application of a belt drive in an automobile is shown in Figure 2–40.

The length of the belt is found by using the following formula:

$$L = \frac{D + d}{2} \times \pi + 2 \times c$$

where L = length of belt in inches
D = diameter of large pulley in inches
d = diameter of small pulley in inches
c = center-to-center distance of pulleys in inches

$$\pi = 3\frac{1}{7}$$

Determine the belt lengths required for each of the 4 belt drives listed in the table in Figure 2–41.

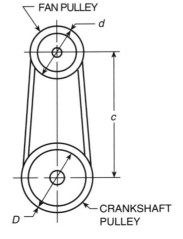

Figure 2–40

	D	d	c	L
a.	$5\frac{1}{2}''$	$3\frac{1}{4}''$	$10\frac{1}{4}''$	
b.	$6\frac{3}{8}''$	$4\frac{1}{8}''$	$11\frac{1}{16}''$	
c.	$4\frac{1}{16}''$	$2\frac{15}{16}''$	$8\frac{5}{8}''$	
d.	$6\frac{13}{16}''$	$5\frac{7}{16}''$	$9\frac{1}{8}''$	

Figure 2–41

4. Hourly paid employees in many companies get time-and-a-half pay for overtime hours worked during their regular workweek. Overtime pay is based on the number of hours worked over the normal workweek hours. The normal workweek varies from company to company, but is usually from 36 to 40 hours. Time-and-a-half means the employee's overtime rate of pay is $1\frac{1}{2}$ times the normal rate of pay or the employee is credited with $1\frac{1}{2}$ hours for each hour worked. Find the total number of hours credited to each employee, A through D, in the table in Figure 2–42.

Total number of hours credited = normal workweek hours + (number of hours worked − normal workweek hours) $\times 1\frac{1}{2}$.

	EMPLOYEE			
	A	B	C	D
Normal workweek hours	38	40	36	40
Hours worked during week	$41\frac{1}{2}$	$46\frac{3}{5}$	$36\frac{3}{4}$	$44\frac{4}{5}$
Total hours credited				

Figure 2–42

5. A tank shown in Figure 2–43 is to be fabricated from steel plate. The specifications call for 3-gauge (approx. $\frac{1}{4}$ inch thick) plate.

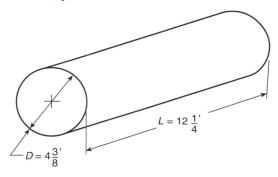

$$L = 12\frac{1}{4}'$$
$$D = 4\frac{3}{8}'$$

Figure 2–43

An engineering aide refers to a metals handbook and finds that 3-gauge steel plate weighs $9\frac{1}{2}$ pounds per square foot. The weight of the tank is computed from the following formula:

$$\text{Weight of tank in pounds} = \pi \times D \times \left(\frac{1}{2} \times D + L\right) \times W$$

where D = diameter of tank in feet
L = length of tank in feet
W = weight of 1 square foot of plate
$\pi = 3\frac{1}{7}$

Find the weight of the tank.

⠿ UNIT EXERCISE AND PROBLEM REVIEW

EQUIVALENT FRACTIONS

Express each of the following fractions as equivalent fractions as indicated.

1. $\dfrac{1}{2} = \dfrac{?}{8}$

2. $\dfrac{7}{12} = \dfrac{?}{36}$

3. $\dfrac{8}{7} = \dfrac{?}{35}$

4. $\dfrac{2}{9} = \dfrac{?}{72}$

5. $\dfrac{3}{16} = \dfrac{?}{256}$

6. $\dfrac{27}{5} = \dfrac{?}{105}$

FRACTIONS IN LOWEST TERMS

Express each of the following as a fraction in lowest terms.

7. $\dfrac{4}{10}$ **9.** $\dfrac{5}{35}$ **11.** $\dfrac{25}{150}$

8. $\dfrac{12}{8}$ **10.** $\dfrac{24}{6}$ **12.** $\dfrac{16}{64}$

MIXED NUMBERS AS FRACTIONS, FRACTIONS AS MIXED NUMBERS

Express each of the following mixed numbers as improper fractions.

13. $1\dfrac{1}{2}$ **15.** $9\dfrac{3}{4}$ **17.** $3\dfrac{63}{64}$

14. $1\dfrac{7}{8}$ **16.** $12\dfrac{2}{5}$ **18.** $53\dfrac{3}{8}$

Express each of the following improper fractions as mixed numbers.

19. $\dfrac{13}{2}$ **21.** $\dfrac{85}{4}$ **23.** $\dfrac{167}{128}$

20. $\dfrac{9}{8}$ **22.** $\dfrac{129}{32}$ **24.** $\dfrac{319}{64}$

LOWEST COMMON DENOMINATORS

Determine the lowest common denominators of the following sets of fractions.

25. $\dfrac{1}{2}, \dfrac{3}{4}, \dfrac{1}{8}$ **26.** $\dfrac{1}{3}, \dfrac{1}{4}, \dfrac{7}{8}, \dfrac{11}{12}$

Determine the lowest common denominators of the following sets of fractions by using prime factors.

27. $\dfrac{1}{15}, \dfrac{5}{6}, \dfrac{8}{9}, \dfrac{3}{10}$ **28.** $\dfrac{5}{12}, \dfrac{11}{14}, \dfrac{3}{16}, \dfrac{3}{7}$

ADDING FRACTIONS

Add the following fractions. Express all answers in lowest terms.

29. $\dfrac{3}{8} + \dfrac{5}{8}$ **31.** $\dfrac{11}{15} + \dfrac{2}{3} + \dfrac{5}{6} + \dfrac{7}{10}$

30. $\dfrac{3}{16} + \dfrac{1}{8} + \dfrac{1}{4}$ **32.** $\dfrac{7}{12} + \dfrac{4}{5} + \dfrac{1}{4} + \dfrac{7}{15}$

ADDING FRACTIONS AND MIXED NUMBERS

Add the following fractions and mixed numbers. Express all answers in lowest terms.

33. $7\dfrac{1}{3} + \dfrac{1}{3}$ **35.** $17\dfrac{17}{25} + 16\dfrac{9}{10} + 25 + 16\dfrac{3}{5}$

34. $9\dfrac{1}{2} + \dfrac{7}{8}$ **36.** $81\dfrac{5}{24} + 19\dfrac{1}{2} + \dfrac{7}{8} + 6\dfrac{1}{6}$

SUBTRACTING FRACTIONS

Subtract the following fractions as indicated. Express the answers in lowest terms.

37. $\dfrac{15}{16} - \dfrac{3}{16}$ **40.** $\dfrac{5}{8} - \dfrac{7}{16}$

38. $\dfrac{47}{100} - \dfrac{9}{25}$ **41.** $\dfrac{13}{32} - \dfrac{7}{64}$

39. $\dfrac{45}{64} - \dfrac{29}{64}$

SUBTRACTING FRACTIONS, WHOLE NUMBERS, MIXED NUMBERS

Subtract the following values as indicated. Express the answers in lowest terms.

42. $12 - \dfrac{1}{4}$ **47.** $78 - \dfrac{8}{25}$

43. $18 - 10\dfrac{5}{8}$ **48.** $17\dfrac{7}{8} - 9\dfrac{9}{16}$

44. $47 - 8\dfrac{7}{8}$ **49.** $47\dfrac{3}{16} - 41\dfrac{9}{32}$

45. $5 - \dfrac{59}{64}$ **50.** $99\dfrac{3}{4} - 99\dfrac{41}{64}$

46. $21 - \dfrac{7}{32}$

MULTIPLYING FRACTIONS

Multiply the following fractions as indicated. Express the answers in lowest terms.

51. $\dfrac{8}{9} \times \dfrac{3}{32}$ **53.** $\dfrac{9}{16} \times \dfrac{2}{3} \times \dfrac{1}{4} \times \dfrac{1}{2}$

52. $\dfrac{3}{5} \times \dfrac{25}{32} \times \dfrac{5}{6}$ **54.** $\dfrac{18}{25} \times \dfrac{5}{9} \times \dfrac{6}{25} \times \dfrac{1}{4}$

MULTIPLYING FRACTIONS, WHOLE NUMBERS, MIXED NUMBERS

Multiply the following values as indicated. Express the answers in lowest terms.

55. $\dfrac{3}{8} \times 16$ **57.** $\dfrac{3}{8} \times 16 \times 2\dfrac{3}{4}$

56. $\dfrac{15}{16} \times 5\dfrac{1}{3}$ **58.** $6\dfrac{4}{5} \times 30 \times 3\dfrac{1}{8}$

DIVIDING FRACTIONS

Divide the following fractions as indicated. Express the answers in lowest terms.

59. $\dfrac{1}{8} \div \dfrac{1}{4}$ **62.** $\dfrac{2}{15} \div \dfrac{11}{60}$

60. $\dfrac{9}{32} \div \dfrac{5}{16}$ **63.** $\dfrac{6}{25} \div \dfrac{3}{5}$

61. $\dfrac{19}{20} \div \dfrac{3}{5}$ **64.** $\dfrac{25}{128} \div \dfrac{5}{32}$

DIVIDING FRACTIONS, WHOLE NUMBERS, MIXED NUMBERS

Divide the following values as indicated. Express the answers in lowest terms.

65. $9 \div \dfrac{1}{3}$

66. $\dfrac{7}{8} \div 6$

67. $150 \div \dfrac{15}{32}$

68. $5\dfrac{3}{8} \div \dfrac{1}{16}$

69. $60 \div 14\dfrac{1}{4}$

70. $60\dfrac{5}{16} \div 10\dfrac{3}{32}$

COMBINED OPERATIONS WITH FRACTIONS, WHOLE NUMBERS, MIXED NUMBERS

Perform the indicated operations.

71. $\dfrac{7}{8} + \dfrac{5}{16} - \dfrac{3}{4}$

72. $\left(\dfrac{5}{9} + \dfrac{2}{3}\right) \div \dfrac{1}{3}$

73. $\dfrac{5}{9} + \dfrac{2}{3} \div \dfrac{1}{3}$

74. $\dfrac{\dfrac{3}{25} - \dfrac{1}{10} + \dfrac{4}{5}}{\dfrac{10}{11}}$

75. $\left(\dfrac{7}{8} \times 14\right) \div \left(\dfrac{3}{8} \times 6\right)$

76. $\dfrac{7\dfrac{1}{8} \times \left(6\dfrac{7}{8} - 4\dfrac{3}{16}\right)}{4\dfrac{3}{4}}$

77. $\dfrac{15}{2\dfrac{2}{3}} + \dfrac{25\dfrac{1}{5}}{4}$

78. $\left(1\dfrac{3}{4} + 3\dfrac{3}{8} \times 6\right) \div 18 + \dfrac{1}{2}$

79. $\dfrac{1}{4} \times 20 \times \dfrac{3}{5} + 120\dfrac{1}{10} \div 14 - 7\dfrac{1}{5}$

80. $10\dfrac{9}{10} \times \left[15\dfrac{7}{16} \times \left(\dfrac{2}{3} + \dfrac{5}{6}\right) - 3\dfrac{3}{10}\right] \div 2\dfrac{1}{2}$

FRACTION PRACTICAL APPLICATION PROBLEMS

81. A baker prepares a cake mix that weighs 100 pounds. The cake mix consists of shortening and other ingredients. The weights of the other ingredients are $20\dfrac{1}{2}$ pounds of flour, $29\dfrac{3}{4}$ pounds of sugar, $18\dfrac{1}{8}$ pounds of milk, 16 pounds of whole eggs, and a total of $5\dfrac{1}{4}$ pounds of flavoring, salt, and baking powder. How many pounds of shortening are used in the mix?

82. Determine dimensions A, B, C, D, and E of the machined part shown in Figure 2–44.

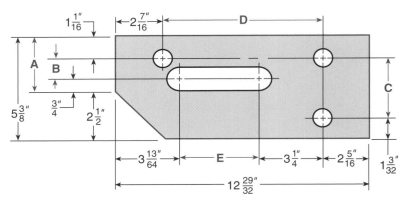

Figure 2–44

83. Before starting a wiring job, an electrician takes an inventory of materials and finds that 4,625 feet of BX cable are in stock. The following lengths of cable are removed from stock for the job: $282\frac{1}{4}$ feet, 482 feet, $56\frac{1}{2}$ feet, $208\frac{1}{2}$ feet, and $157\frac{3}{4}$ feet. Upon completion of the job, $55\frac{1}{2}$ feet are left over and returned to stock. How many feet of cable are in stock after completing the job?

84. A truck is loaded at a structural steel supply house for a delivery to a construction site. The order calls for 125 feet of channel iron, which weighs $3\frac{3}{4}$ tons, 140 feet of I beam, which weighs $4\frac{3}{10}$ tons, and 80 feet of angle iron, which weighs $2\frac{1}{5}$ tons. The maximum legal tonnage permitted to be hauled by the truck is $9\frac{1}{2}$ tons. All of the channel iron and I beam are loaded. Only part of the angle iron is loaded so that the maximum legal tonnage is met but not exceeded. By how many tons of angle iron will the delivery be short of the order?

85. To determine the mathematical ability of an applicant, some firms require the applicant to take a preemployment test. Usually the test is given before the applicant is interviewed. An applicant who fails the test is often not considered for the job. The following problem is taken from a preemployment test given by a large retail firm. What is the total cost of the following items?

a. $6\frac{2}{3}$ boxes of Item A at $3\frac{1}{4}$ per box

b. $\frac{1}{3}$ yard of Item B at $4\frac{1}{2}$ per yard

c. 8 pieces of Item C at $15\frac{3}{4}$ per dozen pieces

d. Find the total cost of the items listed.

86. Horticulture is the science of cultivating plants. A horticultural assistant prepares a soil mixture for house plants, using the materials shown on the chart in Figure 2–45. The amounts indicated on the chart are added to a soilless mix in order to prepare one bushel of potting soil.

a. Determine the amount of each material needed to prepare $1\frac{1}{2}$ bushels of soil.

b. Determine the amount of each material needed to prepare $2\frac{1}{2}$ bushels of soil.

		1 Bushel	a. $1\frac{1}{2}$ Bushels	b. $2\frac{1}{2}$ Bushels
(1)	Ammonium nitrate	$3\frac{1}{2}$ tablespoons		
(2)	Garden fertilizer	6 tablespoons		
(3)	Superphosphate	$2\frac{1}{2}$ tablespoons		
(4)	Ground limestone	9 tablespoons		
(5)	Bone meal	$1\frac{1}{2}$ tablespoons		
(6)	Potassium nitrate	$2\frac{1}{2}$ tablespoons		
(7)	Calcium nitrate	2 tablespoons		

Figure 2–45

87. The fuel oil tank shown in Figure 2–46 is to be constructed by a steel fabricator. The specifications call for a 4-foot diameter tank. The tank must hold a minimum of 2,057 gallons of fuel oil. Determine the required length of the tank.

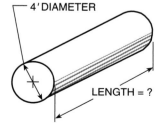

4′ DIAMETER

LENGTH = ?

Figure 2–46

$$L = \frac{G}{\pi \times \dfrac{D}{2} \times \dfrac{D}{2} \times 7\frac{12}{25}}$$

where L = length of tank in feet
G = number of gallons the tank is to hold (capacity)
D = tank diameter in feet

Use $\pi = 3\frac{1}{7}$

88. In estimating the cost of building the table shown in Figure 2–47, a cabinetmaker must find the number of board feet of lumber needed. One board foot of lumber is the equivalent of a piece of lumber 1 foot wide, 1 foot long, and 1 inch thick. Use this formula to find the number of board feet of lumber.

$$\text{bd ft} = \frac{T \times W \times L}{12}$$ where T = thickness in inches
W = width in inches
L = length in feet

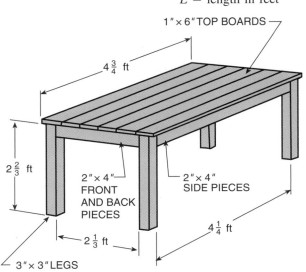

Figure 2–47

Different sizes of lumber are required for the table. These sizes are the measurement of the lumber before finishing. Use these measurements to find the board feet needed.

a. Find the number of board feet of $1'' \times 6''$ lumber needed.

b. Find the number of board feet of $3'' \times 3''$ lumber needed.

c. Find the number of board feet of $2'' \times 4''$ lumber needed.

d. Find the total number of board feet needed.

2–19   Computing with a Calculator: Fractions and Mixed Numbers

Fractions

The fraction key ($\boxed{\text{a}^{\text{b/c}}}$ or $\boxed{\text{A}^{\text{b/c}}}$) is used when entering fractions and mixed numbers in a calculator. The answers to expressions entered as fractions will be given as fractions or mixed numbers with the fraction in lowest terms.

Enter the numerator, press $\boxed{\text{a}^{\text{b/c}}}$, and enter the denominator. The fraction is displayed with the symbol ⌐ between the numerator and denominator.

EXAMPLES •────────────────────────────────

Enter $\dfrac{3}{4}$.

3 $\boxed{\text{a}^{\text{b/c}}}$ 4, 3 ⌐ 4 is displayed.

1. Add. $\dfrac{3}{16} + \dfrac{19}{32}$

Solution.   3 $\boxed{\text{a}^{\text{b/c}}}$ 16 $\boxed{+}$ 19 $\boxed{\text{a}^{\text{b/c}}}$ 32 $\boxed{=}$ 25 ⌐ 32, $\dfrac{25}{32}$ *Ans*

2. Subtract: $\dfrac{7}{8} - \dfrac{5}{64}$

    Solution.   7 $\boxed{a^{b/c}}$ 8 $\boxed{-}$ 5 $\boxed{a^{b/c}}$ 64 $\boxed{=}$ 51 $\lrcorner$ 64, $\dfrac{51}{64}$ *Ans*

3. Multiply: $\dfrac{3}{32} \times \dfrac{11}{16}$

    Solution.   3 $\boxed{a^{b/c}}$ 32 $\boxed{\times}$ 11 $\boxed{a^{b/c}}$ 16 $\boxed{=}$ 33 $\lrcorner$ 512, $\dfrac{33}{512}$ *Ans*

4. Divide. $\dfrac{5}{8} \div \dfrac{13}{15}$

    Solution.   5 $\boxed{a^{b/c}}$ 8 $\boxed{\div}$ 13 $\boxed{a^{b/c}}$ 15 $\boxed{=}$ 75 $\lrcorner$ 104, $\dfrac{75}{104}$ *Ans*

If a calculator does not have an $\boxed{a^{b/c}}$ key, then enter the fraction as a division problem. When the problem involves multiplication or division, parentheses have to be used. Answers are usually given as decimals. Special key combinations have to be used to change from a decimal to a fraction. For example, on the TI-84, press $\boxed{\text{MATH}}$ $\boxed{1}$ $\boxed{\text{ENTER}}$. Pressing the $\boxed{\text{MATH}}$ key produces the screen in Figure 2–48. Notice that item #1 is ▶Frac, which means "change to a fraction." You either need to press $\boxed{1}$ or press the $\boxed{\blacktriangledown}$ key one time (until the 1: is highlighted and then press $\boxed{\text{ENTER}}$).

Figure 2–48

EXAMPLES •————————————————————————————————

1. Add $\dfrac{3}{16} + \dfrac{19}{32}$

    Solution.   3 $\boxed{\div}$ 16 $\boxed{+}$ 19 $\boxed{\div}$ 32 $\boxed{\text{MATH}}$ $\boxed{1}$ $\boxed{\text{ENTER}}$ $\dfrac{25}{32}$ *Ans*

2. Divide $\dfrac{5}{8} \div \dfrac{13}{15}$

    Solution. $\boxed{(}$ 5 $\boxed{\div}$ 8 $\boxed{)}$ $\boxed{\div}$ $\boxed{(}$ 13 $\boxed{\div}$ 15 $\boxed{)}$ $\boxed{\text{MATH}}$ $\boxed{1}$ $\boxed{\text{ENTER}}$ $\dfrac{75}{104}$ *Ans*

Mixed Numbers

Enter the whole number, press $\boxed{a^{b/c}}$, enter the fraction numerator, press $\boxed{a^{b/c}}$, and enter the denominator. Depending on the particular calculator, either the symbol _ or $\lrcorner$ is displayed between the whole number and fraction.

EXAMPLE •————————————————————————————————

Enter $15\dfrac{7}{16}$.

$$15 \boxed{a^{b/c}} 7 \boxed{a^{b/c}} 16$$

Either 15 _ 7 ⌐ 16 or 15 ⌐ 7 ⌐ 16 is displayed.

EXAMPLES •──

The following examples are of mixed numbers with individual arithmetic operations.

1. Add. $7\dfrac{3}{64} + 23\dfrac{5}{8}$

   Solution. $7 \boxed{a^{b/c}} 3 \boxed{a^{b/c}} 64 \boxed{+} 23 \boxed{a^{b/c}} 5 \boxed{a^{b/c}} 8 \boxed{=} 30$ _ 43 ⌐ 64, $30\dfrac{43}{64}$ *Ans*

2. Subtract. $43\dfrac{7}{8} - 36\dfrac{29}{32}$

   Solution. $43 \boxed{a^{b/c}} 7 \boxed{a^{b/c}} 8 \boxed{-} 36 \boxed{a^{b/c}} 29 \boxed{a^{b/c}} 32 \boxed{=} 6$ _ 31 ⌐ 32, $6\dfrac{31}{32}$ *Ans*

3. Multiply. $38\dfrac{5}{6} \times 14\dfrac{13}{16}$

   Solution. $38 \boxed{a^{b/c}} 5 \boxed{a^{b/c}} 6 \boxed{\times} 14 \boxed{a^{b/c}} 13 \boxed{a^{b/c}} 16 \boxed{=} 575$ _ 7 ⌐ 32, $575\dfrac{7}{32}$ *Ans*

4. Divide. $159\dfrac{17}{64} \div 3\dfrac{7}{8}$

   Solution. $159 \boxed{a^{b/c}} 17 \boxed{a^{b/c}} 64 \boxed{\div} 3 \boxed{a^{b/c}} 7 \boxed{a^{b/c}} 8 \boxed{=} 41$ _ 25 ⌐ 248, $41\dfrac{25}{248}$ *Ans*

──•

   If a calculator does not have an $\boxed{a^{b/c}}$, then think of the mixed number as a whole number added to a fraction. Thus, think of $12\frac{8}{15}$ as $12 + \frac{8}{15}$.

EXAMPLE •──

Enter $12\dfrac{8}{15}$ in a calculator without an $\boxed{a^{b/c}}$.

$$12 \boxed{+} 8 \boxed{\div} 15 \boxed{ENTER} \quad 12.5333333 \; Ans$$

or

$$12 \boxed{+} 8 \boxed{\div} 15 \boxed{MATH} \boxed{1} \boxed{ENTER} \quad \dfrac{188}{15} \; Ans$$

──

   Notice that the last answer was expressed as an improper fraction. Calculators without an $\boxed{a^{b/c}}$ key have to be "forced" to think about mixed numbers. First, have the calculator determine the decimal answer, next subtract the whole number, then determine the fraction value for the decimal.

   The next two examples show how to perform arithmetic with mixed numbers on a calculator that does not have an $\boxed{a^{b/c}}$ key.

EXAMPLES •──

1. Add $7\dfrac{3}{64} + 23\dfrac{5}{8}$

   Solution. $7 \boxed{+} 3 \boxed{\div} 64 \boxed{+} 23 \boxed{+} 5 \boxed{\div} 8 \boxed{ENTER}$ The answer is 30.671875.
   30.671875 $\boxed{-} 30 \boxed{MATH} \boxed{1} \boxed{ENTER}$ gives the result $\dfrac{43}{64}$, $30\dfrac{43}{64}$ *Ans*

2. Multiply $38\frac{5}{6} \times 14\frac{13}{16}$

Solution. $\boxed{(}\ 38\ \boxed{+}\ 5 \div 6\ \boxed{)}\ \boxed{\times}\ \boxed{(}\ 14\ \boxed{+}\ 13 \div 16\ \boxed{)}\ \boxed{ENTER}$ gives 575.21875

575.21875 $\boxed{-}$ 575 $\boxed{MATH}\ \boxed{1}\ \boxed{ENTER}$ results in $\frac{7}{32}$, $575\frac{7}{32}$ *Ans*

Practice Exercises, Individual Basic Operations with Fractions and Mixed Numbers

Evaluate the following expressions. The expressions are basic arithmetic operations. Remember to check your answers by estimating the answers and doing each problem twice. The solutions to the problems directly follow the practice exercises. Compare your answers to the given solutions.

1. $\frac{5}{8} + \frac{11}{16}$

2. $\frac{31}{32} - \frac{7}{8}$

3. $\frac{9}{16} \times \frac{5}{8}$

4. $\frac{23}{25} \div \frac{4}{5}$

5. $85\frac{7}{64} + 107\frac{3}{4}$

6. $125\frac{7}{8} - 67\frac{63}{64}$

7. $62\frac{13}{16} \times 47\frac{1}{6}$

8. $785\frac{27}{32} \div 2\frac{3}{4}$

9. $\frac{59}{64} + 46\frac{27}{32}$

10. $37\frac{3}{8} - \frac{45}{64}$

Solutions to Practice Exercises, Individual Basic Operations with Fractions and Mixed Numbers

1. $5\ \boxed{a^{b/c}}\ 8\ \boxed{+}\ 11\ \boxed{a^{b/c}}\ 16\ \boxed{=}\ 1 \_ 5 \lrcorner 16$, $1\frac{5}{16}$ *Ans*

2. $31\ \boxed{a^{b/c}}\ 32\ \boxed{-}\ 7\ \boxed{a^{b/c}}\ 8\ \boxed{=}\ 3 \lrcorner 32$, $\frac{3}{32}$ *Ans*

3. $9\ \boxed{a^{b/c}}\ 16\ \boxed{\times}\ 5\ \boxed{a^{b/c}}\ 8\ \boxed{=}\ 45 \lrcorner 128$, $\frac{45}{128}$ *Ans*

4. $23\ \boxed{a^{b/c}}\ 25\ \boxed{\div}\ 4\ \boxed{a^{b/c}}\ 5\ \boxed{=}\ 1 \_ 3 \lrcorner 20$, $1\frac{3}{20}$ *Ans*

5. $85\ \boxed{a^{b/c}}\ 7\ \boxed{a^{b/c}}\ 64\ \boxed{+}\ 107\ \boxed{a^{b/c}}\ 3\ \boxed{a^{b/c}}\ 4\ \boxed{=}\ 192 \_ 55 \lrcorner 64$, $192\frac{55}{64}$ *Ans*

6. $125\ \boxed{a^{b/c}}\ 7\ \boxed{a^{b/c}}\ 8\ \boxed{-}\ 67\ \boxed{a^{b/c}}\ 63\ \boxed{a^{b/c}}\ 64\ \boxed{=}\ 57 \_ 57 \lrcorner 64$, $57\frac{57}{64}$ *Ans*

7. $62\ \boxed{a^{b/c}}\ 13\ \boxed{a^{b/c}}\ 16\ \boxed{\times}\ 47\ \boxed{a^{b/c}}\ 1\ \boxed{a^{b/c}}\ 6\ \boxed{=}\ 2962 \_ 21 \lrcorner 32$, $2{,}962\frac{21}{32}$ *Ans*

8. $785\ \boxed{a^{b/c}}\ 27\ \boxed{a^{b/c}}\ 32\ \boxed{\div}\ 2\ \boxed{a^{b/c}}\ 3\ \boxed{a^{b/c}}\ 4\ \boxed{=}\ 285 \_ 67 \lrcorner 88$, $285\frac{67}{88}$ *Ans*

9. $59\ \boxed{a^{b/c}}\ 64\ \boxed{+}\ 46\ \boxed{a^{b/c}}\ 27\ \boxed{a^{b/c}}\ 32\ \boxed{=}\ 47 \_ 49 \lrcorner 64$, $47\frac{49}{64}$ *Ans*

10. $37\ \boxed{a^{b/c}}\ 3\ \boxed{a^{b/c}}\ 8\ \boxed{-}\ 45\ \boxed{a^{b/c}}\ 64\ \boxed{=}\ 36 \_ 43 \lrcorner 64$, $36\frac{43}{64}$ *Ans*

Combined Operations

Because the following problems are combined operations expressions, your calculator must have algebraic logic to solve the problems shown. The expressions are solved by entering numbers and operations into the calculator in the same order as the expressions are written. Remember to estimate your answers and to do each problem twice.

EXAMPLES

1. Evaluate. $275\dfrac{17}{32} + \dfrac{7}{8} \times 26\dfrac{3}{4}$

   Solution. 275 [a^b/c] 17 [a^b/c] 32 [+] 7 [a^b/c] 8 [×] 26 [a^b/c] 3 [a^b/c] 4 [=] 298 _ 15 ⌐ 16, $298\dfrac{15}{16}$ Ans

   Because the calculator has algebraic logic, the multiplication operation $\left(\dfrac{7}{8} \times 26\dfrac{3}{4}\right)$ was performed before the addition operation $\left(\text{adding } 275\dfrac{17}{32}\right)$ was performed.

2. Evaluate. $\dfrac{35}{64} - \dfrac{3}{8} + 18 \div 10\dfrac{2}{3}$

   Solution. 35 [a^b/c] 64 [−] 3 [a^b/c] 8 [+] 18 [÷] 10 [a^b/c] 2 [a^b/c] 3 [=] 1 _ 55 ⌐ 64, $1\dfrac{55}{64}$ Ans

3. Evaluate. $380\dfrac{29}{32} - \left(\dfrac{3}{16} + 9\dfrac{15}{64}\right) \times 12$

   As previously discussed, operations enclosed within parentheses are done first. A calculator with algebraic logic performs the operations within parentheses before performing other operations in a combined operations expression. If an expression contains parentheses, enter the expression in the calculator in the order in which it is written. The parentheses keys must be used.

   Solution. 380 [a^b/c] 29 [a^b/c] 32 [−] [(] 3 [a^b/c] 16 [+] 9 [a^b/c] 15 [a^b/c] 64 [)] [×] 12

   [=] 267 _ 27 ⌐ 32, $267\dfrac{27}{32}$ Ans

4. Evaluate. $\dfrac{25\dfrac{47}{64} + 7 \times \dfrac{5}{8}}{\dfrac{3}{16} \times 2 + \dfrac{1}{8}}$

   Recall that for a problem expressed in fractional form, the fraction bar is also used as a grouping symbol. The numerator and denominator are each considered as being enclosed in parentheses.

   Solution. [(] 25 [a^b/c] 47 [a^b/c] 64 [+] 7 [×] 5 [a^b/c] 8 [)] [÷] [(] 3 [a^b/c] 16 [×] 2

   [+] 1 [a^b/c] 8 [)] [=] 60 _ 7 ⌐ 32, $60\dfrac{7}{32}$ Ans

   The expression may also be evaluated by using the [=] key to simplify the numerator without having to enclose the entire numerator in parentheses. However, parentheses must be used to enclose the denominator.

   25 [a^b/c] 47 [a^b/c] 64 [+] 7 [×] 5 [a^b/c] 8 [=] [÷] [(] 3 [a^b/c] 16 [×] 2 [+] 1 [a^b/c] 8 [)]

   [=] 60 _ 7 ⌐ 32, $60\dfrac{7}{32}$ Ans

Practice Exercises, Combined Operations with Fractions and Mixed Numbers

Evaluate the following combined operations expressions. Remember to check your answers by estimating the answers and doing each problem twice. The solutions to the problems directly follow the practice exercises. Compare your answers to the given solutions.

1. $\left(\dfrac{11}{16} + 12\dfrac{31}{32}\right) \div \dfrac{1}{8}$

2. $\dfrac{\dfrac{108}{3}}{\dfrac{3}{8}} - 3\dfrac{5}{64}$

3. $\dfrac{43\dfrac{9}{10} - 17\dfrac{3}{5} + \dfrac{7}{20}}{5}$

4. $120\dfrac{13}{16} + 98\dfrac{5}{8} \times \left(6 - \dfrac{3}{4}\right)$

5. $\dfrac{56\dfrac{3}{4} + 20 \times \dfrac{7}{8}}{4 \times \dfrac{2}{3}}$

6. $\left(\dfrac{25}{32} - \dfrac{3}{4}\right) \div \dfrac{1}{2} \times \dfrac{3}{4}$

7. $50 \times \left(28\dfrac{4}{5} - 17\dfrac{9}{10} + 27\right) \times \dfrac{3}{5}$

8. $\dfrac{40\dfrac{1}{2}}{1\dfrac{1}{8}} - \left(15\dfrac{5}{64} + 8\dfrac{29}{32}\right)$

9. $\dfrac{270 - 175\dfrac{1}{2} \times \dfrac{7}{8}}{\dfrac{1}{64} \times 128}$

10. $\dfrac{\left(426\dfrac{3}{5} - 123\dfrac{9}{25}\right) \times 8}{28\dfrac{4}{5} \div 7\dfrac{1}{5}}$

Solutions to Practice Exercises, Combined Operations with Fractions and Mixed Numbers

1. $\boxed{(}$ 11 $\boxed{a^{b/c}}$ 16 $\boxed{+}$ 12 $\boxed{a^{b/c}}$ 31 $\boxed{a^{b/c}}$ 32 $\boxed{)}$ $\boxed{\div}$ 1 $\boxed{a^{b/c}}$ 8 $\boxed{=}$ 109 _ 1 ⌐ 4, $109\dfrac{1}{4}$ *Ans*

  or 11 $\boxed{a^{b/c}}$ 16 $\boxed{+}$ 12 $\boxed{a^{b/c}}$ 31 $\boxed{a^{b/c}}$ 32 $\boxed{=}$ $\boxed{\div}$ 1 $\boxed{a^{b/c}}$ 8 $\boxed{=}$ 109 _ 1 ⌐ 4, $109\dfrac{1}{4}$ *Ans*

2. 108 $\boxed{\div}$ 3 $\boxed{a^{b/c}}$ 8 $\boxed{-}$ 3 $\boxed{a^{b/c}}$ 5 $\boxed{a^{b/c}}$ 64 $\boxed{=}$ 284 _ 59 ⌐ 64, $284\dfrac{59}{64}$ *Ans*

3. $\boxed{(}$ 43 $\boxed{a^{b/c}}$ 9 $\boxed{a^{b/c}}$ 10 $\boxed{-}$ 17 $\boxed{a^{b/c}}$ 3 $\boxed{a^{b/c}}$ 5 $\boxed{+}$ 7 $\boxed{a^{b/c}}$ 20 $\boxed{)}$ $\boxed{\div}$ 5 $\boxed{=}$ 5 _ 33 ⌐ 100,

  $5\dfrac{33}{100}$ *Ans*

  or 43 $\boxed{a^{b/c}}$ 9 $\boxed{a^{b/c}}$ 10 $\boxed{-}$ 17 $\boxed{a^{b/c}}$ 3 $\boxed{a^{b/c}}$ 5 $\boxed{+}$ 7 $\boxed{a^{b/c}}$ 20 $\boxed{=}$ $\boxed{\div}$ 5 $\boxed{=}$ 5 _ 33 ⌐ 100,

  $5\dfrac{33}{100}$ *Ans*

4. 120 $\boxed{a^{b/c}}$ 13 $\boxed{a^{b/c}}$ 16 $\boxed{+}$ 98 $\boxed{a^{b/c}}$ 5 $\boxed{a^{b/c}}$ 8 $\boxed{\times}$ $\boxed{(}$ 6 $\boxed{-}$ 3 $\boxed{a^{b/c}}$ 4 $\boxed{)}$ $\boxed{=}$ 638 _ 19 ⌐ 32,

  $638\dfrac{19}{32}$ *Ans*

5. $\boxed{(}$ 56 $\boxed{a^{b/c}}$ 3 $\boxed{a^{b/c}}$ 4 $\boxed{+}$ 20 $\boxed{\times}$ 7 $\boxed{a^{b/c}}$ 8 $\boxed{)}$ $\boxed{\div}$ $\boxed{(}$ 4 $\boxed{\times}$ 2 $\boxed{a^{b/c}}$ 3 $\boxed{)}$ $\boxed{=}$ 27 _ 27 ⌐
  32,

  $27\dfrac{27}{32}$ *Ans*

  or 56 $\boxed{a^{b/c}}$ 3 $\boxed{a^{b/c}}$ 4 $\boxed{+}$ 20 $\boxed{\times}$ 7 $\boxed{a^{b/c}}$ 8 $\boxed{=}$ $\boxed{\div}$ $\boxed{(}$ 4 $\boxed{\times}$ 2 $\boxed{a^{b/c}}$ 3 $\boxed{)}$ $\boxed{=}$ 27 _ 27 ⌐ 32,

  $27\dfrac{27}{32}$ *Ans*

6. $\boxed{(}$ 25 $\boxed{a^{b/c}}$ 32 $\boxed{-}$ 3 $\boxed{a^{b/c}}$ 4 $\boxed{)}$ $\boxed{\div}$ 1 $\boxed{a^{b/c}}$ 2 $\boxed{\times}$ 3 $\boxed{a^{b/c}}$ 4 $\boxed{=}$ 3 $\lrcorner$ 64, $\dfrac{3}{64}$ *Ans*

or 25 $\boxed{a^{b/c}}$ 32 $\boxed{-}$ 3 $\boxed{a^{b/c}}$ 4 $\boxed{=}$ $\boxed{\div}$ 1 $\boxed{a^{b/c}}$ 2 $\boxed{\times}$ 3 $\boxed{a^{b/c}}$ 4 $\boxed{=}$ 3 $\lrcorner$ 64, $\dfrac{3}{64}$ *Ans*

7. 50 $\boxed{\times}$ $\boxed{(}$ 28 $\boxed{a^{b/c}}$ 4 $\boxed{a^{b/c}}$ 5 $\boxed{-}$ 17 $\boxed{a^{b/c}}$ 9 $\boxed{a^{b/c}}$ 10 $\boxed{+}$ 27 $\boxed{)}$ $\boxed{\times}$ 3 $\boxed{a^{b/c}}$ 5 $\boxed{=}$ 1137, 1137 *Ans*

8. 40 $\boxed{a^{b/c}}$ 1 $\boxed{a^{b/c}}$ 2 $\boxed{\div}$ 1 $\boxed{a^{b/c}}$ 1 $\boxed{a^{b/c}}$ 8 $\boxed{-}$ $\boxed{(}$ 15 $\boxed{a^{b/c}}$ 5 $\boxed{a^{b/c}}$ 64 $\boxed{+}$ 8 $\boxed{a^{b/c}}$ 29 $\boxed{a^{b/c}}$ 32 $\boxed{)}$ $\boxed{=}$ 12 _ 1 $\lrcorner$ 64, $12\dfrac{1}{64}$ *Ans*

9. $\boxed{(}$ 270 $\boxed{-}$ 175 $\boxed{a^{b/c}}$ 1 $\boxed{a^{b/c}}$ 2 $\boxed{\times}$ 7 $\boxed{a^{b/c}}$ 8 $\boxed{)}$ $\boxed{\div}$ $\boxed{(}$ 1 $\boxed{a^{b/c}}$ 64 $\boxed{\times}$ 128 $\boxed{)}$ $\boxed{=}$ 58 _ 7 $\lrcorner$ 32, $58\dfrac{7}{32}$ *Ans*

or 270 $\boxed{-}$ 175 $\boxed{a^{b/c}}$ 1 $\boxed{a^{b/c}}$ 2 $\boxed{\times}$ 7 $\boxed{a^{b/c}}$ 8 $\boxed{=}$ $\boxed{\div}$ $\boxed{(}$ 1 $\boxed{a^{b/c}}$ 64 $\boxed{\times}$ 128 $\boxed{)}$ $\boxed{=}$ 58 _ 7 $\lrcorner$ 32, $58\dfrac{7}{32}$ *Ans*

10. $\boxed{(}$ $\boxed{(}$ 426 $\boxed{a^{b/c}}$ 3 $\boxed{a^{b/c}}$ 5 $\boxed{-}$ 123 $\boxed{a^{b/c}}$ 9 $\boxed{a^{b/c}}$ 25 $\boxed{)}$ $\boxed{\times}$ 8 $\boxed{)}$ $\boxed{\div}$ $\boxed{(}$ 28 $\boxed{a^{b/c}}$ 4 $\boxed{a^{b/c}}$ 5 $\boxed{\div}$ 7 $\boxed{a^{b/c}}$ 1 $\boxed{a^{b/c}}$ 5 $\boxed{)}$ $\boxed{=}$ 606 _ 12 $\lrcorner$ 25, $606\dfrac{12}{25}$ *Ans*

or $\boxed{(}$ 426 $\boxed{a^{b/c}}$ 3 $\boxed{a^{b/c}}$ 5 $\boxed{-}$ 123 $\boxed{a^{b/c}}$ 9 $\boxed{a^{b/c}}$ 25 $\boxed{=}$ $\boxed{\times}$ 8 $\boxed{)}$ $\boxed{\div}$ $\boxed{(}$ 28 $\boxed{a^{b/c}}$ 4 $\boxed{a^{b/c}}$ 5 $\boxed{\div}$ 7 $\boxed{a^{b/c}}$ 1 $\boxed{a^{b/c}}$ 5 $\boxed{)}$ $\boxed{=}$ 606 _ 12 $\lrcorner$ 25, $606\dfrac{12}{25}$ *Ans*

UNIT 3  Decimal Fractions

OBJECTIVES

After studying this unit you should be able to

- write decimal numbers in word form.

- write numbers expressed in word form as decimal fractions.

- express common fractions as decimal fractions.

- express decimal fractions as common fractions.

- add, subtract, multiply, and divide decimal fractions.

- solve problems using individual operations of addition, subtraction, multiplication, and division of decimal fractions.

- solve practical problems by combining addition, subtraction, multiplication, and division of decimal fractions.

- determine the root of any positive number.

- solve practical problems by using powers and roots.

- solve practical problems by using power and root operations in combination with one or more additional arithmetic operations.

Calculations using decimals are often faster and easier to make than fractional computations. The decimal system of measurement is widely used in occupations where greater precision than fractional parts of an inch is required. Decimals are used to compute to any required degree of precision. Certain industries require a degree of precision to the millionths of an inch.

Most machined parts are manufactured using decimal system dimensions and decimal machine settings. The electrical and electronic industries generally compute and measure using decimals. Computations required for the design of buildings, automobiles, and aircraft are based on the decimal system. Occupations in the retail, wholesale, office, health, transportation, and communication fields require decimal calculations. Finance and insurance companies base their computational procedures on the decimal system. Our monetary system of dollars and cents is based on the decimal system.

A decimal fraction is not written as a common fraction with a numerator and denominator. The decimal fraction is written with a decimal point. Decimal fractions are equivalent to common fractions having denominators that are powers of 10. *Powers of 10* are numbers that are obtained by multiplying 10 by itself a certain number of times. Numbers such as 100; 1,000; 10,000; 100,000; and 1,000,000 are powers of 10.

3–1   Meaning of Fractional Parts

The line segment shown in Figure 3–1 is 1 unit long. It is divided in 10 equal smaller parts. The locations of common fractions and their decimal fraction equivalents are shown on the line.

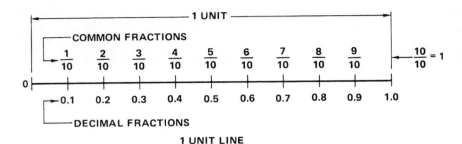

Figure 3–1

3–2   Reading Decimal Fractions

The chart shown in Figure 3–2 gives the names of the parts of a number with respect to their positions from the decimal point.

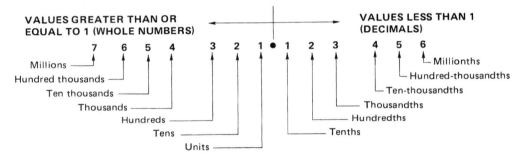

Figure 3–2

To read a decimal, read the number as a whole number. Then say the name of the decimal place of the last digit to the right.

EXAMPLES •

1. 0.43 is read, "forty-three hundredths."
2. 0.532 is read, "five hundred thirty-two thousandths."
3. 0.0028 is read, "twenty-eight ten-thousandths."
4. 0.2800 is read, "two thousand eight hundred ten-thousandths."

To read a mixed decimal (a whole number and a decimal fraction), read the whole number, read the word *and* at the decimal point, and read the decimal.

EXAMPLES •

1. 2.65 is read, "two and sixty-five hundredths."
2. 9.002 is read, "nine and two thousandths."
3. 135.0787 is read, "one hundred thirty-five and seven hundred eighty-seven ten-thousandths."

3–3   Simplified Method of Reading Decimal Fractions

Often a simplified method of reading decimal fractions is used in actual on-the-job applications. This method is generally quicker, easier, and less likely to be misinterpreted. A tool and die maker reads 0.0187 inch as "point zero, one, eight, seven inches." An electronics technician reads 2.125 amperes as "two, point one, two, five amperes."

3–4   Writing Decimal Fractions

A common fraction with a denominator that is a power of 10 can be written as a decimal fraction. Replace the denominator with a decimal point. The decimal point is placed to the left of the first digit of the numerator. There are as many decimal places as there are zeros in the denominator. When writing a decimal fraction, place a zero to the left of the decimal point.

EXAMPLES •——————————————————————————————

Write each common fraction as a decimal fraction.

1. $\dfrac{7}{10} = 0.7$ *Ans*   There is 1 zero in 10 and 1 decimal place in 0.7.

2. $\dfrac{65}{100} = 0.65$ *Ans*   There are 2 zeros in 100 and 2 decimal places in 0.65.

3. $\dfrac{793}{1,000} = 0.793$ *Ans*   There are 3 zeros in 1,000 and 3 decimal places in 0.793.

4. $\dfrac{9}{10,000} = 0.0009$ *Ans*   There are 4 zeros in 10,000 and 4 decimal places in 0.0009. In order to maintain proper place value, 3 zeros are written between the decimal point and the 9.

EXERCISE 3–4

Write the following numbers as words.

1. 0.3	**4.** 0.018	**7.** 15.876	**10.** 351.032
2. 0.03	**5.** 0.0098	**8.** 3.709	**11.** 299.0009
3. 0.175	**6.** 0.00209	**9.** 27.0027	**12.** 158.8008

Write the following words as decimals or mixed decimals.

13. nine tenths

14. six hundredths

15. two ten-thousandths

16. four hundred thirty-five thousandths

17. three hundred one hundred-thousandths

18. seventeen and nine hundredths

19. twelve and one thousandths

20. six and thirty-five ten-thousandths

Each of the following common fractions has a denominator that is a power of 10. Write the equivalent decimal fraction for each.

21. $\dfrac{19}{100}$	**23.** $\dfrac{287}{10,000}$	**25.** $\dfrac{999}{1,000}$	**27.** $\dfrac{8,111}{10,000}$
22. $\dfrac{197}{1,000}$	**24.** $\dfrac{41}{1,000}$	**26.** $\dfrac{7}{10,000}$	**28.** $\dfrac{3}{100,000}$

When working with decimals, the computations and answers may contain more decimal places than are needed. The number of decimal places needed depends on the degree of precision desired. The degree of precision depends on how the obtained decimal value is going to be used. The tools, machinery, equipment, and materials determine the degree of precision that can be obtained.

It is not realistic for a carpenter to attempt to saw a board to a 6.2518-inch length. The 6.2518-inch length is realistic in the machine trades. A surface grinder operator can grind a metal part to four-decimal-place precision.

3–5   Rounding Decimal Fractions

To round a decimal fraction, locate the digit in the number that gives the desired degree of precision. Increase that digit by 1 if the digit that directly follows is 5 or more. Do not change the value of the digit if the digit that follows is less than 5. Drop all digits that follow.

NOTE: The ≈ *sign* means approximately equal to.

EXAMPLES

1. A designer computes a dimension of 0.73862 inch. Three-place precision is needed for the part that is being drawn.

   Locate the digit in the third decimal place. (8)   0.73862 inch ≈ 0.739 inch *Ans*

   The fourth-decimal-place digit, 6, is greater than 5 and increases 8 to 9.

2. In determining rivet hole locations, a sheet metal technician computes a dimension of 1.5038 inches. Two-place precision is needed for laying out the hole locations.

   Locate the digit in the second decimal place. (0)   1.5038 inches ≈ 1.50 inches *Ans*

   The third-decimal-place digit, 3, is less than 5 and does not change the value, 0.

EXERCISE 3–5

Round each of the following numbers to the indicated number of decimal places.

1. 0.837 (2 places)
2. 0.344 (2 places)
3. 0.0072 (3 places)
4. 0.0072 (2 places)
5. 0.8888 (3 places)
6. 0.01497 (4 places)
7. 22.1955 (3 places)
8. 831.40019 (4 places)
9. 89.8994 (3 places)
10. 618.069 (1 place)
11. 722.01010 (3 places)
12. 100.9999 (1 place)

3–6   Expressing Common Fractions as Decimal Fractions

Expressing common fractions as decimal fractions is used in many occupations. A bookkeeper in working a financial statement expresses $16\frac{3}{20}$ as $16.15. In preparing a medication, a nurse may express $1\frac{1}{5}$ ounces of solution as 1.2 ounces.

A common fraction is an indicated division. A common fraction is expressed as a decimal fraction by dividing the numerator by the denominator.

EXAMPLE •————————————————————————

Express $\dfrac{3}{8}$ as a decimal fraction.

$$\begin{array}{r} 0.375 \; Ans \\ 8)\overline{3.000} \end{array}$$

Write $\dfrac{3}{8}$ as an indicated division.

Place a decimal point after the 3 and add zeros to the right of the decimal point.

NOTE: Adding zeros after the decimal point does not change the value of the dividend; 3 has the same value as 3.000.

Place the decimal point for the answer directly above the decimal point in the dividend. Divide.

——•

A common fraction that divides without a remainder is called a *terminating decimal.*
$\dfrac{3}{4} = 0.75, \dfrac{4}{5} = 0.8,$ and $\dfrac{5}{16} = 0.3125$ are examples of terminating decimal fractions.

Repeating or Nonterminating Decimals

A decimal that does not terminate is called a *repeating* or *nonterminating decimal.*
$\dfrac{16}{3} = 5.33333\ldots, \dfrac{7}{6} = 1.16666\ldots,$ and $\dfrac{1}{7} = 0.142857142857\ldots$ are examples of repeating decimals. One way to show that a decimal repeats is to place a bar over the digits that repeat. Using this method, the above three numbers would be written $\dfrac{16}{3} = 5.\overline{3}, \dfrac{7}{6} = 1.1\overline{6},$ and $\dfrac{1}{7} = 0.\overline{142857}.$

EXAMPLE •————————————————————————

Express $\dfrac{2}{3}$ as a decimal.

Write $\dfrac{2}{3}$ as an indicated division.

$$\begin{array}{r} 0.6666\ldots Ans \\ 3)\overline{2.0000} \end{array}$$

Place a decimal point after the 2 and add zeros to the right of the decimal point.

Place the decimal point for the answer directly above the decimal point in the dividend. Divide.

The three dots following the last digit indicate that the digit, 6, continues endlessly. As was mentioned, another way of showing that the digit repeats endlessly is to write a bar above the digit, $\frac{2}{3} = 0.\overline{6}.$

——•

3–7   Expressing Decimal Fractions as Common Fractions

Dimensions given or computed as decimals are often expressed as common fractions for on-the-job measurements. A carpenter expresses 10.625″ as $10\frac{5}{8}″$ when measuring the length to saw a board. In locating bolt holes on a beam, a structural ironworker changes a dimension given as 12′-6.75″ to 12′-$6\frac{3}{4}″$.

To change a decimal fraction to a common fraction, write the number after the decimal point as the numerator of a common fraction. Write the denominator as 1 followed by as many zeros as there are digits to the right of the decimal point. Express the common fraction in lowest terms.

EXAMPLES •——

1. Express 0.7 as a common fraction.

   Write 7 as the numerator.

   Write the denominator as 1 followed by 1 zero. The denominator is 10.

   $\dfrac{7}{10}$ *Ans*

2. Express 0.065 as a common fraction.

   Write 65 as the numerator.

   Write the denominator as 1 followed by 3 zeros. The denominator is 1,000.

   Express the fraction in lowest terms.

   $\dfrac{65}{1,000} = \dfrac{13}{200}$ *Ans*

——•

EXERCISE 3–7 ——

Express each of the following common fractions as decimal fractions. Where necessary, round the answers to four decimal places.

1. $\dfrac{1}{4}$	**4.** $\dfrac{4}{5}$	**7.** $\dfrac{47}{64}$	**10.** $\dfrac{7}{32}$
2. $\dfrac{5}{8}$	**5.** $\dfrac{5}{6}$	**8.** $\dfrac{19}{32}$	**11.** $\dfrac{19}{20}$
3. $\dfrac{13}{32}$	**6.** $\dfrac{3}{25}$	**9.** $\dfrac{1}{16}$	**12.** $\dfrac{29}{64}$

Express each of the following decimal fractions as common fractions. Express the answers in lowest terms.

13. 0.3	**16.** 0.050	**19.** 0.028	**22.** 0.0008
14. 0.42	**17.** 0.005	**20.** 0.0108	**23.** 0.8125
15. 0.325	**18.** 0.903	**21.** 0.999	**24.** 0.03125

3–8   Expressing Decimal Fractions in Practical Applications

EXAMPLE •——

In the circuit shown in Figure 3–3, the total current is the sum of the currents $(I_1 + I_2 + I_3 + I_4)$. What decimal fraction of the total current (amperes) in the circuit shown is received by resistance #2 (R_2)? Round the answer to 3 decimal places.

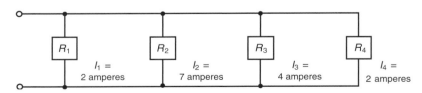

Figure 3–3

Write the common fraction that compares the current (amperes) received by resistance #2 (R_2) with the total current in the circuit.

$$\frac{7 \text{ amperes}}{2 \text{ amperes} + 7 \text{ amperes} + 4 \text{ amperes} + 2 \text{ amperes}} = \frac{7 \text{ amperes}}{15 \text{ amperes}}$$

Express the common fraction as a decimal fraction.

$$\frac{7}{15} = 0.467 \; Ans$$

EXERCISE 3–8

Determine the decimal fraction answers for each of the following problems. Where necessary, round the answers to three decimal places.

1. A building contractor determines the total cost of a job as $54,500. Labor costs are $21,800. What decimal fraction of the total cost is the labor cost?

2. The displacement of an automobile engine is 246 cubic inches. The engine is rebored an additional 4 cubic inches. What decimal fraction of the displacement of the rebored engine is the displacement of the engine before rebore?

3. A mason lays a sidewalk to the dimensions shown in Figure 3–4.

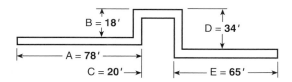

Figure 3–4

   a. What decimal fraction of the total length of sidewalk is distance A?

   b. What decimal fraction of the total length of sidewalk is distance C?

   c. What decimal fraction of the total length of sidewalk is distance E?

   d. What decimal fraction of the total length of sidewalk is distance B plus distance D?

4. The interior walls of a house contain a total area of 3,250 square feet. A painter and decorator paint 1,950 square feet. What decimal fraction of the total wall area is the area that remains to be painted?

5. A hospital dietitian allows 700 calories for a patient's breakfast and 750 calories for lunch. The total daily intake is 2,500 calories.

   a. What decimal fraction of the total daily calorie intake is allowed for breakfast?

   b. What decimal fraction of the total daily calorie intake is allowed for lunch?

   c. What decimal fraction of the total daily calorie intake is allowed for dinner?

6. In pricing merchandise, retail firms sometimes use the following simple formula.

      Retail price = cost of goods + overhead expenses + desired profit

Retail price is the price the customer is charged. Cost of goods is the price the retailer pays the manufacturer or supplier. Refer to the table in Figure 3–5.

Item	Retail Price	Cost of Goods	Overhead Expenses	Desired Profit
A		$325	$105	$28
B		$120	$36	$8
C	$930	$672	$212	

Figure 3–5

   a. What decimal fraction of the retail price is the desired profit of Item A?

   b. What decimal fraction of the retail price is the desired profit of Item B?

c. What decimal fraction of the retail price is the desired profit of Item C?

d. What decimal fraction of the retail price is the cost of goods of Item A?

e. What decimal fraction of the retail price is the overhead expenses of Item B?

3–9   Adding Decimal Fractions

Adding and subtracting decimal fractions are required at various stages in the design and manufacture of products. An estimator in the apparel industry adds and subtracts decimal fractions of an hour in finding cutting and sewing times. Most bakery cost and production calculations are expressed as decimal fractions. A salesclerk adds decimal fractions of dollars when computing sales checks.

To add decimal fractions, arrange the numbers so that the decimal points are directly under each other. The decimal point of a whole number is directly to the right of the last digit. Add each column as with whole numbers. Place the decimal point in the sum directly under the other decimal points.

EXAMPLE •──

Add. 8.75 + 231.062 + 0.7398 + 0.007 + 23

Arrange the numbers so that the decimal points are directly under each other.	8.7500
	231.0620
Add zeros so that all numbers have the same number of places to the right of the decimal point.	0.7398
	0.0070
Add each column of numbers.	+ 23.0000
	263.5588 *Ans*
Place the decimal point in the sum directly under the other decimal points.	

3–10   Subtracting Decimal Fractions

To subtract decimal fractions, arrange the numbers so that the decimal points are directly under each other. Subtract each column as with whole numbers. Place the decimal point in the difference directly under the other decimal points.

EXAMPLE •──

Subtract. 44.6 − 27.368

Arrange the numbers so that the decimal points are directly under each other. Add zeros so that the numbers have the same number of places to the right of the decimal point.	44.600
	−27.368
	17.232 *Ans*
Subtract each column of numbers.	
Place the decimal point in the difference directly under the other decimal points.	

EXERCISE 3–10

Add the following numbers.

1. 0.237 + 0.395

2. 0.836 + 2.917 + 0.02

3. $37.65 + \dfrac{3}{4} + 0.133$

4. 2 + 0.2 + 0.02 + 0.002

5. $0.0009 + 0.03 + 0.1 + 0.005$

6. $0.012 + 0.0075 + 303$

7. $0.063 + 6\dfrac{3}{10} + 630 + 0.63$

8. $0.2073 + 0.209 + 23$

9. $0.313 + 3.032 + 97\dfrac{1}{40} + 0.138$

10. $16.8 + 23.066 + 0.00909 + 45$

Add the following numbers. Round each sum to the indicated number of decimal places.

11. $0.084 + 0.9988$ (3 places)

12. $35.035 + 3 + \dfrac{3}{4}$ (2 places)

13. $43.7 + 0.08 + 0.97$ (1 place)

14. $301.43 + 30.143 + 0.30143$ (3 places)

15. $87.010205 + 36\dfrac{9}{64}$ (4 places)

16. $44.4 + 9.306 + 0.0773$ (2 places)

Subtract the following numbers.

17. $7.932 - 3.107$

18. $0.98 - 0.899$

19. $0.001 - 0.0001$

20. $18\dfrac{3}{8} - 16.027$

21. $45.05 - 44.999$

22. $0.9 - 0.0009$

23. $0.414 - \dfrac{1}{4}$

24. $604.604 - 60.4604$

25. $23.345 - 3.3499$

26. $6\dfrac{91}{1,000} - 0.91$

27. $87.032 - 23.2032$

28. $905.7 - 68.0709$

29. $24.0303 - 20\dfrac{3}{125}$

30. $0.0001 - 0.00001$

Refer to the chart in Figure 3–6 for problems 31–33. Find how much greater value A is than value B.

	A	B
31.	312.067 pounds + 84.12 pounds	107.34 pounds + 172.9 pounds
32.	45.18 meters + 16.25 meters	55.055 meters + 3.25 meters
33.	9.5 inches + 14.66 inches	16.37 inches + 0.878 inch

Figure 3–6

3–11   Adding and Subtracting Decimal Fractions in Practical Applications

Often on-the-job computations require the combination of two or more different operations using decimal fractions. When solving a problem that requires both addition and subtraction operations, follow the procedures for each operation.

EXAMPLE •————————————————————————————————

A part of the structure of a building is shown in Figure 3–7. Girders are large beams under the first floor that carry the ends of the joists. Lally columns support the girders between the foundation walls. Find the length of the lally column in inches. Round the answer to 2 decimal places.

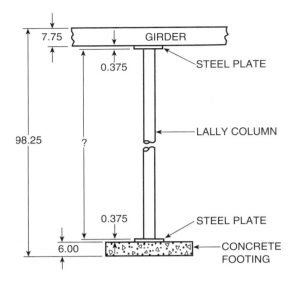

Figure 3–7

Length of lally column $= 98.25'' - (6.00'' + 0.375'' + 0.375'' + 7.75'')$

Add. $6.00'' + 0.375'' + 0.375'' + 7.75'' = 14.500''$

Subtract. $98.25'' - 14.500'' = 83.75''$ *Ans*, rounded to 2 decimal places

EXERCISE 3–11

Solve the following problems.

1. A hardware store clerk bills a cabinetmaker for the following items: nails, $6.85; locks, $13.47; hinges, $5.72; drawer pulls, $4; and cabinet catches, $6.09. What is the total amount of the bill?

2. The following amounts of concrete are delivered to a construction site in one week: 20.5, 32.8, 18.0, 28.75, and 48.3 cubic meters. How many total cubic meters are delivered during the week? Round the answer to 1 decimal place.

3. Find, in inches, each of the following distances on the base plate shown in Figure 3–8.

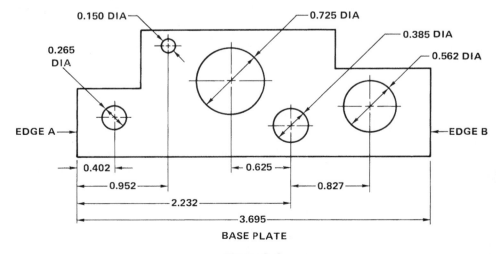

BASE PLATE

Figure 3–8

   a. The horizontal distance between the centers of the 0.265″ diameter hole and the 0.150″ diameter hole.

   b. The horizontal distance between the centers of the 0.385″ diameter hole and the 0.150″ diameter hole.

   c. The distance between edge A and the center of the 0.725″ diameter hole.

   d. The distance between edge B and the center of the 0.385″ diameter hole.

   e. The distance between edge B and the center of the 0.562″ diameter hole.

4. An environmental systems technician measures air pressure at each end of a duct. The first measurement is 0.042 lb/sq in and the second measurement is 0.026 lb/sq in. What is the pressure drop between the ends?

5. An environmental systems technician finds the volume of air flowing through pipes. The amount (volume) of air flowing through a pipe depends on the velocity of the air and the size (diameter) of the pipe. The table in Figure 3–9 lists the number of cubic feet of air flowing at various velocities through different diameter pipes. Round the answers to 1 decimal place.

VELOCITY OF AIR IN FEET PER SECOND	INSIDE DIAMETER OF PIPE IN INCHES			
	1″ Dia	2″ Dia	6″ Dia	10″ Dia
2 ft/s	0.65 cu ft/min	2.62 cu ft/min	23.6 cu ft/min	65.4 cu ft/min
5 ft/s	1.64 cu ft/min	6.55 cu ft/min	59.0 cu ft/min	163.0 cu ft/min
8 ft/s	2.62 cu ft/min	10.50 cu ft/min	94.0 cu ft/min	262.0 cu ft/min
12 ft/s	3.93 cu ft/min	15.70 cu ft/min	141.0 cu ft/min	393.0 cu ft/min

Figure 3–9

   a. At a velocity of 2 ft/s, what is the total volume of air per minute that flows through 4 pipes with diameters of 1″, 2″, 6″, and 10″?

   b. In 1 minute, how much more air flows through a 6″ diameter pipe at 5 ft/s than through a 2″ diameter pipe at 8 ft/s?

   c. In 1 minute, how much more air flows through a 10″ diameter pipe at 5 ft/s than through a 6″ diameter pipe at 12 ft/s?

   d. In 1 minute, how much more air flows through three 1″ diameter pipes at 12 ft/s than through two 2″ diameter pipes at 2 ft/s?

   e. In 1 minute, how much more air flows through one 10″ diameter pipe at 2 ft/s than through three 2″ diameter pipes at 5 ft/s?

6. An automatic screw machine supervisor estimated the setup times for 4 different jobs at a total of 6.25 hours. The jobs actually took 1.75 hours, 0.60 hour, 2.125 hours, and 1.40 hours, respectively. By how many hours was the total of the 4 jobs overestimated? Round the answer to 2 decimal places.

7. During a one-year period, the following kilowatt-hours (kWh) of electricity were consumed in a home.

January	693.75 kWh	May	663.18 kWh	September	668.43 kWh
February	678.24 kWh	June	658.33 kWh	October	659.98 kWh
March	674.83 kWh	July	665.09 kWh	November	671.06 kWh
April	666.05 kWh	August	672.46 kWh	December	682.33 kWh

   a. Find the total number of kilowatt-hours of electricity consumed for the year.

   b. How many more kilowatt-hours of electricity are used during the first 4 months than during the second 4 months of the year?

c. How many more kilowatt-hours of electricity are used during the highest monthly consumption than during the lowest monthly consumption?

8. When overhauling an engine, an automobile mechanic grinds the cylinder walls. After grinding, larger-diameter pistons than the original pistons are installed. On a certain job, the mechanic grinds the cylinder walls, which increases the diameter of the cylinders by 0.0400 inch. The diameters of the cylinders before grinding are 3.6250 inches. A clearance of 0.0025″ is required between the piston and cylinder wall. What size (diameter) pistons are ordered for this job? Refer to Figure 3–10.

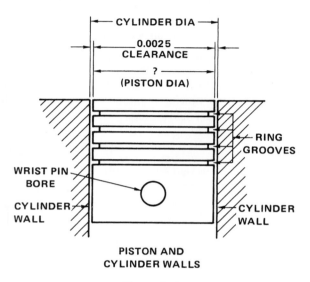

Figure 3–10

3–12   Multiplying Decimal Fractions

A payroll clerk computes the weekly wage of an employee who works 36.25 hours at an hourly rate of $16.75. In preparing a solution, a laboratory technician computes 0.125 of 0.5 liter of acid. A chef finds the total food cost for a banquet for 46 persons at $18.75 per person. A homeowner checks an electricity bill of $33.80 for 650 kilowatt-hours of electricity at $0.052 per kilowatt-hour. Multiplication of decimal fractions is required for these computations.

Recall that multiplication of whole numbers and fractions can be done in any order. Multiplication of decimal fractions can also be done in any order. For example, 4 × 0.3 equals 0.3 × 4.

To multiply decimal fractions, multiply using the same procedure as with whole numbers. Count the number of decimal places in both the multiplier and multiplicand. Begin counting from the last digit on the right of the product and place the decimal point (moving left) the same number of places as there are in both the multiplicand and the multiplier.

EXAMPLE

Multiply. 60.412 × 0.53

Multiply as with whole numbers.	60.412 (3 places)
Beginning at the right of the product, place the decimal point the same number of decimal places as there are in both the multiplicand and the multiplier.	$\times$ 0.53 (2 places)
	1 81236
	30 2060
	32.01836 (5 places)

32.01836 *Ans*

When multiplying certain decimal fractions, the product has a smaller number of digits than the number of decimal places required. For these products, add as many zeros to the left of the product as are necessary to give the required number of decimal places.

EXAMPLE •────────────────────────────────

Multiply. 0.0047 × 0.08. Round the answer to 4 decimal places.

Multiply as with whole numbers.

The product must have 6 decimal places.
Add 3 zeros to the left of the product.

Round to 4 decimal places.

$$\begin{array}{r} 0.0047 \text{ (4 places)} \\ \times\quad 0.08 \text{ (2 places)} \\ \hline 0.000376 \text{ (6 places)} \\ 0.0004 \ Ans \end{array}$$

Multiplying Three or More Factors

When multiplying three or more factors, multiply two factors. Then multiply the product by the third factor. Continue the process until all factors are multiplied.

EXAMPLE •────────────────────────────────

Find the product. $0.74 \times 14 \times 3.8 \times \dfrac{3}{5}$

Multiply. $0.74 \times 14 = 10.36$

Multiply. $10.36 \times 3.8 = 39.368$

Express the common fraction as a decimal $39.368 \times 0.6 = 23.6208 \ Ans$
fraction and multiply.

EXERCISE 3–12A
───

Multiply the following numbers.

1. 0.6×0.9 **5.** 0.053×0.4 **9.** 64.727×6.09

2. 0.42×0.8 **6.** 0.029×0.05 **10.** 124×4.0013

3. 10.25×0.12 **11.** 0.008×0.019

7. $3\dfrac{7}{8} \times 3.66$

4. $2.22 \times \dfrac{3}{4}$ **8.** 0.001×0.01 **12.** $5.077 \times 6\dfrac{3}{25}$

Multiply the following numbers. Round the answers to the indicated number of decimal places.

13. 0.009×0.5 (3 places)

14. 5.26×4.923 (4 places)   **16.** $1.08 \times 2\dfrac{9}{16}$ (3 places)

15. 800.75×10.1 (2 places)   **17.** 0.0304×0.088 (5 places)

18. 0.001×0.006 (5 places)

Multiply the following numbers. Each expression has three or more factors.

19. $0.009 \times 0.09 \times 0.9$ **22.** $0.001 \times 1,000 \times 0.01 \times 100$

20. $2.3 \times \dfrac{1}{4} \times 72.4$ **23.** $3.3 \times 0.33 \times \dfrac{33}{1,000} \times 33$

21. $0.5 \times 0.5 \times 5.5 \times 55$ **24.** $2,817 \times 0.63 \times 78.007$

Multiply the following numbers. Each expression has three or more factors. Round the answers to the indicated number of decimal places.

25. $0.87 \times 3.12 \times 0.06$ (4 places)

26. $14.9 \times 8.25 \times 105 \times 0.4$ (1 place)

27. $0.021 \times 0.376 \times 0.6 \times 42$ (5 places)

28. $88.99 \times 5.4 \times 45 \times 0.46$ (3 places)

Multiplying by Powers of 10

The method of multiplying by powers of 10 is quick and easy to apply. The decimal system is based on groupings of 10. This method of multiplication is based on groupings of 10 place values.

To multiply a number by 10, 100, 1,000, 10,000, and so on, move the decimal point in the multiplicand as many places to the right as there are zeros in the multiplier. If there are not enough digits in the multiplicand, add zeros to the right of the multiplicand.

EXAMPLES

1. $0.085 \times 10 = 0.85$ *Ans*

(1 zero in 10; move 1 place to the right, 0.0 85)

2. $0.085 \times 100 = 8.5$ *Ans*

(2 zeros in 100; move 2 places to the right, 0.08 5)

3. $0.085 \times 1,000 = 85$ *Ans*

(3 zeros in 1,000; move 3 places to the right, 0.085)

4. $0.085 \times 10,000 = 850$ *Ans*

(4 zeros in 10,000; move 4 places to the right, 0.0850)

It is necessary to add 1 zero to the right since the multiplicand 0.085 has only 3 digits.

To multiply a number by 0.1, 0.01, 0.001, 0.0001, and so on, move the decimal point in the multiplicand as many places to the left as there are decimal places in the multiplier. If there are not enough digits in the multiplicand, add zeros to the left of the multiplicand.

EXAMPLES

1. $127.5 \times 0.1 = 12.75$ *Ans*

(1 decimal place in 0.1; move 1 place to the left, 12 7.5)

2. $127.5 \times 0.0001 = 0.01275$ *Ans*

(4 decimal places in 0.0001; move 4 places to the left, 0 0127.5)

It is necessary to add 1 zero to the left since the multiplicand 127.5 has only 3 digits to the left of the decimal point.

EXERCISE 3–12B

Multiply the following numbers. Use the rules for multiplying by a power of 10.

1. 0.72×10

2. $18.7 \times 1,000$

3. 0.005×100

4. $0.039 \times 10,000$

5. $312.88 \times 100,000$

6. 3.7×0.1

7. 0.08×0.01

8. 25.032×0.001

9. 843×0.0001

10. 900.3×0.0001

11. 0.033×100

12. 87.9×0.001

13. $9.35 \times 1,000$

14. $0.0723 \times 10,000$

15. 707×0.0001

3–13   Multiplying Decimal Fractions in Practical Applications

EXAMPLE ●————————————————————————

Large metal parts can be electroplated in a plating bath tank. A plating bath is a solution that contains the plating metal. Parts to be plated are immersed in the bath for a certain period of time. Electroplating provides corrosion protection and often makes a product more attractive. An electroplater finds the number of gallons of plating bath needed for various tank sizes. Find the number of gallons of plating bath needed to provide a $4\frac{1}{4}$-foot bath depth in the tank shown in Figure 3–11. One cubic foot of liquid contains 7.479 gallons. Round the answer to the nearest ten gallons.

$$V = l \times w \times h \qquad \text{where } V = \text{volume}$$
$$l = \text{length}$$
$$w = \text{width}$$
$$h = \text{height (depth of bath)}$$

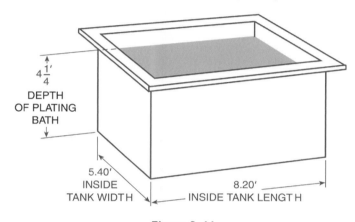

Figure 3–11

1. Compute the volume of the bath: V = 8.20 ft. × 5.40 ft. × 4.25 ft. = 188.19 cu ft

2. Since 1 cubic foot of liquid contains 7.479 gallons, 188.19 cubic feet contain 188.19 × 7.479 gallons.

   188.19 × 7.479 gal = 1,407.473 gal   1,410 gal *Ans,* rounded to the nearest ten gallons

————————————————————————————●

EXERCISE 3–13 ————————————————————————————————————

Solve the following problems.

1. An empty 50-gallon drum weighs 35.75 pounds. What is the weight of the drum when it is filled with tile grout? One gallon of grout weighs 8.50 pounds.

2. A mason loads materials for a job onto a pickup truck. The truck is rated to carry a maximum load of 1.75 tons (1 short ton = 2,000 lb). The following materials are loaded on the truck: 82 four-foot lengths of reinforcing rod, which weigh 0.375 lb/ft; 110 hollow wall tiles, which weigh 21.25 pounds each; and $8\frac{1}{2}$ bags of mortar, which weigh 100 pounds each. How many more pounds of materials can be loaded onto the truck to bring the complete load to 1.75 tons? Round the answer to the nearest hundred pounds.

3. In order to determine selling prices of products, a baker finds the cost of ingredients and adds estimated profit. For large production products, ingredient costs are often broken down to 3-place decimal fractions of a dollar per ounce of ingredients. The total ingredient cost for a cake is $0.065 per ounce (1 pound = 16 ounces). The estimated profit is $1.73. Find the selling price of a cake that weighs 1.625 pounds.

4. An inspector measures various dimensions of a part. Using the front view of the mounting block shown in Figure 3–12, find dimensions A, B, and C in millimeters.

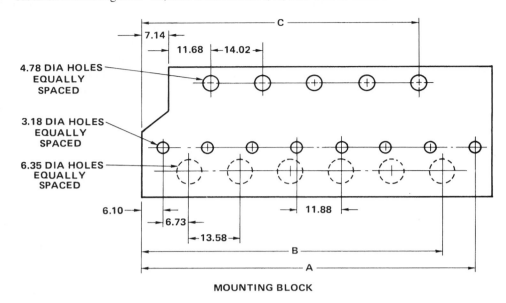

MOUNTING BLOCK

Figure 3–12

NOTE: The 6.35-mm-diameter holes are drawn with broken or hidden lines. Broken lines show that the holes are drilled from the back and do not go through the part. The holes cannot be seen when viewing the part from the front.

5. An air-conditioning technician determines ventilation air requirements of a building by either of the following two methods:

• Total ventilation air required in cubic feet per minute = ventilation required per person in cubic feet per minute × number of persons.

• Total ventilation required in cubic feet per minute = ventilation required per square foot of floor in cubic feet per minute × number of square feet of floor.

The ventilation requirements depend on the use of the building. The ventilation requirements are greater for a hospital than for a store having the same number of persons or the same number of square feet. The table in Figure 3–13 lists different types of buildings with their per person and per square foot of floor requirements.

TYPE OF BUILDING	VENTILATION AIR REQUIREMENTS IN CUBIC FEET PER MINUTE	
	Per Person	Per Square Foot of Floor
Apartment	17.500	0.330
Store	6.250	0.050
Factory	8.750	0.100
Hospital	25.500	0.330
Office	20.250	0.250

Figure 3–13

What is the total ventilation air required in cubic feet per minute for each of the following buildings? Round the answers to the nearest cubic foot.

a. An apartment with 1,250 square feet.

b. A factory with 115 employees.

c. A factory with 12,500 square feet.

d. An office with 15 employees.

e. An apartment with 5 occupants.

f. A hospital with 23,000 square feet.

g. A supermarket with 11,000 square feet.

h. A hospital with an average of 210 patients and 85 employees.

6. A payroll clerk computes the net wages for employees A, B, and C. The net wage is the wage received after all deductions have been made. The gross wage is the wage before any deductions are made. All payroll deductions are based on the gross wage, except health insurance.

The deductions are shown in the table in Figure 3–14. The formula used for computing each deduction or the amount of deduction is given for employees A, B, and C.

Deduction	Formula or Amount
1. Federal Withholding Tax	Employee A: 0.1372 × Gross Wage Employee B: 0.1455 × Gross Wage Employee C: 0.1265 × Gross Wage
2. Social Security and Medicare	0.0765 × Gross Wage (the same rate for all employees)
3. Retirement	0.025 × Gross Wage (the same rate for all employees)
4. Health Insurance	$27.18 (the same amount for all employees)

Figure 3–14

Determine the net wage for each employee in Figure 3–15.

Employee	Number of Hours Worked in Week	Hourly Rate of Pay	Net Wage
A	44.25	$14.73	
B	37.75	$15.28	
C	46.5	$13.62	

Figure 3–15

3–14   Dividing Decimal Fractions

A retailer divides with decimal fractions when computing the unit cost of a product purchased in wholesale quantities. Insurance rates and claim payments are calculated by division of decimal fractions. Division with decimal fractions is used to compute manufacturing time per piece after total production times are determined.

As with division of whole numbers and common fractions, division of decimal fractions is a short method of subtracting a subtrahend a given number of times.

To divide decimal fractions, use the same procedure as with whole numbers. Move the decimal point of the divisor as many places to the right as necessary to make the divisor a whole number. Move the decimal point of the dividend the same number of places to the right. Since division can be expressed as a fraction, the value does not change if both the numerator and denominator are multiplied by the same nonzero number. Add zeros to the dividend if necessary. Place the decimal point in the quotient directly above the decimal point in the dividend. Divide as with whole numbers. Zeros may be added to the dividend to give the degree of precision needed in the quotient.

EXAMPLES

1. Divide. $0.3380 \div 0.52$

Move the decimal point 2 places to the right in the divisor.

Move the decimal point 2 places in the dividend.

Place the decimal point in the quotient directly above the decimal point in the dividend.

Divide.

$$
\begin{array}{r}
0.65 \; Ans \\
0\,52.)\overline{0\,33.80} \\
\underline{31\,2} \\
2\,60 \\
\underline{2\,60}
\end{array}
$$

2. Divide. $11.9 \div 3.072$

Round the answer to 3 decimal places.

Move the decimal point 3 places to the right in the divisor.

Move the decimal point 3 places in the dividend, adding 2 zeros.

Place the decimal point directly above the decimal point in the dividend.

Add 4 zeros to the dividend. One more zero is added than the degree of precision required.

$$
\begin{array}{r}
3.8736 \approx 3.874 \; Ans \\
3\,072.)\overline{11\,900.0000} \\
\underline{9\,216} \\
2\,684\,0 \\
\underline{2\,457\,6} \\
226\,40 \\
\underline{215\,04} \\
11\,360 \\
\underline{9\,216} \\
2\,1440 \\
\underline{1\,8432} \\
3008
\end{array}
$$

EXERCISE 3–14A

Divide the following numbers.

1. $0.6 \div 0.3$

2. $1.2 \div 0.4$

3. $0.72 \div 0.04$

4. $1.875 \div \dfrac{3}{5}$

5. $0.3675 \div \dfrac{1}{4}$

6. $8.024 \div 1.003$

7. $48.036 \div 12$

8. $105 \div 5.25$

9. $10.5 \div 52.5$

10. $0.001 \div 0.125$

11. $0.8 \div 0.02$

12. $12.3 \div 4.1$

13. $0.1875 \div 1.5$

14. $35\dfrac{15}{16} \div 7.1875$

15. $28.747 \div 8.90$

16. $5.948 \div 14.87$

17. $5,948 \div 148.7$

18. $292.1142 \div 18.606$

19. $1.9875 \div \dfrac{5}{16}$

20. $0.0000651 \div 0.0093$

Divide the following numbers. Round the answers to the indicated number of decimal places.

21. $0.613 \div 0.912$ (3 places)

22. $7.059 \div 6.877$ (2 places)

23. $28,900 \div 440,110$ (4 places)

24. $6.998 \div 0.03$ (2 places)

25. $0.4233 \div \dfrac{3}{4}$ (3 places)

26. $5,189.7 \div 6.9$ (1 place)

27. $0.1270 \div 47.60$ (4 places)

28. $9.10208 \div \dfrac{7}{32}$ (5 places)

29. $0.0093 \div 0.979$ (2 places)

30. $15.6309 \div 1.842$ (3 places)

Dividing by Powers of 10

Division is the inverse of multiplication. Dividing by 10 is the same as multiplying by $\frac{1}{10}$ or 0.1. For example, $32.5 \div 10 = 32.5 \times 0.1$. Dividing a number by 10, 100, 1,000, 10,000, and so on, is the same as multiplying a number by 0.1, 0.01, 0.001, 0.0001, and so on. To divide a number by 10, 100, 1,000, 10,000, and so on, move the decimal point in the dividend as many places to the left as there are zeros in the divisor. If there are not enough digits in the dividend, add zeros to the left of the dividend.

EXAMPLES •

1. $732.4 \div 100 = 7.324$ *Ans*   (2 zeros in 100; move 2 places to the left, 7 32.4)

2. $732.4 \div 10,000 = 0.073\ 24$ *Ans*   (4 zeros in 10,000; move 4 places to the left, 0 0732.4)

It is necessary to add 1 zero to the left since the dividend, 732.4, has only 3 digits to the left of the decimal point.

Dividing a number by 0.1, 0.01, 0.001, 0.0001, and so on, is the same as multiplying by 10, 100, 1,000, 10,000, and so on. To divide a number by 0.1, 0.01, 0.001, 0.0001, and so on, move the decimal point in the dividend as many places to the right as there are decimal places in the divisor. If there are not enough digits in the dividend, add zeros to the right of the dividend.

EXAMPLES •

1. $0.065 \div 0.001 = 65$ *Ans*   (3 decimal places in 0.001; move 3 places to the right, 0.065)

2. $0.065 \div 0.0001 = 650$ *Ans*   (4 decimal places in 0.0001; move 4 places to the right, 0.0650)

It is necessary to add 1 zero to the right since the dividend, 0.065, has only 3 digits.

EXERCISE 3–14B

Divide the following numbers. Use the rules for dividing by a power of 10.

1. $0.37 \div 10$

2. $0.09 \div 100$

3. $732 \div 1,000$

4. $1.77 \div 0.001$

5. $0.052 \div 0.0001$

6. $3.288 \div 0.00001$

7. $63.9 \div 1000$

8. $72,086 \div 100,000$

9. $806.58 \div 10,000$

10. $0.901 \div 0.01$

11. $9.08 \div 0.001$

12. $32.7 \div 1,000$

13. $0.087 \div 1,000$

14. $818 \div 10,000$

3–15   Dividing Decimal Fractions in Practical Applications

EXAMPLE •──

Twenty grooves are machined in a plate shown in Figure 3–16. All grooves are equally spaced. Find the center-to-center distance, dimension A, between two consecutive grooves.

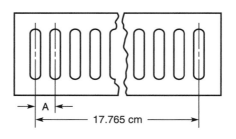

Figure 3–16

There is one less space between grooves (19) than the number of grooves (20). 17.765 cm ÷ 19 = 0.935 cm *Ans*

Actual on-the-job problems often require a combination of different operations in their solutions. A problem must first be thought through to determine how it is going to be solved. After the steps in the solution are determined, apply the procedures for each operation.

EXAMPLE •──

An estimator for a die-casting firm quotes an order for 2,850 castings at a selling price of $3.43 per casting. The materials (molten metals) used to produce the 2,850 castings are listed in the table shown in Figure 3–17. In addition to materials, costs in producing the 2,850 castings are as follows: die cost, $1,750.75; overhead cost, $510.25; labor cost, $625.50.

MATERIAL REQUIRED TO PRODUCE 2,850 CASTINGS		
Material (Molten Metal)	Number of Pounds Required	Cost per Pound
Zinc	5,520.00	$0.52
Aluminum	308.50	$0.43
Copper	180.25	$0.78

Figure 3–17

Find the profit per casting.

Material cost

Cost of zinc	5,520.00 × $0.52 = $2,870.40
Cost of aluminum	308.50 × $0.43 = $132.66
Cost of copper	180.25 × $0.78 = $140.60
Cost of materials	$2,870.40 + $132.66 + $140.60 = $3,143.66
Total cost	$3,143.66 + $1,750.75 + $510.25 + $625.50 = $6,030.16
Cost per casting	$6,030.16 ÷ 2,850 = $2.12
Profit per casting	$3.43 − $2.12 = $1.31 *Ans*

EXERCISE 3–15

Solve the following problems.

1. The amounts of trap rock delivered to a construction site during 1 week are listed in the table in Figure 3–18. Trap rock weighs 1.28 tons per cubic yard. Find the number of cubic yards of trap rock delivered each day. Round answers to 1 decimal place.

Day	Number of Tons Delivered	Number of Cubic Yards Delivered
Monday	21.75	
Tuesday	19.30	
Wednesday	30.80	
Thursday	29.60	
Friday	18.48	

Figure 3–18

2. An offset printer bases prices charged for work on the number of pages per job. The price per page is reduced as the number of pages per job is increased. Find the number of pages (one side of a sheet) printed for each of the following jobs.

    a. Job A: total printing price, $263.70, price per page, $0.045.

    b. Job B: total printing price, $368.76, price per page, $0.042.

    c. Job C: total printing price, $490.20, price per page, $0.038.

3. A floor covering installer is contracted to install vinyl tile in a building. After measuring and finding the building floor area, the installer determines the number of tiles required for the job. Two different size tiles are to be used.

    Ten-inch square tile: Each tile covers an area of 0.694 square foot. A floor area of 3,760 square feet is to be covered with 10-inch square tiles.

    Fourteen-inch square tile: Each tile covers an area of 1.361 square feet. A floor area of 5,150 square feet is to be covered with 14-inch square tiles.

    a. Find the number of 10-inch square tiles needed to the nearest ten tiles. No allowance is made for waste.

    b. Find the number of 14-inch square tiles needed to the nearest ten tiles. No allowance is made for waste.

4. Given the following data, compute the profit per casting.

    Selling price per casting, $2.64   Labor cost, $608.50

    Number of castings, 3,150   Overhead cost, $716.75

    Die cost, $983.25

    Material quantities and costs are listed in the table in Figure 3–19.

MATERIAL REQUIRED TO PRODUCE 3,150 CASTINGS		
Material (Molten Metal)	Number of Kilograms Required	Cost per Kilogram
Zinc	3048.00	$1.14
Aluminum	174.75	$0.95
Copper	102.5	$1.72

Figure 3–19

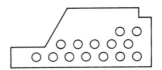

Figure 3–20

5. The bracket shown in Figure 3–20 is part of an aircraft assembly. It is important that the bracket be as light in weight as possible. To reduce weight, equal-sized holes are drilled in the bracket. The bracket weighs 5.20 kilograms before the holes are drilled. After 14 holes are drilled in the bracket, the weight is reduced by 1.26 kilograms. How many more holes of the same size must be drilled to reduce the weight of the bracket to 3.40 kilograms?

6. A home remodeler purchases hardware in the quantities listed in the table in Figure 3–21.

Items	Total Quantity Purchased	Total Cost of Purchased Quantities
Cabinet Hinges	12 boxes	$50.52
Drawer Pulls	15 boxes	$24.15
Cabinet Knobs	20 boxes	$36.60
Hanger Bolts	5 boxes	$14.70
Magnetic Catches	8 boxes	$18.24

Figure 3–21

The remodeler is able to reduce unit costs by quantity purchases. For a certain kitchen remodeling job, the following quantities of hardware are used:

5 boxes of cabinet hinges

$8\frac{1}{2}$ boxes of drawer pulls

7 boxes of cabinet knobs

$1\frac{1}{4}$ boxes of hanger bolts

3 boxes of magnetic catches

Find the total hardware cost that should be charged against this job.

7. Series electrical circuits are shown in Figure 3–22. In a series circuit the total circuit resistance (R_T) equals the sum of the individual resistances.

$$R_T = R_1 + R_2 + R_3 + R_4 + R_5$$

Current in the circuit (I) equals voltage (E) applied to the circuit divided by the total resistance (R_T) of the circuit.

$$I \text{ (amperes)} = \frac{E \text{ (volts)}}{R_T \text{ (ohms)}}$$

Determine the current (amperes) in each of the circuits to 1 decimal place.

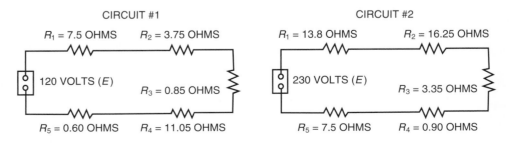

Figure 3–22

3–16   Powers and Roots of Decimal Fractions

Powers of numbers are used to find the area of square surfaces and circular sections. Volumes of cubes, cylinders, and cones are determined by applying the power operation. Determining roots of numbers is used to find the lengths of sides and heights of certain geometric figures. Both powers and roots are required operations in solving many formulas in the electrical, machine, construction, and business occupations.

Meaning of Powers

Two or more numbers multiplied to produce a given number are *factors* of the given number. Two factors of 8 are 2 and 4. The factors of 15 are 3 and 5. A *power* is the product of two or more equal factors. An *exponent* shows how many times a number is taken as a factor. It is written smaller than the number, above the number, and to the right of the number.

EXAMPLES •————————————————————————————

Find the indicated powers for each of the following.

1. 2^5

2^5 means $2 \times 2 \times 2 \times 2 \times 2$; 2 is taken as a factor
5 times. It is read, "two to the fifth power." $2^5 = 32$ *Ans*

2. 0.8^3

0.8^3 means $0.8 \times 0.8 \times 0.8$; 0.8 is taken as a factor 3 times.
It is read, "0.8 to the third power" or "0.8 cubed." $0.8^3 = 0.512$ *Ans*

Use of Parentheses

Parentheses are used as a grouping symbol. When an expression consisting of operations within parentheses is raised to a power, the operations within the parentheses are performed first. The result is then raised to the indicated power.

EXAMPLE •————————————————————————————

Raise to the indicated power. $(1.4 \times 0.3)^2$

Perform the operations within the parentheses first. $(1.4 \times 0.3)^2 = 0.42^2 = 0.1764$ *Ans*
Raise to the indicated power.

Parentheses that enclose a fraction indicate that both the numerator and denominator are raised to the given power.

$$\left(\frac{3}{4}\right)^2 = \frac{3^2}{4^2} = \frac{9}{16} = 0.5625$$

The same answer is obtained by dividing first and squaring second, as by squaring both terms first and dividing second.

$$\left(\frac{3}{4}\right)^2 = 0.75^2 = 0.5625$$

EXERCISE 3–16A

Raise the following numbers to the indicated powers.

1. 0.8^2	**6.** 2^6	**11.** 0.13^3
2. 0.9^3	**7.** 315^2	**12.** 0.2^4
3. 1^7	**8.** 9.6^3	**13.** 3^5
4. 23.25^2	**9.** 0.24^3	**14.** 125.25^2
5. 0.04^3	**10.** 100^3	**15.** 0.009^3

Raise the following expressions to the indicated powers.

16. $\left(\dfrac{1}{2}\right)^2$

17. $\left(\dfrac{1}{2}\right)^3$

18. $(10 \times 1.6)^3$

19. $(0.07 + 2.93)^4$

20. $(33.54 - 21.27)^2$

21. $\left(\dfrac{3}{8}\right)^2$

22. $\left(\dfrac{3}{5}\right)^3$

23. $(14.8 - 11.8)^5$

24. $(9.9 \times 0.01)^2$

25. $(2 + 0.5)^2$

26. $(2 + 0.5)^4$

27. $\left(\dfrac{13}{20}\right)^2$

28. $(187.5 - 186)^3$

29. $(1,000 \times 0.001)^8$

30. $(0.36 \times 18.14)^2$

31. $(175 \times 0.04)^2$

32. $\left(\dfrac{9}{10}\right)^3$

33. $(14.3 - 14.1)^4$

Description of Roots

The *root* of a number is a quantity that is taken two or more times as an equal factor of the number. Finding a root is the opposite or *inverse* operation of finding a power.

The *radical symbol* $\left(\sqrt{}\right)$ is used to indicate a root of a number. The *index* indicates the amount of times that a root is to be taken as an equal factor to produce the given number called the *radicand.* The index is written smaller than the number, above and to the left of the radical symbol. The index 2 is usually omitted. For example, $\sqrt{9}$ means to find the number that can be multiplied by itself and equal 9. In the expression $\sqrt[3]{8}$, the index, 3, indicates the root is taken as a factor 3 times to equal 8. The cube root of 8 is 2.

EXAMPLES •————————————————

Find the indicated roots. The examples have whole number roots that can be determined by observation.

1. $\sqrt{36}$ ⟶ $6 \times 6 = 36$; therefore, $\sqrt{36} = 6$ *Ans*

2. $\sqrt{144}$ ⟶ $12 \times 12 = 144$; therefore, $\sqrt{144} = 12$ *Ans*

3. $\sqrt[3]{8}$ ⟶ $2 \times 2 \times 2 = 8$; therefore, $\sqrt[3]{8} = 2$ *Ans*

4. $\sqrt[3]{125}$ ⟶ $5 \times 5 \times 5 = 125$; therefore, $\sqrt[3]{125} = 5$ *Ans*

5. $\sqrt[4]{81}$ ⟶ $3 \times 3 \times 3 \times 3 = 81$; therefore, $\sqrt[4]{81} = 3$ *Ans*

Expressions Enclosed within the Radical Symbol

The radical symbol is a grouping symbol. An expression consisting of operations within the radical symbol is done using the order of operations. The operations within the radical symbol are performed first. Next find the root.

EXAMPLES •————————————————————————————————

The examples have whole number roots that can be determined by observation.

1. $\sqrt{3 \times 12}$ ———————————→ $\sqrt{36} = \sqrt{6 \times 6}$; therefore, $\sqrt{3 \times 12} = 6$ *Ans*

2. $\sqrt[3]{9.2 + 54.8}$ —————————→ $\sqrt[3]{64} = \sqrt[3]{4 \times 4 \times 4}$; therefore, $\sqrt[3]{9.2 + 54.8} = 4$ *Ans*

3. $\sqrt{148.2 - 27.2}$ —————————→ $\sqrt{121} = \sqrt{11 \times 11}$; therefore, $\sqrt{148.2 - 27.2} = 11$ *Ans*

A radical symbol that encloses a fraction indicates that the roots of both the numerator and denominator are to be taken.

$$\sqrt{\frac{900}{4}} = \frac{\sqrt{900}}{\sqrt{4}}$$

The same answer is obtained by dividing first and extracting the root second as by extracting both roots first and dividing second.

EXAMPLES •————————————————————————————————

1. Dividing first and extracting the root second: $\qquad\qquad \sqrt{\dfrac{900}{4}} = \sqrt{225} = 15$ *Ans*

2. Extracting both roots first and dividing second: $\qquad \dfrac{\sqrt{900}}{\sqrt{4}} = \dfrac{30}{2} = 15$ *Ans*

EXERCISE 3–16B

Determine the indicated whole number roots of the following numbers by observation.

1. $\sqrt{25}$ **5.** $\sqrt[3]{27}$ **9.** $\sqrt[3]{64}$

2. $\sqrt{100}$ **6.** $\sqrt[4]{16}$ **10.** $\sqrt[4]{10,000}$

3. $\sqrt{49}$ **7.** $\sqrt{144}$ **11.** $\sqrt[5]{32}$

4. $\sqrt[3]{1}$ **8.** $\sqrt[3]{125}$ **12.** $\sqrt[6]{64}$

Determine the indicated whole number roots of the following expressions by observation.

13. $\sqrt{3.1 + 12.9}$ **19.** $\sqrt[3]{127.3 - 63.3}$

14. $\sqrt{10.7 - 6.7}$ **20.** $\sqrt{2.5 \times 25.6}$

15. $\sqrt{0.7 \times 70}$

16. $\sqrt[3]{40 \times 0.2}$ **21.** $\sqrt[4]{\dfrac{54.4}{3.4}}$

17. $\sqrt{87.64 + 12.36}$ **22.** $\sqrt{99.03 + 125.97}$

                          23. $\sqrt[3]{101.7 + 23.3}$

18. $\sqrt{\dfrac{360}{2.5}}$ **24.** $\sqrt[4]{6.25 \times 0.16}$

Roots That Are Not Whole Numbers

The root examples and exercises have all consisted of numbers that have whole number roots. These roots are relatively easy to determine by observation.

Most numbers do not have whole number roots. For example, $\sqrt{259} = 16.0935$ (rounded to 4 decimal places) and $\sqrt[3]{17.86} = 2.6139$ (rounded to 4 decimal places). The root of any positive number can easily be computed with a calculator. Calculator solutions to root expressions are given at the end of this unit on pages 120 and 121.

3–17   Decimal Fraction Powers and Roots in Practical Applications

In occupational uses, power and root operations are most often applied in combination with other operations. Before making any computations, think a problem through to determine the steps necessary in its solution.

EXAMPLE •————————————————————————————————————

A construction technician finds the weight of 8 concrete pier footings shown in Figure 3–23. Footings distribute the load (weight) of walls and support columns over a larger area. The concrete used for the pier footings weighs 2,380 kilograms per cubic meter. Determine the weight of the pier footings in metric tons. One metric ton equals 1,000 kilograms. Round the answer to the nearest tenth metric ton.

$$V = s^3 \qquad \text{where } V = \text{volume}$$
$$s = \text{side}$$

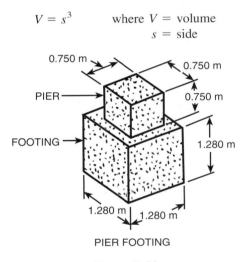

Figure 3–23

Volume of pier: $0.750 \text{ m} \times 0.750 \text{ m} \times 0.750 \text{ m} = 0.421875 \text{ m}^3$

Volume of footing: $1.280 \text{ m} \times 1.280 \text{ m} \times 1.280 \text{ m} = 2.097152 \text{ m}^3$

Total volume: $0.421875 \text{ m}^3 + 2.097152 \text{ m}^3 = 2.519027 \text{ m}^3$

Weight of 1 pier footing:

$$\frac{2.519027 \text{ m}^3}{1} \times \frac{2,380 \text{ kg}}{\text{m}^3} = 5,995.28426 \text{ kg}$$

Weight of 8 pier footings in kilograms:

$$8 \times 5,995.28426 \text{ kg} = 47,962.27408 \text{ kg}$$

Weight of 8 pier footings in metric tons:

$$\frac{47,962.27408 \text{ kg}}{1} \times \frac{1 \text{ metric ton}}{1,000 \text{ kg}} \approx$$

48.0 metric tons *Ans,* rounded to the nearest tenth metric ton

EXERCISE 3–17

*Solve the following problems. The problems that involve computation of roots require calcula-
tor solutions. Refer to pages 126 and 127 for calculator root solutions.*

1. In the table shown in Figure 3–24, the lengths of the sides of squares are given. Find the
   areas of the squares to 2 decimal places.

	Length of Sides (s)	Area (A)
a.	1.25 ft	
b.	0.325 cm	
c.	2.35 yd	
d.	0.66 km	
e.	0.085 ft	

$A = s^2$ where A = area
              s = side

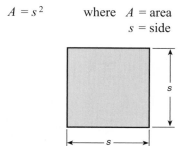

Figure 3–24

2. In the table shown in Figure 3–25, the lengths of the sides of cubes are given. Determine
   the volumes of the cubes to 3 decimal places.

	Length of Sides (s)	Volume (V)
a.	9.705 mm	
b.	3.860 in	
c.	6.600 ft	
d.	1.075 cm	
e.	0.88 ft	

$V = s^3$ where V = volume
              s = side

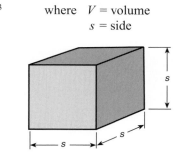

Figure 3–25

3. In the table shown in Figure 3–26, the areas of squares are given. Determine the lengths of
   the sides of the squares to 2 decimal places where necessary.

	Area (A)	Length of Sides (s)
a.	125.0 sq ft	
b.	8.76 km²	
c.	57.75 sq in	
d.	0.88 m²	
e.	2,479 sq ft	

$s = \sqrt{A}$ where s = side
              A = area

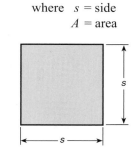

Figure 3–26

4. In the table shown in Figure 3–27, the volumes of cubes are given. Determine the lengths of the sides of the cubes to 2 decimal places where necessary.

	Volume (V)	Length of Sides (s)
a.	18.60 cm³	
b.	143.77 cu ft	
c.	1.896 m³	
d.	0.750 cu yd	
e.	953.25 cu ft	

$s = \sqrt[3]{V}$ where s = side
V = volume

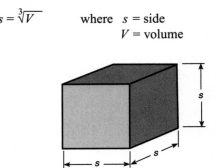

Figure 3–27

5. Find the current in amperes of the circuits listed in the table in Figure 3–28. Express the answer to the nearest tenth ampere when necessary.

$$I = \sqrt{\dfrac{P}{R}}$$ where I = current in amperes
P = power in watts
R = resistance in ohms

Circuit	Power (P)	Resistance (R)	Current (I)
a	288 watts	2.20 ohms	
b	2,320 watts	5.90 ohms	
c	3,050 watts	9.60 ohms	
d	5,240 watts	14.70 ohms	

Figure 3–28

NOTE: Use these formulas for problems 6–11.

$$A = s^2$$ where A = area
$$V = s^3$$ V = volume
$$s = \sqrt{A}$$ s = side
$$s = \sqrt[3]{V}$$

6. Compute the cost of filling a hole 5.50 meters long, 5.50 meters wide, and 5.50 meters deep. The cost of fill soil is $4.75 per cubic meter. Round the answer to the nearest dollar.

7. A paving contractor, in determining the cost of a job, finds the number of square feet (area) to be paved. The shaded area that surrounds a building, as shown in Figure 3–29, is paved. All dimensions are in feet. At $0.87 per square foot, determine the cost of the job to the nearest ten dollars.

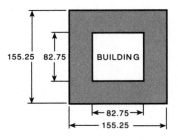

Figure 3–29

8. Keys and keyways have wide applications with shafts and gears. The relationship between shaft diameters and key sizes is shown by the following formula.

$$D = \sqrt{\frac{L \times T}{0.30}}$$

where D = shaft diameter
L = key length
T = key thickness

What is the shaft diameter that would be used with a key where L = 2.70 inches and T = 0.25 inch?

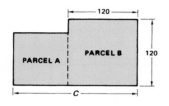

Figure 3–30

9. A plot of land consists of two parcels, A and B, as shown in Figure 3–30. Both parcels are squares. The plot has a total area of 22,500 square meters. Find, in meters, length C. All dimensions are in meters.

10. A steel storage tank as shown in Figure 3–31 is to be fabricated by a welder. The tank is to be made in the shape of a cube capable of holding 2,750 gallons of fuel. One cubic foot contains 7.48 gallons. Find the length of one side of the storage tank. Round the answer to the nearest hundredth foot.

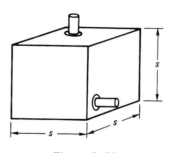

Figure 3–31

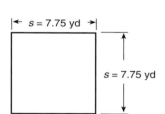

Figure 3–32

11. Find the total cost of carpet and installation for the office floor plan shown in Figure 3–32. The carpet is priced at $38.50 per square yard, a waste allowance of 4.30 square yards is made, and the installation cost is $2.25 per square yard. Round the answer to the nearest ten dollars.

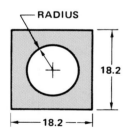

Figure 3–33

12. The plate shown in Figure 3–33 is designed to contain 210 square centimeters or metal after the circular cutout is removed. A designer finds the length of the radius of the cutout to determine the size of the circle to be removed. Find, in centimeters, the length of the required radius to 2 decimal places. All dimensions are in centimeters.

NOTE: A radius is a straight line that connects the center of a circle with a point on the circle.

$$R = \sqrt{\frac{A}{3.1416}}$$

where R = radius
A = area of circle

3–18   Table of Decimal Equivalents

Generally, fractional machine, mechanical, and sheet metal blueprint dimensions are given in multiples of 64ths of an inch. Carpenters, cabinetmakers, and many other woodworkers measure in multiples of 32nds of an inch.

In certain occupations, it is often necessary to express fractional dimensions as decimal dimensions. A machinist is required to express fractional dimensions as decimal equivalents for machine settings in making a part. Decimal dimensions are expressed as fractional dimensions if fractional measuring devices are used. A patternmaker may express decimal dimensions to the nearest equivalent 64th inch.

Using a decimal equivalent table saves time and reduces the chance of error. Decimal equivalent tables are widely used in business and industry. They are posted as large wall charts in work areas and are available as pocket size cards. Skilled workers memorize many of the equivalents after using decimal equivalent tables.

The decimals listed in the table in Figure 3–34 are given to six places. In actual practice, a decimal is rounded to the degree of precision desired for a particular application.

DECIMAL EQUIVALENT TABLE			
1/64	0.015625	33/64	0.515625
1/32	0.03125	17/32	0.53125
3/64	0.046875	35/64	0.546875
1/16	0.0625	9/16	0.5625
5/64	0.078125	37/64	0.578125
3/32	0.09375	19/32	0.59375
7/64	0.109375	39/64	0.609375
1/8	0.125	5/8	0.625
9/64	0.140625	41/64	0.640625
5/32	0.15625	21/32	0.65625
11/64	0.171875	43/64	0.671875
3/16	0.1875	11/16	0.6875
13/64	0.203125	45/64	0.703125
7/32	0.21875	23/32	0.71875
15/64	0.234375	47/64	0.734375
1/4	0.25	3/4	0.75
17/64	0.265625	49/64	0.765625
9/32	0.28125	25/32	0.78125
19/64	0.296875	51/64	0.796875
5/16	0.3125	13/16	0.8125
21/64	0.328125	53/64	0.828125
11/32	0.34375	27/32	0.84375
23/64	0.359375	55/64	0.859375
3/8	0.375	7/8	0.875
25/64	0.390625	57/64	0.890625
13/32	0.40625	29/32	0.90625
27/64	0.421875	59/64	0.921875
7/16	0.4375	15/16	0.9375
29/64	0.453125	61/64	0.953125
15/32	0.46875	31/32	0.96875
31/64	0.484375	63/64	0.984375
1/2	0.5	1	1.0

Figure 3–34

EXAMPLE •────────────────────────────────────

Find the nearer fraction equivalents of the decimal dimensions given on the drawing of the wood pattern shown in Figure 3–35.

Dimension A is between 0.750″ and 0.765625″. Subtract to find the closer dimension.

Dimension A is closer to 0.750″. Find the fraction equivalent for 0.750″.

Dimension B is between 0.96875″ and 0.984375″. Subtract to find the closer dimension.

Dimension B is closer to 0.984375″. Find the fraction equivalent for 0.984375″.

$0.765625″ - 0.757″ = 0.008625″$
$0.757″ - 0.750″ = 0.007″$

$0.757″ \approx \dfrac{3″}{4} \, Ans$

$0.984375″ - 0.978″ = 0.006375″$
$0.978″ - 0.96875″ = 0.00925″$

$0.978″ \approx \dfrac{63″}{64} \, Ans$

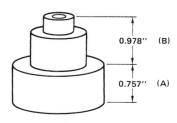

Figure 3–35

EXERCISE 3–18

Find the decimal or fraction equivalents of the following numbers, using the decimal equivalent table.

1. $\dfrac{15}{16}$

2. $\dfrac{11}{32}$

3. $\dfrac{5}{8}$

4. $\dfrac{43}{64}$

5. 0.28125

6. 0.546875

7. 0.078125

8. 0.390625

Determine the nearest fraction equivalents of the following decimals, using the decimal equivalent table.

9. 0.209
10. 0.068
11. 0.351

12. 0.971
13. 0.088
14. 0.243

15. 0.992
16. 0.459
17. 0.148

18. The profile gauge shown in Figure 3–36 is dimensioned fractionally in inches. Use the table of decimal equivalents. Express dimensions A through I in decimal form. Round the answers to 3 decimal places where necessary.

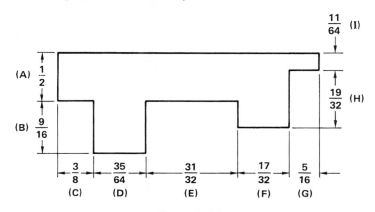

Figure 3–36

19. The hole locations in the bracket shown in Figure 3–37 are dimensioned decimally in inches. Use the table of decimal equivalents. Express dimensions A through H in fractional form. Round the answers to the nearest $\frac{1}{64}$ inch.

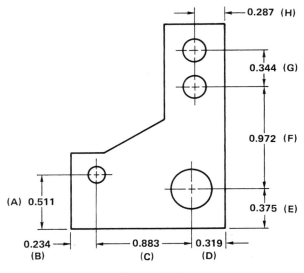

Figure 3–37

3–19   Combined Operations of Decimal Fractions

Combined operation problems are given as arithmetic expressions in this unit. Practical applications problems are based on formulas found in various occupational textbooks, handbooks, manuals, and other related reference materials.

The proper order of operations including powers and roots must be understood to solve expressions made up of any combination of the six arithmetic operations. Study the following order of operations.

- Do all operations within the grouping symbol first. Parentheses, the fraction bar, and the radical symbol are used to group numbers. If an expression contains parentheses within parentheses or brackets, do the work within the innermost parentheses first.
- Do powers and roots next. The operations are performed in the order in which they occur. If a root consists of two or more operations within the radical symbol, perform all the operations within the radical symbol, then extract the root.
- Do multiplication and division next in the order in which they occur.
- Do addition and subtraction last in the order in which they occur.

Again, you can use the memory aid "**P**lease **E**xcuse **M**y **D**ear **A**unt **S**ally" to help remember the order of operations. The **P** in "Please" stands for parentheses, the **E** for exponents or raising to a power, **M** and **D** for multiplication and division, and the **A** and **S** for addition and subtraction.

EXAMPLES •───

1. Find the value of $8.14 + 3.6 \times 0.8 - 1.37$.

Multiply.	$8.14 + 3.6 \times 0.8 - 1.37$
Add.	$8.14 + 2.88 - 1.37$
Subtract.	$11.02 - 1.37$
	9.65 Ans

2. Find the value of $9.6 + \dfrac{18.54 + (12 \times 0.4)^2}{68 \times 0.08 - \sqrt{2.25}}$ to 3 decimal places.

Grouping symbol operations are done first.

a. Perform the work in $[18.54 + (12 \times 0.4)^2]$
Multiply: $(12 \times 0.4) = 4.8$
Square: $4.8^2 = 23.04$
Add: $18.54 + 23.04 = 41.58$

$$9.6 + \dfrac{18.54 + (12 \times 0.4)^2}{68 \times 0.08 - \sqrt{2.25}}$$

b. Perform the work in $(68 \times 0.08 - \sqrt{2.25})$
Extract the square root: $\sqrt{2.25} = 1.5$
Multiply: $68 \times 0.08 = 5.44$
Subtract: $5.44 - 1.5 = 3.94$

Divide.

$$9.6 + 41.58 \div 3.94$$

Add.

$$9.6 + 10.553$$

$$20.153 \; Ans$$

EXERCISE 3–19

Solve the following combined operations expressions. Most expressions that involve roots require calculator solutions. Refer to pages 126 and 127 for calculator root solutions. Round the answers to 2 decimal places.

1. $0.187 + 16.3 \times 1.02$

2. $\dfrac{4.23}{6} - 0.98 \times 0.3$

3. $20 \times 0.86 - 80.4 \div 6$

4. $(13.46 + 18.79) \times 0.3$

5. $(24.78 + 9.07) \times 0.5$

6. $(18.8 - 13.3) \times (2.7 + 9.1)$

7. $40.87 + 16.04 - 3.3^2 \div 6$

8. $40.87 + (16.04 - 3.3^2) \div 6$

9. $(0.73 - 0.37)^2 \times 10.4$

10. $28.39 + (50.6 \div 12 \times 0.8 + 6)^2$

11. $0.051 + 2 \times \sqrt{25} - 6.062$

12. $\left(\dfrac{21.3}{7.1}\right)^3 + 14.4 + 2.2^2$

13. $\dfrac{21.3}{7.2^3} + (14.4 + 2.2)^2$

14. $22.76 \div \sqrt{12.32} + 1.76$

15. $(4.31 \times 0.6)^2 \div (5.96 - 1.05)$

16. $\left(\sqrt{0.23} + 1.06 \times 2.9\right)^2$

17. $(2.39 \times 0.9)^2 \div (1.05 - 0.83)$

18. $2.39 \times (0.9 \div 1.05)^2 - 0.83$

19. $0.360 + 0.112 \times \dfrac{125}{\sqrt{25}}$

20. $0.25 \times \left(\dfrac{\sqrt{49} - 2.4}{3.8}\right) + 0.99$

21. $0.25 \times \dfrac{\sqrt{49} - 2.4 + 0.99}{3.8}$

22. $\dfrac{\sqrt{80.9} \times 3.7}{16.4 \times 1.35} + \left(\dfrac{18.8}{4.7}\right)^2$

23. $23.67 - \sqrt{\dfrac{8.63 \times 5.1}{6.5^2 - 0.59}} \times 0.9$

24. $16.79 + \dfrac{(32.6 \times 0.3)^2}{\sqrt{4.3} + \sqrt{8}} - 2.1$

25. $362.07 - \sqrt[3]{912.6 - 18.532} + 0.763$

26. $(2.36^2 + \sqrt[4]{319.86}) \div 78.230$

27. $123.75 \times \sqrt[3]{13.736} - (86.35 \times 0.94)$

28. $\dfrac{\sqrt[5]{1{,}637} \div 40.07}{0.027 \times 31.023}$

29. $3.86^3 \times (0.875 + 4.63) - (2.032 \div 16.32)^2$

30. $\dfrac{853 - (3.075 \times 0.89^2 + 1.066)}{63.6 - \sqrt{217.95}}$

31. $53.07 - \sqrt[3]{18.35 \times 1.05} - 14.0 \div 6.832$

32. $\dfrac{67.9^2 \div \sqrt{363.74} + 412.36}{2.073 - 14.08 \times 0.065} - \sqrt[3]{360.877}$

3–20   Combined Operations of Decimal Fractions in Practical Applications

EXAMPLE •——————————————————————————————————

A series-parallel electrical circuit is shown in Figure 3–38. The complete circuit consists of 2 minor circuits each connected in parallel. The 2 minor circuits are then connected in series. Use the following formula to compute the total resistance (R_T) of this circuit. Express the answer to the nearest tenth ohm.

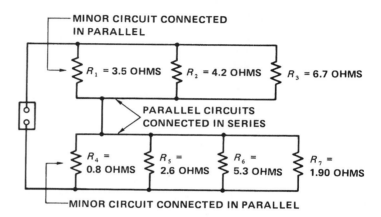

Figure 3–38

$$R_T = \cfrac{1}{\cfrac{1}{R_1} + \cfrac{1}{R_2} + \cfrac{1}{R_3}} + \cfrac{1}{\cfrac{1}{R_4} + \cfrac{1}{R_5} + \cfrac{1}{R_6} + \cfrac{1}{R_7}}$$

$$R_T = \cfrac{1}{\cfrac{1}{3.5} + \cfrac{1}{4.2} + \cfrac{1}{6.7}} + \cfrac{1}{\cfrac{1}{0.8} + \cfrac{1}{2.6} + \cfrac{1}{5.3} + \cfrac{1}{1.9}}$$

$$R_T = 1 \div \left(\frac{1}{3.5} + \frac{1}{4.2} + \frac{1}{6.7}\right) + 1 \div \left(\frac{1}{0.8} + \frac{1}{2.6} + \frac{1}{5.3} + \frac{1}{1.9}\right)$$

$$R_T = 1 \div 0.6731 + 1 \div 2.3496$$

$$R_T = 1.4857 + 0.4256$$

$$R_T = 1.9 \text{ ohms } Ans, \text{ rounded to nearest tenth ohm}$$

EXERCISE 3–20

Solve the following problems. Refer to pages 126 and 127 for calculator root solutions.

1. A surveyor wishes to determine the distance (AB) between two corners (points A and B) of a lot as shown in Figure 3–39. A building between the two corners prevents the taking of a direct measurement. The surveyor makes measurements and locates a stake at point C where distance AC is perpendicular to distance BC. Perpendicular means that AC and BC meet at a 90° angle. Distance AC is measured as 126.50 feet and distance BC is measured as 141.50 feet. Find, to the nearest tenth foot, distance AB.

$$AB = \sqrt{(AC)^2 + (BC)^2}$$

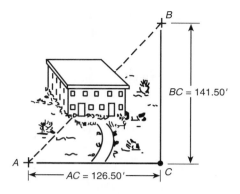

BC = 141.50'

AC = 126.50'

Figure 3–39

2. The bookkeeper for a small trucking firm finds the yearly depreciation of company vehicles. The bookkeeper uses the appraisal method of depreciation. Under this method, yearly depreciation is based on the fraction of the life of the vehicle used in 1 year. The following formula is used to compute yearly depreciation:

Yearly depreciation = (original cost − trade-in value)
 × number of miles driven in one year
 ÷ number of miles of life expectancy

Find the yearly depreciation to the nearest dollar of each of the 4 trucks listed in the table in Figure 3–40.

Truck	Original Cost	Trade-In Value	Number of Miles Driven in One Year	Number of Miles of Life Expectancy	Yearly Depreciation
1	$21,800	$3,000	46,500	250,000	
2	$16,075	$2,650	51,200	200,000	
3	$28,900	$3,800	57,910	300,000	
4	$36,610	$4,350	60,080	350,000	

Figure 3–40

3. Heat transfer by conduction is a basic process in refrigeration. In a refrigeration system a condenser transfers heat by conduction. Refrigerant gas enters a condenser at a high temperature. Heat is absorbed by water surrounding the tubing that contains the gas, and the gas is cooled. Refer to Figure 3–41.

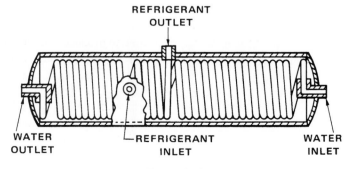

REFRIGERANT OUTLET

WATER OUTLET

REFRIGERANT INLET

WATER INLET

Figure 3–41

Refer to the table in Figure 3–42. Find the rate at which heat is transferred by conduction in each problem. Express the answers to the nearest thousand Btu/min.

$$H = \frac{K \times A \times TD}{60 \times d}$$

	Conductivity of Metal (K)	Surface Area of Metal (A)	Temperature Difference (TD)	Thickness of Metal (d)	Number of Btu Transferred per Minute (H)
a.	2,910	7.50 sq ft	9.00	0.031 in	
b.	1,740	6.30 sq ft	12.00	0.062 in	
c.	408	14.60 sq ft	8.00	0.250 in	
d.	2,910	3.80 sq ft	6.50	0.125 in	
e.	1,740	5.40 sq ft	10.50	0.093 in	

Figure 3–42

4. This problem deals with heat transfer by conduction. A technician wishes to determine the surface area of metal (A) required to transfer 120,000 Btu per minute (H), where $K = 1,740$; $TD = 95°F$; and $d = 0.093$ inch.

$$A = \frac{60 \times H \times d}{K \times TD}$$

Find A in square feet to 1 decimal place.

5. A series-parallel electrical circuit is shown in Figure 3–43.

$$R_T = \frac{1}{\dfrac{1}{R_1} + \dfrac{1}{R_2}} + \frac{1}{\dfrac{1}{R_3} + \dfrac{1}{R_4} + \dfrac{1}{R_5}}$$

Find R_T to 1 decimal place when:

a. $R_1 = 5.20$ ohms, $R_2 = 2.3$ ohms, $R_3 = 0.8$ ohm, $R_4 = 3.4$ ohms, $R_5 = 0.7$ ohm.

b. $R_1 = 0.8$ ohm, $R_2 = 3.4$ ohms, $R_3 = 1.5$ ohms, $R_4 = 5.3$ ohms, $R_5 = 0.9$ ohm.

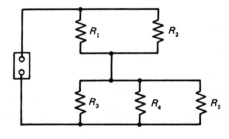

Figure 3–43

6. A flat is to be ground on a 0.750-centimeter-diameter hardened pin. Determine the depth of material to be removed to produce a flat that is 0.325 centimeter long. Express the answer to the nearest thousandth of a centimeter.

$$C = \frac{D}{2} - 0.5 \times \sqrt{4 \times \left(\frac{D}{2}\right)^2 - F^2}$$

where C = depth of material to be removed (depth of cut)
D = diameter of the pin
F = length of the required flat

7. The inside dimensions of a gas tube boiler are given in meters as shown in Figure 3–44. A pipefitter must determine the approximate number of cubic meters (volume) of steam space in the boiler. Steam space is the space above the boiler water line.

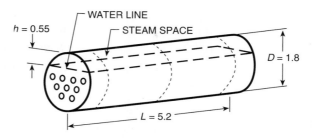

Figure 3–44

$$V = \frac{4 \times h^2}{3} \times \sqrt{\frac{D}{h} - 0.608} \times L$$

where V = number of cubic meters of steam space (volume)
h = height of steam space in meters
D = inside diameter of boiler in meters
L = inside length of boiler in meters

Find the number of cubic meters of steam space in the boiler to the nearest tenth of a cubic meter.

8. A patient often needs to be weaned off some powerful drug like Prednisone. Here is one possible way to wean a patient over a two-week period.

Days Given	Dosage	Times Per Day
Days 1–3	0.02 g	3
Days 4–6	0.01 g	3
Days 7–9	0.005 g	3
Days 10–12	0.0025 g	2
Days 13–14	0.00125 g	2

a. What is the total dosage for the first three days?
b. What is the total dosage for days 7–9?
c. What is the total dosage for day 1?
d. What is the total dosage for day 14?
e. What is the decrease in dosage between day 1 and day 14?
f. What is the total dosage for the 14-day period?

⊞ UNIT EXERCISE AND PROBLEM REVIEW

ROUNDING DECIMAL FRACTIONS

Round each of the following numbers to the indicated number of decimal places.

1. 0.943 (2 places)

2. 0.175 (2 places)

3. 0.0096 (3 places)

4. 0.0073 (1 place)

5. 17.043 (1 place)

6. 34.1355 (3 places)

7. 306.30006 (4 places)

8. 99.999 (2 places)

EXPRESSING COMMON FRACTIONS AS DECIMAL FRACTIONS

Express each of the following common fractions as decimal fractions. Where necessary, round the answers to 3 decimal places.

9. $\dfrac{7}{8}$

10. $\dfrac{5}{9}$

11. $\dfrac{1}{6}$

12. $\dfrac{17}{32}$

13. $\dfrac{33}{64}$

14. $\dfrac{29}{32}$

15. $\dfrac{9}{10}$

16. $\dfrac{8}{15}$

EXPRESSING DECIMAL FRACTIONS AS COMMON FRACTIONS

Express the following decimal fractions as common fractions. Express the answer in lowest terms.

17. 0.6

18. 0.860

19. 0.0625

20. 0.058

21. 0.15625

22. 0.0030

23. 0.998

24. 0.00086

ADDING DECIMAL FRACTIONS

Add the following numbers.

25. 0.413 + 0.033

26. $\dfrac{5}{16}$ + 0.0808 + 0.5909

27. 0.0003 + 0.003 + 0.03

28. 77.77 + 0.31108 + 66

29. 342.0838 + 61 + 0.73012

30. 0.019 + 0.016 + 587

31. $77\dfrac{7}{25}$ + 4.031 + 0.8 + 6.081

32. 494.2063 + 90.631 + 0.2416

SUBTRACTING DECIMAL FRACTIONS

Subtract the following numbers.

33. 0.783 − 0.678

34. 0.95 − 0.3042

35. 0.002 − 0.0009

36. $36\dfrac{1}{8}$ − 36.124

37. 15.1002 − 14.900

38. 71.071 − $68\dfrac{197}{200}$

39. 294.66 − 294.0673

40. 7.003 − 6.9087

MULTIPLYING DECIMAL FRACTIONS

Multiply the following numbers.

41. 0.8×0.7

42. 0.57×0.5

43. 18.13×0.14

44. $62.28 \times \dfrac{1}{4}$

45. 0.024×0.06

46. $4\dfrac{5}{8} \times 4.32$

47. 73.881×1.08

48. 67.022×0.038

Multiply the following numbers. Each expression has three or more factors.

49. $0.13 \times 27 \times 0.9$

50. $0.014 \times 0.913 \times 12$

51. $32.3 \times 6.06 \times 5 \times 0.2$

52. $891 \times 0.77 \times 66.005$

Multiply the following numbers. Round the answers to the indicated number of decimal places.

53. $0.79 \times 8.05 \times 0.07$ (4 decimal places)

54. $218.6 \times 0.89 \times \dfrac{9}{10} \times 27.51$ (3 decimal places)

Multiply the following numbers. Use the rules for multiplying by a power of 10.

55. 0.81×10

56. 0.997×100

57. $16.3 \times 1,000$

58. 0.003×100

59. 17.5×0.01

60. 0.763×0.1

DIVISION OF DECIMAL FRACTIONS

Divide the following numbers.

61. $0.8 \div 0.2$

62. $7.162 \div 0.27$

63. $0.0525 \div 1\dfrac{3}{4}$

64. $90.6059 \div 2.009$

65. $0.00336 \div 0.0016$

66. $0.04875 \div \dfrac{13}{16}$

Divide the following numbers. Round the answers to the indicated number of decimal places.

67. $3.05615 \div 0.009$ (1 place)

68. $8.508 \div 7.971$ (2 places)

69. $0.0046 \div 0.682$ (3 places)

70. $12.21004 \div \dfrac{7}{8}$ (5 places)

Divide the following numbers. Use the rules for dividing by a power of 10.

71. $8.61 \div 100$

72. $79.501 \div 1,000$

73. $358.72 \div 10,000$

74. $29.4 \div 0.001$

75. $0.002 \div 0.01$

76. $4.3921 \div 0.00001$

POWERS AND ROOTS OF DECIMAL FRACTIONS

Raise the following numbers to the indicated powers.

77. 6^2

78. 3.7^2

79. 3.1^3

80. 0.61^3

81. 2.2^3

82. 0.3^4

83. 2^5

84. 207.30^2

85. 0.0008^2

Raise the following expressions to the indicated powers.

86. $(0.6 \times 7)^2$

87. $(0.36 - 0.11)^2$

88. $\left(\dfrac{2}{5}\right)^3$

89. $(28.9 - 19.9)^4$

90. $\left(\dfrac{7}{100}\right)^2$

91. $(5000 \times 0.0002)^6$

Determine by inspection the whole number roots of the following numbers as indicated.

92. $\sqrt{64}$

93. $\sqrt{225}$

94. $\sqrt[3]{8}$

95. $\sqrt[3]{125}$

96. $\sqrt[6]{1}$

97. $\sqrt[3]{27}$

98. $\sqrt{121}$

99. $\sqrt[4]{81}$

100. $\sqrt[5]{32}$

Determine by inspection the whole number roots of the following expressions as indicated.

101. $\sqrt{3 \times 12}$

102. $\sqrt{18.8 - 2.8}$

103. $\sqrt{\dfrac{615.6}{7.6}}$

104. $\sqrt[3]{\dfrac{1.08}{0.04}}$

105. $\sqrt[4]{10.125 \times 8}$

106. $\sqrt{19.09 + 101.91}$

Determine the roots of the following numbers to the indicated number of decimal places. A calculator must be used in their solutions. Refer to pages 126 and 127 for calculator root solutions.

107. $\sqrt{247}$ (2 places)

108. $\sqrt{0.8214}$ (4 places)

109. $\sqrt[3]{9.6234}$ (3 places)

110. $\sqrt[3]{87.705}$ (3 places)

111. $\sqrt[4]{57,376}$ (1 place)

112. $\sqrt[5]{26.204}$ (2 places)

USING THE DECIMAL EQUIVALENT TABLE

Find the decimal or fraction equivalents of the following numbers, using the decimal equivalent table, on page 110.

113. $\dfrac{3}{32}$

114. $\dfrac{9}{16}$

115. $\dfrac{17}{64}$

116. 0.8125

117. 0.296875

118. 0.703125

Determine the nearest fraction equivalents of the following decimals, using the decimal equivalent table, on page 110.

119. 0.070

120. 0.522

121. 0.519

122. 0.205

123. 0.946

124. 0.099

COMBINED OPERATIONS OF DECIMAL FRACTIONS

Solve the following combined operations expressions. Round the answers to 2 decimal places. Some expressions that involve roots require calculator solutions. Refer to pages 126 and 127 for calculator root solutions.

125. $12.08 - 8.74 \times 0.6$

126. $0.98 \times 13 - 14 \div 2.2$

127. $9.34 - 0.7 \times \dfrac{8.08}{15.2}$

128. $1.16 \times (37.81 - 11.02 \times 0.6)$

129. $6.88 + (23.23 - 4.2^2) \div 0.8$

130. $(0.3 - 0.06)^2 \times 12.3$

131. $19.5 - (100 \div 12.5 \times 0.3)^2$

132. $8.18 \div \sqrt{2.85 - 1.06}$

133. $\left(\dfrac{84.4}{21.1}\right)^3 + 16 \div 2.5$

134. $\sqrt{(4.7 + 0.12 + 0.64)^2}$

135. $0.912 - 0.098 \times \dfrac{\sqrt[3]{81.34}}{7.86}$

136. $(6.93 \times 0.5)^2 \div (87.5 - 63.2)$

137. $\sqrt[3]{0.75} \times \left(\dfrac{\sqrt{36} - 3.1}{1.7}\right) + 8.34$

138. $14.33 + \dfrac{(13.1 \times 0.9)^2}{\sqrt{2.6} - 0.07} - 0.88$

DECIMAL FRACTION PRACTICAL APPLICATION PROBLEMS

Solve the following problems. Problems that involve roots require calculator solutions. Refer to pages 126 and 127 for calculator root solutions.

139. The inside width of an air duct is 10.38 inches. The duct is made of 26-gauge metal, which is 0.018 inch thick. Find the outside width of the duct to 2 decimal places.

140. In a parallel circuit, the total circuit current equals the sum of the individual currents. The total circuit current of the parallel circuit shown in Figure 3–45 is 17.50 amperes when all lamps and appliances are operating. Find the current (amperes) of the refrigerator in the parallel circuit shown.

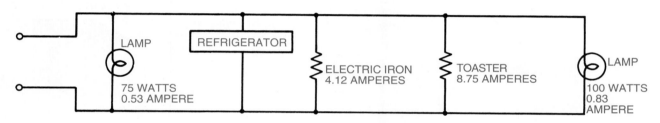

Figure 3–45   Parallel circuit

141. A certain 6-cylinder automobile engine produces 1.07 brake horsepower for each cubic inch of piston displacement. Each piston displaces 28.94 cubic inches. Find the total brake horsepower of the engine to the nearest whole horsepower.

142. The part shown in Figure 3–46, which is dimensioned in centimeters, is to be made by a machinist. Twenty equally spaced holes are drilled along the length of the part. Find, in centimeters, the total length of material needed.

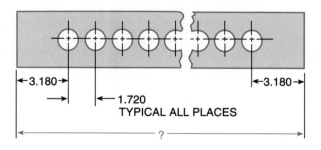

Figure 3–46

143. Plywood sheets are purchased by a carpenter in the quantities and for the costs shown in Figure 3–47 on page 122. Some of the purchased plywood is used on 2 jobs. The following number of sheets are used.

Job A: 12 sheets of $\frac{3}{8}$ inch, 15 sheets of $\frac{1}{2}$ inch, and 8 sheets of $\frac{5}{8}$ inch.

Job B: 14 sheets of $\frac{3}{8}$ inch, 9 sheets of $\frac{1}{2}$ inch, and 5 sheets of $\frac{5}{8}$ inch.

Type of Plywood	Number of Sheets Purchased	Total Cost of Purchased Quantities
$\frac{3}{8}$ inch thick	40 sheets	$260.00
$\frac{1}{2}$ inch thick	25 sheets	$181.25
$\frac{5}{8}$ inch thick	30 sheets	$246.60

Figure 3–47

    a. Find the total cost of plywood that is charged against Job A.

    b. Find the total cost of plywood that is charged against Job B.

144. The lengths of the sides of squares and cubes are given in the table shown in Figure 3–48. Find the area ($A = s^2$) and the volumes ($V = s^3$). Round the answers to 1 decimal place.

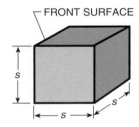

	Lengths of Sides (s)	Areas of Front Surfaces (A)	Volumes of Cubes (V)
a.	1.85 ft		
b.	21.30 mm		
c.	0.80 m		
d.	7.900 in		

FRONT SURFACE

Figure 3–48

145. In the table shown in Figure 3–49, the areas of squares and the volumes of cubes are given. Find the lengths of sides of squares ($s = \sqrt{A}$) and the lengths of sides of cubes ($s = \sqrt[3]{V}$). Round the answers to 2 decimal places.

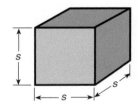

	Areas of Front Surfaces (A)	Volumes of Cubes (V)	Lengths of Sides (s)
a.	56.85 sq ft	—	
b.	—	28.56 m³	
c.	172.9 cm²	—	
d.	—	137.6 cu ft	

Figure 3–49

146. A carton in the shape of a cube is designed to contain 1.25 cubic meters. What is the maximum height of an object that can be packaged in the carton? The object is not tilted. It lies flat on the base of the box. $s = \sqrt[3]{V}$. Round the answer to 2 decimal places.

147. A landscaper is to landscape the shaded area of land around the office building shown in Figure 3–50. The landscaper charges $0.07 per square foot for this job. Find the price charged to complete the job. Round the answer to the nearest hundred dollars. All dimensions are in feet. $A = s^2$.

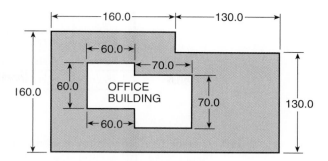

Figure 3–50

148. An inspector checks a 60° groove that has been machined in the fixture shown in Figure 3–51. The groove is checked by placing a pin in the groove and measuring the distance (*H*) between the top of the fixture and the top of the pin. Find *H* to the nearest thousandth inch. All dimensions are in inches.

$$H = 1.5 \times D - 0.866 \times W$$

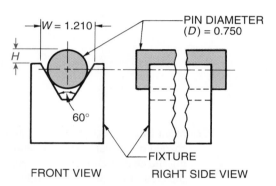

Figure 3–51

149. Four cells are connected in series in an electrical circuit shown in Figure 3–52. A technician finds the amount of current (*I*), in amperes, in the circuit.

$$I = \frac{E \times ns}{r \times ns + R}$$

where E = volts of one cell
ns = number of cells in circuit
r = internal resistance of one cell in ohms
R = external resistance of circuit in ohms

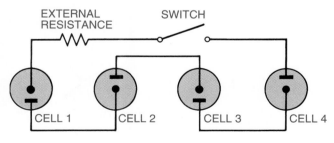

Figure 3–52

Find the amount of current (*I*) in amperes for a, b, and c using the values given on the table in Figure 3–53 on page 124. Express the answers to the nearest tenth ampere.

E	ns	r	R	I
3.25 volts	4 cells	0.85 ohm	2.20 ohms	**a.**
2.50 volts	4 cells	0.67 ohm	1.75 ohms	**b.**
5.75 volts	4 cells	1.13 ohms	2.65 ohms	**c.**

Figure 3–53

150. A steel fabricating firm is contracted to construct the fuel storage tank shown in Figure 3–54. The specifications call for a tank height of 22.00 feet. The tank must hold 25,500 gallons (*G*) of fuel.

$$D = \sqrt{\dfrac{4 \times G}{3.1416 \times H \times 7.479}}$$

Find the diameter to the nearest tenth foot.

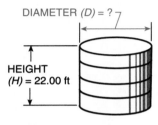

Figure 3–54

151. Main Street, Second Avenue, and Maple Street intersect as shown in Figure 3–55. The shaded triangular portion of land between the streets is to be used as a small park. In finding the cost of converting the parcel of land to a park, a city planning assistant computes the area of the parcel. The sides (*a*, *b*, *c*) of the parcel are measured. Find the area to the nearest ten square meters.

$$\text{Area (number of square meters)} = \frac{b}{2} \times \sqrt{a^2 - \left(\frac{c^2 - a^2 - b^2}{2 \times b}\right)^2}$$

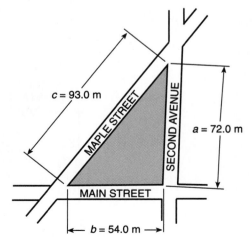

Figure 3–55

3–21   Computing with a Calculator: Decimals

Decimals

The decimal point key ($\boxed{\cdot}$) is used when entering decimal values in a calculator. When entering a decimal fraction, the decimal point key is pressed at the position of the decimal point in the number. For example, to enter the number 0.732, first press $\boxed{\cdot}$ and then enter the digits. To enter the number 567.409, enter 567 $\boxed{\cdot}$ 409.

Performing the four basic operations of addition, subtraction, multiplication, and division with decimals is the same as with whole numbers.

In calculator examples and illustrations of operations with decimals in this text, the decimal key $\boxed{\cdot}$ will *not* be shown to indicate the entering of a decimal point. Wherever the decimal point occurs in a number, it is understood that the decimal point key $\boxed{\cdot}$ is pressed.

Recall that your calculator must have algebraic logic to solve combined operations problems as they are shown in this text. Also recall the procedure for rounding numbers: Locate the digit in the number that gives the desired degree of precision; increase that digit by 1 if the digit immediately following is 5 or more; do not change the value of the digit if the digit immediately following is less than 5. Drop all digits that follow.

Examples of Decimals with Basic Operations of Addition, Subtraction, Multiplication, and Division

1. Add. 19.37 + 123.9 + 7.04
   Solution.   19.36 $\boxed{+}$ 123.9 $\boxed{+}$ 7.04 $\boxed{=}$ 150.31 *Ans*

2. Subtract. 2,876.78 − 405.052
   Solution.   2,876.78 $\boxed{-}$ 405.052 $\boxed{=}$ 2,471.728 *Ans*

3. Multiply. 427.935 × 0.875 × 93.400 (round answer to 1 decimal place)
   Solution.   427.935 $\boxed{\times}$ .875 $\boxed{\times}$ 93.4 $\boxed{=}$ 34 972.988

   34,973.0 *Ans*

   Notice that the two zeros following the 4 are not entered. The final zero or zeros to the right of the decimal point may be omitted.

   Notice that the zero to the left of the decimal point is not entered. The leading zero is omitted.

4. Divide. 813.7621 ÷ 6.466 (round answer to 3 decimal places)
   Solution.   813.7621 $\boxed{\div}$ 6.466 $\boxed{=}$ 125.85247 125.852 *Ans*

Powers

Expressions involving powers and roots are readily computed with a scientific calculator. The *square* key is used to raise a number to the second power (to square a number). Depending on the calculator used, the square of a number is computed in one of the following ways:

Enter the number and press the square key ($\boxed{x^2}$).

EXAMPLE •───────────────────────────────────

To calculate 28.75^2, enter 28.75 and press $\boxed{x^2}$.

Solution.   28.75 $\boxed{x^2}$ 826.5625 *Ans*

NOTE: Upon pressing $\boxed{x^2}$, the answer is displayed. It is not necessary to press $\boxed{=}$ with most calculators.

───

Or, enter the number, press the square key, $\boxed{x^2}$, and press $\boxed{\text{EXE}}$.

EXAMPLE •————————————————————————————————

To calculate 28.75^2, enter 28.75, press $\boxed{x^2}$, and press $\boxed{\text{EXE}}$.

Solution.   28.75 $\boxed{x^2}$ $\boxed{\text{EXE}}$ 826.5625 *Ans*

——•

The universal power key ($\boxed{y^x}$), ($\boxed{x^y}$), or $\boxed{\wedge}$, depending on the calculator used, raises any positive number to a power. To raise a number to a power using the universal power key, do the following:

  Enter the number to be raised to a power (y) or (x).
  Press the universal power key $\boxed{y^x}$, $\boxed{x^y}$, or $\boxed{\wedge}$.
  Enter the power (x) or (y).
  Press the $\boxed{=}$ or $\boxed{\text{EXE}}$ key.

EXAMPLES •———————————————————————————————

1. Calculate 15.72^3. Enter 15.72, press $\boxed{y^x}$, $\boxed{x^y}$, or $\boxed{\wedge}$, enter 3, and press $\boxed{=}$ or $\boxed{\text{EXE}}$.
   Solution.   15.72 $\boxed{x^y}$ 3 $\boxed{=}$ 3884.7012 *Ans*
2. Calculate 0.95^7
   Solution.   .95 $\boxed{x^y}$ 7 $\boxed{=}$ 0.6983373 *Ans*

——•

Roots

To obtain the square root of any positive number, the square root key ($\boxed{\sqrt{x}}$) or ($\boxed{\sqrt{}}$) is used.
  On some calculators the $\sqrt{}$ symbol is above one of the keys. In this case, you need to press the $\boxed{\text{2nd}}$ key and then press the key below the $\sqrt{}$ symbol. Some calculators automatically put a left parenthesis under the $\sqrt{}$ symbol. For example, pressing $\boxed{\text{2nd}}$ $\boxed{\sqrt{}}$ results in $\sqrt{}($ showing on the calculator. While it may not be necessary to put in a right parenthesis, it is a good idea and can help prevent errors.
  Depending on the calculator, the square root of a positive number is computed in one of the following ways.

1. Enter the number and press the square root key ($\boxed{\sqrt{x}}$).

EXAMPLE •————————————————————————————————

Calculate $\sqrt{27.038}$. Enter 27.038 and press $\boxed{\sqrt{x}}$.

Solution.   27.038 $\boxed{\sqrt{x}}$ → 5.199807689 *Ans*

——•

2. Press the square root key ($\boxed{\sqrt{}}$), enter the number, and press $\boxed{\text{EXE}}$, $\boxed{=}$ or $\boxed{\text{ENTER}}$.
   NOTE: The square root is a second function on certain calculators.

EXAMPLE •————————————————————————————————

Calculate $\sqrt{27.038}$. Press $\boxed{\sqrt{}}$, enter 27.038, press $\boxed{\text{EXE}}$.

Solution. $\boxed{\sqrt{}}$ 27.038 $\boxed{\text{EXE}}$ 5.199807689 *Ans*

——•

The root of any positive number can be computed with a calculator. Some calculators have a root key; with other calculators, roots are a second function.
  Depending on the calculator, root calculations are generally performed as follows.

1. Procedure for calculators that have the root key $\boxed{\sqrt[x]{}}$. Enter the root to be taken, press $\boxed{\sqrt[x]{}}$, enter the number whose root is to be taken, press $\boxed{\text{EXE}}$ or $\boxed{=}$.

EXAMPLE •————————————————————————————

Calculate $\sqrt[5]{475.19}$. Enter 5, press $\boxed{\sqrt[x]{}}$, enter 475.19, press $\boxed{\text{EXE}}$ or $\boxed{=}$.

Solution. 5 $\boxed{\sqrt[x]{}}$ 475.19 $\boxed{\text{EXE}}$ 3.430626662 *Ans*

NOTE: Where ($\boxed{\sqrt[x]{}}$) is a second function, press $\boxed{\text{SHIFT}}$ before pressing $\boxed{\sqrt[x]{}}$.

——•

2. On many graphing calculators, such as a TI-83 or TI-84, you find the $\boxed{\sqrt[x]{}}$ key by pressing the $\boxed{\text{MATH}}$ key. Pressing the $\boxed{\text{MATH}}$ key produces the screen in Figure 3–56. Notice that item #5 is $\sqrt[x]{}$. You either need to press $\boxed{5}$ or press the $\boxed{\blacktriangledown}$ key four times (until the 5: is highlighted) and then press $\boxed{\text{ENTER}}$.

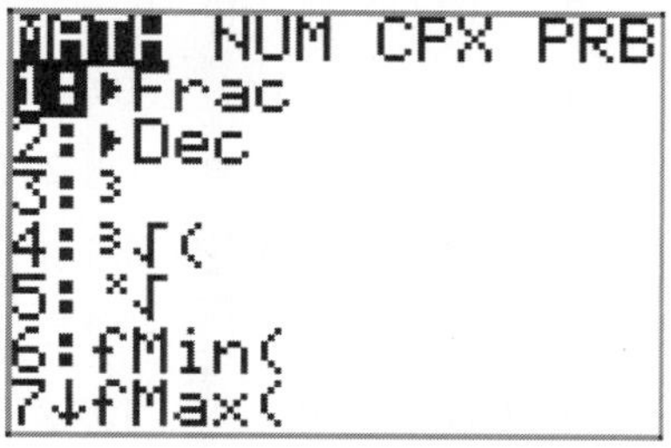

Figure 3–56

EXAMPLE •————————————————————————————

Calculate $\sqrt[4]{389.23}$. on a TI-83 or TI-84 graphing calculator.

Solution. 4 $\boxed{\text{MATH}}$ $\boxed{5}$ $\boxed{(}$ 389.23 $\boxed{)}$ $\boxed{\text{ENTER}}$ 4.441724079 *Ans*

——•

3. Procedure for calculators that do not have the root key $\boxed{\sqrt[x]{}}$ and roots are second functions. The procedures vary somewhat depending on the calculator used. Enter the number you want to find the root for, press $\boxed{\text{2nd}}$, press $\boxed{y^x}$, enter the root to be taken, press $\boxed{=}$.

EXAMPLE •————————————————————————————

Calculate $\sqrt[5]{475.19}$. Enter 475.19, press $\boxed{\text{2nd}}$, press $\boxed{y^x}$, enter 5, press $\boxed{=}$.

Solution. 475.19 $\boxed{\text{2nd}}$ $\boxed{y^x}$ 5 $\boxed{=}$ 3.430626662 *Ans*

or apply the following procedure when $\boxed{x^{1/y}}$ is a second function.

Enter the number for which you are taking the root, press $\boxed{\text{SHIFT}}$, press $\boxed{y^x}$, enter the root to be taken, press $\boxed{=}$.

EXAMPLE •————————————————————————————

Calculate $\sqrt[5]{475.19}$. Enter 475.19, press $\boxed{\text{SHIFT}}$, press $\boxed{x^y}$, enter 5, press $\boxed{=}$.

Solution. 475.19 $\boxed{\text{SHIFT}}$ $\boxed{x^y}$ 5 $\boxed{=}$ 3.4306267 *Ans*

——•

Practice Exercises, Individual Basic Operations

Evaluate the following expressions. The expressions are basic arithmetic operations including powers and roots. Remember to check your answers by estimating answers and doing each problem twice. The solutions to the problems directly follow the practice exercises. Compare your answers to the given solutions. Round each answer to the indicated number of decimal places.

Individual Operations

1. 276.84 + 312.094 (2 places)
2. 16.09 + 0.311 + 5.516 (1 place)
3. 6,704.568 − 4,989.07 (2 places)
4. 0.9244 − 0.0822 (3 places)
5. 43.4967 × 6.0913 (4 places)
6. 8.503 × 0.779 × 13.248 (3 places)

7. 54.419 ÷ 6.7 (1 place)
8. 0.9316 ÷ 0.0877 (4 places)
9. 36.22^2 (2 places)
10. 7.063^5 (1 place)
11. $\sqrt{28.73721}$ (4 places)
12. $\sqrt[5]{1,068.470}$ (3 places)

Solutions to Practice Exercises, Individual Basic Operations

1. 276.84 $\boxed{+}$ 312.094 $\boxed{=}$ 588.934, 588.93 *Ans*
2. 16.09 $\boxed{+}$ .311 $\boxed{+}$ 5.516 $\boxed{=}$ 21.917, 21.9 *Ans*
3. 6704.568 $\boxed{-}$ 4989.07 $\boxed{=}$ 1715.498, 1,715.50 *Ans*
4. .9244 $\boxed{-}$ .0822 $\boxed{=}$ 0.8422, 0.842 *Ans*
5. 43.4967 $\boxed{\times}$ 6.0913 $\boxed{=}$ 264.95145, 264.9515 *Ans*
6. 8.503 $\boxed{\times}$ .779 $\boxed{\times}$ 13.248 $\boxed{=}$ 87.752593, 87.753 *Ans*
7. 54.419 $\boxed{\div}$ 6.7 $\boxed{=}$ 8.1222388, 8.12 *Ans*
8. .9316 $\boxed{\div}$ .0877 $\boxed{=}$ 10.622577, 10.6226 *Ans*
9. 36.22 $\boxed{x^2}$ → 1311.8884, 1,311.89 *Ans*
10. 7.063 $\boxed{y^x}$ 5 $\boxed{=}$ 17577.052, 17,577.1 *Ans*
11. 28.73721 $\boxed{\sqrt{x}}$ → 5.3607098, 5.3607 *Ans*
12. 5 $\boxed{\sqrt[y]{\;}}$ 1068.47 $\boxed{\text{EXE}}$ 4.03415394, 4.034 *Ans*

    or 1068.47 $\boxed{\text{2nd}}$ $\boxed{y^x}$ 5 $\boxed{=}$ 4.03415394, 4.034 *Ans*

    or 1068.47 $\boxed{\text{SHIFT}}$ $\boxed{x^y}$ 5 $\boxed{=}$ 4.03415394, 4.034 *Ans*

Combined Operations

Because the following problems are combined operations expressions, your calculator must have algebraic logic to solve the problems shown. The expressions are solved by entering numbers and operations into the calculator in the same order as the expressions are written.

EXAMPLES •

1. Evaluate. 30.75 + 15 ÷ 4.02 (round answer to 2 decimal places)

   Solution.   30.75 $\boxed{+}$ 15 $\boxed{\div}$ 4.02 $\boxed{=}$ 34.481343, 34.48 *Ans*

2. Evaluate. $51.073 - \dfrac{4}{0.091} + 33.151 \times 2.707$ (round answer to 2 decimal places)

   Solution.   51.073 $\boxed{-}$ 4 $\boxed{\div}$ .091 $\boxed{+}$ 33.151 $\boxed{\times}$ 2.707 $\boxed{=}$ 96.856713, 96.86 *Ans*

3. Evaluate. 46.23 + (5 + 6.92) × (56.07 − 38.5)

   As previously discussed in the order of operations, operations enclosed within parentheses are performed first. A calculator with algebraic logic performs the operations within parentheses before performing other operations in a combined operations expression. If an expression contains parentheses, enter the expression into the calculator in the order in which it is written. The parentheses keys $\boxed{(}$ and $\boxed{)}$ must be used.

   Solution.   46.23 $\boxed{+}$ $\boxed{(}$ 5 $\boxed{+}$ 6.92 $\boxed{)}$ $\boxed{\times}$ $\boxed{(}$ 56.07 $\boxed{-}$ 38.5 $\boxed{)}$ $\boxed{=}$ 255.6644 *Ans*

4. Evaluate. $\dfrac{13.463 + 9.864 \times 6.921}{4.373 + 2.446}$ (round answer to 3 decimal places)

Recall that for problems expressed in fractional form, the fraction bar is also used as a grouping symbol. The numerator and denominator are each considered as being enclosed in parentheses.

$$(13.463 + 9.864 \times 6.921) \div (4.373 + 2.446)$$

Solution. $\boxed{(}$ 13.463 $\boxed{+}$ 9.864 $\boxed{\times}$ 6.921 $\boxed{)}$ $\boxed{\div}$ $\boxed{(}$ 4.373 $\boxed{+}$ 2.446 $\boxed{)}$ $\boxed{=}$ 11.985884, 11.986 *Ans*

The expression may also be evaluated by using the $\boxed{=}$ key to simplify the numerator without having to enclose the entire numerator in parentheses. However, parentheses must be used to enclose the denominator.

Solution.   13.463 $\boxed{+}$ 9.864 $\boxed{\times}$ 6.921 $\boxed{=}$ $\boxed{\div}$ $\boxed{(}$ 4.373 $\boxed{+}$ 2.446 $\boxed{)}$ $\boxed{=}$ 11.985884, 11.986 *Ans*

5. Evaluate. $\dfrac{100.32 - (16.87 + 13)}{111.36 - 78.47}$ (round answer to 2 decimal places)

$$\dfrac{100.32 - (16.87 + 13)}{111.36 - 78.47} = (100.32 - (16.87 + 13)) \div (111.36 - 78.47)$$

Observe these parentheses

To be sure that the complete numerator is evaluated before dividing by the denominator, enclose the complete numerator within parentheses. This is an example of an expression containing parentheses within parentheses.

Solution. $\boxed{(}$ 100.32 $\boxed{-}$ $\boxed{(}$ 16.87 $\boxed{+}$ 13 $\boxed{)}$ $\boxed{)}$ $\boxed{\div}$ $\boxed{(}$ 111.36 $\boxed{-}$ 78.47 $\boxed{)}$ $\boxed{=}$ 2.1419884, 2.14 *Ans*

Using the $\boxed{=}$ key to simplify the numerator:

Solution.   100.32 $\boxed{-}$ $\boxed{(}$ 16.87 $\boxed{-}$ 13 $\boxed{)}$ $\boxed{=}$ $\boxed{\div}$ $\boxed{(}$ 111.36 $\boxed{-}$ 78.47 $\boxed{)}$ $\boxed{=}$ 2.1419884, 2.14 *Ans*

On some calculators, when you press the $\sqrt{}$ key, the screen displays $\sqrt{}($. You need to put in the right parenthesis before you press the $\boxed{\text{ENTER}}$ key. If you fail to do this the calculator will assume that the right parenthesis is at the end of the problem.

6. Evaluate $\sqrt{16} + 9$.

Wrong Solution. $\sqrt{}$ 16 $\boxed{+}$ 9 $\boxed{\text{ENTER}}$ 5

Since the $\boxed{)}$ key was not used, the calculator acted as if the problem was $\sqrt{16 + 9} = \sqrt{25} = 5$.

Solution. $\sqrt{}$ 16 $\boxed{)}$ $\boxed{+}$ 9 $\boxed{\text{ENTER}}$ 13 *Ans*

7. Evaluate. $\dfrac{873.03 + 12.12^3 \times 41}{\sqrt{16.43} - 266.76 \div 107.88}$ (round answer to 2 decimal places)

Solution. $\boxed{(}$ 873.03 $\boxed{+}$ 12.12 $\boxed{y^x}$ 3 $\boxed{\times}$ 41 $\boxed{)}$ $\boxed{\div}$ $\boxed{(}$ 16.43 $\boxed{\sqrt{x}}$ $\boxed{-}$ 266.76 $\boxed{\div}$ 107.88 $\boxed{)}$ $\boxed{=}$ 46732.658, 46,732.66 *Ans*

Using the $\boxed{=}$ key to simplify the numerator:

Solution.   873.03 $\boxed{+}$ 12.12 $\boxed{y^x}$ 3 $\boxed{\times}$ 41 $\boxed{=}$ $\boxed{\div}$ $\boxed{(}$ 16.43 $\boxed{\sqrt{x}}$ $\boxed{-}$ 266.76 $\boxed{\div}$ 107.88 $\boxed{)}$ $\boxed{=}$ 46732.658, 46,732.66 *Ans*

Practice Exercises, Combined Operations

Evaluate the following combined operations expressions. Remember to check your answers by estimating the answer and doing each problem twice. The solutions to the problems directly follow the practice exercises. Compare your answers to the given solutions. Round each answer to the indicated number of decimal places.

1. $503.97 - 487.09 \times 0.777 + 65.14$ (2 places)

2. $27.028 + \dfrac{5}{6.331} - 5.875 \times 1.088$ (3 places)

3. $23.073 \times (0.046 + 5.934 - 3.049) - 17.071$ (3 places)

4. $30.180 \times (0.531 + 12.939 - 2.056) - 60.709$ (3 places)

5. $\dfrac{643.72 - 18.192 \times 0.783}{470.07 - 88.33}$ (2 places)

6. $\dfrac{793.32 - 2.67 \times 0.55}{107.9 + 88.93}$ (1 place)

7. $2{,}446 + 8.917^3 \times 5.095$ (3 places)

8. $679.07 + (36 + 19.973 - 0.887)^2 \times 2.05$ (1 place)

9. $43.71 - \sqrt{256.33 - 107} + 17.59$ (2 places)

10. $\dfrac{\sqrt[5]{14.773} + 93.977 \times \sqrt[3]{282.608}}{3.033}$ (3 places)

11. $\dfrac{1{,}202.03 \div \sqrt[3]{706.8 - 44.317}}{(14.03 \times 0.54 - 2.08)^2} - 2.63$ (1 place)

Solutions to Practice Exercises, Combined Operations

1. 503.97 [−] 487.09 [×] .777 [+] 65.14 [=] 190.64107, *190.64 Ans*

2. 27.028 [+] 5 [÷] 6.331 [−] 5.875 [×] 1.088 [=] 21.425765, *21.426 Ans*

3. 23.073 [×] [(] .046 [+] 5.934 [−] 3.049 [)] [−] 17.071 [=] 50.555963, *50.556 Ans*

4. 30.180 [×] [(] .531 [+] 12.939 [−] 2.056 [)] [−] 60.709 [=] 283.76552, *283.766 Ans*

5. [(] 643.72 [−] 18.192 [×] .783 [)] [÷] [(] 470.07 [−] 88.33 [)] [=] 1.6489644, *1.65 Ans*
   or 643.72 [−] 18.192 [×] 783 [=] [÷] [(] 470.07 [−] 88.33 [)] [=] 1.6489644, *1.65 Ans*

6. [(] 793.32 [−] 2.67 [×] .55 [)] [÷] [(] 107.9 [+] 88.93 [)] [=] 4.0230224, *4.0 Ans*
   or 793.32 [−] 2.67 [×] .55 [=] [÷] [(] 107.9 [+] 88.93 [)] [=] 4.0230224, *4.0 Ans*

7. 2446 [+] 8.917 [y^x] 3 [×] 5.095 [=] 6058.4387, *6,058.4387 Ans*

8. 679.07 [+] [(] 36 [+] 19.973 [−] .887 [)] [x^2] [×] 2.05 [=] 6899.7282, *6,899.7 Ans*

9. 43.71 [−] [(] 256.33 [−] 107 [)] [$\sqrt{x}$] [+] 17.59 [=] 49.079935, *49.08 Ans*
   or 43.71 [−] [$\sqrt{\ }$] [(] 256.33 [−] 107 [)] [+] 17.59 [EXE] 49.079935, *49.08 Ans*

10. [(] 5 [$\sqrt[x]{\ }$] 14.773 [+] 93.977 [×] 3 [$\sqrt[x]{\ }$] 282.608 [)] [÷] 3.033 [EXE] 203.89927, *203.899 Ans*
    or [(] 14.773 [2nd] [y^x] 5 [+] 93.977 [×] 282.608 [2nd] [y^x] 3 [)] [÷] 3.033 [=] 203.89927, *203.899 Ans*

11. [(] 1202.03 [÷] 3 [$\sqrt[x]{\ }$] [(] 706.8 [−] 44.317 [)] [)] [÷] [(] 14.03 [×] .54 [−] 2.08 [)] [x^2] [−] 2.63 [EXE] 1.9345574, *1.9 Ans*
    or [(] 1202.03 [÷] [(] 706.8 [−] 44.317 [)] [2nd] [y^x] 3 [)] [)] [÷] [(] 14.03 [×] .54 [−] 2.08 [)] [x^2] [−] 2.63 [=] 1.9345574, *1.9 Ans*

UNIT 4 ⠿ Ratio and Proportion

OBJECTIVES

After studying this unit you should be able to

• write comparisons as ratios.

• solve applied ratio problems.

• solve for the missing terms of given proportions.

• solve proportion problems by substituting values in formulas.

• analyze problems to determine whether they are direct or inverse proportions, set up proportions, and solve for unknowns.

The ability to solve applied problems using ratio and proportion is a requirement of many occupations. A knowledge of ratio and proportion is necessary in solving many everyday food service occupation problems. Proportions are used to solve many problems in medications in the health care occupations.

Ratio and proportion are widely used in manufacturing applications, such as computing gear speeds and sizes, tapers, and machine cutting times. Electrical resistance, wire sizes, and material requirements are determined by using proportions. The building trades apply ratios in determining roof pitches and pipe capacities. Compression ratios, transmission ratios, and rear axle ratios are commonly used by automobile mechanics. Employees in the business field compute selling price-to-cost ratios, profit-to-cost ratios, and dividend-to-cost ratios. In agricultural applications, fertilizer requirements are often determined by proportions.

4–1 Description of Ratios

Ratio is the comparison of two *like* quantities. For example, the compression ratio of an engine is the comparison between the amount of space in a cylinder when the piston is at the bottom of the stroke and the amount of space when the piston is at the top of the stroke. A compression ratio of 8 to 1 is shown in Figure 4–1.

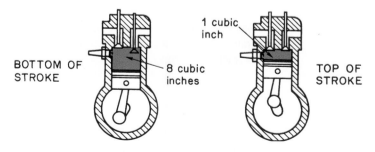

BOTTOM OF STROKE · 8 cubic inches

1 cubic inch · TOP OF STROKE

Figure 4–1

An automobile pulley system is shown in Figure 4–2. The comparison of the fan pulley size to the alternator pulley size is expressed as the ratio of 3 to 4. The comparison of the alternator pulley size to the crankshaft pulley size is expressed as the ratio of 4 to 5. The comparison of the fan pulley size to the crankshaft pulley size is expressed as the ratio of 3 to 5.

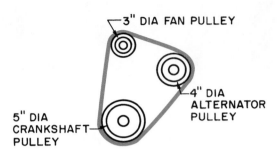

Figure 4–2

The *terms* of a ratio are the two numbers that are compared. **Both terms must be expressed in the same units.** For example, the width and length of the strip of stock shown in Figure 4–3 cannot be compared as a ratio until the 9-centimeter length is expressed as 90 millimeters. Both terms must be in the same units. The width and length are in the ratio of 13 to 90.

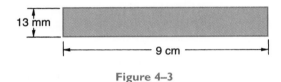

Figure 4–3

It is impossible to express two quantities as ratios if the terms have unlike units that cannot be expressed as like units. For example, inches and pounds as shown in Figure 4–4 cannot be compared as ratios.

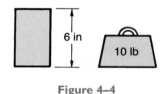

Figure 4–4

Ratios are expressed in the following two ways:

1. With a colon between the two terms, such as 4 : 9. The ratio 4 : 9 is read 4 to 9.

2. With a division sign separating the two numbers, such as 4 ÷ 9 or as a fraction, $\frac{4}{9}$.

4–2   Order of Terms of Ratios

The terms of a ratio must be compared in the order in which they are given. The first term is the numerator of a fraction, and the second term is the denominator. A ratio should be expressed in lowest fractional terms.

$$\text{The ratio 2 to 10} = 2 \div 10 = \frac{2}{10} = \frac{1}{5}$$

$$\text{The ratio 10 to 2} = 10 \div 2 = \frac{10}{2} = \frac{5}{1}$$

Notice that when the ratio of 5 : 1 is written as the fraction $\frac{5}{1}$, we write the denominator of 1.

EXAMPLES •

Express each ratio in lowest terms.

1. $5 : 15 = \dfrac{5}{15} = \dfrac{1}{3}$ *Ans*

2. $21 : 6 = \dfrac{21}{6} = \dfrac{7}{2}$ *Ans*

3. $\dfrac{3}{8} : \dfrac{9}{16} = \dfrac{3}{8} \div \dfrac{9}{16} = \dfrac{3}{8} \times \dfrac{16}{9} = \dfrac{2}{3}$ *Ans*

4. $10 : \dfrac{5}{6} = 10 \div \dfrac{5}{6} = \dfrac{10}{1} \times \dfrac{6}{5} = \dfrac{12}{1}$ *Ans*

EXERCISE 4–2

Express these ratios in lowest fractional form.

1. $3 : 7$

2. $7 : 3$

3. $12 : 24$

4. $24 : 12$

5. $8 : 30$

6. $43 : 32$

7. $12 \text{ in} : 46 \text{ in}$

8. $35 \text{ lb} : 10 \text{ lb}$

9. $9 \text{ cm} : 22 \text{ cm}$

10. $83 \text{ ft} : 100 \text{ ft}$

11. $52 \text{ m} : 16 \text{ m}$

12. $18 \text{ ft}^2 : 288 \text{ ft}^2$

13. $\dfrac{2}{3}$ to $\dfrac{1}{2}$

14. $\dfrac{1}{2}$ to $\dfrac{2}{3}$

15. 8 to $\dfrac{3}{4}$

16. 8 to $\dfrac{4}{3}$

17. 3 in to 3 ft

18. 23 mm to 3 cm

19. 3 cm to 23 mm

20. 3 ft to 2 yd

21. 18 in to 1 yd

22. 16 min to 2 h

23. 20 cm to 0.5 m

24. 6 min to $\dfrac{1}{4}$ h

25. 3 pt to 2 qt

26. $\dfrac{1}{2}$ gal to 5 qt

27. 150 m to 0.45 km

28. 0.4 km to 200 m

29. Refer to the data given in the table in Figure 4–5. Determine the compression ratios of the engines listed.

	AMOUNT OF SPACE WHEN THE PISTON IS AT THE:		Compression Ratio
	Bottom of the Stroke	Top of the Stroke	
a.	27 cubic inches	3 cubic inches	
b.	280 cubic centimeters	35 cubic centimeters	
c.	22 cubic inches	2 cubic inches	
d.	294 cubic centimeters	42 cubic centimeters	

Figure 4–5

30. In the building trades, the terms *pitch, rise, run,* and *span* are used in the layout and construction of roofs. In the gable roof shown in Figure 4–6 on page 134, the span is twice the run. Pitch is the ratio of the rise to the span.

$$\text{Span} = 2 \times \text{run} \qquad \text{Pitch} = \dfrac{\text{rise}}{\text{span}}$$

Determine the pitch of these gable roofs.

a. 12-ft rise, 30-ft span

b. 8-ft rise, 24-ft span

c. 4-m rise, 10-m span

d. 3-m rise, 9-m span

e. 6-m rise, 8-m run

f. 10-ft rise, 12-ft run

g. 15'0"-rise, 17'6"-run

h. 5-m rise, 7.5-m run

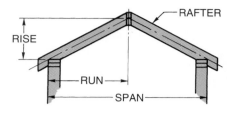

Figure 4–6

31. Refer to the hole locations given for the plate shown in Figure 4–7. Determine each ratio.

a. A to B

b. A to C

c. B to C

d. B to D

e. C to D

f. D to A

g. C to B

h. D to C

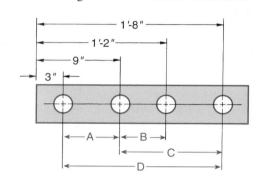

Figure 4–7

4–3   Description of Proportions

A *proportion* is an expression that states the equality of two ratios. Proportions are expressed in the following two ways:

1. $3 : 4 = 6 : 8$, which is read as, "3 is to 4 as 6 is to 8."

2. $\frac{3}{4} = \frac{6}{8}$, which is the equation form. Proportions are generally written as equations in on-the-job applications.

A proportion consists of four terms. The first and the fourth terms are called *extremes,* and the second and third terms are called *means.* In the proportion, $3 : 4 = 6 : 8$, 3 and 8 are the extremes; 4 and 6 are the means. In the proportion, $\frac{4}{5} = \frac{12}{15}$, 4 and 15 are the extremes; 5 and 12 are the means. **The product of the means equals the product of the extremes.** If the terms are cross multiplied, their products are equal.

EXAMPLES •

1. $\dfrac{2}{3} = \dfrac{4}{6}$

Cross multiply.

$$\frac{2}{3} = \frac{4}{6}$$

$$\frac{2}{3} \diagdown\!\!\!\!\diagup \frac{4}{6}$$

$$2 \times 6 = 3 \times 4$$

$$12 = 12$$

2. $\dfrac{a}{b} = \dfrac{c}{d}$

Cross multiply.

$$\dfrac{a}{b} = \dfrac{c}{d}$$

$$\dfrac{a}{b} \diagdown \diagup \dfrac{c}{d}$$

$$ad = bc$$

The method of cross multiplying is used in solving many practical occupational problems. The value of the unknown term can be determined when the values of three terms are known.

EXAMPLES

1. Solve for F.

$$\dfrac{F}{6.2} = \dfrac{9.8}{21.7}$$

$$\dfrac{F}{6.2} = \dfrac{9.8}{21.7}$$

Cross multiply. $\qquad\qquad\qquad 21.7F = 6.2\,(9.8)$

Divide both sides by 21.7. $\qquad \dfrac{21.7F}{21.7} = \dfrac{60.76}{21.7}$

$$F = 2.8 \; Ans$$

Check. $\qquad\qquad\qquad\qquad \dfrac{F}{6.2} = \dfrac{9.8}{21.7}$

Substitute 2.8 for F and divide to obtain the decimal equivalent of each fraction. $\qquad \dfrac{2.8}{6.2} = \dfrac{9.8}{21.7}$

$$0.4516129 = 0.4516129 \; Ck$$

Calculator Application

Solve for F. $\quad \dfrac{F}{6.2} = \dfrac{9.8}{21.7}$

$$F = 6.2 \;\boxed{\times}\; 9.8 \;\boxed{\div}\; 21.7 \;\boxed{=}\; 2.8 \; Ans$$

2. Solve for T.

$$0.7 : T = 3.5 : 2.4$$

Write in fraction form $\qquad\qquad\qquad \dfrac{0.7}{T} = \dfrac{3.5}{2.4}$

Cross multiply. $\qquad\qquad\qquad 0.7 \times 2.4 = 3.5T$

Divide both sides by 3.5 $\qquad\qquad \dfrac{0.7 \times 2.4}{3.5} = \dfrac{3.5T}{3.5}$

$$0.48 = T$$

$$T = 0.48 \; Ans$$

3. This formula relates to the circuit shown in Figure 4–8 on page 136.

$$I = \dfrac{nE}{R + nr} \qquad \text{where } I = \text{circuit current}$$

$$n = \text{number of cells}$$
$$E = \text{voltage of one cell}$$
$$R = \text{external resistance}$$
$$r = \text{internal resistance of one cell}$$

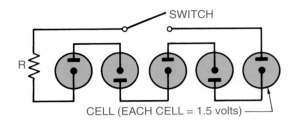

Figure 4–8

There are five 1.5-volt cells connected in series, each with an internal resistance of 1.8 ohms. The circuit current is 0.8 ampere. Find the external resistance in ohms. Round the answer to 1 decimal point.

Substitute the given values for the letters.

$$\frac{0.8}{1} = \frac{5(1.5)}{R + 5(1.8)}$$

Cross multiply.

$$0.8[R + 5(1.8)] = 5(1.5)$$

Remove parentheses.

$$0.8(R + 9) = 7.5$$

Subtract 7.2 from both sides of the equation.

$$0.8R + 7.2 = 7.5$$
$$-7.2 = -7.2$$

Divide both sides of the equation by 0.8.

$$\frac{0.8R}{0.8} = \frac{0.3}{0.8}$$

$$R = 0.375, 0.4 \text{ ohm } Ans$$

Calculator Application

$$R = \frac{nE - Inr}{I} = \frac{5(1.5) - 0.8(5)(1.8)}{0.8}$$

$R = 5 \boxed{\times} 1.5 \boxed{-} .8 \boxed{\times} 5 \boxed{\times} 1.8 \boxed{=} \boxed{\div} .8 \boxed{=} 0.375, 0.4 \, Ans$

EXERCISE 4–3

Solve for the unknown value in each of these proportions. Check each answer. Round the answers to 2 decimal places where necessary.

1. $\dfrac{x}{4} = \dfrac{6}{24}$

2. $\dfrac{3}{A} = \dfrac{15}{30}$

3. $\dfrac{7}{9} = \dfrac{E}{45}$

4. $\dfrac{6}{13} = \dfrac{24}{y}$

5. $\dfrac{20}{c} = \dfrac{5}{9}$

6. $\dfrac{P}{18} = \dfrac{1}{3}$

7. $\dfrac{6}{7} = \dfrac{15}{F}$

8. $\dfrac{12}{H} = \dfrac{4}{25}$

9. $\dfrac{T}{6.6} = \dfrac{7.5}{22.5}$

10. $\dfrac{2.4}{3} = \dfrac{M}{0.8}$

11. $\dfrac{4}{4.1} = \dfrac{8}{L}$

12. $\dfrac{3.4}{y} = \dfrac{-1}{7}$

13. $\dfrac{A}{5} = \dfrac{3.2}{A}$

14. $\dfrac{\frac{3}{8}}{N} = \dfrac{\frac{1}{2}}{4}$

15. $\dfrac{\frac{3}{1}}{2} = \dfrac{5}{F}$

16. $\dfrac{G}{\frac{1}{4}} = \dfrac{\frac{7}{8}}{\frac{3}{8}}$

17. $\dfrac{7}{\frac{-1}{8}} = \dfrac{x}{\frac{9}{16}}$

18. $\dfrac{4}{R} = \dfrac{2R}{12.5}$

Solve these proportion problems. Substitute the known values in the formulas and determine the values of the unknowns.

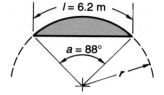

Figure 4–9

19. Compute the radius (*r*) of the circular segment shown in Figure 4–9. Round the answer to 1 decimal place.

$$\alpha = \frac{57.3l}{r}$$

20. A lever is an example of a simple machine. A lever is a rigid bar that is free to turn about its supporting point. The supporting point is called a fulcrum. Levers have a great many practical uses. Scissors, shovels, brooms, and bottle openers are a few common examples of levers. There are three classes of levers. A diagram of a first-class lever is shown in Figure 4–10. F_1 and F_2 are forces, and D_1 and D_2 are distances. Using the given values in the table in Figure 4–11, compute the missing values.

$$\frac{F_1}{F_2} = \frac{D_2}{D_1}$$

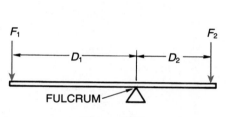

Figure 4–10

	F_1	F_2	D_1	D_2
a.	? lb	200 lb	8 ft	6 ft
b.	72.0 lb	? lb	15.0 ft	2.50 ft
c.	175 lb	1050 lb	? ft	$6\frac{3}{4}$ ft
d.	32.8 lb	393.6 lb	8.4 ft	? ft

Figure 4–11

21. The compression ratio compares the volume of a cylinder at BDC (bottom dead center) to the volume at TDC (top dead center). If the compression ration is 9.3 : 1 and the volume at BDC is 103 cm³, what is the volume at TDC?

22. If the compression ratio of a cylinder is 9.273 : 1 and the volume at TDC is 5.482 in.³, what is the volume at BDC? Round the answer to 2 decimal places.

23. A car was filled with 11.58 gal of gasoline at a cost of $21.89. A van pulled up to the same gas pump and put 17.32 gal in its tank. How much did the owner of the van have to pay for the gasoline?

4–4   Direct Proportions

In actual practice, word statements or other data must be expressed as proportions. When a proportion is set up, the terms of the proportion must be placed in their proper positions. A problem that is set up and solved as a proportion must first be analyzed in order to determine where the terms are placed. Depending on the position of the terms, proportions are either direct or inverse.

Two quantities are *directly proportional* if a change in one produces a change in the other in the same direction. If an increase in one produces an increase in the other, or if a decrease in one produces a decrease in the other, the two quantities are directly proportional. The proportions discussed will be those that change at the same rate. An increase or decrease in one quantity produces the same rate of increase or decrease in the other quantity.

When setting up a direct proportion in fractional form, the numerator of the first ratio must correspond to the numerator of the second ratio. The denominator of the first ratio must correspond to the denominator of the second ratio.

EXAMPLES •——

1. A machine produces 280 pieces in 3.5 hours. How long does it take to produce 720 pieces?

   Analyze the problem. An increase in the number of pieces produced (from 280 to 720) requires an increase in time. Time increases as production increases; therefore, the proportion is direct.

   Set up the proportion. Let t represent the time required to produce 720 pieces.

   $$\frac{280 \text{ pieces}}{720 \text{ pieces}} = \frac{3.5 \text{ hours}}{t}$$

   Notice that the numerator of the first ratio corresponds to the numerator of the second ratio; 280 pieces corresponds to 3.5 hours. The denominator of the first ratio corresponds to the denominator of the second ratio; 720 pieces corresponds to t.

   Solve for t.
   $$\frac{280 \text{ pieces}}{720 \text{ pieces}} = \frac{3.5 \text{ hours}}{t}$$
   $$280t = 3.5 \text{ hours} (720)$$
   $$\frac{280t}{280} = \frac{2\,520 \text{ hours}}{280}$$
   $$t = 9 \text{ hours } \textit{Ans}$$

   Check.
   $$\frac{280 \text{ pieces}}{720 \text{ pieces}} = \frac{3.5 \text{ hours}}{t}$$
   $$\frac{280 \text{ pieces}}{720 \text{ pieces}} = \frac{3.5 \text{ hours}}{9 \text{ hours}}$$
   $$0.3\overline{8} = 0.3\overline{8} \text{ Ck}$$

2. A sheet metal cone is shown in Figure 4–12. The cone is 35 centimeters high with a 38-centimeter-diameter base. Determine the diameter 14 centimeters from the top of the cone.

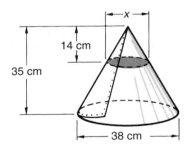

Figure 4–12

   Analyze the problem. As the height of the cone decreases from 35 centimeters to 14 centimeters, the diameter also decreases at the same rate. The proportion is direct.

   Set up the proportion. Let x represent the diameter in centimeters, 14 centimeters from the top.

   $$\frac{14 \text{ centimeters in height}}{35 \text{ centimeters in height}} = \frac{x}{38 \text{ centimeters in diameter}}$$

   Notice that the numerator of the first ratio corresponds to the numerator of the second ratio; the 14-centimeter height corresponds to the x. The denominator of the first ratio corresponds to the denominator of the second ratio; the 35-centimeter height corresponds to the 38-centimeter diameter.

Solve for x.

$$\frac{14\text{ cm}}{35\text{ cm}} = \frac{x}{38\text{ cm}}$$

$$35x = 14\,(38\text{ cm})$$

$$\frac{35x}{35} = \frac{532\text{ cm}}{35}$$

$$x = 15.2\text{ cm } Ans$$

Check.

$$\frac{14\text{ cm}}{35\text{ cm}} = \frac{x}{38\text{ cm}}$$

$$\frac{14\text{ cm}}{35\text{ cm}} = \frac{15.2\text{ cm}}{38\text{ cm}}$$

$$0.4 = 0.4\text{ Ck}$$

Calculator Application

$$\frac{14\text{ cm}}{35\text{ cm}} = \frac{x}{38\text{ cm}}$$

$$14 \boxed{\times} 38 \boxed{\div} 35 \boxed{=} 15.2$$

$$x = 15.2\text{ cm } Ans$$

4–5   Inverse Proportions

Two quantities are *inversely* or *indirectly proportional* if a change in one produces a change in the other in the opposite direction. If an increase in one produces a decrease in the other, or if a decrease in one produces an increase in the other, the two quantities are inversely proportional. For example, if one quantity increases by 4 times its original value, the other quantity decreases by 4 times or is $\frac{1}{4}$ of its original value. Notice 4 or $\frac{4}{1}$ inverted is $\frac{1}{4}$.

When an inverse proportion is set up in fractional form, the numerator of the first ratio must correspond to the denominator of the second ratio. The denominator of the first ratio must correspond to the numerator of the second ratio.

EXAMPLES •────────────────────────────────

1. Five identical machines produce the same parts at the same rate. The 5 machines complete the required number of parts in 1.8 hours. How many hours does it take 3 machines to produce the same number of parts?

   Analyze the problem. A decrease in the number of machines (from 5 to 3) requires an increase in time. Time increases as the number of machines decreases; therefore, the proportion is inverse.

   Set up the proportion. Let x represent the time required by 3 machines to produce the parts.

   $$\frac{5\text{ machines}}{3\text{ machines}} = \frac{x}{1.8\text{ hours}}$$

   Notice that the numerator of the first ratio corresponds to the denominator of the second ratio; 5 machines corresponds to 1.8 hours. The denominator of the first ratio corresponds to the numerator of the second ratio; 3 machines correspond to x.

   Solve for x.

   $$\frac{5}{3} = \frac{x}{1.8\text{ hours}}$$

   $$3x = 5(1.8\text{ hours})$$

   $$\frac{3x}{3} = \frac{9\text{ hours}}{3}$$

   $$x = 3\text{ hours } Ans$$

Check.
$$\frac{5}{3} = \frac{x}{1.8 \text{ hours}}$$

$$\frac{5}{3} = \frac{3 \text{ hours}}{1.8 \text{ hours}}$$

$$1.\overline{6} = 1.\overline{6} \text{ Ck}$$

2. Two gears are in mesh as shown in Figure 4–13. The driver gear has 40 teeth and revolves at 360 revolutions per minute. Determine the number of revolutions per minute of a driven gear with 16 teeth.

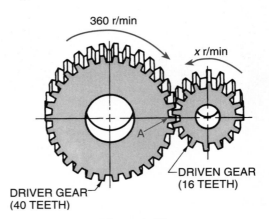

360 r/min

x r/min

A

DRIVEN GEAR
(16 TEETH)

DRIVER GEAR
(40 TEETH)

Figure 4–13

Analyze the problem. When the driver turns one revolution, 40 teeth pass point A. The same number of teeth on the driven gear must pass point A. Therefore, the driven gear turns more than one revolution for each revolution of the driver gear. The gear with 16 teeth (driven gear) revolves at greater revolutions per minute than the gear with 40 teeth (driver gear). A decrease in the number of teeth produces an increase in revolutions per minute. The proportion is inverse.

Set up the proportion. Let x represent the revolutions per minute of the gear with 16 teeth.

$$\frac{40 \text{ teeth}}{16 \text{ teeth}} = \frac{x}{360 \text{ r/min}}$$

Notice that the numerator of the first ratio corresponds to the denominator of the second ratio; the gear with 40 teeth corresponds to 360 r/min. The denominator of the first ratio corresponds to the numerator of the second ratio; the gear with 16 teeth corresponds to x.

Solve for x.
$$\frac{40}{16} = \frac{x}{360 \text{ r/min}}$$

$$16x = 40(360 \text{ r/min})$$

$$\frac{16x}{16} = \frac{14,400 \text{ r/min}}{16}$$

$$x = 900 \text{ r/min } \textit{Ans}$$

Check.
$$\frac{40}{16} = \frac{x}{360 \text{ r/min}}$$

$$\frac{40}{16} = \frac{900 \text{ r/min}}{360 \text{ r/min}}$$

$$14,400 = 14,000 \text{ Ck}$$

EXERCISE 4–5

Analyze each of these problems to determine whether the problem is a direct proportion or an inverse proportion. Set up the proportion and solve.

1. An engine uses 6 gallons of gasoline when it runs for $7\frac{1}{2}$ hours. If it runs at the same speed, how many gallons will be used in 10 hours?

2. In excavating the foundation of a building to a 4-foot depth, 1,800 cubic yards of soil are removed. How many cubic yards are removed when excavating to a 9-foot depth. Round the answer to 1 significant digit.

3. Of the two gears that mesh as shown in Figure 4–14, the one that has the greater number of teeth is called the gear, and the one that has fewer teeth is called the pinion. Refer to the table in Figure 4–15 and determine x in each problem.

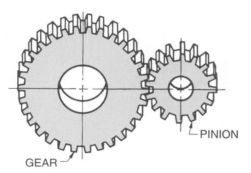

	Number of Teeth on Gear	Number of Teeth on Pinion	Gear (r/min)	Pinion (r/min)
a.	48	20	120.0	x
b.	32	24	x	210.0
c.	35	x	160.0	200.0
d.	x	15	150.0	250.0
e.	54	28	80.00	x

Figure 4–14   Figure 4–15

4. Six bakers take 7 hours to produce the daily bread requirements of a bakery. Working at the same rate, how many bakers are required to produce the same quantity of bread in $5\frac{1}{4}$ hours?

5. A homeowner pays $2,973 in taxes on property assessed at $187,200. After improvements are made, the property is assessed at $226,700. Using the same tax rate, what are the taxes on the $226,700 assessment? Round the answer to the nearest dollar.

6. The tank shown in Figure 4–16 contains 7200.0 liters of water when completely full. How many liters does it contain when filled to these heights (H)?

   a. 2.000 meters   **b.** 3.400 meters   **c.** 1.860 meters

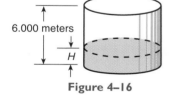

Figure 4–16

7. A balanced lever is shown in Figure 4–17. Observe that the heavier weight is closer to the fulcrum than is the lighter weight. An increase in the distance from the fulcrum produces a decrease in weight required to balance the lever. Refer to the table in Figure 4–18 and determine the unknown values.

	W_1	W_2	D_1	D_2
a.	76.8 lb	24.0 lb	35.0 ft	? ft
b.	? lb	$\frac{3}{4}$ lb	$\frac{1}{4}$ ft	$2\frac{1}{2}$ ft
c.	96.32 lb	60.20 lb	? ft	7.400 ft
d.	175 lb	? lb	$1\frac{1}{2}$ ft	$7\frac{1}{2}$ ft

Figure 4–17   Figure 4–18

8. The crankshaft speed of a car is 2,915 r/min when the car is traveling 55.75 mi/h. What is the crankshaft speed when the car is traveling 42.50 mi/h? Round the answer to 4 significant digits.

9. Two forgings are made of the same stainless steel alloy. A forging that weighs 170 pounds contains 0.80 pound of chromium. How many pounds of chromium does the second forging contain if it weighs 255 pounds?

10. A template is shown in Figure 4–19. A drafter makes an enlarged drawing of the template as shown in Figure 4–20. The original length of 1.80 inches on the enlarged drawing is 3.06 inches as shown. Determine the lengths of A, B, C, and D.

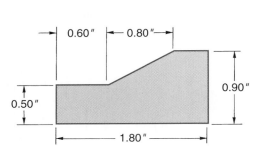

Figure 4–19

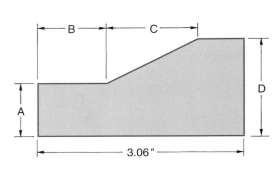

Figure 4–20

UNIT EXERCISE AND PROBLEM REVIEW

EXPRESSING RATIOS IN FRACTIONAL FORM

Express these ratios in lowest fractional form.

1. $15:32$
2. $46:12$
3. $12:46$
4. $27 \text{ mm}:45 \text{ mm}$
5. $21 \text{ ft}:33 \text{ ft}$
6. $45 \text{ in}:27 \text{ in}$
7. $\frac{1}{4}$ to $\frac{1}{2}$
8. 16 to $\frac{2}{3}$

9. 25 cm to 50 mm
10. 2 ft to 8 in
11. $\frac{1}{4} \text{ h}$ to 25 min
12. 3 min to 45 sec
13. 9 in to $\frac{1}{3} \text{ yd}$
14. 0.5 km to 100 m

RATIO PROBLEMS

Solve these ratio problems. Express the answers in lowest fractional form.

15. The cost and selling price of merchandise are listed in the table in Figure 4–21. Determine the cost-to-selling price ratio and the cost-to-profit ratio.

Profit = selling price − cost

	Cost	Selling Price	Ratio of Cost to Selling Price	Ratio of Cost to Profit
a.	$ 60	$ 96	?	?
b.	$105	$180	?	?
c.	$ 18	$ 33	?	?
d.	$204	$440	?	?

Figure 4–21

16. Bronze is an alloy of copper, zinc, and tin with small amounts of other elements. Two types of bronze castings are listed in the table in Figure 4–22 with the percent composition of copper, tin, and zinc in each casting. Determine the ratios called for in the table.

	TYPE OF CASTING	PERCENT COMPOSITION			RATIOS		
		Copper	Tin	Zinc	Copper to Tin	Tin to Zinc	Copper to Zinc
a.	Manganese Bronze	58	1	40	?	?	?
b.	Hard Bronze	86	10	2	?	?	?

Figure 4–22

SOLVING FOR UNKNOWNS IN GIVEN PROPORTIONS

Solve for the unknown value in each of these proportions. Check each answer. Round the answer to 2 decimal places where necessary.

17. $\dfrac{M}{8} = \dfrac{3}{12}$

18. $\dfrac{7}{E} = \dfrac{4}{32}$

19. $\dfrac{5}{20} = \dfrac{C}{96}$

20. $\dfrac{11}{13.2} = \dfrac{88}{T}$

21. $\dfrac{x}{-8.1} = \dfrac{3}{5.4}$

22. $\dfrac{10}{P} = \dfrac{P}{4.9}$

23. $\dfrac{4}{\frac{1}{2}} = \dfrac{B}{7}$

24. $\dfrac{\frac{3}{4}}{\frac{1}{8}} = \dfrac{\frac{1}{2}}{W}$

SOLVING PROPORTIONS GIVEN AS FORMULAS

Solve these proportion problems. Substitute the known values in the formulas and determine the values of the unknowns.

25. The volume of gas decreases as pressure increases. The relationship between pressure and volume of a confined gas is graphed in Figure 4–23 on page 144. Using the given values in the table in Figure 4–24, compute the missing values.

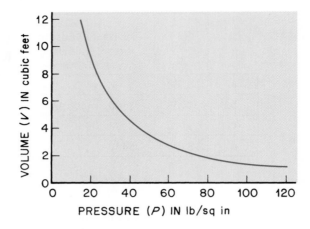

Figure 4–23

	P_1 (lb/sq in)	P_2 (lb/sq in)	V_1 (cu ft)	V_2 (cu ft)
a.	?	45.0	6.0	9.0
b.	15.0	?	12.0	2.0
c.	180.0	60.0	?	3.0
d.	1.8	4.5	1.5	?

Figure 4–24

$$\frac{P_1}{P_2} = \frac{V_2}{V_1}$$

where P_1 = the original pressure
V_1 = the original volume
P_2 = the new pressure
V_2 = the new volume

26. The tool feed (F), in inches per revolution, of a lathe may be computed from this formula.

$$T = \frac{L}{FN}$$

where T = cutting time per cut in minutes
L = length of cut in inches
N = r/min of revolving workpiece

Compute F to 3 decimal places by using the table in Figure 4–25.

	T (min)	L (in)	N (r/min)	F (in/r)
a.	4.8	20	2,100	?
b.	12.5	37	610	?
c.	3	8	335	?
d.	5.2	17	1,200	?

Figure 4–25

SETTING UP AND SOLVING DIRECT AND INVERSE PROPORTIONS

Analyze each of these problems to determine whether the problem is a direct proportion or an inverse proportion. Set up the proportion and solve.

27. An annual interest of $551.25 is received on a savings deposit of $10,500.00. At the same rate, how much annual interest is received on a deposit of $13,090.00.

28. A piece of lumber 2.8 meters long weighs 24.5 kilograms. A piece 0.8 meter long is cut from the 2.8-meter length. Determine the weight of the 0.8-meter piece.

29. Two sump pumps working at the same rate drain a flooded basement in $5\frac{1}{2}$ hours. How long does it take 3 pumps working at the same rate to drain the basement?

30. A solution contains $\frac{1}{4}$ ounce acid and $8\frac{1}{2}$ ounces of water. For the same strength solution, how much acid should be mixed with $12\frac{3}{4}$ ounces of water?

31. A compound gear train is shown in Figure 4–26. Gears B and C are keyed (connected) to the same shaft; therefore, they turn at the same rate. Gear A and Gear C are the driving gears. Gear B and Gear D are the driven gears. Compute the missing values in the table in Figure 4–27. Round the answers to 1 decimal place where necessary.

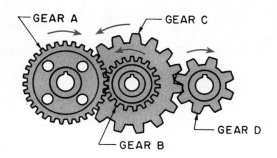

Figure 4–26

	NUMBER OF TEETH				REVOLUTIONS PER MINUTE			
	Gear A	Gear B	Gear C	Gear D	Gear A	Gear B	Gear C	Gear D
a.	80	30	50	20	120.0	?	?	?
b.	60	?	45	?	100.0	300.0	?	450.0
c.	?	24	60	36	144.0	?	?	280.0
d.	55	25	?	15	?	?	175.0	350.0

Figure 4–27

UNIT 5 ::: Percents

OBJECTIVES

After studying this unit you should be able to

- express decimal fractions and common fractions as percents.

- express percents as decimal fractions and common fractions.

- determine the percentage, given the base and rate.

- determine the percent (rate), given the percentage and base.

- determine the base, given the rate and percentage.

- solve more complex percentage problems in which two of the three parts are not directly given.

Each day, people are faced with various kinds of percentage problems to solve. Savings interest, loan payments, insurance premiums, and tax payments are based on percentage concepts.

Percentages are widely used in both business and nonbusiness fields. Merchandise selling prices and discounts, sales commissions, wage deductions, and equipment depreciation are determined by percentages. Business profit and loss are often expressed as percents. Percents are commonly used in making comparisons, such as production and sales increases or decreases over given periods of time. The basic percentage concepts have applications in many areas, including business and finance, manufacturing, agriculture, construction, health, and transportation.

5–1 Definition of Percent

The *percent (%)* indicates the number of hundredths of a whole. The square shown in Figure 5–1 is divided into 100 equal parts. The whole (large square) contains 100 small parts, or 100 percent of the small squares. Each small square is one of the 100 parts or $\frac{1}{100}$ of the large square. Therefore, each small square is $\frac{1}{100}$ of 100 percent or 1 percent.

1 part of 100 parts

$$\frac{1}{100} = 0.01 = 1\%$$

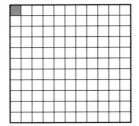

Figure 5–1

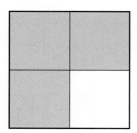

Figure 5–2

EXAMPLE

What percent of the square shown in Figure 5–2 is shaded?

The large square is divided into 4 equal smaller squares. Three of the smaller squares are shaded.

3 parts of 4 parts

$$\frac{3}{4} = 0.75 = \frac{75}{100} = 75\% \ Ans$$

5–2   Expressing Decimal Fractions as Percents

A decimal fraction can be expressed as a percent by moving the decimal point two places to the right and inserting the percent symbol. Moving the decimal point two places to the right is actually multiplying by 100.

EXAMPLES

1. Express 0.0152 as a percent.

   Move the decimal point 2 places to the right. $0.01\,52 = 1.52\% \ Ans$

   Insert the percent symbol.

2. Express 3.876 as a percent.

   Move the decimal point 2 places to the right. $3.87\,6 = 387.6\% \ Ans$

   Insert the percent symbol.

5–3   Expressing Common Fractions and Mixed Numbers as Percents

To express a common fraction as a percent, first express the common fraction as a decimal fraction. Then express the decimal fraction as a percent. If necessary to round, the decimal fraction must be two more decimal places than the desired number of places for the percent.

EXAMPLES

1. Express $\frac{7}{8}$ as a percent.

   Express $\frac{7}{8}$ as a decimal fraction. $\frac{7}{8} = 0.875$

   Express 0.875 as a percent. $0.875 = 87.5\% \ Ans$

2. Express $5\frac{2}{3}$ as a percent to 1 decimal place.

   Express $5\frac{2}{3}$ as a decimal fraction. $5\frac{2}{3} \approx 5.667$

   Express 5.667 as a percent. $5.667 = 566.7\% \ Ans$

Calculator Application

5 [+] 2 [÷] 3 [=] [×] 100 [=] 566.6666667, *566.7 Ans*

or [(] 5 [+] 2 [÷] 3 [)] 100 [ENTER], *566.7 Ans*

EXCERCISE 5–3

Determine the percent of each figure that is shaded.

1. 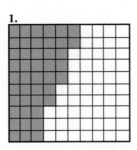   **2.** 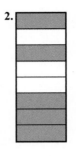   **3.** 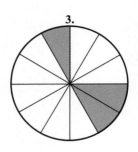   **4.**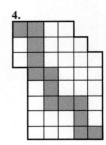

Express each value as a percent.

5. 0.35   **11.** 2.076

6. 0.96   **12.** 0.0639

7. 0.04   **13.** 0.0002

8. 0.062   **14.** 3.005

9. 0.008

10. 1.33   **15.** $\dfrac{1}{4}$

16. $\dfrac{21}{80}$ **19.** $\dfrac{17}{32}$ **22.** $4\dfrac{9}{10}$

17. $\dfrac{3}{20}$ **20.** $\dfrac{1}{250}$ **23.** $14\dfrac{5}{8}$

18. $\dfrac{37}{50}$ **21.** $1\dfrac{59}{100}$ **24.** $3\dfrac{1}{200}$

5–4 Expressing Percents as Decimal Fractions

Expressing a percent as a decimal fraction can be done by dropping the percent symbol and moving the decimal point two places to the left. Moving the decimal point two places to the left is actually dividing by 100.

EXAMPLES •

1. Express $38\dfrac{16}{21}\%$ as a decimal fraction. Round the answer to 4 decimal places.

Express $38\dfrac{16}{21}\%$ as 38.76% $38\dfrac{16}{21}\% = 38.76\% = 0.3876$ *Ans*

Drop the percent symbol and move the decimal point 2 places to the left.

Calculator Application

38 [+] 16 [÷] 21 [=] [÷] 100 [=] 0.387619047, 0.3876 *Ans*

or [(] 38 [+] 16 [÷] 21 [)] [÷] 100 [ENTER], 0.3876 *Ans*

Express each percent as a decimal fraction. Round the answers to 3 decimal places.

1. 0.48% . . . . 0.005 *Ans*

2. $15\dfrac{3}{4}\%$. . . . . 0.158 *Ans*

3. 5% . . . . 0.050 *Ans*

4. 300% . . . . 3.000 *Ans*

5. $1\dfrac{1}{3}\%$ . . . . 0.013 *Ans*

5–5   Expressing Percents as Common Fractions

A percent is expressed as a fraction by first finding the equivalent decimal fraction. The decimal fraction is then expressed as a common fraction.

EXAMPLES

1. Express 37.5% as a common fraction.

   Express 37.5% as a decimal fraction. $37.5\% = 0.375$

   Express 0.375 as a common fraction. $0.375 = \dfrac{375}{1,000} = \dfrac{3}{8}$ *Ans*

Calculator Application

$375 \boxed{a^{b/c}} 1000 \boxed{=} 3 \rfloor 8, \dfrac{3}{8}$ *Ans*

Express each percent as a common fraction.

1. 10% $10\% = 0.10 = \dfrac{10}{100} = \dfrac{1}{10}$ *Ans*

2. 3% $3\% = .03 = \dfrac{3}{100}$ *Ans*

3. $3\dfrac{1}{2}\%$ $3\dfrac{1}{2}\% = 3.5\% = .035 = \dfrac{35}{1,000} = \dfrac{7}{200}$ *Ans*

4. $222\dfrac{1}{2}\%$ $222\dfrac{1}{2}\% = 222.5\% = 2.225 = 2\dfrac{225}{1,000} = 2\dfrac{9}{40}$ *Ans*

5. 0.5% $0.5\% = 0.005 = \dfrac{5}{1,000} = \dfrac{1}{200}$ *Ans*

EXERCISE 5–5

Express each percent as a decimal fraction or mixed decimal.

1. 82%	**6.** 103%	**11.** $\dfrac{3}{4}\%$	**14.** 0.05%
2. 19%	**7.** 224.9%		
3. 3%	**8.** 0.87%	**12.** 0.1%	**15.** $43\dfrac{3}{5}\%$
4. 2.6%	**9.** 4.73%	**13.** $2\dfrac{3}{8}\%$	
5. 27.76%	**10.** $12\dfrac{1}{2}\%$		**16.** $205\dfrac{1}{10}\%$

Express each percent as a common fraction or mixed number.

17. 50%	**20.** 4%	**23.** 190%	**26.** 100.1%
18. 25%	**21.** 16%	**24.** 0.2%	**27.** 0.9%
19. 62.5%	**22.** 275%	**25.** 3.7%	**28.** 0.05%

5–6   Types of Simple Percent Problems

A simple percent problem has three parts. The parts are the rate, the base, and the percentage. In the problem 10% of $80 = $8, the rate is 10%, the base is $80, and the percentage is $8. The *rate* is the percent. The *base* is the number of which the rate or percent is taken. It is the whole or a quantity equal to 100%. The *percentage* is the quantity or part of the percent of the base.

In solving problems, the rate, percentage, and base must be identified. Some people like to use the term "amount" for percentage. This is perfectly correct, however, you will often see percentage used in this book.

In solving percent problems, the words *is* and *of* are often helpful in identifying the three parts. The following descriptions may help you recognize the rate, base, and percentage more quickly.

Base: The total, original, or entire amount. The base usually follows the word "of."

Rate: The number with a % sign. Sometimes it is written as a decimal or fraction.

Percentage: The value that remains after the base and rate have been determined. It is a portion of the base. The percentage is often close to the word "is."

EXAMPLES •────────────────────────────────────

1. What *is* 25% *of* 120? *Is* relates to 25% (the rate) and *of* relates to 120 (the base).

2. What percent *of* 48 *is* 12? *Is* relates to 12 (the percentage) and *of* relates to 48 (the base).

3. 60 *is* 30% *of* what number? *Is* relates to 60 (the percentage) and 30% (the rate). *Of* relates to "what number" (the base).

──•

There are three types of simple percentage problems. The type used depends on which two quantities are given and which quantity must be found. The three types are as follows:

• Finding the percentage, given the rate (percent) and the base.

  A problem of this type is, "What is 15% of 384?"

  If the rate is less than 100%, the percentage is less than the base.

  If the rate is greater than 100%, the percentage is greater than the base.

• Finding the rate (percent), given the base and the percentage.

  A problem of this type is, "What percent of 48 is 12?"

  If the percentage is less than the base, the rate is less than 100%.

  If the percentage is greater than the base, the rate is greater than 100%.

• Finding the base, given the rate (percent) and the percentage.

  A problem of this type is, "Fifty is 30% of what number?"

All three types of percent problems can be handled by the following proportion:

$$\frac{P}{B} = \frac{R}{100} \qquad \text{where}$$

B is the base
P is the percentage or part of the base, and
R is the rate or percent.

Practical applications involve numbers that have units or names of quantities called *denominate numbers.* The base and the percentage have the same unit or denomination. For example, if the base unit is expressed in inches, the percentage is expressed in inches. The rate is not a denominate number; it does not have a unit or denomination. Rate is the part to be taken of the whole quantity, the base.

Finding the Percentage, Given the Base and Rate

In some problems, the base and rate are given and the percentage must be found. First, express the rate (percent) as an equivalent decimal fraction. Then solve with the proportion $\frac{P}{B} = \frac{R}{100}$.

EXAMPLES ●

1. What is 15% of 60?

   Solution. The rate is $15\% = \frac{15}{100}$, so $R = 15$.

   The base, B, is 60. It is the number of which the rate is taken—the whole or a quantity equal to 100%.

   The percentage is to be found. It is the quantity of the percent of the base.

   The proportion is $\dfrac{P}{60} = \dfrac{15}{100}$.

   Now, using cross-products and division.

   $$100P = 15 \times 60$$
   $$100P = 900$$
   $$P = \frac{900}{100} = 9 \; Ans$$

2. Find $56\frac{9}{25}\%$ of $183.76.

   Solution. The rate is $56\frac{9}{25}\%$, so $R = 56\frac{9}{25}$.

   The base, B, is \$183.76.

   The percentage is to be found.

   Express R, $56\frac{9}{25}$ as a decimal, $56\frac{9}{25} = 56.36$.

   The proportion is $\dfrac{P}{\$183.76} = \dfrac{56.36}{100}$.

   Again, using cross-products and division.

   $$100P = 56.36 \times \$183.76$$
   $$100P \approx \$10{,}357$$
   $$P = \frac{\$10{,}357}{100} = \$103.57 \; Ans$$

Calculator Application

56 ⊞ 9 ÷ 25 ⊟ ☒ 183.76 ÷ 100 ⊟ 103.567136, $103.57 *Ans*

If you use the fraction key $\boxed{A^{b/c}}$, it is not necessary to convert R to a decimal.

56 $\boxed{A^{b/c}}$ 9 $\boxed{A^{b/c}}$ 25 ☒ 183.76 ÷ 100 ⊟ 103.567136, $103.57 *Ans*

EXERCISE 5–6

Find each percentage. Round the answers to 2 decimal places where necessary.

1. 20% of 80
2. 2.15% of 80
3. 60% of 200
4. 15.23% of 150
5. 31% of 419.3
6. 7% of 140.34
7. 156% of 65
8. 0.8% of 214
9. 12.7% of 295

10. 122% of 1.68
11. 140% of 280
12. 1.8% of 1240
13. 39% of 18.3
14. 0.42% of 50
15. 0.03% of 424.6
16. $8\frac{1}{2}\%$ of 375

17. $\frac{7}{8}\%$ of 160
18. 296.5% of 81
19. $15\frac{1}{4}\%$ of $35\frac{1}{4}$
20. $\frac{17}{50}\%$ of $139\frac{3}{10}$

5–7   Finding Percentage in Practical Applications

EXAMPLE •————————————————————————————————————

An electrical contractor estimates the total cost of a wiring job as $3,275. Material cost is estimated as 35% of the total cost. What is the estimated material cost to the nearest dollar?
    Think the problem through to determine what is given and what is to be found.

The rate is 35%.

The base, B, is $3,275. It is the total cost or the whole quantity.

The percentage, P, which is the material cost, is to be found.

The proportion is $\dfrac{P}{\$3,275} = \dfrac{35}{100}$.

    Cross multiply $100P = 35 \times \$3,275$

                        $100P = \$114,625$

    Divide $P = \dfrac{\$114,625}{100} = \$1,146.25$ *Ans*

——•

EXERCISE 5–7A

Solve the following problems. Round the answers to 1 decimal place where necessary.

1. A certain automobile cooling system has a capacity of 6.0 gallons. To give protection to −10°F, 40% of the cooling system capacity must be antifreeze. How many gallons of antifreeze should be used?

2. A print shop sells a used cylinder press for 42% of the original cost. If the original cost is $9,255.00, find the selling price of the used press.

3. A machine operator completes a job in 80% of the estimated time. The estimated time is $8\frac{1}{2}$ hours. How long does the job actually take?

4. The horsepower of an engine is increased by 7.8% after an engine is rebored. Find the horsepower increase if the engine is rated at 218.0 horsepower before it is rebored.

5. In an electrical circuit, a certain resistor takes 26% of the total voltage. The total voltage is 115 volts. Find how many volts are taken by the resistor.

6. A nurse computes a dosage of Benadryl for a 4-year-old child at 25% of the adult dosage. The adult dose is 50.0 milligrams. How many milligrams is the child's dose?

7. It is estimated that 37% of an apple harvest is spoiled by an early frost. Before the frost, the expected harvest was 3,800 bushels. How many bushels are estimated to be spoiled? Round the answer to the nearest hundred bushels.

8. A floor area requires 325 board feet of lumber. In ordering material, an additional 12% is allowed for waste. How many board feet are allowed for waste?

Finding the Percent (Rate), Given the Base and Percentage

In some problems, the base and percentage are given, and the percent (rate) must be found.

EXAMPLES •————————————————————————————————————

1. What percent of 12.87 is 9.620? Round the answer to 1 decimal place.

    Since a percent of 12.87 is to be taken, the base or whole quantity equal to 100% is 12.87.

    The percentage or quantity of the percent of the base is 9.620.

    The rate is to be found.

Since the percentage, 9.620, is less than the base, 12.87, the rate must be less than 100%.

The proportion is $\dfrac{9.620}{12.87} = \dfrac{R}{100}$.

Cross multiply $9.620 \times 100 = 12.87R$

$$962 = 12.87R$$

Divide $\dfrac{962}{12.87} = R$

$$R = \dfrac{962}{12.87} = 74.7474 \approx 74.7\% \; Ans$$

Calculator Application

9.62 ⊠ 100 ⊟ 12.87 ⊟ 74.74747475, *74.7% Ans*

2. What percent of 9.620 is 12.87? Round the answer to 1 decimal place.

Notice that although the numbers are the same as in Example 1, the base and percentage are reversed.

Since a percent of 9.620 is to be taken, the base or whole quantity equal to 100% is 9.620.

The percentage or quantity of the percent of the base is 12.87.

Since the percentage, 12.87, is greater than the base, 9.620, the rate must be greater than 100%.

$B = 9.920$ and $P = 12.87$. Thus, the proportion is $\dfrac{12.87}{9.620} = \dfrac{R}{100}$.

$$12.87 \times 100 = 9.620R$$

$$1{,}287 = 9.620R$$

$$\dfrac{1{,}287}{9.620} = R$$

$$R = \dfrac{1{,}287}{9.620} = 133.78 \approx 133.8\% \; Ans \text{ (rounded)}$$

Calculator Application

12.87 ⊠ 100 ⊟ 9.62 ⊟ 133.7837838, *133.8% Ans* (rounded)

EXERCISE 5–7B

Find each percent (rate). Round the answers to 2 decimal places where necessary.

1. What percent of 8 is 4?
2. What percent of 20.7 is 5.6?
3. What percent of 100 is 37?
4. What percent of 84.37 is 70.93?
5. What percent of 70.93 is 84.37?
6. What percent of 318.9 is 63?
7. What percent of 132.7 is 206.3?
8. What percent of 19.5 is 5.5?
9. What percent of 1.25 is 0.5?
10. What percent of 0.5 is 1.25?

11. What percent of $6\dfrac{1}{2}$ is 2?
12. What percent of 134 is $156\dfrac{3}{4}$?
13. What percent of $\dfrac{7}{8}$ is $\dfrac{3}{8}$?
14. What percent of $\dfrac{3}{8}$ is $\dfrac{7}{8}$?
15. What percent of 17.04 is 21.38?
16. What percent of 0.65 is 0.09?

5–8   Finding Percent (Rate) in Practical Applications

EXAMPLE •————————————————————————————————————

An inspector rejects 23 out of a total production of 630 electrical switches. What percent of the total production is rejected? Round the answer to 1 decimal place.

Think the problem through to determine what is given and what is to be found.

Since a percent of the total production of 630 switches is to be found, the base or whole quantity equal to 100% is 630 switches.

The percentage or quantity of the percent of the base is 23 switches.

The rate is to be found.

$$\text{The proportion is } \frac{23 \text{ switches}}{630 \text{ switches}} = \frac{R}{100}.$$

$$23 \text{ switches} \times 100 = 630 \text{ switches} \times R$$

$$2{,}300 \text{ switches} = 630 \text{ switches} \times R$$

$$\frac{2{,}300 \text{ switches}}{630 \text{ switches}} = R$$

$$R = \frac{2{,}300 \text{ switches}}{630 \text{ switches}} = 3.651, 3.7\% \text{ Ans (rounded)}$$

Calculator Application

23 ⊠ 100 ÷ 630 = 3.650793651, 3.7% *Ans* (rounded)

——•

EXERCISE 5–8A
——

Solve the following problems.

1. A garment requires $3\frac{1}{2}$ yards of material. If $\frac{1}{4}$ yard of material is waste, what percent of the required amount is waste?

2. In making a 250-pound batch of bread dough, a baker uses 160 pounds of flour. What percent of the batch is made up of flour?

3. A casting, when first poured, is 17.875 centimeters long. The casting shrinks 0.188 centimeter as it cools. What is the percent shrinkage? Round the answer to 2 decimal places.

4. An electronics technician tests a resistor identified as 130 ohms. The resistance is actually 128 ohms. What percent of the identified resistance is the actual resistance? Round the answer to the nearest whole percent.

5. A small manufacturing plant employs 130 persons. On certain days, 16 employees are absent. What percent of the total number of employees are absent? Round the answer to the nearest whole percent.

6. The total amount of time required to machine a part is 12.5 hours. Milling machine operations take 7.0 hours. What percent of the total time is spent on the milling machine?

7. If 97 acres of 385 acres of timber are cut, what percent of the 385 acres is cut? Round the answer to the nearest whole percent.

8. A road crew resurfaces 12.8 kilometers of a road that is 21.2 kilometers long. What percent of the road is resurfaced? Round the answer to 1 decimal place.

9. A mason ordered 1,850 floor tiles for a commercial job. After the job is completed, 97 tiles remain. What percent of tiles ordered remain?

Determining the Base, Given the Percent (Rate) and the Percentage

In some problems, the percent (rate) and the percentage are given, and the base must be found.

EXAMPLES •————————————————————————————

1. 816 is 68% of what number?

The rate is 68%, so $R = 68$.

Since 816 is the quantity of the percent of the base, the percentage is 816; 816 is 68% of the base.

The base to be found is the whole quantity equal to 100%.

Since the rate, 68% is less than 100%, the base must be greater than the percentage.

$R = 68$ and $P = 816$. The proportion is $\dfrac{816}{B} = \dfrac{68}{100}$.

$$816 \times 100 = 68B$$
$$81{,}600 = 68B$$
$$\dfrac{81{,}600}{68} = B$$
$$B = \dfrac{81{,}600}{68} = 1{,}200 \; Ans$$

2. \$149.50 is $115\frac{2}{3}$% of what value?

The rate is $115\frac{2}{3}$% and $R = 115\frac{2}{3}$.

Since \$149.50 is the quantity of the percent of the base, the percentage is \$149.50; \$149.50 is $115\frac{2}{3}$% of the base.

The base to be found is the whole quality equal to 100%. Since the rate, $115\frac{2}{3}$% is greater than 100%, the percentage must be greater than the base.

Express $115\frac{2}{3}$ as a decimal. $115\frac{2}{3} \approx 115.67$

$R = 115.67$ and $P = \$149.50$ and the proportion is $\dfrac{\$149.50}{B} = \dfrac{115.67}{100}$.

$$\$149.50 \times 100 = 115.67B$$
$$\$14{,}950 = 115.67B$$
$$\dfrac{\$14{,}950}{115.67} = B$$
$$B = \dfrac{\$14{,}950}{115.67} = \$129.25 \; Ans$$

Calculator Application

149.5 $\boxed{\times}$ 100 $\boxed{\div}$ 115.67 $\boxed{=}$ 129.2469958, \$129.25 *Ans*

or, if you want to leave the percent written as a fraction

149.5 $\boxed{\times}$ 100 $\boxed{\div}$ 115 $\boxed{A^{b/c}}$ 2 $\boxed{A^{b/c}}$ 3 $\boxed{=}$ $\boxed{2nd}$ $\boxed{F◄►D}$ 129.2507205, \$129.25 *Ans*

EXERCISE 5–8B ————————————————————————————————————

Find each base. Round the answers to 2 decimal places when necessary.

1. 15 is 10% of what number?

2. 25 is 80% of what number?

3. 80 is 25% of what number?

4. 3.8 is 95.3% of what number?

5. 13.6 is 8% of what number?

6. 123.86 is 88.7% of what number?

7. 312 is 130% of what number?

8. $44\frac{1}{3}$ is 60% of what number?

9. $3\frac{3}{4}$ is 160% of what number?

10. 10 is $6\frac{1}{4}$% of what number?

11. 190.75 is 70.5% of what number?

12. 6.6 is 3.3% of what number?

13. 88 is 205% of what number?

14. 1.3 is 0.9% of what number?

15. $\frac{7}{8}$ is 175% of what number?

16. $\frac{1}{10}$ is $1\frac{1}{5}$% of what number?

5–9   Finding the Base in Practical Applications

EXAMPLE •————————————————————————————————

A motor is said to be 80% efficient if the output (power delivered) is 80% of the input (power received). How many horsepower does a motor receive if it is 80% efficient with a 6.20 horsepower (hp) output?

Think the problem through to determine what is given and what is to be found.

The rate is 80%, so $R = 80$.

Since the output of 6.20 hp is the quantity of the percent of the base, the percentage is 6.20 hp (6.20 hp is 80% of the base).

The base to be found is the input; the whole quantity equal to 100%.

$R = 80$ and $P = 6.20$ hp, so the proportion is $\dfrac{6.20 \text{ hp}}{B} = \dfrac{80}{100}$.

$$6.20 \text{ hp} \times 100 = 80B$$

$$620 \text{ hp} = 80B$$

$$\frac{620 \text{ hp}}{80} = B$$

$$B = \frac{620 \text{ hp}}{80} = 7.75 \text{ hp } Ans$$

———•

EXERCISE 5–9 ——

Solve the following problems.

1. On a production run, 6.5% of the units manufactured are rejected. If 140 units are rejected how many total units are produced?

2. An engine loses 4.2 horsepower through friction. The power loss is 6% of the total rated horsepower. What is the total horsepower rating?

3. This year's earnings of a company are 140% of last year's earnings. The company earned $910,000 this year. How much did the company earn last year?

4. An iron worker fabricates 73.50 feet of railing. This is 28% of a total order. How many feet of railing are ordered?

5. During a sale, 32.8% of a retailer's fabric stock is sold. The income received from the sale is $8,765.00. What is the total retail value of the complete stock? Round the answer to the nearest whole dollar.

6. A pump operating at 70% of its capacity discharges 4,200 liters of water per hour. When the pump is operating at full capacity, how many liters per hour are discharged?

7. How many pounds of mortar can be made with 75 pounds of hydrated lime if the mortar is to contain 15% hydrated lime?

8. The gasoline mileage of a certain automobile is 19.8% greater than last year's model. This represents an increase of 4.80 miles per gallon. Find the mileage per gallon of last year's model. Round the answer to 1 decimal place.

9. With use, a 12-volt battery loses 5.6 ampere-hours, which is 18.0% of its capacity. How many ampere-hours of capacity did the battery originally have? Round the answer to 1 decimal place.

Identifying Rate, Base, and Percentage in Various Types of Problems

In solving simple problems, generally, there is no difficulty in identifying the rate or percent. A common mistake is to incorrectly identify the percentage and the base. There is sometimes confusion as to whether a value is a percentage or a base, the base and percentage are incorrectly interchanged.

The following statements summarize the information that was given when each of the three types of problems was discussed and solved. A review of the statements should be helpful in identifying the rate, percentage, and base.

- The rate (percent) determines the part taken of the whole quantity (base).
- The base is the whole quantity or a quantity that is equal to 100%. It is the quantity of which the rate is taken.
- The percentage is the quantity of the percent that is taken of the base. It is the quantity equal to the percent that is taken of the whole.
- If the rate is 100%, the percentage and the base are the same quantity.

  If the rate is less than 100%, the percentage is less than the base.

  If the rate is greater than 100%, the percentage is greater than the base.
- In practical applications, the percentage and the base have the same unit or denomination. The rate does not have a unit or denomination.
- The word *is* generally relates to the rate or percentage, and the word *of* generally relates to the base.

5–10 More Complex Percentage Practical Applications

In certain percentage problems, two of the three parts are not directly given. One or more additional operations may be required in setting up and solving a problem. Examples of these types of problems follow.

EXAMPLES

1. By replacing high-speed steel cutters with carbide cutters, a machinist increases production by 35%. Using carbide cutters, 270 pieces per day are produced. How many pieces per day were produced with high-speed steel cutters?

Think the problem through. The base (100%) is the daily production using high-speed steel cutters. Since the base is increased by 35%, the carbide cutter production of 270 pieces is 100% + 35% or 135% of the base. Therefore, the rate is 135% and the percentage is 270. The base is to be found.

The proportion is $\dfrac{270 \text{ pieces per day}}{B} = \dfrac{135}{100}$.

$$270 \text{ pieces per day} \times 100 = 135B$$
$$27,000 \text{ pieces per day} = 135B$$
$$\frac{27,000 \text{ pieces per day}}{135} = B$$
$$B = \frac{27,000 \text{ pieces per day}}{135} = 200 \text{ pieces per day } Ans$$

2. A mechanic purchases a set of socket wrenches for $54.94. The purchase price is 33% less than the list price. What is the list price?

Think the problem through. The base (100%) is the list price. Since the base is decreased by 33%, the purchase price, $54.94, is 100% − 33% or 67% of the base. Therefore, the rate is 67% and the percentage is $54.94. The base is to be found.

The proportion is $\dfrac{\$54.94}{B} = \dfrac{67}{100}$.

$$\$54.94 \times 100 = 67B$$
$$\$5,494 = 67B$$
$$\frac{\$5,494}{67} = B$$
$$B = \frac{\$5,494}{67} = \$82 \; Ans$$

3. An aluminum bar measures 137.168 millimeters before it is heated. When heated, the bar measures 137.195 millimeters. What is the percent increase in length? Express the answers to 2 decimal places.

Think the problem through. The base (100%) is the bar length before heating, 137.168 millimeters. The increase in length is 137.195 millimeters − 137.168 millimeters or 0.027 millimeter. Therefore, the percentage is 0.027 millimeter, and the base is 137.168 millimeters. The rate (percent) is to be found.

The proportion is $\dfrac{0.027 \text{ mm}}{137.168 \text{ mm}} = \dfrac{R}{100}$.

$$0.027 \text{ mm} \times 100 = 137.168 \text{ mm } R$$
$$2.7 \text{ mm} = 137.168 \text{ mm } R$$
$$\frac{2.7 \text{ mm}}{137.168 \text{ mm}} = R$$
$$R = \frac{2.7 \text{ mm}}{137.168 \text{ mm}} = 0.019684\%, 0.02\% \; Ans \text{ (rounded)}$$

Calculator Application

137.195 ⊟ 137.168 ▣ ⊡ 137.168 ⊠ 100 ▣ 0.019683891   0.02% *Ans* (rounded)

EXERCISE 5–10

Solve each problem.

1. A contractor is paid $95,000 for the construction of a building. The contractor's expenses are $40,200 for labor, $34,800 for materials, and $6,700 for miscellaneous expenses. What percent profit is made on this job?

2. A vegetable farmer plants 87.4 acres of land this year. This is 15% more than the number of acres planted last year. Find the number of acres planted last year.

3. A laboratory technician usually prepares 245.0 milliliters of a certain solution. The technician prepares a new solution that is 34.75% less than that usually prepared. How many milliliters of new solution are prepared? Round the answer to 1 decimal place.

4. A mason is contracted to lay 230 feet of sidewalk. After laying 158 feet, what percent of the total job remains to be completed? Round the answer to the nearest whole percent.

5. A printer has 782 reams of paper in stock at the beginning of the month. At the end of the first week, 28.0% of the stock is used. At the end of the second week, 50.0% of the stock remaining is used. How many reams of paper remain in stock at the end of the second week? Round the answer to the nearest whole ream.

6. In the electrical circuit shown in Figure 5–3, the total current (total amperes) is equal to the sum of the individual currents. The total current is 10.15 amperes. What percent of the total current is taken by the refrigerator? Round the answer to 1 decimal place.

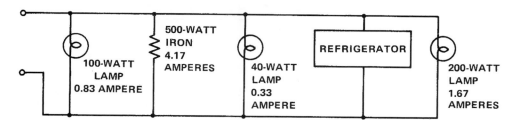

Figure 5–3

7. A welder estimates that 125 meters of channel iron are required for a job. Channel iron is ordered, including an additional 20% allowance for scrap and waste. Actually, 175 meters of channel iron are used for the job. The amount actually used is what percent more than the estimated amount? Round the answer to the nearest whole percent.

8. An alloy of red brass is composed of 85% copper, 5% tin, 6% lead, and zinc. Find the number of pounds of zinc required to make 450 pounds of alloy.

9. The day shift of a manufacturing firm produces 6% defective pieces out of a total production of 1,638 pieces. The night shift produces $4\frac{1}{2}$% defective pieces out of a total of 1,454 pieces. How many more acceptable pieces are produced by the day shift than by the night shift?

10. Two pumps are used to drain a construction site. One pump, with a capacity of pumping 1,500 gallons per hour, is operating at 80% of its capacity. The second pump, with a capacity of pumping 1,800 gallons per hour, is operating at 75% of its capacity. Find the total gallons drained from the site when both pumps operate for 3.15 hours. Round the answer to the nearest hundred gallons.

11. A resistor is rated at 2,500 ohms with a tolerance of ±6%. Tolerance is the amount of variation permitted for a given quantity. The resistor is checked and found to have an actual resistance of 2,320 ohms. By how many ohms is the resistor below the acceptable resistance low limit? Low limit = 2,500 ohms − 6% of 2,500 ohms.

12. A nurse is to prepare a 5% solution of sodium bicarbonate and water. A 5% solution means that 5% of the total solution is sodium bicarbonate. If 1,140 milliliters of water are used in the solution, how many grams of sodium bicarbonate are used?

    NOTE: Use 1 gram equal to 1 milliliter.

13. Forty-two grams of a certain breakfast cereal provide 20% of the United States recommended daily allowance of vitamin D. Find the number of grams of cereal required to provide 90% of the recommended daily allowance of vitamin D. Round the answer to the nearest ten grams.

14. A contractor receives $122,000 for the construction of a building. Total expenses amounted to $110,400. What percent of the $122,000 received is profit. Round the answer to 1 decimal place.

15. A chef for a food catering service prepares 85 pounds of beef. The amount prepared is 15% more than is consumed. Find, to the nearest pound, the number of pounds of beef consumed.

16. The table in Figure 5–4 on page 160 shows the number of pieces of a product produced each day during one week. Also shown are the number of pieces rejected each day by the quality control department. Find the percent rejection for the week's production. Round the answer to 1 decimal place.

	Mon.	Tues.	Wed.	Thur.	Fri.
Number of Pieces Produced	735	763	786	733	748
Number of Pieces Rejected	36	43	52	47	31

Figure 5–4

17. A hot brass casting when first poured in a mold is 9.25 inches long. The shrinkage is 1.38%. What is the length of the casting when cooled? Round the answer to 2 decimal places.

▦ UNIT EXERCISE AND PROBLEM REVIEW

EXPRESSING DECIMALS AND FRACTIONS AS PERCENTS

Express each value as a percent.

1. 0.72   **3.** 2.037   **5.** $\frac{1}{25}$ **7.** 0.0028

2. 0.05   **4.** $\frac{1}{2}$ **8.** 3.1906

                                         6. $1\frac{3}{8}$

EXPRESSING PERCENTS AS DECIMALS AND FRACTIONS

Express each percent as a decimal fraction or mixed decimal.

9. 19%   **11.** 18.09%   **13.** 0.7%   **15.** $\frac{3}{4}\%$

10. 3.4%   **12.** 156%   **16.** $310\frac{3}{10}\%$

                            14. $15\frac{1}{2}\%$

Express each percent as a common fraction or mixed number.

17. 30%   **19.** 140%   **21.** 12.5%   **23.** 0.98%

18. 6%   **20.** 0.9%   **22.** 100.8%   **24.** 0.02%

FINDING PERCENTAGE

Find each percentage. Round the answers to 2 decimal places when necessary.

25. 15% of 60   **31.** 130% of 212   **35.** $10\frac{1}{10}\%$ of $92\frac{1}{5}$

26. 3% of 42.3   **32.** 308% of 6.6

27. 72.8% of 120   **36.** $114\frac{3}{4}\%$ of 84.63

28. 4.93% of 246.8   **33.** $12\frac{1}{2}\%$ of 32

29. 0.7% of 812

30. 42.6% of 53.76   **34.** $\frac{1}{4}\%$ of 627.3

FINDING PERCENT (RATE)

Find each percent (rate). Round the answers to 2 decimal places when necessary.

37. What percent of 10 is 2?   **39.** What percent of 88.7 is 21.9?

38. What percent of 2 is 10?   **40.** What percent of 275 is 108?

41. What percent of 53.82 is 77.63?

42. What percent of 3.09 is 0.78?

43. What percent of $12\frac{1}{4}$ is 3?

44. What percent of 312 is 400.9?

45. What percent of $\frac{3}{4}$ is $\frac{3}{8}$?

46. What percent of $13\frac{4}{5}$ is $6\frac{3}{10}$?

FINDING BASE

Find each base. Round the answers to 2 decimal places when necessary.

47. 20 is 60% of what number?

48. 60 is 20% of what number?

49. 4.1 is 24.9% of what number?

50. 340 is 152% of what number?

51. 50.06 is 67.3% of what number?

52. 9.3 is 238.6% of what number?

53. 0.84 is 2.04% of what number?

54. $20\frac{1}{2}$ is 71% of what number?

55. $\frac{3}{4}$ is 123% of what number?

56. $\frac{4}{5}$ is $3\frac{3}{5}$% of what number?

FINDING PERCENTAGE, PERCENT, OR BASE

Find each percentage, percent (rate), or base. Round the answers to 2 decimal places when necessary.

57. What percent of 24 is 18?

58. What is 40% of 90?

59. What is 123.8% of 12.6?

60. 73 is 82% of what number?

61. What percent of $10\frac{1}{2}$ is 2?

62. __?__ is 48% of 94.82.

63. 72.4% of 212.7 is __?__.

64. What percent of 317 is 388?

65. 51.03 is 88% of what number?

66. 36.5 is __?__% of 27.6.

67. $2\frac{1}{4}$% of 150 is __?__.

68. __?__ is 18% of 120.66.

69. What percent of 36.2 is 45.3?

70. 15.84% of $9\frac{1}{4}$ is __?__.

71. What is 33% of 93.6?

72. 551.23 is __?__% of 357.82.

PRACTICAL APPLICATIONS

Solve the following problems.

73. The carbon content of machine steel for gauges usually ranges from 0.15% to 0.25%. Round the answers for a and b to 2 decimal places.

   a. What is the minimum weight of carbon in 250 kilograms of machine steel?

   b. What is the maximum weight of carbon in 250 kilograms of machine steel?

74. A nautical mile is the unit of length used in sea and air navigation. A nautical mile is equal to 6,076 feet. What percent of a statute mile (5,280 feet) is a nautical mile? Round the answer to 2 decimal places.

75. Air is a mixture composed, by volume, of 78% nitrogen, 21% oxygen, and 1% argon.

   a. Find the number of cubic meters of nitrogen in 25.0 cubic meters of air.

   b. Find the number of cubic meters of oxygen in 25.0 cubic meters of air.

   c. Find the number of cubic meters of argon in 25.0 cubic meters of air.

76. An environmental systems technician estimates that weather-stripping the windows and doors of a certain building decreases the heat loss by 45% or 34,200 British thermal units per hour. Find the heat loss of the building before the weather stripping is installed.

77. A 30-ampere fuse carries a temporary 8% current overload. How many amperes of current flow through the fuse during the overload?

78. To increase efficiency and performance, the frontal area of an automobile is redesigned. The new design results in a decrease of frontal area to 73% of the original design. The frontal area of the new design is 18.25 square feet. What is the frontal area of the original design?

79. A 22-liter capacity radiator requires 6.5 liters of antifreeze to give protection to $-17°C$. What percent of the coolant is antifreeze? Round the answer to the nearest whole percent.

80. A motor is 86.5% efficient; that is, the output is 86.5% of the input. What is the input if the motor delivers 11.50 horsepower? Round the answer to 1 decimal place.

81. When washed, a fabric shrinks $\frac{1}{32}$ inch for each foot of length. What is the percent shrinkage?

82. How many pounds of butterfat are in 125 pounds of cream that is 34% butterfat?

83. An appliance dealer sells a television set at 170% of the wholesale cost. The selling price is $585. What is the wholesale cost?

84. A piece of machinery is purchased for $8,792. In one year, the machine depreciates 14.5%. By how many dollars does the machine depreciate in one year? Round the answer to the nearest dollar.

85. An office remodeling job takes $5\frac{3}{4}$ days to complete. Before working on the job, the remodeler estimated the job would take $4\frac{1}{2}$ days. What percent of the estimated time is the actual time? Round the answer to the nearest whole percent.

86. Engine pistons and cylinder heads are made of an aluminum casting alloy that contains 4% silicon, 1.5% magnesium, and 2% nickel. Round the answers to the nearest tenth kilogram.

    a. How many kilograms of silicon are needed to produce 575 kilograms of alloy?

    b. How many kilograms of magnesium are needed to produce 575 kilograms of alloy?

    c. How many kilograms of nickel are needed to produce 575 kilograms of alloy?

87. Before starting two jobs, a plumber has an inventory of eighteen 15.0-foot lengths of copper tubing. The first job requires 30% of the inventory. The second job requires 25% of the inventory remaining after the first job. How many feet of tubing remain in inventory at the end of the second job? Round the answer to the nearest whole foot.

88. The cost of one dozen milling machine cutters is listed as $525. A multiple discount of 12% and 8% is applied to the purchase of the cutters. Determine the purchase price.

    NOTE: With multiple discounts, the first discount is subtracted from the list price. The second discount is subtracted from the price computed after the first discount is subtracted.

89. A baker usually prepares a 140-pound daily batch of dough. The baker wishes to reduce the batch by 15%. Find the number of pounds of dough prepared.

90. An alloy of stainless steel contains 73.6% iron, 18% chromium, 8% nickel, 0.1% carbon, and sulfur. How many pounds of sulfur are required to make 5,800 pounds of stainless steel? Round the answer to the nearest whole pound.

91. A homeowner computes fuel oil consumption during the coldest 5 months of the year as follows: November, 125 gallons; December, 145 gallons; January, 185 gallons; February, 165 gallons; March, 140 gallons. What percent of the 5-month consumption of fuel oil is January's consumption? Round the answer to 1 decimal place.

92. Two machines together produce a total of 2,015 pieces. Machine A operates for $6\frac{1}{2}$ hours and produces an average of 170 pieces per hour. Machine B operates for 7 hours. What percent of the average hourly production of Machine A is the average hourly production of Machine B? Round the answer to the nearest whole percent.

93. A resistor is rated at 3,200 ohms with a tolerance of ±4%. The resistor is checked and found to have an actual resistance of 3,020 ohms. By what percent is the actual resistance below the low tolerance limit. Round the answer to 1 decimal place.

94. A cosmetologist sanitizes implements with a solution of disinfectant and water. A 4% solution contains 4% disinfectant and 96% water. How many ounces of solution are made with 2 ounces of disinfectant?

UNIT 6 ⠿ Signed Numbers

OBJECTIVES

After studying this unit you should be able to

- express word statements as signed numbers.

- write signed number values using a number scale.

- add and subtract signed numbers.

- multiply and divide signed numbers.

- compute powers and roots of signed numbers.

- solve combined operations of signed number expressions.

- solve signed number problems.

- express decimal numbers in scientific or engineering notation.

- express numbers written in scientific or engineering notation as decimal numbers.

- compute expressions using scientific or engineering notation.

6–1   Meaning of Signed Numbers

Plus and minus signs, which you have worked with so far in this book, have been *signs of operation.* These are the signs used in arithmetic, with the plus sign (+) indicating the operation of addition and the minus sign (−) indicating the operation of subtraction.

In algebra, plus and minus signs are used to indicate both operation and direction from a reference point or zero. A *positive number* is indicated either with no sign or with a plus sign (+) preceding the number. A *negative number* is indicated with a minus sign (−) preceding the number. Positive and negative numbers are called *signed numbers* or *directed numbers.*

Signed numbers are common in everyday use as well as in occupational applications. For example, a Celsius temperature reading of 20 degrees above zero is written as +20°C or 20°C, a temperature reading of 20 degrees below zero is written as −20°C, as shown in Figure 6–1.

Signed numbers are often used to indicate direction and distance from a reference point. Opposites, such as up and down, left and right, north and south, and clockwise and counterclockwise, may be expressed by using positive and negative signs. For example, 100 feet above sea level may be expressed as +100 feet and 100 feet below sea level as −100 feet. Sea level in this case is the zero reference point.

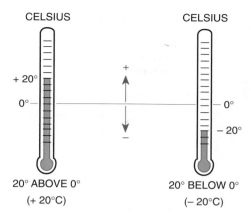

Figure 6–1

The automobile ammeter shown in Figure 6–2 indicates whether a battery is charging (+) or discharging (−).

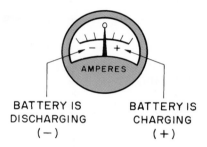

Figure 6–2

In business applications, a profit of $1,000 is expressed as +$1,000, whereas a loss of $1,000 is expressed as −$1,000. Closing prices for stocks are indicated as up (+) or down (−) from the previous day's closing prices.

Signed numbers are used in programming operations for numerical control machines. From a reference point, machine movements are expressed as + and − directions.

EXERCISE 6–1

Express the answer to each of the following problems as a signed number.

1. A speed increase of 12 miles per hour is expressed as +12 mi/h. Express a speed decrease of 8 miles per hour.

2. Traveling 50 kilometers west is expressed as −50 km. How is traveling 75 kilometers east expressed?

3. A wage increase of $25 is expressed at +$25. Express a wage decrease of $18.

4. The reduction of a person's daily calorie intake by 400 calories is expressed as −400 calories. Express a daily intake increase of 350 calories.

5. A company's assets of $383,000 are expressed as +$383,000. Express company liabilities of $167,000.

6. An increase of 30 pounds per square inch of pressure is expressed as +30 lb/sq in. Express a pressure decrease of 28 pounds per square inch.

7. A circuit voltage loss of 7.5 volts is expressed as −7.5 volts. Express a voltage gain of 9 volts.

8. A manufacturing department's production increase of 500 parts per day is expressed as +500 parts. Express a production decrease of 175 parts per day.

9. A savings account deposit of $140 is expressed at +$140. Express a withdrawal of $280.

10. A 0.75 percent contraction of a length of wire is expressed as −0.75%. Express a 1.2 percent expansion.

6–2   The Number Line

The number scale in Figure 6–3 shows the relationship between positive and negative numbers. The scale shows both distance and direction between numbers is referred to as a number line. Considering a number as a starting point and counting to a number to the right represents a positive (+) direction, with numbers increasing in value. Counting to the left represents negative (−) direction, with numbers decreasing in value. The greater of two numbers is the one that is farther to the right on the number line.

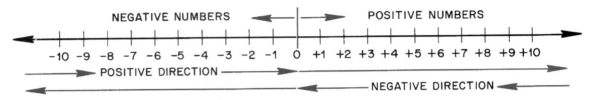

Figure 6–3

EXAMPLES

1. Starting at 0 and counting to the right to +5 represents 5 units in a positive (+) direction; **+5 is 5 units greater than 0**.

2. Starting at 0 and counting to the left to −5 represents 5 units in a negative (−) direction; **−5 is 5 units less than 0**.

3. Starting at −4 and counting to the right to +3 represents 7 units in a positive (+) direction; **+3 is 7 units greater than −4**.

4. Starting at +3 and counting to the left to −4 represents 7 units in a negative (−) direction; **−4 is 7 units less than +3**.

5. Starting at −2 and counting to the left to −10 represents 8 units in a negative (−) direction; **−10 is 8 units less than −2**.

6. Starting at −9 and counting to the right to 0 represents 9 units in a positive (+) direction; **0 is 9 units greater than −9**.

EXERCISE 6–2

Refer to the number line shown in Figure 6–4. Give the direction (+ or −) and the number of units counted going from the first to the second number.

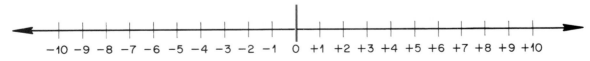

Figure 6–4

1. 0 to +6	**6.** −7 to +1	**11.** +10 to −10	**16.** −8 to −3
2. 0 to −6	**7.** +8 to −3	**12.** −10 to +10	**17.** +4 to +10
3. −2 to 0	**8.** +6 to −6	**13.** +6 to −5	**18.** +7 to +2
4. +2 to 0	**9.** −10 to −4	**14.** −9 to +8	**19.** −4 to +7
5. −4 to +6	**10.** +9 to +1	**15.** −3 to −7	**20.** +6 to −4

Refer to the number line shown in Figure 6–5. Give the direction (+ or −) and the number of units counted going from the first number to the second.

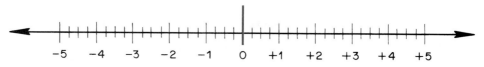

Figure 6–5

21. -2 to $+3\frac{1}{2}$

22. -5 to $-4\frac{1}{4}$

23. $+1.5$ to -2

24. -2.75 to 0

25. $+3$ to -4.25

26. $+4\frac{1}{2}$ to $+1$

27. $+3\frac{1}{4}$ to $+\frac{1}{2}$

28. -4.25 to -4.5

29. -0.25 to -4

30. $+2\frac{1}{4}$ to $-3\frac{3}{4}$

31. $+4.75$ to 0.5

32. -1.5 to $+4.25$

Select the greater of each of the two signed numbers and indicate the number of units by which it is greater.

33. $+3, -2$

34. $-6, 0$

35. $-7, +3$

36. $-12, -4$

37. $-28, -73$

38. $+18.62, +14.08$

39. $-18.62, -14.08$

40. $+10.57, -12.85$

41. $-2.5, +2.5$

42. $-86\frac{3}{4}, 0$

43. $+16.17, -21.86$

44. $-23\frac{1}{4}, -15\frac{3}{8}$

45. $+1\frac{1}{16}, -1\frac{7}{8}$

46. $-50.23, -41.76$

47. $+\frac{3}{16}, -\frac{9}{32}$

List each set of signed numbers in order of increasing value, starting with the smallest value.

48. $+14, -25, +25, 0, +7, -7, -10$

49. $0, -18, +4, -22, -1, +2, +16$

50. $-2, -19, -21, +13, +27, 0, -5$

51. $-15, +17, +3, -3, +15, 0, -8$

52. $-11.6, 0, +10, -4.3, +1, -10.8$

53. $+17.8, +2.3, -1, +1, -1.1, -0.4$

54. $+4\frac{3}{8}, -6, -12\frac{7}{8}, 0, -12\frac{13}{16}, -\frac{1}{4}$

55. $0, -1\frac{1}{2}, -6\frac{5}{32}, -6\frac{1}{4}, -1\frac{15}{32}, -1\frac{7}{16}$

6–3 Operations Using Signed Numbers

In order to solve problems in algebra, you must be able to perform basic operations using signed numbers. The operations of addition, subtraction, multiplication, division, powers, and roots with both positive and negative numbers are presented.

6–4 Absolute Value

The procedures for performing certain operations of signed numbers are based on an understanding of absolute value. The *absolute value* of a number is the distance from the number 0. The absolute value of a number is indicated by placing the number between a pair of vertical bars. The absolute values of $+8$ and -8 are written as follows.

$$|+8| = 8 \text{ because } +8 \text{ is 8 units from 0}$$
$$|-8| = 8 \text{ because } -8 \text{ is 8 units from 0}$$

The absolute values of $+8$ and -8 are the same. The absolute values of -15 and 5 are written as follows.

$$|-15| = 15 \qquad |5| = 5$$

The absolute value of -15 is 10 greater than the absolute value of 5. The number 5 is the same as the number $+5$. Positive numbers do not require a positive sign as a prefix.

EXERCISE 6–4

Express each of the pairs of signed numbers as absolute values. Subtract the smaller absolute value from the larger absolute value.

1. $+15, -10$
2. $-15, +10$
3. $-6, +2$
4. $-14, 14$
5. $-9, 0$
6. $+9, 0$

7. $-23, +22$
8. $+18, -18$
9. $+18, 18$
10. $-18, -18$
11. $-6\frac{1}{4}, -8\frac{3}{4}$

12. $-3\frac{1}{2}, -12\frac{7}{8}$
13. $+15.9, -10.7$
14. $10.54, -12.46$
15. $-0.03, -0.007$

6–5 Addition of Signed Numbers

Procedure for Adding Two or More Numbers with the Same Signs

- Add the absolute values of the numbers.
- If all the numbers are positive, then the sum is positive. A positive sign is not needed as a prefix.
- If all the numbers are negative, prefix a negative sign to the sum.

EXAMPLES

1. $4 + 8 = 12$ *Ans*
2. $\left(+25\frac{1}{2}\right) + (+10) = 35\frac{1}{2}$ *Ans*
3. $9 + 5.6 + 2.1 = 16.7$ *Ans*
4. $6 + (+2) + 7 = 15$ *Ans*
5. Add -6 and -14.

   The absolute value of -6 is 6.

   The absolute value of -14 is 14.

   Add. $6 + 14 = 20$

   Prefix a negative sign to the sum. $(-6) + (-14) = -20$ *Ans*
6. $(-2) + (-5) + (-8) + (-10) = -25$ *Ans*
7. $(-4) + (-12) = -16$ *Ans*
8. $-2.5 + (-3) + (-0.2) + (-5.8) = -11.5$ *Ans*

Calculator Application

Pressing the change sign key, $\boxed{+/-}$, instructs the calculator to change the sign of the displayed value. Calculations involving negative numbers can be made by using the change sign key. To enter a negative number, enter the absolute value of the number, then press the change sign key.

EXAMPLES •——————————————————————————————————

1. Add. $-25.873 + (-138.029)$

25.873 $\boxed{+/-}$ $\boxed{+}$ 138.029 $\boxed{+/-}$ $\boxed{=}$ -163.902 *Ans*

2. Add. $-6.053 + (-0.072) + (-15.763) + (-0.009)$

6.053 $\boxed{+/-}$ $\boxed{+}$ .072 $\boxed{+/-}$ $\boxed{+}$ 15.763 $\boxed{+/-}$ $\boxed{+}$ .009 $\boxed{+/-}$ $\boxed{=}$ -21.897 *Ans*

———•

Certain more advanced calculators permit direct entry of negative values. These calculators do *not* have the change sign key, $\boxed{+/-}$. The subtraction key, $\boxed{-}$ or negative key, $\boxed{(-)}$, is used to enter negative values. The negative sign is entered before the number is entered. A negative value is displayed. To determine if your calculator has this capability, press $\boxed{-}$ or, $\boxed{(-)}$, and enter a number. The display will show a negative value. For example, -125.87 is entered directly as $\boxed{-}$ or, $\boxed{(-)}$ 125.87. The value displayed is -125.87.

Be careful! On some calculators the $\boxed{-}$ key is used for subtraction only and the $\boxed{(-)}$ key for negative numbers. On these calculators you will get an error message if the wrong key is used. For example, on a Texas Instruments TI-30, the following message is displayed:

$$\boxed{\begin{array}{c} \text{SYNTAX} \\ \text{Error} \end{array}}$$

EXAMPLES •——————————————————————————————————

1. Add. $-25.873 + (-138.029)$

$\boxed{-}$ or $\boxed{(-)}$ 25.873 $\boxed{+}$ $\boxed{-}$ or $\boxed{(-)}$ 138.029 $\boxed{\text{EXE}}$ -163.902 *Ans*

2. Add. $-6.053 + (-0.072) + (-15.763) + (-0.009)$

$\boxed{-}$ or $\boxed{(-)}$ 6.053 $\boxed{+}$ $\boxed{-}$ or $\boxed{(-)}$ .072 $\boxed{+}$ $\boxed{-}$ or $\boxed{(-)}$ 15.763 $\boxed{+}$ $\boxed{-}$ or $\boxed{(-)}$ .009 $\boxed{\text{EXE}}$
-21.897 *Ans*

———•

Procedure for Adding a Positive Number and a Negative Number

• Subtract the smaller absolute value from the larger absolute value.
• The answer has the sign of the number having the larger absolute value.

EXAMPLES •——————————————————————————————————

1. Add $+10$ and -4.

The absolute value of $+10$ is 10.

The absolute value of -4 is 4.

Subtract. $10 - 4 = 6$

Prefix the positive sign to the difference. $(+10) + (-4) = +6$ or 6 *Ans*

2. Add -10 and $+4$

The absolute value of -10 is 10.

The absolute value of $+4$ is 4.

Subtract. $10 - 4 = 6$

Prefix the negative sign to the difference. $(-10) + (+4) = -6$ *Ans*

3. $15.8 + (-2.4) = +13.4$ or 13.4 *Ans*

4. $6\dfrac{1}{4} + \left(-10\dfrac{3}{4}\right) = -4\dfrac{1}{2}$ *Ans*

5. $-20 + (+20) = 0$ *Ans*

Calculator Application

Add. $16\dfrac{17}{32} + \left(-23\dfrac{29}{64}\right)$

16 $\boxed{a^{b/c}}$ 17 $\boxed{a^{b/c}}$ 32 $\boxed{+}$ 23 $\boxed{a^{b/c}}$ 29 $\boxed{a^{b/c}}$ 64 $\boxed{+/-}$ $\boxed{=}$ $-6 \, \_ \, 59 \, \_\rfloor \, 64$

$-6\dfrac{59}{64}$ *Ans*

or 16 $\boxed{a^{b/c}}$ 17 $\boxed{a^{b/c}}$ 32 $\boxed{-}$ 23 $\boxed{a^{b/c}}$ 29 $\boxed{a^{b/c}}$ 64 $\boxed{\text{EXE}}$ $-6 \, \_ \, 59 \, \_\rfloor \, 64$

$-6\dfrac{59}{64}$ *Ans*

Calculators that do not have an $\boxed{a^{b/c}}$ fraction key require you to remember that a mixed number is the sum of a whole number and a fraction.

EXAMPLE •

Enter the mixed number $-24\frac{3}{4}$ in a calculator that does not have an $\boxed{a^{b/c}}$ key.

$\boxed{(-)}$ $\boxed{(}$ 24 $\boxed{+}$ 3 $\boxed{÷}$ 4 $\boxed{)}$

Procedure for Adding Combinations of Two or More Positive and Negative Numbers

• Add all the positive numbers.
• Add all the negative numbers.
• Add their sums, following the procedure for adding signed numbers.

EXAMPLES •

1. $-12 + 7 + 3 + (-5) + 20 = 30 + (-17) = +13$ or 13 *Ans*

2. $6 + (-10) + (-5) + 8 + 2 + (-7) = 16 + (-22) = -6$ *Ans*

Calculator Application

Add. $15.86 + (-5.07) + (-8.95) + 23.56 + (-0.92)$

15.86 $\boxed{+}$ 5.07 $\boxed{+/-}$ $\boxed{+}$ 8.95 $\boxed{+/-}$ $\boxed{+}$ 23.56 $\boxed{+}$ .92 $\boxed{+/-}$ $\boxed{=}$ 24.48 *Ans*

or 15.86 $\boxed{+}$ $\boxed{-}$ or $\boxed{(-)}$ 5.07 $\boxed{+}$ $\boxed{-}$ or $\boxed{(-)}$ 8.95 $\boxed{+}$ 23.56 $\boxed{+}$ $\boxed{-}$ or $\boxed{(-)}$ .92 $\boxed{\text{EXE}}$

24.48 *Ans*

EXERCISE 6–5

Add the signed numbers as indicated.

1. $+6 + (+9)$

2. $15 + 8$

3. $8 + 36$

4. $7 + (+18) + 2$

5. $0 + 25$

6. $-12 + (-7)$

7. $-8 + (-15)$

8. $0 + (-16)$

9. $-14 + (-4) + (-11)$

10. $-3 + (-6) + (-17)$

11. $9 + (-4)$

12. $18 + (-26)$

13. $-20 + 17$

14. $46 + (-14)$

15. $-23 + 17$

16. $25 + (-3)$

17. $-25 + 3$

18. $-18 + (-25)$

19. $-4 + (-31)$

20. $27 + (-27)$

21. $-15.3 + (-3.5)$

22. $-15.3 + 3.5$

23. $-3.8 + (-14.7)$

24. $37.96 + (-40.38)$

25. $-9\frac{1}{4} + \left(-3\frac{3}{4}\right)$

26. $18\frac{31}{32} + \left(-36\frac{11}{64}\right)$

27. $-13\frac{1}{8} + \left(-7\frac{13}{32}\right)$

28. $-13\frac{1}{8} + 7\frac{13}{32}$

29. $-4.25 + (-7) + (-3.22)$

30. $18.07 + (-17.64)$

31. $16 + (-4) + (-11)$

32. $-36.33 + (-4.20) + 26.87$

33. $30.88 + (-0.95) + 1.32$

34. $-12.77 + (-9) + (-7.61) + 0.48$

35. $2.53 + 16.09 + (-54.05) + 21.37$

36. $-12\frac{1}{4} + 2\frac{7}{32} + \left(-23\frac{9}{16}\right) + \left(-19\frac{37}{64}\right)$

6–6 Subtraction of Signed Numbers

Procedure for Subtracting Signed Numbers

- Change the sign of the number being subtracted (subtrahend) to the opposite sign.
- Add the resulting signed numbers.

NOTE: When the sign of the subtrahend is changed, the problem becomes one in addition. Therefore, subtracting a negative number is the same as adding a positive number. Subtracting a positive number is the same as adding a negative number.

EXAMPLES

1. Subtract 8 from 5. $\qquad\qquad\qquad\qquad\qquad\qquad$ $5 - 8$

Change the sign of the subtrahend to the opposite sign.

The subtrahend is 8. Change 8 (or $+8$) to -8.

Add the resulting signed numbers. $\qquad\qquad\qquad\qquad$ $5 + (-8) = -3$ *Ans*

2. Subtract -10 from 4. $\qquad\qquad\qquad\qquad\qquad\qquad$ $4 - (-10)$

The subtrahend is -10. Change the sign of the subtrahend from -10 to 10.

Add the resulting signed numbers. $\qquad\qquad\qquad$ $4 + (10) = 14$ *Ans*

3. $-7 - (-12) = -7 + (+12) = +5$ or 5 *Ans*

4. $10.6 - (-7.2) = 10.6 + (+7.2) = +17.8$ or 17.8 *Ans*

5. $-10.6 - (+7.2) = -10.6 + (-7.2) = -17.8$ *Ans*

6. $0 - (-14) = 0 + (+14) = +14$ or 14 *Ans*

7. $0 - (+14) = 0 + (-14) = -14$ *Ans*

8. $-20 - (-20) = -20 + (+20) = 0$ *Ans*

9. $(18 - 4) - (-20 + 3) = 14 - (-17) = 14 + 17 = 31$ *Ans*

NOTE: Following the proper order of operations, the operations enclosed within parentheses must be done first as shown in example 9.

Calculator Applications

1. Subtract. $-163.94 - (-150.65)$

   163.94 ⌈+/−⌉ ⌈−⌉ 150.65 ⌈+/−⌉ ⌈=⌉ -13.29 *Ans*

   or ⌈−⌉ or ⌈(−)⌉ 163.94 ⌈−⌉ ⌈−⌉ or ⌈(−)⌉ 150.65 ⌈EXE⌉ -13.29 *Ans*

2. Subtract. $-27.55 - (-8.64 + 0.74) - (-53.41)$

   27.55 ⌈+/−⌉ ⌈−⌉ ⌈(⌉ 8.64 ⌈+/−⌉ ⌈+⌉ .74 ⌈)⌉ ⌈−⌉ 53.41 ⌈+/−⌉ ⌈=⌉ 33.76 *Ans*

   or ⌈−⌉ or ⌈(−)⌉ 27.55 ⌈−⌉ ⌈(⌉ ⌈−⌉ or ⌈(−)⌉ ⌈−⌉ 8.64 ⌈+⌉ .74 ⌈)⌉ ⌈−⌉ ⌈−⌉ or ⌈(−)⌉ 53.41 ⌈EXE⌉

   33.76 *Ans*

EXERCISE 6–6

Subtract the signed numbers as indicated.

1. $-10 - (-8)$
2. $10 - 8$
3. $5 - (-13)$
4. $5 - (+13)$
5. $-22 - (-14)$
6. $+15 - (+9)$
7. $3 - (-19)$
8. $26 - 31$
9. $+40 - (+40)$
10. $-40 - (-40)$
11. $-40 - (+40)$
12. $-25 - 0$
13. $0 - 7$
14. $0 - (-7)$
15. $36 - 41$
16. $-52 - (-8)$
17. $-7 - (-46)$
18. $34 - (+17)$
19. $16.5 - (+14.3)$
20. $16.5 - (-14.3)$
21. $-16.5 - (-14.3)$

22. $-50.23 - 51.87$
23. $+50.23 - (-51.87)$
24. $-0.003 - 0.05$
25. $10\frac{1}{2} - \left(+7\frac{1}{4}\right)$
26. $-10\frac{1}{2} - \left(-7\frac{1}{4}\right)$
27. $27\frac{3}{8} - \left(-13\frac{27}{64}\right)$
28. $\frac{9}{64} - \left(+\frac{11}{64}\right)$
29. $(6 + 10) - (-7 + 8)$
30. $(-14 + 8) - (-5 + 8)$
31. $(-14 + 5) - (2 - 10)$
32. $(2.7 - 5.6) - (18.4 - 6.3)$
33. $(7.23 - 6.81) - (-10.73)$
34. $\left(9\frac{3}{8} + 1\frac{1}{2}\right) - \left(8 - 9\frac{1}{4}\right)$
35. $[3 - (-7)] - [14 - (-6)]$
36. $[-8.76 + (-5.83)] - [12.06 - (-0.97)]$

6–7   Multiplication of Signed Numbers

Procedure for Multiplying Two or More Signed Numbers

- Multiply the absolute values of the numbers.
- If all numbers are positive, the product is positive.
- Count the number of negative signs.
  - If there is an odd number of negative signs, the product is negative.
  - If there is an even number of negative signs, the product is positive.

It is not necessary to count the number of positive values in an expression consisting of both positive and negative numbers. Count only the number of negative values to determine the sign of the product.

EXAMPLES •

1. Multiply. 3(−5)

   Multiply the absolute values.

   There is one negative sign.

   Since one is an odd number, the product is negative.   3(−5) = −15 *Ans*

2. Multiply. −3(−5)

   Multiply the absolute values.

   There are two negative signs.

   Since two is an even number, the product is positive.   −3(−5) = +15 or 15 *Ans*

3. (−3)(−1)(−2)(−3)(−2)(−1) = +36 or 36 *Ans*

4. (−3)(+1)(−2)(−3)(−2)(−1) = −36 *Ans*

5. (3)(−1)(−2)(3)(2)(−1) = −36 *Ans*

6. (−3)(−1)(−2)(3)(−2)(1) = +36 or 36 *Ans*

7. (3)(1)(2)(3)(2)(1) = +36 or 36 *Ans*

NOTE: The product of any number or numbers and 0 equals 0. For example, 0(4) = 0; 0(−4) = 0; (7)(−4)(0)(3) = 0.

Calculator Application

Multiply. (−8.61)(3.04)(−1.85)(−4.03)(0.162). Round the answer to 1 decimal place.

8.61 $\boxed{+/-}$ $\boxed{\times}$ 3.04 $\boxed{\times}$ 1.85 $\boxed{+/-}$ $\boxed{\times}$ 4.03 $\boxed{+/-}$ $\boxed{\times}$ .162 $\boxed{=}$ −31.61320475

                                                                    −31.6 (rounded) *Ans*

or $\boxed{-}$ or $\boxed{(-)}$ 8.61 $\boxed{\times}$ 3.04 $\boxed{\times}$ $\boxed{-}$ or $\boxed{(-)}$ 1.85 $\boxed{\times}$ $\boxed{-}$ or $\boxed{(-)}$ 4.03 $\boxed{\times}$ .162 $\boxed{EXE}$

−31.61320475

−31.6 (rounded) *Ans*

EXERCISE 6–7

Multiply the signed numbers as indicated. Where necessary, round the answers to 2 decimal places.

1. (−4)(6)
2. 4(−6)
3. (−4)(−6)
4. (4)(6)
5. (+12)(−3)
6. (−10)(−2)
7. (−5)(7)
8. (−2)(−14)
9. 0(16)
10. 0(−16)
11. (6.5)(−2)
12. (−3.2)(−0.1)
13. (8.23)(−1.46)
14. (−0.06)(−0.60)
15. $\left(1\frac{1}{2}\right)\left(-\frac{1}{2}\right)$
16. $\left(-2\frac{1}{8}\right)\left(-1\frac{1}{2}\right)$

17. $\frac{1}{4}(0)$
18. $\left(-\frac{1}{4}\right)(0)$
19. (−2)(−2)(−2)
20. (−2)(+2)(+2)
21. (−2)(+2)(−2)
22. (−3.86)(−2.1)(27.85)(−32.56)
23. (4)(−3.86)(0.7)(−1)
24. (−6.3)(−0.35)(2)(−1)(0.05)
25. (1)(1)(−1)(1)(1)(1)(1)
26. (1)(1)(−1)(1)(1)(−1)(1)
27. (−4.03)(−0.25)(−3)(−0.127)
28. (−0.02)(−120)(−0.20)
29. $\left(+\frac{1}{4}\right)(-8)\left(\frac{1}{2}\right)\left(2\frac{3}{8}\right)$
30. $\left(-\frac{1}{4}\right)(-8)\left(\frac{1}{2}\right)\left(-2\frac{3}{8}\right)$

6–8 Division of Signed Numbers

Procedure for Dividing Signed Numbers

- Divide the absolute values of the numbers.
- Determine the sign of the quotient.
  - If both numbers have the same sign (both negative or both positive), the quotient is positive.
  - If the two numbers have unlike signs (one positive and one negative), the quotient is negative.

NOTE: Recall that zero divided by any number equals zero. For example, $0 \div (+9) = 0$, $0 \div (-9) = 0$.

Dividing by zero is *not* possible. For example, $+9 \div 0$ and $-9 \div 0$ are not possible.

EXAMPLES •─────────────────────────

1. Divide -20 by -4.

   Divide the absolute values.

   There are two negative signs.

   The quotient is positive. $-20 \div (-4) = +5$ or 5 *Ans*

2. Divide 24 by -8.

   Divide the absolute values.

   The signs are unlike.

   The quotient is negative. $24 \div (-8) = -3$ *Ans*

3. $30 \div 15 = +2$ or 2 *Ans*

4. $-24 \div 3 = -8$ *Ans*

Calculator Application

Divide. $31.875 \div (-56.625)$. Round the answer to 3 decimal places.

$31.875 \;\boxed{\div}\; 56.625 \;\boxed{+/-}\; \boxed{=} \; -0.562913907, -0.563$ (rounded) *Ans*

or $31.875 \;\boxed{\div}\; \boxed{-}$ or $\boxed{(-)} \; 56.625 \;\boxed{\text{EXE}}\; -0.562913907, -0.563$ (rounded) *Ans*

─── •

EXERCISE 6–8

Divide the signed numbers as indicated. Round the answers to 2 decimal places where necessary.

1. $-10 \div (-5)$
2. $-10 \div (+5)$
3. $10 \div (-5)$
4. $+18 \div (+9)$
5. $-21 \div 3$
6. $12 \div (-4)$
7. $-27 \div (-9)$
8. $+48 \div (-6)$
9. $-48 \div (-6)$
10. $-35 \div 7$
11. $\dfrac{-16}{-4}$

12. $\dfrac{0}{-10}$
13. $\dfrac{30}{-10}$
14. $\dfrac{-40}{-8}$
15. $\dfrac{-36}{6}$
16. $\dfrac{39}{13}$
17. $\dfrac{-60}{-0.5}$

18. $\dfrac{-20}{-2.5}$
19. $\dfrac{+6.4}{-4}$
20. $\dfrac{-17.92}{3.28}$
21. $0.562 \div (-0.821)$
22. $-\dfrac{1}{2} \div \left(-\dfrac{1}{2}\right)$
23. $-15 \div 1\dfrac{1}{4}$
24. $-6 \div \dfrac{3}{8}$

25. $4\frac{1}{3} \div \left(-2\frac{2}{3}\right)$ **27.** $-29.96 \div 5.35$ **29.** $-20.47 \div 0.537$

28. $-4.125 \div (-1.5)$ **30.** $-44.876 \div (-7.836)$

26. $0 \div \left(-\frac{7}{8}\right)$

6–9 Powers of Signed Numbers

Procedure for Determining Values with Positive Exponents

• Apply the procedure for multiplying signed numbers to raising signed numbers to powers.

EXAMPLES •——————————————————————————————————

1. $2^3 = (2)(2)(2) = +8$ or 8 *Ans*

2. $2^4 = (2)(2)(2)(2) = +16$ or 16 *Ans*

3. $(-4)^2 = (-4)(-4) = +16$ or 16 *Ans*

4. $(-4)^3 = (-4)(-4)(-4) = -64$ *Ans*

5. $(-2)^4 = (-2)(-2)(-2)(-2) = +16$ or 16 *Ans*

6. $(-2)^5 = (-2)(-2)(-2)(-2)(-2) = -32$ *Ans*

7. $\left(-\frac{2}{3}\right)^3 = \left(-\frac{2}{3}\right)\left(-\frac{2}{3}\right)\left(-\frac{2}{3}\right) = -\frac{8}{27}$ or -0.296 (rounded) *Ans*

NOTE:

• A positive number raised to any power is positive.

• A negative number raised to an even power is positive.

• A negative number raised to an odd power is negative.

Calculator Application

As presented in Unit 3, the universal power key, $\boxed{y^x}$, $\boxed{x^y}$, or $\boxed{\wedge}$ raises any *positive* number to a power.

Solve. 2.073^5. Round the answer to 2 decimal places.

   $2.073 \boxed{y^x} 5 \boxed{=} 38.28216674$, 38.28 (rounded) *Ans*

The universal power key can also be used to raise a *negative* number to a power. Most calculators are capable of directly raising a negative number to a power. Use the change sign key $\boxed{+/-}$ or the negative key $\boxed{(-)}$ or $\boxed{-}$ and the universal power key $\boxed{y^x}$, $\boxed{x^y}$, or $\boxed{\wedge}$.

Be careful when either writing or raising a negative number to a power. The number $-(3)^4$ is not the same as -3^4. The number -3^4 is shorthand for $-1(3)^4 = -81$. But, $(-3)^4 = 81$.

EXAMPLES •——————————————————————————————————

Round the answers to 4 significant digits.

1. Solve. $(-3.874)^4$

   $3.874 \boxed{+/-} \boxed{x^y} 4 \boxed{=} 225.236342$, 225.2 *Ans*

   or $\boxed{(}\ \boxed{-}$ or $\boxed{(-)}\ 3.874\ \boxed{)}\ \boxed{x^y}$ or $\boxed{\wedge}\ 4\ \boxed{EXE}\ 225.236342$, 225.2 *Ans*

   or $\boxed{(}\boxed{(-)}\ 3.874\ \boxed{)}\ \boxed{\wedge}\ 4\ \boxed{ENTER}\ 225.236342$, 225.2 *Ans*

NOTE: -3.874 must be enclosed within parentheses.

2. Solve. $(-3.874)^5$

3.874 $\boxed{+/-}$ $\boxed{x^y}$ 5 $\boxed{=}$ -872.565589, -872.6 *Ans*

or $\boxed{(}$ $\boxed{-}$ or $\boxed{(-)}$ 3.874 $\boxed{)}$ $\boxed{x^y}$ or $\boxed{\wedge}$ 5 $\boxed{EXE}$ -872.565589, -872.6 *Ans*

or $\boxed{(}$ $\boxed{(-)}$ 3.874 $\boxed{)}$ $\boxed{\wedge}$ 5 $\boxed{ENTER}$ -872.565589, -872.6 *Ans*

Negative Exponents

Two numbers whose product is 1 are *multiplicative inverses* or *reciprocals* of each other. For example, x or $\frac{x}{1}$ and $\frac{1}{x}$ are reciprocals of each other:

$$\frac{x}{1} \times \frac{1}{x} = 1.$$

A number with a negative exponent is equal to the reciprocal of the number with a positive exponent:

$$\frac{x^{-n}}{1} = \frac{1}{x^n}$$

Procedure for Determining Values with Negative Exponents

* Write the reciprocal of the number (invert the number) changing the negative exponent to a positive exponent.
* Simplify.

EXAMPLES

1. Find the value of 5^{-2}.

Write the reciprocal of 5^{-2} (invert 5^{-2}) and change the negative exponent (-2) to a positive exponent (2).

$$5^{-2} = \frac{5^{-2}}{1}$$

$$\frac{5^{-2}}{1} = \frac{1}{5^2}$$

Simplify.

$$\frac{1}{5^2} = \frac{1}{(5)(5)} = \frac{1}{25} \text{ or } 0.04 \text{ Ans}$$

2. $2^{-3} = \dfrac{1}{2^3} = \dfrac{1}{(2)(2)(2)} = \dfrac{1}{8}$ or 0.125 *Ans*

3. $(-5)^{-2} = \dfrac{1}{(-5)^2} = \dfrac{1}{(-5)(-5)} = \dfrac{1}{+25} = \dfrac{1}{25}$ or 0.04 *Ans*

4. $(-4)^{-3} = \dfrac{1}{(-4)^3} = \dfrac{1}{(-4)(-4)(-4)} = \dfrac{1}{-64} = -\dfrac{1}{64}$ or -0.015625 *Ans*

Calculator Applications

Depending on the calculator used, a *negative* exponent is entered with the change sign key $\boxed{+/-}$ or the negative key $\boxed{(-)}$ or $\boxed{-}$. The rest of the procedure is the same as used with positive exponents.

EXAMPLES

Round the answers to 3 decimal places.

1. Calculate. 3.162^{-3}

3.162 $\boxed{y^x}$ 3 $\boxed{+/-}$ $\boxed{=}$ 0.0316311078, 0.032 *Ans*

or 3.162 $\boxed{x^y}$ or $\boxed{\wedge}$ $\boxed{-}$ 3 $\boxed{EXE}$ 0.0316311078, 0.032 *Ans*

2. Calculate. $(-3.162)^{-3}$

The solutions shown are with calculators capable of directly raising a negative number to a power.

3.162 $\boxed{+/-}$ $\boxed{y^x}$ 3 $\boxed{+/-}$ $\boxed{=}$ -0.031631108, -0.032 *Ans*

or $\boxed{(}$ $\boxed{-}$ 3.162 $\boxed{)}$ $\boxed{x^y}$ or $\boxed{\wedge}$ $\boxed{(}$ $\boxed{(}$ $\boxed{(-)}$ 3 $\boxed{)}$ $\boxed{\text{EXE}}$ -0.031631108, -0.032 *Ans*

If this exponent is negative or 10 or more, then it is a good idea to place parentheses around the exponnet.

EXAMPLE •

Calculate $(-1.25)^{-12}$

$\boxed{(}$ $\boxed{(-)}$ 1.25 $\boxed{)}$ $\boxed{\wedge}$ $\boxed{(}$ $\boxed{(-)}$ 12 $\boxed{)}$ $\boxed{\text{ENTER}}$ 0.0687194767, 0.0687 *Ans*

EXERCISE 6–9

Raise each signed number to the indicated power. Round answers to 2 decimal places where necessary.

1. 2^2	**10.** $(-4.70)^2$	**19.** $\left(-\dfrac{1}{2}\right)^2$	**25.** 4^{-2}
2. $(-2)^2$	**11.** 4.70^3		**26.** $(-3)^{-3}$
3. 2^3	**12.** $(-50.87)^1$	**20.** $\left(\dfrac{1}{2}\right)^3$	**27.** $(-5)^{-2}$
4. $(-2)^3$	**13.** 0.93^6		**28.** $(-4.07)^{-3}$
5. $(-4)^3$	**14.** $(-1.58)^2$	**21.** $\left(-\dfrac{3}{4}\right)^2$	**29.** 4.98^{-2}
6. 3^3	**15.** $(-0.85)^3$		**30.** $(-1.038)^{-5}$
7. 2^4	**16.** 0.73^3	**22.** $\left(-\dfrac{3}{4}\right)^3$	**31.** 17.66^{-2}
8. $(-2)^4$	**17.** $(-0.58)^3$		**32.** $(-0.83)^{-3}$
9. $(-2)^5$	**18.** 2.36^5	**23.** $(-2)^{-2}$	**33.** $(-6.087)^{-4}$
		24. $(-2)^{-1}$	

6–10 Roots of Signed Numbers

A *root* of a number is a quantity that is taken two or more times as an equal factor of the number. The expression $\sqrt[3]{64}$ is called a radical. A *radical* is an indicated root of a number. The symbol $\sqrt{\ }$ is called a *radical sign* and indicates a root of a number. The digit 3 is called the index. An *index* indicates the number of times that a root is to be taken as an equal factor to produce the given number. The given number 64 is called a *radicand.*

When either a positive number or a negative number is squared, a positive number results. For example, $3^2 = 9$ and $(-3)^2 = 9$. Therefore, every positive number has two square roots, one positive root and one negative root. The square roots of 9 are $+3$ and -3. The expression $\sqrt{9}$ is used to indicate the positive or *principal root,* $+3$ or 3. The expression $-\sqrt{9}$ is used to indicate the negative root, -3. The expression $\pm\sqrt{9}$ indicates both the positive and negative square roots, ±3. The principal cube root of 8 is 2. The principal cube root of -8 is -2.

$$\overset{\text{index}}{\underset{\text{radical sign}}{\sqrt[3]{8}}} = 2 \qquad \sqrt[3]{-8} = -2 \quad \text{radicand}$$

The square root of a negative number has no solution in the real number system. For example, $\sqrt{-4}$ has no solution; $\sqrt{-4}$ is not equal to $\sqrt{(-2)(-2)}$ nor is it equal to $\sqrt{(+2)(+2)}$.

Any even root (even index) of a negative number has no solution in the real number system. For example, $\sqrt[4]{-16}$ and the $\sqrt[6]{-64}$ have no solution.

Calculations in this book will involve only principal roots unless otherwise specified, such as square roots of quadratic equations in Unit 18. The table in Figure 6–6 shows the sign of the principal root depending on the sign of the number and whether the index is even or odd.

INDEX	RADICAND	ROOT
Even	Positive (+)	Positive (+)
Even	Negative (−)	No Solution
Odd	Positive (+)	Positive (+)
Odd	Negative (−)	Negative (−)

Figure 6–6

EXAMPLES

1. $\sqrt{36} = \sqrt{(6)(6)} = 6$ *Ans*

    Even index (2), positive radicand (36), positive root (6)

2. $\sqrt[4]{-81} \ne \sqrt[4]{(3)(3)(3)(3)} \ne \sqrt[4]{(-3)(-3)(-3)(-3)}$. No solution. *Ans*

    Even index (4), negative radicand (−81)

3. $\sqrt[5]{32} = \sqrt[5]{(2)(2)(2)(2)(2)} = 2$ *Ans*

    Odd index (5), positive radicand (32), positive root (2)

4. $\sqrt[3]{-27} = \sqrt{(-3)(-3)(-3)} = -3$ *Ans*

    Odd index (3), negative radicand (−27), negative root (−3)

5. $\sqrt[3]{\dfrac{-8}{27}} = \sqrt[3]{\left(-\dfrac{2}{3}\right)\left(-\dfrac{2}{3}\right)\left(-\dfrac{2}{3}\right)} = -\dfrac{2}{3}$ or -0.67 (rounded) *Ans*

    Odd index (3), negative radicand $\left(-\dfrac{8}{27}\right)$, negative number $\left(-\dfrac{2}{3}\right)$

Calculator Applications

As presented in Unit 3, not all calculators have the root key $\boxed{\sqrt[x]{}}$, $\boxed{\sqrt[x]{y}}$, or $\boxed{x^{1/y}}$. Two basic methods of calculating roots of *positive* numbers were shown. The following examples show each method.

1. **Solve.** $\sqrt[4]{562.824}$. The procedure shown is used with calculators that have the root key $\boxed{\sqrt[x]{}}$.

    Solution.   4 $\boxed{\sqrt[x]{}}$ 562.824 $\boxed{\text{EXE}}$ 4.870719863 *Ans*

2. **Solve.** $\sqrt[4]{562.824}$. The procedures shown are used with calculators that do not have the root key and where roots are second functions. The procedures vary somewhat depending on the calculator used.

    Solution.   562.824 $\boxed{\text{2nd}}$ $\boxed{y^x}$ 4 $\boxed{=}$ 4.870719863 *Ans*

    or 562.824 $\boxed{\text{SHIFT}}$ $\boxed{x^y}$ 4 $\boxed{=}$ 4.870719863 *Ans*

3. **Solve.** $\sqrt[4]{562.824}$. Many graphing calculators, such as a TI-83 or TI-84, access the $\boxed{\sqrt[x]{}}$ key by pressing the $\boxed{\text{MATH}}$ key to get the screen in Figure 6–7. Item #5 is $\sqrt[x]{}$. You either need to press $\boxed{5}$ or press the $\boxed{\blacktriangledown}$ key four times (until the 5: is highlighted, and then press $\boxed{\text{ENTER}}$.

Figure 6–7

Solution. 4 MATH 5 ENTER (562.824) ENTER 4.870719863 *Ans*

or 4 MATH ▼ ▼ ▼ ▼ ENTER (562.824) ENTER

Most calculators are capable of directly computing roots of *negative* numbers. The following examples show the procedures for calculating roots of *negative* numbers depending on the make and model of the calculator.

1. Solve. $\sqrt[5]{-85.376}$. The procedure shown is used with a calculator that has the root key $\boxed{\sqrt[x]{}}$.

Solution. 5 $\boxed{\sqrt[x]{}}$ − 85.376 EXE −2.433700665 *Ans*

or 5 $\boxed{\sqrt[x]{}}$ 85.376 +/− = −2,433700665 *Ans*

2. Solve. $\sqrt[5]{-85.376}$. The procedures shown are used with calculators that do not have the root key and where roots are second functions.

Solution. 85.376 +/− 2nd y^x 5 = −2.433700665 *Ans*

or 85.376 +/− SHIFT x^y 5 = −2.433700665 *Ans*

3. Solve. $\sqrt[5]{-85.376}$. The following can be used with graphing calculators, such as a TI-83 or TI-84.

Solution. 5 MATH 5 ENTER ((−) 85.376) ENTER −2.433700665 *Ans*

or 5 MATH ▼ ▼ ▼ ▼ ENTER ((−) 85.376) ENTER

−2.433700665 *Ans*

With a calculator that is *not* capable of directly computing roots of negative numbers, enter the absolute value of the negative number and calculate as a positive number. Assign a positive sign or a negative sign to the displayed calculator answer, following the procedure for signs of roots of negative numbers.

EXERCISE 6–10

Use observation to determine the indicated root of each signed number. Round the answers to 3 decimal places where necessary.

1. $\sqrt{9}$	**7.** $\sqrt[4]{81}$	**13.** $\sqrt{1}$
2. $\sqrt[3]{64}$	**8.** $\sqrt[5]{-32}$	**14.** $\sqrt[3]{1}$
3. $\sqrt[3]{-64}$	**9.** $\sqrt[3]{-125}$	**15.** $\sqrt[3]{-1}$
4. $\sqrt{64}$	**10.** $\sqrt[3]{125}$	**16.** $\sqrt[5]{-1}$
5. $\sqrt[3]{-27}$	**11.** $\sqrt[7]{-128}$	**17.** $\sqrt[6]{1}$
6. $\sqrt[3]{-1000}$	**12.** $\sqrt[5]{+32}$	**18.** $\sqrt[9]{-1}$

19. $\sqrt[4]{81}$ **21.** $\sqrt[4]{\dfrac{1}{16}}$ **23.** $\dfrac{\sqrt[3]{-27}}{8}$

20. $\sqrt[3]{-216}$ **22.** $\sqrt[5]{\dfrac{-1}{32}}$ **24.** $\dfrac{27}{\sqrt[3]{-8}}$

Use a calculator to determine the indicated root of each signed number. Round the answers to 3 decimal places.

25. $\sqrt{28.073}$ **28.** $\sqrt[5]{-73.091}$ **30.** $\dfrac{\sqrt[3]{-89.096}}{-17.323}$

26. $\sqrt[3]{-236.539}$

27. $\sqrt[5]{424.637}$ **29.** $\sqrt[3]{\dfrac{-97.326}{123.592}}$

6–11   Combined Operations of Signed Numbers

Expressions consisting of two or more operations of signed numbers are solved using the same order of operations as in arithmetic.

Order of Operations

- Do all operations within the grouping symbol first. Parentheses, the fraction bar and the radical symbol are used to group numbers. If an expression contains parentheses within parentheses or brackets, do the work within the innermost parentheses first.
- Do powers and roots next. The operations are performed in the order in which they occur. If a root consists of two or more operations within the radical symbol, perform all the operations within the radical symbol, then extract the root.
- Do multiplication and division next in the order in which they occur.
- Do addition and subtraction last in the order in which they occur.

The memory aid "**P**lease **E**xcuse **M**y **D**ear **A**unt **S**ally" can be again used to help remember the order of operations. Remember that **P** in "Please" stands for parentheses, the **E** for exponents or raising to a power, **M** and **D** for multiplication and division, and the **A** and **S** for addition and subtraction.

EXAMPLES •————————————————————————————————

1. Find the value of $50 + (-2)[6 + (-2)^3(4)]$.

Perform the operations within brackets in the proper order. $50 + (-2)[6 + (-2)^3(4)]$

    Raise to a power.
    $-2^3 = -8$ $50 + (-2)[6 + (-8)(4)]$

    Multiply.
    $(-8)(4) = -32$ $50 + (-2)[6 + (-32)]$

    Add.
    $6 + (-32) = -26$ $50 + (-2)(-26)$

    Multiply.
    $(-2)(-26) = +52$ $50 + 52$

    Add.
    $50 + 52 = 102$ *102 Ans*

2. Find the values of

$$\frac{-37 - \sqrt{b^3 - (-8)c + (-7)}}{-3a^2 - b}$$

when $a = -2$, $b = -4$, and $c = 10$.

Substitute for a, b, and c in the expression.

Perform the operations within the radical sign.

    Raise to a power.

    $(-4)^3 = -64$

    Multiply.

    $(-8)(10) = -80$

    Subtract.

    $-64 - (-80) = 16$

    Add.

    $16 + (-7) = 9$

    Take the square root.

    $\sqrt{9} = 3$

Complete the operations in the numerator

$-37 - 3 = -40$

Complete the operations in the denominator.

    Raise to a power

    $(-2)^2 = 4$

    Multiply.

    $-3(4) = -12$

    Subtract.

    $-12 - (-4) = -8$

Divide.

$-40 \div -8 = +5$ or 5

$$\frac{-37 - \sqrt{(-4)^3 - (-8)(10) + (-7)}}{-3(-2)^2 - (-4)}$$

$$\frac{-37 - \sqrt{-64 - (-8)(10) + (-7)}}{-3(-2)^2 - (-4)}$$

$$\frac{-37 - \sqrt{-64 - (-80) + (-7)}}{-3(-2)^2 - (-4)}$$

$$\frac{-37 - \sqrt{16 + (-7)}}{-3(-2)^2 - (-4)}$$

$$\frac{-37 - \sqrt{9}}{-3(-2)^2 - (-4)}$$

$$\frac{-37 - 3}{-3(-2)^2 - (-4)}$$

$$\frac{-40}{-3(-2)^2 - (-4)}$$

$$\frac{-40}{-3(4) - (-4)}$$

$$\frac{-40}{-12 - (-4)}$$

$$\frac{-40}{-8}$$

5 Ans

Calculator Application

1. Solve. $\sqrt{38.44} - (-3)[8.2 - (5.6)^3(-7)]$

38.44 $\boxed{\sqrt{x}}$ $\boxed{-}$ 3 $\boxed{+/-}$ $\boxed{\times}$ $\boxed{(}$ 8.2 $\boxed{-}$ 5.6 $\boxed{y^x}$ 3 $\boxed{\times}$ 7 $\boxed{+/-}$ $\boxed{)}$ $\boxed{=}$ 3,718.736 *Ans*

or $\boxed{\sqrt{}}$ 38.44 $\boxed{-}$ $\boxed{(-)}$ or $\boxed{-}$ 3 $\boxed{\times}$ $\boxed{(}$ 8.2 $\boxed{-}$ 5.6 $\boxed{x^y}$ 3 $\boxed{\times}$ $\boxed{(-)}$ or $\boxed{-}$ 7 $\boxed{)}$ $\boxed{EXE}$

                                                           3,718.736 *Ans*

or $\boxed{\sqrt{}}$ 38.44 $\boxed{)}$ $\boxed{-}$ $\boxed{(-)}$ 3 $\boxed{(}$ 8.2 $\boxed{-}$ 5.6 $\boxed{\wedge}$ 3 $\boxed{\times}$ $\boxed{(-)}$ 7 $\boxed{)}$ $\boxed{ENTER}$ 3,718.736 *Ans*

2. Solve. $18.32 - (-4.52) + \dfrac{\sqrt[4]{93.724 - 6.023}}{-1.236^3}$. Round the answer to 2 decimal places.

The solutions shown are with calculators capable of directly computing powers of negative numbers.

NOTE: The universal power key is a second function on certain calculators.

18.32 $\boxed{-}$ 4.52 $\boxed{+/-}$ $\boxed{+}$ $\boxed{(}$ 93.724 $\boxed{-}$ 6.023 $\boxed{)}$ $\boxed{2nd}$ $\boxed{y^x}$ 4 $\boxed{\div}$ 1.236 $\boxed{+/-}$ $\boxed{y^x}$ 3 $\boxed{=}$
21.21932578, 21.22 *Ans*

or 18.32 $\boxed{-}$ 4.52 $\boxed{+/-}$ $\boxed{+}$ $\boxed{(}$ 93.724 $\boxed{-}$ 6.023 $\boxed{)}$ $\boxed{SHIFT}$ $\boxed{x^y}$ 4 $\boxed{\div}$ 1.236 $\boxed{+/-}$ $\boxed{x^y}$ 3 $\boxed{=}$
21.21932578, 21.22 *Ans*

or 18.32 $\boxed{-}$ $\boxed{(-)}$ or $\boxed{-}$ 4.52 $\boxed{+}$ 4 $\boxed{\sqrt[x]{\,}}$ $\boxed{(}$ 93.724 $\boxed{-}$ 6.023 $\boxed{)}$ $\boxed{\div}$ $\boxed{(}$ $\boxed{(-)}$ or $\boxed{-}$ 1.236 $\boxed{)}$ $\boxed{x^y}$ or $\boxed{\wedge}$ 3 $\boxed{\text{EXE}}$ 21.21932578, 21.22 *Ans*

EXERCISE 6–11

Solve each of the combined operation signed number exercises. Use the proper order of operations. Round the answers to 2 decimal places where necessary.

1. $6(-5) + (2)(7) - (-3)(-4)$

2. $\dfrac{2(-1)(-3) - (6)(5)}{3(7) - 9}$

3. $[4 + (2)(-5)]^2 + (-3)$

4. $4(-2) + 3(10-4)$

5. $\dfrac{5(6 - 3)}{3(2 + 9) - 27}$

6. $[4^2 + (2)(5)(-3)]^2 + 2(-3)^3$

7. $(-2.87)^3 + \sqrt{15.93} - (5.63)(4)(-5.26)^3$

8. $\dfrac{2(-5.16)^2}{3.07(4.98)} - \dfrac{(-4.66)^3}{18.37 + (-2.02)}$

9. $(-2.46)^3 + \sqrt[3]{(-3.86)(-10.41) - (-6.16)}$

10. $10.78^{-2} + [43.28 + (9)(-0.563)]^{-3}$

Substitute the given numbers for letters in these expressions and solve. Use the proper order of operations. Round the answers to 2 decimal places where necessary.

11. $6xy + 15 - xy; x = -2, y = 5$

12. $\dfrac{-3ab - 2bc}{abc - 35}; a = -3, b = 10, c = -4$

13. $rst(2r - 5st); r = -1, s = 4, t = 6$

14. $(x - y)(3x - 2y); x = -5, y = -7$

15. $\dfrac{p(2m - 2w)}{m(p + w) + 8}; m = 9, p = -6, w = -3$

16. $\dfrac{d^3 + 4f + fh}{h^2 - (2 + d)}; d = -2, f = -3, h = 4$

17. $\dfrac{x^2}{n} - \dfrac{21 + y^3}{xy}; n = 5.31, x = -5.67, y = -1.87$

18. $\sqrt{6(ab - 6)} - (c)^3; a = -6.07, b = -2.91, c = 1.56$

19. $5\sqrt[3]{e} + (ef - d) - (d)^3; d = -10.55, e = 8.26, f = -7.09$

20. $\dfrac{\sqrt[4]{(mpt + pt + 19)}}{t^2 + 2p - 7}; m = 2, p = -2.93, t = -5.86$

6–12   Scientific Notation

In scientific applications and certain technical fields, computations with very large and very small numbers are required. The numbers in their regular or standard form are inconvenient to read, write, and use in computations. For example, a coulomb (unit of electrical charge) equals 6,241,960,000,000,000,000 electrical charges. Copper expands 0.00000900 per unit of length per degree Fahrenheit. Scientific notation simplifies reading, writing, and computing with large and small numbers.

In scientific notation, a number is written as a whole number or decimal between 1 and 10 multiplied by 10 with a suitable exponent. For example, a value of 325,000 is written in scientific notation as 3.25×10^5.

The effect of multiplying a number by 10 is to shift the position of the decimal point. Changing a number from the standard decimal form to scientific notation involves counting the number of decimal places the decimal point must be shifted.

Expressing Decimal (Standard Form) Numbers in Scientific Notation

A positive or negative number whose absolute value is 10 or greater has a positive exponent when expressed in scientific notation.

EXAMPLES

Express the following values in scientific notation.

1. 146,000
   a. Write the number as a value between 1 and 10: 1.46
   b. Count the number of places the decimal point is shifted to determine the exponent of 10: $1\,46,000..$ The decimal point is shifted 5 places. The exponent of 10 is 5: 10^5.
   c. Multiply. 1.46×10^5
      $146,000 = 1.46 \times 10^5$ *Ans*

2. $6\,3,150,000. = 6.315 \times 10^7$ *Ans*
   Shift 7 places

3. $-9\,7.856 = -9.785\,6 \times 10^1$ *Ans*
   Shift 1 place

A positive or negative number whose absolute value is less than 1 has a negative exponent when expressed in scientific notation.

EXAMPLES

Express the following values in scientific notation.

1. $0.02\,89 = 2.89 \times 10^{-2}$ *Ans*
   Shift 2 places. Observe that the decimal point is shifted to the right, resulting in a negative exponent.

2. $0.00003\,18 = 3.18 \times 10^{-5}$ *Ans*
   Shift 5 places

3. $-0.8\,59 = -8.59 \times 10^{-1}$ *Ans*
   Shift 1 place

EXERCISE 6–12A

Rewrite the following standard form numbers in scientific notation.

1. 625	5. 959,100	9. 0.0000095	13. −0.104
2. 67,000	6. 1258.67	10. −415,000	14. 793,200
3. −3789	7. −0.00063	11. 2,030,000	15. 0.0083
4. 0.037	8. 59,300	12. 10.109	16. −0.0000276

The following are science and technology unit conversions. Rewrite each in scientific notation.

17. 1 kilowatt-hour (kWh) = 3,600,000 joules (J)
18. 1 dyne (dyn) = 0.00001 newton (N)
19. 1 foot-pound-force (ft-lb) = 0.000000377 kilowatt-hour (kWh)

20. 1 joule (J) = 0.00094845 British thermal unit (Btu)

21. 1 light year = 9,460,550,000,000 kilometers (km)

Expressing Scientific Notation as Decimal (Standard Form) Numbers

To express a number given in scientific notation as a decimal number, shift the decimal point in the reverse direction and attach required zeros. Move the decimal point according to the exponent of 10. With positive exponents, the decimal point is moved to the right; with negative, it is moved to the left.

EXAMPLES •

Express the following values in decimal form.

1. $4.3 \times 10^3 = 4\ 300.$ *Ans*

Shift right 3 places. Attach required zeros.

2. $8.907 \times 10^5 = 8\ 90{,}700.$ *Ans*

Shift right 5 places. Attach required zeros.

3. $-3.8 \times 10^{-4} = -0.000\ 3\ 8.$ *Ans*

Shift left 4 places. Attach required zeros

EXERCISE 6–12B

Write the following scientific notation numbers in decimal (standard) form.

1. 3×10^3	**5.** 5.093×10^{-5}	**9.** 4.0052×10^6	**13.** 7.771×10^{-8}
2. 8.5×10^5	**6.** -9.667×10^4	**10.** 4.0052×10^{-6}	**14.** -1.019×10^6
3. -4.73×10^{-2}	**7.** -2.008×10^{-7}	**11.** 4.983×10^5	**15.** 6.107×10^7
4. 2.028×10^7	**8.** 7.106×10^9	**12.** 8.818×10^{-4}	**16.** -3.202×10^{-9}

The following are science and technology unit conversions. Rewrite each in decimal (standard) form.

17. 1 inch (in) = 2.54×10^8 Angstroms (Å)

18. 1 degree per minute = 4.629629×10^{-5} revolutions per second (r/s)

19. 1 ampere-hour (Ah) = 3.6×10^3 coulombs (C)

20. 1 atmosphere (atm) = 1.03323×10^7 grams per square meter (g/m²)

21. 1 foot-pound-force per hour (ft-lb/hr) = 5.050×10^{-7} horsepower (hp)

Multiplication and Division Using Scientific Notation

Scientific notation is used primarily for multiplication and division operations. The procedures presented in this unit for the algebraic operations of multiplication and division are applied to operations involving scientific notation.

EXAMPLES •

Compute the following expressions.

1. $(2.8 \times 10^3) \times (3.5 \times 10^5)$

   a. Multiply the decimals:

      $2.8 \times 3.5 = 9.8$

    b. The product of the 10s equals 10 raised to a power, which is the sum of the exponents:

$$10^3 \times 10^5 = 10^{3+5} = 10^8$$

    c. Combine both parts (9.8 and 10^8) as a product:

$$(2.8 \times 10^3) \times (3.5 \times 10^5) = 9.8 \times 10^8 \; Ans$$

2. $340,000 \times 7,040,000$

Rewrite the numbers in scientific notation and solve: $340,000 \times 7,040,000 = (3.4 \times 10^5) \times (7.04 \times 10^6) = 23.936 \times 10^{11}$. Notice that the decimal part is greater than 10. Rewrite the decimal part and solve: $23.936 = 2.3936 \times 10^1$.

$$(2.3936 \times 10^1) \times 10^{11} = 2.3936 \times 10^{12} \; Ans$$

3. $-840,000 \div 0.0006$

$$
\begin{aligned}
-840,000 \div 0.0006 &= (-8.4 \times 10^5) \div (6 \times 10^{-4}) \\
&= (-8.4 \div 6) \times (10^5 \div 10^{-4}) \\
&= -1.4 \times 10^{5-(-4)} \\
&= -1.4 \times 10^9 \; Ans
\end{aligned}
$$

4. $\dfrac{(3.4 \times 10^{-8}) \times (7.9 \times 10^5)}{(-2 \times 10^6)}$

$$3.4 \times 7.9 \div -2 = -13.43$$

$$10^{-8} \times 10^5 \div 10^6 = 10^{-8+5-6} = 10^{-9}$$

$$-13.43 \times 10^{-9} = (-1.343 \times 10^1) \times 10^{-9}$$

$$(-1.343 \times 10^1) \times 10^{-9} = -1.343 \times 10^{1-9} = -1.343 \times 10^{-8} \; Ans$$

Calculator Applications

With 10-digit calculators, the number shown in the calculator display is limited to 10 digits. Calculations with answers that are greater than 9,999,999,999 or less than 0.000000001 are automatically expressed in scientific notation.

EXAMPLES •————————————————————————————

1. 80000000 ⊠ 400000 ⊟ 3.2 ^{13} (Answer displayed as 3.2 ^{13})

There is some variation among calculators as to how the answer is displayed. Many calculators display the answer as shown, with a space between the 3.2 and the 13 and the 13 smaller in size than the 3.2.

The display shows the number (mantissa) and the exponent of 10; it does *not* show the 10. The displayed answer of 3.2 ^{13} does *not* mean that 3.2 is raised to the thirteenth power. The display 3.2 ^{13} means 3.2×10^{13}; $80,000,000 \times 400,000 = 3.2 \times 10^{13}$.

Some calculators, such as graphing calculators, display the answer with a small capital E between the mantissa and the exponent. For the problem $80,000,000 \times 400,000$, the answer is displayed as 3.2 E 13.

2. .0000007 ⊠ .000002 ⊟ 1.4 $^{-12}$ (Answer displayed as 1.4 $^{-12}$)

The display 1.4 $^{-12}$ means 1.4×10^{-12}; $0.0000007 \times 0.000002 = 1.4 \times 10^{-12} \; Ans$

Numbers in scientific notation can be directly entered in a calculator. For calculations whose answer does *not exceed* the number of digits in the calculator display, the answer is displayed in standard decimal form.

The answer is displayed in decimal (standard) form with certain calculators with the exponent entry key, $\boxed{\text{EE}}$, or exponent key, $\boxed{\text{EXP}}$.

EXAMPLE •——

Solve. $(3.86 \times 10^3) \times (4.53 \times 10^4)$

  3.86 $\boxed{\text{EE}}$ 3 $\boxed{\times}$ 4.53 $\boxed{\text{EE}}$ 4 $\boxed{=}$ 174858000 *Ans*

  or 3.86 $\boxed{\text{EXP}}$ 3 $\boxed{\times}$ 4.53 $\boxed{\text{EXP}}$ 4 $\boxed{\text{EXE}}$ or $\boxed{=}$ 174858000 *Ans*

  The answer is displayed in standard form.

——•

For calculations with answers that *exceed* the number of digits in the calculator display, the answer is displayed in scientific notation. Both calculators with the $\boxed{\text{EE}}$ key or $\boxed{\text{EXP}}$ key display the answer in scientific notation.

EXAMPLE •——

Solve. $\dfrac{(-1.96 \times 10^7) \times (2.73 \times 10^5)}{8.09 \times 10^{-4}}$

1. Using the $\boxed{\text{EE}}$ key:

  1.96 $\boxed{+/-}$ $\boxed{\text{EE}}$ 7 $\boxed{\times}$ 2.73 $\boxed{\text{EE}}$ 5 $\boxed{\div}$ 8.09 $\boxed{\text{EE}}$ 4 $\boxed{+/-}$ $\boxed{=}$ $-6.614091471\ ^{15}$,
  $-6.614091471 \times 10^{15}$ *Ans*

  or $\boxed{(-)}$ 1.96 $\boxed{\text{EE}}$ 7 $\boxed{\times}$ 2.73 $\boxed{\text{EE}}$ 5 $\boxed{\div}$ 8.09 $\boxed{\text{EE}}$ $\boxed{(-)}$ 4 $\boxed{\text{ENTER}}$ $-6.614091471\text{E}15$,
  $-6.614091471 \times 10^{15}$ *Ans*

2. Using the $\boxed{\text{EXP}}$ key:

  1.96 $\boxed{+/-}$ $\boxed{\text{EXP}}$ 7 $\boxed{\times}$ 2.73 $\boxed{\text{EXP}}$ 5 $\boxed{\div}$ 8.09 $\boxed{\text{EXP}}$ 4 $\boxed{+/-}$ $\boxed{=}$ $-6.614091471\ ^{15}$,
  $-6.614091471 \times 10^{15}$ *Ans*

  or $\boxed{(-)}$ or $\boxed{-}$ 1.96 $\boxed{\text{EXP}}$ 7 $\boxed{\times}$ 2.73 $\boxed{\text{EXP}}$ 5 $\boxed{\div}$ 8.09 $\boxed{\text{EXP}}$ $\boxed{(-)}$ or $\boxed{-}$ 4 $\boxed{\text{EXE}}$
  $-6614091471\ ^{15}$,
  $-6.614091471 \times 10^{15}$ *Ans*

——•

EXERCISE 6–12C

In problems 1 through 14, the numbers are in scientific notation. Solve and leave answers in scientific notation. Round the answers (mantissas) to 2 decimal places

1. $(2.50 \times 10^3) \times (5.10 \times 10^5)$

2. $(4.08 \times 10^{-4}) \times (6.10 \times 10^{-5})$

3. $(-7.60 \times 10^4) \times (1.90 \times 10^5)$

4. $(2.43 \times 10^{-6}) \div (7.60 \times 10^3)$

5. $(8.51 \times 10^7) \div (6.30 \times 10^{-5})$

6. $(9.16 \times 10^5) \times (-5.36 \times 10^{-4})$

7. $(3.53 \times 10^{-4}) \times (6.46 \times 10^{-6})$

8. $(8.26 \times 10^{-6}) \times (4.35 \times 10^7)$

9. $\dfrac{(1.25 \times 10^4) \times (6.30 \times 10^5)}{(7.83 \times 10^3)}$

10. $\dfrac{(8.76 \times 10^{-5}) \times (1.05 \times 10^9)}{(6.37 \times 10^3)}$

11. $\dfrac{(5.50 \times 10^4) \times (-6.00 \times 10^6)}{(6.92 \times 10^{-3})}$

12. $\dfrac{(8.46 \times 10^{-5})}{(3.90 \times 10^7) \times (6.77 \times 10^{-3})}$

13. $\dfrac{(2.73 \times 10^{-3}) \times (4.08 \times 10^6)}{(1.05 \times 10^4) \times (7.55 \times 10^{-6})}$

14. $\dfrac{(5.48 \times 10^{-3}) \times (9.72 \times 10^{-5})}{(-6.35 \times 10^6) \times (-3.03 \times 10^{-4})}$

In problems 15 through 28, the numbers are in decimal (standard) form. Rewrite the numbers in scientific notation, calculate and give answers in scientific notation. Round the answers (mantissas) to 2 decimal places.

15. $1510 \times 30,500$

16. 0.000300×0.00210

17. $-81,300 \times 902,000$

18. $82.10 \div 0.00000605$

19. $61,770 \times 53,100$

20. $0.0000821 \div -315$

21. $38,400 \times 851,000$

22. $0.0000430 \times 1,230,000$

23. $\dfrac{-0.00623 \times 742,000}{651,000}$

24. $\dfrac{70,800 \times 423,000}{0.0984}$

25. $\dfrac{0.0000276 \times 207,000,000}{0.00892}$

26. $\dfrac{-0.000829}{405,000 \times 0.00312}$

27. $\dfrac{0.00503 \times 0.000406}{0.00423 \times 577,000}$

28. $\dfrac{518,000 \times 0.00612}{37,400 \times 0.0000830}$

Solve the following science and technology problems.

29. The wave length of radio waves (ℓ) is calculated by the following formula:

$$\ell = \frac{V}{n}$$ where V = velocity of radio waves in km/s
   n = frequency in cycles/s

Calculate the wave length (ℓ) in km/cycle when $V = 3.0 \times 10^5$ km/s and $n = 1.02 \times 10^6$ cycles/s. Give the answer in decimal (standard) form and round to 2 significant digits.

30. When a uranium (U^{235}) nucleus splits, energy is released. The change in energy is given by the following statement:

Change in energy (in ergs) = change in mass (in grams) $\times$ (velocity of light (in cm/s))^2. Determine the change in energy (ergs) when 352.500 grams of U^{235} are split and 352.155 grams remain. The velocity of light is 3.00×10^{10} cm/s. Give the answer in scientific notation. Round the mantissa to 3 significant digits.

31. The amount of expansion of metal when heated is computed as follows:

$$\text{Expansion} = \begin{pmatrix}\text{original}\\\text{length}\end{pmatrix} \times \begin{pmatrix}\text{linear expansion per unit of}\\\text{length per degree Fahrenheit}\end{pmatrix} \times \begin{pmatrix}\text{temperature}\\\text{change}\end{pmatrix}$$

Calculate the amount of expansion for the metals shown in Figure 6–8. Give the answers in decimal (standard) form to 3 significant digits.

	Metal	Original Length of Metal	Linear Expansion per Unit of Length per Degree Fahrenheit	Original Temperature	Temperature to Which Heated
a.	Aluminum	6.7520 in	1.244×10^{-5}	68.0°F	225.0°F
b.	Copper	35.7520 ft	9.000×10^{-6}	35.0°F	97.0°F
c.	Carbon Steel	3.0950 in	6.330×10^{-6}	84.0°F	743.0°F

Figure 6–8

32. The bending of light when it passes from one substance to another is called refraction.

$$\text{Index of refraction} = \frac{\text{Velocity of light in air}}{\text{Velocity of light in substance}}$$

The velocity of light in air is 1.86×10^5 miles per second (mi/sec). The velocity of light though glass is 1.23×10^5 miles per second (mi/sec). Determine the index of refraction of glass. Give answer in decimal (standard) form to 3 significant digits.

33. A formula in electricity for determining power when current and resistance are known is:

$$P = I^2 \times R \qquad \text{where } P = \text{power in watts (W)}$$
$$I = \text{current in amperes (A)}$$
$$R = \text{resistance in ohms } (\Omega)$$

Determine the power in watts (W) when $I = 3.80 \times 10^{-5}$ A and $R = 2.90 \times 10^5\,\Omega$. Give the answer in scientific notation and round the mantissa to 2 significant digits.

6–13   Engineering Notation

Engineering notation is very similar to scientific notation. In engineering notation, the exponents of the 10 are always multiples of three. The main advantage of engineering notation is when SI (metric) units are used. In the metric system, the most widely used prefixes are for every third power of 10.

Expressing Decimal (Standard Form) Numbers in Engineering Notation

A positive or negative number with an absolute value of 1000 or greater has a positive exponent when expressed in engineering notation.

EXAMPLES ●

1. 25,700

    a. Write the number as a value between 1 and 1000: 25.7

    b. Count the number of places the decimal point is shifted to determine the exponent of 10: 25 700. The decimal point is shifted 3 places. The exponent of 10 is 3: 10^3.

    c. Multiply. 25.7×10^3

        $25,700 = 25.7 \times 10^3$ *Ans*

2. 92 500,000,000. $= 92.5 \times 10^9$ *Ans*

        └─ Shift 9 places

3. $-152\,000,000. = -152 \times 10^6$ *Ans*

        └─ Shift 6 places

A positive or negative number with an absolute value less than 1 has a negative exponent when expressed in engineering notation.

EXAMPLES ●

1. $-0.35 = 0.350 = 350 \times 10^{-3}$ *Ans*

        └─ Shift 3 places. Notice that the decimal point is shifted to the right, and so the exponent is negative. Notice that a 0 had to be added at the right.

2. $-0.0000024 = 0.000002\,4 = 2.4 \times 10^{-6}$ *Ans*

        └─ Shift 6 places.

3. $-0.000000073 = -0.000\,000\,073 = -73 \times 10^{-9}$ *Ans*

Shift 9 places.

EXERCISE 6–13A

Rewrite the following standard form numbers in engineering notation.

1. 625   **5.** 959,100   **9.** 0.000000058   **13.** 930,000.5

2. 67,500   **6.** 3278.94   **10.** −4,710,000,000   **14.** 35,700,000.5

3. −3789   **7.** −0.00063   **11.** 723,000,000,000,000   **15.** 0.00000375

4. 0.037   **8.** 59,3$\overline{0}$0   **12.** 13.569   **16.** −0.00000092

The following are science and technology unit conversions. Rewrite each in engineering notation.

17. 1 horsepower-hour (hp-h) $= 2\,647\,768$ joules (J)

18. 1 square mile $(\text{mi}^2) = 4\,014\,490$ square inches (in.^2)

19. 1 erg $= 0.00000007375616$ foot-pound-force

20. 1 second $= 0.0002777778$ hour

21. 1 parsec $= 30\,837\,400\,000\,000$ kilometer

Expressing Engineering Notation as Decimal (Standard Form) Numbers

To express a number given in engineering notation as a decimal number, shift the decimal point in the reverse direction and attach any required zeros. Move the decimal place according to the exponent of 10. With positive exponents, the decimal point is moved to the right; with negative exponents, it is moved to the left.

EXAMPLES •

Express the following values in decimal form.

1. $5.7 \times 10^3 = 5\,700. = 5,700$ *Ans*

Shift 3 places to the right.

2. $-35.2 \times 10^{-9} = -0.000\,000\,035\,2 = -0.0000000352$ *Ans*

Shift 9 places to the left. Attach required zeros.

EXERCISE 6–13B

Write the following engineering notation numbers in decimal (standard) form.

1. 5×10^3 **3.** -31.24×10^{-3} **5.** 135.07×10^{-6} **7.** -7.85×10^6

2. 1.75×10^6 **4.** 5.026×10^{-6} **6.** -531.2×10^9 **8.** -1.24×10^{-9}

The following are science and technology unit conversions. Rewrite each in decimal (standard) form.

9. 1 astronomical unit is 149.598×10^9 meters

10. 1 atmosphere is 98.06×10^3 newtons per square meter (N/m^2)

11. 1 electron volt (eV) is 160.21×10^{-21} joules

12. 1 cubic inch (in^3) is 16.38706×10^{-6} cubic meter (m^3)

Multiplication and Division Using Engineering Notation

Like scientific notation, engineering notation is useful to give short expressions for very large and very small numbers. Computation with engineering notation is used primarily for multiplication and division operations. The procedures are exactly the same as the ones for scientific notation.

EXAMPLES •————————————————————————————————

Compute the following expressions.

1. $(5.7 \times 10^3) \times (3.2 \times 10^9)$

   a. Multiply the decimals:

      $5.7 \times 3.2 = 18.24$

   b. The product of the 10s equals 10 raised to a power, which is the sum of the exponents:

      $10^3 \times 10^9 = 10^{3+9} = 10^{12}$

   c. Combine both parts (18.24 and 10^{12}) as a product:

      $(5.7 \times 10^3) \times (3.2 \times 10^9) = 18.24 \times 10^{12}$ *Ans*

2. $25{,}000{,}000 \times 0.0000000135$

   Rewrite the numbers in engineering notation and solve:

   $25{,}000{,}000 \times 0.0000000135 = (25 \times 10^6) \times (13.5 \times 10^{-9}) = 337.5 \times 10^{-3}.$

3. $-0.0000000135 \div 0.0000015$

   $$-0.000000013\,5 \div 0.0000015 = (-13.5 \times 10^{-9}) \div (1.5 \times 10^{-6})$$
   $$= (-13.5 \div 1.5) \times (10^{-9} \div 10^{-6})$$
   $$= -9 \times (10^{-9-(-6)})$$
   $$= -9 \times (10^{-3})\ Ans$$

EXERCISE 6–13C

In problems 1–12, the numbers are in engineering notation. Solve and leave answers in engineering notation. Round the answers (mantissas) to 2 decimal places.

1. $(7.5 \times 10^3) \times (6.2 \times 10^{12})$

2. $(3.08 \times 10^6) \times (8.2 \times 10^9)$

3. $(-5.3 \times 10^9) \times (20.4 \times 10^{12})$

4. $(-59.75 \times 10^6) \times (32.5 \times 10^{-6})$

5. $(350 \times 10^9) \div (7 \times 10^{-6})$

6. $(2.75 \times 10^{-9}) \div (110 \times 10^{-3})$

7. $(163.2 \times 10^{-12}) \div (3.84 \times 10^6)$

8. $(9.59 \times 10^6) \div (255 \times 10^{12})$

9. $\dfrac{(83.2 \times 10^9) \times (643.5 \times 10^{12})}{1.74 \times 10^{12}}$

10. $\dfrac{(472.5 \times 10^{-6}) \times (72.37 \times 10^9)}{352 \times 10^6}$

11. $\dfrac{(34.2 \times 10^{-12}) \times (543.6 \times 10^6)}{(26.25 \times 10^{-9}) \times (15.2 \times 10^9)}$

12. $\dfrac{(678 \times 10^{12}) \times (23.75 \times 10^6)}{(21.3 \times 10^{-6}) \times (42.3 \times 10^{-9})}$

In problems 13–20, the numbers are in decimal (standard) form. Rewrite the numbers in engineering notation, calculate, and give answers in engineering notation. Round the answers (mantissas) to 2 decimal places.

13. $43{,}500 \times 27{,}250$

14. $0.000050 \times 0.0000035$

15. $0.00000043 \times -12{,}300{,}000$

16. $-15{,}200{,}000 \times 275{,}000$

17. $47,200,000 \div 0.00000000589$

18. $-0.000000000058 \div 12,300,000$

19. $\dfrac{73,200,000 \times -0.000000923}{0.0000087 \times 963,000,000,000,000}$

20. $\dfrac{-0.0000000063 \times 0.000000785}{23,000 \times -158,000,000}$

Solve the following science and technology problems.

21. The rest energy, E, of an electron with rest mass, m, is given by Einstein's equation $E = mc^2$ where c is the speed of light. Find E if $m \approx 911 \times 10^{-33}$ kg and $c \approx 299.8 \times 10^6$ m/s.

22. The mass of the Earth is about 5.975×10^{24} kg and its volume is about 1.083×10^{21} m³. Density is defined as mass divided by volume. What is the density of the Earth?

UNIT EXERCISE AND PROBLEM REVIEW

WORD EXPRESSIONS AS SIGNED NUMBERS

Express the answer to each of these problems as a signed number.

1. A business profit of $15,000 is expressed as +$15,000. Express a business loss of $20,000.

2. A force of 600 pounds that pulls an object to the left is expressed as −600 pounds. Express a force of 780 pounds that pulls the object to the right.

3. A certain decrease of the cost of wholesale merchandise to a retailer is expressed as −15%. Express a wholesale cost increase of 18%.

4. A circuit current increase of 8.6 amperes is expressed as +8.6 amperes. Express a current decrease of 7.3 amperes.

THE NUMBER LINE

5. Refer to the number scale shown in Figure 6–9. Give the direction (+ or −) and the number of units counted going from the first to the second number.

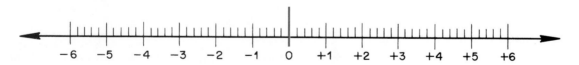

Figure 6–9

    a. −3 to +6   **c.** +4.8 to −4.4   **e.** 0.8 to −3.2   **g.** −2.7 to +2.7

    b. −5.4 to 0   **d.** −0.6 to +0.2   **f.** +3.4 to +0.6   **h.** +2.6 to −3.2

6. List each set of signed numbers in order of increasing value starting with the smallest value.

    a. −4.6, −20.9, +6.3, +1.7, −16.4, +18.3

    b. −7.5, 0, 7.5, −2.3, +0.5, −0.3

    c. 21.3, 0, 20.6, −4.6, −7, −23.4

    d. $+3\frac{1}{2}, -3, +6\frac{3}{4}, -6\frac{7}{8}, -6\frac{13}{16}$

ADDITION AND SUBTRACTION OF SIGNED NUMBERS

Add or subtract the signed numbers as indicated.

7. $-6 + (-13)$

8. $14 + (-6)$

9. $-25 + 18$

10. $43 + (-29)$

11. $21 + (-21)$

12. $-14.7 + (-3.4)$

13. $-7.2 + 2.5$

14. $-10\frac{3}{8} + \left(-2\frac{3}{4}\right)$

15. $+18 - (+7)$

16. $5 - (-22)$

17. $-37 - (-31)$

18. $-23 - 0$

19. $0 - (-23)$

20. $-30.7 - 5.5$

21. $0.923 - (-10.631)$

22. $-12\frac{1}{2} - \left(-4\frac{1}{8}\right)$

23. $15 + (-8) + (-15) + (-10)$

24. $-20.73 + 12.87 + (-3.08) + 36.77$

25. $-3.91 + 1.87 + 3.22 + 7.50$

26. $2\frac{1}{2} + \left(-3\frac{3}{4}\right) + \left(-\frac{1}{8}\right) + \frac{3}{8}$

27. $(3 + 18) - (-8 + 5)$

28. $(-13.72 - 6.06) - (4.54 + 7.82)$

29. $(6.48 - 5.32) - (4 - 8.31)$

30. $\left(8 - 10\frac{1}{2}\right) - \left(9\frac{5}{8} + 2\frac{1}{4}\right)$

MULTIPLICATION AND DIVISION OF SIGNED NUMBERS

Multiply or divide the signed numbers as indicated.

31. $(-5)(3)$

32. $(10)(-7)$

33. $(-30)(-15)$

34. $(-16)(0)$

35. $(5.6)(-3)$

36. $(-1.2)(-2.1)$

37. $(-0.5)(+0.3)$

38. $(-3)(-3)(-3)$

39. $(-3)(-3)(-3)(-3)$

40. $\left(1\frac{1}{4}\right)(-2)$

41. $\left(-3\frac{1}{2}\right)\left(-2\frac{1}{4}\right)$

42. $\left(-\frac{1}{4}\right)\left(-\frac{1}{4}\right)\left(\frac{1}{4}\right)$

43. $(-5.36)(0.28)(3)(-1.87)$

44. $(-0.01)(50.62)(-2)(0.32)$

45. $-8 \div (-2)$

46. $12 \div (-3)$

47. $\dfrac{-24}{-3}$

48. $\dfrac{-21}{7}$

49. $\dfrac{-4.8}{0.8}$

50. $\dfrac{0.86}{-0.19}$

51. $0 \div 12\frac{1}{2}$

52. $-2\frac{1}{2} \div \left(-1\frac{1}{4}\right)$

53. $\dfrac{-16.86}{4.17}$

54. $-3.03 \div (-6.86)$

POWERS AND ROOTS OF SIGNED NUMBERS

Raise to a power or determine a root as indicated. Round the answers to 2 decimal places where necessary.

55. $(-4)^2$

56. $(-4)^3$

57. $(+4)^3$

58. $(-2)^4$

59. $(-2)^5$

60. $(-6.7)^2$

61. $(-0.2)^3$

62. $\left(-\frac{1}{4}\right)^2$

63. $\left(\frac{1}{2}\right)^3$

64. $\left(-\frac{1}{2}\right)^3$

65. 10^{-2}

66. $(-2)^{-3}$

67. $(-57.93)^3$

68. $(15.05)^{-2}$

69. $\sqrt[3]{-27}$

70. $\sqrt[5]{-32}$

71. $\sqrt[4]{85.62}$

72. $\sqrt[5]{-387.63}$

73. $\sqrt[3]{-83.732}$

74. $\sqrt[3]{\dfrac{+27}{-64}}$

75. $\sqrt[5]{\dfrac{1}{+32}}$

76. $\dfrac{\sqrt[3]{-84.27}}{5.19}$

COMBINED OPERATIONS OF SIGNED NUMBERS

Solve each combined operation signed number exercise. For exercises 83–88, substitute given numbers for letters, then solve. Use the proper order of operations. Round the answers to 2 decimal places where necessary.

77. $8(-4) + (-1)(6) - (-2)(3)$

78. $\dfrac{6(9 + 3)}{7(11 - 6) - 5}$

79. $[5^2 - (3)(-1)(-4)]^2 + 3(-2)^3$

80. $(-3)^2 + \sqrt[3]{-27} - (3)(-4)(2)$

81. $\sqrt{(-5) - (-6)(+5)} + (4)^2 - (-4)^3$

82. $\sqrt{67.2 - (-8)(-6.32)} - (2.86)^{-2}$

83. $abc(3a - 4bc); a = -2, b = 4, c = 6$

84. $(m - p)(4m - 3p); m = -5, p = -8$

85. $\dfrac{x^3 + 2y - ys}{s^2 - (x + 2)}; x = -2, y = -3, s = 4$

86. $b^3 + \sqrt{8 + (ab - 4)}; a = -8.73, b = -4.08$

87. $(-3)\sqrt[3]{d} + (fh - d) - (h)^3; d = 8.65, f = -10.94, h = -3.51$

88. $\dfrac{\sqrt[3]{(17 + 2xyt - 2t)}}{2y + t^3 - 2}; x = -3.44, y = 5.87, t = -1.66$

SIGNED NUMBER PROBLEMS

Express the answers as signed numbers for each problem.

89. The daily closing price net changes of a certain stock for one week are shown in Figure 6–10. What is the average daily net change in price?

DAY	MON.	TUES.	WED.	THUR.	FRI.
Net Change	$-1\frac{1}{4}$	$+\frac{1}{4}$	$+\frac{7}{8}$	$-\frac{1}{4}$	$+1\frac{5}{8}$

Figure 6–10

90. An hourly temperature report in degrees Celsius is given in Figure 6–11.

TIME	TEMPERATURE	TIME	TEMPERATURE
12 Noon	−1.0°C	6 P.M.	0.4°C
1 P.M.	1.2°C	7 P.M.	−1.0°C
2 P.M.	3.6°C	8 P.M.	−3.8°C
3 P.M.	5.8°C	9 P.M.	−4.4°C
4 P.M.	3.2°C	10 P.M.	−6.6°C
5 P.M.	0.0°C	11 P.M.	−8.0°C

Figure 6–11

a. What is the temperature change for each time period listed?

(1) Noon to 3 P.M.   (3) 4 P.M. to 9 P.M.

(2) 2 P.M. to 6 P.M.   (4) 3 P.M. to 11 P.M.

    b. What is the average temperature to 1 decimal place during each time period listed?

       (1) Noon to 5 P.M.   **(3)** 6 P.M. to 11 P.M.

       (2) 4 P.M. to 9 P.M.   **(4)** Noon to 11 P.M.

91. During a 7-year period a company experiences profits some years and losses other years. Company annual profits ($+$) and losses ($-$) are shown in Figure 6–12.

YEAR	1994	1995	1996	1997	1998	1999	2000
Profit (+) or loss (–)	+$580,000	+$493,000	–$103,000	–$267,000	–$179,000	–$86,000	+$319,000

Figure 6–12

    a. What is the total dollar change for the years listed?

       (1) 1994 to 1995   **(3)** 1996 to 1997   **(5)** 1998 to 1999

       (2) 1995 to 1996   **(4)** 1997 to 1998   **(6)** 1999 to 2000

    b. What is the average annual profit or loss for the years listed? Round the answers to the nearest thousand dollars.

       (1) 1994 to 1998   **(2)** 1997 to 1999   **(3)** 1994 to 2000

92. Holes are drilled in a plate as shown in Figure 6–13. The holes are drilled in the sequence shown; that is, hole 1 is drilled first, hole 2 is drilled second, and so on. Movement to the left from one hole to the next is expressed as the negative ($-$) direction. Movement to the right from one hole to the next is expressed as the positive ($+$) direction. Express the distance and direction ($+$ or $-$) when moving from the holes listed.

    a. Hole 1 to hole 2

    b. Hole 2 to hole 3

    c. Hole 3 to hole 4

    d. Hole 4 to hole 5

    e. Hole 5 to hole 6

    f. Hole 3 to hole 6

    g. Hole 6 to hole 3

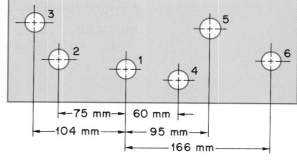

Figure 6–13

SCIENTIFIC NOTATION

Rewrite the following numbers in scientific notation.

93. 976,000   **94.** 0.015   **95.** 0.00039

Rewrite the following scientific notation values in decimal form.

96. 1.6×10^5 **97.** 5.09×10^6 **98.** 4.03×10^{-4}

Solve the following expressions given in scientific notation. Give the answers in scientific notation. Round the answers (mantissas) to 2 decimal places.

99. $(3.54 \times 10^{-7}) \times (6.03 \times 10^{-4})$

100. $(6.19 \times 10^6) \div (9.42 \times 10^{-5})$

101. $\dfrac{(7.30 \times 10^{-5}) \times (6.18 \times 10^4)}{(-3.77 \times 10^6)}$

102. $\dfrac{(-1.67 \times 10^6) \times (9.18 \times 10^{-3})}{(2.07 \times 10^4) \times (-5.55 \times 10^{-7})}$

The following expressions are given in decimal (standard) form. Rewrite the numbers in scientific notation, calculate, and give the answers in scientific notation. Round the answers (mantissas) to 2 decimal places.

103. $43,600 \times 753,000$

104. $0.000421 \div (-1640)$

105. $\dfrac{0.00712 \times 471,000}{507,000}$

106. $\dfrac{429,000 \times 0.0916}{0.00000194 \times 40,500}$

ENGINEERING NOTATION

Rewrite the following numbers in engineering notation.

107. $1,850,000$ **108.** 0.0000357 **109.** -0.000000618

Rewrite the following engineering notation values in decimal form.

110. 43.2×10^3 **111.** -571×10^{-9} **112.** 12.3×10^{-6}

Solve the following expression given in engineering notation. Give the answers in engineering notation. Round the answers (mantissas) to 2 decimal places.

113. $(43.2 \times 10^3) \times (-571 \times 10^{-9})$

114. $(3.72 \times 10^9) \div (971 \times 10^6)$

115. $\dfrac{(-12.6 \times 10^{-9}) \times (157 \times 10^3)}{-5.22 \times 10^9}$

116. $\dfrac{(5.31 \times 10^9) \times (94.8 \times 10^6)}{(-2.55 \times 10^{-9}) \times (-94.2 \times 10^{-6})}$

The following expressions are given in decimal (standard) form. Rewrite the numbers in engineering notation, calculate, and give the answers in engineering notation. Round the answers (mantissas) to 2 decimal places.

117. $5,700,000 \times 89,300$

118. $0.0000258 \div (-8,390)$

119. $\dfrac{0.00527 \times 359,300}{0.000209}$

120. $\dfrac{0.000739 \times 37,000}{0.000000048 \times 23,400}$

SECTION 11 ⠿

Measurement

UNIT 7 ⣿ Precision, Accuracy, and Tolerance

OBJECTIVES

After studying this unit you should be able to

• determine the degree of precision of any measurement number.

• round sums and differences of measurement numbers to proper degrees of precision.

• determine the number of significant digits of measurement numbers.

• round products and quotients of measurement numbers to proper degrees of accuracy.

• compute absolute and relative error between true and measured values.

• compute maximum and minimum clearances and interferences of bilateral and unilateral tolerance-dimensioned parts (customary and metric).

• compute total tolerance, maximum limits, and minimum limits of customary and metric unit lengths.

• solve practical applied problems involving tolerances and limits (customary and metric).

Measurement is used to communicate size, quantity, position, and time. Without measurement, a building could not be built nor a product manufactured.

The ability to measure with tools and instruments and to compute with measurements is required in almost all occupations. In the construction field, measurements are calculated and measurements are made with tape measures, carpenters squares, and transits. The manufacturing industry uses a great variety of measuring instruments, such as micrometers, calipers, and gauge blocks. Measurement calculations from engineering drawings are requirements. Electricians and electronics technicians compute circuit measurements and read measurements on electrical meters. Environmental systems occupations require heat load and pressure calculations and make measurements with instruments such as manometers and pressure gauges.

7–1 Exact and Approximate (Measurement) Numbers

If a board is cut into 6 pieces, 6 is an exact number; exactly 6 pieces are cut. If 150 bolts are counted, 150 bolts is an exact number; exactly 150 bolts are counted. These are examples of counting numbers and are exact.

However, if the length of a board is measured as $7\frac{3}{8}$ inches, $7\frac{3}{8}$ inches is not exact. If the diameter of a bolt is measured as 12.5 millimeters, 12.5 millimeters is not exact. The $7\frac{3}{8}$ inches and 12.5 millimeters are approximate values. Measured values are always approximate.

Measurement is the comparison of a quantity with a standard unit. For example, a linear measurement is a means of expressing the distance between two points; it is the measurement of lengths. A linear measurement has two parts: a unit of length and a multiplier.

$$\text{multiplier} \underset{\llcorner}{\overset{2.5 \text{ inches}}{\overset{\urcorner}{}}} \text{unit of length} \qquad \text{multiplier} \underset{\llcorner}{\overset{15\frac{1}{4} \text{ miles}}{\overset{\urcorner}{}}} \text{unit of length}$$

The measurements 2.5 inches and $15\frac{1}{4}$ miles are examples of denominate numbers. A *denominate number* is a number that refers to a special unit of measure. A *compound denominate number* consists of more than one unit of measure, such as 7 feet 2 inches.

7–2   Degree of Precision of Measuring Instruments

The exact length of an object cannot be measured. All measurements are approximations. By increasing the number of graduations on a measuring instrument, the degree of precision is increased. Increasing the number of graduations enables the user to get closer to the exact length. The precision of a measurement depends on the measuring instrument used. The *degree of precision* of a measuring instrument depends on the smallest graduated unit of the instrument.

The degree of precision necessary in different trades varies. In building construction, generally $\frac{1}{16}$-inch or 2-millimeter precision is adequate. Sheet metal technicians often work to $\frac{1}{32}$-inch or 1-millimeter precision. Machinists and automobile mechanics usually work to 0.001-inch or 0.02-millimeter precision. In the manufacture of some products, very precise measurements to 0.00001 inch or 0.0003 millimeter and 0.000001 inch or 0.00003 millimeter are sometimes required. For example, the dial indicator in Figure 7–1 can be used to measure to the nearest 0.001″ while the one in Figure 7–2 measures to the nearest 0.0005″.

Figure 7–1   (Courtesy of S-T Industries)   **Figure 7–2**   (Courtesy of Chicago Dial Indicator Co.)

Various measuring instruments have different limitations on the degree of precision possible. The accuracy achieved in measurement does not depend only on the limitations of the measuring instrument. Accuracy can also be affected by errors of measurement. Errors can be caused by defects in the measuring instruments and by environmental changes such as differences in temperature. Perhaps the greatest cause of error is the inaccuracy of the person using the measuring instrument.

7–3   Common Linear Measuring Instruments

Tape Measure. Tape measures are commonly used by garment makers and tailors. Customary tape measures are generally 5 feet long with $\frac{1}{8}$ inch the smallest graduation. Therefore, the degree of precision is $\frac{1}{8}$ inch. Metric tape measures are generally 2 meters long with 1 millimeter the smallest graduation. The degree of precision for a metric tape measure is 1 millimeter.

Folding Rule. Folding rules are used by construction workers such as carpenters, cabinet-makers, electricians, and masons. Customary rules are generally 6 feet long and fold to 6 inches. The smallest graduation is usually $\frac{1}{16}$ inch. The smallest graduation on metric rules is generally 1 millimeter. Customary units and metric units are available on the same rule. The customary units are on one side of the rule, and the metric units are on the opposite side.

Steel Tape. Steel tapes are used by contractors, construction workers, and surveyors. Customary steel tapes are available in 25-foot, 50-foot, and 100-foot lengths. Generally, the smallest graduation is $\frac{1}{8}$ inch. Metric tapes are available in 10-meter, 15-meter, 20-meter, and 30-meter lengths. Generally, the smallest graduation is 1 millimeter. Customary units and metric units are also available on the same tape. Customary units are on one side of the tape, and metric units are on the opposite side.

Steel Rules. Steel rules are widely used in manufacturing industries by machine operators, machinists, and sheet metal technicians. Customary steel rules are available in sizes from 1 inch to 144 inches; 6 inches is the most common length. Customary rules are available in both fractional and decimal-inch graduations. The smallest graduation on fractional rules is $\frac{1}{64}$ inch; the smallest graduation on a decimal-inch is 0.01 inch. Metric measure steel rules are available in a range from 150 millimeters to 1,000 millimeters (1 meter) in length. The smallest graduation is 0.5 millimeter.

Vernier and Dial Calipers. Vernier calipers and dial calipers are widely used in the metal trades. The most common customary unit lengths are 6 inches and 12 inches, although calipers are available in up to 72-inch lengths. The smallest unit that can be read is 0.001 inch. Metric measure rules are available in lengths from 150 millimeters to 600 millimeters. The smallest unit that can be read is 0.02 millimeter. Some vernier calipers are designed with both customary and metric unit scales on the same instrument.

Micrometers. Micrometers are used by tool and die makers, automobile mechanics, and inspectors. Micrometers are used when relatively high precision measurements are required. There are many different types and sizes of micrometers. Customary outside micrometers are available in sizes from 0.5 inch to 60 inches. The smallest graduation is 0.0001 inch with a vernier attachment. Metric outside micrometers are available in sizes up to 600 millimeters. The smallest graduation is 0.002 millimeter with a vernier attachment.

7–4   Degree of Precision of a Measurement Number

The degree of precision of a measurement number depends on the number of decimal places used. The number becomes more precise as the number of decimal places increases. For example, 4.923 inches is more precise than 4.92 inches. The range includes all of the values that are represented by the number.

EXAMPLES •————————————————————————————————

1. What is the degree of precision and the range for 2 inches?

    The degree of precision of 2 inches is to the nearest inch as shown in Figure 7–3. The range of values includes all numbers equal to or greater than 1.5 inches or less than 2.5 inches.

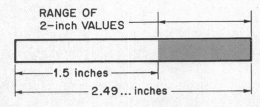

Figure 7–3

2. What is the degree of precision and the range for 2.0 inches?

The degree of precision of 2.0 inches is to the nearest 0.1 inch as shown in Figure 7–4. The range of values includes all numbers equal to or greater than 1.95 inches and less than 2.05 inches.

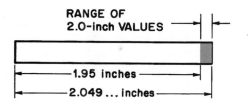

**RANGE OF
2.0-inch VALUES**

1.95 inches

2.049 ... inches

Figure 7–4

3. What is the degree of precision and the range for 2.00 inches?

The degree of precision of 2.00 inches is to the nearest 0.01 inch as shown in Figure 7–5. The range of values includes all numbers equal to or greater than 1.995 inches and less than 2.005 inches.

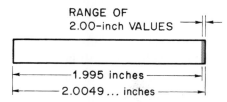

**RANGE OF
2.00-inch VALUES**

1.995 inches

2.0049 ... inches

Figure 7–5

4. What is the degree of precision and the range for 2.000 inches?

The degree of precision of 2.000 inches is to the nearest 0.001 inch. The range of values includes all numbers equal to or greater than 1.9995 inches and less than 2.0005 inches.

EXERCISE 7–4

For each measurement find:

a. the degree of precision   *the # of places behind decimal*
b. the range   *x by ½ then ⊕ and ⊖*

7-12

1. 3.6 in	**4.** 7.08 mm	**7.** 12.002 in	**10.** 23.00 in
2. 1.62 in	**5.** 15.885 in	**8.** 36.0 mm	**11.** 9.1 mm
3. 4.3 mm	**6.** 9.1837 in	**9.** 7.01 mm	**12.** 14.01070 in

7–5 Degrees of Precision in Adding and Substracting Measurement Numbers

When adding or subtracting measurements, all measurements must be expressed in the same kind of units. Often measurement numbers of different degrees of precision are added or subtracted. When adding and subtracting numbers, there is a tendency to make answers more precise than they are. An answer that has more decimal places than it should gives a false degree of precision. A sum or difference cannot be more precise than the least precise measurement number used in the computations. Round the answer to the least precise measurement number used in the computations.

EXAMPLES •

1. 15.63 in + 2.7 in + 0.348 in = 18.678 in, 18.7 in *Ans*

Since the least precise number is 2.7 in, round the answer to 1 decimal place.

2. 3.0928 cm − 0.532 cm = 2.5608 cm, 2.561 cm *Ans*

Since the least precise number is 0.532 cm, round the answer to 3 decimal places.

 3. 73 ft + 34.21 ft = 107.21 ft, 107 ft *Ans*

Since the least precise number is 73 ft, round the answer to the nearest whole number.

4. 73.0 ft + 34.21 ft = 107.21 ft, 107.2 ft *Ans*

Notice that this example is identical to Example 3, except the first measurement is 73.0 ft instead of 73 ft. Since the least precise measurement is 73.0 ft, round the answer to 1 decimal place.

EXERCISE 7–5

Your answer should be the lowest # behind the decimal.

#7-12

Add or subtract the following measurement numbers. Round answers to the degree of precision of the least precise number.

1. 2.69 in + 7.871 in

2. 14.863 mm − 5.0943 mm

3. 80.0 ft + 7.34 ft

4. 0.0009 in + 0.001 in

5. 2,256 mi − 783.7 mi

6. 31.708 cm² − 27.69 cm²

7. 18.005 in − 10.00362 in

8. 0.0187 cm³ + 0.70 cm³

9. 33.92 gal + 27 gal

10. 6.01 lb + 15.93 lb + 18.0 lb

11. 26.50 sq in − 26.49275 sq in

12. 84.987 mm + 39.01 mm − 77 mm

7–6   Significant Digits

It is important to understand what is meant by significant digits and to apply significant digits in measurement calculations. A measurement number has all of its digits significant if all digits, except the last, are exact and the last digit has an error of no more than half the unit of measurement in the last digit. For example, 6.28 inches has 3 significant digits when measured to the nearest hundredth of an inch.

The following rules are used for determining significant digits:

1. All nonzero digits are significant.

2. Zeros between nonzero digits are significant.

3. Final zeros in a decimal or mixed decimal are significant.

4. Zeros used as place holders are *not* significant unless they are identified as significant. Usually a zero is identified as significant by tagging it with a bar above it.

EXAMPLES

The next seven items are examples of significant digits. They represent measurement (approximate) numbers.

1. 812   3 significant digits, all nonzero digits are significant

2. 7.139   4 significant digits, all nonzero digits are significant

3. 14.3005   6 significant digits, zeros between nonzero digits are significant

4. 9.300   4 significant digits, final zeros of a decimal are significant

5. 0.008   1 significant digit, zeros used as place holders are *not* significant

6. 23,000   2 significant digits, zeros used as place holders are *not* significant

7. 23,0̄00   3 significant digits, a zero tagged is significant

In addition to the previous examples, study the following examples. The number of signifi-cant digits, shown in parentheses, is given for each number.

1. 3.905 (4)	**5.** 147.500 (6)	**9.** 1.00187 (6)	**13.** 8,600 (2)
2. 3.950 (4)	**6.** 7.004 (4)	**10.** 8.020 (4)	**14.** 0.01040 (4)
3. 83.693 (5)	**7.** 0.004 (1)	**11.** 0.020 (2)	**15.** 95,080.7 (6)
4. 147.005 (6)	**8.** 0.00187 (3)	**12.** 8,603.0 (5)	**16.** 90,00$\overline{0}$ (5)

EXERCISE 7–6

Write Rule also

 # 1-20

Determine the number of significant digits for the following measurement (approximate) numbers.

1. 2.0378	**6.** 9,700	**11.** 0.00095	**16.** 87,195
2. 0.0378	**7.** 12.090	**12.** 385.007	**17.** 66,08$\overline{0}$
3. 126.10	**8.** 137.000	**13.** 4,353.0	**18.** 87,200.00
4. 0.020	**9.** 137,0$\overline{0}$0	**14.** 1.040	**19.** 6.010
5. 9,709.3	**10.** 8.005	**15.** 0.0370	**20.** 4,000,100

7–7 Accuracy

The number of significant digits in a measurement number determines its accuracy. The greater the number of significant digits, the more accurate the number. For example, consider the measurements of 8 millimeters and 126 millimeters. Both measurements are equally precise; they are both measured to the nearest millimeter. The two measurements are not equally accurate. The greatest error in both measurements is 0.5 millimeter. However, the error in the 8 millimeter measurement is more serious than the error in the 126 millimeter measurement. The 126 millimeter measurement (3 significant digits) is more accurate than the 8 millimeter measurement (1 significant digit).

Examples of the Accuracy of Measurement Numbers.

1. 2.09 is accurate to 3 significant digits

2. 0.1250 is accurate to 4 significant digits

3. 0.0087 is accurate to 2 significant digits

4. 50,000 is accurate to 1 significant digit

5. 68.9520 is accurate to 6 significant digits

When measurement numbers have the same number of significant digits, the number that begins with the largest digit is the most accurate. For example, consider the measurement numbers 3,700; 4,100; and 2,900. Although all 3 numbers have 2 significant digits, 4,100 is the most accurate.

EXERCISE 7–7

Greatest number of S.D.

#9-16

For each of the following groups of measurement numbers, identify the number that is most accurate.

Ans Pg 810

1. 5.05; 4.9; 0.002	**9.** 70,108; 69.07; 8.09
2. 18.6; 1.860; 0.186	**10.** 0.930; 0.0086; 5.31
3. 1,000; 29; 173	**11.** 917; 43.08; 0.0936
4. 0.0009; 0.004; 0.44	**12.** 86,000; 9,300; 435
5. 123.0; 9,460; 36.7	**13.** 0.0002; 0.0200; 0.0020
6. 0.27; 50,720; 52.6	**14.** 5.0003; 5.030; 5.003
7. 4.16; 5.16; 8.92	**15.** 636.0; 818.0; 727.0
8. 39.03; 436; 0.0235	**16.** 0.1229; 7.063; 20.125

7–8   Accuracy in Multiplying and Dividing Measurement Numbers

Care must be taken to maintain proper accuracy when multiplying and dividing measurement numbers. There is a tendency to make answers more accurate than they actually are. An answer that has more significant digits than it should gives a false impression of accuracy. A product or quotient cannot be more accurate than the least accurate measurement number used in the computations.

Examples of Multiplying and Dividing Measurement Number.

1. 3.896 in × 63.6 = 247.7856 in, 248 in *Ans*
   Since the least accurate number is 63.6, round the answer to 3 significant digits.

2. 7,500 mi × 2.250 = 16,875 mi, 17,000 mi *Ans*
   Since the least accurate number is 7,500, round the answer to 2 significant digits.

3. 0.009 mm ÷ 0.4876 = 0.018457752 mm, 0.02 mm *Ans*
   Since the least accurate number is 0.009, round the answer to 1 significant digit.

4. 802,000 lb ÷ 430.78 × 1.494 = 2,781.438321 lb, 2,780 lb *Ans*
   Since the least accurate number is 802,000, round the answer to 3 significant digits.

5. If a machined part weighs 0.1386 kilogram, what is the weight of 8 parts? Since 8 is a counting number and is exact, only the number of significant digits in the measurement number, 0.1386, is considered.
   8 × 0.1386 kg = 1.1088 kg, 1.109 kg *Ans*, rounded to 4 significant digits.

EXERCISE 7–8

Smallest
S.D.
#9-16

Multiply or divide the following measurement numbers. Round answers to the same number of significant digits as the least accurate number.

1. 18.9 mm × 2.373
2. 1.85 in × 3.7
3. 8,900 ÷ 52.861
4. 9.085 cm ÷ 1.07
5. 33.08 mi × 0.23
6. 51.9 ÷ 0.97623
7. 0.007 × 0.852
8. 830.367 × 9.455

9. 6.80 ÷ 9.765 × 0.007
10. 71,200 × 19.470 × 0.168
11. 5.00017 × 16.874 × 0.12300
12. 30,000 ÷ 154.9 ÷ 80.03
13. 0.00956 × 34.3 × 0.75
14. 15.635 × 0.415 × 10.07
15. 270.001 ÷ 7.100 × 19.853
16. 52.3 × 6.890 × 0.0073

7–9   Absolute and Relative Error

Absolute error and relative error are commonly used to express the amount of error between an actual or true value and a measured value.

Absolute error is the difference between a true value and a measured value. Since the measured value can be either a smaller or larger value than the true value, subtract the smaller value from the larger value.

$$\text{Absolute Error} = \text{True Value} - \text{Measured Value}$$

or

$$\text{Absolute Error} = \text{Measured Value} - \text{True Value}$$

Relative error is the ratio of the absolute error to the true value. It is expressed as a percent.

$$\text{Relative Error} = \frac{\text{Absolute Error}}{\text{True Value}} \times 100$$

NOTE: 100 is an exact number

Examples of Absolute and Relative Error.

1. The actual or true value of the diameter of a shaft is 1.7056 inches. The shaft is measured as 1.7040 inches. Compute the absolute error and the relative error.

   The true value is larger than the measured value, therefore:

   Absolute Error = True Value − Measured Value

   Absolute Error = 1.7056 in − 1.7040 in = 0.0016 in *Ans*

   $$\text{Relative Error} = \frac{\text{Absolute Error}}{\text{True Value}} \times 100$$

   $$\text{Relative Error} = \frac{0.0016 \text{ in}}{1.7056 \text{ in}} \times 100 \approx 0.094\% \text{ } Ans, \text{ rounded to 2 significant digits}$$

Calculator Application

.0016 $\boxed{\div}$ 1.7056 $\boxed{\times}$ 100 = 0.09380863, 0.094% *Ans*, rounded

2. In an electrical circuit, a calculated or measured value calls for a resistance of 98 ohms. What are the absolute error and the relative error in using a resistor that has an actual or true value of 91 ohms?

   The measured value is larger than the true value, therefore:

   Absolute Error = Measured Value − True Value

   Absolute error = 98 ohms − 91 ohms = 7 ohms *Ans*

   $$\text{Relative error} = \frac{7 \text{ ohms}}{91 \text{ ohms}} \times 100 \approx 8\% \text{ } Ans, \text{ rounded to 1 significant digit}$$

EXERCISE 7–9

Compute the absolute error and the relative error of each of the values given in the table in Figure 7–6. Round answers to the proper number of significant digits.

#7-12

	Actual or True Value	Measured Value		Actual or True Value	Measured Value
1.	3.872 in	3.870 in	7.	105 ohms	102 ohms
2.	0.53 mm	0.52 mm	8.	0.9347 in	0.9341 in
3.	12.7 lb	12.9 lb	9.	1.005 m²	1.015 m²
4.	485 mi	482 mi	10.	27.2 ft	26.9 ft
5.	23.86 cm	24.00 cm	11.	1 827.6 m	1 830.2 m
6.	6 056 kg	6 100 kg	12.	0.983 cu ft	1.000 cu ft

Figure 7–6

7–10   Tolerance (Linear)

Tolerance (linear) is the amount of variation permitted for a given length. Tolerance is equal to the difference between the maximum and minimum limits of a given length.

EXAMPLES •————————————————————————————

1. The maximum permitted length (limit) of the tapered shaft shown in Figure 7–7 is 134.2 millimeters. The minimum permitted length (limit) is 133.4 millimeters. Find the tolerance.

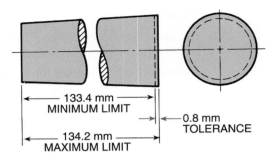

133.4 mm
MINIMUM LIMIT

0.8 mm
TOLERANCE

134.2 mm
MAXIMUM LIMIT

Figure 7–7   Tapered shaft.

The tolerance equals the maximum limit minus the minimum limit.
134.2 mm − 133.4 mm = 0.8 mm *Ans*

2. The maximum permitted depth (limit) of the dado joint shown in Figure 7–8 is $\frac{21}{32}$ inch. The tolerance is $\frac{1}{16}$ inch. Find the minimum permitted depth (limit).

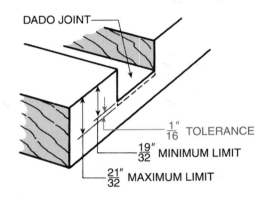

DADO JOINT

$\frac{1}{16}''$ TOLERANCE

$\frac{19}{32}''$ MINIMUM LIMIT

$\frac{21}{32}''$ MAXIMUM LIMIT

Figure 7–8   Dado joint.

The minimum limit equals the maximum limit minus the tolerance.

$$\frac{21''}{32} - \frac{1''}{16} = \frac{19''}{32} \ Ans$$

EXERCISE 7–10

Refer to the tables in Figures 7–9 and 7–10 and determine the tolerance, maximum limit, or minimum limit as required for each problem.

	Tolerance	Maximum Limit	Minimum Limit
1.		$5\frac{7}{16}''$	$5\frac{13}{32}''$
2.		$7'-9\frac{1}{16}''$	$7'-8\frac{15}{16}''$
3.	0.02"	16.76"	
4.	0.007"		0.904"
5.		1.7001"	1.6998"
6.	0.0025"		3.069"

	Tolerance	Maximum Limit	Minimum Limit
7.		50.7 mm	49.8 mm
8.		26.8 cm	26.6 cm
9.	0.04 mm		258.03 mm
10.	0.12 mm	80.09 mm	
11.	0.006 cm		12.731 cm
12.		4.01 mm	3.98 mm

Figure 7–9   Figure 7–10

7–11 Unilateral and Bilateral Tolerance with Clearance and Interference Fits

A *basic dimension* is the standard size from which the maximum and minimum limits are made.

Unilateral tolerance means that the total tolerance is taken in only one direction from the basic dimension, such as:

$$2.6856 \quad \begin{matrix} + 0.0000 \\ - 0.0020. \end{matrix}$$

Bilateral tolerance means that the tolerance is divided partly above (+) and partly below (−) the basic dimension, such as 2.6846 ± 0.0010.

When one part is to move within another there is a *clearance* between the parts. A shaft made to turn in a bushing is an example of a clearance fit. The shaft diameter is less than the bushing hole diameter. When one part is made to be forced into the other, there is *interference* between parts. A pin pressed into a hole is an example of an interference fit. The pin diameter is greater than the hole diameter.

EXAMPLES ●──

1. This is an illustration of a clearance fit between a shaft and a hole using unilateral tolerancing. Refer to Figure 7–11 and determine the following:

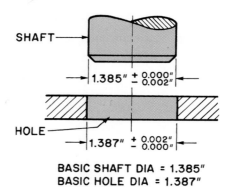

BASIC SHAFT DIA = 1.385"
BASIC HOLE DIA = 1.387"

Figure 7–11

    a. Maximum shaft diameter
        1.385" + 0.000" = 1.385" *Ans*
    b. Minimum shaft diameter
        1.385" − 0.002" = 1.383" *Ans*
    c. Maximum hole diameter
        1.387" + 0.002" = 1.389" *Ans*
    d. Minimum hole diameter
        1.387" − 0.000" = 1.387" *Ans*
    e. Maximum clearance equals maximum hole diameter minus minimum shaft diameter
        1.389" − 1.383" = 0.006" *Ans*
    f. Minimum clearance equals minimum hole diameter minus maximum shaft diameter
        1.387" − 1.385" = 0.002" *Ans*

2. This is an illustration of an interference fit between a pin and a hole using bilateral toleranc-ing. Refer to Figure 7–12 and determine the following:

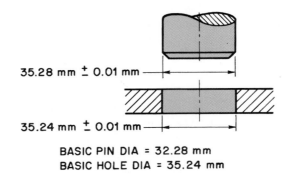

BASIC PIN DIA = 32.28 mm
BASIC HOLE DIA = 35.24 mm

Figure 7–12

a. Maximum pin diameter

35.28 mm + 0.01 mm = 35.29 mm *Ans*

b. Minimum pin diameter

35.28 mm − 0.01 mm = 35.27 mm *Ans*

c. Maximum hole diameter

35.24 mm + 0.01 mm = 35.25 mm *Ans*

d. Minimum hole diameter

35.24 mm − 0.01 mm = 35.23 mm *Ans*

e. Maximum interference equals maximum pin diameter minus minimum hole diameter

35.29 mm − 35.23 mm = 0.06 mm *Ans*

f. Minimum interference equals minimum pin diameter minus maximum hole diameter

35.27 mm − 35.25 mm = 0.02 mm *Ans*

EXERCISE 7–11

Problems 1 through 5 involve clearance fits between a shaft and hole using unilateral toleranc-ing. Given diameters A *and* B, *compute the missing values in the table.*

Refer to Figure 7–13 to determine the table values in Figure 7–14. The values for problem 1 are given.

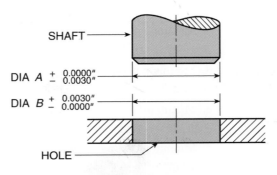

Figure 7–13

		Basic Dimension (inches)	Maximum Diameter (inches)	Minimum Diameter (inches)	Maximum Clearance (inches)	Minimum Clearance (inches)
1.	DIA A	1.4580	1.4580	1.4550	0.0090	0.0030
	DIA B	1.4610	1.4640	1.4610		
2.	DIA A	0.9345				
	DIA B	0.9365				
3.	DIA A	2.1053				
	DIA B	2.1078				
4.	DIA A	0.4961				
	DIA B	0.4970				
5.	DIA A	0.9996				
	DIA B	1.0007				

Figure 7–14

#1-10

Problems 6 through 10 involve interference fits between a pin and hole using bilateral toler-ancing. Given diameters A *and* B, *compute the missing values in the table.*

Refer to Figure 7–15 to determine the table values in Figure 7–16. The values for problem 6 are given.

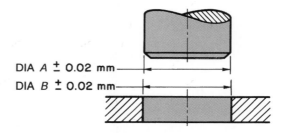

DIA A ± 0.02 mm
DIA B ± 0.02 mm

Figure 7–15

		Basic Dimension (millimeters)	Maximum Diameter (millimeters)	Minimum Diameter (millimeters)	Maximum Clearance (millimeters)	Minimum Clearance (millimeters)
6.	DIA A	20.73	20.75	20.71	0.09	0.01
	DIA B	20.68	20.70	20.66		
7.	DIA A	32.07				
	DIA B	32.01				
8.	DIA A	10.82				
	DIA B	10.75				
9.	DIA A	41.91				
	DIA B	41.85				
10.	DIA A	26.73				
	DIA B	26.65				

Figure 7–16

⠿ UNIT EXERCISE AND PROBLEM REVIEW

DEGREE OF PRECISION OF NUMBERS

For each measurement find:

a. *the degree of precision*

b. *the range*

1. 5.3 in

2. 2.78 mm

3. 1.834 in

4. 12.9 mm

5. 19.001 in

6. 28.35 mm

7. 29.0 mm

8. 6.1088 in

DEGREES OF PRECISION IN ADDING AND SUBTRACTING MEASUREMENT NUMBERS

Add or subtract the following measurement numbers. Round answers to the degree of precision of the least precise number.

9. 26.954 mm − 6.0374 mm

10. 0.0008 in + 0.003 in

11. 3,343 mi − 894.5 mi

12. 28.609 cm + 19.73 cm

13. 27.004 in − 13.00727 in

14. 0.0263 cm² + 0.80 cm²

15. 16 in + 6.93 in + 18.0 in

16. 96.823 mm + 43.06 mm + 52 mm

SIGNIFICANT DIGITS

Determine the number of significant digits for the following measurement numbers.

17. 9.8350

18. 0.0463

19. 8,604.3

20. 0.00086

21. 27.005

22. 89,100

23. 94,126.0

24. 70,000

ACCURACY

For each of the following groups of measurement numbers, identify the number which is most accurate.

25. 6.07; 3.2; 0.005

26. 0.0004; 0.006; 0.56

27. 16.3; 13.0; 48,070

28. 41.02; 364; 0.0384

29. 0.870; 0.0091; 4.22

30. 71,000; 4,200; 593

31. 3.0006; 2.070; 9.001

32. 0.0007; 0.0600; 0.0030

ACCURACY IN MULTIPLYING AND DIVIDING MEASUREMENT NUMBERS

Multiply or divide the following measurement numbers. Round answers to the same number of significant digits as the least accurate number.

33. 2.76 × 4.9

34. 9.043 × 1.02

35. 0.005 × 0.973

36. 55,000 ÷ 767

37. 82,400 × 21.503 × 0.203

38. 30,000 ÷ 127.8 ÷ 86.07

39. 0.00827 × 43.2 × 0.66

40. 360.002 ÷ 8.200 × 15.107

ABSOLUTE AND RELATIVE ERROR

Compute the absolute error and the relative error of each of the values in the table in Figure 7–17. Round answers to the proper number of significant digits.

	Actual or True Value	Measured Value		Actual or True Value	Measured Value
41.	5.963 in	5.960 in	44.	107 ohms	99 ohms
42.	392 mm	388 mm	45.	0.8639 cm	0.8634 cm
43.	5,056 lb	4,998 lb	46.	71.3 ft	70.6 ft

Figure 7–17

TOLERANCE

Refer to the tables in Figures 7–18 and 7–19 and determine the tolerance, maximum limit, or minimum limit as required for each exercise.

	Tolerance	Maximum Limit	Miminum Limit
47.		$7\frac{3}{16}''$	$7\frac{1}{8}''$
48.	$\frac{1}{64}''$	$18\frac{1}{4}''$	
49.	0.006″		2.775″
50.		0.3064″	0.3051″

Figure 7–18

	Tolerance	Maximum Limit	Maximum Limit
51.		40.3 mm	39.7 mm
52.	0.008 cm		6.502 cm
53.	0.18 cm	78.84 mm	
54.		34.02 mm	33.95 cm

Figure 7–19

UNILATERAL TOLERANCE

These exercises require computation with unilateral tolerance clearance fits between mating parts. Given dimensions A *and* B, *compute the missing values in the tables.*

Refer to Figure 7–20 to determine the table values in Figure 7–21.

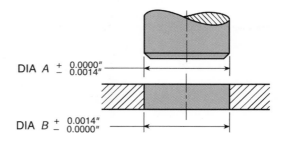

DIA $A \pm {}^{+\,0.0000''}_{-\,0.0014''}$

DIA $B \pm {}^{+\,0.0014''}_{-\,0.0000''}$

Figure 7–20

		Basic Dimension (inches)	Maximum Diameter (inches)	Minimum Diameter (inches)	Maximum Clearance (inches)	Minimum Clearance (inches)
55.	DIA A	1.7120				
	DIA B	1.7136				
56.	DIA A	0.2962				
	DIA B	0.2970				
57.	DIA A	2.8064				
	DIA B	2.8075				

Figure 7–21

BILATERAL TOLERANCE

These exercises require computations with bilateral tolerances of mating parts with interference fits. Given dimensions A *and* B, *compute the missing values in the tables.*

Refer to Figure 7–22 to determine the table values in Figure 7–23.

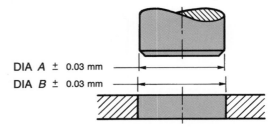

DIA *A* ± 0.03 mm

DIA *B* ± 0.03 mm

Figure 7–22

		Basic Dimension (millimeters)	Maximum Diameter (millimeters)	Minimum Diameter (millimeters)	Maximum Clearance (millimeters)	Minimum Clearance (millimeters)
58.	DIA *A*	78.78				
	DIA *B*	78.70				
59.	DIA *A*	9.94				
	DIA *B*	9.85				
60.	DIA *A*	130.03				
	DIA *B*	129.96				

Figure 7–23

PRACTICAL APPLIED PROBLEMS

61. A cabinetmaker saws a board as shown in Figure 7–24. What are the maximum and minimum permissible values of length *A*?

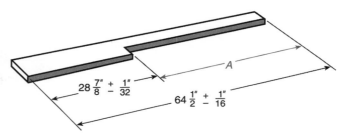

$28 \frac{7}{8}'' {}^{+}_{-} \frac{1}{32}''$

$64 \frac{1}{2}'' {}^{+}_{-} \frac{1}{16}''$

A

Figure 7–24

62. A sheet metal technician lays out a job to the dimensions and tolerances shown in Figure 7–25. Determine the maximum permissible value of length *A*.

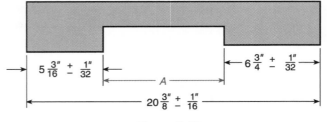

$5 \frac{3}{16}'' {}^{+}_{-} \frac{1}{32}''$

$6 \frac{3}{4}'' {}^{+}_{-} \frac{1}{32}''$

A

$20 \frac{3}{8}'' {}^{+}_{-} \frac{1}{16}''$

Figure 7–25

63. Determine the maximum and minimum permissible wall thickness of the steel sleeve shown in Figure 7–26.

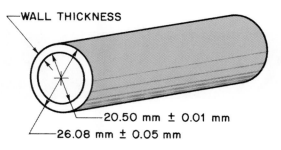

WALL THICKNESS

20.50 mm ± 0.01 mm

26.08 mm ± 0.05 mm

Figure 7–26

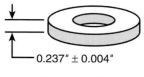

0.237″ ± 0.004″

Figure 7–27

64. Spacers are manufactured to the dimension and tolerance shown in Figure 7–27. An inspector measures 10 bushings and records the following thicknesses:

0.243″	0.239″	0.236″	0.242″	0.234″
0.231″	0.241″	0.238″	0.240″	0.232″

Which spacers are defective (above the maximum limit or below the minimum limit)?

65. The drawing in Figure 7–28 gives the locations with tolerances of 6 holes that are to be drilled in a length of angle iron. An ironworker drills the holes, then checks them for proper locations from edge A. The actual locations of the drilled holes are shown in Figure 7–29. Which holes are drilled out of tolerance (located incorrectly)?

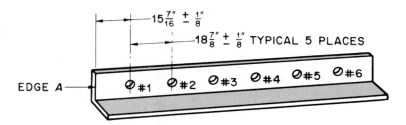

$15\frac{7}{16}'' \pm \frac{1}{8}''$

$18\frac{7}{8}'' \pm \frac{1}{8}''$ TYPICAL 5 PLACES

EDGE A

#1 #2 #3 #4 #5 #6

Figure 7–28

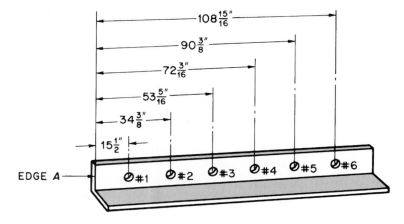

$108\frac{15}{16}''$

$90\frac{3}{8}''$

$72\frac{3}{16}''$

$53\frac{5}{16}''$

$34\frac{3}{8}''$

$15\frac{1}{2}''$

EDGE A

#1 #2 #3 #4 #5 #6

Figure 7–29

UNIT 8 ⁙ Customary Measurement Units

OBJECTIVES

After studying this unit you should be able to

- express lengths as smaller or larger customary linear compound numbers.

- perform arithmetic operations with customary linear units and compound numbers.

- express given customary length, area, and volume measures in larger and smaller units.

- express given customary capacity and weight units as larger and smaller units.

- solve practical applied customary length, area, volume, capacity, and weight problems.

- express customary compound unit measures as equivalent compound unit measures.

- solve practical applied compound unit measures problems.

The United States uses two systems of weights and measures, the American customary system and the SI metric system. The American or U.S. customary system is based on the English system of weights and measures. The International System of Units called the SI metric system is used by all but a few nations.

The American customary length unit, yard, is defined in terms of the metric length base unit, meter. The American customary mass (weight) unit, pound, is defined in terms of the SI metric mass base unit, kilogram.

Throughout this book, the American customary units are called "customary" units and the SI metric units called "metric" units. It is important that you have the ability to measure and compute with both the customary and metric systems. This chapter will examine the customary measurement system and the next chapter will look the metric system.

Linear Measure

A linear measurement is a means of expressing the distance between two points; it is the measurement of lengths. Most occupations require the ability to compute linear measurements and to make direct length measurements.

A drafter computes length measurements when drawing a machined part or an architectural floor plan, an electrician determines the amount of cable required for a job, a welder calculates the length of material needed for a weldment, a printer "figures" the number of pieces that can be cut from a sheet of stock, a carpenter calculates the total length of baseboard required for a building, and an automobile technician computes the amount of metal to be removed for a cylinder re-bore.

8–1 Customary Linear Units

The yard is the standard unit of linear measure in the customary system. From the yard, other units such as the inch and foot are established. The smallest unit is the inch. Customary units of linear measure are shown in Figure 8–1.

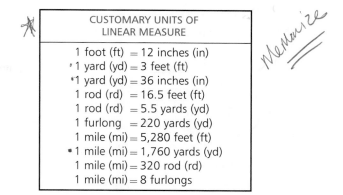

Figure 8–1

Notice that most of the symbols, ft for foot, mi for mile, yd for yard, do not have periods at the end. That is because they are symbols and not abbreviations. The one exception is in. for inch. Many people prefer in. because the period helps you know that they do not mean the word "in."

8–2 Expressing Equivalent Units of Measure

When expressing equivalent units of measure, either of two methods can be used. Throughout Unit 8, examples are given using either of the two methods. Many examples show how both methods are used in expressing equivalent units of measure.

METHOD I

This is a practical method used for many on-the-job applications. It is useful when simple unit conversions are made.

METHOD 2

This method is called the *unity fraction method.* The unity fraction method eliminates the problem of incorrectly expressing equivalent units of measure. Using this method removes any doubt as to whether to multiply or divide when making a conversion. The unity fraction method is particularly useful in solving problems that involve a number of unit conversions.

This method multiplies the given unit of measure by a fraction equal to one. The unity fraction contains the given unit of measure and the unit of measure to which the given unit is to be converted. The unity fraction is set up in such a way that the original unit cancels out and the unit you are converting to remains. Recall that cancelling is the common term used when a numerator and a denominator are divided by a common factor.

Expressing Smaller Customary Units of Linear Measure as Larger Units

To express a smaller unit of length as a larger unit of length, divide the given length by the number of smaller units contained in one of the larger units.

EXAMPLE •————————————————————————————

Express 76.53 inches as feet.

METHOD I

Since 12 inches equal 1 foot, divide 76.53 by 12.

$$76.53 \div 12 = 6.3775$$
$$76.53 \text{ inches} \approx 6.378 \text{ feet } Ans$$

METHOD 2

Since 76.53 inches is to be expressed as feet,

multiplying by the unity fraction $\frac{1\ ft}{12\ in}$

permits the inch unit to be canceled and the foot unit to remain.

$$76.53\ \cancel{in} \times \frac{1\ ft}{12\ \cancel{in}} = \frac{76.53\ ft}{12} \approx 6.378\ ft\ \textit{Ans}$$

In the numerator and denominator, divide by the common factor, 1 inch.

Divide 76.53 ft by 12.

Calculator Application

76.53 $\div$ 12 $=$ 6.3775

6.378 ft *Ans* rounded to 4 significant digits

EXERCISE 8–2A

Express each length as indicated. Round each answer to the same number of significant digits as in the original quantity. Customary units of linear measure are given in the table in Figure 8–1.

1. 51.0 inches as feet
2. 272.5 inches as feet
3. 21.25 feet as yards
4. 67.8 feet as yards
5. 6,300 feet as miles
6. 404.6 inches as yards
7. 44.4 inches as feet
8. 4,928 yards as miles
9. 56.8 feet as yards
10. 53.25 feet as yards
11. 216 rods as miles
12. 6.05 furlongs as miles

Expressing Smaller Units as a Combination of Units

For actual on-the-job applications, smaller units are often expressed as a combination of larger and smaller units (compound denominate numbers).

EXAMPLE

A carpenter wants to express $134\frac{7}{8}$ inches as feet and inches as shown in Figure 8–2.

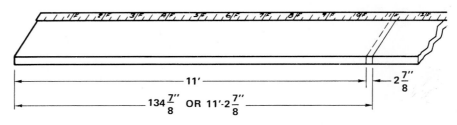

Figure 8–2

Since 12 inches equal 1 foot, divide $134\frac{7}{8}$ by 12. There are 11 feet plus a remainder of $2\frac{7}{8}$ inches in $134\frac{7}{8}$ inches.

The carpenter uses 11 feet $2\frac{7}{8}$ inches as an actual on-the-job measurement. *Ans*

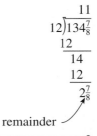

EXERCISE 8–2B

Express each length as indicated. Customary units of linear measure are given in the table in Figure 8–1.

1. 75 inches as feet and inches

2. 40 inches as feet and inches

3. 2,420 yards as miles and yards

4. 15,000 feet as miles and feet

5. $127\frac{1}{2}$ inches as feet and inches

6. 63.2 feet as yards and feet

7. $1,925\frac{1}{3}$ yards as miles and yards

8. $678\frac{3}{4}$ rods as miles and rods

Expressing Larger Customary Units of Linear Measure as Smaller Units

To express a larger unit of length as a smaller unit of length, multiply the given length by the number of smaller units contained in one of the larger units.

EXAMPLE •

Express 2.28 yards as inches.

METHOD 1

Since 36 inches equal 1 yard, multiply 2.28 by 36.

$2.28 \times 36 = 82.08$

2.28 yards ≈ 82.1 in *Ans*

 METHOD 2

Multiply 2.28 yards by the unity fraction.

$$\frac{36 \text{ in}}{1 \text{ yd}}$$

$$2.28 \text{ yd} \times \frac{36 \text{ in}}{1 \text{ yd}} = 2.28 \times 36 \text{ in} \approx 82.1 \text{ in } Ans$$

Divide the numerator and denominator by the common factor, 1 yard

EXERCISE 8–2C

Express each length as indicated. Round each answer to the same number of significant digits as in the original quantity. Customary units of linear measure are given in the table in Figure 8–1.

1. 6.0 feet as inches

2. 0.75 yard as inches

3. 16.30 yards as feet

4. 0.122 mile as yards

5. 1.350 miles as feet

6. 9.046 feet as inches

7. 4.25 yards as feet

8. 2.309 miles as yards

9. 0.250 mile as feet

10. 3.20 yards as inches

11. 1.45 miles as rods

12. 3.6 miles as furlongs

Expressing Larger Units as a Combination of Units

Larger units are often expressed as a combination of two different smaller units.

EXAMPLE •————————————————————————

Express 2.3 yards as feet and inches.

METHOD 1

Express 2.3 yards as feet.	$2.3 \times 3 = 6.9$
Multiply 2.3 by 3.	2.3 yd = 6.9 ft
Express 0.9 foot as inches.	$0.9 \times 12 = 10.8$
Multiply 0.9 by 12.	0.9 ft = 10.8 in
Combine feet and inches.	2.3 yd = 6 ft 10.8 in *Ans*

 METHOD 2

Multiply 2.3 yards by the unity fraction $\dfrac{3 \text{ ft}}{1 \text{ yd}}$. $2.3 \text{ yd} \times \dfrac{3 \text{ ft}}{1 \text{ yd}} = 2.3 \times 3 \text{ ft} = 6.9 \text{ ft}$

Multiply 0.9 feet by the unity fraction $\dfrac{12 \text{ in}}{1 \text{ ft}}$. $0.9 \text{ ft} \times \dfrac{12 \text{ in}}{1 \text{ ft}} = 0.9 \times 12 \text{ in} = 10.8 \text{ in}$

Combine feet and inches.   2.3 yd = 6 ft 10.8 in *Ans*

——•

EXERCISE 8–2D

Express each length as indicated.

1. $6\dfrac{1}{2}$ yards as feet and inches

2. 8.250 yards as feet and inches

3. 0.0900 mile as yards and feet

4. $\dfrac{5}{12}$ mile as yards and feet

5. 2.180 miles as rods and yards

6. $8\dfrac{7}{32}$ yards as feet and inches

7. 0.90 yard as feet and inches

8. 0.3700 mile as yards and feet

 8–3 Arithmetic Operations with Compound Numbers

Basic arithmetic operations with compound numbers are often required for on-the-job applications. For example, a material estimator may compute the stock requirements of a certain job by adding 16 feet $7\frac{1}{2}$ inches and 9 feet 10 inches. An ironworker may be required to divide a 14-foot 10-inch beam in three equal parts.

The method generally used for occupational problems is given for each basic operation.

Addition of Compound Numbers

To add compound numbers, arrange like units in the same column, then add each column. When necessary, simplify the sum.

EXAMPLE •————————————————————————

Determine the amount of stock, in feet and inches, required to make the welded angle bracket shown in Figure 8–3 on page 219.

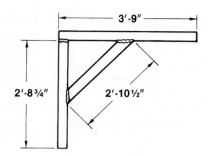

Figure 8–3

Arrange like units in the same column.	3 ft 9 in
	2 ft 10$\frac{1}{2}$ in
Add each column.	2 ft 8$\frac{3}{4}$ in
	7 ft 28$\frac{1}{4}$ in

Simplify the sum. Divide 28$\frac{1}{4}$ by 12 to express 28$\frac{1}{4}$ inches as 2 feet 4$\frac{1}{4}$ inches.

28$\frac{1}{4}$ inches = 2 feet 4$\frac{1}{4}$ inches

Add. 7 feet + 2 feet 4$\frac{1}{4}$ inches = 9 feet 4$\frac{1}{4}$ inches *Ans*

Subtraction of Compound Numbers

To subtract compound numbers, arrange like units in the same column, then subtract each column starting from the right. Regroup as necessary.

EXAMPLES

1. Determine length A of the pipe shown in Figure 8–4.

Arrange like units in the same column.	15 ft 8$\frac{1}{2}$ in
	7 ft 3$\frac{1}{4}$ in
Subtract each column.	8 ft 5$\frac{1}{4}$ in *Ans*

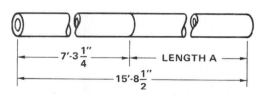

Figure 8–4

2. Subtract 8 yards 2 feet 7 inches from 12 yards 1 foot 3 inches.

Arrange like units in the same column.

$$\begin{array}{r} 12 \text{ yd } 1 \text{ ft }\ \ 3 \text{ in} \\ -8 \text{ yd } 2 \text{ ft }\ \ 7 \text{ in} \\ \hline \end{array}$$

Subtract each column.

Since 7 inches cannot be subtracted from 3 inches, subtract 1 foot from the foot column (leaving 0 feet). Express 1 foot as 12 inches; then add 12 inches to 3 inches.

$$\begin{array}{r} 12 \text{ yd } 0 \text{ ft } 15 \text{ in} \\ -8 \text{ yd } 2 \text{ ft }\ \ 7 \text{ in} \\ \hline \end{array}$$

Since 2 feet cannot be subtracted from 0 feet, subtract 1 yard from the yard column (leaving 11 yards). Express 1 yard as 3 feet; then add 3 feet to 0 feet.

$$\begin{array}{r} 11 \text{ yd } 3 \text{ ft } 15 \text{ in} \\ -8 \text{ yd } 2 \text{ ft }\ \ 7 \text{ in} \\ \hline \end{array}$$

Subtract each column.

3 yd 1 ft   8 in *Ans*

EXERCISE 8–3A

Add. Express each answer in the same units as those given in the exercise. Regroup the answer when necessary.

1. 5 ft 6 in + 7 ft 3 in

2. 3 ft 9 in + 4 ft 8 in

3. 6 ft $3\frac{3}{8}$ in + 4 ft $1\frac{1}{2}$ in + 8 ft $10\frac{1}{4}$ in

4. 5 yd 2 ft + 2 yd $\frac{1}{2}$ ft + 7 yd $\frac{1}{4}$ ft

5. 3 yd 2 ft + 5 yd $\frac{1}{4}$ ft + 9 yd $2\frac{3}{4}$ ft

6. 9 yd 2 ft 3 in + 2 yd 0 ft 6 in

7. 12 yd 2 ft 8 in + 10 yd 2 ft 7 in

8. 4 yd 1 ft $3\frac{1}{2}$ in + 7 yd 0 ft 9 in + 4 yd 2 ft 0 in

9. 3 rd 4 yd + 2 rd 1 yd

10. 6 rd $3\frac{1}{4}$ yd + 8 rd + 4 yd

11. 1 mi 150 rd + 1 mi 285 rd

12. 3 mi 75 rd 2 yd + 2 mi 150 rd $3\frac{1}{4}$ yd + 1 mi 200 rd 5 yd

Subtract. Express each answer in the same units as those given in the exercise. Regroup the answer when necessary.

13. 6 ft 7 in − 2 ft 4 in

14. 15 ft 3 in − 12 ft 9 in

15. 10 ft $1\frac{3}{8}$ in − 7 ft $8\frac{7}{16}$ in

16. 8 yd $1\frac{1}{2}$ ft − 4 yd $2\frac{3}{4}$ ft

17. 14 yd 2 ft − 11 yd 1.5 ft

18. 7 yd 1 ft 9 in − 2 yd 2 ft 11 in

19. 16 yd 2 ft 2.15 in − 14 yd 2 ft 4.25 in

20. 23 yd 1 ft 0 in − 3 yd 0 ft $6\frac{5}{8}$ in

21. 5 rd 3 yd 2 ft − 4 rd 2 yd 1 ft

22. 2 rd 5 yd $1\frac{1}{3}$ ft − 1 rd 0 yd $1\frac{2}{3}$ ft

23. 7 mi 240 rd − 3 mi 310 rd

24. 4 mi 150 rd 4 yd − 1 mi 175 rd 5 yd

Multiplication of Compound Numbers

To multiply compound numbers, multiply each unit of the compound number by the multiplier. When necessary, simplify the product.

EXAMPLE •——

A plumber cuts 5 pieces of copper tubing. Each piece is 8 feet $9\frac{3}{4}$ inches long. Determine the total length of tubing required.

Multiply each unit by 5.	$8 \text{ ft } 9\frac{3}{4} \text{ in}$
	$\underline{\phantom{8 \text{ ft } 9\frac{3}{4}} 5}$
Simplify the product.	$40 \text{ ft } 48\frac{3}{4} \text{ in}$
Divide $48\frac{3}{4}$ by 12 to express $48\frac{3}{4}$ inches as 4 feet $\frac{3}{4}$ inch.	$48\frac{3}{4} \text{ inches} = 4 \text{ feet } \frac{3}{4} \text{ inches}$
Add.	$40 \text{ feet} + 4 \text{ feet } \frac{3}{4} \text{ inches} = 44 \text{ feet } \frac{3}{4} \text{ inches } Ans$

——•

Division of Compound Numbers

To divide compound numbers, divide each unit by the divisor starting at the left. If a unit is not exactly divisible, express the remainder as the next smaller unit and add it to the given number of smaller units. Continue the process until all units are divided.

EXAMPLE •——

The 4 holes in the I beam shown in Figure 8–5 are equally spaced. Determine the distance between 2 consecutive holes.

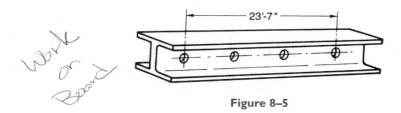

Figure 8–5

Since there are 3 spaces between holes, divide 23 feet 7 inches by 3.

Divide 23 feet by 3.	23 ft ÷ 3 = 7 ft. (quotient) and a 2 ft remainder.
Express the 2-foot remainder as 24 inches.	2 ft = 2 × 12 in = 24 in
Add 24 inches to the 7 inches given in the problem.	24 in + 7 in = 31 in
Divide 31 inches by 3.	$31 \text{ in} \div 3 = 10\frac{1}{3} \text{ in (quotient)}$
Collect quotients.	$7 \text{ ft } 10\frac{1}{3} \text{ in } Ans$

EXERCISE 8–3B

Multiply. Express each answer in the same units as those given in the exercise. Regroup the answer when necessary.

1. 7 ft 3 in × 2

2. 4 ft 5 in × 3

3. 12 ft 3 in × 5.5

4. 6 yd $\frac{1}{2}$ ft × 4

5. 16 yd 2 ft × 8

6. 5 yd 1.25 ft × 4.8

7. 9 yd 2 ft 3 in × 2

8. 11 yd 1 ft 7$\frac{3}{4}$ in × 3

9. 10 yd 2 ft 9 in × $\frac{1}{2}$

10. 6 rd 4 yd × 5

11. 5 mi 210 rd × 1.4

12. 3 mi 180 rd 5 yd × 2

Divide. Express each answer in the same units as those given in the exercise.

13. 9 ft 6 in ÷ 3

14. 7 ft 4 in ÷ 2

15. 18 ft 3.9 in ÷ 4

16. 16 yd 2 ft ÷ 8

17. 21 yd 1 ft ÷ 1$\frac{1}{2}$

18. 4 yd 3.75 ft ÷ 3

19. 14 yd 2 ft 6 in ÷ 2

20. 17 yd 1 ft 10 in ÷ 5

21. 6 yd 2 ft 3$\frac{1}{4}$ in ÷ 0.5

22. 5 rd 2 yd ÷ 4

23. 3 mi 150 rd ÷ 1$\frac{1}{2}$

24. 4 mi 310 rd 4 yd ÷ 3

8–4   Customary Linear Measure Practical Applications

EXAMPLE •

The electrical conduit in Figure 8–6 is made from $\frac{5}{8}$-inch diameter tubing. What is the total length of the straight tubing used for the conduit? Give the answer in feet and inches.

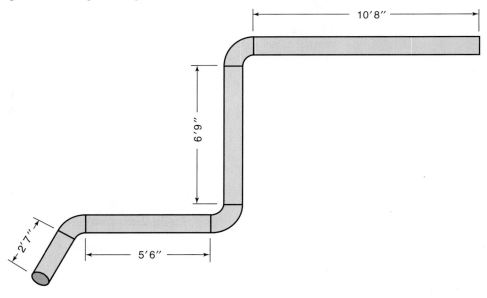

Figure 8–6

Arrange like units in the same column.   10 ft 8 in

6 ft 9 in

5 ft 6 in

Add each column.   2 ft 7 in
                                                   ―――――――
                                                   23 ft 30 in

Simplify the sum. Divide 30 inches by 12 to change 30 inches to 2 feet 6 inches. Add 23 ft + 2 ft 6 in = 25 ft 6 in *Ans*

EXERCISE 8–4

Solve the following problems.

1. The first-floor plan of a house is shown in Figure 8–7. Find distances A, B, C, and D in feet and inches.

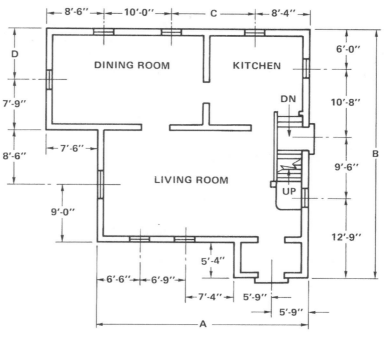

Figure 8–7

2. A survey subdivides a parcel of land in 5 lots of equal width as shown in Figure 8–8. Find the number of feet in distances A and B. Round the answers to 3 significant digits.

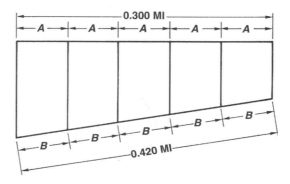

Figure 8–8

3. A bolt (roll) contains 70 yards 2 feet of fabric. The following lengths of fabric are sold from the bolt: 4 yards 2 feet, 5 yards $1\frac{1}{4}$ feet, 7 yards $2\frac{1}{2}$ feet. Find the length of fabric left on the bolt. Express the answer in yards and feet.

4. A building construction assistant lays out the stairway shown in Figure 8–9 on page 224.

   a. Find, in feet and inches, the total run of the stairs.

   b. Find, in feet and inches, the total rise of the stairs.

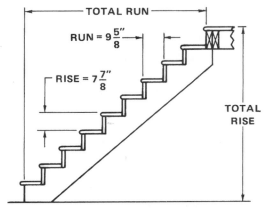

Figure 8–9

5. A structural steel fabricator cuts 4 equal lengths from a channel iron shown in Figure 8–10. Allow $\frac{1}{8}''$ waste for each cut. Find the length, in feet and inches, of the remaining piece.

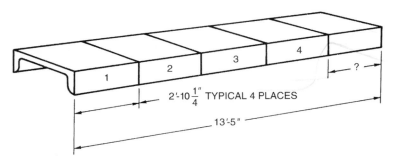

Figure 8–10

6. The floor of a room shown in Figure 8–11 is to be covered with oak flooring. The flooring is $2\frac{1}{4}$ inches wide. Allow 320 linear feet for waste. Find the total number of linear feet of oak flooring, including waste, needed for the floor.

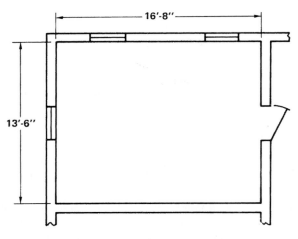

Figure 8–11

7. A concrete beam shown in Figure 8–12 is $12\frac{1}{3}$ yards long. Find distances A, B, and C in feet and inches.

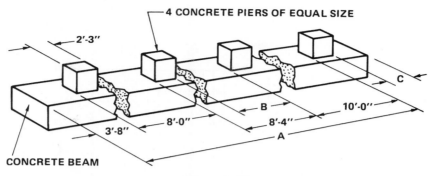

Figure 8–12

8. A carpet installer contracts to supply and install carpeting in the hallways of an office building. The locations of the hallways are shown in Figure 8–13. The hallways are $4\frac{1}{2}$ feet wide. The price charged for both carpet cost and installation is $38.75 per running (linear) yard. Find the total cost of the installation job.

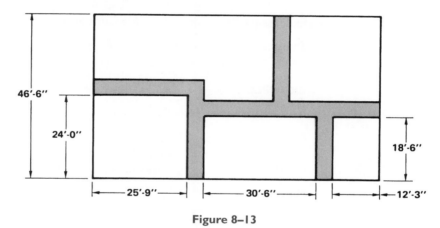

Figure 8–13

9. An apparel maker must know the fabric cost per garment. Find the material cost of a garment that requires 68 inches of fabric at $4.75 per yard and 52 inches of lining at $2.20 per yard.

8–5 Customary Units of Surface Measure (Area)

The ability to compute areas is necessary in many occupations. In agriculture, crop yields and production are determined in relation to land area. Fertilizers and other chemical requirements are computed in terms of land area. In the construction field, carpenters are regularly involved with surface measure, such as floor and roof areas.

A surface is measured by determining the number of surface units contained in it. A surface is two-dimensional. It has length and width, but no thickness. Both length and width must be expressed in the same unit of measure. Area is computed as the product of two linear measures and is expressed in square units. For example, 2 inches × 4 inches = 8 square inches.

The surface enclosed by a square that is 1 foot on a side is 1 square foot. The surface enclosed by a square that is 1 inch on a side is 1 square inch. Similar meanings are attached to square yard, square rod, and square mile. For our uses, the statement "area of the surface enclosed by a figure" is shortened to "area of a figure." Therefore, areas will be referred to as the area of a rectangle, area of a triangle, area of a circle, and so forth.

Look at the reduced drawing in Figure 8–14 on page 226 showing a square inch and a square foot. Observe that 1 linear foot equals 12 linear inches, but 1 square foot equals 12 inches times 12 inches or 144 square inches.

The table shown in Figure 8–15 lists common units of surface measure. Other than the unit acre, surface measure units are the same as linear measure units with the addition of the term *square*.

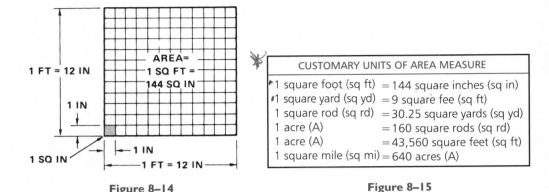

Figure 8–14   **Figure 8–15**

Expressing Customary Area Measure Equivalents

To express a given customary unit of area as a larger customary unit of area, divide the given area by the number of square units contained in one of the larger units.

EXAMPLE ●

Express 728 square inches as square feet.

METHOD I

Since 144 sq in = 1 sq ft, divide 728 by 144.
728 ÷ 144 ≈ 5.06; 728 sq in ≈ 5.06 sq ft *Ans*

METHOD 2

$$728 \text{ sq in} \times \frac{1 \text{ sq ft}}{144 \text{ sq in}} = \frac{728 \text{ sq ft}}{144} \approx 5.06 \text{ sq ft } \textit{Ans}$$

To express a given customary unit of area as a smaller customary unit of area, multiply the given area by the number of square units contained in one of the larger units.

EXAMPLE ●

Express 0.612 square yard as square inches.

Multiply 0.612 square yard by the unity fractions $\dfrac{9 \text{ sq ft}}{1 \text{ sq yd}}$ and $\dfrac{144 \text{ sq in}}{1 \text{ sq ft}}$.

$$0.612 \text{ sq yd} \times \frac{9 \text{ sq ft}}{1 \text{ sq yd}} \times \frac{144 \text{ sq in}}{1 \text{ sq ft}} \approx 793 \text{ sq in } \textit{Ans}$$

Calculator Application
.612 ⊠ 9 ⊠ 144 ▣ 793.152
793 sq in (rounded to 3 significant digits) *Ans*

EXERCISE 8–5

Express each areas as indicated. Round each answer to the same number of significant digits as in the original quantity. Customary units of area measure are given in the table in Figure 8–15.

1. 196 square inches as square feet
2. 1,085 square inches as square feet
3. 45.8 square feet as square yards
4. 2.02 square feet as square yards

5. 1,600 acres as square miles
6. 192 acres as square miles
7. 120,000 square feet as acres
8. 122.5 square yards as square rods
9. 17,400 square feet as acres
10. 2,300 square inches as square yards
11. 871,000 square feet as square miles
12. 2,600 square feet as square rods
13. 2.35 square feet as square inches
14. 0.624 square foot as square inches

15. 4.30 square yards as square feet
16. 0.59 square yard as square feet
17. 3.8075 square miles as acres
18. 0.462 square mile as acres
19. 2.150 acres as square feet
20. 0.25 acre as square feet
21. 5.45 square rods as square yards
22. 0.612 square yard as square inches
23. 0.0250 square mile as square feet
24. 1.75 square rods as square feet

8–6 Customary Area Measure Practical Applications

EXAMPLE

A sheet of aluminum that contains 18.00 square feet is sheared into 38 strips of equal size. What is the area of each strip in square inches?

Since 1 square foot equals 144 square inches, multiply 18.00 by 144.

Divide 2,592 square inches by the number of strips (38).

$18.00 \times 144 = 2{,}592$
18.00 square feet = 2,592 square inches

$2{,}592 \div 38 \approx 68.21$ rounded to 4 significant digits
The area of each strip is 68.21 square inches. *Ans*

Calculator Application

$18 \boxed{\times} 144 \boxed{\div} 38 \boxed{=} 68.21052632$
68.21 square inches (rounded to 4 significant digits) *Ans*

EXERCISE 8–6

Solve the following problems.

1. How many strips, each having an area of 48.00 square inches, can be sheared from a sheet of steel that measures 18.00 square feet?

2. A contractor estimates the cost of developing a 0.3000-square-mile parcel of land at $1,200 per acre. What is the total cost of developing this parcel?

3. A painter and decorator compute the total interior wall surface of a building as 220 square yards after allowing for windows and doors. Two coats of paint are required for the job. If 1 gallon of paint covers 500 square feet, how many gallons of paint are required? Give the answer to the nearest gallon.

4. A bag of lawn food sells for $16.50 and covers 12,500 square feet. What is the cost to the nearest dollar to cover $1\frac{1}{2}$ acres of lawn?

5. A basement floor that measures 875 square feet is to be covered with floor tiles. Each tile measures 100 square inches. Make an allowance of 5% for waste. How many tiles are needed? Give answer to the nearest 10 tiles.

6. A land developer purchased 0.200 square mile of land. The land was subdivided into 256 building lots of approximately the same area. What is the average number of square feet per building lot? Give answer to the nearest 100 square feet.

7. A total of 180 square yards of the interior walls of a building are to be paneled. Each panel is 32 square feet. Allowing for 15% waste, how many whole panels are required?

8–7   Customary Units of Volume (Cubic Measure)

A solid is measured by determining the number of cubic units contained in it. A solid is three-dimensional; it has length, width, and thickness or height. Length, width, and thickness must be expressed in the same unit of measure. Volume is the product of three linear measures and is expressed in cubic units. For example, 2 inches × 3 inches × 5 inches = 30 cubic inches.

The volume of a cube having sides 1 foot long is 1 cubic foot. The volume of a cube having sides 1 inch long is 1 cubic inch. A similar meaning is attached to the cubic yard.

A reduced illustration of a cubic foot and a cubic inch is shown in Figure 8–16. Observe that 1 linear foot equals 12 linear inches, but 1 cubic foot equals 12 inches × 12 inches × 12 inches, or 1,728 cubic inches.

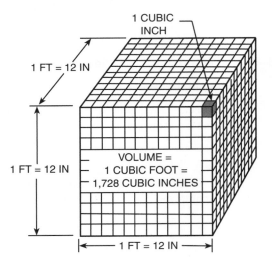

1 CUBIC INCH

1 FT = 12 IN

1 FT = 12 IN

VOLUME =
1 CUBIC FOOT =
1,728 CUBIC INCHES

1 FT = 12 IN

Figure 8–16

The table in Figure 8–17 lists common units of volume measure with their abbreviations. Volume measure units are the same as linear unit measures with the addition of the term *cubic.*

CUSTOMARY UNITS OF VOLUME MEASURE
1 cubic foot (cu ft) = 1,728 cubic inches (cu in)
1 cubic yard (cu yd) = 27 cubic feet (cu ft)

Figure 8–17

Expressing Customary Volume Measure Equivalents

To express a given unit of volume as a larger unit, divide the given volume by the number of cubic units contained in one of the larger units.

EXAMPLE •

Express 4,300 cubic inches as cubic feet.

METHOD 1

Since 1,728 cu in = 1 cu ft, divide 4,300 by 1,728.

4,300 ÷ 1,728 ≈ 2.5; 4,320 cu in ≈ 2.5 cu ft *Ans*

METHOD 2

$$4,300 \text{ cu in} \times \frac{1 \text{cu ft}}{1,728 \text{ cu in}} \approx 2.5 \text{ cu ft} \quad Ans$$

To express a given unit of volume as a smaller unit, multiply the given volume by the number of cubic units contained in one of the larger units.

EXAMPLE •————————————————————————————

Express 0.0197 cubic yards as cubic inches.

Multiply 0.0197 cubic yard by the unity fractions $\dfrac{27 \text{ cu ft}}{1 \text{ cu yd}}$ and $\dfrac{1728 \text{ cu in}}{1 \text{ cu ft}}$.

$$0.0197 \text{ cu yd} \times \frac{27 \text{ cu ft}}{1 \text{ cu yd}} \times \frac{1728 \text{ cu in}}{1 \text{ cu ft}} \approx 919 \text{ cu in} \quad Ans$$

Calculator Application

.0197 ☒ 27 ☒ 1728 ═ 919.1212; 919 cu in (rounded to 3 significant digits) *Ans*

EXERCISE 8–7

Express each volume as indicated. Round each answer to the same number of significant digits as the original quantity. Customary units of volume measure are given in the table in Figure 8–17.

1. 4,320 cu in = ? cu ft
2. 860 cu in = ? cu ft
3. 117 cu ft = ? cu yd
4. 187 cu ft = ? cu yd
5. 12,900 cu in = ? cu ft
6. 18,000 cu in = ? cu yd
7. 73 cu ft = ? cu yd
8. 124.7 cu ft = ? cu yd
9. 562 cu in = ? cu ft
10. 51,000 cu in = ? cu yd

11. 1.650 cu ft = ? cu in
12. 0.325 cu ft = ? cu in
13. 16.4 cu yd = ? cu ft
14. 243.0 cu yd = ? cu ft
15. 0.273 cu ft = ? cu in
16. 0.09 cu yd = ? cu in
17. 113.4 cu yd = ? cu ft
18. 0.36 cu yd = ? cu ft
19. 0.55 cu ft = ? cu in
20. 0.1300 cu yd = ? cu in

8–8   Customary Volume Practical Applications

EXAMPLE •————————————————————————————

Castings are to be made from 2.250 cubic feet of molten metal. If each casting requires 15.8 cubic inches of metal, how many castings can be made?

Since 1,728 cubic inches equal 1 cubic foot, multiply 1,728 by 2.250.

$$1{,}728 \times 2.250 = 3{,}888$$

Therefore, 2.250 cubic feet = 3,888 cubic inches.

Divide.

$$3{,}888 \div 15.8 \approx 246$$

246 castings can be made. *Ans*

Calculator Application

1728 × 2.25 ÷ 15.8 ═ 246.0759494; 246 castings *Ans*

EXERCISE 8–8

Solve the following problems.

1. Each casting requires 8.500 cubic inches of bronze. How many cubic feet of molten bronze are required to make 2,500 castings?

2. A cord is a unit of measure of cut and stacked fuel wood equal to 128 cubic feet. Wood is burned at the rate of $\frac{1}{2}$ cord per week. How many weeks will a stack of wood measuring $21\frac{1}{3}$ cubic yards last?

3. Common brick weighs 112.0 pounds per cubic foot. How many cubic yards of brick can be carried by a truck whose maximum carrying load is rated as 8 gross tons? One gross ton = 2,240 pounds.

4. Hot air passes through a duct at the rate of 550 cubic inches per second. Find the number of cubic feet of hot air passing through the duct in 1 minute. Give the answer to the nearest cubic foot.

5. For lumber that is 1 inch thick, 1 board foot of lumber has a volume of $\frac{1}{12}$ cubic foot. Seasoned white pine weighs $31\frac{1}{2}$ pounds per cubic foot. Find the weight of 1,400 board feet of seasoned white pine.

6. Concrete is poured for a building foundation at the rate of $8\frac{1}{2}$ cubic feet per minute. How many cubic yards are pumped in $\frac{3}{4}$ hour? Give the answer to the nearest cubic yard.

8–9   Customary Units of Capacity

Capacity is a measure of volume. The capacity of a container is the number of units of material that the container can hold.

In the customary system, there are three different kinds of measures of capacity. *Liquid measure* is for measuring liquids. For example, it is used for measuring water and gasoline and for stating the capacity of fuel tanks and reservoirs. *Dry measure* is for measuring fruit, vegetables, grain, and the like.

Apothecaries' fluid measure is for measuring drugs and prescriptions.

The most common units of customary liquid and fluid measure are listed in the table shown in Figure 8–18. Also listed are common capacity–cubic measure equivalents.

CUSTOMARY UNITS OF CAPACITY MEASURE
16 ounces (oz) = 1 pint (pt)
2 pints (pt)   = 1 quart (qt)
4 quarts (qt) = 1 gallon (gal)
COMMONLY USED CAPACITY–CUBIC MEASURE EQUIVALENTS
1 gallon (gal)   = 231 cubic inches (cu in)
7.5 gallons (gal) = 1 cubic foot (cu ft)

Figure 8–18

Expressing Equivalent Customary Capacity Measures

It is often necessary to express given capacity units as either larger or smaller units. The procedure is the same as used with linear, square, and cubic units of measure.

EXAMPLE •——————————————————————————————————

1. Express 20.5 ounces as pints.

   METHOD I

   Since 16 ounces = 1 pint, divide 20.5 by 16. $20.5 \div 16 = 1.28125$

   $20.5 \text{ oz} \approx 1.28 \text{ pt}$ (rounded to 3 significant digits) *Ans*

METHOD 2

Multiply 20.5 oz by the unity

fraction $\dfrac{1 \text{ pt}}{16 \text{ oz}}$.

$$20.5 \text{ oz} \times \frac{1 \text{ pt}}{16 \text{ oz}} = \frac{20.5 \text{ pt}}{16} \approx 1.28 \text{ pt (rounded to}$$
3 significant
digits) *Ans*

EXERCISE 8–9

Express each unit of measure as indicated. Round each answer to the same number of significant digits as in the original quantity. Customary and metric units of capacity measure are given in the table in Figure 8–17.

1. 6.52 pt = ? qt
2. 38 oz = ? pt
3. 9.25 gal = ? qt
4. 0.35 pt = ? oz
5. 35 qt = ? gal
6. 17.75 qt = ? pt
7. 3.07 gal = ? cu in

8. 53.8 gal = ? cu ft
9. 46 cu in = ? gal
10. 0.90 cu ft = ? gal
11. 62 oz = ? qt
12. 0.22 gal = ? pt
13. 1.6 qt = ? oz
14. 43 pt = ? gal

8–10 Customary Capacity Practical Applications

EXAMPLE

How many ounces of solution are contained in a $\frac{3}{4}$-quart container when full?

METHOD 1

Since 16 oz = 1 pt and 2 pt = 1 qt, there are 16 × 2 or 32 oz in 1 qt.

Multiply 32 by $\frac{3}{4}$. $32 \times \dfrac{3}{4} = 24; \dfrac{3}{4}$ qt = 24 oz *Ans*

METHOD 2

Multiply $\dfrac{3}{4}$ quart by unity fractions $\dfrac{2 \text{ pt}}{1 \text{ qt}}$

and $\dfrac{16 \text{ oz}}{1 \text{ pt}}$.

$$\frac{3}{4} \text{ qt} \times \frac{2 \text{ pt}}{1 \text{ qt}} \times \frac{\overset{4}{\cancel{16}} \text{ oz}}{1 \text{ pt}} = 24 \text{ oz } Ans$$

EXERCISE 8–10 omit

Solve the following problems.

1. In planning for a reception, a chef estimates that eighty 4-ounce servings of tomato juice are required. How many quarts of juice must be ordered?

2. When mixing oil with gasoline for an outboard engine, 1 part of oil is added to 10 parts of gasoline. How many pints of oil are added to 4.5 gal of gasoline?

3. A water tank has a volume of 4,550 cubic feet. The tank is $\frac{9}{10}$ full. How many gallons of water are contained in the tank? Round answer to the nearest hundred gallons.

4. Automotive cooling systems require 4 qt of coolant for each 10 qt of capacity to provide protection to −13°F. A cooling system has a capacity of 17 qt. How many quarts and pints of coolant are required?

5. An empty fuel oil tank has a volume of 575 cubic feet. Oil is pumped into the tank at the rate of 132 gallons per minute. How long does it take to fill the tank? Round answer to the nearest minute.

6. A small oil can holds $\frac{1}{5}$ pt. How many times can it be refilled from a 1.25 gal can of oil?

7. A cutting lubricant requires 8.75 oz of concentrate for 1 qt of water. How much concentrate will be required for 3.6 gal of water?

8. A solution contains 12.5% acid and 87.5% water. How many gallons of solution are made with 27 ounces of acid? Round answer to the nearest tenth gallon.

9. A water storage tank has a volume of 165 cubic yards. When the tank is $\frac{1}{4}$ full of water, how many gallons of water are required to fill the tank? Round the answer to the nearest one hundred gallons.

8–11   Customary Units of Weight (Mass)

Weight is a measure of the force of attraction of the earth on an object. *Mass* is a measure of the amount of matter contained in an object. The weight of an object varies with its distance from the earth's center. The mass of an object remains the same regardless of its location in the universe.

Scientific applications dealing with objects located other than on the earth's surface are *not* considered in this book. Therefore, the terms *weight* and *mass* are used interchangeably.

As with capacity measures, the customary system has three types of weight measures. *Troy weights* are used in weighing jewels and precious metals such as gold and silver. *Apothecaries' weights* are for measuring drugs and prescriptions. *Avoirdupois* or *commercial weights* are used for all commodities except precious metals, jewels, and drugs.

The most common units of customary weight measure are listed in the table shown in Figure 8–19.

CUSTOMARY UNITS OF WEIGHT MEASURE	
16 ounces (oz)	=1 pound (lb)
2,000 pounds (lb)	=1 net or short ton
2,240 pounds (lb)	=1 gross or long ton

Figure 8–19

The long ton is seldom used. Originally, it was used to measure the weight of anthracite coal in Pennsylvania, bulk amounts of certain iron and steel products, and the deadweight tonnage of ships.

Expressing Equivalent Customary Weight Measures

In the customary system, apply the same procedures that are used with other measures. The following example shows the method of expressing given weight units as larger or smaller units.

EXAMPLE •————————————————————————

Express 0.28 pound as ounces.

METHOD 1

Since 16 oz = 1 lb, multiply 0.28 by 16.

$0.28 \times 16 = 4.48$

$0.28 \text{ lb} \approx 4.5 \text{ oz } Ans$

METHOD 2

Multiply 0.28 by the unity fraction $\dfrac{16 \text{ oz}}{1 \text{ lb}}$.

$0.28 \, \cancel{\text{lb}} \times \dfrac{16 \text{ oz}}{1 \, \cancel{\text{lb}}} \approx 4.5 \text{ oz } Ans$

EXERCISE 8–11

Express each unit of weight as indicated. Round each answer to the same number of significant digits as in the original quantity. Customary and metric units of weight are given in the table in Figure 8–18.

1. 35 oz = ? lb
2. 0.6 lb = ? oz
3. 2.4 long tons = ? lb
4. 3.1 short tons = ? lb
5. 5,300 lb = ? short tons

6. 7,850 lb = ? long tons
7. 43.5 oz = ? lb
8. 0.120 lb = ? oz
9. 0.12 short ton = ? lb
10. 720 lb = ? short tons

8–12 Customary Weight Practical Applications

EXAMPLE

A truck delivers 4 prefabricated concrete wall sections to a job site. Each wall section has a volume of 1.65 cubic yards. One cubic yard of concrete weight 4040 pounds. How many tons are carried on this delivery?

Find the volume of 4 wall sections	$4 \times 1.65 \text{ yd}^3 = 6.6 \text{ yd}^3$
Find the weight of 6.6 cubic yards	$6.6 \times 4040 \text{ lb} = 25{,}664 \text{ lb}$
Find the number of tons	$25{,}664 \text{ lb} \div 2000 \text{ lb/t} \approx 13.3 \text{ tons } Ans$

Calculator Application

4 ☒ 1.65 ☒ 4040 ⊡ 2000 ▣ 13.332, 13.3 tons *Ans* (rounded to 3 significant digits.)

EXERCISE 8–12   Omit

Solve each problem.

1. What is the total weight, in pounds, of 1 gross (144) 12.0-ounce cans of fruit?
2. 250 identical strips are sheared from a sheet of steel that weighs 42.25 lb. Find the weight, in ounces, of each strip.
3. One cubic foot of stainless steel weighs 486.9 pounds. How many pounds and ounces does 0.175 cubic foot of stainless steel weigh? Give answer to the nearest whole ounce.
4. A technician measures the weights of some objects as 12.3 oz, 9.6 oz, 7.4 oz, 11.6 oz, 9.2 oz, 13.8 oz, and 8.4 oz. Find the total weight in pounds and ounces.
5. A force of 760 tons (short tons) is exerted on the base of a steel support column. The base has a cross-sectional area of 160 square inches. How many pounds of force are exerted per square inch of cross-sectional area?
6. An assembly housing weighs 8.25 pounds. The weight of the housing is reduced to 6.50 pounds by drilling holes in the housing. Each drilled hole removes 0.80 ounce of material. How many holes are drilled?
7. A cubic foot of water weighs 62.42 pounds at 40 degrees Fahrenheit and 61.21 pounds at 120 degrees Fahrenheit. Over this temperature change, what is the average decrease in weight in ounces per cubic foot for each degree increase in temperature? Give answer to the nearest hundredth ounce.

8–13 Compound Units

In this unit, you have worked with basic units of length, area, volume, capacity, and weight. Methods of expressing equivalent measures between smaller and larger basic units of measure have been presented.

In actual practice, it is often required to use a combination of the basic units in solving problems. *Compound units* are the products or quotients of two or more basic units. Many quantities or rates are expressed as compound units. Following are examples of compound units.

EXAMPLES

1. Speed can be expressed as miles per hour. *Per* indicates division. Miles per hour can be written as mi/hr, $\frac{mi}{hr}$, or mph.

2. Pressure can be expressed as pounds per square inch. Pounds per square inch can be written as lb/sq in, $\frac{lb}{sq\ in}$, or psi. Since square inch may be written as in^2, pounds per square inch may be written as lb/in^2.

Expressing Simple Compound Unit Equivalents

Compound unit measures are converted to smaller or larger equivalent compound unit measures using unity fractions. Following are examples of simple equivalent compound unit conversions.

EXAMPLE

Express 2,870 pounds per square foot as pounds per square inch.

Multiply 2,870 lb/sq ft by unity fraction $\frac{1\ sq\ ft}{144\ sq\ in}$.

$$\frac{2,870\ lb}{sq\ ft} \times \frac{1\ sq\ ft}{144\ sq\ in} \approx 19.9\ lb/sq\ in\ Ans$$

EXERCISE 8–13A

Express each simple compound unit measure as indicated. Round each answer to the same number of significant digits as in the original quantity.

1. 61.3 mi/hr = ? mi/min
2. 7.025 lb/sq in = ? lb/sq ft
3. 2,150 rev/min = ? rev/sec
4. $1.69/gal = ? $/qt

5. 0.260 ft/sec = ? ft/hr
6. 0.500 lb/cu in = ? lb/cu ft
7. 0.16 short ton/sq in = ? lb/sq in
8. 538 cu in/sec = ? cu ft/sec

Expressing Complex Compound Unit Equivalents

Exercise 8–13A involves the conversion of simple compound units. The conversion of only one unit is required; the other unit remains the same. In problem 1, the hour unit is converted to a minute unit; the mile unit remains the same. In problem 2, the square inch unit is converted to a square foot unit; the pound unit remains the same.

Problems often involve the conversion of more than one unit in their solutions. It is necessary to convert both units of a compound unit quantity to smaller or larger units. Following is an example of complex compound unit conversions in which both units are converted.

EXAMPLE •———————————————————————————————————————

Express 0.283 short tons per square foot as pounds per square inch.

Multiply 0.283 ton by unity fractions $\dfrac{2,000\ lb}{1\ ton}$

and $\dfrac{1\ sq\ ft}{144\ sq\ in}$.

$$\frac{0.283\ \cancel{ton}}{1\ sq\ \cancel{ft}} \times \frac{2,000\ lb}{1\ \cancel{ton}} \times \frac{1\ sq\ \cancel{ft}}{144\ sq\ in} \approx 3.93\ lb/sq\ in\ \textit{Ans}$$

Calculator Application

.283 ⊠ 2000 ÷ 144 = 3.930555556, 3.93 lb/sq in (rounded to 3 significant digits) *Ans*

———•

✓ **EXERCISE 8–13B**
——

Express each complex compound unit measure as indicated. Round each answer to the same number of significant digits as in the original quantity.

1. 73.9 lb/sq in = ? short tons/sq ft
2. 4,870 ft/min = ? mi/hr
3. $2.32/gal = ? cents/pt

4. 3.5 short ton/sq ft = ? lb/sq in
5. 63.8 cu in/sec = ? cu ft/hr
6. 53.6 mi/hr = ? ft/sec

8–14   Compound Units Practical Applications

EXAMPLE •———————————————————————————————————————

Aluminum weighs 0.0975 pound per cubic inch. What is the weight, in short tons, of 26.8 cubic feet of aluminum?

Find the weight in tons per 1 cubic foot. Multiply 0.0975 lb/cu in by unity fractions $\dfrac{1,728\ cu\ in}{1\ cu\ ft}$ and $\dfrac{1\ ton}{2,000\ lb}$.

$$\frac{0.0975\ \cancel{lb}}{\cancel{cu\ in}} \times \frac{1,728\ \cancel{cu\ in}}{1\ cu\ ft} \times \frac{1\ ton}{2,000\ \cancel{lb}} = 0.08424\ \frac{ton}{cu\ ft}$$

Find the weight of 26.8 cubic feet. Multiply 0.08424 ton/cu ft by 26.8 cu ft.

$$\frac{0.08424\ ton}{\cancel{cu\ ft}} \times 26.8\ \cancel{cu\ ft} \approx 2.26\ tons\ \textit{Ans}$$

Calculator Application

.0975 ⊠ 1728 ÷ 2000 ⊠ 26.8 = 2.257632, 2.26 tons (rounded to 3 significant digits) *Ans*

———•

EXERCISE 8–14   omit
——

Solve the following problems.

1. Unbroken anthracite coal weighs 93.6 lb/ft³. Find the weight of 3.5 cubic yards of coal in
   (a) short tons and
   (b) long tons.

   Round each answer to 2 significant digits.

2. A British thermal unit (Btu) is the amount of heat required to raise the temperature of one pound of water one degree Fahrenheit. A building has 12 air-conditioning units. Each unit

removes 24,000 British thermal units of heat per hour (Btu/hr). How many Btu are removed from the building in 10 minutes?

3. With a concrete curb machine, a two-person crew can install 47.5 linear feet of edging per hour. Each linear foot takes 0.25 ft^3 of concrete.

   (a) How many cubic feet of concrete will be needed to edge 1,217 linear feet of land?

   (b) If each cubic foot of wet concrete weighs 194.8 lb, what is the total weight of the wet concrete used for this curbing? Round the answer to the nearest hundred pounds.

   (c) How long will it take the two-person crew to complete the job?

   (d) If the price for natural gray curb is $3.50/lf with an average cost of $0.40/lf for materials, how much will this curbing cost?

4. An oil spill needs to be treated with a bacterium culture at the rate of 1 oz of culture per 100 cubic feet of oil. If the spill is 786,000 barrels, how much culture will be needed? 1 barrel is 31.5 gal. Give the answer in pounds rounded to three decimal places.

5. Hot air passes through a duct at the rate of 675 cubic inches per second. Find the number of cubic feet of hot air passing through the duct in 18.5 minutes. Round the answer to the nearest cubic foot.

6. The speed of sound in air is 1,090 feet per second at 32°F. How many miles does sound travel in 0.225 hour? Round the answer to the nearest mile.

7. Materials expand when heated. Different materials expand at different rates. Mechanical and construction technicians must often consider material rates of expansion. The amount of expansion for short lengths of material is very small. Expansion is computed if a product is made to a high degree of precision and is subjected to a large temperature change. Also, expansion is computed for large structures because of the long lengths of structural members used. The table in Figure 8–20 lists the expansion per inch for each Fahrenheit degree rise in temperature for a 1-inch length of material. For example, the expansion of a 1-inch length of aluminum is expressed as $0.00001244 \frac{in/in}{1°F}$.

Material	Expansion of One Inch of Material in One Fahrenheit Degree Temperature Increase
Aluminum	0.00001244 inch
Copper	0.00000900 inch
Structural Steel	0.00000722 inch
Brick	0.00000300 inch
Concrete	0.00000800 inch

Figure 8–20

Find the total expansion for each of the materials listed in the table in Figure 8–21. Express the answer to 3 significant digits.

	Material	Length Before Temperature Increase	Fahrenheit Temperature Change	Total Expansion in Inches
a.	Copper Cable	315.2 ft	From 35.3°F to 92.5°F	
b.	Steel I Beam	50.7 ft	From 5.27°F to 90.3°F	
c.	Brick Wall	224.4 ft	From 20.8°F to 81.6°F	
d.	Aluminum Wire	415.0 ft	From 43.2°F to 105.7°F	
e.	Concrete	120.5 ft	From 15.3°F to 89.2°F	

Figure 8–21

UNIT EXERCISE AND PROBLEM REVIEW

EQUIVALENT CUSTOMARY UNITS OF LINEAR MEASURE

Express each length as indicated. Round the answers to 3 decimal places when necessary.

1. $25\frac{1}{2}$ inches as feet

2. 16.25 feet as yards

3. 3,960 feet as miles

4. 78 inches as feet and inches

5. 47 feet as yards and feet

6. $7\frac{1}{4}$ feet as inches

7. $8\frac{1}{6}$ yards as feet

8. 0.6 mile as feet

9. $2\frac{1}{4}$ yards as inches

10. $5\frac{1}{2}$ yards as feet and inches

11. $\frac{1}{12}$ mile as yards and feet

12. 6.2 yards as feet and inches

ARITHMETIC OPERATIONS WITH CUSTOMARY COMPOUND NUMBERS

Perform the indicated arithmetic operation. Express the answer in the same units as those given in the exercise. Regroup the answer when necessary.

13. 5 ft 7 in + 6 ft 8 in

14. 4 yd 2 ft + 5 yd 2 ft

15. 6 ft $9\frac{1}{2}$ in + 3 ft $4\frac{1}{4}$ in

16. 10 ft 7 in − 3 ft 4 in

17. 14 yd $\frac{1}{2}$ ft − 11 yd 2 ft

18. 6 ft $1\frac{3}{4}$ in − 4 ft $8\frac{1}{4}$ in

19. 5 ft 3 in × 5

20. 6 ft $8\frac{1}{2}$ in × 3

21. 3 yd 2 ft 7 in × 4

22. 20 ft 10 in ÷ 5

23. 11 yd 2 ft ÷ 3

24. 3 yd 2 ft 4 in ÷ 2

EQUIVALENT CUSTOMARY UNITS OF AREA MEASURE

Express each customary area measure in the indicated unit. Round each answer to the same number of significant digits as in the original quantity.

25. 504 sq in as sq ft

26. 128 sq ft as sq yd

27. 4.08 sq ft as sq in

28. 2,480 acres as sq mi

29. 217,800 sq ft as acres

30. 0.2600 acres as sq ft

31. 5.33 sq yd as sq ft

32. 0.275 sq mi as acres

33. 0.080 sq yd as sq in

34. 0.0108 sq mi as sq ft

EQUIVALENT CUSTOMARY UNITS OF VOLUME MEASURE

Express each volume in the unit indicated. Round each answer to the same number of significant digits as in the original quantity.

35. 4,700 cu in as cu ft

36. 215 cu ft as cu yd

37. 0.712 cu ft as cu in

38. 12.34 cu yd as cu ft

39. 0.5935 cu ft as cu in

40. 19.80 cu ft as cu yd

41. 20,000 cu in as cu yd

42. 0.030 cu yd as cu in

EQUIVALENT CUSTOMARY UNITS OF CAPACITY MEASURE

Express each capacity in the unit indicated. Round each answer to the same number of significant digits as in the original quantity.

43. 15.3 gal as qt

44. 31 oz as pt

45. 6.5 pt as qt

46. 6.2 qt as gal

47. 1.04 gal as cu in

48. 84 cu ft as gal

49. 0.20 gal as pt

50. 1.40 qt as oz

EQUIVALENT CUSTOMARY UNITS OF WEIGHT MEASURE

Express each weight in the unit indicated. Round each answer to the same number of significant digits as in the original quantity.

51. 34 oz as lb

52. 0.060 lb as oz

53. 48,400 lb as long tons

54. 7,800 lb as short tons

55. 0.660 short tons as lb

56. 1.087 long tons as lb

SIMPLE COMPOUND UNIT MEASURES

Express each simple compound unit measure as indicated. Round each answer to the same number of significant digits as in the original quantity.

57. 8.123 lb/sq in = ? lb/sq ft

58. 50.7 mi/hr = ? mi/min

59. 618 cu ft/sec = ? cu ft/hr

60. 2,090 rev/min = ? rev/sec

COMPLEX COMPOUND UNIT MEASURES

Express each complex compound unit measure as indicated. Round each answer to the same number of significant digits as in the original quantity.

61. 5,190 ft/min = ? mi/hr

62. $3.81/gal = ? cents/pt

63. 57.2 cu in/sec = ? cu ft/hr

64. 62.9 mi/hr = ? ft/sec

PRACTICAL APPLICATIONS PROBLEMS

Solve the following problems.

65. The first-floor plan of a ranch house is shown in Figure 8–22. Determine distances A, B, C, and D in feet and inches.

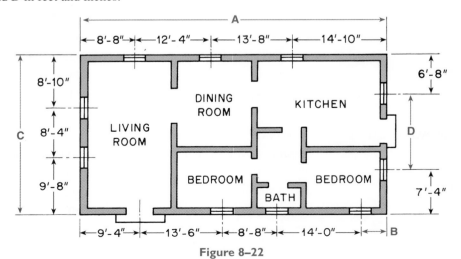

Figure 8–22

66. A bolt of fabric contains $80\frac{1}{2}$ yards of fabric. The following lengths of fabric are sold: 2 lengths each 5 yards 2 feet long, 3 lengths each 8 yards $1\frac{1}{2}$ feet long, and 5 lengths each 6 yards 2 feet long. What length of fabric does the bolt now contain? Express the answer in yards and feet.

67. How many strips, each having an area of 36 square inches, can be sheared from a sheet of aluminum that measures 6 square feet?

68. A painter computes the total interior wall surface of a building as 330 square yards after allowing for windows and doors. Two coats of paint are required for the job. One gallon of paint covers 500 square feet. How many gallons of paint are required? Round answer to the nearest gallon.

69. Hot air passes through a duct at the rate of 830 cubic inches per second. Compute the number of cubic feet of hot air that passes through the duct in 2.5 minutes. Round answer to 2 significant digits.

70. A cord is a unit of measure of cut fuel wood equal to 128 cubic feet. If wood is burned at the rate of $\frac{1}{2}$ cord per week, how many weeks would a stack of wood measuring 12 cubic yards last? Round answer to one decimal place.

71. Common brick weighs 112 pounds per cubic foot. How many cubic yards of brick can be carried by a truck whose maximum carrying load is rated at 12 short tons? Round answer to 2 significant digits.

72. A solution contains 5% acid and 95% water. How many quarts of the solution can be made with 3.8 ounces of acid? Round answer to 1 decimal place.

73. A water tank that has a volume of 4,550 cubic feet is $\frac{3}{4}$ full. How many gallons of water are contained in the tank? Round answer to the nearest hundred gallons.

74. Carpet is installed in the hallways of a building. The cost of carpet and installation is $32.50 per square yard. A total length of 432.0 feet of carpet 5.0 feet wide is required. What is the total cost of carpet and installation? (Number of square feet = length in feet × width in feet)

75. The interior walls of a building are to be covered with plasterboard. After making allowances for windows and doors, a contractor estimates that 438 square yards of wall area are to be covered. One sheet of plasterboard has a surface area of 24.0 square feet. Allowing 15% for waste, how many sheets of plasterboard are required for this job? Round answer to the nearest whole sheet.

UNIT 9 ⠿ Metric Measurement Units

OBJECTIVES

After studying this unit you should be able to

- express lengths as smaller or larger metric linear numbers.

- perform arithmetic operations with metric linear units.

- select appropriate linear metric units in various applications.

- express given metric length, area, and volume measures in larger and smaller units.

- express given metric capacity and weight units as larger and smaller units.

- solve practical applied metric length, area, volume, capacity, and weight problems.

- express metric compound unit measures as equivalent compound unit measures.

- solve practical applied compound unit measures.

- convert between customary measures and metric measures.

The International System of Units, called the SI metric system, is the primary measurement system used by all countries except the United States, Liberia, and Myanmar. In the United States, the customary units tend to be used in areas such as construction, real estate transactions, and retail trade. Other areas, such as automotive maintenance, nursing, other health care areas, and biotechnology use the metric system.

Linear Measure

A linear measurement is a means of expressing the distance between two points; it is the measurement of lengths. Most occupations require the ability to compute linear measurements and to make direct length measurements.

A drafter computes length measurements when drawing a machined part or an architectural floor plan, an electrician determines the amount of cable required for a job, a welder calculates the length of material needed for a weldment, a printer "figures" the number of pieces that can be cut from a sheet of stock, a carpenter calculates the total length of baseboard required for a building, and an automobile technician computes the amount of metal to be removed for a cylinder re-bore.

9–1 Metric Units of Linear Measure

An advantage of the metric system is that it allows easy, fast computations. Since metric system units are based on powers of 10, figuring is simplified. To express a certain metric unit as a larger or smaller metric unit, all that is required is to move the decimal point a proper number of places to the left or right. Metric system units are also easy to learn.

The metric system does not require difficult conversions as in the customary system. It is easier to remember that 1000 meters equal 1 kilometer than to remember that 1760 yards equal 1 mile.

Many occupations require working with metric units of linear measure. A manufacturing technician may measure and compute using millimeters. An architectural drafter and a construction technician may use meters and centimeters. Kilometers are used to measure relatively long distances such as those traveled by a vehicle per unit of time, such as kilometers per hour.

Millimeters are used to measure small lengths often requiring a high degree of precision. The thickness of a spoon, fork, and compact disc are approximately 1 millimeter. The thickness of your pen or pencil is a little less than 1 centimeter; this book is about 3 centimeters thick. Most home kitchen counters are roughly 1 meter high and doors are 2 meters high. The length of 10 football fields is about 1 kilometer.

EXAMPLES

For each of the following, an estimate of length with the appropriate unit is given.

1. The length of your pen is about 15 centimeters.
2. The thickness of each of your fingernails is less than one millimeter.
3. The length of most automobiles is approximately 5 meters.
4. The thickness of a saw blade is about 1 millimeter.
5. The thickness of a brick is about 6 centimeters.
6. The room ceiling height in a typical house is between 2 and 2.5 meters.
7. Most automobiles are capable of exceeding 160 kilometers per hour.

EXERCISE 9–1A

Select the linear measurement unit most appropriate for each of the following. Identify as millimeter, centimeter, meter, or kilometer.

1. The height of a drinking glass.
2. The distance from New York City to Chicago.
3. The length of a bus.
4. The thickness of a photographic print.
5. The length of your index finger.
6. The thickness of a blade of grass.
7. The length of the Mississippi River.
8. The width of a house.

EXERCISE 9–1B

For each of the following, write the most appropriate metric unit; identify it as millimeter, centimeter, meter, or kilometer.

1. Most handheld calculators are about 15 ? long.
2. The Empire State Building is approximately 442 ? high.

3. The thickness of a hardcover book is about 3 ? thick.

4. Many young men have an 80 ? waist.

5. The diameter of a large safety pin is approximately 0.5 ?.

6. The speed of light is roughly 300 000 ? per second.

7. My driveway is about 45 ? long.

8. Computer monitor screens are often about 28 ? wide.

9. Most kitchen counter tops are about 30 ? thick.

10. The handle of a hammer is about 20 ? long.

Prefixes and Symbols for Metric Units of Length

The following metric power-of-10 prefixes are based on the meter:

milli means one thousandth (0.001)   *deka* means ten (10)

centi means one hundredth (0.01)   *hecto* means hundred (100)

deci means one tenth (0.1)   *kilo* means thousand (1 000)

The table shown in Figure 9–1 lists the metric units of length with their symbols. These units are based on the meter. Observe that each unit is 10 times greater than the unit directly above it.

METRIC UNITS OF LINEAR MEASURE	
1 millimeter (mm) = 0.001 meter (m)	1 000 millimeters (mm) = 1 meter (m)
1 centimeter (cm) = 0.01 meter (m)	100 centimeters (cm) = 1 meter (m)
1 decimeter (dm) = 0.1 meter (m)	10 decimeters (dm) = 1 meter (m)
1 meter (m) = 1 meter (m)	1 meter (m) = 1 meter (m)
1 dekameter (dam) = 10 meters (m)	0.1 dekameter (dam) = 1 meter (m)
1 hectometer (hm) = 100 meters (m)	0.01 hectometer (hm) = 1 meter (m)
1 kilometer (km) = 1 000 meters (m)	0.001 kilometers (km) = 1 meter (m)

Figure 9–1

The most frequently used metric units of length are the kilometer (km), meter (m), centimeter (cm), and millimeter (mm). In actual applications, the dekameter (dam) and hectometer (hm) are not used. The decimeter (dm) is seldom used. The metric prefixes for very large and very small numbers will be studied in Unit 9–17.

To make numbers easier to read they may be divided into groups of three, separated by spaces (or thin spaces), as in 12 345, but not commas or points. This applies to digits on both sides of the decimal marker (0.901 234 56). Numbers with four digits may be written either with the space (5 678) or without it (5678).

This practice not only makes large numbers easier to read, but also allows all countries to keep their custom of using either a point or a comma as decimal marker. For example, engine size in the United States is written as 3.2 L and in Germany as 3,2 L. The space prevents possible confusion and sources of error.

9–2   Expressing Equivalent Units within the Metric System

To express a given unit of length as a larger unit, move the decimal point a certain number of places to the left. To express a given unit of length as a smaller unit, move the decimal point a certain number of places to the right. The exact procedure of moving decimal points is shown in the following examples. Refer to the metric units of linear measure table shown in Figure 9–1.

EXAMPLES •————————————————————————————————

1. Express 65 decimeters as meters.

   Since a meter is the next larger unit to a decimeter, move the decimal point 1 place to the left.   65.
                                                                                                   ↖

   In moving the decimal point 1 place to the left, you are actually dividing by 10.

   65 dm = 6.5 m *Ans*

2. Express 0.28 decimeter as centimeters.

   Since a centimeter is the next smaller unit to a decimeter, move the decimal point 1 place to the right.   0.28
                                                                                                   ↗

   In moving the decimal point 1 place to the right, you are actually multiplying by 10.

   0.28 dm = 2.8 cm *Ans*

3. Express 0.378 meter (m) as millimeters (mm).

   Expressing meters as millimeters involves 3 steps.

   0.378 m = 3.78 dm = 37.8 cm = 378 mm
          ①   ②   ③

   Since a millimeter is 3 smaller units from a meter, move the decimal point 3 places to the right. In moving the decimal point 3 places to the right, you are actually multiplying by 10^3 ($10 \times 10 \times 10$), or 1 000.

   0.378

   0.378 m = 378 mm *Ans*

4. Express 2 700 centimeters as meters.

   Since a meter is 2 larger units from a centimeter, move the decimal point 2 places to the left   2 700.
                                                                                                   ↖

   2 700 cm = 27 m *Ans*

   Notice the answer is 27 meters, not 27.00 meters. Because 2 700 centimeters has 2 significant digits, the answer is rounded to 2 significant digits.

——•

    Moving the decimal point a certain number of places to the left or right is the most practical way of expressing equivalent metric units. However, the unity fraction method can also be used in expressing equivalent metric units.

EXAMPLES •————————————————————————————————

1. Express 0.378 kilometer as meters.

   Multiply 0.378 kilometer by the unity fraction $\dfrac{1\,000 \text{ m}}{1 \text{ km}}$. $0.378 \text{ k\!m} \times \dfrac{1\,000 \text{ m}}{1 \text{ k\!m}} = 378 \text{ m}$ *Ans*

2. Express 237 millimeters as decimeters.

   Since a decimeter is 10×10 or 100 times greater than a millimeter, multiply 237 millimeters by the unity fraction $\dfrac{1 \text{ dm}}{100 \text{ mm}}$. $237 \text{ m\!m} \times \dfrac{1 \text{ dm}}{100 \text{ m\!m}} = 2.37 \text{ dm}$ *Ans*

——•

EXERCISE 9–2

Express these lengths in meters. Metric units of linear measure are given in the table in Figure 9–1.

1. 34 decimeters
2. 4,320 millimeters
3. 0.05 kilometers
4. 2.58 dekameters

5. 335 millimeters
6. 95.6 centimeters
7. 0.84 hectometers
8. 402 decimeters

9. 1.05 kilometers
10. 56.9 millimeters
11. 14.8 dekameters
12. 2,070 centimeters

Express each value as indicated.

13. 7 decimeters as centimeters
14. 28 millimeters as centimeters
15. 5 centimeters as millimeters
16. 0.38 meter as dekameters
17. 2.4 kilometers as hectometers
18. 27 dekameters as meters
19. 310.6 decimeters as meters
20. 3.9 hectometers as kilometers

21. 735 millimeters as decimeters
22. 8.5 meters as centimeters
23. 616 meters as kilometers
24. 404 dekameters as decimeters
25. 0.08 kilometers as decimeters
26. 8,975 millimeters as dekameters
27. 0.06 hectometers as centimeters
28. 302 decimeters as kilometers

9–3   Arithmetic Operations with Metric Lengths

Arithmetic operations are performed with metric denominate numbers the same as with customary denominate numbers. Compute the arithmetic operations, then write the proper metric unit of measure.

EXAMPLES •

1. 3.2 m + 5.3 m = 8.5 m
2. 20.65 mm − 16.32 mm = 4.33 mm
3. 7.225 cm × 10.60 = 76.59 cm, rounded to 4 significant digits
4. 24.8 km ÷ 4.625 = 5.36 km, rounded to 3 significant digits

As with the customary system, only like units can be added or subtracted.

9–4   Metric Linear Measure Practical Applications

EXAMPLE •

A structural steel fabricator cuts 25 pieces each 16.2 centimeters long from a 6.35-meter length of channel iron. Allowing 4 millimeters waste for each piece, find the length in meters of channel iron left after all 25 pieces have been cut.

Express each 16.2-centimeter piece as meters.	16.2 cm = 0.162 m
Find the total length of 25 pieces.	25 × 0.162 m = 4.05 m
Express each 4 millimeters of waste as meters.	4 mm = 0.004 m
Find the total amount of waste.	25 × 0.004 m = 0.1 m
Find the amount of channel iron left.	6.35 m − (4.05 m + 0.1 m) = 2.2 m *Ans*

Calculator Application

6.35 $\boxed{-}$ $\boxed{(}$ 25 $\boxed{\times}$ .162 $\boxed{+}$ 25 $\boxed{=}$ .004 $\boxed{)}$ $\boxed{=}$ 2.2

2.2 m of channel iron are left *Ans*

EXERCISE 9–4

Solve the following problems.

1. Three pieces of stock measuring 3.2 decimeters, 9 centimeters, and 7 centimeters in length are cut from a piece of fabric 0.6 meter long. How many centimeters long is the remaining piece?

2. Find, in meters, the total length of the wall section shown in Figure 9–2.

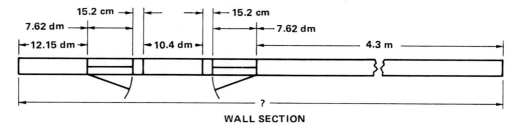

WALL SECTION

Figure 9–2

3. Preshrunk fabric is shrunk by the manufacturer. A length of fabric measures 150 meters before shrinking. If the shrinkage is 7 millimeters per meter of length, what is the total length of fabric after shrinking?

4. Find dimensions A and B, in centimeters, of the pattern shown in Figure 9–3.

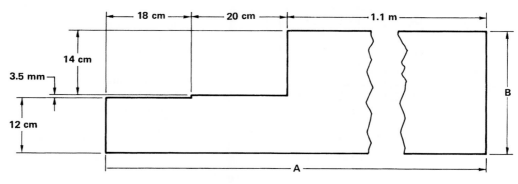

Figure 9–3

5. Three different parts, each of a different material, are made in a manufacturing plant. Refer to the table in Figure 9–4. Compute the cost of material per piece and the cost of a production run of 2,500 pieces of each part including a 15% waste and scrap allowance.

Part Number	Length of Each Piece in Millimeters	Cost per Meter of Material	Cost per Piece	Cost per 2500 Pieces including 15% Allowance (Round to Nearer Dollar)
105-AD	86.00	$2.20		
106-AD	51.00	$2.80		
107-AD	90.00	$3.03		

Figure 9–4

9–5   Metric Units of Surface Measure (Area)

The method of computing surface measure is the same in the metric system as in the customary system. The product of two linear measures produces square measure. The only difference is in the use of metric rather than customary units. For example, 2 centimeters × 4 centimeters = 8 square centimeters.

Surface measure symbols are expressed as linear measure symbols with an exponent of 2. For example, 4 square meters is written as 4 m^2, and 25 square centimeters is written as 25 cm^2.

The basic unit of area is the square meter. The surface enclosed by a square that is 1 meter on a side is 1 square meter. The surface enclosed by a square that is 1 centimeter on a side is 1 square centimeter. Similar meanings are attached to the other square units of measure.

A reduced drawing of a square decimeter and a square meter is shown in Figure 9–5. Observe that 1 linear meter equals 10 linear decimeters, but 1 square meter equals 10 decimeters × 10 decimeters or 100 square decimeters.

The table in Figure 9–6 shows the units of surface measure with their symbols. These units are based on the square meter. Notice that each unit in the table is 100 times greater than the unit directly above it.

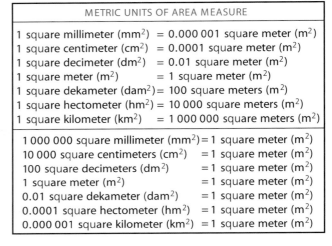

METRIC UNITS OF AREA MEASURE	
1 square millimeter (mm^2)	= 0.000 001 square meter (m^2)
1 square centimeter (cm^2)	= 0.0001 square meter (m^2)
1 square decimeter (dm^2)	= 0.01 square meter (m^2)
1 square meter (m^2)	= 1 square meter (m^2)
1 square dekameter (dam^2)	= 100 square meters (m^2)
1 square hectometer (hm^2)	= 10 000 square meters (m^2)
1 square kilometer (km^2)	= 1 000 000 square meters (m^2)
1 000 000 square millimeter (mm^2)	= 1 square meter (m^2)
10 000 square centimeters (cm^2)	= 1 square meter (m^2)
100 square decimeters (dm^2)	= 1 square meter (m^2)
1 square meter (m^2)	= 1 square meter (m^2)
0.01 square dekameter (dam^2)	= 1 square meter (m^2)
0.0001 square hectometer (hm^2)	= 1 square meter (m^2)
0.000 001 square kilometer (km^2)	= 1 square meter (m^2)

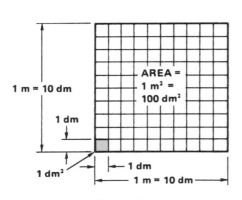

Figure 9–5

Figure 9–6

Expressing Metric Area Measure Equivalents

To express a given metric unit of area as the next larger metric unit of area, move the decimal point two places to the left. Moving the decimal point two places to the left is actually a shortcut method of dividing by 100.

EXAMPLE •

Express 840.5 square decimeters (dm^2) as square meters (m^2).

Since a square meter is the next larger unit to a square decimeter, move the decimal point 2 places to the left:

8 40.5

840.5 dm^2 = 8.405 m^2 *Ans*

In moving the decimal point 2 places to the left, you are actually dividing by 100.

To express a given metric unit of area as the next smaller metric unit of area, move the decimal point two places to the right. Moving the decimal point two places to the right is actually a shortcut method of multiplying by 100.

EXAMPLES •——————————————————————————————————————

1. Express 46 square centimeters (cm²) as square millimeters (mm²).

   Since a square millimeter is the next smaller unit to a square centimeter, move the decimal point 2 places to the right.

   $$46.00$$

   $$46 \text{ cm}^2 = 4600 \text{ mm}^2 \text{ } Ans$$

   In moving the decimal point 2 places to the right, you are actually multiplying by 100.

2. Express 0.08 square kilometer (km²) as square meters (m²).

   Since a square meter is 3 units smaller than a square kilometer, the decimal point is moved 3×2 or 6 places to the right.

   $$0.080\,000$$

   $$0.08 \text{ km}^2 = 80\,000 \text{ m}^2 \text{ } Ans$$

   In moving the decimal point 6 places to the right, you are actually multiplying by $100 \times 100 \times 100$ or $1\,000\,000$.

EXERCISE 9–5 ——————————————————————————————————————

Express each area as indicated. Metric units of area measure are given in the table in Figure 9–6.

1. 500 square millimeters as square centimeters
2. 82 square decimeters as square meters
3. 4900 square centimeters as square decimeters
4. 15.6 square hectometers as square kilometers
5. 10000 square millimeters as square decimeters
6. 7300 square centimeters as square meters
7. 350000 square millimeters as square meters
8. 2700000 square meters as square kilometers
9. 8 square meters as square decimeters
10. 23 square centimeters as square millimeters
11. 0.48 square meter as square centimeters
12. 0.06 square meter as square millimeters
13. 2.08 square decimeters as square centimeters
14. 0.009 square kilometer as square meters
15. 0.044 square kilometer as square decimeters

9–6   Arithmetic Operations with Metric Area Units

Arithmetic operations are performed with metric area denominate numbers the same as with customary area denominate numbers. Compute the arithmetic operations, then write the proper metric unit of surface measure.

EXAMPLES •——————————————————————————————————————

1. $42.87 \text{ cm}^2 + 16.05 \text{ cm}^2 = 58.92 \text{ cm}^2 \text{ } Ans$ 3. $6.15 \times 30.8 \text{ mm}^2 \approx 189 \text{ mm}^2 \text{ } Ans$

2. $7.62 \text{ m}^2 - 4.06 \text{ m}^2 = 3.56 \text{ m}^2 \text{ } Ans$ 4. $12.95 \text{ km}^2 \div 4.233 \approx 3.059 \text{ km}^2 \text{ } Ans$

   As with the customary system, only like units can be added or subtracted.

9–7   Metric Area Measure Practical Applications

EXAMPLE •————————————————————————————————

A flooring contractor measures the floor of a room and calculates the area as 42.2 square meters. The floor is to be covered with tiles. Each tile measures 680.0 square centimeters. Allowing 5% for waste, determine the number of tiles needed.

Express each 680.0 square centimeter tile as square meters.	$680.0 \text{ cm}^2 = 0.0680 \text{ m}^2$
Find the number of tiles needed to cover 42.2 m².	$42.2 \text{ m}^2 \div 0.0680 \text{ m}^2 \approx 621$ 621 tiles
Find the number of tiles needed allowing 5% for waste.	$1.05 \times 621 \text{ tiles} \approx 652 \text{ tiles } Ans$

Calculator Application

42.2 ÷ .068 × 1.05 = 651.6176471, 652 tiles *Ans*

EXERCISE 9–7

Solve the following problems.

1. How many pieces, each having an area of 450 square centimeters, can be cut from an aluminum sheet that measures 2.7 square meters?

2. A state purchases 3 parcels of land that are to be developed into a park. The respective areas of the parcels are 16 000 square meters, 21 000 square meters, and 23 000 square meters. How many square kilometers are purchased for the park?

3. Acid soil is corrected (neutralized) by liming. A soil sample shows that 0.4 metric ton of lime per 1 000 square meters of a certain soil is required to correct an acid condition. How many metric tons of lime are needed to neutralize 0.3 square kilometer of soil?

4. An assembly consists of 4 metal plates. The respective areas of the plates are 500 square centimeters, 700 square centimeters, 18 square decimeters, and 0.15 square meter. Find the total surface measure of the 4 plates in square meters.

5. A roll of gasket material has a surface measure of 2.25 square meters. Gaskets, each requiring 1,200 square centimeters of material, are cut from the roll. Allow 20% for waste. Find the number of gaskets that can be cut from the roll.

9–8   Metric Units of Volume (Cubic Measure)

The method of computing volume measure is the same in the metric system as in the customary system. The product of three linear measures produces cubic measure. The only difference is in the use of metric rather than customary units. For example, 2 centimeters × 3 centimeters × 5 centimeters = 30 cubic centimeters.

Volume measure symbols are expressed as linear measure symbols with an exponent of 3. For example, 6 cubic meters is written as 6 m³, and 45 cubic decimeters is written as 45 dm³.

The basic unit of volume is the cubic meter. The volume of a cube having sides 1 meter long is 1 cubic meter. The volume of a cube having sides 1 decimeter long is 1 cubic decimeter. Similar meanings are attached to the cubic centimeter and cubic millimeter.

A reduced illustration of a cubic meter and a cubic decimeter is shown in Figure 9–7 on page 249. Observe that 1 linear meter equals 10 linear decimeters, but 1 cubic meter equals 10 decimeters × 10 decimeters × 10 decimeters or 1 000 cubic decimeters.

The table in Figure 9–8 shows the units of volume measure with their symbols. These units are based on the cubic meter. Notice that each unit in the table is 1 000 times greater than the unit directly above it.

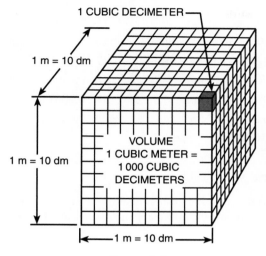

METRIC UNITS OF VOLUME MEASURE	
1 cubic millimeter (mm³) = 0.000 000 001 cubic meter (m³)	
1 cubic centimeter (cm³) = 0.000 001 cubic meter (m³)	
1 cubic decimeter (dm³) = 0.001 cubic meter (m³)	
1 cubic meter (m³) = 1 cubic meter (m³)	
1 000 000 000 cubic millimeters (mm³)	= 1 cubic meter (m³)
1 000 000 cubic centimeters (cm³)	= 1 cubic meter (m³)
1 000 cubic decimeters (dm³)	= 1 cubic meter (m³)
1 cubic meter (m³)	= 1 cubic meter (m³)

Figure 9–7

Figure 9–8

Expressing Metric Volume Measure Equivalents

To express a given unit of volume as the next larger unit, move the decimal point three places to the left. Moving the decimal point three places to the left is actually a shortcut method of dividing by 1 000.

EXAMPLES •————————————————————————————

1. Express 1 450 cubic millimeters (mm³) as cubic centimeters (cm³).

Since a cubic centimeter is the next larger unit to a cubic millimeter, move the decimal point 3 places to the left.

$$1\,450.$$

$$1450 \text{ mm}^3 = 1.45 \text{ cm}^3 \text{ } Ans$$

In moving the decimal point 3 places to the left, you are actually dividing by 1,000.

2. Express 27 000 cubic centimeters (cm³) as cubic meters (m³).

Since a cubic meter is 2 units larger than a cubic centimeter, the decimal point is moved 2×3 or 6 places to the left.

$$027\,000.$$

$$27000 \text{ cm}^3 = 0.027 \text{ m}^3 \text{ } Ans$$

In moving the decimal point 6 places to the left, you are actually dividing by $1\,000 \times 1\,000$ or $1\,000\,000$.

——•

To express a given unit of volume as the next smaller unit, move the decimal point three places to the right. Moving the decimal point three places to the right is actually a shortcut method of multiplying by 1 000.

EXAMPLE •————————————————————————————

Express 12.6 cubic meters (m³) as cubic decimeters (dm³).

Since a cubic decimeter is the next smaller unit to a cubic meter, move the decimal point 3 places to the right.

$$12.600$$

$$12.6 \text{ m}^3 = 12\,600 \text{ dm}^3 \text{ } Ans$$

In moving the decimal point 3 places to the right, you are actually multiplying by 1 000.

——•

EXERCISE 9–8

Express each volume as indicated. Metric units of volume measure are given in the table in Figure 9–8.

1. $2\,700 \text{ mm}^3 = ? \text{ cm}^3$
2. $4\,320 \text{ cm}^3 = ? \text{ dm}^3$
3. $940 \text{ dm}^3 = ? \text{ m}^3$
4. $80 \text{ cm}^3 = ? \text{ dm}^3$
5. $48\,000 \text{ mm}^3 = ? \text{ dm}^3$
6. $650 \text{ cm}^3 = ? \text{ dm}^3$
7. $150\,000 \text{ dm}^3 = ? \text{ m}^3$

8. $20 \text{ mm}^3 = ? \text{ cm}^3$
9. $70\,000 \text{ mm}^3 = ? \text{ dm}^3$
10. $120\,000 \text{ cm}^3 = ? \text{ m}^3$
11. $5 \text{ dm}^3 = ? \text{ cm}^3$
12. $38 \text{ cm}^3 = ? \text{ mm}^3$
13. $0.8 \text{ m}^3 = ? \text{ cm}^3$
14. $0.075 \text{ dm}^3 = ? \text{ cm}^3$

15. $5.23 \text{ cm}^3 = ? \text{ mm}^3$
16. $0.94 \text{ m}^3 = ? \text{ dm}^3$
17. $1.03 \text{ dm}^3 = ? \text{ cm}^3$
18. $0.096 \text{ m}^3 = ? \text{ cm}^3$
19. $0.106 \text{ dm}^3 = ? \text{ mm}^3$
20. $0.006 \text{ m}^3 = ? \text{ cm}^3$

9–9 Arithmetic Operations with Metric Volume Units

Arithmetic operations are performed with metric volume denominate numbers the same as with customary volume denominate numbers. Compute the arithmetic operations, and then write the proper metric unit of volume.

EXAMPLES

1. $4.37 \text{ m}^3 + 11.52 \text{ m}^3 = 15.89 \text{ m}^3$ *Ans*
2. $280.6 \text{ cm}^3 - 102.9 \text{ cm}^3 = 177.7 \text{ cm}^3$ *Ans*
3. $0.590 \times 1400 \text{ mm}^3 \approx 830 \text{ mm}^3$ *Ans*
4. $126 \text{ dm}^3 \div 6.515 \approx 19.3 \text{ dm}^3$ *Ans*

As with the customary system, only like units can be added or subtracted.

9–10 Metric Volume Practical Applications

EXAMPLE

A total of 340 pieces are punched from a strip of stock that has a volume of 12.8 cubic centimeters. Each piece has a volume of 22.5 cubic millimeters. What percent of the volume of stock is wasted?

Express each 22.5 cubic millimeter piece as cubic centimeters. $22.5 \text{ mm}^3 = 0.0225 \text{ cm}^3$

Find the total volume of 340 pieces. $0.0225 \text{ cm}^3 \times 340 = 7.65 \text{ cm}^3$

Find the amount of stock wasted. $12.8 \text{ cm}^3 - 7.65 \text{ cm}^3 = 5.15 \text{ cm}^3$

Find the percent of stock wasted. $\dfrac{5.15 \text{ cm}^3}{12.8 \text{ cm}^3} \approx 0.402$

$0.402 = 40.2\%, 40.2\%$ waste *Ans*

Calculator Application

.0225 $\boxed{\times}$ 340 $\boxed{=}$ = 7.65

$\boxed{(}$ 12.8 $\boxed{-}$ 7.65 $\boxed{)}$ $\boxed{\div}$ 12.8 $\boxed{=}$ 0.402 343 75, 0.402 rounded 40.2% waste *Ans*

EXERCISE 9–10

Solve each volume exercise.

1. Thirty concrete support bases are required for a construction job. Eighty-two cubic decimeters of concrete are used for each base. Find the total number of cubic meters of concrete needed for the bases. Round answer to 2 significant digits.

2. Before machining, an aluminum piece has a volume of 3.75 cubic decimeters. Machining operations remove 50.0 cubic centimeters from the top. There are 6 holes drilled. Each hole removes 30.0 cubic centimeters. There are 4 grooves milled. Each groove removes 2,500 cubic millimeters of stock. Find the volume of the piece, in cubic decimeters, after the machining operations.

3. Anthracite coal weighs 1.50 kilograms per cubic decimeter. Find the weight of a 5.60 cubic meter load of coal.

4. A total of 620 pieces are punched from a strip of stock that has a volume of 38.6 cubic centimeters. Each piece has a volume of 45.0 cubic millimeters. How many cubic centimeters of strip stock are wasted after the pieces are punched?

5. A magnesium alloy contains the following volumes of each element: 426.5 cubic decimeters of magnesium, 38.7 cubic decimeters of aluminum, 610 cubic centimeters of manganese, and 840 cubic centimeters of zinc. Find the percent composition by volume of each element in the alloy. Round answers to the nearest hundredth percent.

9–11   Metric Units of Capacity

Capacity is a measure of volume. The capacity of a container is the number of units of material that the container can hold.

The metric system uses only one kind of capacity measure; the units are standardized for all types of measure.

In the metric system, the liter is the standard unit of capacity. Measures made in gallons in the customary system are measured in liters in the metric system. In addition to the liter, the milliliter is used as a unit of capacity measure. Liters and milliliters are used for fluids (gases and liquids) and for dry ingredients in recipes.

The relationship between the liter and milliliter is shown in the table in Figure 9–9. Also listed are common metric capacity–cubic measure equivalents.

METRIC UNITS OF CAPACITY MEASURE
1 000 milliliters (mL) = 1 liter (L)
COMMONLY USED CAPACITY–CUBIC MEASURE EQUIVALENTS
1 milliliter (mL) = 1 cubic centimeter (cm³) 1 liter (L)   = 1 cubic decimeter (dm³) 1 liter (L)   = 1 000 cubic centimeters (cm³) 1 000 liters (L) = 1 cubic meter (m³)

Figure 9–9

Expressing Equivalent Metric Capacity Measures

It is often necessary to express given capacity units as either larger or smaller units. The procedure is the same as used with linear, square, and cubic units of measure.

EXAMPLE •—————————————————————————

Express 0.714 liters as milliliters.

Since 1 liter = 1 000 milliliters, multiply 1 000 by 0.714.

1 000 × 0.714 = 714; 0.714 L = 714 mL *Ans*

EXERCISE 9–11

Express each unit of measure as indicated. Round each answer to the same number of significant digits as in the original quantity. Customary and metric units of capacity measure are given in the table in Figure 9–9.

1. $3670 \text{ mL} = ? \text{ L}$
2. $1.2 \text{ L} = ? \text{ mL}$
3. $23.6 \text{ mL} = ? \text{ cm}^3$
4. $3.9 \text{ L} = ? \text{ cm}^3$
5. $5300 \text{ cm}^3 = ? \text{ L}$
6. $218 \text{ cm}^3 = ? \text{ mL}$
7. $0.08 \text{ m}^3 = ? \text{ L}$
8. $650 \text{ L} = ? \text{ m}^3$
9. $83 \text{ dm}^3 = ? \text{ L}$
10. $0.63 \text{ L} = ? \text{ mL}$
11. $7.3 \text{ dm}^3 = ? \text{ L}$
12. $478 \text{ mL} = ? \text{ L}$
13. $29000 \text{ mL} = ? \text{ L}$
14. $0.75 \text{ m}^3 = ? \text{ L}$

9–12   Metric Capacity Practical Applications

EXAMPLE •

An automobile gasoline tank holds 72 liters. Gasoline weighs 0.803 g/cm^3. What is the weight of the gasoline in a full tank? Give the answer in kilograms rounded off to the nearest tenth kilogram.

$$72 \, \cancel{L} \times \frac{1000 \, \cancel{mL}}{1 \, \cancel{L}} \times \frac{1 \, \cancel{cm^3}}{1 \, \cancel{mL}} \times \frac{0.803 \, \cancel{g}}{1 \, \cancel{cm^3}} \times \frac{1 \text{ kg}}{1000 \, \cancel{g}} = 57.816 \text{ kg}, \; 57.8 \text{ kg } Ans$$

EXERCISE 9–12

Solve the following problems.

1. In planning for a banquet, the chef estimates that 150 glasses of orange juice will be needed. If each glass holds 120 mL of juice, how many liters of orange juice should be ordered?

2. The liquid intake of a hospital patient during a specified period of time is 300 mL, 250 mL, 125 mL, 275 mL, 350 mL, 150 mL, and 200 mL. Find the total liter intake of liquid for this time period.

3. An oil-storage tank has a volume of 300,000 barrels (bbl). (1 bbl is 119.25 L.) However, there are 30,000 bbl of sludge at the bottom of the tank. The rest of the tank is full of usable oil. How many liters of oil are in the tank? Round answer to the nearest thousand liters.

4. An automobile engine originally has a displacement of 2.300 liters. The engine is rebored an additional 150.0 cubic centimeters. What is the engine displacement in liters after it is rebored?

5. An empty underground storage tank for unleaded gasoline has a volume of 37.85 m^3. The tank is being filled at the rate of 500 liters per minute. How long will it take to fill the tank? Round the answer to the nearest minute.

6. A bottle contains 2.250 liters of solution. A laboratory technician takes 28 samples from the bottle. Each sample contains 35.0 milliliters of solution. How many liters of solution remain in the bottle?

7. An engine running at a constant speed uses 120 milliliters of gasoline in one minute. How many liters of gasoline are used in 5 hours?

8. A solution contains 17.5 % acid and 82.5 % water. How many liters of the solution can be made with 750 mL of acid? Round the answer to the nearest tenth liter.

9. A water-storage tank has a volume of 135 cubic meters. When the tank is $\frac{1}{3}$ full of water, how many liters of water are required to fill the tank?

9–13 Metric Units of Weight (Mass)

Weight is a measure of the force of attraction of the earth on an object. *Mass* is a measure of the amount of matter contained in an object. The weight of an object varies with its distance from the earth's center. The mass of an object remains the same regardless of its location in the universe.

Scientific applications dealing with objects located other than on the earth's surface are *not* considered in this book. Therefore, the terms *weight* and *mass* are used interchangeably.

In the metric system, the kilogram is the standard unit of mass. Objects that are measured in pounds in the customary system are measured in kilograms in the metric system.

The most common units of metric weight (mass) are listed in the table shown in Figure 9–10.

METRIC UNITS OF WEIGHT (MASS) MEASURE
1000 milligrams (mg) = 1 gram (g)
1000 grams (g)   = 1 kilogram (kg)
1000 kilograms (kg) = 1 metric ton (t)

Figure 9–10

Expressing Equivalent Metric Weight Measures

In both the customary and metric systems, apply the same procedures that are used with other measures. The following examples show the method of expressing given weight units as larger or smaller units.

EXAMPLE •————————————————————————

Express 657 grams as kilograms.

Since 1000 g = 1 kg, divide 657 by 1000.   657 g = 0.657 kg *Ans*
Move the decimal point 3 places to the left.

EXERCISE 9–13

Express each unit of weight as indicated. Round each answer to the same number of significant digits as in the original quantity. Metric units of weight are given in the table in Figure 9–10.

1. 1.72 g = ? mg
2. 890 mg = ? g
3. 2.6 metric tons = ? kg
4. 1230 g = ? kg
5. 2700 kg = ? metric tons
6. 0.6 kg = ? g
7. 0.04 g = ? mg
8. 900 kg = ? metric tons
9. 23 000 mg = ? g
10. 95 g = ? kg

9–14   Metric Weight Practical Applications

EXAMPLE •—————————————————————————————————

A truck delivers 4 prefabricated concrete wall sections to a job site. Each wall section has a volume of 1.28 cubic meters. One cubic meter of concrete weighs 2350 kilograms. How many metric tons are carried on this delivery?

Find the volume of 4 wall sections.	$4 \times 1.28 \text{ m}^3 = 5.12 \text{ m}^3$
Find the weight of 5.12 cubic meters.	$5.12 \times 2350 \text{ kg} = 12032 \text{ kg}$
Find the number of metric tons.	$12032 \text{ kg} \div 1000 \approx 12.0$ metric tons *Ans*

Calculator Application

$4 \boxed{\times} 1.28 \boxed{\times} 2350 \boxed{\div} 1000 \boxed{=} 12.03712$, 12.0 metric tons (rounded to 3 significant digits) *Ans*

———•

EXERCISE 9–14

Solve each problem.

1. What is the total weight, in kilograms, of 3 cases of 425-g cans of peas if each case has 48 cans?

2. An analytical balance is used by laboratory technicians in measuring the following weights: 750 mg, 600 mg, 920 mg, 550 mg, and 870 mg. Find the total measured weight in grams.

3. Three hundred identical strips are sheared from a sheet of steel. The sheet weighs 16.5 kilograms. Find the weight, in grams, of each strip.

4. One cubic meter of aluminum weighs 2.707 metric tons. How many kilograms does 0.155 cubic meter of aluminum weigh? Give answer to the nearest whole kilogram.

5. A force of 760 metric tons is exerted on the base of a steel support column. The base has a cross-sectional area of 12100 cm². How many kilograms of force are exerted per square centimeter of cross-sectional area?

6. A piece of steel weighed 3.75 kg. Its weight was reduced to 3.07 kg by drilling holes in the steel. Each drilled hole removed 42.5 g of steel. How many holes were drilled?

7. A liter of water weighs 1.032 kg at 4°C and 0.988 kg at 50°C. Over this temperature change, what is the average weight decrease in grams per liter for each degree increase in temperature? Give answers to the nearest hundredth gram.

9–15   Compound Units

In this unit, you have worked with basic units of length, area, volume, capacity, and weight. Methods of expressing equivalent measures between smaller and larger basic units of measure have been presented.

In actual practice, it is often required to use a combination of the basic units in solving problems. *Compound units* are the products or quotients of two or more basic units. Many quantities or rates are expressed as compound units. Following are examples of compound units.

EXAMPLES •—————————————————————————————————

1. Speed can be expressed as miles per hour. *Per* indicates division. Kilometers per hour can be written as km/h, $\dfrac{\text{km}}{\text{h}}$, or kph.

2. A bending force or torque is expressed as a newton·meter, which can be written as N·m.

3. Volume flow can be expressed as cubic centimeters per second. Cubic centimeters per second is written as cm^3/s.

Expressing Simple Compound Unit Equivalents

Compound unit measures are converted to smaller or larger equivalent compound unit measures using unity fractions. Following are examples of simple equivalent compound unit conversions.

EXAMPLE

Express 4 700 liters per hour as liters per second.

Multiply 4 700 L/h by unity fraction $\dfrac{1 \text{ hr}}{3\,600 \text{ s}}$.

$$\frac{4\,700 \text{ L}}{\cancel{h}} \times \frac{1\,\cancel{h}}{3\,600 \text{ s}} \approx 1.3 \text{ L/s } Ans$$

EXERCISE 9–15A

Express each simple compound unit measure as indicated. Round each answer to the same number of significant digits as in the original quantity.

1. $129 \text{ g/cm}^2 = ? \text{ g/mm}^2$

2. $53 \text{ km/hr} = ? \text{ km/min}$

3. $0.128 \text{ dm}^3/\text{sec} = ? \text{ m}^3/\text{sec}$

4. $31\,520 \text{ kg/m}^2 = ? \text{ kg/dm}^2$

5. $87.0 \text{ hp/L} = ? \text{ hp/cm}^3$

6. $930 \text{ mg/mm}^2 = ? \text{ g/mm}^2$

7. $\$9.03/\text{kg} = ? \$/\text{g}$

8. $510 \text{ cm/sec} = ? \text{ m/sec}$

Expressing Complex Compound Unit Equivalents

Exercise 9–15A involves the conversion of simple compound units. The conversion of only one unit is required; the other unit remains the same. In problem 1, the square centimeter unit is converted to a square millimeter unit; the gram unit remains the same. In problem 2, the hour unit is converted to a minute unit; the kilometer unit remains the same.

Problems often involve the conversion of more than one unit in their solutions. It is necessary to convert both units of a compound unit quantity to smaller or larger units. Following is an example of complex compound unit conversions in which both units are converted.

EXAMPLE

Express 62.35 kilometers per hour as meters per minute.

Multiply 62.35 km/hr by unity fractions $\dfrac{1\,000 \text{ m}}{1 \text{ km}}$ and $\dfrac{1 \text{ hr}}{60 \text{ min}}$.

$$\frac{62.35 \cancel{\text{km}}}{\cancel{hr}} \times \frac{1\,000 \text{ m}}{1 \cancel{\text{km}}} \times \frac{1 \cancel{hr}}{60 \text{ min}} \approx 1\,039 \text{ m/min } Ans$$

Calculator Application

62.35 ☒ 1000 ÷ 60 ▣ 1039.166667, 1 039 m/min (rounded to 4 significant digits) *Ans*

EXERCISE 9–15B

Express each complex compound unit measure as indicated. Round each answer to the same number of significant digits as in the original quantity.

1. 67 km/hr = ? m/sec
2. 0.43 kg/cm² = ? g/mm²
3. 12.66 m/sec = ? km/min
4. 0.88 g/mm² = ? mg/cm²
5. $4.77/L = ? cents/mL
6. 0.06 kg/cm³ = ? metric tons/m³

9–16   Compound Units Practical Applications

EXAMPLE •

A heating installation consists of 5 air ducts of equal size. Hot air passes through each duct at the rate of 12.8 cubic decimeters per second. Compute the total number of cubic meters of hot air that passes through the 5 ducts in 3.25 minutes.

Find the number of cubic meters per minute of air flowing through 1 duct. Multiply 12.8 dm³/sec by unity fractions $\dfrac{1 \text{ m}^3}{1\,000 \text{ dm}^3}$ and $\dfrac{60 \text{ sec}}{1 \text{ min}}$.

$$12.8 \;\frac{\cancel{dm^3}}{\cancel{sec}} \times \frac{1 \text{ m}^3}{1\,000 \;\cancel{dm^3}} \times \frac{60 \;\cancel{sec}}{1 \text{ min}} = 0.768 \text{ m}^3/\text{min}$$

Find the total volume of air passing through 5 ducts in 3.25 minutes. Multiply 0.768 m³/min by 5 and by 3.25 min.

$$0.768 \;\frac{\text{m}^3}{\cancel{min}} \times 5 \times 3.25 \;\cancel{min} \approx 12.5 \text{ m}^3 \; Ans$$

Calculator Application

12.8 $\div$ 1000 $\times$ 60 $\times$ 5 $\times$ 3.25 $=$ 12.48, 12.5 m³ (rounded to 3 significant digits) *Ans*

EXERCISE 9–16

Solve the following problems.

1. Anthracite coal weighs 1.5 kilograms per cubic decimeter. Find the weight, in metric tons, of a 3.5 cubic meter load of coal. Round the answer to 2 significant digits.

2. A joule (J) is the amount of energy required to raise the temperature of 0.24 g of water from 0°C to 1°C. A building has 18 air-conditioning units. Each unit removes 25 300 000 J per hour (J/h). How many joules are removed from the building in 20 minutes?

3. A 0.15-square-kilometer tract of land is to be seeded with Kentucky Bluegrass. At a seeding rate of 7.5 grams per square meter of land, how many kilograms of seed are required for the complete tract? Round the answer to 2 significant digits.

4. In drilling a piece of stock, a drill makes 360 revolutions per minute with a feed of 0.50 millimeter. Feed is the distance the drill advances per revolution. How many seconds are required to cut through a steel plate which is 3.5 centimeters thick? Round the answer to the nearest second.

5. Air passes through a duct at the rate of 12 000 cm³/s. Find the number of cubic meters of air passing through the duct in 27.75 minutes. Round the answer to the nearest cubic meter.

6. The speed of sound in air is 332 meters per second (mps) at 0 degrees Celsius. At this speed, how many kilometers does sound travel in 0.345 hour? Round the answer to the nearest kilometer.

7. Materials expand when heated. Different materials expand at different rates. Mechanical and construction technicians must often consider material rates of expansion. The amount of expansion for short lengths of material is very small. Expansion is computed if a product is made to a high degree of precision and is subjected to a large temperature change. Also, expansion is computed for large structures because of the long lengths of structural member used. The table in Figure 9–11 lists the expansion per millimeter for each Celsius degree rise in temperature for a 1-mm length of material. For example, the expansion of a 1-mm length of aluminum is expressed as $0.000024 \dfrac{mm/mm}{1°C}$ or $0.000024 \dfrac{mm}{mm \cdot 1°C}$.

Material	Expansion of One Millimeter of Material in One Celsius Degree Temperature Increase
Aluminum	0.000 024 mm
Copper	0.000 019 8 mm
Structural Steel	0.000 015 8 mm
Brick	0.000 006 6 mm
Concrete	0.000 017 5 mm

Figure 9–11

Find the total expansion for each of the materials listed in the table in Figure 9–12. Express the answer to 3 significant digits.

		Length Before Temperature Increase	Celsius Temperature Change	Total Expansion in mm
a.	Copper Cable	107.5 m	From 2.3°C to 25.4°C	
b.	Steel I-Beam	17.25 m	From −23.6°C to 26.4°C	
c.	Brick Wall	87.25 m	From −7.2°C to 21.9°C	
d.	Aluminum Wire	143.6 m	From 4.6°C to 41.2°C	
e.	Concrete Foundation	38.5 m	From −15.7°C to 23.5°C	

Figure 9–12

9–17   Metric Prefixes Applied to Very Large and Very Small Numbers

Electronics and physics often involve applications and computations with very large and very small numbers. Biotechnology uses very small numbers. Computers process data in the central processing unit (CPU). Small silicon wafers called chips contain integrated circuits other processing circuitry. A single chip possesses an enormous amount of computing power. Data transformation operation takes place at extremely high speeds. Some computers can process millions of functions in a fraction of a second.

Data transmission can be measured by the number of bits per second. A bit is the on or off state of a single circuit represented by the binary digits 1 and 0. Coaxial cables can transmit data at the rate of ten million bits per second for short distances. Transmission speeds using high-frequency radio waves (microwaves) can send signals at fifty million bits per second. Optical fibers using laser technologies can transmit with speeds of one billion bits per second.

NOTE: The binary system is presented on pages 326–329.

The prefixes most commonly used with very large and very small numbers and their corresponding values are listed in the table in Figure 9–13. Notice that each value is 1000 or 10^3 times larger or smaller than the value it directly precedes.

PREFIXES USED FOR LARGE AND SMALL VALUES				
	Symbol	Meaning	Factor Value	Power of 10
tera	T	one trillion	1 000 000 000 000	10^{12}
giga	G	one billion	1 000 000 000	10^{9}
mega	M	one million	1 000 000	10^{6}
kilo	k	one thousand	1 000	10^{3}
—	—	—	1	10^{0}
milli	m	one-thousandth	0.001	10^{-3}
micro	μ	one-millionth	0.000 001	10^{-6}
nano	n	one-billionth	0.000 000 001	10^{-9}
pico	p	one-trillionth	0.000 000 000 001	10^{-12}

Figure 9–13

Some quantities with their definitions and units, which are commonly used in electronics/computer technology, are listed in the table in Figure 9–14.

SOME COMMONLY USED ELECTRONICS/COMPUTER QUANTITIES		
Quantity and (Symbol)	Definition	Unit and (Symbol)
Current (I)	The transfer of electrical charge through materials.	ampere (A)
Frequency (F)	The number of complete cycles in a unit of time.	hertz (Hz)
Resistance (R)	The opposition a material has to current flow.	ohm (Ω)
Power (P)	The rate at which energy is generated or dissipated.	watt (W)
Voltage (E)	Electronmotive force or electrical pressure.	volt (V)
Capacitance (C)	The property of a capacitor that permits the storage of electrostatic energy.	farad (F)

Figure 9–14

Expressing Electrical/Computer Measure Unit Equivalents

In expressing equivalent units either of the following two methods can be used.

Method 1. The decimal point is moved to the proper number of decimal places to the left or right.

Method 2. The unity fraction method multiplies the given unit of measure by a fraction equal to one.

EXAMPLES

Refer to Figure 9–13 for the following three examples.

1. Express 5 300 000 bits (b) as megabits (Mb).

METHOD 1. Move the decimal point 6 places to the left.

5 300 000.

5 300 000 b = 5.3 Mb *Ans*

METHOD 2. Multiply 5 300 000 b

by the unity fraction $\dfrac{1\,\text{Mb}}{1\,000\,000\,\text{b}}$

$5\,300\,000\,\text{b} \times \dfrac{1\,\text{Mb}}{1\,000\,000\,\text{b}} = 5.3\,\text{Mb } Ans$

2. Express 4.6 amperes (A) as milliamperes (mA)

METHOD 1. Move the decimal point 3 places to the right.

4.600   4.6 A = 4600 mA *Ans*

METHOD 2. Multiply 4.3 A by the

unity fraction $\dfrac{1000\,\text{mA}}{1\,\text{A}}$

$4.6\,\text{A} \times \dfrac{1000\,\text{mA}}{1\,\text{A}} = 4600\,\text{mA } Ans$

3. Express 6 500 microseconds (μs) as milliseconds (ms)

6 500 μs = 6.5 ms *Ans*

or $6\,500\,\text{μs} \times \dfrac{1\,\text{ms}}{1,000\,\text{μs}} = 6.5\,\text{ms } Ans$

EXERCISE 9–17A

Express each of the following values as the indicated unit value.

1. 15 200 milliamperes (mA) as amperes (A)
2. 0.26 second (s) as microseconds (μs)
3. 750 watts (W) as kilowatts (kW)
4. 0.097 megaohm (MΩ) as ohms (Ω)
5. 8.2×10^9 bits (b) as gigabits (Gb)
6. 414 kilohertz (kHz) as hertz (Hz)
7. 380 milliseconds (ms) as seconds (s)
8. 4.4×10^6 microamperes (μA) as amperes (A)
9. 1.68 terabits per second (Tbps) as bits per second (bps)
10. 350 000 microfarads (μF) as farads (F)
11. 270 watts (W) as milliwatts (mW)
12. 0.03 second (s) as nano seconds (ns)
13. 5.8×10^5 hertz (Hz) as megahertz (MHz)
14. 120 picofarads (pF) as farads (F)
15. 2600 milliamperes (mA) as microamperes (μA)
16. 97 000 kilobits (kb) as gigabits (Gb)

Expressing Biotechnology Measure Unit Equivalents

The unity fraction method is the most useful procedure to use when converting these metric units in biotechnology.

EXAMPLES •———————————————————————————————————————

1. Express 4500 pmol as nmol.

   Multiply 4500 pmol by the unity fraction $\dfrac{1 \text{ nmol}}{1000 \text{ pmol}}$.

   $4500 \text{ pmol} \times \dfrac{1 \text{ nmol}}{1000 \text{ pmol}} = 4.5 \text{ nmol } Ans$

2. Express 27.5 micrograms as picograms.

   Since one microgram is 10^{-6} g and one picogram is 10^{-12} g the unity fraction is $\dfrac{10^{6} \text{pg}}{1 \text{ } \mu g}$.

   Thus, $27.5 \text{ } \mu g \times \dfrac{10^{6} \text{pg}}{1 \mu g} = 27.5 \times 10^{6} \text{ pg} = 27\,500\,000 \text{ pg } Ans$

EXERCISE 9–17B

Express each of the following values as the indicated unit value. Express the answer to 3 significant digits.

1. 25.3 centimeters as nanometers
2. 595 nanometers as picometers
3. 172.5 nanograms as micrograms
4. 38.75 picograms as micrograms
5. 23.6 microliters as nanoliters
6. 2.4 picomoles as nanomoles

Arithmetic Operations

Arithmetic operations are performed the same way as with any other metric value. Compute the arithmetic operations, then write the appropriate unit of measure. Remember, when values are added or subtracted they must be in the same units.

EXAMPLES •———————————————————————————————————————

1. Express answer as V: $18.60 \text{ V} + 410.0 \text{ mV} = 18.60 \text{ V} + 0.4100 \text{ V} = 19.01 \text{ V } Ans$
2. Express answer as μs: $510 \text{ } \mu s - 12000 \text{ ns} = 510 \text{ } \mu s - 12 \text{ } \mu s = 498 \text{ } \mu s \, Ans$
3. Express answer as A: $(2.4 \times 10^{4}) \text{ } \mu A \times 375 = (9 \times 10^{6}) \text{ } \mu A = 9 \text{ A } Ans$
4. Express answer as Hz: $0.75 \text{ MHz} \div 970 \approx 0.000773 \text{ MHz} \approx 773 \text{ Hz } Ans$

EXERCISE 9–17C

Compute each of the following values. Express each answer as the indicated unit value.

1. $13.0 \text{ V} + 810 \text{ mV} = ? \text{ V}$
2. $0.35 \text{ MW} + 6500 \text{ W} = ? \text{ kW}$
3. $0.04 \text{ A} - 1400 \text{ } \mu A = ? \text{ mA}$

4. 15.8×0.018 W = ? mW

5. 3.96 MHz $\div 1.32$ = ? kHz

6. 870 pF + 0.002 μF = ? pF

7. 96.5 Ω − 0.025 kΩ = ? Ω

8. (1.2×10^4) ns × (3×10^6) = ? s

9. 5400 b × (2.5×10^7) = ? Gb

10. 0.93 MHz $\div 31$ = ? Hz

11. 440 mA + $260\,000$ μA = ? A

12. (8.42×10^3) mF − (7.15×10^5) μF = ? F

13. 8000 ks + 360 Ms = ? Gs

14. (1.7×10^6) W × (5×10^4) = ? GW

15. 20 ns $\div (5 \times 10^{-10})$ = ? s

16. (3.6×10^5) μA + (5.0×10^2) mA = ? A

17. $52\,000$ nF − 14 μF = ? mF

9–18 Conversion Between Metric and Customary Systems

In technical work it is sometimes necessary to change from one measurement system to the other. Use the following metric–customary conversions for the length of an object. Because the length of an inch is defined in terms of a centimeter, some of these conversions are exact.

Metric–Customary Length Conversions
1 in = 2.54 cm
1 ft = 30.48 cm
1 yd = 0.9144 m
1 mi ≈ 1.6093 km

To convert from one system to the other, you can either use unity fractions or multiply by the conversion factor given in the table.

EXAMPLES

1. Convert 3.7 ft to centimeters.

METHOD 1

Since 1 ft = 30.48 cm, multiply 3.7 by 30.48.
$$3.7 \times 30.48 = 112.8$$
$$3.7 \text{ ft} = 112.8 \text{ cm } Ans$$

METHOD 2

Since 3.7 ft is to be expressed in centimeters, multiply by the unity fraction $\dfrac{30.48 \text{ cm}}{1 \text{ ft}}$.

$$3.7 \text{ ft} = 3.7 \text{ ft} \times \frac{30.48 \text{ cm}}{1 \text{ ft}} = 112.8 \text{ cm } Ans$$

2. Convert 5.4 km to miles.

METHOD 1

Since 1 mi ≈ 1.6093 km, divide 5.4 by 1.6093.
$$5.4 \div 1.6093 = 3.4$$
$$5.4 \text{ km} \approx 3.4 \text{ mi } Ans$$

METHOD 2

Since 5.4 km is to be expressed in miles, multiply by the unity fraction $\dfrac{1 \text{ mi}}{1.6093 \text{ km}}$.

$$3.7 \text{ ft} = 5.4 \cancel{\text{km}} \times \frac{1 \text{ mi}}{1.6093 \cancel{\text{km}}} \approx 3.4 \text{ mi } \textit{Ans}$$

There will be times when more than one unity fraction will have to be used.

3. Convert 7.36 in to millimeters.

There is no inch–millimeter conversion in the table. So, we must use two conversions. First, convert inches to centimeters and then convert centimeters to millimeters. The unity fractions are $\dfrac{2.54 \text{ cm}}{1 \text{ in}}$ and $\dfrac{10 \text{ mm}}{1 \text{ cm}}$.

$$7.36 \text{ in} = 7.36 \cancel{\text{in}} \times \frac{2.54 \cancel{\text{cm}}}{1 \cancel{\text{in}}} \times \frac{10 \text{ mm}}{1 \cancel{\text{cm}}} = 186.9 \text{ mm } \textit{Ans}$$

EXERCISE 9–18A

Express each length as indicated. Round each answer to the same number of significant digits as in the original quantity.

1. 12.0 in as centimeters
2. 25.3 in as millimeters
3. 3.25 ft as millimeters
4. 12.65 ft as centimeters
5. 1.20 m as feet

6. 4.2 m as yards
7. 36.75 mi as kilometers
8. 152.6 km as miles
9. 115.2 yd as centimeters
10. 8 ft $7\frac{1}{2}$ in as meters

Dual Dimensions

Companies involved in international trade may use "dual dimensioning" on technical drawings and specifications. Dual dimensioning means that both metric and customary dimensions are given, as shown in Figure 9–15.

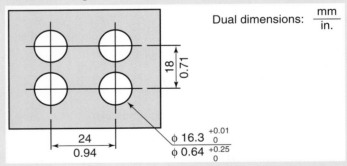

Figure 9–15

The metric measurements are supposed to be written on top of the fraction bar, but it is a good idea to look for a key. Diameter dimensions are marked with a ϕ.

Metric–Customary Area Conversions

$$1 \text{ square inch (sq in. or in}^2) = 6.4516 \text{ cm}^2$$
$$1 \text{ square foot (sq ft or ft}^2) \approx 0.0929 \text{ m}^2$$
$$1 \text{ square yard (sq yd or yd}^2) \approx 0.8361 \text{ m}^2$$
$$1 \text{ acre} \approx 0.4047 \text{ ha}$$
$$1 \text{ square mile (sq mi or mi}^2) \approx 2.59 \text{ km}^2$$

EXAMPLE

Convert 12.75 ft² to square centimeters.

Since 12.75 ft² is to be expressed in square centimeters, multiply by the unity fractions $\dfrac{0.0929 \text{ m}^2}{1 \text{ ft}^2}$ and $\dfrac{10000 \text{ cm}^2}{1 \text{ m}^2}$.

$$12.75 \text{ ft}^2 = 12.75 \text{ ft}^2 \times \frac{0.0929 \text{ m}^2}{1 \text{ ft}^2} \times \frac{10000 \text{ cm}^2}{1 \text{ m}^2} = 11844.75 \text{ cm}^2, \; 11840 \text{ cm}^2 \; Ans$$

EXERCISE 9–18B

Express each area as indicated. Round each answer to the same number of significant digits as in the original quantity.

1. 18.5 ft² as square centimeters
2. 47.75 in² as square millimeters
3. 3.9 yd² as square meters
4. 12$\overline{0}$ ft² as square meters

5. 65 ha as acres
6. 50$\overline{0}$ ha as square miles
7. 18.75 m² as square feet
8. 18.75 m² as square yards

Metric–Customary Volume Conversions

1 cubic inch (cu in. or in³) = 16.387 cm³

1 fluid ounce (fl oz) ≈ 29.574 cm³

1 teaspoon (tsp) ≈ 4.929 mL

1 tablespoon (tbsp) ≈ 14.787 mL

1 cup ≈ 236.6 mL

1 quart (qt) ≈ 0.9464 L

1 gallon (gl) ≈ 3.785 L

EXAMPLES

1. Convert 2.5 gal to liters.

$$2.5 \text{ gal} = 2.5 \text{ gal} \times \frac{3.785 \text{ L}}{1 \text{ gal}} = 9.4625 \text{ L}, \; 9.5 \text{ L} \; Ans$$

2. Convert 17.25 liters to quarts.

$$17.25 \text{ L} = 17.25 \text{ L} \times \frac{1 \text{ qt}}{0.9464 \text{ L}} \approx 18.226965, \; 18.23 \text{ qt} \; Ans$$

EXERCISE 9–18C

Express each volume as indicated. Round each answer to the same number of significant digits as in the original quantity.

1. 278.5 cubic inches as
   (a) cubic centimeters and
   (b) liters

2. $1\frac{1}{2}$ cups as milliliters

3. 25.75 fluid ounces as milliliters

4. 6.5 quarts as liters

5. 42.75 liters as gallons

6. 2.4 liters as quarts

7. 15.8 milliliters as fluid ounces

8. 135.4 milliliters as cubic inches

Metric–Customary Weight Conversions

$$1 \text{ ounce (oz)} = 28.35 \text{ g}$$

$$1 \text{ pound (lb)} \approx 0.4536 \text{ kg}$$

$$1 \text{ (short) ton} \approx 907.2 \text{ kg}$$

EXAMPLE •

Convert 75.2 grams to pounds.

$$75.2 \text{ g} = 75.2 \text{ g} \times \frac{1 \text{ oz}}{28.35 \text{ g}} \times \frac{1 \text{ lb}}{16 \text{ oz}} = 0.165784 \text{ lb},\ 0.166 \text{ lb } Ans$$

EXERCISE 9–18D

Express each weight as indicated. Round each answer to the same number of significant digits as in the original quantity.

1. 165 pounds as kilograms

2. 5.25 tons as metric tons

3. 43.76 ounces as grams

4. 70.5 grams as ounces

5. 23.8 kilograms as pounds

6. 2 759 kilograms as (short) tons

⠿ UNIT EXERCISE AND PROBLEM REVIEW

METRIC UNITS OF LINEAR MEASURE

For each of the following write the most appropriate metric unit.

1. Most audio compact discs are about 12 ? in diameter.

2. Some large trees grow over 30 ? high.

3. Slices of cheese are usually between 1 and 2?

4. Many home refrigerators are about 0.8 ? wide.

5. My desktop computer keyboard is 48 ? long.

EQUIVALENT METRIC UNITS OF LINEAR MEASURE

Express each value in the unit indicated.

6. 30 mm as cm   **9.** 23 m as cm   **12.** 0.014 m as mm

7. 8 cm as mm   **10.** 650 m as km   **13.** 12.2 cm as mm

8. 2,460 mm as m   **11.** 0.8 km as m   **14.** 372.5 m as km

ARITHMETIC OPERATIONS WITH METRIC LENGTHS

Solve each exercise. Express the answers in the unit indicated.

15. 6.3 cm + 13.6 mm = ? mm

16. 1.7 m − 92 cm = ? cm

17. 20.8 × 31.0 m = ? m

18. 8.46 dm ÷ 6.27 = ? dm

19. 0.264 km + 37.9 m + 21 hm = ? m

20. 723.2 cm − 5.1 m = ? cm

21. 70.6 dm + 127 mm + 4.7 m = ? dm

22. 41.8 cm + 4.3 dm + 77.7 mm + 0.03 m = ? cm

EQUIVALENT METRIC UNITS OF AREA MEASURE

Express each metric area measure in the indicated unit.

23. 532 mm^2 as cm^2 **28.** 1.96 m^2 as cm^2

24. 23.6 dm^2 as m^2 **29.** 0.009 km^2 as m^2

25. 14,660 cm^2 as m^2 **30.** 173,000 m^2 as km^2

26. 53 cm^2 as mm^2 **31.** 28,000 mm^2 as m^2

27. 6 m^2 as dm^2 **32.** 0.7 dm^2 as mm^2

EQUIVALENT METRIC UNITS OF VOLUME MEASURE

Express each volume in the unit indicated.

33. 2,400 mm^3 as cm^3 **37.** 4.6 m^3 as dm^3

34. 1,700 cm^3 as dm^3 **38.** 420 dm^3 as m^3

35. 7 dm^3 as cm^3 **39.** 60,000 cm^3 as m^3

36. 15 cm^3 as mm^3 **40.** 0.0048 dm^3 as mm^3

EQUIVALENT METRIC UNITS OF CAPACITY MEASURE

Express each metric capacity in the unit indicated.

41. 1.3 L as mL   **45.** 618 L as m^3

42. 2,100 mL as L   **46.** 3.17 dm^3 as L

43. 93.4 mL as cm^3 **47.** 0.06 L as mL

44. 5,210 cm^3 as L   **48.** 19,000 mL as L

EQUIVALENT METRIC UNITS OF WEIGHT MEASURE

Express each metric weight in the unit indicated.

49. 1,880 g as kg

50. 730 mg as g

51. 2.7 metric tons as kg

52. 4.75 g as mg

53. 0.21 kg as g

54. 310,000 kg as metric tons

SIMPLE COMPOUND UNIT MEASURES

Express each simple compound unit measure as indicated. Round each answer to the same number of significant digits as in the original quantity.

55. 58 km/h = ? km/min

56. 148 g/cm^2 = ? g/mm^2

57. 32 040 kg/m^2 = ? kg/dm^2

58. 94.0 hp/L = ? hp/cm^3

COMPLEX COMPOUND UNIT MEASURES

Express each complex compound unit measure as indicated. Round each answer to the same number of significant digits as in the original quantity.

59. 10.58 m/sec = ? km/min

60. 42 km/hr = ? m/sec

61. 0.39 kg/cm^2 = ? g/mm^2

62. 0.90 g/mm^2 = ? mg/cm^2

METRIC PREFIXES

Express each of the following values as the indicated unit value.

63. 940 watts (W) as kilowatts (kW)

64. 7.3×10^8 bits (b) as gigabits (Gb)

65. 0.005 second (s) as nanoseconds (ns)

66. 4.9×10^5 hertz (Hz) as megahertz (MHz)

67. 1,780 milliamperes (mA) as microamperes (μA)

68. 63×10^{-11} farads (F) as picofarads (pF)

69. 1 294 picometers as nanometers

70. 95.73 microliters as nanoliters

CONVERSION BETWEEN METRIC AND CUSTOMARY SYSTEMS

Convert each unit to the indicated unit. Round each answer to the same number of significant digits as in the original quantity.

71. Convert 147.4 in. to meters

72. Convert 6.75 ft to centimeters

73. Convert 47.35 mm to inches

74. Convert 853.25 kilometers to miles

75. Convert 23.6 square yards to square meters

76. Convert 125 hectars to acres

77. Convert 47.25 cubic feet to cubic meters

78. Convert 17.6 gallons to liters
79. Convert 482.3 milliliters to quarts
80. Convert 369.5 cubic centimeters to fluid ounces
81. Convert 621.8 pounds to kilograms
82. Convert 16 ounces to fluid grams

PRACTICAL APPLICATIONS PROBLEMS

Solve the following problems.

83. A car travels from Town A to Town C by way of Town B. The car travels 135 kilometers. The trip takes 2.25 hours. It takes 0.8 hour to get from Town A to Town B. Assuming the same speed is maintained for the entire trip, how many kilometers apart are Town A and Town B?

84. Determine the total length, in meters, of the wall section shown in Figure 9–16.

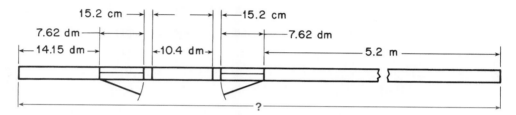

Figure 9–16

85. An assembly consists of 5 metal plates. The respective areas of the plates are 650 cm², 800 cm², 16.3 dm², 12 dm², and 0.12 m². Determine the total surface measure, in square meters, of the 5 plates.

86. A roll of fabric has a surface measure of 12.0 square meters. How many pieces, each requiring 1,800 square centimeters of fabric, can be cut from the roll? Make an allowance of 10% for waste.

87. A total of 325 pieces are punched from a strip of stock that has a volume of 11.6 cubic centimeters. Each piece has a volume of 22.4 cubic millimeters. How many cubic centimeters of strip stock are wasted after the pieces are punched? Round answer to 3 significant digits.

88. Twenty concrete support bases are required for a construction job. Ninety-five cubic decimeters of concrete are used for each base. Compute the total number of cubic meters of concrete required for the 20 bases.

89. A truck is to deliver 8 prefabricated concrete wall sections to a job site. Each wall section has a volume of 0.620 cubic meter. One cubic meter of concrete weighs 2,350 kilograms. How many metric tons are carried on this delivery? Round answer to 3 significant digits.

90. The liquid intake of a hospital patient during a specified period of time is as follows: 275 mL, 150 mL, 325 mL, 275 mL, 175 mL, 200 mL, and 300 mL. What is the total liter intake of liquid for the time period?

UNIT 10 ⠿ Steel Rules and Vernier Calipers

O B J E C T I V E S

After studying this unit you should be able to

• read measurements on a customary rule graduated in 32nds and 64ths.

• read measurements on a customary rule graduated in 50ths and 100ths.

• read measurements on a metric rule with 1-millimeter and 0.5-millimeter graduations.

• read customary and metric vernier caliper settings.

10–1   Types of Steel Rules

Steel rules are widely used in the metal trades and in certain woodworking occupations. There are many different types of rules designed for specific job requirements. Steel rules are available in various customary and metric graduations. Rules can be obtained in a wide range of lengths, widths, and thicknesses. Two of the many types of steel rules are shown in Figures 10–1 and 10–2.

Figure 10–1   Customary rule with graduations in 32nds and 64ths. (The L. S. Starrett Company)

Figure 10–2   Customary rule with decimal graduations in 50ths and 100ths. (The L. S. Starrett Company)

10–2   Reading Fractional Measurements

An enlarged customary rule is shown in Figure 10–3 on page 269. The top scale is graduated in 64ths of an inch. The bottom scale is graduated in 32nds of an inch. The staggered graduations are for halves, quarters, eighths, sixteenths, and thirty-seconds.

Measurements can be read on a rule by noting the last complete inch unit and counting the number of fractional units past the inch unit. For actual on-the-job uses, shortcut methods for reading measurements are used. Refer to the enlarged customary rule with graduations in 32nds and 64ths shown in Figure 10–4 on page 269 for these examples. Two methods for reading measurements are shown.

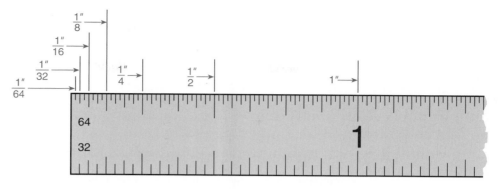

Figure 10–3

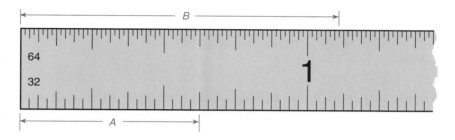

Figure 10–4

EXAMPLES

1. Read the measurement of length A in Figure 10–4.

METHOD 1

Observe the number of 1-inch graduations.

$0 \times 1'' = 0$

Length A falls on a $\frac{1}{8}$-inch graduation. Count the number of 8ths from zero.

$$5 \times \frac{1''}{8} = \frac{5''}{8}$$

$$A = \frac{5''}{8} \ Ans$$

METHOD 2

Length A is one $\frac{1}{8}$-inch graduation more than $\frac{1}{2}$ inch.

$$A = \frac{1''}{2} + \frac{1''}{8} = \frac{4''}{8} + \frac{1''}{8} = \frac{5''}{8} \ Ans$$

2. Read the measurement of length B in Figure 10–4.

METHOD 1

Observe the number of 1-inch graduations. $1 \times 1'' = 1''$

Length B falls on a $\frac{1}{64}$-inch graduation. Count the number of 64ths from 1 inch.

$$7 \times \frac{1''}{64} = \frac{7''}{64}$$

$$B = 1'' + \frac{7''}{64} = 1\frac{7''}{64} \ Ans$$

METHOD 2

Length *B* is one $\frac{1}{64}$-inch graduation less than $1\frac{1}{8}''$.

$$B = 1\frac{1''}{8} - \frac{1''}{64} = 1\frac{8''}{64} - \frac{1''}{64} = 1\frac{7''}{64} \; Ans$$

10–3 Measurements that Do Not Fall on Rule Graduations

Often the end of the object being measured does not fall on a rule graduation. In these cases, read the closer rule graduation. Refer to the enlarged customary rule shown in Figure 10–5 for these examples.

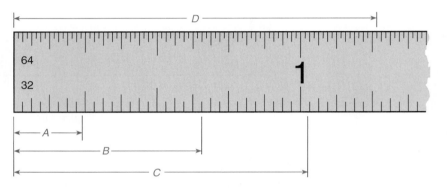

Figure 10–5 Enlarged customary rule with graduations in 32nds and 64ths.

EXAMPLES •

1. Read the measurement of length *A*.

   The measurement is closer to $\frac{1}{4}$ inch than $\frac{7}{32}$ inch.

   $$A = \frac{1''}{4} \; Ans$$

2. Read the measurement of length *B*.

   The measurement is closer to $\frac{21}{32}$ inch than $\frac{11}{16}$ inch.

   $$B = \frac{21''}{32} \; Ans$$

3. Read the measurement of length *C*.

   The measurement is closer to $1\frac{1}{32}$ inches than 1 inch.

   $$C = 1\frac{1''}{32} \; Ans$$

4. Read the measurement of length *D*.

   The measurement is closer to $1\frac{17}{64}$ inches than $1\frac{9}{32}$ inches.

   $$D = 1\frac{17''}{64} \; Ans$$

EXERCISE 10–3

Read measurements a *through* p *on the enlarged customary rule with graduations in 32nds and 64ths shown in Figure 10–6.*

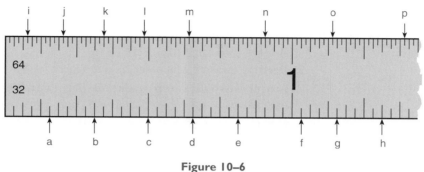

Figure 10–6

10–4 Reading Decimal-Inch Measurements

An enlarged customary rule is shown in Figure 10–7. The top scale is graduated in 100ths of an inch (0.01 inch). The bottom scale is graduated in 50ths of an inch (0.02 inch). The staggered graduations are for halves, tenths, and fiftieths.

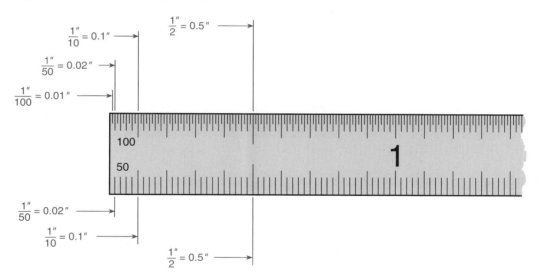

Figure 10–7 Enlarged customary rule with decimal graduations in 50ths and 100ths.

Refer to the enlarged customary rule with decimal graduations in 50ths and 100ths shown in Figure 10–8 for these examples. One method for reading the measurements is shown.

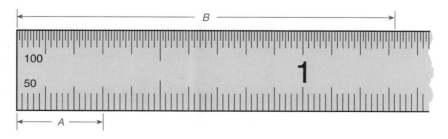

Figure 10–8

EXAMPLES •————————————————————————————————————

1. Read the measurement of length *A* in Figure 10–8.

   Observe the number of 1-inch graduations.
   $0 \times 1'' = 0$

   Length *A* falls on a 0.1-inch graduation. Count the number of 10ths from zero.
   $3 \times 0.1'' = 0.3''$.

   $A = 0.3''$ *Ans*

2. Read the measurement of length *B* in Figure 10–8.

   Length *B* is one 1-inch graduation plus three 0.1-inch graduations plus two 0.01-inch graduation.

   $B = (1 \times 1'') + (3 \times 0.1'') + (2 \times 0.01'') = 1.32''$ *Ans*

—— •

EXERCISE 10–4 ———————————————————————————————————————

Read measurements a *through* p *on the enlarged customary rule with decimal graduations in 50ths and 100ths shown in Figure 10–9.*

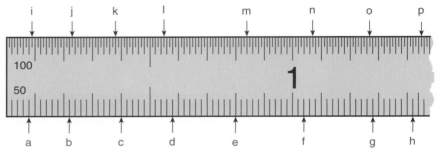

Figure 10–9

10–5 Reading Metric Measurements

An enlarged metric rule is shown in Figure 10–10. The top scale is graduated in one-half millimeters (0.5 mm). The bottom scale is graduated in millimeters (1 mm). Refer to the enlarged metric rule shown for these examples.

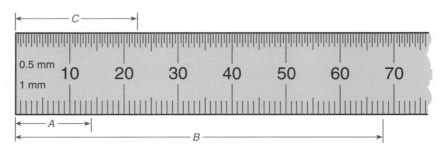

Figure 10–10   Enlarged metric rule with 1-mm and 0.5-mm graduations.

EXAMPLES •

1. Read the measurement of length *A* in Figure 10–10.

   Length *A* is 10 millimeters plus 4 millimeters.

   A = 10 mm + 4 mm = 14 mm *Ans*

2. Read the measurement of length *B* in Figure 10–10.

   Length *B* is 2 millimeters less than 70 millimeters.

   B = 70 mm − 2 mm = 68 mm *Ans*

3. Read the measurement of length *C* in Figure 10–10.

   Length *C* is 20 millimeters plus two 1-millimeter graduations plus one 0.5-millimeter graduation.

   C = 20 mm + 2 mm + 0.5 mm = 22.5 mm *Ans*

EXERCISE 10–5

Read measurements a *through* p *on the enlarged metric rule with 1-millimeter and 0.5-millimeter graduations shown in Figure 10–11.*

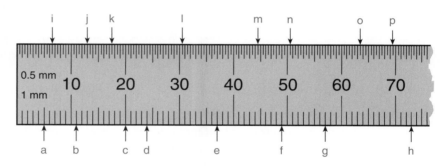

Figure 10–11

10–6   Vernier Calipers: Types and Description

Vernier calipers are widely used in the manufacturing occupations. They are used for many different applications where precision to thousandths of an inch or hundredths of a millimeter is required. Vernier calipers are commonly used for measuring lengths of objects, determining distances between holes in parts, and measuring inside and outside diameters of cylinders.

Vernier calipers are available in a wide range of lengths with different types of jaws and scale graduations.

There are two basic parts of a vernier caliper. One part is the main scale, which is similar to a steel rule with a fixed jaw. The other part is a sliding jaw with a vernier scale. The vernier scale slides parallel to the main scale and provides a degree of precision to 0.001 inch.

The front side of a commonly used customary vernier caliper is shown in Figure 10–12 on page 274. The parts are identified. The main scale is divided into inches, and the inches are divided into 10 divisions each equal to 0.1 inch. The 0.1-inch divisions are divided into four parts, each equal to 0.025 inch. The vernier scale has 25 divisions in a length equal to the length on the main scale, which has 24 divisions, as shown in Figure 10–13. The difference between a main scale division and a vernier scale division is $\frac{1}{25}$ of 0.025 inch or 0.001 inch.

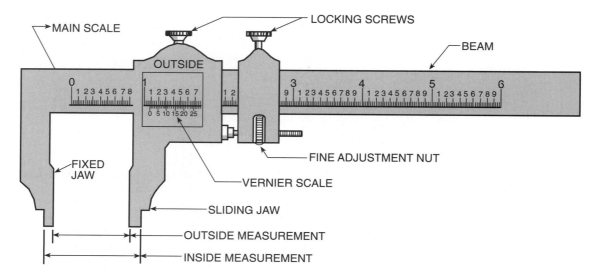

Figure 10–12

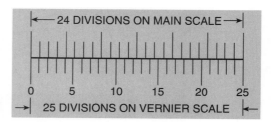

Figure 10–13

The front side of the customary vernier caliper (25 divisions) is used for outside measurements as shown in Figure 10–14. The reverse or back side is used for inside measurements as shown in Figure 10–15 on page 275.

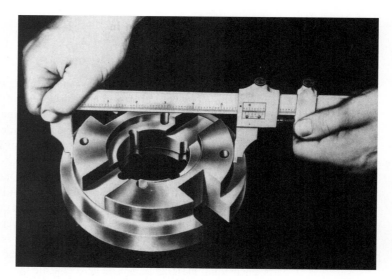

Figure 10–14 Measuring an outside diameter. The measurement is read on the front side of the caliper. (The L. S. Starrett Company)

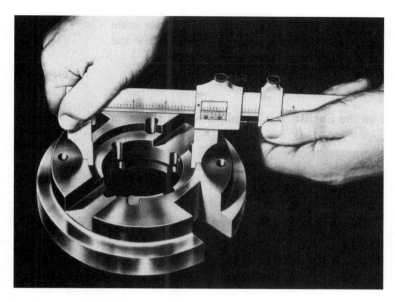

Figure 10–15 Measuring an inside diameter. The measurement is read on the back side of the caliper. (The L. S. Starrett Company)

The accuracy of a measurement obtainable with a vernier caliper depends on the user's ability to align the caliper with the part being measured. The line of measurement must be parallel to the beam of the caliper and must lie in the same plane as the caliper. Care must be used to prevent too loose or too tight a caliper setting.

10–7   Reading Measurements on a Customary Vernier Caliper

A measurement is read by adding the thousandths reading on the vernier scale to the reading from the main scale.

On the main scale, read the number of 1-inch divisions, 0.1-inch divisions, and 0.025-inch divisions that are to the left of the zero graduation on the vernier scale. On the vernier scale, find the graduation that most closely coincides with a graduation on the main scale. This vernier graduation indicates the number of thousandths that are added to the main scale reading.

EXAMPLES •————————————————————————————————

1. Read the measurement set on the customary vernier caliper scales shown in Figure 10–16.

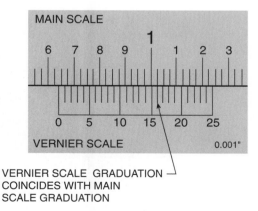

VERNIER SCALE GRADUATION
COINCIDES WITH MAIN
SCALE GRADUATION

Figure 10–16

To the left of the zero graduation on the vernier scale, read the main scale reading: zero 1-inch division, six 0.1-inch divisions, and one 0.025-inch division.
($0 \times 1'' + 6 \times 0.1'' + 1 \times 0.025'' = 0.625''$)

Observe which vernier scale graduation most closely coincides with a main scale graduation. The sixteenth vernier scale graduation coincides. Add 0.016 inch to the main scale reading.
($0.625'' + 0.016'' = 0.641''$)

Vernier caliper reading: 0.641″ *Ans*

2. Read the measurement set on the customary vernier caliper scales shown in Figure 10–17.

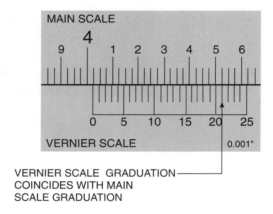

Figure 10–17

To the left of the zero graduation on the vernier scale, read the main scale reading: four 1-inch divisions, zero 0.1-inch division, and zero 0.025-inch division.
($4 \times 1'' + 0 \times 0.1'' + 0 \times 0.025'' = 4''$)

Observe which vernier scale graduation most closely coincides with a main scale graduation. The twenty-first vernier scale graduation coincides. Add 0.021 inch to the main scale reading.
($4'' + 0.021'' = 4.021''$)

Vernier caliper reading: 4.021″ *Ans*

EXERCISE 10–7

Read the customary vernier caliper measurements for these settings.

1.
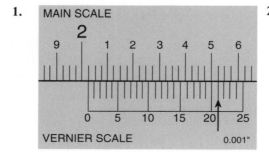

2.
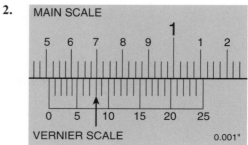

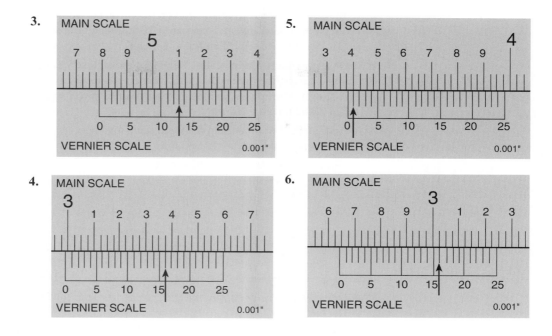

10–8   Reading Measurements on a Metric Vernier Caliper

The same principles are used in reading and setting metric vernier calipers as for customary vernier calipers. The main scale is divided in 1–millimeter divisions. Each millimeter division is divided in half or 0.5-millimeter divisions. A graduation is numbered every ten millimeters in the sequence: 10 mm, 20 mm, 30 mm, and so on. The vernier scale has 25 divisions. Each division is $\frac{1}{25}$ of 0.5 millimeter or 0.02 millimeter.

A measurement is read by adding the 0.02-millimeter reading on the vernier scale to the reading from the main scale. On the main scale, read the number of millimeter divisions and 0.5-millimeter divisions that are to the left of the zero graduation on the vernier scale. On the vernier scale, find the graduation that most closely coincides with a graduation on the main scale. Multiply the graduation by 0.02 millimeter and add the value obtained to the main scale reading.

EXAMPLE •———————————————————————————————

Read the measurement set on the metric scales in Figure 10–18.

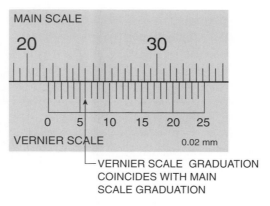

Figure 10–18

To the left of the zero graduation on the vernier scale, read the main scale reading: twenty-one 1-millimeter divisions and one 0.5-millimeter division.
(21 × 1 mm + 1 × 0.5 mm = 21.5 mm)

Observe which vernier scale graduation most closely coincides with a main scale graduation. The sixth vernier scale graduation coincides. Each vernier scale graduation represents 0.02 mm. Multiply to find the number of millimeters represented by 6 divisions.
(6 × 0.02 mm = 0.12 mm)

Add the 0.12 millimeter to the main scale reading.
(21.5 mm + 0.12 mm = 21.62 mm)

Vernier caliper reading: 21.62 mm *Ans*

EXERCISE 10–8

Read the metric vernier caliper measurements for the following settings.

1.
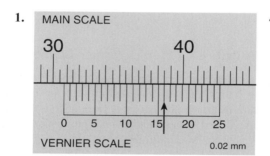

4.
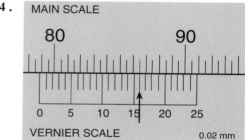

2.
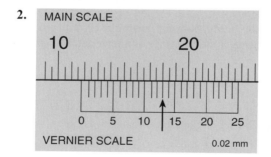

5.
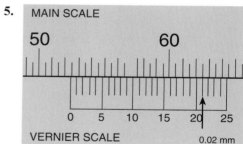

3.
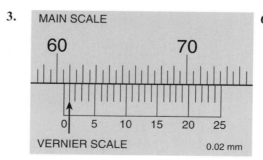

6.
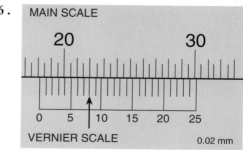

::: UNIT EXERCISE AND PROBLEM REVIEW

READING FRACTIONAL–INCH MEASUREMENTS ON A CUSTOMARY RULE

1. Read measurements a through p on the enlarged English rule graduated in 32nds and 64ths in Figure 10–19.

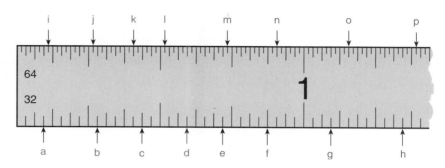

Figure 10–19

READING DECIMAL–INCH MEASUREMENTS ON A CUSTOMARY RULE

2. Read measurements a through p on the enlarged English rule graduated in 50ths and 100ths shown in Figure 10–20.

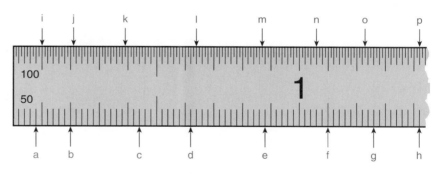

Figure 10–20

READING MEASUREMENTS WITH A METRIC RULE

3. Read measurements a through p on the enlarged metric rule with 1-mm and 0.5-mm graduations in Figure 10–21.

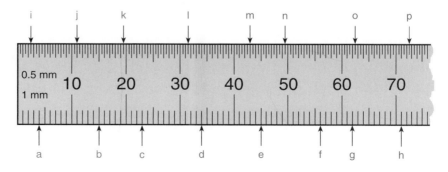

Figure 10–21

READING CUSTOMARY AND METRIC VERNIER CALIPER SETTINGS

Read the vernier caliper measurements for these settings.

4. Customary measurements

a.
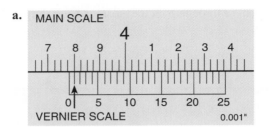

c.
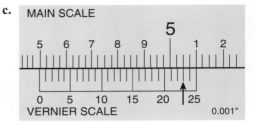

b.
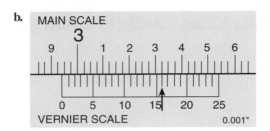

d.
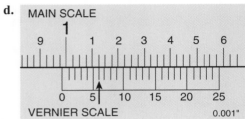

5. Metric measurements

a.
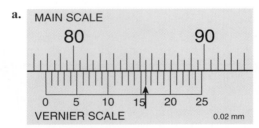

c.
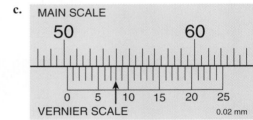

b.
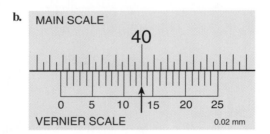

d.
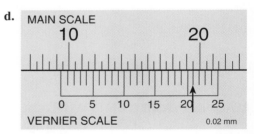

UNIT 11 ⠿ Micrometers

OBJECTIVES

After studying this unit you should be able to

• read settings on 0.001 decimal-inch micrometer scales.

• read settings on 0.0001 decimal-inch vernier micrometer scales.

• read settings on 0.01-millimeter metric micrometer scales.

• read settings on 0.002-millimeter metric vernier micrometer scales.

Micrometers are basic measuring instruments that are widely used in the manufacture and inspection of products. Occupations in various technical fields require making measurements with a number of different types of micrometers. Micrometers are commonly used by machinists, pattern makers, sheet metal technicians, inspectors, and automobile mechanics.

Micrometers are available in a wide range of sizes and types. Outside micrometers are used to measure lengths between parallel surfaces of objects. Other types of micrometers, such as depth micrometers, inside micrometers, screw-thread micrometers, and wire micrometers have specific applications.

The micrometer descriptions and procedures for reading measurements that are presented only apply to conventional non-digital micrometers. Digital micrometers are also widely used. Measurements are read directly as a five-digit LCD display.

11–1   Description of a Customary Outside Micrometer

Figure 11–1 shows a customary outside micrometer graduated in thousandths of an inch (0.001″). The principal parts are labeled.

The part is placed between the anvil and the spindle. The barrel of a micrometer consists of a scale that is 1 inch long.

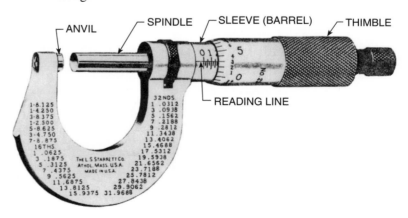

Figure 11–1   A customary outside micrometer. (The L. S. Starrett Company)

Refer to the barrel and thimble scales in Figure 11–2. The 1-inch barrel scale length is divided into 10 divisions each equal to 0.100 inch. The 0.100-inch divisions are further divided into four divisions each equal to 0.025 inch.

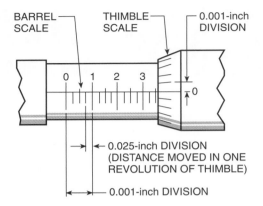

Figure 11–2   Enlarged barrel and thimble scales.

The thimble scale is divided into 25 parts. One revolution of the thimble moves 0.025 inch on the barrel scale. A movement of one graduation on the thimble equals $\frac{1}{25}$ of 0.025 inch or 0.001 inch along the barrel.

11–2   Reading a Customary Micrometer

A micrometer is read by observing the position of the bevel edge of the thimble in reference to the scale on the barrel. The user observes the greatest 0.100-inch division and the number of 0.025-inch divisions on the barrel scale. To this barrel reading, add the number of the 0.001-inch divisions on the thimble that coincide with the horizontal line (reading line) on the barrel scale.

Procedure for Reading a Micrometer

• Observe the greatest 0.100-inch division on the barrel scale.

• Observe the number of 0.025-inch divisions on the barrel scale.

• Add the thimble scale reading (0.001-inch division) that coincides with the horizontal line on the barrel scale.

EXAMPLES •————————————————————————————————

1. Read the customary micrometer setting shown in Figure 11–3.

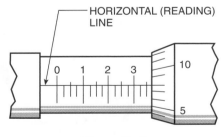

Figure 11–3

Observe the greatest 0.100-inch division on the barrel scale.
(three 0.100-inch divisions = 0.300 inch)

Observe the number of 0.025-inch divisions between the 0.300-inch mark and the thimble. (two 0.025-inch divisions = 0.050 inch)

Add the thimble scale reading that coincides with the horizontal line on the barrel scale. (eight 0.001-inch divisions = 0.008 inch)

Micrometer reading:
0.300″ + 0.050″ + 0.008″ = 0.358″ Ans

2. Read the customary micrometer setting shown in Figure 11–4.

On the barrel scale, two 0.100-inch divisions = 0.200 inch.

On the barrel scale, zero 0.025-inch division = 0 inch.

On the thimble scale, twenty-three 0.001-inch divisions = 0.023 inch.

Micrometer reading:
0.200″ + 0.023″ = 0.223″ Ans

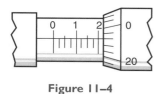

Figure 11–4

EXERCISE 11–2

Read the settings on these customary micrometer scales graduated in 0.001″.

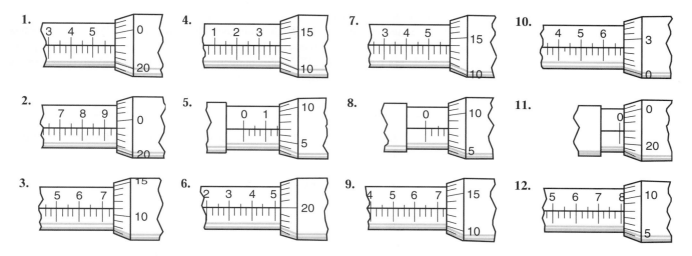

1.

2.

3.

4.

5.

6.

7.

8.

9.

10.

11.

12.

11–3 The Customary Vernier Micrometer

The addition of a vernier scale on the barrel of a 0.001-inch micrometer increases the degree of precision of the instrument to 0.0001 inch. The barrel scale and the thimble scale of a vernier micrometer are identical to that of a 0.001-inch micrometer.

Figure 11–5 shows the relative positions of the barrel scale, thimble scale, and vernier scale of a 0.0001-inch micrometer.

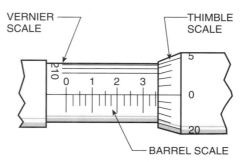

Figure 11–5

The vernier scale consists of 10 divisions. Ten vernier divisions on the circumference of the barrel are equal in length to nine divisions of the thimble scale. The difference between one vernier division and one thimble division is 0.0001 inch. Figure 11–6 shows a flattened view of a vernier scale and a thimble scale.

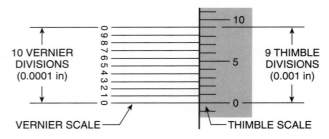

Figure 11–6

11–4   Reading a Customary Vernier Micrometer

Reading a customary vernier micrometer is the same as reading a 0.001-inch micrometer except for the addition of reading the vernier scale. A particular vernier graduation coincides with a thimble scale graduation. The vernier graduation gives the number of 0.0001-inch divisions that are added to the barrel and thimble scale readings. A vernier division may not line up exactly with a thimble scale graduation. When this happens, select the vernier division that comes the *nearest* to matching a thimble marking.

EXAMPLES •——————————————————————————————

1. A flattened view of a customary vernier micrometer setting is shown in Figure 11–7. Read this setting.

Read the barrel scale reading.
$(3 \times 0.100'' + 3 \times 0.025'' = 0.375'')$

Read the thimble scale.
$(9 \times 0.001'' = 0.009'')$

Read the vernier scale.
$(4 \times 0.0001'' = 0.0004'')$

Vernier micrometer reading:
$0.375'' + 0.009'' + 0.0004'' = 0.3844''$ *Ans*

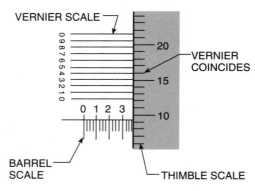

Figure 11–7

2. A flattened view of a customary vernier micrometer setting is shown in Figure 11–8. Read this setting.

On the barrel scale read 0.200 inch.

On the thimble scale read 0.020 inch.

On the vernier scale read 0.0008 inch.

Vernier micrometer reading:
0.200″ + 0.020″ + 0.0008″ = 0.2208″ *Ans*

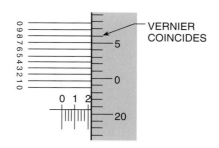

Figure 11–8

EXERCISE 11–4

Read the settings on these customary vernier micrometer scales graduated in 0.0001″.

1.

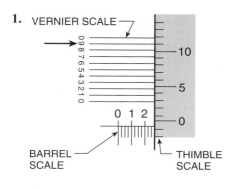

2.

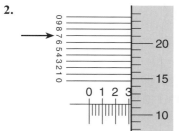

3.

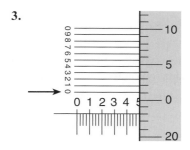

4.

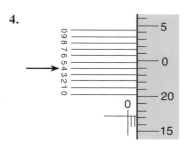

5.

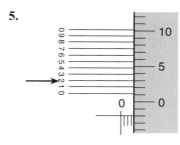

6.

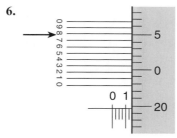

7.

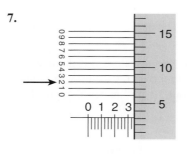

8.

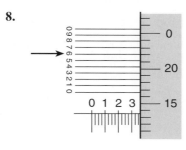

9.

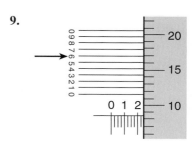

10.

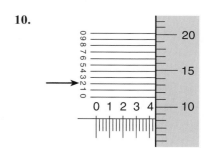

11.

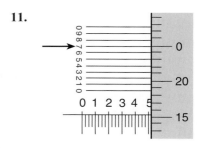

12.

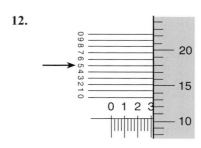

11–5   Description of a Metric Micrometer

Figure 11–9 shows a 0.01-millimeter outside micrometer.

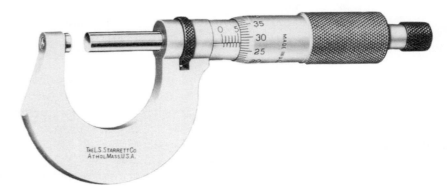

Figure 11–9   A metric outside micrometer. (The L. S. Starrett Company)

The barrel of a 0.01-millimeter micrometer consists of a scale that is 25 millimeters long. Refer to the barrel and thimble scales in Figure 11–10. The 25-millimeter barrel scale length is divided into 25 divisions each equal to 1 millimeter. Every fifth millimeter is numbered from 0 to 25 (0, 5, 10, 15, 20, 25). On the lower part of the barrel scale, each millimeter is divided in half (0.5 mm).

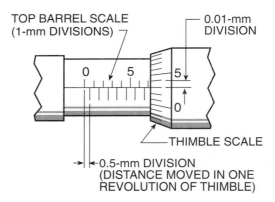

Figure 11–10

The thimble has a scale that is divided into 50 parts. One revolution of the thimble moves 0.5 millimeter on the barrel scale. A movement of one graduation on the thimble equals $\frac{1}{50}$ of 0.5 millimeter or 0.01 millimeter along the barrel.

11–6   Reading a Metric Micrometer

Procedure for Reading a 0.01-Millimeter Micrometer

- Observe the number of 1-millimeter divisions on the barrel scale.
- Observe the number of 0.5-millimeter divisions (either 0 or 1) on the lower part of the barrel scale.
- Add the thimble scale reading (0.01 division) that coincides with the horizontal line on the barrel scale.

EXAMPLES •

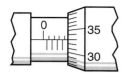

Figure 11–11

1. Read the metric micrometer setting shown in Figure 11–11.

   Observe the number of 1-millimeter divisions on the barrel scale.
   (4 × 1 mm = 4 mm)

   Observe the number of 0.5-millimeter divisions on the lower barrel scale.
   (0 × 0.5 mm = 0)

   Add the thimble scale reading that coincides with the horizontal line on the barrel scale.
   (33 × 0.01 mm = 0.33 mm)

   Micrometer reading:
   4 mm + 0.33 mm = 4.33 mm *Ans*

2. Read the metric micrometer setting shown in Figure 11–12.

   On the barrel scale read 17 millimeters.

   On the lower barrel scale read 0.5 millimeter.

   On the thimble scale read 0.26 millimeter.

   Micrometer reading:
   17 mm + 0.5 mm + 0.26 mm = 17.76 mm *Ans*

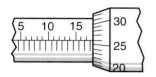

Figure 11–12

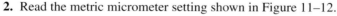

EXERCISE 11–6

Read the settings on these metric micrometer scales graduated in 0.01 mm.

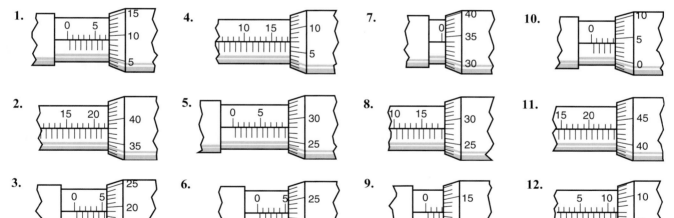

11–7 The Metric Vernier Micrometer

The addition of a vernier scale on the barrel of a 0.01-millimeter micrometer increases the degree of precision of the instrument to 0.002 millimeter. The barrel scale and the thimble scale of a vernier micrometer are identical to that of a 0.01-millimeter micrometer.

Figure 11–13 on page 288 shows the relative positions of the barrel scale, thimble scale, and vernier scale of a 0.002-millimeter micrometer.

The vernier scale consists of five divisions. Each division equals one-fifth of a thimble division or $\frac{1}{5}$ of 0.01 millimeter or 0.002 millimeter. Figure 11–14 on page 288 shows a flattened view of a vernier scale and a thimble scale.

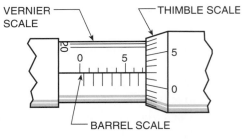

Figure 11–13

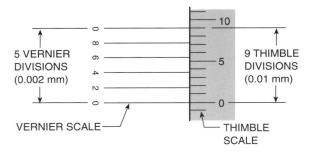

Figure 11–14

11–8   Reading a Metric Vernier Micrometer

Reading a metric vernier micrometer is the same as reading a 0.01-millimeter micrometer except for the addition of reading the vernier scale. Observe which division on the vernier scale coincides with a division on the thimble scale. If the vernier division that coincides is marked 2, add 0.002 millimeter to the barrel and thimble scale reading. Add 0.004 millimeter for a coinciding vernier division marked 4, add 0.006 millimeter for a division marked 6, and add 0.008 millimeter for a division marked 8.

EXAMPLES •——

1. A flattened view of a metric vernier micrometer is shown in Figure 11–15. Read this setting.

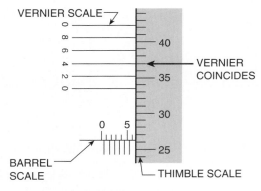

Figure 11–15

Read the barrel scale.
(6 × 1 mm + 0 × 0.5 mm = 6 mm)

Read the thimble scale.
(26 × 0.01 mm = 0.26 mm)

Read the vernier scale.
(0.004 mm)

Vernier micrometer reading:
6 mm + 0.26 mm + 0.004 mm = 6.264 mm *Ans*

2. A flattened view of a metric vernier micrometer is shown in Figure 11–16. Read this setting.

On the barrel scale read 9.5 millimeters.

On the thimble scale read 0.43 millimeter.

On the vernier scale read 0.008 millimeter.

Vernier micrometer reading:
9.5 mm + 0.43 mm + 0.008 mm = 9.938 mm *Ans*

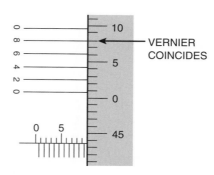

Figure 11–16

EXERCISE 11–8

Read the settings on these metric vernier micrometer scales graduated in 0.002 mm. In each exercise the arrow shows where the vernier division matches a thimble scale graduation.

1.

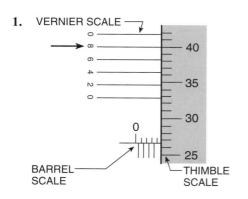

3.

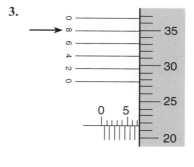

5.

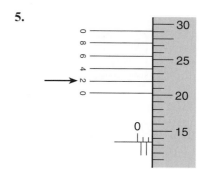

2.

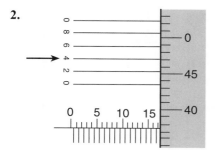

4.

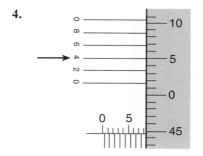

6.

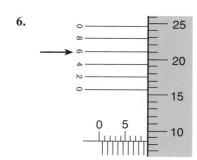

7.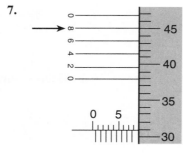

8.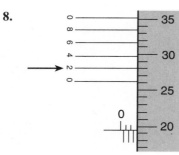

9.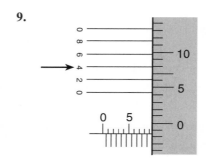

10.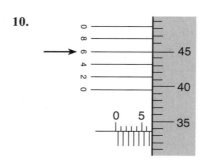

11.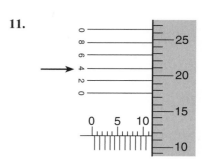

12.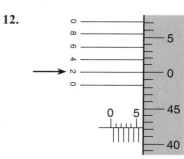

▦ UNIT EXERCISE AND PROBLEM REVIEW

CUSTOMARY MICROMETER (0.001″)

1. Read the settings on these customary micrometer scales graduated in 0.001″.

a.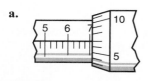

b.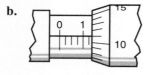

c.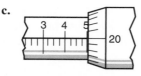

d.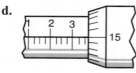

e.

f.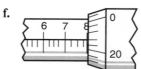

g.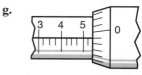

h.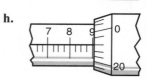

CUSTOMARY VERNIER MICROMETER (0.0001″)

2. Read the settings on these customary vernier micrometer scales graduated in 0.0001″. *In each exercise the arrow shows where the vernier division matches a thimble scale graduation.*

a.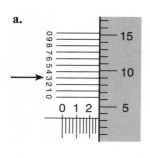

b.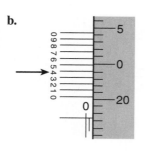

c.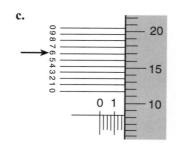

d.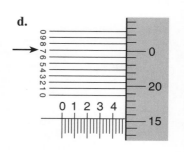

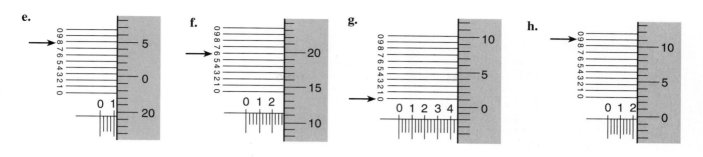

METRIC MICROMETER (0.01 mm)

3. Read the settings on these metric micrometer scales graduated in 0.01 mm.

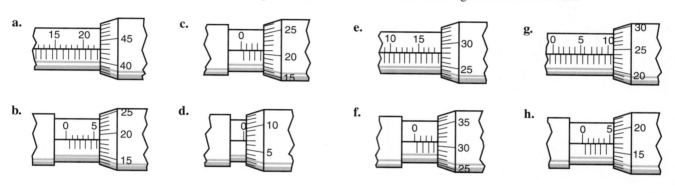

METRIC VERNIER MICROMETER (0.002 mm)

4. Read the settings on these metric vernier micrometer scales graduated in 0.002 mm. *In each exercise the arrow shows where the vernier division matches a thimble scale graduation.*

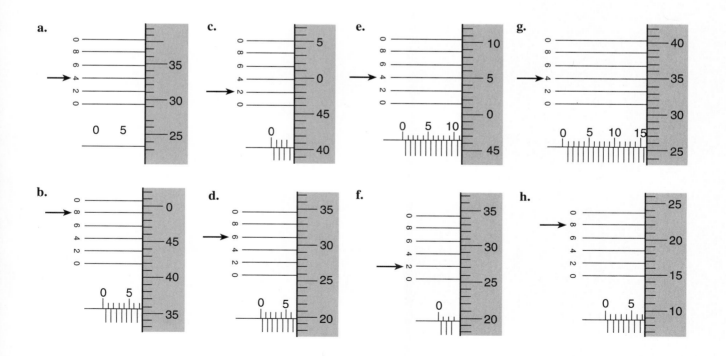

SECTION III ⠿

Fundamentals of Algebra

UNIT 12 ⁘ Introduction to Algebra

OBJECTIVES

After studying this unit you should be able to

- express word statements as algebraic expressions.
- express diagram values as algebraic expressions.
- evaluate algebraic expressions by substituting given numbers for letter values.
- solve formulas by substituting numbers for letters, word statements, and diagram values.

Algebra is a branch of mathematics that uses letters to represent numbers; *algebra* is an extension of arithmetic. The rules and procedures that apply to arithmetic also apply to algebra.

By the use of letters, general rules called *formulas* can be stated mathematically. The expression, $°C = \frac{5(°F - 32°)}{9}$ is an example of a formula that is used to express degrees Fahrenheit as degrees Celsius. Many operations in shop, construction, and industrial work are expressed as formulas. Business, finance, transportation, agriculture, and health occupations require the employee to understand and apply formulas.

A knowledge of algebra fundamentals is necessary in a wide range of occupations. Algebra is often used in solving on-the-job geometry and trigonometry problems. The basic principles of algebra presented in this text are intended to provide a practical background for diverse occupational applications.

12–1 Symbolism

Symbols are the language of algebra. Both arithmetic and literal numbers are used in algebra.

Arithmetic numbers or *constants* are numbers that have definite numerical values, such as 2, 8.5, and $\frac{1}{4}$.

Literal numbers or *variables* are letters that represent arithmetic numbers, such as x, y, a, A, and T. Depending on how it is used, a literal number can represent one particular arithmetic number, a wide range of numerical values, or all numerical values.

Customarily, the multiplication sign ($\times$) is not used in algebra because it can be misinterpreted as the letter x. When a variable (letter) is multiplied by a numerical value, or when two or more variables are multiplied, no sign of operation is required. For example, $2 \times a$ is written $2a$; $b \times c$ is written bc, $4 \times L \times W$ is written $4LW$. When two or more arithmetic numbers are multiplied, parentheses ( ) are used in place of the multiplication sign. For example, 3 times 5 can be written as (3)5, 3(5), or (3)(5). A raised dot may also be used to indicate multiplication. Here, 3 times 5 would be written $3 \cdot 5$.

12–2 Algebraic Expressions

An *algebraic expression* is a word statement put into mathematical form by using variables, arithmetic numbers, and signs of operation. Generally, part of a word statement contains an unknown quantity. The unknown quantity is indicated by a symbol. The symbol usually used is a single letter, such as x, y, a, V, or P.

A variety of words and phrases indicate mathematical operations in word statements. Some of the many words and phrases that indicate the mathematical operations of addition, subtraction, multiplication, and division are listed.

Addition. The operation symbol (+) is substituted for words and phrases such as add, sum, plus, increase, greater than, heavier than, larger than, exceeded by, and gain of.

Subtraction. The operation symbol (−) is substituted for words and phrases such as subtract, minus, decreased by, less than, lighter than, smaller than, shorter than, reduced by, and loss of.

Multiplication. No sign of operation is required for the product of all variables or the product of a variable and a numerical value. Otherwise, the symbol ( ) or · is substituted for words such as multiply, times, and product of.

Division. The operation symbols are the division sign, ÷, a horizontal bar, —, and a slash, /. The horizontal bar and slash are used in fractional forms of algebraic expressions, as in $\frac{a}{b}$ and a/b. Any of the three symbols can be substituted for words and phrases such as "divide by" or "quotient of".

Examples of Algebraic Expressions

1. The statement "add 5 to x" is expressed algebraically as **$x + 5$**.
2. The statement "12 is decreased by b" is expressed algebraically as **$12 - b$**.
3. The statement "x is subtracted from 10" is written algebraically as **$10 - x$**.
4. The cost, in dollars, of 1 pound of grass seed is d. The cost of 6 pounds of seed is expressed as **$6d$**.
5. The weight, in pounds, of 10 gallons of gasoline is W. The weight of 1 gallon is expressed as **$W \div 10$** or **$\frac{W}{10}$**.
6. The length of a spring, in millimeters, is l. The spring is stretched to 3 times its original length plus 0.4 millimeter. The stretched spring length is expressed as **$3l + 0.4$**.
7. A patio is shown in Figure 12–1. Length A is expressed in feet as x. Length B is $\frac{1}{2}$ of Length A or $\frac{1}{2}x$. Length C is twice Length A or $2x$. The total length of the patio is expressed as **$x + \frac{1}{2}x + 2x$**.
8. A plate with 2 drilled holes is shown in Figure 12–2. The total length of the plate is 14 centimeters. The distance, in centimeters, from the left edge of the plate to the center of hole 1 is c. The distance, in centimeters, from the right edge of the plate to hole 2 is b. The distance between holes, in centimeters, is expressed as **$14 - c - b$**, or **$14 - (c + b)$**.

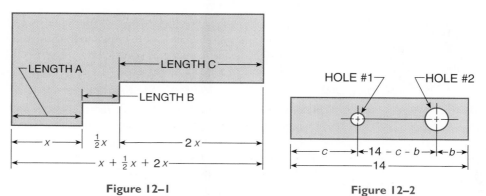

Figure 12–1   Figure 12–2

9. Perimeter (P) is the distance around an object. The perimeter of a rectangle equals twice its length (l) plus twice its width (w). The perimeter of a rectangle expressed as a formula is **$P = 2l + 2w$**.

EXERCISE 12–2

Express each exercise as an algebraic expression.

1. Add 3 to a
2. Subtract 7 from d
3. Subtract d from 7
4. Multiply 8 times m
5. Multiply x times y
6. Divide 25 by b
7. Divide b by 25
8. Square x
9. Increase e by 12
10. The product of r and s
11. Multiply 102 by x
12. Reduce 75 by y
13. Reduce y by 75
14. The sum of e and f reduced by g
15. Increase a by the square of b
16. The square root of x plus 6.8
17. Three times V minus 12
18. The product of c and d increased by e
19. One-half x minus four times y
20. The sum of $4\frac{1}{4}$ and b reduced by c
21. The product of 9 and m increased by the product of 2 and n
22. The square root of m divided by the cube root of n
23. Divide x by the product of 25 and y
24. Take the square root of r, add s, and subtract the product of 2 and t

Express each problem as an algebraic expression.

25. Refer to Figure 12–3. All values are in inches.

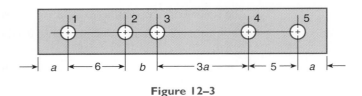

Figure 12–3

    a. Express the distance from the left edge of the part to the center of hole 4.

    b. Express the distance from the center of hole 1 to the center of hole 5.

    c. Express the distance from the center of hole 3 to the center of hole 5.

    d. Express the distance from the right edge of the part to the center of hole 1.

26. A machine produces P parts per hour. Express the number of parts produced in h hours.

27. The length of a board, in meters, is L. The board is cut into N number of equal pieces. Express the length of each piece.

28. A cross-sectional view of a pipe is shown in Figure 12–4.

    a. The pipe wall thickness (T) is equal to the difference between the outside diameter (D) and the inside diameter (d) divided by 2. Express the wall thickness.

    b. The inside diameter (d) is equal to the outside diameter (D) minus twice the wall thickness (T). Express the inside diameter.

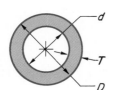

Figure 12–4

29. A person has a checkbook balance represented as B. A check is made out for an amount represented as C. The amount deposited in the account is represented as D. Express the new account balance.

30. A series circuit is shown in Figure 12–5. The total resistance (R_T) of the circuit is equal to the sum of the individual resistances R_1, R_2, and R_3. The circuit has a total resistance of 150 ohms. Express the resistance of R_1.

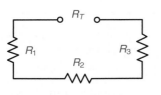

Figure 12–5

31. The total piston displacement of an engine is determined by computing the product of 0.7854 times the square of the cylinder bore (D) times the length of the piston stroke (L) times the number of cylinders (N). Express the total piston displacement.

32. A stairway is shown in Figure 12–6. The actual number of steps is shown.

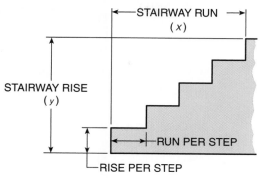

Figure 12–6

a. The stairway run is *x*. Express the run per step.

b. The stairway rise is *y*. Express the rise per step.

33. Impedance (*Z*) of the circuit shown in Figure 12–7 is computed by adding the square of resistance (*R*) to the square of reactance (*X*), then taking the square root of the sum. Express the circuit impedance.

34. Refer to Figure 12–8. Express the distances between the following points.

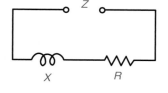

Figure 12–7

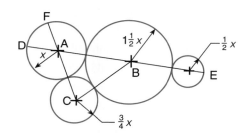

Figure 12–8

a. Point A to point B   **c.** Point B to point C

b. Point F to point C   **d.** Point D to point E

12–3   Evaluation of Algebraic Expressions

Certain problems in this text involve the use of formulas. Some problems require substituting numerical values for letter values. The problems are solved by applying the order of operations of arithmetic. Review the order of operations before proceeding to solve the exercises and problems that follow.

Order of Operations for Combined Operations of Addition, Subtraction, Multiplication, Division, Powers, and Roots

• **Do all the work in parentheses first**. Parentheses are used to group numbers. In a problem expressed in fractional form, two or more numbers in the dividend (numerator) and/or divisor (denominator) may be considered as being enclosed in parentheses. For example, $\frac{4.87 + 0.34}{9.75 - 8.12}$ may be considered as $(4.87 + 0.34) \div (9.75 - 8.12)$. If an algebraic expression contains parentheses within parentheses or brackets, such as $[5.6 \times (7 - 0.09) + 8.8]$, do the work within the innermost parentheses first.

- **Do powers and roots next**. These operations are performed in the order in which they occur from left to right. If a root consists of two or more operations within the radical sign, perform all the operations within the radical sign, then extract the root.
- **Do multiplication and division next**. These operations are performed in the order in which they occur from left to right.
- **Do addition and subtraction last**. These operations are performed in the order in which they occur from left to right.

Once again, the memory aid "**P**lease **E**xcuse **M**y **D**ear **A**unt **S**ally" can be used to help remember the order of operations. The **P** in "Please" stands for parentheses, the **E** for exponents or raising to a power, the **M** and **D** for multiplication and division, and the **A** and **S** for addition and subtraction.

EXAMPLES •

1. What is the value of the expression $53.8 - x(xy - m)$, where $x = 8.7$, $y = 3.2$, and $m = 22.6$? Round the answer to 1 decimal place.

   Substitute the numerical values for x, y, and m. $53.8 - 8.7 [8.7(3.2) - 22.6]$

   Perform the operations in the proper order.

   a. Perform the operations within parentheses or brackets. Inside the brackets, the multiplication is performed first.

$8.7 (3.2) = 27.84$	$53.8 - 8.7 (27.84 - 22.6)$
$27.84 - 22.6 = 5.24$	$53.8 - 8.7 (5.24)$

   b. Perform the multiplication.

$8.7 (5.24) = 45.588$	$53.8 - 45.588$

   c. Perform the subtraction.

$53.8 - 45.588 = 8.212$	8.2 *Ans* (rounded)

   Calculator Application

   $53.8 \boxed{-} 8.7 \boxed{\times} \boxed{(} \boxed{(} 8.7 \boxed{\times} 3.2 \boxed{-} 22.6 \boxed{)} \boxed{)} \boxed{=}$ 8.212, 8.2 *Ans* (rounded)

2. The total resistance (R_T) of the circuit shown in Figure 12–9 is computed from formula:

   $$R_T = R_1 + \frac{R_2 R_3}{R_2 + R_3}$$

   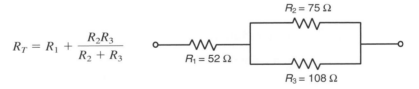

   Figure 12–9

   The values of the individual resistances (R_1, R_2, R_3) are given in the figure. Determine the total resistance (R_T) of the circuit to the nearest ohm. The symbol for ohm is Ω.

   Substitute the numerical values for R_1, R_2, and R_3.

   $$R_T = 52\ \Omega + \frac{75\ \Omega\ (108\ \Omega)}{75\ \Omega + 108\ \Omega}$$

   Perform the operations in the proper order.

   a. Consider the numerator and the denominator as being enclosed within parentheses. Perform the operation within parentheses.

$75\ \Omega\ (108\ \Omega) = 8,100\ \Omega^2$	$R_T = 52\ \Omega + \dfrac{8,100\ \Omega^2}{183\ \Omega}$
$75\ \Omega + 108\ \Omega = 183\ \Omega$	

b. Perform the division.

$8{,}100 \ \Omega^2 \div 183 \ \Omega \approx 44.3 \ \Omega$ $R_T \approx 52 \ \Omega + 44.3 \ \Omega$

c. Perform the addition.

$52 \ \Omega + 44.3 \ \Omega = 96.3 \ \Omega$ $R_T \approx 96 \ \Omega \ Ans$

Calculator Application

$R_T = 52 + \dfrac{75(108)}{75 + 108}$

$R_T = 52 \boxed{+} \boxed{(} 75 \boxed{\times} 108 \boxed{)} \boxed{\div} \boxed{(} 75 \boxed{+} 108 \boxed{)} \boxed{=}$ 96.26229508

$R_T \approx 96 \ \Omega \ Ans$

3. To determine the center-to-center hole distance (c) shown in Figure 12–10, an inspector uses the formula $c = \sqrt{a^2 + b^2}$. Compute the value of c to 2 decimal places.

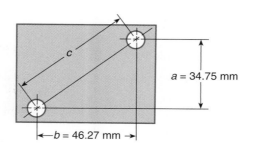

Figure 12–10

Substitute the length values for a and b. $c = \sqrt{(34.75 \text{ mm})^2 + (46.27 \text{ mm})^2}$

Perform the operations in the proper order.

a. Perform the operations within the radical sign.

$(34.75 \text{ mm})^2 \approx 1\,207.563 \text{ mm}^2$ $c \approx \sqrt{1\,207.563 \text{ mm}^2 + 2\,140.913 \text{ mm}^2}$

$(46.27 \text{ mm})^2 \approx 2\,140.913 \text{ mm}^2$

$1\,207.563 \text{ mm}^2 + 2\,140.913 \text{ mm}^2$ $c \approx \sqrt{3\,348.476 \text{ mm}^2}$
$= 3\,348.476 \text{ mm}^2$

b. Extract the square root.

$\sqrt{3\,348.476 \text{ mm}^2} \approx 57.87 \text{ mm}$ $c \approx 57.87 \text{ mm } Ans$

Calculator Application

$c = \sqrt{34.75^2 + 46.27^2}$

$c = \boxed{(} 34.75 \boxed{x^2} \boxed{+} 46.27 \boxed{x^2} \boxed{)} \boxed{\sqrt{x}}$ 57.86601248 $c \approx 57.87 \text{ mm } Ans$

or $c = \boxed{\sqrt{}} \boxed{(} 34.75 \boxed{x^2} \boxed{+} 46.27 \boxed{x^2} \boxed{)} \boxed{\text{EXE}}$ 57.86601248 $c \approx 57.87 \text{ mm } Ans$

EXERCISE 12–3

Substitute the given numbers for letters and compute the value of each expression. Where necessary, round the answers to 2 decimal places.

1. If $a = 5$ and $b = 3$, find

a. $4a - 2$ **d.** $b(a + b)$

b. $5 + b - a$ **e.** $3a - (2 + a)$

c. $6a \div b$

2. If $x = 6.2$ and $y = 2.5$, find

   a. $2xy - y$

   b. $(x + y)(x - y)$

   c. $\dfrac{x + y}{x - y}$

   d. $2x - x \div 3$

   e. $2x + xy - 4y$

3. If $e = 8$ and $f = 4$, find

   a. $3ef + 9$

   b. $5f + ef$

   c. $5e + f\left(\dfrac{e}{4}\right)$

   d. $\dfrac{10e - 6f}{8}$

   e. $12e \div (2f + 2)$

4. If $m = 8.3$, $s = 4.1$, and $t = 2$, find

   a. $\dfrac{m}{s} + t - 1$

   b. $ms(5 + 2s - 3t)$

   c. $12s(m + 5 - t)$

   d. $\dfrac{3m - s + 4t}{22 - st}$

   e. $\dfrac{12s}{t} - [3m - (s + t) + 4]$

5. If $x = 12$, $y = 8$, and $w = 15$, find

   a. $20 - \dfrac{w}{y} + 12x$

   b. $(x + w) \div (2y - x)$

   c. $\dfrac{xy + 4}{2x - 2y}$

   d. $\sqrt{x + 5y - (w + 3)}$

   e. $\dfrac{4x^2}{\sqrt{6w - 8}}$

6. If $p = 5.1$, $h = 4.3$, and $k = 3.2$, find

   a. $p + ph^2 - k^3$

   b. $(h + 2)^2(p - k)^2$

   c. $\left[\dfrac{(hk)^2}{2} - hk\right] + p^3$

   d. $\dfrac{h^3 + 3h - 12}{p^2 + 15}$

   e. $\dfrac{k^3}{3h - 9} + p^2(ph - 6k)^2$

Each problem requires working with formulas. Substitute numerical values for letters and solve.

7. A drill revolving at 300 revolutions per minute has a feed of 0.025 inch per revolution. Determine the cutting time required to drill through a workpiece 3.60 inches thick. Use this formula for finding cutting time. Round the answer to 1 decimal place.

$$T = \frac{L}{FN}$$
where T = time in minutes
L = length of cut in the workpiece in inches
F = feed in inches per revolution
N = speed in revolutions per minute

8. The resistance of an aluminum wire is 10 ohms. The constant value for the resistance of a circular-mil foot of aluminum wire at 75°F is 17.7 ohms. Compute the wire diameter, to the nearest whole mil, for a wire 500 feet long.

$$d = \sqrt{\frac{KL}{R}}$$
where d = diameter in mils
K = constant (17.7 Ω/CM-ft)
L = length in feet
R = resistance in ohms

9. Express 75°F as degrees Celsius using this formula. Round the answer to the nearest whole degree.

$$°C = \frac{5(°F - 32°)}{9}$$

10. Express 12°C as degrees Fahrenheit using this formula. Round the answer to the nearest whole degree.

$$°F = \frac{9}{5}(°C) + 32°$$

11. An original principal of $8,750.00 is deposited in a compound interest savings account. The money is left on deposit for 2 compounding periods with an interest rate of 4.12% per period. Determine the amount of money in the account at the end of the 2 periods. Use this formula and express the answer to the nearest cent.

$$A = P(1 + R)^n$$ where A = accumulated principal
P = original principal
R = rate per period
n = number of periods

12. An engine is turning at the rate of 1,525 revolutions per minute. The piston has a diameter (d) of 3 inches and a stroke length of 4.00 inches. The mean effective pressure on the piston is 60.0 lb/sq in. Find the horsepower developed by this engine. Round the answer to 1 decimal place.

$$hp = \frac{PLAN}{33,000}$$ where P = mean effective pressure in pounds per square inch
L = length of stroke in inches
A = piston cross-sectional area in square inches $(0.7854d^2)$
N = number of revolutions per minute

13. Find the area (A) of the plot of land shown in Figure 12–11 using this formula.

$$A = \frac{(H + h)b + ch + aH}{2}$$

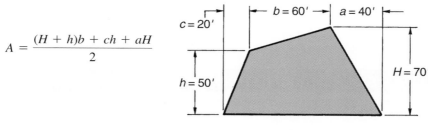

Figure 12–11

14. A cabinetmaker cuts a piece of plywood to the form and dimensions shown in Figure 12–12. Use this formula to determine the radius (r) of the circle from which the piece is cut.

$$r = \frac{c^2 + 4h^2}{8h}$$

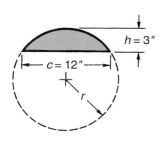

Figure 12–12

15. Pulley dimensions are given in Figure 12–13. Compute the length (L) of the belt required using this formula. Round the answer to 1 decimal place.

$$L = 2c + \frac{11D + 11d}{7} + \frac{(D - d)^2}{4c}$$

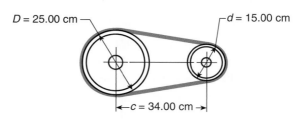

Figure 12–13

16. The total resistance (R_T) of the parallel circuit shown in Figure 12–14 is computed using this formula. Compute the total circuit resistance in ohms.

$$R_T = \cfrac{1}{\cfrac{1}{R_1} + \cfrac{1}{R_2} + \cfrac{1}{R_3} + \cfrac{1}{R_4} + \cfrac{1}{R_5}}$$

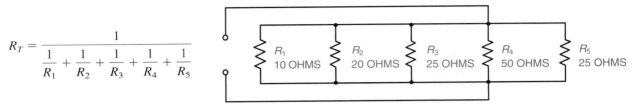

Figure 12–14

17. An elliptical platform is shown in Figure 12–15. Compute the perimeter (P) of the platform using this formula. Round the answer to 1 decimal place.

$$P = 3.14\sqrt{2(a^2 + b^2)}$$

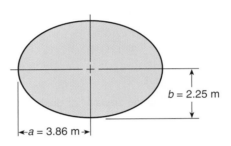

Figure 12–15

⠿ UNIT EXERCISE AND PROBLEM REVIEW

WRITING ALGEBRAIC EXPRESSIONS

Express each of these exercises as an algebraic expression.

1. Add 12 to six times x

2. The sum of a and b minus c

3. One-quarter m times R

4. The square of V reduced by the product of 3 and P

5. Divide d by the product of 14 and f

6. The square of y increased by the cube root of x

7. Twice M decreased by one-third R

8. The sum of a and b divided by the difference between a and b

9. Square F, add G, and divide the sum by H

10. Multiply x and y, take the square root of the product, and subtract 5

WRITING ALGEBRAIC EXPRESSIONS

Express each of these problems as an algebraic expression.

11. A car averages C miles per gallon of gasoline. Express the number of gallons of gasoline used when the car travels M miles.

12. Refer to the template shown in Figure 12–16. All values are given in millimeters.

a. Express the distance from point A to point C.

b. Express the distance from point B to point F.

c. Express the distance from point D to point E.

d. Express the distance from point A to point E.

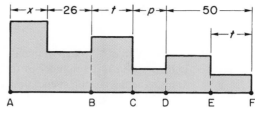

Figure12–16

13. Three pumps are used to drain water from a construction site. Each pump discharges G gallons of water per hour. How much water is drained from the site in H hours?

14. A piece of property is going to be enclosed by a fence with 2 gates. The property is in the shape of a square with each side S feet long. Each gate is G feet in length. Express the total number of feet of fencing required.

SUBSTITUTING VALUES IN EXPRESSIONS

Substitute the given numbers for letters and compute the value of each expression. Round answers to 2 decimal places where necessary.

15. If $x = 12$ and $y = 9$, find

   a. $23 + x - y$

   b. $y(3 + x)$

   c. $\frac{1}{2}x - (y - 6)$

   d. $x + y\left(\frac{x}{3}\right)$

16. If $c = 3.2$ and $d = 1.8$, find

   a. $4c - 5d \div 2$

   b. $12cd + (c + d)$

   c. $(c - d)(c + d)$

   d. $\dfrac{cd + 14}{5c - 5d}$

17. If $h = 6.7$, $m = 3.9$, and $s = 7.8$, find

   a. $hm(2s + 1 + 0.5h)$

   b. $(3s - 2m) \div (6m - 2h)$

   c. $s^2 + 5m^2 - h^2$

   d. $\dfrac{2m^3}{0.5s^2 - 5h + 7}$

18. If $x = 2$, $y = 4$, and $t = 5$, find

   a. $95.6 - [7t + (2x + y)]$

   b. $\dfrac{(xy)^2}{y} + xyt - \dfrac{2t^2}{5x}$

   c. $\sqrt{2xy(xy-t)}$

   d. $y\left(\dfrac{x^2t}{y}\right) + \sqrt{2x^4 + t - 12}$

SOLVING FORMULAS

Each problem requires working with formulas. Substitute numerical values for letters and solve.

19. Three cells are connected as shown in Figure 12–17. In each cell the internal resistance is 2.2 ohms and the voltage is 1.5 volts. The resistance of the external circuit is 2.5 ohms. Use this formula to determine the circuit current to the nearest tenth ampere.

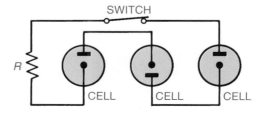

Figure 12–17

$$I = \frac{En}{rn + R}$$

where I = current in amperes
E = voltage of each cell in volts
n = number of cells
r = internal resistance of each cell in ohms
R = external resistance in ohms

20. A shaft with a 2.76-inch diameter is turned in a lathe at 200.0 revolutions per minute. The cutting speed is the number of feet that the shaft travels past the cutting tool in 1 minute. Determine the cutting speed, to the nearest foot per minute, using this formula.

$$C = \frac{3.1416DN}{12}$$

where C = cutting speed in feet per minute
D = diameter in inches
N = revolutions per minute

21. A tapered pin is shown in Figure 12–18. Compute the length of the side, to the nearest one-tenth millimeter, using this formula.

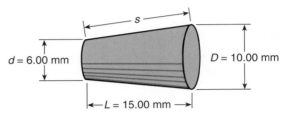

Figure 12–18

$$s = \sqrt{\left(\frac{D}{2} - \frac{d}{2}\right)^2 + L^2}$$

where s = side in millimeters
L = length in millimeters
D = diameter of larger end in millimeters
d = diameter of smaller end in millimeters

UNIT 13 ⠿ Basic Algebraic Operations

OBJECTIVES

After studying this unit you should be able to

- add and subtract single and multiple literal terms.
- multiply and divide single and multiple literal terms.
- compute powers of single and multiple literal terms.
- compute roots of single literal terms.
- remove parentheses that are preceded by plus or minus signs.
- simplify combined operations of literal term expressions.
- solve literal term problems.
- express decimal numbers as binary numbers.
- express binary numbers as decimal numbers.

A knowledge of basic operations is required in order to solve certain algebraic expressions. In solving trade applied problems, it is sometimes necessary to perform operations with literal or letter values. Formulas given in trade handbooks cannot always be used directly as given, but must be rearranged. Operations are performed to rearrange a formula so that it can be used for a particular occupational application.

13–1   Definitions

It is important that you understand the following definitions in order to apply procedures that are required for solving problems involving basic operations.

A *term* of an algebraic expression is that part of the expression that is separated from the rest by a plus or minus sign. There are five terms in this expression.

$$6x + \frac{xy}{2} - 20 + 2a^2b - dhx^3\sqrt{c}$$

A *factor* is one of two or more literal and/or numerical values of a term that are multiplied. For example, 6 and x are each factors of $6x$; 2, a^2 and b are each factors of $2a^2b$; d, h, x^3, and $\sqrt{c}$ are each factors of $dhx^3\sqrt{c}$.

It is absolutely essential that you distinguish between factors and terms.

A *numerical coefficient* is the number factor of a term. The letter factors of a term are called *literal factors*. For example, in the term $7xy$, 7 is the numerical coefficient; x and y are the literal factors.

Like terms are terms that have identical literal factors, including exponents. The numerical coefficients do not have to be the same. For example, $3a$ and $10a$ are like terms; $2x^2y$ and $5x^2y$ are like terms.

Unlike terms are terms that have different literal factors or exponents. For example, $8x$ and $8y$ are unlike terms. The terms $12xy$, $4x^2y$, and $5xy^2$ are unlike terms. Although the literal factors are x and y in each of the terms, they are raised to different powers.

13–2   Addition

Only like terms can be added. The addition of unlike terms can only be indicated. As in arithmetic, like things can be added, but unlike things cannot be added. For example, 2 inches and 3 inches = 5 inches. Both values are like units of measure. Both are inches, therefore, they can be added. It can be readily seen that 2 inches and 3 pounds cannot be added. Inches and pounds are unlike units of measure.

Procedure for Adding Like Terms

• Add the numerical coefficients, applying the procedure for addition of signed numbers.

• Leave the literal factors unchanged.

NOTE: If a term does not have a numerical coefficient, the coefficient 1 is understood. For example, $x = 1x$; $axy = 1axy$; $c^3dy^2 = 1c^3dy^2$.

EXAMPLES •————————————————————————————————

1. Add $5x$ and $10x$.

   Both terms have the identical literal factor, x. These terms are like terms.

   Add the numerical coefficients.

   $5 + 10 = 15$

   Leave the literal factor unchanged. $5x + 10x = 15x$ *Ans*

2. $R + (-12R) = -11R$ *Ans*

3. $-7xy + 7xy = 0$ *Ans*

4. $-14a^2b^3 + (-6a^2b^3) = -20a^2b^3$ *Ans*

5. $3CD + (-5CD) + 8CD = 6CD$ *Ans*

——

Procedure for Adding Unlike Terms

The addition of unlike terms can only be indicated.

EXAMPLES •————————————————————————————————

1. Add 13 and x.

   The literal factors are not identical. These terms are unlike. Indicate the addition.

   $13 + x$ *Ans*

2. Add $12M$ and $8P$.

   $12M + 8P$ *Ans*

3. Add $4W$ and $-9W^2$.

   $4W + (-9W^2)$ *Ans*

4. Add $-7a$, $-5b$, and $10ab$.

   $-7a + (-5b) + 10ab$ *Ans*

——

Procedure for Adding Expressions that Consist of Two or More Terms

- Group like terms in the same column.
- Add like terms and indicate the addition of the unlike terms.

EXAMPLES

1. Add $7x + (-xy) + 5xy^2$ and $-2x + 3xy + (-6xy^2)$.

Group like terms in the same column.	$7x + (-xy) + \quad 5xy^2$
Add the like terms.	$-2x + \quad 3xy + (-6xy^2)$
Indicate the addition of the unlike terms.	$5x + \quad 2xy + \quad (-xy^2)$ Ans

2. Add $[3c + (-4d)]$, $[c + 10d + (-4cd)]$, and $[-12c + (-5cd) + cd^2]$.

$$3c + (-4d)$$
$$c + \quad 10d + (-4cd)$$
$$-12c \quad\quad\quad + (-5cd) + cd^2$$
$$\overline{-8c + \quad 6d + (-9cd) + cd^2} \;\; Ans$$

EXERCISE 13–2

These expressions consist of groups of single terms. Add these terms.

1. $4a, 7a$
2. $6b, 12b$
3. $-7x, 3x$
4. $7x, -3x$
5. $20y, y$
6. $15xy, 7xy$
7. $-21xy, -13xy$
8. $25m^2, -m^2$
9. $-5x^2y, 5x^2y$
10. $4c^3, 0$
11. $-7pt, -pt$
12. $0.4x, -0.8x$
13. $8.3a^2b, 6.9a^2b$
14. $-0.05y, 0.006y$

15. $\frac{1}{2}xy, \frac{3}{4}xy$
16. $2\frac{7}{8}c^2d, -3\frac{1}{8}c^2d$
17. $-2.06gh^3, -0.85gh^3$
18. $-50.6abc, 50.5abc$
19. $9P, -14P, P, 5P$
20. $-0.3dt^2, -1.7dt^2, -dt^2$
21. $\frac{1}{4}xy, \frac{7}{8}xy, xy, -4xy$
22. $20.06D, -19.97D, -0.9D$
23. $6M, 0.6M, 0.06M, 0.006M$
24. $-\frac{3}{8}C, -C, 2C, -\frac{1}{16}C$

These expressions consist of groups of two or more terms. Group like terms and add.

25. $(9x + 7y), (12x + 12y)$
26. $(2a + 6b + 3c), (a + 5b + 4c)$
27. $(x + 4xy + 3y), (9x + 3xy + y)$
28. $[6a + (-10ab)], [(-a) + 12ab]$
29. $[3x + (-9xy) + y], [x + 8xy + (-y)]$
30. $[(-8cd) + 7c^2d + 14cd^2], [7cd + (-12c^2d) + (-17cd^2)]$
31. $[3x^2y + 4xy^2 + (-15x^2y^2)], [(-2x^2y) + (-5xy^2)]$
32. $[1.3M + (-3N)], [(-8M) + 0.5N], [20M + (-0.7N)]$
33. $[c + 3.6cd + (-5.7d)], [(-1.4c) + 8.6d]$
34. $[0.5T + (-2.8T^2) + (-T^3)], [5.5T^2 + 0.7T^3]$

35. $[b^4 + 4b^3c + 3b^2c], [5b^4 + (-4b^3c) + (-9b^2c)]$

36. $\left[1\frac{1}{2}P + \left(-\frac{1}{2}V\right) + \left(\frac{1}{4}PV\right)\right], \left[\frac{3}{4}P + \frac{3}{4}V + (-2PV) + (-P^2V)\right]$

Determine the literal value answers for these problems.

37. Six stamping machines produce the same product. The number of pieces per hour produced by each machine is shown in Figure 13–1. What is the total number of pieces produced per hour by all six machines?

MACHINE	#1	#2	#3	#4	#5	#6
Number of Pieces Produced per Hour	0.9x	1.2x	1.6x	0.7x	1.4x	0.8x

Figure 13–1

38. The total voltage (E_t), in volts, of the circuit shown in Figure 13–2 is represented as x. The amount of voltage taken by each of the six resistors is given as E_{R_1} through E_{R_6}. What is the sum of the voltage taken by the resistors listed?

a. R_1 and R_2

b. R_3 and R_4

c. R_4, R_5, and R_6

d. All 6 resistors

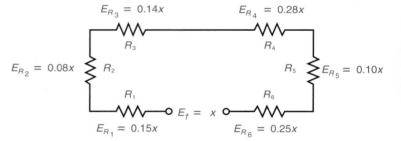

Figure 13–2

39. A checking account has a balance represented as B. The following deposits are made: $\frac{1}{2}B$, $\frac{3}{4}B$, $\frac{1}{4}B$, and $2B$. No checks are issued during this time. What is the new balance?

13–3 Subtraction

As in addition, only like terms can be subtracted. The subtraction of unlike terms can only be indicated. The same principles apply in arithmetic. For example, 7 meters − 5 meters = 2 meters. Both values are like units of measure. Both are meters; therefore, they can be subtracted. The values 7 meters and 5 liters cannot be subtracted. Meters and liters are unlike units of measure.

Procedures for Subtracting Like Terms

• Subtract the numerical coefficients applying the procedure for subtraction of signed numbers.

• Leave the literal factors unchanged.

EXAMPLES •————————————————————————————————

1. $14xy - (-6xy)$

Both terms have identical literal factors, xy. These terms are like terms.

Subtract the numerical coefficients.

$14 - (-6) = 20$

Leave the literal factor unchanged. $14xy - (-6xy) = 20xy$ *Ans*

2. $16H - 13H = 3H$ *Ans*

3. $ab - 10ab = -9ab$ *Ans*

4. $-4L^2 - 7L^2 = -11L^2$ *Ans*

5. $-15x^2y - (-15x^2y) = 0$ *Ans*

Procedure for Subtracting Unlike Terms

• The subtraction of unlike terms can only be indicated.

EXAMPLES

1. Subtract $3b$ from $4a$.

   The literal factors are not identical. These terms are unlike.

   Indicate the subtraction. $4a - 3b$ *Ans*

2. Subtract $5xy^2$ from $-16xy$.

   $-16xy - 5xy^2$ *Ans*

3. Subtract $-2P$ from $-PT$.

   $-PT - (-2P)$ or $-PT + 2P$ *Ans*

Procedure for Subtracting Expressions that Consist of Two or More Terms

• Group like terms in the same column.

• Subtract like terms and indicate the subtractions of the unlike terms.

NOTE: Each term of the subtrahend is subtracted following the procedure for subtraction of signed numbers.

EXAMPLES

1. Subtract $6a + 8b - 5c$ from $9a - 13b + 7c$.

   Group like terms in the same column.

   $$\begin{array}{r} 9a - 13b + 7c \\ -\ (6a + 8b - 5c) \end{array}$$

   Change the sign of each term in the subtrahend and follow the procedure for addition of signed numbers.

   $$\begin{array}{r} 9a - 13b + 7c \\ +\ (-6a - 8b + 5c) \\ \hline 3a - 21b + 12c \ \textit{Ans} \end{array}$$

2. Subtract $(4x^2 + 6x - 15xy) - (9x^2 - x - 2y + 5y^2)$

   $$\begin{array}{rcl} 4x^2 + 6x - 15xy & = & 4x^2 + 6x - 15xy \\ -\ (9x^2 - x -2y + 5y^2) & = & -9x^2 + x + 2y - 5y^2 \\ \hline & & -5x^2 + 7x - 15xy + 2y - 5y^2 \ \textit{Ans} \end{array}$$

EXERCISE 13–3

These expressions consist of groups of single terms. Subtract terms as indicated.

1. $5x - 3x$

2. $5x - (-3x)$

3. $-7a - a$

4. $-6ab - (-4ab)$

5. $3y^2 - 11y^2$

6. $16xy - (-16xy)$

7. $-10xy - (-10xy)$

8. $-12c^2 - c^2$

9. $-12c^2 - (-c^2)$

10. $0 - 16M$

11. $0 - (-16M)$

12. $0.5x^2y - 1.2x^2y$

13. $-18.7P - (-12.6P)$

14. $-12ax^4 - 7ax^4$

15. $0.025D - (-0.075D)$

16. $8.12n - 8.82n$

17. $\frac{1}{2}c^2d - \left(-\frac{1}{2}c^2d\right)$

18. $-\frac{1}{2}c^2d - \left(-\frac{1}{2}c^2d\right)$

19. $1\frac{3}{4}H - \frac{3}{8}H$

20. $\frac{3}{8}H - 1\frac{3}{4}H$

21. $-2.7xy - 3.4xy$

22. $2.03F - (-0.08F)$

23. $0 - (-g^2h)$

24. $-3\frac{3}{16}G - 5\frac{5}{8}G$

These expressions consist of groups of two or more terms. Group like terms and subtract.

25. $(3P^2 - 2P) - (6P^2 - 7P)$

26. $(5x + 9xy) - (2x + 6xy)$

27. $(10y^2 + 2) - (10y^2 - 2)$

28. $(10y^2 - 2) - (10y^2 - 2)$

29. $(3N + 12NS) - (4N + NS)$

30. $(ab - a^2b + ab^2) - 0$

31. $0 - (ab - a^2b + ab^2)$

32. $(0.5y - 0.7y^2) - (y - 0.2y^2)$

33. $(-3x^3 + 7x^2 - x) - (-12x^3 + 8x)$

34. $(15L - 12H) - (-12L + 6H - 4)$

35. $\left(-\frac{1}{2}x + \frac{1}{4}x^2 - \frac{1}{8}x^3\right) - \left(-\frac{1}{4}x + \frac{1}{4}x^3\right)$

36. $\left(\frac{3}{8}R - \frac{1}{8}D + 2\frac{1}{4}\right) - \left(\frac{3}{8}D - 3\frac{5}{8}\right)$

37. $(11.09e + 14.76f) - (e - f - 10.03)$

38. $(-20T + 8.5T^2 - 0.3T^3) - (T^3 + 4.4)$

Determine the literal value answers for these problems.

39. The support bracket dimensions shown in Figure 13–3 are given, in inches, in terms of x. Determine dimensions A through F.

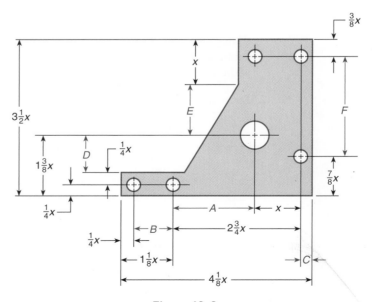

Figure 13–3

40. An employee earns a gross wage represented as *W*. The employee's payroll deductions are shown in Figure 13–4. What is the employee's net wage?

TYPE OF DEDUCTION	FEDERAL INCOME TAX	SOCIAL SECURITY	HEALTH AND ACCIDENT INSURANCE	RETIREMENT	MISCELLANEOUS
Amount of Deduction	$0.22W$	$0.076W$	$0.065W$	$0.05W$	$0.042W$

Figure 13–4

13–4   Multiplication

It was shown that unlike terms could not be added or subtracted. In multiplication, the exponents of the literal factors do not have to be the same to multiply the values. For example, x^2 can be multiplied by x^4. The term x^2 means $(x)(x)$. The term x^4 means $(x)(x)(x)(x)$.

$$(x^2)(x^4) = \underbrace{(x)(x)}_{x^2}\ \underbrace{(x)(x)(x)(x)}_{x^4} = x^{2+4} = x^6$$

Notice that this behaves the same as when numbers are multiplied.

$$(3^2)(3^4) = \underbrace{(3)(3)}_{3^2}\ \underbrace{(3)(3)(3)(3)}_{3^4} = 3^{2+4} = 3^6$$

In general, we write $a^m a^n = a^{m+n}$.

Area units of measure can be multiplied by linear units of measure. One side of the cube shown in Figure 13–5 has an area of 9 cm². The volume of the cube is determined by multiplying the area of a side (9 cm²) by the side length (3 cm).

$$\text{Volume} = (9 \text{ cm}^2)(3 \text{ cm}) = 27 \text{ cm}^3$$

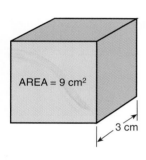

AREA = 9 cm²

3 cm

Figure 13–5

Procedure for Multiplying Two or More Terms

- Multiply the numerical coefficients, following the procedure for multiplication of signed numbers.
- Add the exponents of the same literal factors.
- Show the product as a combination of all numerical and literal factors.

EXAMPLES •

1. Multiply. $(2xy^2)(-3x^2y^3)$

Multiply the numerical coefficients following the procedure for multiplication of signed numbers. $(2)(-3) = -6$

Add the exponents of the same literal factors.

$(x^1)(x^2) = x^{1+2} = x^3$

$(y^2)(y^3) = y^{2+3} = y^5$

Show the product as a combination of all numerical and literal factors.

$(2xy^2)(-3x^2y^3) = -6x^3y^5\ Ans$

2. $(-4a^2b^3)(-5a^2b^4) = (-4)(-5)(a^{2+2})(b^{3+4}) = 20a^4b^7\ Ans$

3. $(-2)(3a)(-5b^2c^2)(-2ac^3d^3) = (-2)(3)(-5)(-2)(a^{1+1})(b^2)(c^{2+3})(d^3)$

$= -60a^2b^2c^5d^3\ Ans$

It is sometimes necessary to multiply expressions that consist of more than one term within an expression, such as $3a(6 + 4a)$ and $(2x - 4y)(-x + 5y)$.

Procedure for Multiplying Expressions that Consist of More than One Term

- Multiply each term of one expression by each term of the other expression.
- Combine like terms.

Before applying the procedure to algebraic expressions, two examples are given to show that the procedure is consistent with arithmetic.

EXAMPLES •————————————————————————————————

1. Multiply $5(7 + 4)$
   Multiply each term of one expression by each term of the other expression. (Distributive property)

   $$5(7 + 4) = 5 \cdot 7 + 5 \cdot 4$$
   $$\text{Combine} \qquad = 35 + 20$$
   $$= 55$$

2. Multiply $(6 + 3)(-4 + 2)$
   Multiply each term of one expression by each term of the other expression. (Distributive property)

   $$(6 + 3)(-4 + 2) = 6(-4) + 6(5) + 3(-4) + 3(5)$$
   $$\text{Combine} \qquad = -24 + 30 + -12 + 15$$
   $$= 9$$

3. Multiply $3a(6 + 2a^2)$
   Multiply each term of one expression by each term of the other expression. (Distributive property)

   $$3a(6 + 2a^2) = 3a(6) + 3a(2a^2) = 18a + 6a^3$$

Combine like terms. Since $18a$ and $6a^3$ are unlike terms, they cannot be combined. The answer is $18a + 6a^3$.

Foil Method

EXAMPLE •————————————————————————————————

Multiply. $(3c + 5d^2)(4d^2 - 2c)$

This is an example in which both expressions have two terms. The solution illustrates a shortcut of the distributive property called the FOIL method.

FOIL Method

Find the sum of the products of:

1. the first terms: **F**
2. the outer terms: **O**
3. the inner terms: **I**
4. the last terms: **L**

Then combine like terms.

$$(3c + 5d^2)(4d^2 - 2c) = 3c(4d^2) + 3c(-2c) + 5d^2(4d^2) + 5d^2(-2c)$$
$$12cd^2 + (-6c^2) + 20d^4 + (-10cd^2)$$

Product of **First** terms	Product of **Outer** terms	Product of **Inner** terms	Product of **Last** terms
F	O	I	L

Combine like terms.

┌────── COMBINE ──────┐

$$12cd^2 + (-6c^2) + 20d^4 + (-10cd^2)$$
$$= 2cd^2 + (-6c^2) + 20d^4$$
$$= 2cd^2 - 6c^2 + 20d^4 \text{ Ans}$$

EXERCISE 13–4

These expressions consist of single terms. Multiply these terms as indicated.

1. $(x)(2x^2)$

2. $(4ab)(6a^2b^2)$

3. $(9c^3)(-6c^4)$

4. $(-10x)(-7x^2)$

5. $(9ab^2c)(3a^2bc^4)$

6. $(-5x^2y^2)(x^3y^3)$

7. $(6cd^2)(2c^4d)$

8. $(-15M)(0)$

9. $(-12P^5N^4)(-P^3)$

10. $(-12P^5N^4)(-P^3)(-1)$

11. $(2.5x^4y)(0.5y^3)$

12. $(0.2ST^2)(-0.3S^4)$

13. $(15V^2)(0)(-2V)$

14. $\left(\dfrac{1}{2}x^3\right)\left(\dfrac{1}{4}x^2y\right)$

15. $\left(-\dfrac{3}{4}x^2y\right)\left(-\dfrac{5}{8}x^3y^2\right)$

16. $(-6)(-LW^2)(W)$

17. $(a^2b)(bc)(d)$

18. $(-a^2b)(bc)(d)$

19. $(0.6F)(3F^2G)(G^2)$

20. $(-4.2m^2n)(-5m^3)(-m)$

21. $(-5x^2y)(4xy^4)(-3n)$

22. $\left(-\dfrac{1}{3}B\right)\left(\dfrac{1}{2}C^2D\right)(-BD^2)$

23. $(x^3y^2)(-x^2y)(-x)$

24. $(0.06H^2L^2)(-1)(5L)$

These expressions consist of groups of two or more terms. Multiply as indicated and combine like terms where possible.

25. $2x(3x + y)$

26. $-a^3(a^2 + b + b^3)$

27. $3M(M^2 - MN)$

28. $-6x^3(-x - x^2y)$

29. $10c^2d(3cd^3 - 4d)$

30. $-2(PV + V^2 - 6)$

31. $r^2t^3s(-r^2s^2 + s - r^3t)$

32. $-0.3L^2H(-0.4H^3 - L + 4H^2)$

33. $-1(f^2g - 9fg^2 + 12fh)$

34. $-1(-f^2g + 9fg^2 - 12fh)$

35. $(x + y)(x + y)$

36. $(x - y)(x + y)$

37. $(x - y)(x - y)$

38. $(A + F)(A - 20)$

39. $(7a + 12)(a^4 + 3)$

40. $(2m^2 - n^3)(-3m^3 + 6n)$

41. $(3ax^2 + cx)(7a^2x^3 - c^2)$

42. $(-0.1S^3T - 10)(0.5T^2 - 0.2S^2)$

43. $(4x^2y^3 - 6xy)(4x^2y^3 + 6xy)$

44. $(-4x^2y^3 + 6xy)(-4x^2y^3 + 6xy)$

Determine the literal value answers for these problems.

45. A gear speed is measured in revolutions per minute (r/min). The speed of a gear is represented by N.

   a. How many revolutions does the gear turn in 20 minutes?

   b. When the gear speed is reduced by 35 revolutions per minute, the speed is $N - 35$ r/min. How many revolutions does the gear turn in 4.25 minutes at the reduced speed?

46. A rectangle is shown in Figure 13–6. The area of a rectangle equals its length times its width.

  a. What is the area of the rectangle?

  b. What is the area of the rectangle if the width is increased by 3 inches?

  c. What is the area of the rectangle if the width is increased by 5 inches and the length is decreased by 6 inches?

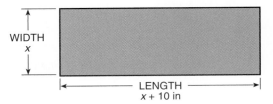

WIDTH
x

LENGTH
$x + 10$ in

Figure 13–6

47. Power (watts) is equal to the product of current (amperes) and voltage (volts). In the electrical circuit shown in Figure 13–7, the current received by each of the resistors R_1 through R_5 is $3y$ and the voltage is $0.2x$. The amperes received by each of the resistors R_6 through R_9 is $5y$, and the voltage is $0.25x$.

  a. What is the power, in watts, of each of the resistors R_1 through R_5?

  b. What is the power, in watts, of each of the resistors R_6 through R_9?

  c. What is the total power, in watts, of resistors R_1 through R_5?

  d. What is the total power, in watts, of resistors R_6 through R_9?

  e. What is the total power, in watts, of the complete circuit?

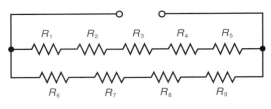

Figure 13–7

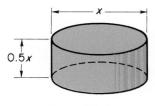

Figure 13–8

48. The diameter, in meters, of the storage tank shown in Figure 13–8 is x and the height is $0.5x$. The approximate volume (number of cubic meters) of the storage tank is computed by multiplying 0.7854 by the square of the diameter by the height.

  a. Determine the volume (number of cubic meters) of the tank when the tank is full.

  b. Determine the volume (number of cubic meters) in the tank when the tank is filled to a height 6 meters below the top of the tank.

13–5   Division

As with multiplication, the exponents of the literal factors do not have to be the same to divide the values. For example, x^4 can be divided by x.

$$\frac{x^4}{x} = \frac{(x)(x)(x)(x)}{x} = x^{4-1} = x^3$$

This behaves the same as when numbers are divided.

$$\frac{3^7}{3^2} = \frac{3 \cdot 3 \cdot 3 \cdot 3 \cdot 3 \cdot \cancel{3} \cdot \cancel{3}}{\cancel{3} \cdot \cancel{3}}$$

$$= 3^{7-2} = 3^5$$

In general,

$$\frac{a^m}{a^n} = a^{m-n}, \quad \text{if } a \neq 0$$

Volume units of measure can be divided by linear units of measure. The container shown in Figure 13–9 has a volume of 480 cm³ and a height of 6 cm. The area of the container top is determined by dividing the volume by the height.

$$\text{Top area} = \frac{480 \text{ cm}^3}{6 \text{ cm}} = 80 \text{ cm}^2$$

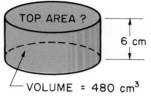

TOP AREA ?

6 cm

VOLUME = 480 cm³

Figure 13–9

Procedure for Dividing Two Terms

• Divide the numerical coefficients following the procedure for division of signed numbers.
• Subtract the exponents of the literal factors of the divisor from the exponents of the same literal factors of the dividend.
• Combine numerical and literal factors.

EXAMPLES •————————————————————————————

1. Divide. $-12a^3 \div 3a$

   Divide the numerical coefficients following the procedure for signed numbers.

   $-12 \div 3 = -4$

   Subtract the exponents of the literal factors in the divisor from the exponents of the same literal factors in the dividend.

   $a^3 \div a = a^{3-1} = a^2$

   Combine the numerical and literal factors.

   $\dfrac{-12a^3}{3a} = -4a^2 \; Ans$

2. Divide. $(-20a^3x^5y^2) \div (-2ax^2)$

   Divide the numerical coefficients following the procedure for signed numbers.

   $-20 \div -2 = 10$

   Subtract the exponents of the literal factors in the divisor from the exponents of the same literal factors in the dividend.

   $a^{3-1} = a^2$
   $x^{5-2} = x^3$
   $y^2 = y^2$

   Combine. $\dfrac{-20a^3x^5y^2}{-2ax^2} = 10a^2x^3y^2 \; Ans$

——•

In arithmetic, any number except 0 divided by itself equals 1. For example, $5 \div 5 = 1$. Applying the division procedure, $5 \div 5 = 5^{1-1} = 5^0$. Therefore, $5^0 = 1$. Any number except 0 raised to the zero power equals 1.

EXAMPLES •————————————————————————————————

1. $\dfrac{6^2}{6^2} = 6^{2-2} = 6^0 = 1 \; Ans$

2. $\dfrac{a^3b^2c}{a^3b^2c} = (a^{3-3})(b^{2-2})(c^{1-1}) = a^0b^0c^0 = (1)(1)(1) = 1 \; Ans$

3. $\dfrac{4P}{P} = 4P^{1-1} = 4P^0 = 4(1) = 4 \; Ans$

Procedure for Dividing when the Dividend Consists of More than One Term

- Divide each term of the dividend by the divisor, following the procedure for division of signed numbers.
- Combine terms.

Before the procedure is applied to algebraic expressions, an example is given to show that the procedure is consistent with arithmetic.

ARITHMETIC EXAMPLE •————————————————————————

From arithmetic:

$(12 + 8) \div 4 = 20 \div 4 = 5 \; Ans$

From algebra:

Divide each term of the dividend by the divisor.

$\dfrac{12 + 8}{4} = \dfrac{12}{4} + \dfrac{8}{4} = 3 + 2$

Combine terms. $3 + 2 = 5 \; Ans$

ALGEBRA EXAMPLE •————————————————————————————

Divide. $\dfrac{-16x^2y + 8x^3y^2 + 24x^5y^3z}{-8x^2y}$

Divide each term of the dividend by the divisor.

$-16x^2y \div -8x^2y = 2x^{2-2}y^{1-1} = 2x^0y^0 = 2(1)(1) = 2$

$+8x^3y^2 \div -8x^2y = -1x^{3-2}y^{2-1} = -1xy = -xy$

$+24x^5y^3z \div -8x^2y = -3x^{5-2}y^{3-1}z^{1-0} = -3x^3y^2z$

Combine. $2 - xy - 3x^3y^2z \; Ans$

EXERCISE 13–5

These exercises require division of single terms. Divide terms as indicated.

1. $4x^2 \div 2x$

2. $21a^4 \div 7a$

3. $12T^3 \div 4T^3$

4. $-16a^4b^5 \div 4ab^3$

5. $25x^3y^4 \div 5x^2$

6. $FS^2 \div (-FS^2)$

7. $-FS^2 \div (-FS^2)$

8. $0 \div 14mn$

9. $(-42a^5d^2) \div (-7a^2d^2)$

10. $(-3.6H^2P) \div (0.6HP)$

11. $DM^2 \div (-1)$

12. $-DM^2 \div (-1)$

13. $-8.4ab \div ab$

14. $0.8PV^2 \div (-0.2V)$

15. $1\frac{1}{4}c^2d^3 \div \frac{1}{4}cd^2$

16. $\left(-\frac{1}{2}x^3y^3\right) \div \frac{1}{8}x^3$

17. $-6g^3h^2 \div \left(-\frac{3}{4}gh\right)$

18. $-24x^2y^5 \div (-0.5x^2y^4)$

19. $x^2y^3z^4 \div xy^3z$

20. $18a^2bc^2y \div (-a^2)$

21. $\frac{1}{4}P^2V \div \frac{1}{16}$

22. $-0.08xy \div 0.4y$

23. $-9.6x^2yz \div (-1.2x)$

24. $\frac{3}{4}FS^3 \div (-3S)$

These exercises consist of expressions in which the dividends have two or more terms. Divide as indicated.

25. $(6x^3 + 10x^2) \div 2x$

26. $(6x^3 - 10x^2) \div (-2x)$

27. $(28x^3y^2 - 14x^2y) \div 7xy$

28. $(35a^2 + 15a^3) \div (-a)$

29. $(14M - 12MN) \div (-1)$

30. $(21b - 24c) \div 3$

31. $(-36a^2b^5 - 27a^3b^4) \div (-9ab^4)$

32. $(15TF^2 - 45T^2F + 30) \div (-15)$

33. $(0.6x^4y^5 + 0.2x^3y^4) \div 2x^3y^2$

34. $(-4.5D^3H + 0.3D^2 - 7.5D^4) \div 1.5D^2$

35. $(-x^3y^3z + x^2z^4 + x^3y) \div (-x^2)$

36. $(5MN^3P^2 - 2M^3N^3P) \div 0.01MN^2P$

37. $\left(\frac{1}{2}a^2c - \frac{3}{4}a^3c^2 - ac^3\right) \div \frac{1}{8}ac$

38. $(-2.5e^2f - 0.5ef^2 + e^2f^2) \div 0.5f$

39. $\left(\frac{3}{10}EG^2 + \frac{7}{10}E^2G^3 - \frac{9}{10}E^3GH\right) \div \left(-\frac{1}{10}EG\right)$

40. $(0.8x^2y^3z - 0.6xy^2z^2 + 0.4y^2) \div (0.2y^2)$

Determine the literal value answers for the following problems.

Figure 13–10

41. A concrete slab is shown in Figure 13–10. The thickness of a slab is determined by dividing the slab volume by the face area. The face area is determined by dividing the volume by the thickness. Dimensions of six slabs of different sizes are given in the table in Figure 13–11. Determine either the thickness or face area required.

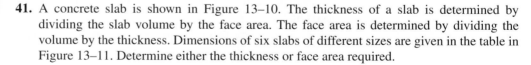

	VOLUME (cubic meters)	FACE AREA (square meters)	THICKNESS (meters)		VOLUME (cubic feet)	FACE AREA (square feet)	THICKNESS (feet)
a.	$1\,000x^3$	$500x^2$	?	**d.**	$2,400x^3$	?	$0.25x$
b.	$750x^3$	$1\,500x^2$	?	**e.**	$2,100x^3$	?	$1.4x$
c.	$1\,920x^3$	$1\,600x^2$	?	**f.**	$360x^3$	?	$0.09x$

Figure 13–11

42. Twenty loaves of bread are made from a batch of dough. The weight, in pounds, of the dough is represented by x.

   a. What is the weight of each of the 20 loaves?

b. If the batch of dough is increased by an additional 5 pounds of dough, what is the weight of each of the 20 loaves?

c. If the batch of dough is decreased by 0.2x, what is the weight of each of the 20 loaves?

43. The daily sales amounts of a company for one week are shown in the chart in Figure 13–12. Determine the average daily sales amount when x represents a number of dollars.

DAY	MONDAY	TUESDAY	WEDNESDAY	THURSDAY	FRIDAY
Amount	8x + $360	10x + $240	6x + $400	10x	6x + $500

Figure 13–12

44. Refer to the drill jig shown in Figure 13–13 and determine, in millimeters, distances A, B, C, and D. The values of x and y represent a number of millimeters.

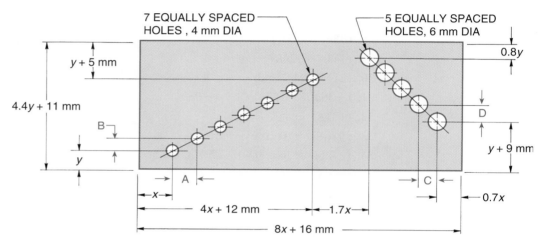

Figure 13–13

13–6   Powers

Procedure for Raising a Single Term to a Power

- Raise the numerical coefficients to the indicated power following the procedure for powers of signed numbers.
- Multiply each of the literal factor exponents by the exponent of the power to which it is raised.
- Combine numerical and literal factors.

Algebraically, this is written $(a^m)^n = a^{mn}$.

ARITHMETIC EXAMPLE •————————————————————

Raise to the indicated power. $(2^2)^3$

From arithmetic:

$$(2^2)^3 = (4)^3 = (4)(4)(4) = 64 \ Ans$$

From algebra:

$$(2^2)^3 = 2^{2(3)} = 2^6 = (2)(2)(2)(2)(2)(2) = 64 \ Ans$$

———•

ALGEBRA EXAMPLES •————————————————————————————————

1. Raise to the indicated power. $(5x^3)^2$

Raise the numerical coefficient to the indicated power following the procedure for powers of signed numbers.

$5^2 = 25$

Multiply each literal factor exponent by the exponent of the power to which it is to be raised.

$(x^3)^2 = x^{3(2)} = x^6$

Combine numerical and literal factors. $(5x^3)^2 = 25x^6$ *Ans*

NOTE: $(x^3)^2$ is not the same as $x^3 x^2$.

$(x^3)^2 = (x^3)(x^3) = (x)(x)(x)(x)(x)(x) = x^6$

$x^3 x^2 = (x)(x)(x)(x)(x) = x^5$

2. Raise to the indicated power. $(-4x^2 y^4 z)^3$

Raise the numerical coefficient to the indicated power.

$(-4)^3 = (-4)(-4)(-4) = -64$

Multiply the exponents of the literal factors by the indicated power.

$(x^2 y^4 z)^3 = x^{2(3)} y^{4(3)} z^{1(3)} = x^6 y^{12} z^3$

Combine. $(-4x^2 y^4 z)^3 = -64x^6 y^{12} z^3$ *Ans*

3. Raise to the indicated power. $\left[-\dfrac{1}{4} a^3 (bc^2)^3 d^4 \right]^2$

Perform the correct order of operations.

Remove the innermost parentheses first. $(bc^2)^3 = b^3 c^6$ $\left[-\dfrac{1}{4} a^3 (bc^2)^3 d^4 \right]^2 = \left[-\dfrac{1}{4} a^3 b^3 c^6 d^4 \right]^2$

Apply the power procedure.

$\left(-\dfrac{1}{4} \right)^2 = \dfrac{1}{16}$

$(a^3)^2 = a^6$

$(b^3)^2 = b^6$

$(c^6)^2 = c^{12}$

$(d^4)^2 = d^8$

Combine. $\dfrac{1}{16} a^6 b^6 c^{12} d^8$ *Ans*

——

Procedure for Raising Two or More Terms to a Power

• Apply the procedure for multiplying expressions that consist of more than one term.

EXAMPLE •——

Solve. $(3a + 5b^3)^2$ Apply the FOIL method.

	F	O	I	L
	Step 1	Step 2	Step 3	Step 4

$(3a + 5b^3)(3a + 5b^3) = 3a(3a) + 3a(5b^3) + 5b^3(3a) + 5b^3(5b^3)$

$ 9a^2 \quad + \quad 15ab^3 + 15ab^3 \quad + \quad 25b^6$

$ \underline{\text{COMBINE}}$

$ 9a^2 + 30ab^3 + 25b^6$ *Ans*

EXERCISE 13–6

These exercises consist of expressions with single terms. Raise terms to the indicated powers.

1. $(ab)^2$

2. $(DF)^3$

3. $(3ab)^2$

4. $(-4xy)^2$

5. $(2x^2y)^3$

6. $(3a^3b^2)^3$

7. $(-3c^3d^2e^4)^3$

8. $(2MS^2)^4$

9. $(-7x^4y^5)^2$

10. $(-3N^2P^2T^3)^4$

11. $(a^2bc^3)^3$

12. $(-2a^2bc^3)^3$

13. $(-x^4y^5z)^3$

14. $(9C^4F^2H)^2$

15. $(0.3x^4y^2)^3$

16. $(-0.5c^2d^3e)^3$

17. $(3.2M^3NP^2)^2$

18. $\left(\dfrac{1}{2}x^3y^2z\right)^2$

19. $\left(\dfrac{3}{4}abc^3\right)^3$

20. $\left[-8(a^2b^3)^2c\right]^2$

21. $\left[-3x^2(y^2)^2z^3\right]^3$

22. $\left[0.6d^3(ef^2)^3\right]^2$

23. $\left[\dfrac{5}{8}(a^2bc^3)^2\right]^2$

24. $\left[(-2x^2y)^2(xy^2)^2\right]^3$

These exercises consist of expressions of more than one term. Raise these expressions to the indicated powers and combine like terms where possible.

25. $(a + b)^2$

26. $(a - b)^2$

27. $(x^2 + y)^2$

28. $(D^3 + G^4)^2$

29. $(3x^3 - 2y^2)^2$

30. $(6m^2 - 5n^2)^2$

31. $(x^2y^3 + xy^2)^2$

32. $(0.9c^3e^2 - 0.2e)^2$

33. $(4.5M^2P - P^4)^2$

34. $\left(\dfrac{3}{4}d^2h + \dfrac{1}{4}dh^2\right)^2$

35. $\left[(a^2)^3 - (b^3)^2\right]^2$

36. $\left[(-x^4y)^2 + (x^2y)^3\right]^2$

Determine the literal value answers for the following problems.

37. The volume of a cube shown in Figure 13–14 is computed by cubing the length of a side ($V = s^3$). Express the volumes of cubes for each of the side lengths given in Figure 13–15.

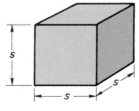

Figure 13–14

	LENGTH OF SIDE (s)	VOLUME (V)
a.	$5x$	?
b.	$0.2x$	?
c.	$10\frac{1}{2}x$	?
d.	$1.8x$	?

Figure 13–15

38. The approximate area of the circle shown in Figure 13–16 is computed by multiplying 3.14 by the square of the radius ($A = 3.14R^2$). Express the areas of circles for each radius given in Figure 13–17. Round answers to 2 decimal places where necessary.

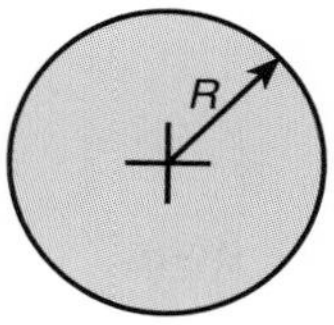

	LENGTH OF RADIUS (R)	AREA (A)
a.	5x	?
b.	0.3x	?
c.	8.3x	?
d.	2x + 4 ft	?
e.	x + 1.5 mm	?

Figure 13–16

Figure 13–17

39. The approximate volume of the sphere shown in Figure 13–18 is computed by multiplying 0.52 by the cube of the diameter ($V = 0.52D^3$). Express the volumes of spheres for each diameter given in Figure 13–19. Round answers to 2 decimal places where necessary.

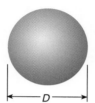

	LENGTH OF DIAMETER (D)	VOLUME (V)
a.	2x	?
b.	0.5x	?
c.	1.1x	?
d.	x	?

Figure 13–18

Figure 13–19

13–7   Roots

Procedures for Extracting the Root of a Term

- Determine the root of the numerical coefficient following the procedure for roots of signed numbers.
- The roots of the literal factors are determined by dividing the exponent of each literal factor by the index of the root.
- Combine the numerical and literal factors.

ARITHMETIC EXAMPLE •————————————————————

$\sqrt{2^6}$

From arithmetic:

$$\sqrt{2^6} = \sqrt{(2)(2)(2)(2)(2)(2)} = \sqrt{64} = 8 \, Ans$$

From algebra:

$\sqrt{2^6}$ has an exponent of 6 and an index of 2. Divide the exponent by the index.

$$\sqrt{2^6} = 2^{6 \div 2} = 2^3 = (2)(2)(2) = 8 \, Ans$$

ALGEBRA EXAMPLES •————————————————————

1. Solve. $\sqrt{36x^4y^2z^6}$

Determine the root of the numerical coefficient. $\sqrt{36} = 6$

Determine the roots of the literal factors by dividing the exponents of each literal factor by the index of the root.

$$\sqrt{x^4} = x^{4 \div 2} = x^2$$
$$\sqrt{y^2} = y^{2 \div 2} = y$$
$$\sqrt{z^6} = z^{6 \div 2} = z^3$$

Combine. $\sqrt{36x^4y^2z^6} = 6x^2yz^3$ *Ans*

2. Solve. $\sqrt[3]{-27ab^6c^2}$

Determine the cube root of -27.

$$\sqrt[3]{-27} = -3$$

Divide the exponent of each literal factor by the index 3.

$$\sqrt[3]{a} = \sqrt[3]{a}$$
$$\sqrt[3]{b^6} = b^{6 \div 3} = b^2$$
$$\sqrt[3]{c^2} = \sqrt[3]{c^2}$$

Combine. $\sqrt[3]{-27ab^6c^2} = -3b^2\sqrt[3]{ac^2}$ *Ans*

NOTE: Roots of expressions that consist of two or more terms *cannot* be extracted by this procedure. For example, $\sqrt{x^2 + y^2}$ consists of two terms and does *not* equal $\sqrt{x^2} + \sqrt{y^2}$. This mistake, commonly made by students, must be avoided. This fact is consistent with arithmetic.

$$\sqrt{3^2 + 4^2} = \sqrt{9 + 16} = \sqrt{25} = 5 \ Ans$$
$$\sqrt{3^2 + 4^2} \neq \sqrt{3^2} + \sqrt{4^2}$$
$$\sqrt{3^2 + 4^2} \neq 3 + 4$$
$$\sqrt{3^2 + 4^2} \neq 7$$

Fractional Exponents

Fractional exponents can be used to indicate roots.

- $a^{1/n} = \sqrt[n]{a}$
- $a^{m/n} = \left(\sqrt[n]{a}\right)^m = \sqrt[n]{a^m}$

EXAMPLES •

1. Solve $49^{1/2}$.

$$49^{1/2} = \sqrt{49} = 7$$

2. Evaluate $8^{1/3}$.

$$8^{1/3} = \sqrt[3]{8} = 2.$$

3. Simplify $(81y^6)^{1/2}$.

$$(81y^6)^{1/2} = \sqrt{81y^6}$$
$$81^{1/2} = \sqrt{81} = 9$$
$$(y^6)^{1/2} = \sqrt{y^{6/2}} = y^3$$

Combine. $(81y^6)^{1/2} = 9y^3$

4. Solve $(125x^6y^9)^{1/3}$

$$125^{1/3} = \sqrt[3]{125} = 5$$
$$(x^6)^{1/3} = x^{6/3} = x^2$$
$$(y^9)^{1/3} = y^{9/3} = y^3$$

Combine. $(125x^6y^9)^{1/3} = 5x^2y^3$

EXERCISE 13–7

Determine the roots of these terms.

1. $\sqrt{4a^2b^2c^2}$

2. $\sqrt{4x^2y^4}$

3. $\sqrt{16c^2d^6}$

4. $\sqrt[3]{64x^3y^9}$

5. $\sqrt[3]{-64x^3y^9}$

6. $\sqrt{m^4n^2s^6}$

7. $(25f^2g^8)^{1/2}$

8. $(81x^8y^6)^{1/2}$

9. $\sqrt{49c^2d^6e^{10}}$

10. $\sqrt[3]{8p^6t^3w^9}$

11. $\sqrt[3]{-125x^9y^3}$

12. $\sqrt{0.36x^2y^6}$

13. $(0.64a^6c^8f^2)^{1/2}$

14. $(25ab^2)^{1/2}$

15. $\sqrt{100xy}$

16. $\sqrt[3]{64ab^3}$

17. $\sqrt{144mp^4s}$

18. $\sqrt{\dfrac{4}{9}a^2b^4c^6}$

19. $\sqrt{\dfrac{1}{16}xy^4}$

20. $\sqrt[3]{\dfrac{8}{27}m^3n^6}$

21. $\sqrt[3]{-64d^6t^9}$

22. $\sqrt[3]{-8ef^2}$

23. $\sqrt[3]{27d^2ef^6}$

24. $\sqrt[4]{16x^4y^2}$

25. $\sqrt[5]{32h^{10}}$

26. $\sqrt[5]{-32a^2b^5}$

27. $(c^6dt^9)^{1/3}$

28. $(-n^2xy^3)^{1/3}$

29. $\sqrt{\dfrac{4}{9}a^4bc^6}$

30. $\sqrt[3]{-\dfrac{1}{64}x^2y^3z}$

Determine the literal value answers for these problems.

31. The length of a side of the cube shown in Figure 13–20 is computed by taking the cube root of the volume $\left(s = \sqrt[3]{V}\right)$. Express the lengths of sides for each of the cube volumes given in Figure 13–21.

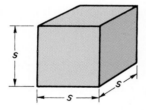

Figure 13–20

	VOLUME (V)	LENGTH OF EACH SIDE (s)
a.	$64x^3$	?
b.	$0.027x^3$	?
c.	$\dfrac{8}{27}x^3$	?
d.	$\dfrac{27}{64}x^3$	?

Figure 13–21

32. The approximate radius of the circle shown in Figure 13–22 is computed by dividing the area by 3.14 and taking the square root of the quotient $\left(R = \sqrt{\dfrac{A}{3.14}}\right)$. Express the radius for each of the circle areas given in Figure 13–23.

Figure 13–22

	AREA (A)	RADIUS (R)
a.	$12.56x^2$	?
b.	$50.24x^2$	?
c.	$0.1256x^2$	?
d.	$0.2826x^2$	?

Figure 13–23

33. The approximate diameter of the sphere shown in Figure 13–24 is computed by multiplying the cube root of the volume by 1.24 $\left(D = 1.24\sqrt[3]{V}\right)$. Express the diameters of each of the volumes given in Figure 13–25 on page 324.

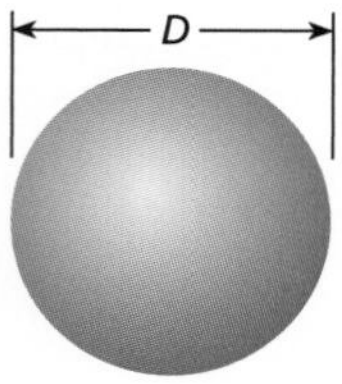

	VOLUME (*V*)	DIAMETER (*D*)
a.	$64x^3$	?
b.	$1{,}000x^3$	?
c.	$0.027x^3$	?
d.	$0.125x^3$	?

Figure 13–24   Figure 13–25

13–8   Removal of Parentheses

In certain expressions, terms are enclosed within parentheses or other grouping symbols such as brackets, [ ], braces, { }, or absolute values, | |, that are preceded by a plus or minus sign. In order to combine like terms, it is necessary to first remove the grouping symbols.

Procedure for Removal of Parentheses Preceded by a Plus Sign

• Remove the parentheses without changing the signs of any terms within the parentheses.
• Combine like terms.

EXAMPLE •───────────────────────────────

$3x + (5x - 8 + 6y) = 3x + 5x - 8 + 6y = 8x - 8 + 6y$ *Ans*

Procedure for Removal of Parentheses Preceded by a Minus Sign

• Remove the parentheses and change the sign of each term within the parentheses.
• Combine like terms.

EXAMPLES •───────────────────────────────

1. $8x - (4y - 6a + 7) = 8x - 4y + 6a - 7$ *Ans*
2. $-(7a^2 + b - 3) + 12 - (-b + 5)$
$\quad = -7a^2 - b + 3 + 12 + b - 5$
$\quad = -7a^2 + 10$ *Ans*

EXERCISE 13–8

Remove parentheses and combine like terms where possible.

1. $4a + (3a - 2a^2 + a^3)$ **9.** $-10c^3 - (-8c^3 - d + 12)$
2. $4a - (3a - 2a^2 + a^3)$ **10.** $-10c^3 + (-8c^3 - d + 12)$
3. $9b - (15b^2 - c + d)$ **11.** $-(16 + xy - x) + (-x)$
4. $15 + (x^2 - 10)$ **12.** $18 - (r^3 + r^2) + (r^2 - 11)$
5. $8y^2 - (y^2 + 12)$ **13.** $-(a^2 + b^2) + (a^2 + b^2)$
6. $-7m + (3m + m^2)$ **14.** $-(3x + xy - 6) + 12 + (x + xy)$
7. $xy^2 - (xy + xy^2)$ **15.** $20 + (cd - c^2d + d) + 14 - (cd + d)$
8. $-(ab + a^2b - 6a)$ **16.** $20 - (cd - c^2d + d) - 14 + (cd + d)$

13–9   Combined Operations

Expressions that consist of two or more different operations are solved by applying the proper order of operations.

Order of Operations

- First, do all operations within grouping symbols. Grouping symbols are parentheses ( ), brackets [ ], braces { }, and absolute values | |.
- Second, do powers and roots.
- Next, do multiplication and division operations in order from left to right.
- Last, do addition and subtraction operations in order from left to right.

EXAMPLES •────────────────────────────

1. Simplify. $5b + 4b(5 + a - 2b^2)$

   Multiply. $\hspace{4cm}$ $5b + 4b(5 + a - 2b^2)$

   $4b(5 + a - 2b^2) = 20b + 4ab - 8b^3$ $\hspace{1cm}$ $5b + 20b + 4ab - 8b^3$

   Combine like terms.

   $5b + 20b = 25b$ $\hspace{3cm}$ $25b + 4ab - 8b^3$ *Ans*

2. Simplify. $18 + \sqrt{\dfrac{32x^3}{2x}} + 3(7x + 6y) - 2y$

   Perform the division under the
   radical symbol. $\hspace{3cm}$ $18 + \sqrt{\dfrac{32x^3}{2x}} + 3(7x + 6y) - 2y$

   $\dfrac{32x^3}{2x} = 16x^2$ $\hspace{3cm}$ $18 + \sqrt{16x^2} + 3(7x + 6y) - 2y$

   Take the square root.

   $\sqrt{16x^2} = 4x$ $\hspace{3cm}$ $18 + 4x + 3(7x + 6y) - 2y$

   Multiply.

   $3(7x + 6y) = 21x + 18y$ $\hspace{2cm}$ $18 + 4x + 21x + 18y - 2y$

   Combine like terms.

   $4x + 21x = 25x$

   $18y - 2y = 16y$ $\hspace{3cm}$ $18 + 25x + 16y$ *Ans*

3. Simplify. $15a^6b^3 + (2a^2b)^3 - \dfrac{a^7(b^3)^2}{ab^3}$

   Raise to the indicated powers. $\hspace{1cm}$ $15a^6b^3 + (2a^2b)^3 - \dfrac{a^7(b^3)^2}{ab^3}$

   $(2a^2b)^3 = 8a^6b^3$

   $a^7(b^3)^2 = a^7b^6$ $\hspace{3cm}$ $15a^6b^3 + 8a^6b^3 - \dfrac{a^7b^6}{ab^3}$

   Divide.

   $\dfrac{a^7b^6}{ab^3} = a^6b^3$ $\hspace{3cm}$ $15a^6b^3 + 8a^6b^3 - a^6b^3$

   Combine like terms.

   $15a^6b^3 + 8a^6b^3 - a^6b^3 = 22a^6b^3$ $\hspace{1cm}$ $22a^6b^3$ *Ans*

These expressions consist of combined operations. Simplify.

1. $14 + 4(3a) + a$

2. $23 - 5(-2x) - 4x$

3. $20 - 2(5xy)^2 + x^2y^2$

4. $7 + 4(M^2N^2) - (MN)^2$

5. $6(c^2 - d) + c^2 - 3d$

6. $(10 - P^2)(4 + P^2) + P$

7. $(2 + H^2)(3 + H^2) - 4H^4$

8. $-7(x^2 + 5)^2 + (10x)^2$

9. $(xy \div x) - (x^2y \div x) + 5x$

10. $(15 - 20L - 9L^2) \div 2 + 7L^2$

11. $(8ab^8) \div (4ab^2) - (b^2)^3 + 12$

12. $\sqrt{36x^4y^2} - 5y(2x)^2$

13. $\sqrt{(64D^6)} \div 4 \div D^2$

14. $(20x^6 + 12x^4) \div (2x)^2 + x^4$

15. $5a[-6 + (ab^2)^3 - 10]$

16. $[9 - (xy^2)^2]^2$

17. $-c[-12 + (c^2d^2) - 4c] + 8c$

18. $(M + 3N)^2 + (M - 3N)^2 - 5M^2$

19. $x(x - 4y)^2 - x(x + 4y)^2$

20. $(10f^6 + 12f^4h) \div \sqrt{4f^4}$

13–10 Basic Structure of the Binary Numeration System

Many different kinds of data (numeric, text, audio-visual, physical) are stored and processed in computers. Each computer circuit assumes only one of two states: ON (binary system digit 1) or OFF (binary system digit 0). The two values, 1 and 0, are binary digits or bits. Bytes, as well as bits, are used in computer data transmission. A byte is group of bits representing a single character or digit of data. The typical computer uses codes that assign either seven or eight bits per character.

Many personal computer systems use the American Standard Code for Information Interchange (ASCII). Each ASCII character has a corresponding 7-bit code. The code can display 2^7 or 128 characters. An eighth bit may be added for a parity check, otherwise the eighth bit is 0. Various patterns of 1 and 0 in a binary code are used to represent alphanumeric (text) and control characters. For example, the pattern 01000101 represents the letter E. The pattern 00111101 represents the equal sign (=).

As previously stated, the binary system digits 1 and 0 are the building blocks for the binary code and are used to represent data and program instructions for computers. Therefore, it is important to have a basic understanding of the binary system. An understanding of the structure of the decimal system is helpful in discussing the binary system.

Structure of the Decimal System

The elements of a mathematical system are the base of the system, the particular digits used, and the locations of the digits with respect to the decimal point (place value). In the decimal system, all numbers are combinations of the digits 0–9. The decimal system is built on powers of the base 10. Each place value is ten times greater than the place value directly to its right. Since any nonzero number with an exponent of 0 equals 1, 10^0 equals 1.

An analysis of the number 64,216 (Figure 13–26) shows this structure.

6	4	2	1	6	Number
$10^4 = 10,000$	$10^3 = 1000$	$10^2 = 100$	$10^1 = 10$	$10^0 = 1$	Place Value
$6 \times 10^4 =$	$4 \times 10^3 =$	$2 \times 10^2 =$	$1 \times 10^1 =$	$6 \times 10^0 =$	Value
$6 \times 10,000 =$	$4 \times 1000 =$	$2 \times 100 =$	$1 \times 10 =$	$6 \times 1 =$	
60,000	4000	200	10	6	
60,000 $+$	4000 $+$	200 $+$	10 $+$	6 $=$	64,216

Figure 13 –26

EXAMPLES

Analyze the following numbers.

1. $16 = 1(10^1) + 6(10^0) = 10 + 6 \, Ans$

2. $216 = 2(10^2) + 1(10^1) + 6(10^0) = 200 + 10 + 6 \, Ans$

3. $4216 = 4(10^3) + 2(10^2) + 1(10^1) + 6(10^0) = 4000 + 200 + 10 + 6 \, Ans$

4. $64{,}216 = 6(10^4) + 4(10^3) + 2(10^2) + 1(10^1) + 6(10^0)$
$= 60{,}000 + 4000 + 200 + 10 + 6 \, Ans$

The same principles of structure hold true for numbers that are less than one. A number less than one can be expressed by using negative exponents. Recall, as represented in Unit 6, that a number with a negative exponent is equal to the reciprocal of the number with a positive exponent. When the number is inverted and the negative exponent changed to a positive exponent, the result is as follows.

$$10^{-1} = \frac{1}{10^1} = 0.1$$

$$10^{-2} = \frac{1}{10^2} = \frac{1}{100} = 0.01$$

$$10^{-3} = \frac{1}{10^3} = \frac{1}{1000} = 0.001$$

$$10^{-4} = \frac{1}{10^4} = \frac{1}{10{,}000} = 0.0001$$

An analysis of the number 0.8502 (Figure 13–27) shows this structure.

• 8	5	0	2	Number
$10^{-1} = 0.1$	$10^{-2} = 0.01$	$10^{-3} = 0.001$	$10^{-4} = 0.0001$	Place Value
$8 \times 10^{-1} =$ $8 \times 0.1 =$ 0.8	$5 \times 10^{-2} =$ $5 \times 0.01 =$ 0.05	$0 \times 10^{-3} =$ $0 \times 0.001 =$ 0	$2 \times 10^{-4} =$ $2 \times 0.0001 =$ 0.0002	Value
0.8 +	0.05 +	0 +	0.0002 =	0.8502

Figure 13–27

Structure of the Binary System

The same principles of structure apply to the binary system as to the decimal system. The binary system is built upon the base 2 and uses only the digits 0 and 1. Numbers are shown as binary numbers by putting a 2 to the right and below the number (subscript) as shown: 11_2, 100_2, 1_2, 10001_2 are binary numbers. As with the decimal system, the elements that must be considered are the base, the particular digits used, and the place value of the digits. The binary system is built on the powers of the base 2; each place value is twice as large as the place value directly to its right. As in the decimal system, the zero is also a place marker in the binary system. See Figure 13–28.

PLACE VALUES OF BINARY NUMBERS												
2^7	2^6	2^5	2^4	2^3	2^2	2^1	2^0		2^{-1}	2^{-2}	2^{-3}	2^{-4}
128	64	32	16	8	4	2	1	•	0.5	0.25	0.125	0.0625

Figure 13–28

Expressing Binary Numbers as Decimal Numbers

Numbers in the decimal system are usually shown without a subscript. It is understood the number is in the decimal system. In certain instances, for clarity, decimal numbers are shown with the subscript 10. The following examples, going from left to right, show a method of expressing binary numbers as equivalent decimal numbers. There is more than one way of converting binary numbers to decimal numbers. The method shown is clearly related to the place value structure of the binary system. Remember that 0 and 1 are the only digits in the binary system.

EXAMPLES •———

Express each binary number as an equivalent decimal number.

1. $11_2 = 1(2^1) + 1(2^0) = 2 + 1 = 3_{10}$ *Ans*
2. $111_2 = 1(2^2) + 1(2^1) + 1(2^0) = 4 + 2 + 1 = 7_{10}$ *Ans*
3. $11101_2 = 1(2^4) + 1(2^3) + 1(2^2) + 0(2^1) + 1(2^0) = 16 + 8 + 4 + 0 + 1 = 29_{10}$ *Ans*
4. In the ASCII code, the bit pattern for the upper case letter R is 01010010.
   $$01010010_2 = 0(2^7) + 1(2^6) + 0(2^5) + 1(2^4) + 0(2^3) + 0(2^2) + 1(2^1) + 0(2^0)$$
   $$= 0 + 64 + 0 + 16 + 0 + 0 + 2 + 0 = 82_{10}$$ *Ans*
5. $101.11_2 = 1(2^2) + 0(2^1) + 1(2^0) + 1(2^{-1}) + 1(2^{-2})$
   $$= 4 + 0 + 1 + 0.5 + 0.25 = 5.75_{10}$$ *Ans*

Expressing Decimal Numbers as Binary Numbers

The following examples show a method of expressing decimal numbers as equivalent binary numbers. As with converting binary numbers to decimal numbers, there are various ways of converting decimal numbers to binary numbers. The method shown is clearly related to the structure of the binary system.

EXAMPLE 1 •———

Express 25_{10} as an equivalent binary number.

Determine the largest power of 2 in 25; $2^4 = 16$. There is one 2^4.

Subtract 16 from 25;

$$25 - 16 = 9.$$

Determine the largest power of 2 in 9; $2^3 = 8$. There is one 2^3.

Subtract 8 from 9;

$$9 - 8 = 1.$$

Determine the largest power of 2 in 1; $2^0 = 1$. There is one 2^0.

Subtract 1 from 1;

$$1 - 1 = 0.$$

There are no 2^2 and 2^1. The place positions for these values must be shown as zeros.

$$25_{10} = 1(2^4) + 1(2^3) + 0(2^2) + 0(2^1) + 1(2^0)$$
$$25_{10} = \quad 1 \quad\quad 1 \quad\quad 0 \quad\quad 0 \quad\quad 1$$
$$25_{10} = 11001_2 \; Ans$$

EXAMPLE 2

Express 11.625_{10} as an equivalent binary number.

$$2^3 = 8; 11.625 - 8 = 3.625$$
$$2^1 = 2; 3.625 - 2 = 1.625$$
$$2^0 = 1; 1.625 - 1 = 0.625$$
$$2^{-1} = 0.5; 0.625 - 0.5 = 0.125$$
$$2^{-3} = 0.125; 0.125 - 0.125 = 0$$

There are no 2^2 and 2^{-2}.

$$11.625_{10} = 1(2^3) + 0(2^2) + 1(2^1) + 1(2^0) \; . \; + 1(2^{-1}) + 0(2^{-2}) + 1(2^{-3})$$
$$11.625_{10} = \quad 1 \qquad 0 \qquad 1 \qquad 1 \; . \; \quad 1 \qquad 0 \qquad 1$$
$$11.625_{10} = 1011.101_2 \; Ans$$

EXERCISE 13–10

Analyze the following numbers.

1. 265   **4.** 0.802   **7.** 4751.107
2. 2855   **5.** 23.023   **8.** 3006.0204
3. 90,500   **6.** 105.009   **9.** 163.0643

Express the following binary numbers as decimal numbers.

10. 10_2 **17.** 1011_2 **24.** 0.1011_2
11. 1_2 **18.** 11000_2 **25.** 11.11_2
12. 100_2 **19.** 10101_2 **26.** 11.01_2
13. 101_2 **20.** 101010_2 **27.** 10.000_2
14. 1101_2 **21.** 110101_2 **28.** 1111.11_2
15. 1111_2 **22.** 111010_2 **29.** 1001.0101_2
16. 10100_2 **23.** 0.1_2 **30.** 10011.0101_2

In the ASCII code, the following binary bit patterns are given for various characters. Determine the equivalent decimal numbers.

31. Uppercase letter G is represented by 01000111.
32. Lowercase letter h is represented by 01101000.
33. Percent sign (%) is represented by 00100101.
34. Greater than symbol (>) is represented by 00111110.
35. Number sign (#) is represented by 00100011.
36. Lowercase letter z is represented by 01111010.
37. Dollar sign ($) is represented by 00100100.

Express the following decimal numbers as binary numbers.

38. 14   **45.** 98   **52.** 0.375
39. 100   **46.** 1   **53.** 10.5
40. 87   **47.** 6   **54.** 81.75
41. 23   **48.** 51   **55.** 19.0625
42. 43   **49.** 270   **56.** 101.25
43. 4   **50.** 0.5   **57.** 1.125
44. 105   **51.** 0.125   **58.** 163.875

UNIT EXERCISE AND PROBLEM REVIEW

ADDITION OF SINGLE TERMS

These expressions consist of groups of single terms. Add these terms.

1. $-8x, 5x$
2. $10m^2, -m^2$
3. $7MP, MP$
4. $0.09xy, 0.04xy$
5. $7.4a^2c, 7.3a^2c$
6. $-0.07F, -0.02F$
7. $\frac{1}{4}x^2y^3, -\frac{7}{8}x^2y^3$

8. $3\frac{1}{2}V, -3\frac{1}{2}V$
9. $5x, -6x, -x, 8x$
10. $-0.4HL, -3.6HL, -0.3HL$
11. $\frac{1}{4}ab, -\frac{3}{8}ab, ab, \frac{1}{2}ab$
12. $5.5N, 0.55N, -0.055N$

ADDITION OF GROUPS OF TWO OR MORE TERMS

These expressions consist of groups of two or more terms. Group like terms and add.

13. $(5a + 6b), (4a + 8b)$
14. $(9x + 2y + 17), (x + 4y + 14)$
15. $[7m + (-3mn)], [(-m) + 6mn]$
16. $[8P + (-7PT)], [P + (-5PT) + (-T)]$
17. $[2.4F + (-4G)], [(-7.6F) + 0.3G], [0.9F + (-1.2G)]$
18. $\left[1\frac{3}{4}x + \left(-\frac{1}{4}y\right) + \frac{3}{8}xy\right], \left[\frac{1}{2}x + \frac{7}{8}y + (-3xy) + (-xy^2)\right]$

SUBTRACTION OF SINGLE TERMS

These expressions consist of groups of single terms. Subtract terms as indicated.

19. $1P - 7P$
20. $9ab - (-3ab)$
21. $-14x^2y - (-14x^2y)$
22. $0 - (-15E)$
23. $0.7x^2 - 1.4x^2$
24. $-cd - 5.7cd$
25. $0.09H - (-0.15H)$

26. $\frac{1}{4}fg^2 - \frac{3}{16}fg^2$
27. $-\frac{5}{16}B - \left(-\frac{5}{8}B\right)$
28. $4.90M - 6.04M$
29. $0 - cd^2$
30. $-4\frac{1}{8}x^2y - 3\frac{7}{16}x^2y$

SUBTRACTION OF GROUPS OF TWO OR MORE TERMS

These expressions consist of groups of two or more terms. Group like terms and subtract.

31. $(6R - 5R^2) - (4R - 8R^2)$
32. $(12x^2 - 3) - (12x^2 - 3)$
33. $(9T - 4TW) - (2T + TW)$
34. $0 - (x^2y - xy^2 + x^2y^2)$
35. $(-y^3 - y^2 - y) - (-5y^2 + 3y)$
36. $(7.5M + 9.6N) - (3.4M - N + 3.2)$
37. $\left(\frac{1}{2}c - \frac{1}{4}d + 1\frac{3}{4}\right) - \left(\frac{7}{8}c - 2\frac{1}{2} + cd\right)$
38. $(-15L + 6.1L^2) - (9.3L + 0.6L^2 - L^3)$

MULTIPLICATION OF SINGLE TERMS

These expressions consist of single terms. Multiply these terms as indicated.

39. $(6xy)(10x^2y^2)$

40. $(-12a)(-5a^2)$

41. $(5c^2de^2)(-4cd^2e^3)$

42. $(-7M^4N)(-N^2)$

43. $(4.3x^4y)(0.6y^2)$

44. $(0.06S^3T)(4S^2)$

45. $\left(-\dfrac{3}{4}a^2b\right)\left(-\dfrac{3}{8}a^2b^2c\right)$

46. $(-8)(-FH^4)(F^3)$

47. $(c^2d)(b^2c)(d^2)$

48. $(-5.6M^2N)(-M^3)(-MN^2)$

49. $\left(\dfrac{1}{5}P\right)\left(\dfrac{1}{2}PS^3\right)\left(-P^2\right)$

50. $(0.1x^3y)(-y)(-3x^2z)$

MULTIPLICATION OF TWO OR MORE TERMS

These expressions consist of groups of two or more terms. Multiply as indicated and combine like terms where possible.

51. $7D^2(D^3 - DH)$

52. $-5x^2(-x - x^2y)$

53. $a^2b^3c^4(-a^2c^2 + b - cb^2)$

54. $-0.9A^2E(-0.2E^2 - A + 3A^2E)$

55. $(4m^2 - 3n^3)(-5m^2 + 7n^3)$

56. $(9.8h^2y - ax)(0.3b^4 - x^3)$

57. $(-0.8P^3S - 12)(-0.8S^2 - 0.5P)$

58. $(6e^2f^3 - 8ef)(-7ef + 3e^2f^3)$

DIVISION OF SINGLE TERMS

These exercises require division of single terms. Divide terms as indicated.

59. $16y^2 \div 8y$

60. $-10a^5b^4 \div 2a^2b^3$

61. $C^2D \div (-CD)$

62. $0 \div x^2y$

63. $(-35m^5n^2) \div (-5m^2n^2)$

64. $0.6BF^2 \div (-0.2F^2)$

65. $5.2a^3b \div a$

66. $-\dfrac{1}{4}x^4y^3 \div \dfrac{1}{8}x^4y$

67. $-9E^3F^2 \div \left(-\dfrac{3}{4}E^2F\right)$

68. $0.06x^5y^3 \div (-0.5x^3y^2)$

69. $\dfrac{5}{16}a^2b^2 \div 5a^2$

70. $-10.5NS^4 \div (-0.2S)$

DIVISION OF TWO OR MORE TERMS

These exercises consist of expressions in which the dividends have two or more terms. Divide as indicated.

71. $(12x^3 - 8x^2) \div 4x$

72. $(25a^5b^3 - 10a^3b^2) \div 5a^2b$

73. $(-40C^2D^5 - 32C^3D^4) \div (-8CD)$

74. $(25xy^2 - 35x^2y + 50) \div (-5)$

75. $(0.8F^3G^4 + 0.4F^4G^3) \div 4F^2G^2$

76. $(-3.5c^2d - 0.5cd^2 + c^2d^2) \div (0.5c)$

77. $\left(\dfrac{1}{2}x^2y - \dfrac{3}{4}x^3y^2 - 2xy^3\right) \div \dfrac{1}{4}xy$

78. $\left(\dfrac{3}{5}HM^3 - \dfrac{1}{5}H^2M^2 + \dfrac{4}{5}H^2M^3\right) \div (-2M^2)$

RAISING SINGLE TERMS TO POWERS

These exercises consist of expressions with single terms. Raise terms to the indicated powers.

79. $(5ab)^2$

80. $(-7xy)^2$

81. $(3M^4P^2)^2$

82. $(-2a^3b^2c)^3$

83. $(-3M^3P^2T^4)^4$

84. $(0.4x^2y^3)^3$

85. $(6.1d^3fh^2)^2$

86. $\left(\dfrac{3}{4}a^3bc^2\right)^2$

87. $[-7(x^2b^3)^2c]^2$

88. $[-3d^2(e^2)^2f^3]^3$

89. $[0.4m^3(ns^2)^3]^2$

90. $[(-2C^2D)^3(CD^2)^2]^2$

RAISING EXPRESSIONS CONSISTING OF MORE THAN ONE TERM TO POWERS

These exercises consist of expressions of more than one term. Raise these expressions to the indicated powers and combine like terms where possible.

91. $(x^2 + y)^2$

92. $(E^2 - F^3)^2$

93. $(5a^2 + 4b^2)^2$

94. $(c^2d^2 - cd)^2$

95. $(0.8P^2T^3 - 0.4T)^2$

96. $\left(\dfrac{1}{2}x^2y + \dfrac{1}{2}xy^2\right)^2$

97. $[(F^3)^2 - (H^2)^3]^2$

98. $[(ab^2)^3 + (-a^2b)^3]^2$

EXTRACTING ROOTS OF TERMS

Determine the roots of these terms.

99. $\sqrt{9x^2y^4z^2}$

100. $\sqrt{25a^4b^2c^6}$

101. $\sqrt[3]{8M^3P^6T^9}$

102. $(-27d^6e^3f^3)^{1/3}$

103. $(0.16F^4H^2)^{1/2}$

104. $\sqrt{0.36a^4b^8c^2}$

105. $\sqrt{121x^2y}$

106. $\sqrt{\dfrac{1}{4}C^2D^6}$

107. $(-64d^2e)^{1/3}$

108. $(81x^4y^8z^2)^{1/4}$

109. $\sqrt[5]{-32a^{10}b^5c^2}$

110. $\sqrt[3]{-\dfrac{8}{27}G^3HL^2}$

REMOVING PARENTHESES

Remove parentheses and combine like terms where possible.

111. $-(3a^2 + b - c^2)$

112. $+(18x^2y - y^2 - x)$

113. $x^2y - (xy + x^2y)$

114. $7C^2 + (-8C^2 - D + 4)$

115. $20 - (P^2 + P) + (P^2 - 7)$

116. $(E^3 + F^2) - (2E^3 + 3F^2)$

117. $17 + (mr - m^2r + r) + 8 - (mr - r)$

118. $17 - (mr - m^2r + r) - 8 + (mr - r)$

COMBINED OPERATIONS

These expressions consist of combined operations. Simplify.

119. $18 + 5(6x) + x$

120. $17 - 4(-4a^2) - 3a^2$

121. $4(M^2 - P) + P - 2M^2$

122. $-10(C^2 + 4)^2 + (5C^2)^2$

123. $(12 - 8D + 6D^2) \div 2 + D$

124. $\sqrt{25f^2g^4} - 3f(4f)^2$

125. $-x[-6 + (x^2y)^2 - 2x] + 4y$

126. $[7 - (a^2b)^2]^2$

127. $m(m - 3t)^2 - m(m + 3t)^2$

128. $(20R^4 - 24R^2T) \div \sqrt{16R^4}$

EXPRESSING BINARY NUMBERS AS DECIMAL NUMBERS

Express the following binary numbers as decimal numbers.

129. 100_2

130. 1101_2

131. 11011_2

132. 10101_2

133. 0.1001_2

134. 10.011_2

135. 1001.1010_2

136. 11011.0101_2

EXPRESSING DECIMAL NUMBERS AS BINARY NUMBERS

Express the following decimal numbers as binary numbers.

137. 16

138. 93

139. 117

140. 136

141. 0.25

142. 0.375

143. 12.875

144. 109.0625

EIGHT-BIT ASCII CODE CHARACTERS

In the ASCII code, the following binary bit patterns are given for various characters. Determine the equivalent decimal numbers.

145. Lowercase letter r is represented by 01110010.

146. Less than symbol (<) is represented by 00111100.

147. Uppercase letter M is represented by 01001101.

148. Question mark (?) is represented by 00111111.

LITERAL PROBLEMS

Determine the literal value answers for these problems.

149. The total voltage (E_t), in volts, of the circuit shown in Figure 13–29 is represented as x. The amount of voltage taken by 5 of the 6 resistors is given as E_{R_1} through E_{R_5}.

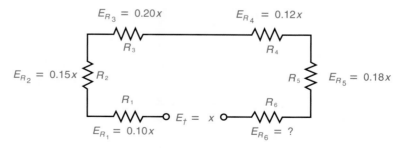

$E_{R_3} = 0.20x$ $E_{R_4} = 0.12x$

R_3 R_4

$E_{R_2} = 0.15x$ R_2 R_5 $E_{R_5} = 0.18x$

R_1 R_6

$E_t = x$

$E_{R_1} = 0.10x$ $E_{R_6} = ?$

Figure 13–29

a. What is the sum of the voltage taken by R_1, R_2, R_3, R_4, and R_5?

b. What is the average resistor voltage for resistors R_1 through R_5?

c. What voltage is taken by R_6?

150. A corporation's annual profits and losses for a six-year period are shown in the graph in Figure 13–30.

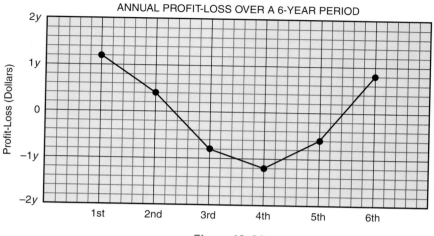

ANNUAL PROFIT-LOSS OVER A 6-YEAR PERIOD

Figure 13–30

    a. What is the increase or decrease in profits for the years listed?
        (1) the 1st to the 2nd year
        (2) the 2nd to the 4th year
        (3) the 3rd to the 5th year
        (4) the 4th to the 6th year
    b. What is the total dollar amount lost by the corporation from the 3rd through the 5th year?
    c. What is the net profit or loss for the 6-year period?
    d. What is the average annual profit or loss for the 6-year period?

151. Refer to the bracket shown in Figure 13–31 and determine distances A through G.

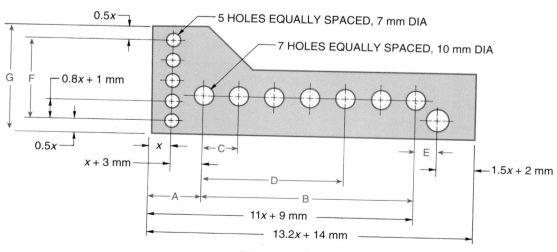

Figure 13–31

152. The area of the triangle shown in Figure 13–32 equals one-half the product of the base and height $\left(A = \frac{1}{2}bh\right)$.

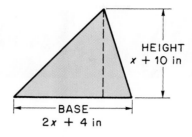

Figure 13–32

a. What is the area of the triangle?

b. What is the area of the triangle if the base is increased by 2 inches?

c. What is the area of the triangle if the height is decreased by 4 inches?

d. What is the area of the triangle if both the base and height are doubled?

UNIT 14 ⠿ Simple Equations

After studying this unit you should be able to

- express word problems as equations.
- express problems given in graphic form as equations.
- solve equations using the six fundamental principles of equality.
- solve problems by writing equations and determining the values of unknowns.
- substitute values in formulas and solve for the unknowns.

14–1 Expression of Equality

An *equation* is a mathematical statement of equality between two or more quantities. An equation always contains an equal sign (=). The value of all the quantities on the left side of the equal sign equals the value of all quantities on the right side of the equal sign. A *formula* is a particular type of equation that states a mathematical rule.

Because it expresses the equality of the quantities on the left and on the right of the equal sign, an equation is a balanced mathematical statement. An equation may be considered similar to a balanced scale as illustrated in Figure 14–1. The total weight on the left side of the scale equals the total weight on the right side.

$$3 \text{ lb} + 5 \text{ lb} + 2 \text{ lb} = 4 \text{ lb} + 6 \text{ lb}$$

$$10 \text{ lb} = 10 \text{ lb} \text{ (The scale is balanced.)}$$

Figure 14–2 shows that when the 2-pound weight is removed from the scale, the scale is no longer in balance.

$$3 \text{ lb} + 5 \text{ lb} \neq 4 \text{ lb} + 6 \text{ lb}$$

$$8 \text{ lb} \neq 10 \text{ lb} \text{ (The scale is } not \text{ balanced.)}$$

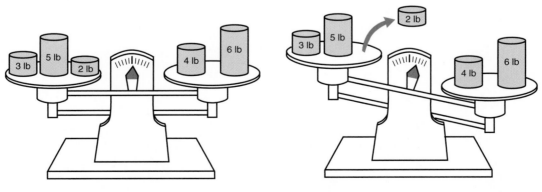

Figure 14–1   **Figure 14–2**

In general, an equation is solved to determine the value of an unknown quantity. Although any letter or symbol can be used to represent the unknown quantity, the letter x is commonly used.

The first letter of the unknown quantity is sometimes used to represent a quantity. These are some common letter designations used to represent specific quantities.

L to represent length	P to represent pressure	V to represent volume
A to represent area	W to represent weight	h to represent height
t to represent time	D to represent diameter	

14–2   Writing Equations from Word Statements

It is important to develop the ability to express word statements as mathematical symbols or equations. A problem must be fully understood before it can be written as an equation.

Whether the word problem is simple or complex, a definite logical procedure should be followed to analyze the problem. A few or all of the following steps may be required, depending on the complexity of the particular problem.

• Carefully read the entire problem, several times if necessary.
• Break the problem down into simpler parts.
• It is sometimes helpful to draw a simple picture as an aid in visualizing the various parts of the problem.
• Identify and list the unknowns. Give each unknown a letter name, such as x.
• Decide where the equal sign should be, and group the parts of the problem on the proper side of the equal sign.
• Check. Are the statements on the left equal to the statements on the right of the equal sign?
• Solve for the unknown.
• Check the answer against the original problem, step by step.

The following examples illustrate the method of writing equations from given word statements. After each equation is written, the value of the unknown quantity is obtained. No specific procedures are given at this time in solving for the unknown. The unknown quantity values are determined by logical reasoning.

EXAMPLES •———————————————————————————————

1. What weight must be added to a 12-pound weight so that it will be in balance with a 20-pound weight?

To help visualize the problem, a picture is shown in Figure 14–3.

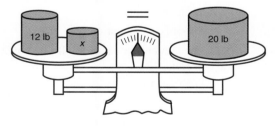

Figure 14–3

Identify the unknown.	Let x = the number of pounds of the unknown weight
Write the equation.	$12 + x = 20$
The number 8 must be added to 12 to equal 20.	
$12 + 8 = 20$	$x = 8$
	8 pounds *Ans*

Check the answer by substituting 8 for *x* in the equation.

$$12 + 8 = 20$$

Perform the indicated operation.

$$12 + 8 = 20 \qquad\qquad 20 = 20 \text{ Ck}$$

The equation is balanced since the left side of the equation equals the right side. Figure 14–4 shows that 8 pounds must be added to 12 pounds to equal 20 pounds and make the scale balance.

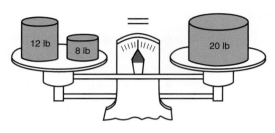

Figure 14–4

2. In a series circuit, the sum of the individual voltages equals the applied line voltage. A certain series circuit with 3 resistors has an applied line voltage (E_t) of 120 volts. The voltage across $R_2(E_{R_2})$ is twice the voltage across $R_1(E_{R_1})$. The voltage across $R_3(E_{R_3})$ is 3 times the voltage across R_1. What is the voltage across each resistor?

Identify the unknown.   Let x = number of volts across R_1

$2x$ = number of volts across R_2

$3x$ = number of volts across R_3

A picture of the problem is shown in Figure 14–5.

Figure 14–5

Write the equation $x + 2x + 3x = 120$

Simplify the equation. $6x = 120$

The number 6 must be multiplied by 20 to equal 120.

$6(20) = 120$ $x = 20$

$2x = 40$

$3x = 60$

$E_{R_1} = 20$ volts *Ans*

$E_{R_2} = 40$ volts *Ans*

$E_{R_3} = 60$ volts *Ans*

Check the answers by substituting the answers. $20 + 40 + 60 = 120$

Perform the indicated operations.

$20 + 40 + 60 = 120$ $120 = 120$ Ck

The equation is balanced since the left side of the equation equals the right side.

Figure 14–6 shows the resistor voltages equal the applied voltage.

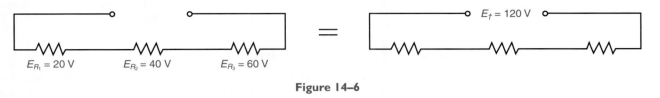

$E_{R_1} = 20$ V $E_{R_2} = 40$ V $E_{R_3} = 60$ V $E_t = 120$ V

Figure 14–6

14–3   Checking the Equation

In the final step in each of the preceding examples, the value found for the unknown was substituted in the original equation to prove that it was the correct value. If an equation is properly written and if both sides of the equation are equal, the equation is balanced and the solution is correct. If the solution to an equation is a rounded value, the check may result in a very small difference between both sides.

It is important that you check your computations. When working with equations for actual on-the-job applications, checking your work is essential. Errors in computation can often be costly in terms of time, labor, and materials.

EXERCISE 14–3

Express each of these word problems as an equation. Let the unknown number equal x *and, by logical reasoning, solve for the value of the unknown. Check the equation by comparing it to the word problem. Does the equation state mathematically what the problem states in words? Check whether the equation is balanced by substituting the value of the unknown in the equation.*

1. A number plus 20 equals 35.
2. A number less 12 equals 18.
3. Four times a number equals 40.
4. A number divided by 2 equals 14.
5. Twenty divided by a number equals 5.
6. A number plus twice the number equals 45.
7. Five times a number, plus the number, equals 36.
8. Four times a number, minus the number, equals 21.
9. Six times a number divided by 3 equals 12.
10. Twice a number, plus 3 times the number, equals 20.
11. Ten times a number, minus 4 times the number, equals 42.
12. Three subtracted from 3 times a number, plus twice the number, equals 27.
13. Four added to 5 times a number, minus twice the number, equals 34.
14. Sixty multiplied by 3 times a number equals 360.
15. A length of lumber 12 feet long is cut into 2 unequal lengths. One piece is 3 times as long as the other.
    a. Find the length of the shorter piece.
    b. Find the length of the longer piece.
16. A building contractor estimates labor cost for a job will be $25,000 more than material cost. The total labor and material cost is $95,000.
    a. What is the estimated material cost?
    b. What is the estimated labor cost?

17. A plot of land that has an area of 65,000 square feet is subdivided into 4 building lots. Lot 2 and lot 3 each have twice the area of lot 1. Lot 4 is 5,000 square feet greater in area than lot 1. Find the area (number of square feet) of each lot listed.

   a. Lot 1   **c.** Lot 3

   b. Lot 2   **d.** Lot 4

18. In 2 cuts, the total amount of stock milled off an aluminum casting is 7.5 millimeters. The first cut (roughing cut) is 6.5 millimeters greater than the second cut (finish cut). How much stock is removed on the finish cut?

19 A hospital dietician computes the daily calorie intake for a patient as 2,550 calories. The calories allowed for lunch are twice the calories allowed for breakfast. The calories allowed for dinner are one and one-half the calories allowed for lunch. The allowance for an evening snack is 150 calories. How many calories are allowed for each meal listed?

   a. Breakfast   **b.** Lunch   **c.** Dinner

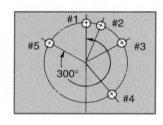

Figure 14–7

20. Five holes are drilled in a steel plate as shown in Figure 14–7. There are 300 degrees between hole 1 and hole 5. The number of degrees between any two consecutive holes doubles in going from hole 1 to hole 5. Find the number of degrees between the holes listed.

   a. Hole 1 and hole 2   **c.** Hole 3 and hole 4

   b. Hole 2 and hole 3   **d.** Hole 4 and hole 5

For each of these problems, refer to the corresponding figure. Write an equation and solve for x. Check the equation.

21. In Figure 14–8, let x = a given number of inches.

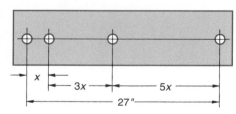

Figure 14–8

23. In Figure 14–10, let x = a given number of inches.

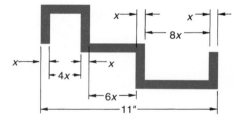

Figure 14–10

22. In Figure 14–9, let x = a given number of degrees.

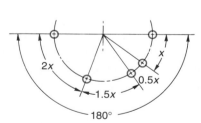

Figure 14–9

24. In Figure 14–11, let x = a given number of degrees.

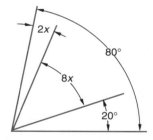

Figure 14–11

For each of these problems, write an equation. Solve for the unknown and check the equation.

25. In the series circuit shown in Figure 14–12, the sum of the voltages across each of the individual resistors is equal to the applied line voltage of 120 volts. Find the number of volts for each voltage listed.

Let x = volts across R_1.

a. E_{R_1}

b. E_{R_2}

c. E_{R_3}

d. E_{R_4}

e. E_{R_5}

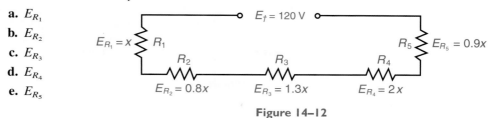

Figure 14–12

26. Refer to Figure 14–13. Determine the distances between these holes.

Let l = a given number of millimeters.

a. Hole 1 to hole 2

b. Hole 2 to hole 3

c. Hole 3 to hole 4

d. Hole 4 to hole 5

e. Hole 1 to hole 5

f. Hole 2 to hole 6

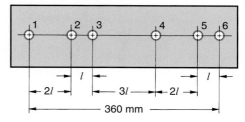

Figure 14–13

27. Refer to Figure 14–14. Determine the distance between these points.

Let h = a given number of inches.

a. A and B

b. C and D

c. E and F

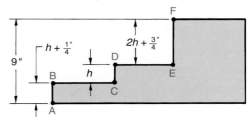

Figure 14–14

14–4 Principles of Equality

In actual practice, equations cannot always be solved by inspection or common sense. There are specific procedures for solving equations using the fundamental principles of equality. The principles of equality that will be presented are those of addition, subtraction, multiplication, division, powers, and roots. Equations are solved directly and efficiently by the application of these principles.

14–5 Solution of Equations by the Subtraction Principle of Equality

The subtraction principle of equality states that if the same number is subtracted from both sides of an equation, the sides remain equal and the equation remains balanced. The subtraction principle is used to solve an equation in which a number is added to the unknown, such as $x + 4 = 10$.

Figure 14–15 shows a balanced scale. If 4 pounds are removed from the left side only, the scale is not in balance, as shown in Figure 14–16. If 4 pounds are removed from both the left and right sides, the scale remains in balance, as shown in Figure 14–17.

$$6 \text{ lb} + 4 \text{ lb} = 10 \text{ lb}$$
$$6 \text{ lb} + 4 \text{ lb} - 4 \text{ lb} \neq 10 \text{ lb}$$
$$6 \text{ lb} \neq 10 \text{ lb}$$
$$6 \text{ lb} + 4 \text{ lb} - 4 \text{ lb} = 10 \text{ lb} - 4 \text{ lb}$$
$$6 \text{ lb} = 6 \text{ lb}$$

6 lb + 4 lb = 10 lb

| 6 lb | 4 lb | | 10 lb |

Figure 14–15

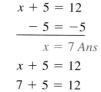

Figure 14–16   **Figure 14–17**

Procedure for Solving an Equation in which a Number Is Added to the Unknown

- Subtract the number that is added to the unknown from both sides of the equation.
- Check.

EXAMPLES •

1. Solve for x. See Figures 14–18, 14–19, and 14–20.

$x + 5 = 12$ (Figure 14–18)

In the equation, the number 5 is added to x. To solve the equation, subtract 5 from both sides of the equation (Figure 14–19).

$$\begin{aligned} x + 5 &= 12 \\ -5 &= -5 \\ \hline x &= 7 \; Ans \end{aligned}$$

Check. Substitute 7 for x in the original equation (Figure 14–20).

$$\begin{aligned} x + 5 &= 12 \\ 7 + 5 &= 12 \\ 12 &= 12 \; Ck \end{aligned}$$

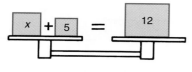

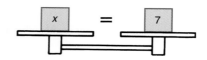

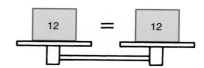

Figure 14–18   **Figure 14–19**   **Figure 14–20**

2. Solve for T.

$-12°C = T + 4°C$

In the equation, 4°C is added to T.

To solve the equation, subtract 4°C from both sides of the equation.

$$\begin{aligned} -12°C &= T + 4°C \\ -4°C &= \quad -4°C \\ \hline -16°C &= T \; Ans \end{aligned}$$

Check.

$$\begin{aligned} -12°C &= T + 4°C \\ -12°C &= -16°C + 4°C \\ -12°C &= -12°C \; Ck \end{aligned}$$

3. Determine dimension D of the plate shown in Figure 14–21.

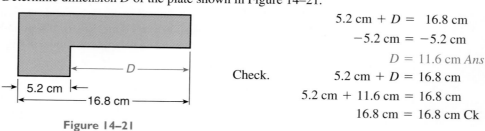

Figure 14–21

$$\begin{aligned} 5.2 \text{ cm} + D &= 16.8 \text{ cm} \\ -5.2 \text{ cm} &= -5.2 \text{ cm} \\ \hline D &= 11.6 \text{ cm } Ans \end{aligned}$$

Check.

$$\begin{aligned} 5.2 \text{ cm} + D &= 16.8 \text{ cm} \\ 5.2 \text{ cm} + 11.6 \text{ cm} &= 16.8 \text{ cm} \\ 16.8 \text{ cm} &= 16.8 \text{ cm } Ck \end{aligned}$$

Write the equation.
Subtract 5.2 cm from both sides of the equation.

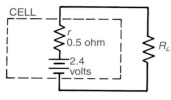

Figure 14–22

4. The 2.4-volt cell shown in Figure 14–22 has an internal resistance (r) of 0.5 ohm. The total circuit resistance (R) is 1.8 ohms. The external resistance is expressed as R_L. Using the formula $R = r + R_L$, determine the external resistance.

Substitute the given values in the formula.

$$R = r + R_L$$
$$1.8 \; \Omega = 0.5 \; \Omega + R_L$$

Subtract $0.5 \; \Omega$ from both sides of the equation.

$$\underline{-0.5 \; \Omega = -0.5 \; \Omega}$$
$$1.3 \; \Omega = R_L \; Ans$$

Check.

$$1.8 \; \Omega = 0.5 \; \Omega + R_L$$
$$1.8 \; \Omega = 0.5 \; \Omega + 1.3 \; \Omega$$
$$1.8 \; \Omega = 1.8 \; \Omega \; Ck$$

EXERCISE 14–5

Solve each of these equations using the subtraction principle of equality. Check all answers.

1. $P + 15 = 25$

2. $x + 18 = 27$

3. $M + 24 = 43$

4. $y + 50 = 82$

5. $13 = T + 9$

6. $37 = D + 2$

7. $55 = a + 19$

8. $y + 16 = 15$

9. $C + 34 = 12$

10. $x + 6 = -11$

11. $y + 30 = -23$

12. $x + 63 = 17$

13. $10 + R = 44$

14. $51 = 48 + E$

15. $-18.3 = 2.6 + x$

16. $H + 7.6 = 15.2$

17. $22.5 = L + 3.7$

18. $-36.2 = y + 6.2$

19. $86.04 = x + 61.95$

20. $F + 0.007 = 1.006$

21. $T + 9.07 = 9.07$

22. $H + 3\frac{1}{4} = 7\frac{1}{2}$

23. $-\frac{7}{16} = x - \frac{9}{32}$

24. $20\frac{3}{16} = A + 17\frac{1}{8}$

25. $39\frac{5}{8} = y + 42\frac{7}{8}$

26. $1\frac{7}{16} = W + \frac{9}{16}$

27. $x + 13\frac{1}{8} = -10$

28. $0.015 = 1.009 + H$

29. $-14.067 = 3.034 + x$

30. $20.863 = D + 25.942$

For each of the problems (Figures 14–23 through 14–28) write an equation. Solve for the unknown and check.

31.

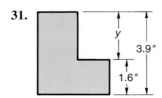

Figure 14–23

32.

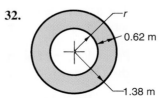

Figure 14–24

33.

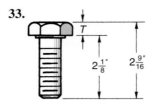

Figure 14–25

34.

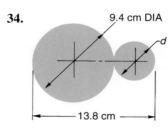

9.4 cm DIA

13.8 cm

Figure 14–26

35.

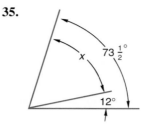

$73\frac{1}{2}°$

x

$12°$

Figure 14–27

36.

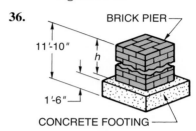

BRICK PIER

11'-10"

h

1'-6"

CONCRETE FOOTING

Figure 14–28

37. The retail selling price of merchandise is $367.50. The retail markup is $172.75. What is the cost (C) of the merchandise to the retailer?

38. A shaft rotates in a bearing. The bearing is 0.3750 inch in diameter. The total clearance between the shaft and bearing is 0.0008 inch. Find the diameter (d) of the shaft.

39. Three holes are drilled in a housing. The center distance between the first hole and the second hole is 17.54 centimeters. The center-to-center distance between the first hole and the third hole is 24.76 centimeters. Find the distance (D) between the second hole and the third hole.

40. A savings account balance (amount) is $3,835. The principal deposited is $3,572. What is the amount of earned interest (I)?

For each of these problems, substitute the given values in the formula and solve for the unknown. Check.

41. Using the formula $P_T = P_1 + P_2$, determine P_1 when $P_2 = 135$ watts and $P_T = 250$ watts.

42. On a company's balance sheet, assets equal liabilities plus net worth. $A = L + NW$. Determine L when $NW = \$56,200$ and $A = \$127,370$.

43. One of the formulas used in computing spur gear dimensions is $D_0 = D + 2a$. Determine D when $a = 0.1429$ inch and $D_0 = 4.7144$ inches.

44. A sheet metal formula used in computing the size of a stretchout is $LS = 4s + W$. Determine W when $s = 3$ inches and $LS = 12\frac{1}{8}$ inches.

45. A formula used to compute the dimensions of a ring is $D = d + 2T$. Determine d when $D = 5.20$ centimeters and $T = 0.86$ centimeters.

14–6 Solution of Equations by the Addition Principle of Equality

The addition principle of equality states that if the same number is added to both sides of an equation, the sides remain equal and the equation remains balanced. The addition principle is used to solve an equation in which a number is subtracted from the unknown, such as $x - 6 = 24$.

Procedure for Solving an Equation in which a Number Is Subtracted from the Unknown

• Add the number, which is subtracted from the unknown, to both sides of the equation.

• Check.

EXAMPLES

1. Solve for y. See Figures 14–29, 14–30, 14–31, and 14–32.

$y - 7 = 10$

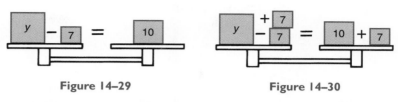

| Figure 14–29 | Figure 14–30 |

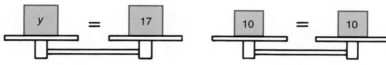

| Figure 14–31 | Figure 14–32 |

In the equation, the number 7 is subtracted from y (Figure 14–29).

To solve the equation, add 7 to both sides of the equation (Figure 14–30).

(Figure 14–31)

(Figure 14–32)

$$y - 7 = 10$$
$$\underline{+ 7 = +7}$$
$$y = 17 \; Ans$$

Check.
$$y - 7 = 10$$
$$17 - 7 = 10$$
$$10 = 10 \; Ck$$

2. A $5\frac{1}{4}$-inch length is cut from a block as shown in Figure 14–33. The remaining block is $6\frac{3}{4}$ inches high. What was the original height, H, of the block?

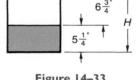

Figure 14–33

Write the equation.

$$H - 5\frac{1''}{4} = 6\frac{3''}{4}$$

Add $5\frac{1}{4}''$ to both sides of the equation.

$$\underline{+ 5\frac{1''}{4} = + 5\frac{1''}{4}}$$
$$H = 12'' \; Ans$$

Check.
$$H - 5\frac{1''}{4} = 6\frac{3''}{4}$$
$$12'' - 5\frac{1''}{4} = 6\frac{3''}{4}$$
$$6\frac{3''}{4} = 6\frac{3''}{4} \; Ck$$

EXERCISE 14–6

Solve each of these equations using the addition principle of equality. Check all answers.

1. $T - 12 = 28$

2. $x - 9 = -19$

3. $B - 4 = 9$

4. $P - 50 = 87$

5. $y - 23 = -20$

6. $16 = M - 12$

7. $-35 = E - 21$

8. $47 = R - 36$

9. $h - 8 = 12$

10. $T - 19 = -6$

11. $-22 = x - 31$

12. $39 = F - 39$

13. $W - 15.75 = 83.00$

14. $N - 2.4 = 6.9$

15. $A - 0.8 = 0.5$

16. $x - 10.09 = -13.78$

17. $4.93 = r - 3.07$

18. $-30.003 = x - 29.998$

19. $91.96 = L - 13.74$

20. $P - 0.02 = 0.07$

21. $G - 59.875 = 49.986$

22. $x - 7.63 = -12.06$

23. $D - \dfrac{1}{2} = \dfrac{1}{2}$

24. $y - \dfrac{7}{8} = -\dfrac{3}{4}$

25. $15\dfrac{5}{8} = H - 2\dfrac{7}{8}$

26. $-46\dfrac{3}{32} = x - 29\dfrac{15}{16}$

27. $C - 5\dfrac{7}{16} = -5\dfrac{7}{16}$

28. $W - 10.0039 = 9.0583$

29. $-14\dfrac{15}{32} = y - 14\dfrac{7}{16}$

30. $E - 29.8936 = 18.3059$

For each of these problems, write an equation and solve for the unknown. Check.

31. The bushing shown in Figure 14–34 has a body diameter of 44.17 millimeters. The body diameter is 14.20 millimeters less than the head diameter. What is the size of the head diameter (*D*)?

32. A series circuit is shown in Figure 14–35. R_2 is 50 ohms. R_2 is 125 ohms less than the total circuit resistance, R_T. What is total circuit resistance R_T?

33. A wall section is shown in Figure 14–36. The distance from the left edge of the wall to the center of a window is 8′4″. This is 9′6″ less than the distance from the right edge of the wall to the center of the window (*x*). What is the distance from the right edge of the wall to the center of the window?

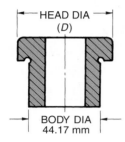

HEAD DIA (D)

BODY DIA 44.17 mm

Figure 14–34

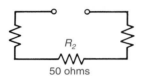

R_2

50 ohms

Figure 14–35

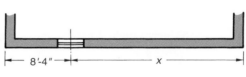

8′-4″   x

Figure 14–36

34. The flute length of the reamer shown in Figure 14–37 is 2.858 centimeters, which is 8.572 centimeters less than the shank length. How long is the shank (*x*)?

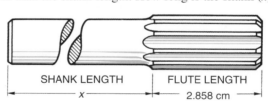

SHANK LENGTH   FLUTE LENGTH

x   2.858 cm

Figure 14–37

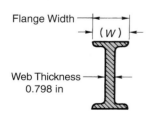

Flange Width (W)

Web Thickness 0.798 in

Figure 14–38

35. A cross-sectional view of an I beam is shown in Figure 14–38. The web thickness is 0.798 inch. The web thickness is 7.250 inches less than the flange width. What is the width (*W*) of the flange?

For each of these problems, substitute the given values in the formula and solve for the unknown. Check.

36. A retailer's net profit (*NP*) on the sale of merchandise is determined by subtracting overhead (*O*) expenses from the merchandise markup (*M*): $NP = M - O$. Determine *M* when $NP = \$122$ and $O = \$74$.

37. The total taper of a shaft equals the diameter of the large end minus the diameter of the small end: $T = D - d$. Determine *D* when $T = 22.5$ millimeters and $d = 30.8$ millimeters.

38. A formula often used for computing total weekly salaries of salespeople is total salary (*TS*) minus flat salary (*FS*) equals commission (*C*): $TS - FS = C$. Determine the total salary of a salesperson whose flat salary is $225 and whose commission is $387.

39. The net change of stock equals the closing price of stock on the day of the stock quotation minus the closing price of the stock on the previous day: $NC = CP - PCP$. The closing price of the stock on the previous day is $27\frac{1}{4}$ ($27.25). If the net change is $-1\frac{1}{2}$ ($-$1.50), what is the closing price on the day of the stock quotation?

40. The spur gear formula is $D_R = D - 2d$. Compute the pitch diameter (D) when the root diameter (D_R) = 3.0118 inches and the dedendum (d) = 0.1608 inch.

41. A sheet metal formula is $W = LS - 4S$. Determine the length size (LS) when $W = 38.2$ centimeters and $S = 10.6$ centimeters.

42. The formula for computing straight-line depreciation of equipment book value is book value (BV) equals original cost (OC) minus the number of years depreciated (n) times the annual depreciation (AD): $BV = OC - n(AD)$. Determine the original cost of the equipment when $BV = $2,400$, $n = 3$, and $AD = 350.

14–7 Solution of Equations by the Division Principle of Equality

The division principle of equality states that if both sides of an equation are divided by the same number, the sides remain equal and the equation remains balanced. The division principle is used to solve an equation in which a number is multiplied by the unknown, such as $3x = 18$.

Procedure for Solving an Equation in which the Unknown Is Multiplied by a Number

- Divide both sides of the equation by the number that multiplies the unknown.
- Check.

EXAMPLES •————————————————————————

1. Solve for x. See Figures 14–39, 14–40, and 14–41.

$$6x = 30$$

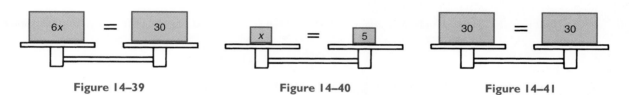

Figure 14–39   Figure 14–40   Figure 14–41

In the equation, x is multiplied by 6 (Figure 14–39).

To solve the equation, divide both sides by 6 (Figure 14–40).

(Figure 14–41)

$$\frac{6x}{6} = \frac{30}{6}$$

$$x = 5 \ Ans$$

Check. $6x = 30$

$$6(5) = 30$$

$$30 = 30 \ Ck$$

2. A plate with equally spaced holes is shown in Figure 14–42. Find the center-to-center distance between the holes as represented by x.

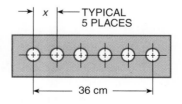

Figure 14–42

Write the equation. $5x = 36$ cm

Divide both sides of the equation by 5. $\dfrac{5x}{5} = \dfrac{36 \text{ cm}}{5}$

                                                                    $x = 7.2$ cm *Ans*

Check. $5x = 36$ cm

                                                                    $5(7.2 \text{ cm}) = 36$ cm

                                                                    36 cm $= 36$ cm Ck

EXERCISE 14–7

Solve each of these equations using the division principle of equality. Check all answers.

1. $4D = 32$

2. $7x = -21$

3. $15M = 60$

4. $54 = 9P$

5. $-27 = 3y$

6. $30 = 6x$

7. $10y = 0.80$

8. $18T = 41.4$

9. $-11.07x = 40.93$

10. $-x = 19$

11. $0 = 7H$

12. $-5C = 0$

13. $7.1E = 42.6$

14. $0.6L = 12$

15. $-2.7x = 23.76$

16. $0.21y = -0.0126$

17. $13.2W = 0$

18. $-x = -19.75$

19. $0.125P = 0.875$

20. $9.37R = 103.07$

21. $-0.66x = 4.752$

22. $\dfrac{1}{4}D = 8$

23. $24 = \dfrac{3}{8}B$

24. $-\dfrac{1}{2}y = 36$

25. $1\dfrac{5}{8}L = 8\dfrac{1}{8}$

26. $-48\dfrac{3}{8} = 10\dfrac{3}{4}x$

27. $-\dfrac{7}{16} = -\dfrac{7}{16}y$

28. $50.98W = 10.196$

29. $-0.002x = 4.938$

30. $-\dfrac{3}{16} = -1\dfrac{1}{16}y$

For each of these problems (Figures 14–43, 14–44, and 14–45) write an equation and solve for the unknown. Check.

31.

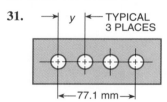

Figure 14–43

32.

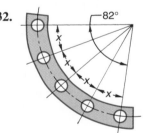

Figure 14–44

33.

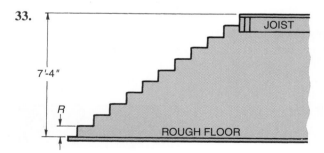

Figure 14–45

34. The feed of a drill is the depth of material that the drill penetrates in one revolution. The total depth of penetration equals the product of the number of revolutions and the feed. A drill is shown in Figure 14–46. Compute the feed (F) of a drill that cuts to a depth of 2.400 inches while turning 400 revolutions.

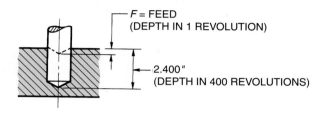

Figure 14–46

For each of these problems substitute the given values in the formula and solve for the unknown. Check.

35. The perimeter of a square equals 4 times the length of a side of the square, $p = 4s$. Determine s when $p = 63.2$ feet.

36. The circumference of a circle equals π (approximately 3.14) times the diameter of the circle: $C = \pi d$. Determine d when $C = 3.768$ meters.

37. Power, in watts, is equal to the product of current, in amperes, and voltage, in volts: $P_W = IE$. Determine E when $P_W = 4{,}600$ watts and $I = 20.0$ amperes.

38. Power, in watts, is also equal to the product of the square of the current, in amperes, and the total line resistance, in ohms, $P_W = I^2 R$. Determine R when $P_W = 100.0$ watts and $I = 20.0$ amperes.

39. Power, in kilowatts, is equal to 0.001 times the voltage, in volts, times the current, in amperes: $P_{kW} = 0.001EI$. Determine I when $P_{kW} = 2.415$ kilowatts and $E = 230.0$ volts.

40. The length of cut, in inches, of a workpiece in a lathe is equal to the product of the cutting time, in minutes, the tool feed, in inches per revolution, and the number of revolutions per minute of the workpiece: $L = TFN$. Determine N when $L = 18$ inches, $T = 3.0$ minutes, and $F = 0.050$ inch per revolution.

41. Simple annual interest earned on money loaned is equal to the product of the principal, the rate of interest, and the time, in years: $i = prt$. Determine t when $i = \$1{,}260$, $p = \$4{,}500$, and $r = 8\%$.

14–8   Solution of Equations by the Multiplication Principle of Equality

The multiplication principle of equality states that if both sides of an equation are multiplied by the same number, the sides remain equal and the equation remains balanced. The multiplication principle is used to solve an equation in which the unknown is divided by a number; for example, $\frac{y}{7} = 42$.

Procedure for Solving an Equation in which the Unknown Is Divided by a Number

• Multiply both sides of the equation by the number that divides the unknown.
• Check.

EXAMPLES •————————————————————————

1. Solve for y. See Figures 14–47, 14–48, and 14–49 on page 350.

$$\frac{y}{3} = 5$$

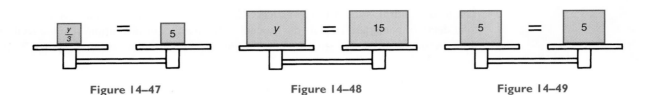

Figure 14–47   **Figure 14–48**   **Figure 14–49**

In the equation, y is divided by 3 (Figure 14–47). To solve the equation, multiply both sides by 3 (Figure 14–48).

$$\frac{3}{1}\left(\frac{y}{3}\right) = \left(\frac{5}{1}\right)\frac{3}{1}$$

$$y = 15 \; Ans$$

(Figure 14–49)   Check.

$$\frac{y}{3} = 5$$

$$\frac{15}{3} = 5$$

$$5 = 5 \; Ck$$

2. Six strips of equal width are sheared from the piece of flat stock shown in Figure 14–50. Each strip is $1\frac{3}{4}$ inches wide. The original width (W) of the piece of stock can be expressed by the equation $\frac{W}{6} = 1\frac{3}{4}''$. Solve for W.

Multiply both sides of the equation by 6.

$$\frac{6}{1}\left(\frac{W}{6}\right) = \left(1\frac{3''}{4}\right)\frac{6}{1}$$

$$W = 10\frac{1''}{2} \; Ans$$

Check.

$$\frac{W}{6} = 1\frac{3''}{4}$$

$$\frac{10\frac{1''}{2}}{6} = 1\frac{3''}{4}$$

$$1\frac{3''}{4} = 1\frac{3''}{4} \; Ck$$

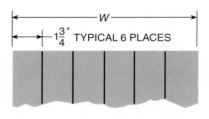

$\longleftarrow$ W $\longrightarrow$

$\longleftarrow 1\frac{3}{4}''$ TYPICAL 6 PLACES

Figure 14–50

EXERCISE 14–8

Solve each of these equations using the multiplication principle of equality. Check all answers.

1. $\dfrac{P}{5} = 4$

2. $\dfrac{M}{12} = 5$

3. $D \div 9 = 7$

4. $3 = L \div 6$

5. $3 = W \div 9$

6. $\dfrac{N}{12} = -2$

7. $\dfrac{C}{14} = 0$

8. $\dfrac{x}{-10} = 7$

9. $\dfrac{E}{-2} = -18$

10. $13 = y \div (-4)$

11. $\dfrac{F}{3.6} = 5$

12. $\dfrac{A}{-0.5} = 24$

13. $S \div (7.8) = 3$

14. $x \div (-0.4) = 16$

15. $-20 = \dfrac{y}{0.3}$

16. $\dfrac{T}{-1.8} = 2.4$

17. $0 = H \div (-3.8)$

18. $M \div 9.5 = -10$

19. $\dfrac{y}{-0.1} = -0.01$

20. $\dfrac{R}{12.6} = 0.002$

21. $1.04 = \dfrac{H}{0.08}$

22. $\dfrac{B}{\frac{3}{4}} = 12$

23. $V \div 1\frac{1}{4} = 3$

24. $\dfrac{x}{\frac{3}{8}} = -\dfrac{1}{2}$

25. $D \div \left(-\dfrac{1}{16}\right) = -32$

26. $4 = y \div \left(-\dfrac{7}{8}\right)$

27. $1.020 = \dfrac{T}{7.350}$

28. $H \div (-2) = 7\dfrac{9}{16}$

29. $\dfrac{M}{0.009} = 100$

30. $x \div (6.004) = -0.17$

For each of these problems write an equation. Solve for the unknown and check.

31. See Figure 14–51.

32. The depth of an American Standard thread divided by 0.6495 is equal to the pitch. Compute the depth of the thread shown in Figure 14–52.

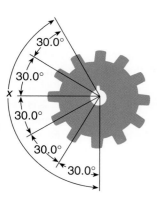

Figure 14–51

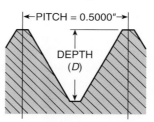

Figure 14–52

33. The width of a rectangle is equal to its area divided by its length. Compute the area of the rectangular patio shown in Figure 14–53.

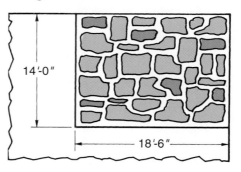

Figure 14–53

34. The grade of a road is expressed as a percent. It is equal to the amount of rise divided by the horizontal distance. Compute the rise of the road shown in Figure 14–54.

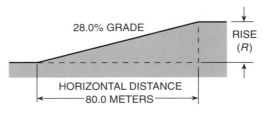

28.0% GRADE

RISE (*R*)

HORIZONTAL DISTANCE
80.0 METERS

Figure 14–54

For each of these problems, substitute the given values in the formula and solve for the unknown. Check.

35. The average rate of speed (*R*) of a vehicle, in miles per hour, is computed by dividing distance (*D*), in miles, by time (*t*), in hours: $R = \frac{D}{t}$. Determine *D* when *t* = 0.80 hour and *R* = 55 miles per hour.

36. The pitch (*P*) of a spur gear equals the number (*N*) of gear teeth divided by the pitch diameter (*D*): $P = \frac{N}{D}$. Determine *N* when *P* = 5 and *D* = 5.6000 inches.

37. Ohm's law states that current, in amperes (*I*), equals voltage, in volts (*E*), divided by resistance, in ohms (*R*): $I = \frac{E}{R}$. Determine *E* when *I* = 1.5 amperes and *R* = 8.0 ohms.

38. The diameter of a circle equals the circle circumference divided by 3.1416: $D = \frac{C}{3.1416}$. Determine *C* when *D* = 17.06 centimeters. Round the answer to 4 significant digits.

39. In mechanical energy applications, force, in pounds (*F*), equals work in foot-pounds (*W*), divided by distance, in feet (*D*): $F = \frac{W}{D}$. Determine *W* when *F* = 150.5 pounds and *D* = 7.53 feet. Round the answer to 3 significant digits.

40. In electrical energy applications, power, in watts (*W*), equals energy, in watt-hours (*Wh*), divided by time, in hours (*T*): $W = \frac{Wh}{T}$. Determine *Wh* when *W* = 250 watts and *T* = 3.2 hours.

41. The number of gallons (*g*) in a gasoline tank is the number of liters (*L*) multiplied by 0.2642 gallon/liter: *g* = 0.2642*L*. Determine the number of liters that a gas tank can hold if its capacity is 37.4 gal.

42. A grinding wheel can have an allowable surface speed (*S*) in feet per minute where *S* is the product of π(3.1416), the diameter of the wheel (*D*) and the revolutions per minute, rpm, (*R*) of the grinder spindle divided by 12: that is, $S = \frac{\pi DR}{12}$. Determine the rpm of the grinder spindle if the surface speed in 6,280 feet per minute and the diameter of the wheel is 12 inches.

14–9   Solution of Equations by the Root Principle of Equality

The root principle of equality states that if the same root of both sides of an equation is taken, the sides remain equal and the equation remains balanced. The root principle is used to solve an equation that contains an unknown that is raised to a power; for example, $s^2 = 36$.

Procedure for Solving an Equation in which the Unknown Is Raised to a Power

- Extract the root of both sides of the equation that leaves the unknown with an exponent of 1.
- Check.

EXAMPLES

1. Solve for *x*. $x^2 = 25$. See Figures 14–55, 14–56, and 14–57.

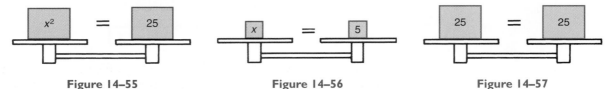

Figure 14–55   **Figure 14–56**   **Figure 14–57**

In the equation, x is squared (Figure 14–55). To solve the equation, extract the square root of both sides (Figure 14–56).

$$x^2 = 25$$
$$\sqrt{x^2} = \sqrt{25}$$
$$x = 5 \; Ans$$

(Figure 14–57)

Check.

$$x^2 = 25$$
$$5^2 = 25$$
$$25 = 25 \; Ck$$

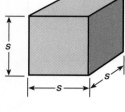

Figure 14–58

2. The volume of the cube shown in Figure 14–58 equals 164.3 cubic centimeters. Expressed as an equation $s^3 = 164.3 \; cm^3$. Solve for s.

$$s^3 = 164.3 \; cm^3$$
$$\sqrt[3]{s^3} = \sqrt[3]{164.3 \; cm^3}$$

Extract the cube root of both sides of the equation.

Calculator Application

3 $\boxed{\sqrt[x]{\;}}$ 164.3 $\boxed{EXE}$ 5.477039266, $s \approx 5.477$ cm Ans
or 164.3 $\boxed{2nd}$ $\boxed{y^x}$ 3 $\boxed{=}$ 5.477039266, $s \approx 5.477$ cm Ans
or 164.3 $\boxed{SHIFT}$ $\boxed{x^y}$ 3 $\boxed{=}$ 5.477039266, $s \approx 5.477$ cm Ans
or 164.3 $\boxed{\wedge}$ $\boxed{(}$ 1 $\boxed{\div}$ 3 $\boxed{)}$ $\boxed{ENTER}$ 5.477039266, $s \approx 5.477$ cm Ans
Check. $(5.477039266 \; cm)^3 \approx 164.3$ cm

Calculator Application

5.477039266 $\boxed{x^y}$ or $\boxed{y^x}$ or $\boxed{\wedge}$ 3 $\boxed{=}$ 164.3 (rounded)
164.3 $cm^3 = 164.3 \; cm^3 \; Ck$

EXERCISE 14–9

Solve each of these equations using the root principle of equality. Check all answers. Problems 23 through 30 require calculator solutions. Round the answers to 3 significant digits where necessary.

1. $S^2 = 16$
2. $P^2 = 36$
3. $81 = M^2$
4. $49 = B^2$
5. $D^3 = 27$
6. $x^3 = -27$
7. $144 = F^2$
8. $-64 = y^3$
9. $L^3 = 125$
10. $T^3 = 0$
11. $10,000 = L^2$
12. $-125 = x^3$
13. $\dfrac{4}{9} = W^2$

14. $C^2 = \dfrac{1}{16}$
15. $P^2 = \dfrac{9}{25}$
16. $M^3 = \dfrac{1}{8}$
17. $-\dfrac{1}{8} = y^3$
18. $D^3 = \dfrac{1}{-64}$
19. $G^3 = \dfrac{64}{125}$

20. $x^3 = \dfrac{-64}{125}$
21. $E^2 = 0.09$
22. $0.64 = H^2$
23. $W^2 = 2.753$
24. $0.0017 = R^2$
25. $N^3 = 87.25$
26. $-0.609 = x^3$
27. $8.794 = F^4$
28. $T^2 = 361.9$
29. $y^5 = -7852.6$
30. $17.004 = y^3$

31. The volume of a cube equals the length of a side (s) cubed. See Figure 14–59. Given the volumes of cubes in Figure 14–60, write an equation for each, solve for s, and check.

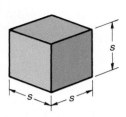

Figure 14–59

	VOLUME	EQUATION	s
a.	125 cubic centimeters		
b.	0.064 cubic yard		
c.	$\frac{8}{27}$ cubic foot		
d.	0.027 cubic meter		
e.	$\frac{1}{8}$ cubic inch		

Figure 14–60

32. The cross-section area of a square pipe is the length of a side (s) squared. If the cross-section area of a square pipe is 67 in²,

    a. Write an equation for the length of each side

    b. Find the length of s to the nearest 16th on an inch

    c. Check your answer

14–10   Solution of Equations by the Power Principle of Equality

The power principle of equality states that if both sides of an equation are raised to the same power, the sides remain equal and the equation remains balanced. The power principle is used to solve an equation that contains a root of the unknown; for example, $\sqrt{A} = 8$.

Procedure for Solving an Equation Which Contains a Root of the Unknown

• Raise both sides of the equation to the power that leaves the unknown with an exponent of 1.

• Check.

EXAMPLES •───────────────────────────────

1. Solve for y. See Figures 14–61, 14–62, and 14–63.

$$\sqrt{y} = 3$$

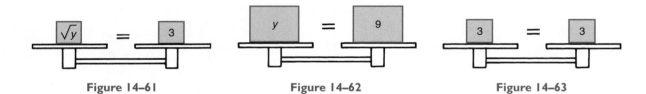

Figure 14–61   Figure 14–62   Figure 14–63

In the equation, y is expressed as a square root (Figure 14–61). $\sqrt{y} = 3$

To solve the equation, square both sides (Figure 14–62). $(\sqrt{y})^2 = 3^2$

     $y = 9$ *Ans*

(Figure 14–63)   Check. $\sqrt{y} = 3$

     $\sqrt{9} = 3$

     $3 = 3$ Ck

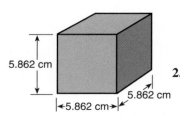

5.862 cm

5.862 cm

5.862 cm

Figure 14–64

2. The length of each side of the cube shown in Figure 14–64 equals 5.862 centimeters. The volume (V) of the cube can be expressed as the equation $\sqrt[3]{V} = 5.862$ cm.

Solve for V. $\sqrt[3]{V} = 5.862$ cm

Cube both sides of the equation. $(\sqrt[3]{V})^3 = (5.862 \text{ cm})^3$

Calculator Application

5.862 $\boxed{x^y}$ or $\boxed{y^x}$ 3 $\boxed{=}$ 201.4361639, $V \approx 201.4$ cm^3 *Ans*

or 5.862 $\boxed{\wedge}$ 3 $\boxed{\text{EXE}}$ 201.4361639, $V \approx 201.4$ cm^3 *Ans*

Check.

Calculator Application

NOTE: $\boxed{\sqrt[x]{}}$ and $\boxed{\sqrt[x]{y}}$ are second functions on certain calculators.

3 $\boxed{\sqrt[x]{}}$ 201.4361639 $\boxed{\text{EXE}}$ or $\boxed{=}$ 5.862

or 201.4361639 $\boxed{\text{2nd}}$ $\boxed{y^x}$ 3 = 5.862

or 201.4361639 $\boxed{\text{SHIFT}}$ $\boxed{x^y}$ $\boxed{=}$ 5.862

5.862 cm = 5.862 cm Ck

EXERCISE 14–10

Solve each of these equations using the power principle of equality. Check all answers. Round answers to 3 significant digits where necessary.

1. $\sqrt{C} = 8$

2. $\sqrt{T} = 12$

3. $\sqrt{P} = 1.2$

4. $0.7 = \sqrt{M}$

5. $0.82 = \sqrt{F}$

6. $\sqrt[3]{V} = 3$

7. $\sqrt[3]{H} = 2.3$

8. $\sqrt[3]{x} = -4$

9. $-0.1 = \sqrt[3]{y}$

10. $\sqrt[4]{M} = 2$

11. $\sqrt{A} = 0$

12. $\sqrt[5]{N} = 0.9$

13. $-1.72 = \sqrt[5]{y}$

14. $0.2 = \sqrt[4]{D}$

15. $\sqrt[3]{x} = -0.6$

16. $\sqrt[4]{P} = 0.1$

17. $0.1 = \sqrt[3]{B}$

18. $\dfrac{1}{2} = \sqrt{A}$

19. $\sqrt{R} = \dfrac{3}{8}$

20. $\sqrt[3]{V} = \dfrac{2}{3}$

21. $\sqrt[4]{F} = \dfrac{1}{2}$

22. $-\dfrac{3}{5} = \sqrt[3]{y}$

23. $\dfrac{5}{8} = \sqrt{H}$

24. $\sqrt{P} = 1\dfrac{1}{4}$

25. $\sqrt[3]{B} = 2.86$

26. $\sqrt[5]{x} = -1.09$

27. $7.83 = \sqrt[3]{y}$

28. $0.364 = \sqrt[3]{y}$

29. $\sqrt[4]{x} = 12.04$

30. $\sqrt[5]{x} = -2.96$

Solve these problems using the power principle of equality.

31. The length of a side of a square equals the square root of the area (*A*). See Figure 14–65. Given the length of the sides of squares in Figure 14–66, write an equation for each, solve for area (*A*), and check. Round answers to 4 significant digits where necessary.

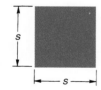

Figure 14–65

	LENGTH OF SIDE (s)	EQUATION	AREA (A)
a.	3.6 centimeters		
b.	12.92 meters		
c.	$5\dfrac{1}{2}$ inches		
d.	$\dfrac{3}{4}$ foot		
e.	0.087 meter		

Figure 14–66

⠿ UNIT EXERCISE AND PROBLEM REVIEW

EQUATIONS INVOLVING ADDITION OR SUBTRACTION

Solve each of these equations. Check all answers.

1. $M + 7 = 22$

2. $16 = P + 12$

3. $T - 18 = 9$

4. $y - 7 = -20$

5. $x + 13 = -20$

6. $25 = D - 8$

7. $25 = D + 8$

8. $-17 = T - 15$

9. $y - 36 = -18$

10. $x + 12.9 = -7.3$

11. $H + 9.3 = 14.8$

12. $A - 0.6 = 0.4$

13. $-22 = L - 26.1$

14. $2.6 = F + 3.7$

15. $M + 0.03 = 2.05$

16. $R - 4.52 = 5.48$

17. $29.197 = x + 31.093$

18. $y - 14.69 = -10.05$

19. $D - \dfrac{3}{4} = -\dfrac{3}{4}$

20. $H + 2\dfrac{1}{2} = 6\dfrac{1}{4}$

21. $27\dfrac{3}{8} = B + 19\dfrac{1}{8}$

22. $y - \dfrac{5}{8} = -\dfrac{1}{2}$

23. $6\dfrac{9}{16} = F - 8\dfrac{11}{16}$

24. $x + 14\dfrac{13}{16} = -10\dfrac{7}{8}$

EQUATIONS INVOLVING MULTIPLICATION OR DIVISION

Solve each of these equations. Check all answers.

25. $5L = 55$

26. $9x = -63$

27. $\dfrac{D}{7} = 42$

28. $H \div 4 = 8$

29. $50 = -10x$

30. $\dfrac{A}{-6} = 54$

31. $\dfrac{B}{-3} = -27$

32. $9M = 0$

33. $-12E = -48$

34. $-y = -43.6$

35. $26 = \dfrac{H}{2.5}$

36. $0.66 = 0.3F$

37. $-7.5x = 24$

38. $\dfrac{W}{19.60} = 0.015$

39. $\dfrac{y}{0.1} = -0.06$

40. $8.73B = 29.682$

41. $-20 = M \div 18.85$

42. $P \div (-5.6) = -10.3$

43. $-0.073x = 0.584$

44. $\dfrac{1}{2}E = 6$

45. $V \div 2\dfrac{3}{4} = 8$

46. $\dfrac{x}{\frac{3}{8}} = -\dfrac{1}{4}$

47. $-\dfrac{5}{16} = -2\dfrac{1}{2}F$

48. $y \div \left(-\dfrac{3}{4}\right) = 3\dfrac{1}{2}$

EQUATIONS INVOLVING POWERS OR ROOTS

Solve each of these equations. Check all answers. Round answers to 3 significant digits where necessary.

49. $B^2 = 64$

50. $x^3 = 64$

51. $x^3 = -64$

52. $\sqrt{M} = 11$

53. $15 = \sqrt{D}$

54. $H^3 = 125$

55. $\sqrt[3]{V} = 6$

56. $\sqrt{P} = 0$

57. $-8 = y^3$

58. $\sqrt{E} = 0.7$

59. $\sqrt[4]{L} = 0.1$

60. $x^3 = -0.125$

61. $0.2 = \sqrt[3]{F}$

62. $0.09 = B^2$

63. $\sqrt[3]{y} = -0.86$

64. $x^3 = 0.0312$

65. $T^2 = 43.56$

66. $\sqrt[5]{M} = -1$

67. $E^2 = 265.8$

68. $G^3 = -1.089$

69. $\sqrt{A} = \dfrac{5}{3}$

70. $\sqrt[3]{V} = \dfrac{2}{5}$

71. $y^3 = \dfrac{-8}{125}$

72. $\sqrt[3]{x} = -2\dfrac{3}{4}$

PROBLEMS: WRITING AND SOLVING EQUATIONS

For each of the following problems, write an equation and solve, by logical reasoning, for the unknown. Check the equation by comparing it to the word problem. Does the equation state mathematically what the problem states in words? Check whether the equation is balanced by substituting the value of the unknown in the equation.

73. Five times a number, minus the number, equals 24.

74. Three times a number, divided by 6, equals $\frac{1}{2}$.

75. Two added to 4 times a number, minus 3 times the number, equals 5.

76. See Figure 14–67.   **77.** See Figure 14–68.

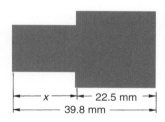

Figure 14–67

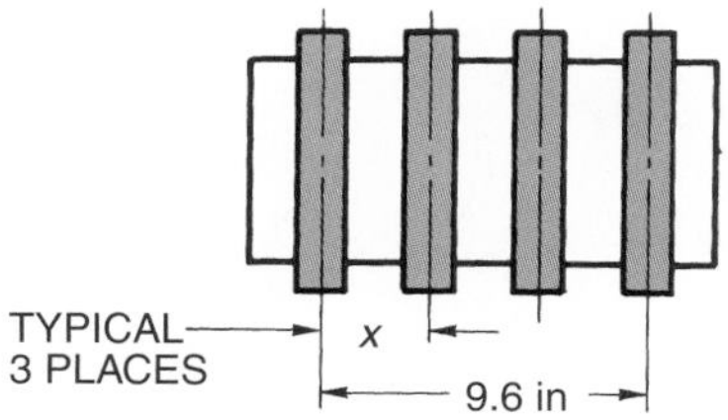

Figure 14–68

78. A length of tubing 40.0 inches long is cut into 2 unequal lengths. One piece is 4 times as long as the other.

   a. Find the length of the shorter piece.

   b. Find the length of the longer piece.

79. A company's profit for the second half year is $150,000 greater than the profit for the first half year. The total annual profit is $850,000. What is the profit for the first half year?

80. Four holes are drilled along a straight line in a plate. The distance between hole 1 and hole 4 is 35.0 centimeters. The distance between hole 2 and hole 3 is twice the distance between hole 1 and hole 2. The distance between hole 3 and hole 4 is the same as the distance between hole 2 and hole 3. What is the distance between hole 1 and hole 3?

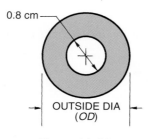

0.8 cm

OUTSIDE DIA
(OD)

Figure 14–69

81. In a series circuit, the sum of the voltages across each of the individual resistors is equal to the applied line voltage. In a 120-volt circuit with 3 resistors, the voltage across R_2 is 20 volts greater than the voltage across R_1. The voltage across R_3 is 20 volts greater than the voltage across R_2. What is the voltage across R_3?

82. The inside diameter of the washer shown in Figure 14–69 is 0.8 centimeter, which is 1.7 centimeters less than the outside diameter. What is the outside diameter (OD)?

83. Power, in watts, is equal to the product of the square of the current, in amperes, and the total line resistance, in ohms. Determine the line resistance (R) when power equals 400 watts and current equals 40 amperes. Round the answer to 1 decimal place.

84. The book value of equipment using the straight-line depreciation method is equal to the original cost minus the product of number of years depreciated and the annual depreciation. Determine the original cost (OC) when the book value equals $3,200, the annual depreciation is $400, and the depreciation is over a 5-year period.

PROBLEMS: SUBSTITUTING VALUES IN FORMULAS

For each of these problems, substitute the given values in the formula and solve for the unknown. Check.

85. The spur gear formula is $D_R = D - 2d$. Determine the pitch diameter (D) when the root diameter (D_R) = 76.50 millimeters and the dedendum (d) = 4.08 millimeters.

86. Power, in kilowatts (P_{kW}), is equal to 0.001 times voltage, in volts (E), times current, in amperes (I): $P_{kW} = 0.001EI$. Determine E when $P_{kW} = 3.6$ kilowatts and $I = 30.0$ amperes.

87. The area of a triangle is equal to $\frac{1}{2}$ times the base times the height: $A = \frac{1}{2}bh$. Determine h when $A = 3\frac{1}{2}$ square feet and $b = 1\frac{3}{4}$ feet.

88. The volume of a cube equals the length of a side cubed: $V = s^3$. Determine s when $V = 307.3$ cubic centimeters. Round the answer to 3 significant digits.

UNIT 15 ⠿ Complex Equations

OBJECTIVES

After studying this unit you should be able to

- solve equations consisting of combined operations.
- substitute values in formulas and solve for unknowns.
- rearrange formulas to solve for any letter value.

15–1 Equations Consisting of Combined Operations

Often in actual occupational applications, formulas are used and equations are developed that are complex. These equations require the use of two or more principles of equality for their solutions. For example, the equation $0.13x - 4.73(x + 6.35) = 5.06x - 2.87$ requires a definite step-by-step procedure for determining the value of x. Use of proper procedure results in the unknown standing alone on one side of the equation with its value on the other.

Procedure for Solving Equations Consisting of Combined Operations

> It is essential that the steps used in solving an equation be taken in this order. Some or all of these steps may be used, depending on the equation.
>
> - Remove parentheses.
> - Combine like terms on each side of the equation.
> - Apply the addition and subtraction principle of equality to get all unknown terms on one side of the equation and all known terms on the other side.
> - Combine like terms.
> - Apply the multiplication and division principles of equality.
> - Apply the power and root principles of equality.

NOTE: Always solve for a positive unknown. A positive unknown may equal a negative value, but a negative unknown is not a solution. For example, $x = -10$ is correct, but $-x = 10$ is incorrect. When solving equations where the unknown remains a negative value, multiply both sides of the equation by -1. Multiplying a negative unknown by -1 results in a positive unknown. For example, multiplying both sides of $-x = 10$ by -1 gives $(-1)(-x) = (-1)(10), x = -10$.

EXAMPLES •————————————————————————————

The following are examples of equations consisting of combined operations.

1. Solve for x.

$8x + 12 = 52$

In the equation, the operations involved are multiplication and addition. First subtract 12 from both sides of the equation.

$$8x + 12 = 52$$

$$\underline{-12 = -12}$$

Next divide both sides of the equation by 8.

$$\frac{8x}{8} = \frac{40}{8}$$

$$x = 5 \ Ans$$

Check.

$$8x + 12 = 52$$
$$8(5) + 12 = 52$$
$$40 + 12 = 52$$
$$52 = 52 \ Ck$$

2. Solve for *D*.

$$-18D + 4D = 3D - 5D + 19 + 5$$

Combine like terms on each side of the equation.

Add 2*D* to both sides of the equation.

Divide both sides of the equation by −12.

$$-18D + 4D = 3D - 5D + 19 + 5$$
$$-14D = -2D + 24$$

$$\underline{+ \ 2D = +2D}$$

$$\frac{-12D}{-12} = \frac{24}{-12}$$

$$D = -2 \ Ans$$

Check.

$$-18D + 4D = 3D - 5D + 19 + 5$$
$$-18(-2) + 4(-2) = 3(-2) - 5(-2) + 19 + 5$$
$$36 + (-8) = -6 - (-10) + 19 + 5$$
$$28 = 28 \ Ck$$

3. Solve for *y*.

$$14y - 6(y - 3) = 22$$

Remove the parentheses.

Combine like terms.
Subtract 18 from both sides of the equation.

$$14y - 6(y - 3) = 22$$
$$14y - 6y + 18 = 22$$
$$8y + 18 = 22$$
$$\underline{-18 = -18}$$

Divide both sides of the equation by 8.

$$\frac{8y}{8} = \frac{4}{8}$$

$$y = \frac{1}{2} \quad Ans$$

Check.

$$14y - 6(y - 3) = 22$$

$$14\left(\frac{1}{2}\right) - 6\left(\frac{1}{2} - 3\right) = 22$$

$$14\left(\frac{1}{2}\right) - 6\left(-2\frac{1}{2}\right) = 22$$

$$7 + 15 = 22$$

$$22 = 22 \ Ck$$

4. Solve for x.

$$\frac{x^2}{6} - 36.5 = -35$$

	$\dfrac{x^2}{6} - 36.5 = -35$
Add 36.5 to both sides of the equation.	$\underline{+ 36.5 = +36.5}$
Multiply both sides of the equation by 6.	$\dfrac{6}{1}\left(\dfrac{x^2}{6}\right) = (1.5)6$
Simplify.	$\dfrac{\cancel{6}}{1}\left(\dfrac{x^2}{\cancel{6}}\right) = (1.5)6$
	$x^2 = 9$
Extract the square root of both sides of the equation.	$\sqrt{x^2} = \sqrt{9}$
	$x = 3 \qquad Ans$

$$\text{Check.} \qquad \frac{x^2}{6} - 36.5 = -35$$

$$\frac{3^2}{6} - 36.5 = -35$$

$$\frac{9}{6} - 36.5 = -35$$

$$1.5 - 36.5 = -35$$

$$-35 = -35 \text{ Ck}$$

5. Solve for R.

$$4 = \frac{130}{R + 20}$$

Observe that R appears in the denominator of a fractional expression. Whenever the unknown appears in the denominator, an operation is performed to remove the *complete* denominator. Multiply both sides of the equation by the *complete* denominator.

Multiply both sides of the equation by $R + 20$.	$4(R + 20) = \dfrac{130(R + 20)}{R + 20}$
Simplify.	$4(R + 20) = \dfrac{130(\cancel{R + 20})}{\cancel{R + 20}}$
	$4(R + 20) = 130$
Remove parentheses.	$4R + 80 = 130$
Subtract 80 from both sides of the equation.	$\underline{\quad -80 \quad\; -80 \quad}$
Divide both sides of the equation by 4.	$\dfrac{4R}{4} = \dfrac{50}{4}$
	$R = 12.5 \; Ans$

$$\text{Check.} \qquad 4 = \frac{130}{R + 20}$$

$$4 = \frac{130}{12.5 + 20}$$

$$4 = \frac{130}{32.5}$$

$$4 = 4 \text{ Ck}$$

6. Solve for P.

$$6\sqrt[3]{P} = 4\left(\sqrt[3]{P} + 1.5\right)$$

Remove parentheses.

Subtract $4\sqrt[3]{P}$ from both sides of the equation.

Divide both sides of the equation by 2.

Raise both sides of the equation to the third power.

$$6\sqrt[3]{P} = 4\left(\sqrt[3]{P} + 1.5\right)$$
$$6\sqrt[3]{P} = 4\sqrt[3]{P} + 6$$
$$\underline{-4\sqrt[3]{P} = -4\sqrt[3]{P}}$$
$$\frac{2\sqrt[3]{P}}{2} = \frac{6}{2}$$
$$\sqrt[3]{P} = 3$$
$$\left(\sqrt[3]{P}\right)^3 = (3)^3$$

$$P = 27 \; Ans$$

Check.

$$6\sqrt[3]{P} = 4\left(\sqrt[3]{P} + 1.5\right)$$
$$6\sqrt[3]{27} = 4\left(\sqrt[3]{27} + 1.5\right)$$
$$6(3) = 4(3 + 1.5)$$
$$6(3) = 4(4.5)$$
$$18 = 18 \; Ck$$

EXERCISE 15–1

Solve for the unknown and check each of these combined operations equations. Round answers to 2 decimal places where necessary.

1. $5x - 33 = 7$

2. $10M + 5 + 4M = 89$

3. $8E - 14 = 2E + 28$

4. $4B - 7 = B + 14$

5. $7T - 14 = 0$

6. $6N + 4 = 84 + N$

7. $10.8A - 12.3 = -23.4 - 16.9$

8. $12 - (-x + 8) = 18$

9. $3H + (2 - H) = 20$

10. $6 = -(2 + C) - (4 + 2C)$

11. $5(R + 6) = 10(R - 2)$

12. $0.29E = 9.39 - 0.01E$

13. $7.2F + 5(F - 8.1) = -0.06F + 15.18$

14. $\dfrac{P}{7} + 8 = 5.9$

15. $\dfrac{1}{4}W + (W - 8) = \dfrac{3}{4}$

16. $\dfrac{1}{8}D - 3(D - 7) = 5\dfrac{1}{8}D - 3$

17. $0.58y = 18.78 - 0.02y$

18. $2H^2 - 20 = (H + 4)(H - 4)$

19. $4A^2 + 3A + 36 = 8A^2 + 3A$

20. $x(3 + x) + 20 = x^2 - (x - 5)$

21. $\dfrac{b^3}{2} + 34 = 42$

22. $3F^2 + F(F + 8) = 8F + F^2 + 81$

23. $\dfrac{18}{x + 7} = 10$

24. $\dfrac{1}{4}(2B - 12) + B^2 = \dfrac{1}{2}B + 22$

25. $\dfrac{14.6}{1.8x} = \dfrac{-8.3}{6x + 0.9}$

26. $14\sqrt{x} = 6\left(\sqrt{x} + 8\right) + 16$

27. $8.12P^2 + 6.83P + 5.05 = 16.7P^2 + 6.83P$

28. $7.3\sqrt{x} = 3\left(\sqrt{x} + 8.06\right) - 4.59$

29. $\sqrt{B^2} - 2.53B = -2.53(B - 3.95)$

30. $(2y)^3 - 2.80(5.89 + 3y) = -23.87 - 8.40y$

15–2   Solving for the Unknown in Formulas

Occupational applications often require solving formulas in which all but one numerical value of the letter values are known. The unknown letter value can appear anywhere within the formula. There are two ways of solving for the unknown.

1. Substituting known number values for letter values directly in the given formula and solving for the unknown.

2. First rearranging the given formula so the unknown stands alone on one side of the formula, then substituting known number values for letter values, and solving for the unknown.

Both ways of solving for the unknown are shown.

15–3   Substituting Values Directly in Given Formulas

To determine the numerical value of the unknown, write the original formula, substitute the known number values for their respective letter values, and simplify. Then follow the procedure given for solving equations consisting of combined operations.

EXAMPLES •

1. An open belt pulley system is shown in Figure 15–1. The number of inches between the pulley centers is represented by x. The larger pulley diameter (D) is 6.25 inches, and the smaller pulley diameter (d) is 4.25 inches. The belt length (L) is 56.0 inches. This formula is found in a trade handbook.

$$L = 3.14(0.5D + 0.5d) + 2x$$

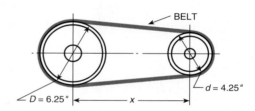

Figure 15–1

Use the formula to compute x.

Write the formula.	$L = 3.14(0.5D + 0.5d) + 2x$
Substitute the known numerical values for their respective letter values and simplify.	$56.0 \text{ in} = 3.14[0.5(6.25 \text{ in}) + 0.5(4.25 \text{ in})] + 2x$ $56.0 \text{ in} = 3.14(5.25 \text{ in}) + 2x$ $56.0 \text{ in} = 16.485 \text{ in} + 2x$

Subtract 16.485 inches from both sides.

$$\begin{array}{rcl} 56.0 \text{ in} &=& 16.485 \text{ in} + 2x \\ -16.485 \text{ in} &=& -16.485 \text{ in} \\ \hline \end{array}$$

Divide both sides by 2.

$$\frac{39.515 \text{ in}}{2} = \frac{2x}{2}$$

$$19.7575 \text{ in} = x, x \approx 19.8 \text{ in } Ans$$

Check.

Calculator Application

$L = 3.14 (0.5D + 0.5d) + 2x$

$56 = 3.14 \boxed{\times} \boxed{(} .5 \boxed{\times} 6.25 \boxed{+} .5 \boxed{\times} 4.25 \boxed{)} \boxed{+} 2 \boxed{\times} 19.7575 \boxed{=} 56$

$56 = 56 \text{ Ck}$

2. A slot is cut in the circular piece shown in Figure 15–2. The piece has a radius (R) of 97.60 millimeters. The number of millimeters in the width is represented by W. <u>Dimension A is</u> 20.20 millimeters. The formula is found in a trade handbook is $A = R - \sqrt{R^2 - 0.2500W^2}$. Find dimension W.

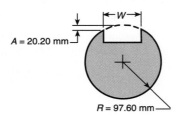

Figure 15–2

Write the formula.	$A = R - \sqrt{R^2 - 0.2500W^2}$
Substitute the known numerical values for their respective letter values and simplify.	$20.20 \text{ mm} = 97.60 \text{ mm} - \sqrt{(97.60 \text{ mm})^2 - 0.2500W^2}$ $20.20 \text{ mm} = 97.60 \text{ mm} - \sqrt{9525.76 \text{ mm}^2 - 0.2500W^2}$
Subtract 97.60 mm from both sides.	$\dfrac{-97.60 \text{ mm} = -97.60 \text{ mm}}{-77.40 \text{ mm} = -\sqrt{9{,}525.76 \text{ mm}^2 - 0.2500W^2}}$

Observe that both terms, 9,525.76 mm^2 and 0.2500W^2, are enclosed within the radical sign. Neither term can be removed until the radical sign is eliminated.

Square both sides.	$(-77.40 \text{ mm})^2 = \left(-\sqrt{9{,}525.76 \text{ mm}^2 - 0.2500W^2}\right)^2$ $5{,}990.76 \text{ mm}^2 = 9{,}525.76 \text{ mm}^2 - 0.2500W^2$
Subtract 9,525.76 mm^2 from both sides.	$\dfrac{-9{,}525.76 \text{ mm}^2 = -9{,}525.76 \text{ mm}^2}{-3{,}535.00 \text{ mm}^2 = -0.2500W^2}$
Divide both sides by -0.2500.	$\dfrac{-3{,}535.00 \text{ mm}^2}{-0.2500} = \dfrac{-0.2500W^2}{-0.2500}$ $14{,}140 \text{ mm}^2 = W^2$
Take the square root of both sides.	$\sqrt{14{,}140 \text{ mm}^2} = \sqrt{W^2}$ $118.911732 \text{ mm} = W, \; W \approx 118.9 \text{ mm } Ans$
Check.	

Calculator Application

$A = R - \sqrt{R^2 - 0.2500W^2}$

$20.20 = 97.6 \;\boxed{-}\;\boxed{(}\; 97.6 \;\boxed{x^2}\;\boxed{-}\; .25 \;\boxed{\times}\; 118.911732 \;\boxed{x^2}\;\boxed{)}\;\boxed{\sqrt{x}}\;\boxed{=}\; 20.20000001$

or $20.20 = 97.6 \;\boxed{-}\;\boxed{\sqrt{}}\;\boxed{(}\; 97.6 \;\boxed{x^2}\;\boxed{-}\; .25 \;\boxed{\times}\; 118.911732 \;\boxed{x^2}\;\boxed{)}\;\boxed{\text{EXE}}\; 20.20000001$

$20.20 = 20.20$ Ck

EXERCISE 15–3

The formulas for this set of problems have been taken from various technical fields. Substitute the given numerical values for letter values and solve for the unknown.

1. This is the formula for finding the area of a triangle.

$$A = \frac{ab}{2}$$ where A = area
   a = altitude
   b = base

   Solve for a when A = 24 sq ft and b = 8 ft.

2. The inductive reactance is found from this formula.

$$X_L = 2\pi f L$$ where X_L = inductive reactance, in ohms
   π = 3.14
   f = frequency in hertz
   L = inductance in henrys

   Solve for L when f = 65 Hz and X_L = 82 ohms. Round the answer to 2 significant digits.

3. In a power transformer, the ratio of the primary and the secondary voltages is directly proportional to the number of turns on the primary and the secondary.

$$\frac{E_p}{E_s} = \frac{T_p}{T_s}$$

   Solve for T_s when E_p = 440 volts, E_s = 2,200 volts, and T_p = 150 turns.

4. This formula is used to express a temperature in degrees Celsius as a temperature in degrees Fahrenheit.

$$°F = \frac{9}{5}(°C) + 32°$$

   Solve for °C when the temperature is 28°F. Round the answer to 2 significant digits.

5. The area of a trapezoid is found from this formula.

$$A = \frac{h}{2}(b + b')$$ where A = area
   h = height
   b = first base
   b' = second base

   Solve for b when A = 108 sq ft, h = 8.65 ft, and b' = 12.50 ft. Round the answer to 3 significant digits.

6. The accumulated amount in a savings account is found from this formula. Round the answer to 3 significant digits.

$$A = p + prt$$ where A = amount at end of period
   p = principal invested
   r = rate
   t = time in years

   Solve for t when p = $1,800, r = 7%, and A = $3,312.

7. The sum of an arithmetic progression is found from this formula.

$$S = \frac{n}{2}(a + l)$$

where S = sum
n = number of terms
a = first term
l = last term

Solve for l when $S = 150$, $n = 6$, and $a = 5$.

8. There are 4 cells connected in a series circuit. The internal resistance of each cell is 5.0 ohms. The external circuit resistance is 8 ohms. The current in the circuit is 4.0 amperes. Find the voltage of each cell.

$$I = \frac{nE}{R + nr}$$

where I = current in amperes
n = number of cells
E = voltage of one cell
r = internal resistance of one cell
R = resistance of external circuit

9. This is a formula for finding horsepower.

$$\text{hp} = \frac{IE(\text{Eff})}{746}$$

where hp = horsepower
E = voltage in volts
I = current in amperes
Eff = efficiency

Solve for I when hp = 10.5 horsepower, $E = 220$ volts, and Eff = 85%. Round the answer to the nearest whole ampere.

10. This formula is used to find the impedance of a circuit.

$$Z = \sqrt{R^2 + X^2}$$

where Z = impedance in ohms
R = resistance in ohms
X = reactance in ohms

Solve for X when $R = 12$ ohms and $Z = 65$ ohms. Round the answer to the nearest whole ohm.

15–4   Rearranging Formulas

A formula that is used to find a particular value must sometimes be rearranged to solve for another value. Consider the letter to be solved for as the unknown term and the other letters in the formula as the known values. The formula must be rearranged so that the unknown term is on one side of the equation and all other values are on the other side. A formula is rearranged by using the same procedure that is used for solving equations consisting of combined operations.

Problems are often solved more efficiently by first rearranging formulas than by directly substituting values in the original formula and solving for the unknown. This is particularly true in solving more complex formulas which involve many operations. Also, it is sometimes necessary to solve for the same unknown after a formula has been rearranged using different known values. Since the formula has been rearranged in terms of the specific unknown, solutions are more readily computed.

First rearranging formulas and then substituting known values enables you to solve for the unknown using a calculator for continuous operations. This is illustrated in example 3.

EXAMPLES •——————————————————————————

1. A screw thread is checked with a micrometer and 3 wires as shown in Figure 15–3. The measurement is checked by using this formula.

$$M = D - 1.5155p + 3w$$ where M = measurement over the wires
D = major diameter
p = pitch
w = wire size

Solve the formula for w.

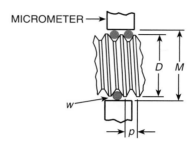

Figure 15–3

Subtract D from both sides of the equation.	$M = D - 1.5155p + 3w$
	$-D = -D$
Add $1.5155p$ to both sides of the equation.	$M - D = -1.5155p + 3w$
	$+1.5155p = +1.5155p$
Divide both sides of the equation by 3.	$\dfrac{M - D + 1.5155p}{3} = \dfrac{3w}{3}$

$$w = \frac{M - D + 1.5155p}{3} \quad Ans$$

Figure 15–4

2. The area of a ring shown in Figure 15–4 is computed from this formula.

$$A = \pi(R^2 - r^2)$$ where A = area
π = pi
R = outside radius
r = inside radius

Solve for R.

Remove parentheses.	$A = \pi(R^2 - r^2)$
	$A = \pi R^2 - \pi r^2$
Add πr^2 to both sides of the equation.	$+\pi r^2 = \qquad + \pi r^2$
Divide both sides of the equation by π.	$\dfrac{A + \pi r^2}{\pi} = \dfrac{\pi R^2}{\pi}$
Take the square root of both sides.	$\sqrt{\dfrac{A + \pi r^2}{\pi}} = \sqrt{R^2}$

$$R = \sqrt{\frac{A + \pi r^2}{\pi}} \quad Ans$$

3. This example is the same problem as example 2 on page 364, which was solved by first substituting known number values in the given formula. Compare both methods of solution.

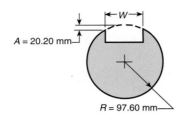

A = 20.20 mm

R = 97.60 mm

Figure 15–5

A slot is cut in the circular piece shown in Figure 15–5. The piece has a radius (R) of 97.60 millimeters. The number of millimeters in the width is represented by W. Dimension A is 20.20 millimeters. This formula is found in a trade handbook.

$$A = R - \sqrt{R^2 - 0.2500W^2}$$

Solve for W.

Subtract R from both sides of the equation.

$$\begin{array}{l} A = R - \sqrt{R^2 - 0.2500W^2} \\ \underline{-R = -R} \\ A - R = -\sqrt{R^2 - 0.2500W^2} \end{array}$$

Square both sides of the equation.

$$(A - R)^2 = \left(-\sqrt{R^2 - 0.2500W^2}\right)^2$$
$$(A - R)^2 = R^2 - 0.2500W^2$$

Subtract R^2 from both sides of the equation.

$$\begin{array}{l} (A - R)^2 = R^2 - 0.2500W^2 \\ \underline{-R^2 \quad -R^2} \end{array}$$

Divide both sides of the equation by -0.2500.

$$\frac{(A - R)^2 - R^2}{-0.2500} = \frac{-0.2500W^2}{-0.2500}$$

Take the square root of both sides.

$$\sqrt{\frac{(A - R)^2 - R^2}{-0.2500}} = \sqrt{W^2}$$

$$\sqrt{\frac{(A - R)^2 - R^2}{-0.2500}} = W$$

Substitute the given numerical values for their respective letter values and find dimension W.

$$W = \sqrt{\frac{(20.20 - 97.60)^2 - 97.60^2}{-0.2500}}$$

Calculator Application

$W = \boxed{(}\,\boxed{(}\, 20.2 \,\boxed{-}\, 97.6 \,\boxed{)}\, \boxed{x^2} \,\boxed{-}\, 97.6 \,\boxed{x^2} \,\boxed{=}\, \boxed{\div}\, .25 \,\boxed{+/-}\, \boxed{)}\, \boxed{\sqrt{x}} \rightarrow 118.911732$

or $W = \boxed{\sqrt{}}\, \boxed{(}\, \boxed{(}\, \boxed{(}\, 20.2 \,\boxed{-}\, 97.6 \,\boxed{)}\, \boxed{x^2} \,\boxed{-}\, 97.6 \,\boxed{x^2} \,\boxed{)}\, \boxed{\div}\, \boxed{(-)}$ or $\boxed{-}\, .25 \,\boxed{)}\, \boxed{EXE}$

118.911732 $\;\llcorner$ or .25 $\boxed{+/-}\, \boxed{)}\, \boxed{=}$

$W \approx 118.9$ mm *Ans*

Check. Substitute numerical values in the *original* formula.

$20.20 = 97.6 \,\boxed{-}\, \boxed{(}\, 97.6 \,\boxed{x^2}\, \boxed{-}\, .25 \,\boxed{\times}\, 118.911732 \,\boxed{x^2}\, \boxed{)}\, \boxed{\sqrt{x}}\, \boxed{=}\, 20.20000001$

or $20.20 = 97.6 \,\boxed{-}\, \boxed{\sqrt{}}\, \boxed{(}\, 97.6 \,\boxed{x^2}\, \boxed{-}\, .25 \,\boxed{\times}\, 118.911732 \,\boxed{x^2}\, \boxed{)}\, \boxed{EXE}$

or $\boxed{=}\, 20.20000001$

$20.20 = 20.20$ Ck

EXERCISE 15–4

The formulas for this set of problems are found in various occupational handbooks and manuals. Rearrange each and solve for the designated letter.

1. Solve for b.

$A = ab$

2. Solve for l.

$p = 2l + 2w$

3. a. Solve for E.

    b. Solve for R.

$$I = \frac{E}{R}$$

4. Solve for w.

$V = lwh$

5. Solve for $\angle C$.

$\angle A + \angle B + \angle C = 180°$

6. Solve for M.

$hp = 0.000016MN$

7. Solve for T.

$S = T - \dfrac{1.732}{N}$

8. Solve for r.

$E = I(R + r)$

9. Solve for C.

$C_a = S(C - F)$

10. Solve for $P.F.$

$P = 2EI(P.F.)$

11. Solve for I.

$P = \dfrac{EI}{1,000}$

12. a. Solve for K.

b. Solve for d.

$R = \dfrac{KL}{d^2}$

13. a. Solve for E_x.

b. Solve for R.

$I = \dfrac{E_x - E_c}{R}$

14. a. Solve for D.

b. Solve for c.

$CM = \dfrac{KIND}{c}$

15. a. Solve for E.

b. Solve for R.

$P = \dfrac{E^2}{R}$

16. Solve for I.

$kW = \dfrac{2IE(PE)}{1000}$

17. Solve for C.

$D_o = 2C - d + 2a$

18. Solve for P.

$M = D - 1.5155P + 3W$

19. a. Solve for P.

b. Solve for R.

$I = \sqrt{\dfrac{P}{R}}$

20. Solve for F.

$C_x = B_y(F - 1)$

21. Solve for E.

$M = E - 0.866P + 3W$

22. Solve for D.

$hp = \dfrac{D^2N}{2.5}$

23. Solve for d.

$L = 3.14(0.5D + 0.5d) + 2x$

24. Solve for D.

$C = \dfrac{0.7854D^2L}{231}$

25. a. Solve for R.

b. Solve for r.

$I = \dfrac{nE}{R + nr}$

26. A differential pulley is shown in Figure 15–6. This formula is used to find the pulling force.

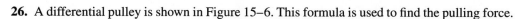

$$P = \dfrac{W(R - r)}{2R}$$

where P = pulling force
R = larger pulley
r = smaller pulley
W = weight

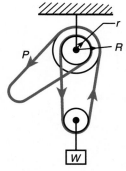

a. Solve for W.

b. Solve for r.

Figure 15–6

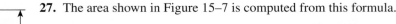

For problems 27 through 30, rearrange each formula for the designated letter and solve.

27. The area shown in Figure 15–7 is computed from this formula.

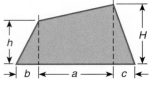

Figure 15–7

$$A = \frac{a(H + h) + bh + cH}{2}$$

  a. Solve for b when $A = 124.26$ m², $a = 12.5$ m, $c = 3.2$ m, $h = 6.0$ m, $H = 9.6$ m.

  b. Solve for a when $A = 265.8$ sq ft, $b = 5.1$ ft, $c = 4.8$ ft, $h = 9.2$ ft, $H = 13.4$ ft.

28. The horsepower of an electric motor is found with this formula.

$$hp = \frac{6.2832\ T(rpm)}{33,000}$$

  where hp = horsepower
  T = torque in pound feet (lb ft)
  rpm = revolutions per minute

  Solve for T when hp = 2.25 and rpm = 2,170. Round the answer to 3 significant digits.

29. This problem deals with heat transfer by conduction.

$$A = \frac{60Hd}{K(TD)}$$

  where A = surface area of metal (sq ft)
  d = thickness of metal (in)
  K = conductivity of metal
  TD = temperature difference (°F)
  H = measure of heat (Btu/min)

  a. Determine H when $A = 5.40$ sq ft, $d = 0.093$ in, $K = 1,740$, and $TD = 10.50$°F. Round the answer to 3 significant digits.

  b. Determine TD when $A = 3.80$ sq ft, $H = 9,500$ Btu/min, $d = 0.125$ in, and $K = 2,910$. Round the answer to 2 significant digits.

30. The total equivalent resistance of two parallel resistances is found with this formula.

$$R_T = \frac{R_1 \times R_2}{R_1 + R_2}$$

  where R_T = total resistance in ohms (Ω)
  R_1 and R_2 = individual resistances in ohms (Ω)

  Solve for R_1 when $R_T = 5$ Ω and $R_2 = 17$ Ω. Round the answer to the nearest whole ohm.

⦂ UNIT EXERCISE AND PROBLEM REVIEW

SOLVING COMBINED OPERATIONS EQUATIONS

Solve for the unknown and check each of these combined operations equations.

1. $8F - 7 + 5F = 45$

2. $12P - 9 = 39$

3. $3.6M - 6 = 26.2 - M$

4. $20 - (-B + 14) = 32$

5. $-14 = 3 - C + (9 + 5C)$

6. $15(E - 3) = -10(E + 2)$

7. $0.34T = 3.45 - 0.12T$

8. $6.8R + 2(R - 3.1) = 0.3R + 32.9$

9. $\dfrac{M}{10} + 7.6 = -5.8$

10. $\dfrac{1}{2}D + 2\left(D - \dfrac{1}{4}\right) = 5\dfrac{1}{2}$

11. $\dfrac{3}{8}N + 5(N - 6) = 1\dfrac{7}{8}N - 2$

12. $6A^2 - 54 = (A + 3)(A - 3)$

13. $28 + x^2 = (x - 6)(x - 2)$

14. $10\sqrt{B} = 4\left(\sqrt{B} + 12\right)$

15. $W + 5\sqrt{W} = -7\sqrt{W} + (W + 6)$

16. $3.7(4y - 2) + y^3 = 14.8y + 0.6$

SUBSTITUTING VALUES AND SOLVING FORMULAS

These formulas have been taken from various technical fields. Substitute the given values for letter values and solve for the unknown. Check each answer.

17. This is the formula for finding the circumference of a circle.

$$C = 2\pi r \qquad \text{where } C = \text{circumference}$$
$$\pi = 3.1416$$
$$r = \text{radius}$$

Solve for *r* when *C* = 13.53 inches. Round the answer to 4 significant digits.

18. Use this formula and solve for °C when the temperature is −8.6°F. Round the answer to the nearest degree.

$$°F = \frac{9}{5}(°C) + 32°$$

19. The current for a motor is found from this formula.

$$I = \frac{E_x}{R} - \frac{E_c}{R} \qquad \text{where } I = \text{current in amperes}$$
$$R = \text{resistance in ohms}$$
$$E_x = \text{impressed voltage}$$
$$E_c = \text{counter voltage}$$

Solve for E_c when *I* = 6.50 amperes, *R* = 0.350 ohm, and E_x = 250.0 volts. Round the answer to 3 significant digits.

20. This is the formula for finding the hypotenuse of a right triangle. Round the answer to 4 significant digits.

$$c = \sqrt{a^2 + b^2} \qquad \text{where } c = \text{hypotenuse}$$
$$a = \text{first side}$$
$$b = \text{second side}$$

Solve for *b* when *c* = 50.00 mm and *a* = 16.00 mm.

REARRANGING FORMULAS

These formulas are found in various occupational handbooks and manuals. Rearrange each and solve for the designated letter in problems 21 through 24.

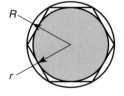

Figure 15–8

21. Figure 15–8 shows the inscribed and circumscribed circles for a regular hexagon.

    a. Solve for *r*.

        $R = 1.155r$

    b. Solve for *R*.

        $A = 2.598R^2$

22. This formula is used to find the area of the cross-section of the I-beam shown in Figure 15–9.

    $A = dt + 2a(s + n)$

    a. Solve for *t*.

    b. Solve for *n*.

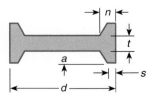

Figure 15–9

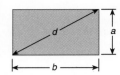

Figure 15–10

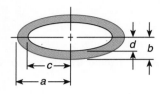

Figure 15–11

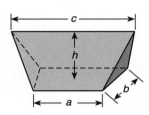

Figure 15–12

23. This formula is used to find the length of the diagonal of the rectangle shown in Figure 15–10.

$$d = \sqrt{a^2 + b^2}$$

 a. Solve for a.

 b. Solve for b.

24. This formula is used to find the area of the elliptical ring shown in Figure 15–11.

$$A = \pi(ab - cd)$$

 a. Solve for b.

 b. Solve for d.

For problems 25 and 26, rearrange each formula for the designated letter and solve.

25. This formula is used to find the volume of the wedge shown in Figure 15–12.

$$V = \frac{bh(2a + c)}{6}$$

 a. Solve for c when $a = 8.10$ in, $b = 6.24$ in, $h = 7.13$ in, and $V = 152$ cu in. Round the answer to 3 significant digits.

 b. Solve for a when $b = 11.03$ cm, $h = 14.17$ cm, $c = 20.15$ cm, and $V = 1630$ cm^3. Round the answer to 4 significant digits.

26. A tapered pin is shown in Figure 15–13.

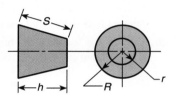

Figure 15–13

 a. Solve for h when $R = 2.38$ cm, $r = 1.46$ cm, and $V = 69.5$ cm^3. Round the answer to 3 significant digits.

$$V = 1.05(R^2 + Rr + r^2)h$$

 b. Solve for h when $S = 0.875$ in, $R = 0.420$ in, and $r = 0.200$ in. Round the answer to 3 significant digits.

$$S = \sqrt{(R - r)^2 + h^2}$$

UNIT 16 ⁘ The Cartesian Coordinate System and Graphs of Linear Equations

OBJECTIVES

After studying this unit you should be able to

• locate points in the Cartesian coordinate system.

• graph linear equations.

• determine linear equations given the slope and y-intercept.

• determine linear equations given two points.

A graph is a picture that shows the relationship between sets of quantities. Graphs are widely used in business, industry, government, and scientific and technical fields. Newspapers, magazines, books, and manuals often contain graphs. Since they are used in both occupations and everyday living, it is important to know how to interpret and construct basic types of graphs.

In this unit, Cartesian graphs will be studied. This is a system for locating points on a grid.

16–1 Description of the Cartesian (Rectangular) Coordinate System

An intersecting horizontal and vertical line are drawn on graph paper. The lines are used as reference lines. The horizontal line is called the *x axis;* the vertical line is called the *y axis.* The point of intersection of the *x* and *y* axes is called the *origin* (0).

The *x* axis and *y* axis divide the plane (graph paper) into four sections called *quadrants.* The quadrants are numbered I, II, III, and IV, starting with the upper, right-hand quadrant and going in a counterclockwise direction.

By means of this system, the location of a point can be determined with reference to the two axes. Points located to the right of the *y* axis have positive (+) *x* values. Points to the left of the *y* axis have negative (−) *x* values. Points above the *x* axis have positive (+) *y* values. Points below the *x* axis have negative (−) *y* values.

The line spacings of the graph can be numbered along the two axes. Starting at the origin (0), numbers are positive values when going to the right along the *x* axis; numbers are negative values when going to the left. Starting at the origin (0), numbers are positive values when going upward along the *y* axis; numbers are negative values when going downward.

The Cartesian (rectangular) coordinate system is shown in Figure 16–1 on page 374.

Locating Points

To locate or plot a point, we measure its distance from each axis. For example, consider a point that is located 5 units to the right of the origin and 3 units upward from the *x* axis. The two numbers, 5 and 3, *taken in that order*, will give a definite location of the point. The numbers, 5 and 3, are called an *ordered pair* because it makes a difference which number comes first. The two numbers in the ordered pair are called *coordinates* of the point and the numbers are enclosed in parentheses, such as (5,3). The first number is called the *x-coordinate* or *abscissa* ; the second number is called the *y-coordinate* or *ordinate.*

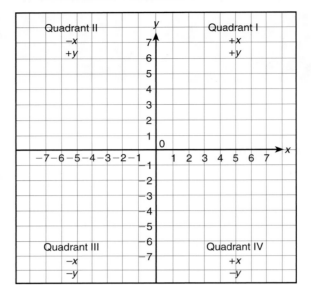

Figure 16–1

The following points and their coordinates are shown on the graph in Figure 16–2.

Point A (5,3)

Point B (0,4)

Point C (3,0)

Point D (−6,6)

Point E (−7,0)

Point F (−5,−3)

Point G (2,−5)

Point H (0,−6)

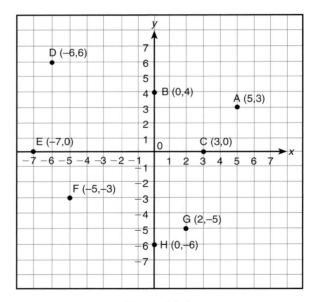

Figure 16–2

16–2   Graphing a Linear Equation

The graph of a straight line (linear) equation has an infinite number of ordered pairs of numbers that satisfy the equation. For example, the graph of the equation $x + y = 4$ contains all points whose coordinates have a sum of 4. A few of the infinite ordered pairs of numbers or coordinates are shown in the following table. Observe that the sum of each ordered pair is 4.

x	−3	−1	0	4	5	7
y	7	5	4	0	−1	−3

The coordinates in the table are plotted in Figure 16–3.

NOTE: The x and y axes are not numbered; each space is equal to one unit. The straight line that connects the points is the graph of the equation $x + y = 4$.

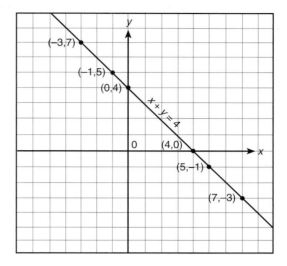

Figure 16–3

Procedure for Graphing an Equation

1. Write the equation in terms of either variable (x or y).

2. a. Make a table of values. Head one column x and the other column y.

   b. Select at least three convenient values for one variable and find their corresponding values for the second variable by substituting the first variable values in the equation. These values (coordinates) will be plotted on the graph.

NOTE: Although two points determine a straight line, at least one extra point should be plotted as a check. The coordinates selected should result in points on the graph that are not too close together to assure greater accuracy when the points are connected by a line.

3. Draw and label the x axis and the y axis on the graph.

4. Plot the points (coordinates) on the graph. If a coordinate value is a fraction or decimal, estimate its location on the graph. Make the points small, distinct, and as accurate as you can.

5. Carefully connect the points with a thin straight line. *The straight line is the graph of the equation.* The straight line should extend through the points and not end at the plotted points; the line is infinitely long. If the points cannot be connected with a straight line, a mistake has been made. Check the table values and the accuracy of the plotted points.

EXAMPLE •————————————————————————————————

Graph $3x + 2y = 5$.

Write the equation in terms of either x or y.

The equation has been written in terms of x by solving it for y. $y = \dfrac{5 - 3x}{2}$

Make a table of values. Select at least three values for x and find the corresponding value of y. The x-values of -2, 3, and 5 are chosen.

x	y
-2	5.5
3	-2
5	-5

Draw a label the x and y axes on the graph.

Plot the points. The location of the point $(-2, 5.5)$ is estimated.

Connect the points with a straight line, extending the line past the points.

The graph of $3x + 2y = 5$ is shown in Figure16–4.

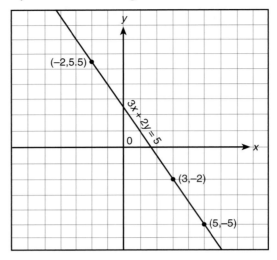

Figure 16–4

It is traditional to solve an equation for y (in terms of x) as in the previous example. One could just as easily have written the equation in terms of y. This is shown in the next example.

EXAMPLE •——

Graph $4x - 5y = 6$.

Write the equation in terms of either x or y.

The equation is written in terms of y by solving it for x. $x = \dfrac{6 + 5y}{4}$

Make a table of values. Select at least three values for y and find the corresponding values of x; y values of -6, 0, and 4 are selected.

x	y
-6	-6
1.5	0
6.5	4

Draw and label the x and y axes on the graph.

Plot the points on the graph. The location of x values of 1.5 and 6.5 are estimated.

Connect the points with a straight line, extending the line past the points.

The graph of $4x - 5y = 6$ is shown in Figure 16–5.

NOTE: The coordinates for this example are shown on the graph. Generally, it is not necessary to show the coordinates on a graph.

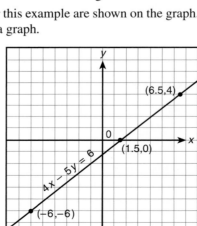

Figure 16–5

EXERCISE 16–2

Graph the following linear equations.

1. $x = y + 2$	**9.** $3x - 5y = 0$	**17.** $6x + 3y = 9$
2. $y = x + 1$	**10.** $4x - y = -4$	**18.** $2x - y = -4$
3. $x + y = 4$	**11.** $3x + 4y = 12$	**19.** $10x - 4y = 40$
4. $x - y = 5$	**12.** $4x + 2y = 15$	**20.** $8x = 2y$
5. $y = x$	**13.** $2y + 5x = 1$	**21.** $9x - 12y = 15$
6. $y = -x$	**14.** $2x - 3y = -12$	**22.** $8y = 30 - x$
7. $4x = 2y$	**15.** $2x + 3y = 12$	**23.** $2x - 7y = 5$
8. $x - 2y = 0$	**16.** $5x - 3y = 0$	**24.** $16x - 4y = 20$

16–3 Slope of a Linear Equation

The slope of a linear equation (a straight line) refers to its steepness or inclination. The *slope* is the ratio of a change in y to a change in x as we move from one point to another point along the line. The change in y is called the *rise;* the change in x is called the *run.*

$$\text{slope} = \frac{\text{change in } y}{\text{change in } x} = \frac{\text{rise}}{\text{run}}$$

If a line rises moving from left to right, it has a positive slope.

If a line falls moving from left to right, it has a negative slope.

A horizontal line has a zero slope.

A vertical line does not have a slope.

A positive slope is shown in Figure 16–6.

$$\text{slope} = \frac{\text{change in } y}{\text{change in } x} = \frac{6 - 2}{7 - 1} = \frac{4}{6} = \frac{2}{3}$$

Notice that the slope rises going from left to right. The slope is positive (a positive change in x and a positive change in y).

A negative slope is shown in Figure 16–7.

$$\text{slope} = \frac{\text{change in } y}{\text{change in } x} = \frac{-7 - (-1)}{8 - 3} = \frac{-6}{5} = -\frac{6}{5}$$

Notice that the slope falls going from left to right. The slope is negative (a positive change in x and a negative change in y).

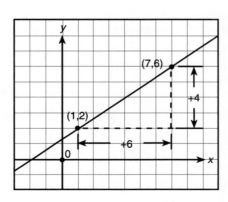

Figure 16–6 Positive slope $\left(\dfrac{2}{3}\right)$

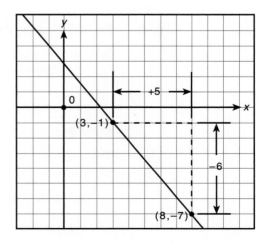

Figure 16–7 Negative slope $\left(-\dfrac{6}{5}\right)$

EXERCISE 16–3

Plot each of the pairs of points (coordinates) and draw a straight line through them. Then find the slope of each line.

1. (4,2); (3,5)	**6.** (−7,5); (−8,2)	**11.** (−7,−3); (7,3)
2. (5,3); (4,6)	**7.** (4,−2); (7,−9)	**12.** (4,0); (0,4)
3. (7,0); (3,2)	**8.** (−3,−3); (5,5)	**13.** (0,−3); (−1,0)
4. (1,−3); (4,−6)	**9.** (−6,4); (−7,8)	**14.** (8,6); (2,1)
5. (0,−5); (−3,8)	**10.** (−2,1); (−6,−6)	**15.** (−6,5); (−5,−6)

16–4   Slope Intercept Equation of a Straight Line

If the slope and the y-intercept of a straight line are known, the equation of the line can easily be determined. The *y-intercept* is the ordinate (y-coordinate) where the line intercepts (crosses) the y axis; the value of the x is 0 at this point.

The general equation for the slope-intercept form of a straight line is:

$$y = mx + b \text{ where } m \text{ is the } slope \text{ and } b \text{ is the } y\text{-}intercept.$$

EXAMPLE •————————————————————————————————

The slope (m) of a straight line is $\frac{3}{4}$ and the y-intercept (b) is 6. Write the equation of the line.

Substitute $\frac{3}{4}$ for m and 6 for b in $y = mx + b$

The equation of the straight line is $y = \dfrac{3}{4}x + 6 \, Ans$

EXERCISE 16–4

Write the equation of the line, given the slope (m) and the y intercept (b).

1. $m = \dfrac{1}{2}, b = 5$	**6.** $m = -3, b = -5$	**10.** $m = -\dfrac{3}{7}, b = -\dfrac{1}{2}$
2. $m = -\dfrac{3}{4}, b = 12$	**7.** $m = \dfrac{1}{5}, b = 12$	**11.** $m = \dfrac{5}{8}, b = 0$
3. $m = \dfrac{3}{2}, b = 10$	**8.** $m = \dfrac{6}{5}, b = -7\dfrac{1}{2}$	**12.** $m = -\dfrac{10}{3}, b = 0$
4. $m = 3, b = -5$	**9.** $m = 2, b = 4\dfrac{1}{4}$	
5. $m = -3, b = 5$		

16–5   Point-Slope Equation of a Straight Line

If the slope of a line and one point on the line are known, the equation of the line can be determined.

The general equation for the point-slope equation of a line is

$$y - y_1 = m(x - x_1)$$

Where m is the slope and (x_1, y_1) is a point on the line.

EXAMPLE •————————————————————————————————

The slope (m) of a line is $\frac{2}{5}$ and the line passes through the point (3,7). Write the equation of the line.

Substitute $\dfrac{2}{5}$ for m, 3 for x_1, and 7 for y_1.

The equation of the line is $y - 7 = \dfrac{2}{5}(x-3) \, Ans$

EXERCISE 16–5

Write the equation of the line given the slope (m) and the point (x_1, y_1) on the line.

1. $m = 3$, point: $(5,4)$

2. $m = 7$, point: $(-4,1)$

3. $m = \dfrac{1}{2}$, point: $(-2,-4)$

4. $m = -\dfrac{2}{3}$, point: $(12,-3)$

5. $m = -\dfrac{2}{5}$, point: $(-3.25,5.1)$

6. $m = \dfrac{7}{5}$, point: $(2.73,-7.6)$

7. $m = 0$, point: $(-5.7,3.4)$

8. $m = 0$, point: $(-9.7,-2.5)$

16–6 Determining an Equation, Given Two Points

If two points on a straight line are known, an equation can be determined by finding the slope of the line, and then finding the *y*-intercept.

EXAMPLES

1. Find the equation of a straight line that passes through two points whose coordinates are $(1,2)$ and $(5,7)$.

Find the slope. Notice that either coordinate can be subtracted from the other.

$$\text{Slope } (m) = \frac{\text{change in } y}{\text{change in } x} = \frac{2 - 7}{1 - 5} = \frac{-5}{-4} = \frac{5}{4} \text{ or } \frac{7 - 2}{5 - 1} = \frac{5}{4}$$

Find the *y*-intercept. Since both coordinates lie on the line, choose either one of the coordinates and substitute the values for *m*, *x*, and *y* in $y = mx + b$. The coordinate $(1,2)$ is chosen.

$$y = mx + b$$

$$2 = \frac{5}{4}(1) + b$$

$$\frac{3}{4} = b$$

The equation of the line is $y = \dfrac{5}{4}x + \dfrac{3}{4}$ *Ans*

2. Find the equation of a straight line that passes through two points whose coordinates are $(-2,-1)$ and $(1,-3)$.

$$\text{Slope } (m) = \frac{\text{change in } y}{\text{change in } x} = \frac{-1 - (-3)}{-2 - 1} = \frac{-1 + 3}{-3} = \frac{2}{-3} = -\frac{2}{3}$$

$$\text{or } \frac{-3 - (-1)}{1 - (-2)} = \frac{-3 + 1}{1 + 2} = \frac{-2}{3} = -\frac{2}{3}$$

Find the *y*-intercept. Choose either coordinate and substitute values for *m*, *x*, and *y*. The coordinate $(1,-3)$ is chosen.

$$y = mx + b$$

$$-3 = -\frac{2}{3}(1) + b$$

$$-3 + \frac{2}{3} = b$$

$$-2\frac{1}{3} = b$$

The equation of the line is $y = -\dfrac{2}{3}x - 2\dfrac{1}{3}$ *Ans*

EXERCISE 16–6

Find the slope and the y-intercept of the straight line that passes through each of the following pairs of points (coordinates). Then, write the equation of each line.

1. $(3,2)$; $(5,3)$	**6.** $(-8,3)$; $(8,6)$	**11.** $(7,0)$; $(0,7)$
2. $(6,4)$; $(2,7)$	**7.** $(1,-2)$; $(7,-5)$	**12.** $(0,-1)$; $(-1,0)$
3. $(0,5)$; $(6,1)$	**8.** $(-2,-2)$; $(6,6)$	**13.** $(7,6)$; $(-7,-6)$
4. $(2,-5)$; $(4,-6)$	**9.** $(-4,3)$; $(-5,7)$	**14.** $(3,-1)$; $(-1,3)$
5. $(-6,7)$; $(0,3)$	**10.** $(3,0)$; $(-4,-4)$	**15.** $(-7,-4)$; $(-8,-2)$

16–7   Describing a Straight Line

A straight line is often described in terms of its slope and y-intercept. If the equation (regardless of its form) of the straight line is known, the slope and y-intercept are easily found. The equation is rearranged in the slope-intercept form, and the slope and y-intercept are identified. The following examples in the chart in Figure 16–8 show how to determine the slope and y-intercept of any linear equation.

	Given Linear Equation	Rearranging Equation to $y = mx + b$	Line Description	
			Slope (m)	y-Intercept (b)
1.	$8x + 4y = 9$	$4y = -8x + 9$ Divide the equation by 4 $y = -2x + 2\frac{1}{4}$	-2	$2\frac{1}{4}$
2.	$7x - 5y - 20 = 0$	$-5y = -7x + 20$ Divide the equation by -5 $y = \frac{7}{5}x + (-4)$	$\frac{7}{5}$	-4
3.	$-5x = -\frac{2}{3}y$	$-\frac{2}{3}y = -5x + 0$ Multiply the equation by $-\frac{3}{2}$ $y = \frac{15}{2}x + 0$	$\frac{15}{2}$	0

Figure 16–8

Graphing a Linear Equation When Slope and y-Intercept Are Known

A linear equation can be quickly graphed when its slope and y-intercept are known.

EXAMPLE •────────────────────────────

Graph $5x + 7y = 28$ using only the slope and y-intercept.

Rearrange equation to $y = mx + b$

$$5x + 7y = 28$$
$$7y = -5x + 28$$
$$y = -\frac{5}{7}x + 4$$
$$m = -\frac{5}{7}, b = 4$$

Refer to Figure 16–9.

Locate the y-intercept at the point $(0,4)$.

The slope is $-\frac{5}{7}$. Think of $-\frac{5}{7}$ as $\frac{-5}{7}$.

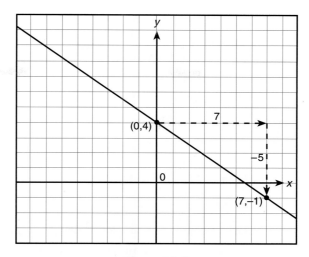

Figure 16–9

From (0,4) move 7 units to the right and 5 units down. Locate the second point. Draw a line through the two points. This line is the graph of $5x + 7y = 28$.

Because $-\frac{5}{7}$ can also be written as $\frac{5}{-7}$, another point can be found on the same line by starting at (0,4) and moving 7 units to the left and 5 units up, as shown in Figure 16–10.

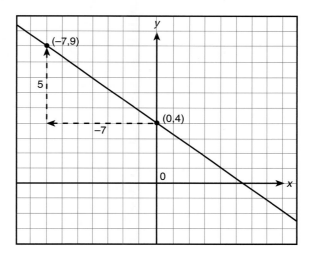

Figure 16–10

EXERCISE 16–7A

Rearrange equations 1 through 15 as y = mx + b. *Then, find the slope (*m*) and y-intercept (*b*) of each.*

1. $2x + 6y = -20$

2. $4x + 7y = 18$

3. $x - 3y = -20$

4. $-4x - 5y = -28$

5. $5x - y - 21 = 0$

6. $6x - 4y + 16 = 0$

7. $3x + 6y = -27$

8. $2x + 5y = 16$

9. $4x - 2y + 6 = 0$

10. $3x - 2y = 14$

11. $-2x + \frac{1}{4}y = 0$

12. $7y - 5x = 2$

13. $5x - 3y = 2$

14. $5x = -7y$

15. $x + 3y - 7 = 0$

Graph equations 16 through 24, using only the slope and y-intercept.

16. $x + 2y = 6$

17. $2x - 5y = 15$

18. $-8x + 3y = 21$

19. $6x + 5y = -30$

20. $4x - 7y + 21 = 0$

21. $7x + 4y = 22$

22. $-x + \dfrac{1}{2}y = -3$

23. $7x - \dfrac{1}{3}y = 0$

24. $3x = 4y$

Graphing a Linear Equation When Slope and One Point Are Known

If the slope and one point on a line are known, the line can be graphed.

EXAMPLE •

Graph the line with the slope $\frac{2}{7}$ that passes through the point $(-4,3)$.

Refer to Figure 16–11.

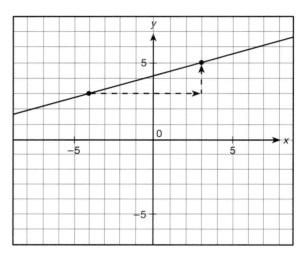

Figure 16–11

Locate the given point $(-4,3)$.

The slope is $\frac{2}{7}$. From $(-4,3)$ move 7 units to the right and 2 units up. Locate the second point, $(3,5)$, and draw a line through the two points. This line is the graph of $y - 3 = \frac{2}{7}(x + 4)$.

It is often easier to compare equations for lines if they are all written in the same form. Probably the most popular form is the slope-intercept form.

EXAMPLE •

Rewrite $y - 3 = \frac{2}{7}(x + 4)$ in the slope-intercept form, $y = mx + b$.

$$y - 3 = \frac{2}{7}(x + 4)$$

$$y - 3 = \frac{2}{7}x + \frac{8}{7}$$

$$y = \frac{2}{7}x + \frac{8}{7} + 3$$

$$y = \frac{2}{7}x + \frac{29}{7} \text{ or } y = \frac{2}{7}x + 4\frac{1}{7}$$

EXERCISE 16–7B

*For each problem, (**a**) graph the line with the given slope that passes through the given point, (**b**) write the point-slope equation of the line, and (**c**) write the slope-intercept equation for the line.*

1. $m = \dfrac{1}{3}$; point: $(3,-2)$

2. $m = -\dfrac{2}{5}$; point: $(-1,4)$

3. $m = -\dfrac{3}{4}$; point: $(-5,-2)$

4. $m = 2$; point: $(-3,-3)$

5. $m = 3$; point: $(-5,-2)$

6. $m = \dfrac{4}{3}$; point: $(2,-6)$

7. $m = 2\dfrac{1}{2}$; point: $(1,3)$

8. $m = -3\dfrac{1}{3}$; point: $(4,2)$

UNIT EXERCISE AND PROBLEM REVIEW

GRAPHING EQUATIONS

Graph the following linear equations.

1. $y = 5 - x$

2. $x - y = 5$

3. $3x - 2y = 0$

4. $5x - 2y = 20$

5. $-3y + 2x = -12$

6. $4x = 15 - y$

FINDING THE SLOPE OF A STRAIGHT LINE BY GRAPHING

Plot each of the pairs of points (coordinates) and draw a straight line through them. Then, find the slope of each line.

7. $(3,5)$; $(5,6)$

8. $(-8,1)$; $(-9,3)$

9. $(-9,-4)$; $(9,4)$

10. $(8,0)$; $(4,7)$

11. $(-5,2)$; $(-7,-8)$

12. $(5,0)$; $(0,-6)$

WRITING EQUATIONS OF LINES

Write the equation of the line, given the slope (m) and the y-intercept (b)

13. $m = \dfrac{1}{2}, b = 6$

14. $m = 5, b = 9$

15. $m = -\dfrac{3}{4}, b = -12$

16. $m = \dfrac{5}{3}, b = 0$

17. $m = -3, b = 12$

18. $m = -\dfrac{10}{7}, b = -15$

Write the equation of the line give the slope (m) and the point (x_1,y_1) on the line.

19. $m = -5$; point: $(7.1,-6.5)$

20. $m = \dfrac{3}{4}$; point: $(-4.2, -8.3)$

21. $m = \dfrac{12}{5}$; point: $\left(\dfrac{3}{4}, -\dfrac{5}{8}\right)$

22. $m = -\dfrac{7}{4}$; point: $\left(2\dfrac{1}{2}, 4\dfrac{1}{3}\right)$

FINDING SLOPES AND y-INTERCEPTS AND WRITING EQUATIONS

Find the slope and the y-intercept of the straight line that passes through each of the following pairs of points (coordinates). Then, write the equation of each line.

23. $(5,2)$; $(3,6)$

24. $(0,7)$; $(6,4)$

25. $(-2,1)$; $(8,-5)$

26. $(-9,0)$; $(-5,-5)$

27. $(7,-2)$; $(-8,-5)$

28. $(-3,-7)$; $(-5,-9)$

Rearrange equations 29 through 34 as $y = mx + b$. Then, find the slope (m) and the y-intercept (b) of each.

29. $5x + 7y = -23$

30. $-6x - 5y = -32$

31. $-3x - \dfrac{1}{3}y = 17$

32. $-9x + 2y = 15$

33. $4x + 8y = 0$

34. $5x - 2y - 10 = 0$

GRAPHING EQUATIONS USING SLOPE AND y-INTERCEPT

Graph each of the following equations using only the slope and y-intercept.

35. $x + 3y = 5$

36. $5x + 10y = -12$

37. $-6x + y = 3$

38. $7x + 5y - 25 = 0$

39. $-4x = 5y$

40. $2x + \dfrac{1}{3}y - 1 = 0$

GRAPHING LINES USING SLOPE AND ONE POINT

Graph each of the lines with the given slope that passes through the given point.

41. $m = -\dfrac{1}{3}$; point: (4,2)

42. $m = -\dfrac{5}{4}$; point: (6,3)

43. $m = \dfrac{2}{7}$; point: (−5,1)

44. $m = 1.5$; point: (−4,−5)

UNIT 17 ▦ Systems of Equations

OBJECTIVES

After studying this unit you should be able to

- solve systems of equations graphically.
- solve systems of equations by the substitution method.
- solve systems of equations by the addition and subtraction method.
- determine whether systems of equations are consistent, inconsistent, or dependent.
- evaluate a determinant.
- solve systems of equations using Cramer's rule.
- write and solve systems of equations from word problems.

As previously discussed, one equation with two unknowns (variables) has an infinite number of solutions. When making the table of coordinates for graphing an equation with two variables, you could have selected any value for one variable and find the corresponding value for the other variable. A single equation with two variables is called an *indeterminate* equation—there are an infinite number of pairs of numbers (ordered pairs) that satisfy the equation.

If a problem involves two variables, two equations are required to find a single value for each of the two variables. If a problem involves three variables, three equations are required in its solution. The sets of equations are called *systems of equations.* Only sets of two equations will be presented.

Systems of equations can have one solution, no solution, or an infinite number of solutions.

If the lines intersect in one point, the system has but one solution. If the two lines overlap, that is, they are two equations for the same line, then there are an infinite number of solutions. Two lines that are parallel (and do not overlap) have no solution.

Systems of equations with one solution (a single ordered pair) will be discussed first. Systems with no solutions or infinite solutions will be discussed later in this unit.

Systems of equations can be solved by the following methods.

Graphing the sets of equations

Elimination by substitution

Elimination by addition or subtraction

Elimination by comparison

Determinants

Graphing systems of equations will be shown first. The algebraic methods of substitution and addition or subtraction follow.

17–1   Graphical Method of Solving Systems of Equations

The primary purpose for graphing systems of equations is to find the coordinates at the point of intersection of the two graphed equations. The point of intersection is where two straight lines

385

cross. Straight lines that are not parallel can only intersect at one point. The point of intersection is the only point that lies on both lines; it is the only point where both equations have common x and y coordinates. Therefore, the common x and y coordinates are the only pair of values that satisfies both equations.

It is important to keep in mind that graphs and information read from graphs are often approximate. The solution to a system of equations may not be exact. Check the solution by substituting the x and y coordinates in the original equations. The values on each side of the equal sign should be the same or approximately the same.

EXAMPLE •

Graph and solve the following systems of equations.

$$2x - 5y = -3$$
$$x - 6y = 2$$

Rewrite each equation in slope-intercept form.

$$2x - 5y = -3 \qquad\qquad x - 6y = -3$$
$$y = \frac{2}{5}x + \frac{3}{5} \qquad\qquad y = \frac{1}{6}x - \frac{1}{3}$$

Since the y-intercepts are fractions, we will set up a table of values for each equation and use the points in the table to graph the lines.

x	y
-6	-1.8
1	1
6	3

x	y
-10	-2
2	0
8	1

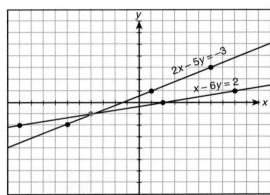

Figure 17–1

The equations are shown in Figure 17–1. The point of intersection is $(-4, -1)$. *Ans*

Check. Substitute $x = -4$ and $y = -1$ in the original equations.

$$2x - 5y = 3 \qquad\qquad\qquad x - 6y = 2$$
$$2(-4) - 5(-1) = -3 \qquad\qquad -4 - 6(-1) = 2$$
$$-3 = -3 \text{ Ck} \qquad\qquad\qquad 2 = 2 \text{ Ck}$$

EXERCISE 17–1

Graph and solve each of the following systems of equations. Check the solutions.

1. $x = 2y$
   $x - y = 2$

2. $x - y = 1$
   $x + y = 11$

3. $-2x - 3y = 4$
   $2x - y = 0$

4. $x + 4y = 0$
   $4x + 6y = 20$

5. $x - y = 7$
   $2x - y = 10$

6. $x - y = 8$
   $4x + 6y = 12$

7. $-x - 2y = 0$
   $x - y = -6$

8. $3x + y = 0$
   $2x + y = 2$

9. $x + y = 6$
   $x - y = -10$

10. $x + 2y = 2$
    $2x + y = -5$

11. $2x - 3y = -8$
    $5x + y = -3$

12. $4x - 2y = 0$
    $3x + 2y = -7$

13. $x - 3y = 1$
    $3x + y = -7$

14. $2x - y = -8$
    $3x - y = -12$

15. $x - 2y = -9$
    $x + 2y = 15$

16. $4x - 3y = 9$
    $4x + 3y = 15$

17. $2x + 3y = 13$
    $4x + 3y = 5$

18. $3x + 2y = -6$
    $3x + 8y = -6$

17–2   Substitution Method of Solving Systems of Equations

The substitution method is one of the algebraic ways of solving systems of equations. Precise solutions are obtained by algebraic methods. Algebraic methods of solving two equations for the values of the two variables involve the *elimination* of one variable. One equation with one variable is obtained.

The substitution method is an efficient way of solving systems of equations in which one equation has a variable with a numerical coefficient of one.

The following is an example of a system of equations that is readily solved by the substitution method.

$$x = 2y$$
$$4x + 3y = 15$$

Procedure for Solving Systems of Equations by Substitution

- Solve one of the equations for one of the variables.
- Substitute the solution from Step 1 into the other equation and solve for the remaining variable.
- Substitute the value from Step 2 into the equation from Step 1 and solve for the other variable.
- Check by substituting the values for both variables in the two original equations.

EXAMPLES

1. **Solve.** $2x + y = 14 , 4x - 2y = -12$

   $\qquad\qquad\qquad\qquad\qquad\qquad$ $2x + y = 14$
   $\qquad\qquad\qquad\qquad\qquad\qquad$ $4x - 2y = -12$

   Solve $2x + y = 14$ for y. $\qquad\qquad\qquad$ $y = 14 - 2x$

   Substitute $14 - 2x$ for y in the second $\qquad$ $4x - 2(14 - 2x) = -12$
   equation and solve for x. $\qquad\qquad\qquad$ $4x - 28 + 4x = -12$
   $\qquad\qquad\qquad\qquad\qquad\qquad\qquad\qquad$ $8x = 16$
   $\qquad\qquad\qquad\qquad\qquad\qquad\qquad\qquad$ $x = 2$

   Substitute 2 for x in the first equation $\qquad$ $2x + y = 14$
   and solve for y. $\qquad\qquad\qquad\qquad\qquad$ $2(2) + y = 14$
   $\qquad\qquad\qquad\qquad\qquad\qquad\qquad\qquad$ $y = 10$
   $\qquad\qquad\qquad\qquad\qquad\qquad$ $x = 2, y = 10$ *Ans*

   Check. Substitute the values $\qquad$ $2x + y = 14$ $\qquad\qquad$ $4x - 2y = -12$
   for x and y in both of the $\qquad$ $2(2) + 10 = 14$ $\qquad$ $4(2) - 2(10) = -12$
   original equations. $\qquad\qquad\qquad\qquad$ $14 = 14$ Ck $\qquad\qquad$ $-12 = -12$ Ck

2. Solve. $2x + 3y = 7, 3x - 2y = 4$

$$2x + 3y = 7$$
$$3x - 2y = 4$$

Solve $2x + 3y = 7$ for x.

Substitute $\dfrac{7 - 3y}{2}$ for x in the second

equation and solve for y.

$$x = \frac{7 - 3y}{2}$$
$$3x - 2y = 4$$
$$3\left(\frac{7 - 3y}{2}\right) - 2y = 4$$
$$10\frac{1}{2} - 4\frac{1}{2}y - 2y = 4$$
$$-6\frac{1}{2}y = -6\frac{1}{2}$$
$$y = 1$$

Substitute 1 for y in the first equation
and solve for x.

$$2x + 3y = 7$$
$$2x + 3(1) = 7$$
$$x = 2$$
$$x = 2, y = 1 \ \ Ans$$

Check. Substitute the values for x and
y in both of the original equations.

$$2x + 3y = 7$$
$$2(2) + 3(1) = 7$$
$$7 = 7 \ \text{Ck}$$

$$3x - 2y = 4$$
$$3(2) - 2(1) = 4$$
$$4 = 4 \ \text{Ck}$$

EXERCISE 17–2

Solve each of the following systems of equations by the substitution method. Check the answers.

1. $x = y + 4$
  $3x + 7y = -18$

2. $x = y - 1$
  $3x + 2y = -13$

3. $x = 2y$
  $7y - 4x = -3$

4. $x = 4y$
  $x - 5y = 2$

5. $x = y + 3$
  $2y - x = -5$

6. $y = 3x + 2$
  $9x - 7y = 22$

7. $x = 4y - 5$
  $5y - x = 4$

8. $y = 8 - 3x$
  $4x - 5y = -2$

9. $y = x + 2$
  $2x - y = 1$

10. $y = -3x$
  $2x - 3y = 22$

11. $y = 3x$
  $2x - 3y = -7$

12. $y = -2x$
  $6x - 4y = -70$

13. $x = 3 - y$
  $4x + 7y = 18$

14. $4x = 2y$
  $5x - 3y = -3$

15. $5y = 2x$
  $3x - 2y = 22$

16. $4x - 3y = -31$
  $5x - 3y = -35$

17. $3x = 6y$
  $6x - 5y = 9$

18. $6x + 5y = 13$
  $3x + 4y = 14$

19. $2x + 5y = 9$
  $3x - 5y = 1$

20. $5x - 3y = 19$
  $4x + 6y = 32$

21. $4x + 3y = -1.25$
  $1.5x - y = -1$

22. $\dfrac{2}{3}x = y$
  $\dfrac{3}{5}x - \dfrac{2}{5}y = 1$

23. $0.25x + 1.75y = 3$
  $1.5x + 2y = 1$

24. $2x - 5y = -9$
  $3x + 1.25y = 4$

17–3   Addition or Subtraction Method of Solving Systems of Equations

The most convenient way of solving certain systems of equations is by the addition or subtraction method. Some systems of equations are more readily solved by this method than by the substitution method.

Procedure for Solving Systems of Equations by Addition or Subtraction

- Eliminate a variable by adding or subtracting the equations so that one equation with one variable remains. It may be necessary to rearrange the equations to align like terms.
- Solve the resulting equation for the remaining variable.
- Substitute the value of this variable in either of the two original equations and solve for the other variable.
- Check by substituting the values of both variables in the two original equations.

Solutions Not Requiring Multiplication

If either of the two variables in both equations has the same coefficient (number factor), one of the variables can be directly eliminated by adding or subtracting equations.

EXAMPLES

1. Solve. $5x + 2y = 19, 3x - 2y = 5$

The y variables in both equations have the same coefficient (2). Add the equations to eliminate the y variables. Solve for x.

$$5x + 2y = 19$$
$$\underline{3x - 2y = 5}$$
$$8x = 24$$
$$x = 3$$

Substitute the value of x in either of the original equations, and solve for y.

$$5x + 2y = 19$$
$$5(3) + 2y = 19$$
$$15 + 2y = 19$$
$$y = 2$$
$$x = 3, y = 2 \; Ans$$

Check. Substitute the values for x and y in both of the original equations.

$$5x + 2y = 19 \qquad 3x - 2y = 5$$
$$5(3) + 2(2) = 19 \qquad 3(3) - 2(2) = 5$$
$$19 = 19 \; Ck \qquad 5 = 5 \; Ck$$

2. Solve. $5y = -7x + 53, 4x = 41 - 5y$

Rearrange both equations to align like terms. Terms must be in the same order.

$$7x + 5y = 53$$
$$4x + 5y = 41$$

The y variables in both equations have the same coefficient (5). Subtract the second equation from the first to eliminate y. Solve for x.

$$7x + 5y = 53$$
$$\underline{-(4x + 5y = 41)}$$
$$3x = 12$$
$$x = 4$$

Substitute the value of x in either of the original equations and solve for y.

$$7x + 5y = 53$$
$$7(4) + 5y = 53$$
$$5y = 25$$
$$y = 5$$
$$x = 4, y = 5 \; Ans$$

Check. Substitute the values for x and y in both of the original equations.

$$5y = -7x + 53 \qquad 4x = 41 - 5y$$
$$5(5) = -7(4) + 53 \qquad 4(4) = 41 - 5(5)$$
$$25 = 25 \; Ck \qquad 16 = 16 \; Ck$$

EXERCISE 17–3A

Solve each of the following systems of equations by the addition or subtraction method. Check the answers.

1. $x + y = 5$
  $x - y = 1$

2. $x = y - 4$
  $x + y = 8$

3. $2x - y = 5$
  $6x + y = 27$

4. $y = -3x + 11$
  $4x + y = 14$

5. $6x - 3y = 27$
  $x + 3y = 8$

6. $4x + 7y = 70$
  $6x - 7y = 0$

7. $-2x - 3y = -13$
  $6y = 2x + 14$

8. $x + 6y = 19$
  $x - 2y = -13$

9. $9x = 32 - 10y$
  $-3x + 10y = 56$

10. $6x + 4y = 72$
  $7x + 4y = 80$

11. $5x - 3y = -9$
  $7x - 3y = -23$

12. $-7x = -2 - 8y$
  $-5x - 8y = 26$

13. $5x - 4y = 0$
  $8x - 4y = 12$

14. $-x + 7y = 21$
  $x + 3y = 9$

15. $5x = -2y - 19$
  $-2y = -5 - 3x$

16. $-5x + 9y = 40$
  $2x + 9y = 47$

17. $x - 8y = -35$
  $-5x - 8y = -65$

18. $-5.5x + 4.5y = 41.4$
  $-5.5x - 2.6y = 7.32$

Solutions Requiring Multiplication

The previous examples and exercises had the same coefficient for one of the variables in both equations. The variable was directly eliminated by adding or subtracting the equations. If neither of the variables have the same coefficient, it is necessary to multiply one or both equations by a factor or factors that will make the coefficient of one of the variables equal.

EXAMPLES •────────────────────────────

1. Solve. $8x + 2y = -18, 10x - 6y = -14$

Neither the x variables nor the y variables have the same coefficients. If both sides of the first equation are multiplied by 3, the y terms will have the same coefficients and can be eliminated.

Add the equations to eliminate the y variable. Solve for x.

Substitute the value of x in either of the original equations. Solve for y.

$$8x + 2y = -18$$
$$10x - 6y = -14$$
$$\overline{3(8x + 2y) = 3(-18)}$$
$$24x + 6y = -54$$

$$24x + 6y = -54$$
$$10x - 6y = -14$$
$$\overline{34x = -68}$$
$$x = -2$$

$$8x + 2y = -18$$
$$8(-2) + 2y = -18$$
$$-16 + 2y = -18$$
$$y = -1$$

$$x = -2, y = -1 \; Ans$$

Check. Substitute the values for x and y in both of the original equations.

$$8x + 2y = -18$$
$$8(-2) + 2(-1) = -18$$
$$-18 = -18 \text{ Ck}$$

$$10x - 6y = -14$$
$$10(-2) - 6(-1) = -14$$
$$-14 = -14 \text{ Ck}$$

2. Solve. $15x + 10y = 40, 9x + 8y = 32$

Both equations must be multiplied by factors in order to eliminate one of the variables. Either the x or y variable can be eliminated. The x variable will be eliminated.

Multiply the first equation by -3.

Multiply the second equation by 5.

$$-3(15x + 10y = 40)$$
$$5(9x + 8y = 32)$$
$$-45x - 30y = -120$$
$$\underline{45x + 40y = 160}$$

Add the equations to eliminate the x variable. Solve for y.

$$10y = 40$$
$$y = 4$$

Substitute the value of y in either of the original equations. Solve for x.

$$15x + 10y = 40$$
$$15x + 10(4) = 40$$
$$15x = 0$$
$$x = 0$$
$$x = 0, y = 4 \; Ans$$

Check. Substitute the values for x and y in both of the original equations.

$15x + 10y = 40$	$9x + 8y = 32$
$15(0) + 10(4) = 40$	$9(0) + 8(4) = 32$
$40 = 40 \; Ck$	$32 = 32 \; Ck$

3. Solve. $0.5x - 2y = -2, 0.04x - 0.08y = 0.16$

If one or both of the equations have variables with decimal coefficients, multiply the equation or equations by a multiple of 10 to clear the variables of decimal coefficients. Then, solve the system of equations using the same procedure as with the previous examples.

$$0.5x - 2y = -2$$
$$0.04x - 0.08y = 0.16$$

Multiply the first equation by 10.

Multiply the second equation by 100.

$$5x - 20y = -20$$
$$4x - 8y = 16$$

Multiply both equations by factors so x can be eliminated.

Multiply the first equation by -4.

Multiply the second equation by 5.

$$-4(5x - 20y = -20)$$
$$5(4x - 8y = 16)$$
$$-20x + 80y = 80$$
$$\underline{20x - 40y = 80}$$

Add the equations to eliminate x. Solve for y.

$$40y = 160$$
$$y = 4$$

$$-20x + 80y = 80$$
$$-20x + 80(4) = 80$$

Substitute the value of y in either equation. Solve for x.

$$-20x = -240$$
$$x = 12$$
$$x = 12, y = 4 \; Ans$$

Check. Substitute the values for x and y in both of the original equations.

$0.5x - 2y = -2$	$0.04x - 0.08y = 0.16$
$0.5(12) - 2(4) = -2$	$0.04(12) - 0.08(4) = 0.16$
$6 - 8 = -2$	$0.48 - 0.32 = 0.16$
$-2 = -2 \; Ck$	$0.16 = 0.16 \; Ck$

EXERCISE 17–3B

Solve each of the following systems of equations by the addition or subtraction method. Check the answers.

1. $2x + y = 7$
$4x - 2y = -6$

2. $x - y = 5$
$2x + 4y = 10$

3. $6x = y$
$3x - 5y = 27$

4. $7x + 4y = 10$
$x + 3y = 16$

5. $3x - 2y = -9$
$12x - 4y = 0$

6. $5x = -y + 34$
$2x = 3y$

7. $2x + 2y = 12$
$4x + 6y = -4$

8. $10x - 4y = -8$
$6x + 8y = 68$

9. $x = y + 3$
$5x + 3y = 39$

10. $-8x + 3y = 20$
$x + 2y = 7$

11. $3x + 6y = -10$
$9x - 2y = 10$

12. $3x - 2y = 1$
$3y = 5 - 2x$

13. $4x - 6y = 2$
$9x - 5y = 13$

14. $8x - 2y = 8$
$5x - 3y = 12$

15. $-7x + 3y = 6$
$-6x + 4y = 2$

16. $-7y = -9x + 68$
$11x - 4y = 74$

17. $0.7x + 0.4y = 5.1$
$0.6x - 0.5y = 1.0$

18. $0.06x + 0.08y = 0.6$
$0.12x + 0.02y = 0.78$

19. $1.5x - 0.6y = 3.3$
$1.2x + 0.4y = 4.4$

20. $3x + 0.2y = 12.4$
$6x - 0.4y = 11.2$

21. $0.15x + 0.25y = 125$
$0.2x + 0.2y = 140$

17–4   Types of Systems of Equations

The examples and exercises presented have all had one solution (a single ordered pair). Some systems of equations have no solutions; other systems have an infinite number of solutions.

The Three Types of Systems of Equations

1. One Solution. These are the types of systems that you have solved graphically and by the algebraic methods of substitution and addition or subtraction. The solution is a single ordered pair; the graphs of the equations are two intersecting straight lines. Such a system is said to be *consistent* and *independent.*

2. No Solution. Such a system is said to be *inconsistent.*

The following example shows what happens when solving a set of inconsistent equations.

EXAMPLE •——————————————————————————————

$$3x + 5y = 15$$
$$6x + 10y = 8$$

Multiply the first equation by 2 and subtract the second equation.

$$6x + 10y = 30$$
$$\underline{-(6x + 10y = 8)}$$
$$0 = 22?$$

Both variables have been eliminated and an *incorrect* statement remains. There is no solution.

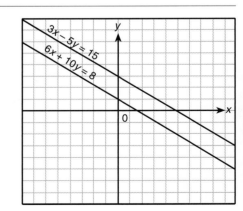

Figure 17–2

The graph of this system is shown in Figure 17–2. The graphs of the equations are two parallel lines.

The lines cannot intersect; therefore, there is no solution. *Ans*

3. An Infinite Number of Solutions. Such a system is said to be $dependent.$ Two dependent equations can be reduced to one; the two equations are actually the same. There is no unique solution set. Any set of values for the variables that satisfies one equation will also satisfy the other equation.

The following example shows what happens when solving a set of dependent equations.

EXAMPLE •────────────────────────────────────

$$6x + 8y = 16$$
$$12x + 16y = 32$$

Multiply the first equation by 2 and subtract the second equation.

$$12x + 16y = 32$$
$$\underline{-(12x + 16y = 32)}$$
$$0 = 0$$

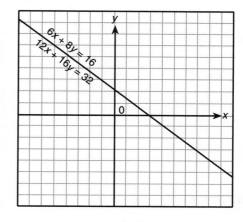

Figure 17–3

The result, $0 = 0$, is a correct statement, but all terms have been eliminated. The graph of this system is shown in Figure 17–3.

The two equations are one and the same. Ans

•──•

EXERCISE 17–4 ───

Solve the following systems of equations. Identify each system as to whether it is consistent and independent (one solution), inconsistent (no solution set), or dependent (infinite solution sets.)

1. $3x - 6y = 21$
$\quad 2x - 4y = 14$

2. $4x + 6y = 28$
$\quad 10x + 15y = 61$

3. $3x = 7 + 5y$
$\quad 9x - 15y = 16$

4. $4x = 3y - 7$
$\quad 7x = -4 - 3y$

5. $8x + 4y = 6$
$\quad 20x + 10y = 15$

6. $18y + 6x = -36$
$\quad 5x = -15y - 30$

7. $12x - 9y = 21$
$\quad 10x - 6y = 10$

8. $6x = 2y - 23$
$\quad -5y = -15x - 18$

9. $5x + 9y = 13$
$\quad 20x + 36y = 47$

10. $12y = 10x + 42$
$\quad 42y = 35x + 147$

11. $16x - 4y = -36$
$\quad 12x - 3y = -27$

12. $0.3x + 0.7y = -1.5$
$\quad 0.45x = 0.2y + 1.5$

13. $-0.2x + y = -1.8$
$\quad -0.5x + 2.25y = -4.2$

14. $0.3x - 0.6y = 2.8$
$\quad 0.75x - 1.5y = 7$

15. $4.5x + 0.3y = 18.6$
$\quad 9x - 1.2y = 33.6$

17–5 Determinants

A third method to solve systems of equations is called Cramer's rule and uses determinants. A determinant is a number. If a, b, c, and d are any four numbers, then the symbol

$$\begin{vmatrix} a & b \\ c & d \end{vmatrix}$$

is called a **2 × 2 determinant.** The numbers a, b, c, and d are called the elements or entries of the determinant. The value of a determinant is $ad - bc$, which gives

$$\begin{vmatrix} a & b \\ c & d \end{vmatrix} = ad - bc$$

EXAMPLES •

1. Evaluate the determinant $\begin{vmatrix} 7 & 5 \\ 2 & 6 \end{vmatrix}$

$$\begin{vmatrix} 7 & 5 \\ 2 & 6 \end{vmatrix} = 7(6) - 2(5) = 42 - 10 = 32 \ Ans$$

2. Evaluate the determinant $\begin{vmatrix} -6 & -2 \\ 3 & 4 \end{vmatrix}$

$$\begin{vmatrix} -6 & -2 \\ 3 & 4 \end{vmatrix} = -6(4) - (3)(-2) = -24 - (-6) = -18 \ Ans$$

EXERCISE 17–5

Evaluate each of the following determinants.

1. $\begin{vmatrix} 5 & 9 \\ 3 & 7 \end{vmatrix}$

2. $\begin{vmatrix} 12 & 5 \\ 2 & -6 \end{vmatrix}$

3. $\begin{vmatrix} -9 & 7 \\ 12 & -5 \end{vmatrix}$

4. $\begin{vmatrix} 6 & -3 \\ -11 & -14 \end{vmatrix}$

5. $\begin{vmatrix} 3.5 & 1.7 \\ 4.6 & 9.1 \end{vmatrix}$

6. $\begin{vmatrix} -7.2 & 3.4 \\ 6.2 & -11.9 \end{vmatrix}$

7. $\begin{vmatrix} 15.3 & 18.4 \\ 9.4 & -6.7 \end{vmatrix}$

8. $\begin{vmatrix} 32.6 & -21.8 \\ 14.5 & -14.5 \end{vmatrix}$

9. $\begin{vmatrix} 5\frac{1}{3} & 6\frac{1}{4} \\ -9\frac{1}{6} & 3\frac{1}{4} \end{vmatrix}$

10. $\begin{vmatrix} -4\frac{1}{4} & 5\frac{1}{3} \\ 7\frac{1}{5} & -6\frac{1}{2} \end{vmatrix}$

17–6   Cramer's Rule

Systems of linear equations can be solved using Cramer's rule. To use Cramer's rule on a 2 × 2 linear system requires three determinants. Consider the linear system

$$ax + by = k_1$$
$$cx + dy = k_2$$

Set up three determinants: $D = \begin{vmatrix} a & b \\ c & d \end{vmatrix}$, $D_x = \begin{vmatrix} k_1 & b \\ k_2 & d \end{vmatrix}$, and $D_y = \begin{vmatrix} a & k_1 \\ c & k_2 \end{vmatrix}$. First, calculate the determinant D. If $D = 0$, then the system is either inconsistent or dependent. If $D \neq 0$, then the system is independent and the solution is

$$x = \frac{D_x}{D} \qquad\qquad y = \frac{D_y}{D}$$

EXAMPLES •

1. Use Cramer's rule to solve the system

$$x + 2y = 5$$
$$3x + 5y = 13$$

Begin by setting up and evaluating determinant D. $D = \begin{vmatrix} 1 & 2 \\ 3 & 5 \end{vmatrix} = 1(5) - 3(2) = -1$

$D \neq 0$, so the system is independent and we continue.

Set up and evaluate D_x and D_y.

$$D_x = \begin{vmatrix} 5 & 2 \\ 13 & 5 \end{vmatrix} = 5(5) - 13(2) = -1$$

$$D_y = \begin{vmatrix} 1 & 5 \\ 3 & 13 \end{vmatrix} = 1(13) - 3(5) = -2$$

Solve for x.

$$x = \frac{D_x}{D} = \frac{-1}{-1} = 1.$$

Solve for y.

$$y = \frac{D_y}{D} = \frac{-2}{-1} = 2.$$

$$x = 1, y = 2 \; Ans$$

2. Use Cramer's rule to solve the system

$$2x = 3y - 7$$
$$5x - 2y = 10$$

Begin by rewriting the equations so the terms with the variables are on one side of the equal signs.

$$2x - 3y = -7$$
$$5x - 2y = 10$$

Next, set up and evaluate determinant D. $D = \begin{vmatrix} 2 & -3 \\ 5 & -2 \end{vmatrix} = 2(-2) - 5(-3) = 11$

$D \neq 0$, so the system is independent and we continue.

Set up and evaluate D_x and D_y.

$$D_x = \begin{vmatrix} -7 & -3 \\ 10 & -2 \end{vmatrix} = -7(-2) - 10(-3) = 44$$

$$D_y = \begin{vmatrix} 2 & -7 \\ 5 & 10 \end{vmatrix} = 2(10) - 5(-7) = 55$$

Solve for x.

$$x = \frac{D_x}{D} = \frac{44}{11} = 4.$$

Solve for y.

$$y = \frac{D_y}{D} = \frac{55}{11} = 5.$$

$$x = 4, y = 5 \; Ans$$

EXERCISE 17–6

Use Cramer's rule to solve each system or determine if it cannot be solved.

1. $2x + y = 7$
   $3x - 2y = -7$

2. $3x + y = 3$
   $2x - 3y = 13$

3. $4x + 3y = 4$
   $3x - 2y = -14$

4. $5x - 2y = -30$
   $3x + 6y = 18$

5. $\dfrac{1}{2}x - \dfrac{2}{3}y = \dfrac{3}{4}$
   $\dfrac{1}{3}x + 2y = \dfrac{5}{6}$

6. $\dfrac{3}{4}x + \dfrac{1}{3}y = \dfrac{7}{12}$
   $-\dfrac{2}{3}x + \dfrac{1}{2}y = \dfrac{-1}{48}$

7. $1.3x - 0.8y = 2.9$
   $1.7x - 0.7y = 0.4$

8. $6x + 2.5y = 8.2$
   $13.8x + 5.75y = 18.86$

9. $4.3x - 2.7y = -4$
   $3.5x + 4.2y = -1$

10. $3.5x - 6.5y = 22.45$
    $5.5x + 3.3y = -1.21$

17–7 Writing and Solving Systems of Equations from Word Statements, Number Problems, and Practical Applications

It is important to develop the ability to express word statements as algebraic expressions and equations. Translating word statements into mathematical form using literal numbers, arithmetic numbers, and signs of operation was presented in Unit 12. Writing single equations with one variable from word statements was presented in Unit 14. Basically, the same procedure of writing equations from word statements is used with systems of equations. Instead of translating sentences into one equation with one variable, you will translate sentences into two equations, each with two variables.

Writing and solving systems of equations from word statements has many practical applications. Problems, which otherwise would be difficult to solve, are often easily solved with systems of equations.

A problem must be fully understood before it can be expressed in equation form. Some or all of the following steps may be required in translating word problems into systems of equations and solving for the variables.

Solving Word Problems with Systems of Equations

- Carefully read the entire problem, several times if necessary.
- Analyze the problem; break the problem down into simpler parts.
- It is sometimes helpful to make a simple sketch or a table of values as an aid in visualizing and organizing the various parts of the problem.
- Identify the variables. Represent one of the variables by a letter, such as x; represent the other variable by another letter, such as y.
- Write the two equations, each with the same two variables. Check. Are the expressions on the left equal to the expressions on the right of the equal sign?
- Solve using either the substitution method or the addition or subtraction method.
- Check the equations by substituting values for the variables.
- Check the answer against the original word problem. Does the answer fit the problem statements?

The following examples show the method of writing and solving systems of equations from word statements. Checking the answers is left up to the student. Check by substituting the values in the equations and check the answers against the word statements.

NUMBER PROBLEM EXAMPLE •

The difference between two numbers is 6, and twice their sum is 24. Find the two numbers.

Identify the variables. Let one number equal x and the other number equal y. Write the two equations.

a. The difference between the two numbers is 6. $x - y = 6$

b. Twice the sum of the two numbers is 24. $2(x + y) = 24$ or

                                                                                   $2x + 2y = 24$

Rearrange the first equation and solve by the $x = 6 + y$
substitution method. $2(6 + y) + 2y = 24$

                                                                                         $y = 3$

Substitute 3 for y in the first original equation. $x - 3 = 6$

                                                                                         $x = 9$

The two numbers are 3 and 9 *Ans*

PRACTICAL APPLICATION EXAMPLES •

1. Before modifications, two machines that produce at different rates, together produced 7,800 pieces per day. Modifications to improve both machines were made to increase production.

After modifications, the lower producing machine's production increased by 45%; the higher producing machine's production increased by 35%. The total production of both machines increased to 10,800 pieces per day. How many pieces per day were produced by each machine before the modifications were made?

Identify the variables:

Let the lower producing machine's production before modification equal x. Let the higher producing machine's production before modification equal y.

Write the two equations:

a. Before modifications $x + y = 7,800$

b. After modifications $1.45x + 1.35y = 10,800$

NOTE: The lower producing machine's production increased by 45% or 145% of production before modification. The higher producing machine's production increased by 35% or 135% of production before modification.

Rearrange the first equation, and solve by the substitution method.

$$x = 7,800 - y$$
$$1.45(7,800 - y) + 1.35y = 10,800$$
$$11,310 - 1.45y + 1.35y = 10,800$$
$$-0.1y = -510$$
$$y = 5,100$$

Substitute 5,100 for y in the first equation.

$$x + 5,100 = 7,800$$
$$x = 2,700$$

Before modifications, one machine produced 2,700 pieces and the other machine produced 5,100 pieces. *Ans*

2. A manufacturer wishes to make 3,500 pounds of a new product, which is a mixture of two different ingredients. One ingredient costs $1.60 a pound; the other ingredient costs $3.90 a pound. The cost of the new product is to be priced at $2.40 a pound. How many pounds of each ingredient should be used to make 3,500 pounds of the new product?

Systems of equations are used in a great variety of mixture applications. In solving these problems, one equation is written in terms of amount and the other equation is written in terms of cost or value.

Making a table with amounts and costs, such as the table in Figure 17–4, is often helpful in writing the equations. The table has all of the values that are needed to write the two equations.

Identify the variables:

Let x = amount of the first ingredient and y = the amount of the second ingredient.

	Cost per lb	Amount (No. of lbs)	Total Cost or Value (Cost per lb × No. of lbs)
First ingredient	$1.60	x	$1.60x
Second ingredient	$3.90	y	$3.90y
Mixture	$2.40	3,500 lbs	$2.40(3,500) or $8,400

Figure 17–4

Refer to the table and write the two equations.

a. Amount of mixture $x + y = 3,500$

b. Cost or value of mixture $1.60x + 3.90y = 8,400$

Solve by the substitution method.

Rearrange the first equation. $x = 3{,}500 - y$

Multiply the second equation by $10(1.60x + 3.90y = 8{,}400)$
10 to remove decimals.

Solve for y. $16x + 39y = 84{,}000$

$16(3{,}500 - y) + 39y = 84{,}000$

$56{,}000 - 16y + 39y = 84{,}000$

$23y = 28{,}000$

$y \approx 1{,}217$

Substitute 1,217 for y in the first $x + 1{,}217 \approx 3{,}500$
equation, and solve for x. $x \approx 2{,}283$

2,283 pounds of the $1.60 a pound ingredient and 1,217 pounds of the $3.90 a pound ingredient should be used. *Ans*

3. A medical assistant has two different solutions of acid and water. One is a 16.0% acid solution, and the other is a 35.0% acid solution. How many cubic centimeters of each solution
should be used to make 150.0 cubic centimeters of a 28.0% acid solution?

NOTE: A 28.0% acid solution means 28.0% of the solution is acid and 72.0% is water.

Systems of equations are often used to solve problems involving solutions. A solution is a
liquid mixture of two or more substances; often one substance is water. One equation is written in terms of the amount of solution; the other equation is written in terms of the amount
of one substance in the solution.

The table in Figure 17–5 has all of the values that are needed to write the two equations.

Identify the variables:

Let x = amount of 16% acid solution and y = amount of 35% solution.

Decimal Fraction of Acid in Solution	Amount of Solution (Cubic Centimeters)	Amount of Acid in Solution (Cubic Centimeters)
0.16	x	0.16x
0.35	y	0.35y
0.28	150	0.28(150) or 42

Figure 17–5

Refer to the table and write the two equations.

a. Amount of solution $x + y = 150$

b. Amount of acid $0.16x + 0.35y = 42$

Solve by the addition method. $x + y = 150$

Multiply the second equation by 100 to $16x + 3y = 4{,}200$
remove decimals.

Multiply each term of the first equation $-16x - 16y = -2{,}400$
by -16 and add the equations. $\underline{16x + 35y = 4{,}200}$

$19y = 1{,}800$

$y \approx 94.7$

Substitute 94.7 for y in the first equation. $x + 94.7 \approx 150$

$x \approx 55.3$

55.3 cubic centimeters of the 16.0% acid solution and 94.7 cubic centimeters of the 35.0%
acid solution should be used. *Ans*

4. A small business owner wishes to invest $38,000 of company profits. Part of the $38,000 is to be invested in a safe investment at 4.2% annual interest and the rest at a less safe investment at 7.3% annual interest. If the owner wishes to receive a total annual interest income of $2,000, how much should be invested at each rate?

Systems of equations are used in business and finance applications. This is an example of a practical investment problem that is easily solved using systems of equations. One equation is written in terms of the amount invested; the other equation is written in terms of income received from the investment.

The table in Figure 17–6 has all the values that are needed to write the two equations.

Identify the variables:

Let x = the amount invested at 4.2% and y = amount invested at 7.3%.

	Rate of Interest	Amount Invested	Income (Rate × Amount)
Safe Investment (4.2%)	0.042	x	$0.042x$
Less safe investment (7.3%)	0.073	y	$0.073y$
Totals		$38,000	$2,000

Figure 17–6

Refer to the table and write the two equations.

a. Amount invested

$$x + y = 38,000$$

b. Income from investment

$$0.042x + 0.073y = 2,000$$

Solve by the substitution method.

Rearrange the first equation. The second equation can be multiplied by 1,000 to remove decimals. It can also be left in decimal form and easily solved using a calculator. This example will be left in decimal form.

$$x = 38,000 - y$$
$$0.042x + 0.073y = 2,000$$
$$0.042(38,000 - y) + 0.073y = 2,000$$
$$1,596 - 0.042y + 0.073y = 2,000$$
$$0.031y = 404$$
$$y = 13,032$$
$$x = 38,000 - y$$
$$x = 38,000 - 13,032$$
$$x = 24,968$$

$24,968 should be invested in the safe investment (4.2%) and $13,032 should be invested in the less safe investment (7.3%). *Ans*

EXERCISE 17–7

Write a system of equations for the following number problems and practical applications and solve. It may be helpful to make a table of values for certain problems.

1. The difference between two numbers is 7 and three times their sum is 75. Find the two numbers.

2. The sum of two numbers is 18. Four times one number is equal to five times the other number. Find the two numbers.

3. The total cost of two different items is $19. The difference in cost of the two items is $5. Find the cost of each item.

4. The difference between two numbers is 4. Twice the larger number is 3 less than three times the smaller number. Find the two numbers.

5. Part A is 7.5 pounds heavier than part B. The difference between twice part A's weight and three times part B's weight is 2.0 pounds. Find the weights of part A and part B.

6. Two resistors in series, R_1 and R_2, have a total resistance of 1,200 ohms. The resistance of R_1 is 150 ohms more than twice R_2. Find the resistances of R_1 and R_2.

7. A metal strip 42 inches long is to be cut in two pieces. The smaller piece is to be 3 inches more than half the larger piece. Find the length of each piece.

8. A rectangular display panel is designed so the length is 14 inches shorter than twice the width. the perimeter of the panel is 206 inches. Find the length and width.

    NOTE: Perimeter equals twice the length plus twice the width.

9. How much of an 18% acid solution and how much of a 30% acid solution should be mixed together to make 15.0 liters of a 25% acid solution?

10. How many ounces of 74% pure silver and 87% pure silver when combined will make 20.0 ounces of 83% pure silver? Round the answers to one decimal place.

11. Two pumps, each operating at partial capacity and pumping at different rates, together discharge 8,450 liters of water per hour. At full capacity, the smaller pump's rate of discharge is increased by 55% and the larger pump's rate of discharge is increased by 60%. The two pumps, operating at full capacity, together discharge 13,345 liters per hour. How many liters per hour were discharged by each pump when operating at partial capacity?

12. A total of $54,25\overline{0}$ is to be invested, part at 5.370% annual interest and the rest at 6.720% annual interest. How much is invested at each rate to provide an annual interest income of $3,10\overline{0}$? Round the answer to the nearest dollar.

13. A mason estimates the cost of both cement and lime required for a job at $381. Between the cement and lime, a total of 98 bags are required. Cement costs $4.20 a bag and lime costs $2.50 a bag. How many bags of cement and how many bags of lime are to be used on the job?

14. A building supply company has a quantity sale of paint brushes. Seven 2-inch brushes and four 3-inch brushes sell for a total of $32.50. Nine 2-inch brushes and three 3-inch brushes sell for a total of $33.75. Find the cost of one 2-inch and one 3-inch brush.

15. A chemist has one solution that tests 12.80% acid and another solution that tests 26.20% acid. How much of each should be combined to make 275.0 cubic centimeters of solution testing 15.50% acid. Round the answers to one decimal place.

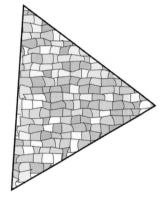

Figure 17–7

16. A triangular patio, like the one in Figure 17–7, has two sides of equal length. The third side is 12 feet longer than $1\frac{1}{4}$ times the length of each of the two equal sides. The perimeter of the patio is $83\frac{1}{2}$ feet. Perimeter equals the sum of the lengths of the sides. Find the length of each of the three sides.

17. A landscaper mixed two chemicals to make 75.0 gallons of insecticide. One chemical costs $8.60 a gallon; the other costs $6.75 a gallon. The cost of the insecticide is $7.40 a gallon. How many gallons of each chemical are used? Round the answers to one decimal place.

18. Part of savings is invested at 4.30% annual interest and part at 5.90% annual interest. An annual interest income of $765 is received from the total investment of $15,870. How much is invested at each rate? Round theanswers to the nearest dollar.

19. In a certain triangle, like the one in Figure 17–8, angle A is 15° less than half of the sum of angle B and angle C. Angle B is equal to angle C. The sum of the three angles of a triangle equals 180°. Find angle A.

20. A 4.80% investment brings an annual return of $94 less than a 6.20% investment. The total amount invested is $21,600. How much is invested at each rate?

21. Ten gallons of a 15.0% alcohol solution are to be mixed with a 28.0% alcohol solution to make a 20.0% alcohol solution.

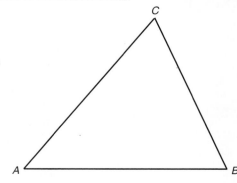

Figure 17–8

a. How many gallons of a 28.0% alcohol solution must be used?

b. How many gallons of a 20.0% alcohol solution are made?

22. Two manufacturing production lines, each making the same part, together produced an average of 1,750 defective parts per week. Improvements to both production lines are made resulting in both lines together producing an average of 700 defective parts per week. After improvements, the line that produced the most defective parts reduced the number of defective parts by 62%. The line that produced the least defective parts reduced the number of defective parts by 56%.

a. How many defective parts per week were produced by each production line before improvements?

b. How many defective parts per week are produced by each production line after improvements?

UNIT EXERCISE AND PROBLEM REVIEW

GRAPHING SYSTEMS OF EQUATIONS

Graph and solve each of the following systems of equations. Check the solutions.

1. $2x + 3y = 10$
   $x + 4y = 0$

2. $2x - 3y = -7$
   $3x = y$

3. $4x - 2y = -16$
   $9x - 3y = -36$

4. $-2x - y = 0$
   $3x - 2y = -35$

5. $x - 5y = -4$
   $x - 4y = -5$

6. $9x - 7y = 22$
   $-3x + y = 2$

SOLVING SYSTEMS OF EQUATIONS BY SUBSTITUTION

Solve each of the following systems of equations by the substitution method. Check the answers.

7. $x = 9 - 3y$
   $x - 7y = -21$

8. $-y = -2x + 5$
   $6x + y = 27$

9. $x = -3y + 8$
   $2x - y = 9$

10. $-2x = -5y$
    $-3x + 2y = -22$

11. $x + 3y = 8$
    $6x - 3y = 27$

12. $-4x + 7y = 41$
    $x + 3y = 42$

13. $0.6x - 0.7y = 0$
    $0.4x - 0.7y = 7$

14. $-0.5x - 1.5y = 4.5$
    $x - 7y = -21$

15. $0.09x - 0.05y = -0.22$
    $0.06x - 0.05y = -0.28$

SOLVING SYSTEMS OF EQUATIONS BY ADDITION OR SUBTRACTION

Solve each of the following systems of equations by the addition or subtraction method. Check the answers.

16. $x + y = 5$
    $x - y = -3$

17. $x + y = 6$
    $x - y = 0$

18. $y + 5x = 43$
    $12x + y = 99$

19. $2x + 5y = 14$
    $4x - 5y = -2$

20. $3y = 30 - 5x$
    $3x + 3y = 18$

21. $4x + 7y = 40$
    $-8x + 7y = 4$

22. $-6x + 3y = -24$
    $2x = 40 + 9y$

23. $-2x - 3y = -17$
    $5y = 27 - 3x$

24. $x - 0.8y = 0.2$
    $0.4x + 0.6y = 1$

25. $0.5x + 0.2y = 1.4$
    $0.6x + 1.8y = 4.8$

26. $0.06x - 0.1y = -0.26$
    $0.08x + 0.06y = 0.04$

27. $0.9x - 0.4y = 3$
    $0.06x + 0.14y = -0.3$

SOLVING CONSISTENT, INCONSISTENT, AND DEPENDENT SYSTEMS OF EQUATIONS

Solve the following systems of equations. Identify each system as to whether it is consistent and independent (one solution set), inconsistent (no solution set), or dependent (infinite solution sets).

28. $18x = 70 - 6y$
    $12x + 4y = 22$

29. $15x - 9y = 42$
    $3y = 5x - 14$

30. $4x - 8y = -12$
    $7x - 14y = -21$

31. $x = 8 - 3y$
$-6x + 3y = -27$

32. $8x - 2y = -9$
$28x - 7y = -28$

33. $8y = 16 - 12x$
$15x + 10y = 20$

34. $-0.21x - 0.12y = 1.42$
$0.07x + 0.04y = -0.48$

35. $0.3x + 0.2y = 3$
$0.7x + 0.4y = 8$

36. $0.9x - 0.6y = 1.3$
$2.25x - 1.5y = 3.25$

DETERMINANTS

Evaluate each of the following determinants.

37. $\begin{vmatrix} -7 & 4 \\ 12 & -6 \end{vmatrix}$

38. $\begin{vmatrix} 4.5 & -5.4 \\ 3.8 & 5.1 \end{vmatrix}$

39. $\begin{vmatrix} \frac{2}{3} & -\frac{1}{5} \\ -\frac{3}{4} & \frac{3}{5} \end{vmatrix}$

40. $\begin{vmatrix} -6.7 & -4.5 \\ 7.4 & 8.1 \end{vmatrix}$

CRAMER'S RULE

Use Cramer's rule to solve each of the following systems of linear equations.

41. $-7x + 4y = -42.8$
$12x - 6y = 69$

42. $5x - 7y = 12$
$-10x - 14y = 6$

43. $5x - 8y = 16$
$-7.5x + 12y = -6$

44. $6.5x - 3.2y = 2$
$3.25x + 1.6y = 8$

WRITING AND SOLVING SYSTEMS OF EQUATIONS FROM WORD STATEMENTS

Write a system of equations for each problem and solve.

45. The difference between two numbers is 5. Three times the larger number is equal to four times the smaller number plus 6. Find the two numbers.

46. Two air ducts together remove 580 cubic feet of air per minute. The larger duct removes 50 cubic feet of air less than twice the amount of air removed by the smaller duct. How many cubic feet of air per minute are removed by each duct?

47. How much of a 20.3% acid solution and how much of a 35.6% acid solution should be mixed together to make 25.5 liters of a 30.5% acid solution?

48. A total of $38,350 is to be invested at 4.78% annual interest and the rest at 5.93% annual interest. How much is invested at each rate to provide an annual income of $2,100? Round the answers to the nearest dollar.

49. A sidewalk is to be made around a rectangular field that has a perimeter of 1,930 feet. The width of the field is 20 feet greater than half the length. Find the length and width.

50. A landscaper mixes rye grass seed worth $1.85 a pound with blue grass seed worth $4.10 a pound. The mixture weighs 75.0 pounds and is worth $2.40 a pound. How many pounds of each kind of seed are used? Round the answers to one decimal place.

51. Two branches of a company produce the same product. Both branches together produced 12,400 items per day. The manufacturing process at the lower producing branch was changed, which resulted in a 25% increase in production. The process at the other branch was not changed and the rate of production remained the same. The total production of both branches together is increased to 13,550 items per day. How many items per day were produced by each branch before the lower producing branch's process was changed?

52. A contractor originally estimated the total cost of a job including labor and materials at $57,000. In reviewing the costs, a new estimate of $48,000 is made, reducing the labor cost by 15% and the material cost by 18%.

a. What was each cost, labor and materials, of the original estimate?

b. What is each cost, labor and materials, of the new estimate?

UNIT 18 ⁘ Quadratic Equations

OBJECTIVES

After studying this unit you should be able to

- solve incomplete quadratic equations.
- solve practical problems as incomplete quadratic equations.
- solve complete quadratic equations using the quadratic formula.
- solve practical problems with given equations using the quadratic formula.
- solve practical problems in word form without given equations using the quadratic formula.

With a few exceptions, only linear equations have been presented in this book. Recall that with linear equations the unknown or variable is raised to the first power only. The following are examples of linear equations typical of those that you have solved: $3x + 5 = 20$, $8x = 3x - 24$, and $16 - (y - 4) = 9 + 3y$.

A quadratic equation with one unknown or variable has at least one term raised to the second power with no terms of a higher power. Observe that an equation may also contain the unknown in the first power as well as the second power. The following are examples of quadratic equations: $7x^2 + x = 0$, $15y^2 - 12y = 93$, and $45x^2 = 108$.

Quadratic equations have wide practical applications in physics and in a variety of technological fields. Only a few examples of the many fields that use quadratic equations are given. Construction occupations apply quadratic equations in determining material sizes and checking the squareness of structural members. Currents, voltages, and times related to variable voltages and currents are computed with quadratic equations in electronics and electrical technologies. Design and manufacturing technologies apply quadratic equations in calculating dimensions of two- and three-dimensional objects. In physics, quadratic equations are used to compute times related to distances traveled by fired projectiles.

18–1 General or Standard Form of Quadratic Equations

The general or standard form of a quadratic equation is:

$$ax^2 + bx + c = 0 \quad (a \text{ cannot equal } 0)$$

where a is the coefficient of the x^2 term
b is the coefficient of the x term
c is the constant term that does not contain x in any form

All quadratic equations *must have* a term raised to the second power (x^2). A quadratic equation does not have to have a term raised to the first power (x). A quadratic equation does not have to have a constant term (c).

403

A quadratic equation has two roots. The *roots* of a quadratic equation are the values that satisfy the equation or make the equation true. The two roots are sometimes called a solution set and are enclosed in braces { }.

NOTE: In solving some practical applications with quadratic equations, two solutions result, but only one solution may be correct when applied to a real situation. For example, a practical problem using a quadratic equation dealing with building materials may result in solutions of $+10$ feet and -10 feet. Although both solutions may be mathematically correct, -10 feet in length is ridiculous and must be eliminated as a possible solution. The $+10$-foot length is the only reasonable solution.

18–2 Incomplete Quadratic Equations ($ax^2 = c$)

A quadratic equation that does not contain a first power variable (x term) is called an *incomplete or pure quadratic equation.* Some examples are: $x^2 = 64$, $3x^2 = 43$, $64x^2 - 92 = 0$, $3x^2 + 5 = 65$, and $4x^2 - 3 = 7x^2 - 51$. Notice that the last two equations have to be re-arranged to be in the $ax^2 = c$ form.

As previously stated, all quadratic equations have two roots. All positive numbers have two square roots; a positive and a negative square root. Recall that the square root of a negative number has no solution in the real number system. For example, $\sqrt{-4}$ has no solution.

The procedure for solving incomplete quadratic equations is similar to the procedure for solving linear equations.

Procedure for Solving Incomplete Quadratic Equations

- Isolate the term containing x^2 on one side of the equation.

  NOTE: It may be necessary to remove parentheses, combine like terms, or apply the principles of equality.

- If x^2 has a coefficient other than 1, divide both sides of the equation by the coefficient.

- Take the square root of both sides of the equation. Write a $\pm$ sign before the square root quantity.

- Check. Substitute each root in the original equation. Recall that if the solution to an equation is a rounded value, the check may result in a very small difference between both sides of the equation because of rounding.

EXAMPLES •────────────────────────────────

1. $6x^2 = 54$. Solve for x.

Divide both sides of the equation by 6.

$$\frac{6x^2}{6} = \frac{54}{6}$$

$$x^2 = 9$$

Take the square root of both sides of the equation.

$$\sqrt{x^2} = \sqrt{9}$$

$$x = \pm\sqrt{9} = \pm3 \text{ or}$$

$$x = 3 \text{ or } -3 \text{ } Ans$$

Check. Substitute the x values in the original equation and simplify.

For $x = 3$	For $x = -3$
$6(3^2) = 54$	$6(-3^2) = 54$
$6(9) = 54$	$6(9) = 54$
$54 = 54$	$54 = 54$

2. $4x^2 - 3 = 7x^2 - 51$. Solve for x.

Solution

Subtract $7x^2$ from both sides.

Add 3 to both sides of the equation.

Divide both sides of the equation by -3.

$$
\begin{aligned}
4x^2 - 3 &= 7x^2 - 51 \\
\underline{-7x^2} \qquad & \quad \underline{-7x^2} \\
-3x^2 - 3 &= \qquad -51 \\
\underline{+3} &= \qquad \underline{+ \ 3} \\
-3x^2 &= \qquad -48
\end{aligned}
$$

$$\frac{-3x^2}{-3} = \frac{-48}{-3}$$

$$x^2 = 16$$

Take the square root of both sides of the equation.

$$\sqrt{x^2} = \sqrt{16}$$

$$x = \pm\sqrt{16} = \pm 4 \text{ or}$$

$$x = 4 \text{ or } -4 \ \textit{Ans}$$

3. A landscaper, in designing a circular patio which is to contain 275.0 square feet, must compute the patio diameter. The following formula is used:

$$A = 0.7854d^2 \qquad \text{where } A = \text{area}$$
$$d = \text{diameter}$$

Compute the diameter to 4 significant digits.

Substitute known number values for letter values.

$$275.0 \text{ ft}^2 = 0.7854d^2$$

Divide both sides of the equation by 0.7854.

$$\frac{275.0 \text{ ft}^2}{0.7854} = \frac{0.7854d^2}{0.7854}$$

$$350.140 \text{ ft}^2 \approx d^2$$

Take the square root of both sides.

$$\sqrt{350.140 \text{ ft}^2} \approx \sqrt{d^2}$$

$$\pm \ 18.71 \text{ ft} \approx d$$

-18.71 ft makes no sense and is eliminated.

A patio cannot have a negative diameter, so the solution is the diameter $= 18.71$ ft (rounded to 4 significant digits).

Check. Substitute the d value in the original equation and simplify.

$$A = 0.7854d^2$$
$$275.0 \text{ ft}^2 \approx 0.7854 \ (18.71 \text{ ft})^2$$
$$275.0 \text{ ft}^2 \approx 0.7854 \ (350.06 \text{ ft}^2)$$
$$275.0 \text{ ft}^2 \approx 274.9 \text{ ft}^2$$

Observe the small difference between both sides of the equation because of rounding.

EXERCISE 18–2A

Solve and check the following incomplete quadratic equation problems. Round the answers to 2 decimal places where necessary.

1. $x^2 = 144$

2. $x^2 = 529$

3. $x^2 = 51.84$

4. $6x^2 = 150$

5. $4x^2 = 20$

6. $x^2 - 225 = 0$

7. $6x^2 - 84 = 0$

8. $x^2 + 2x^2 = 48$

9. $8x^2 - 3x^2 = 320$

10. $5x^2 - 2x^2 = 214$

11. $3.6x^2 + 4.8x^2 = 19.2$

12. $0.9x^2 + 0.6x^2 = 1.06$

13. $4x^2 = 15.8 - 1.5x^2$

14. $18.32x^2 = 26.12 + 16.50x^2$

15. $3x^2 = 156 - 2x^2$

16. $10x^2 - 70 = 28 - 12x^2$

17. $0.3x^2 - 0.95 = 0.08 - 0.5x^2$

18. $14.75x^2 - 18.32 = -15.25x^2 - 6.62$

19. $\dfrac{x}{4} = \dfrac{16}{x}$

20. $\dfrac{5.6}{x} = \dfrac{x}{19.8}$

21. $\dfrac{x - 7}{2} = \dfrac{8}{x + 7}$

22. $\dfrac{7.82 + x}{5.3} = \dfrac{6}{7.82 - x}$

23. $(x + 6)(x - 6) = 24$

24. $(x + 0.9)(x - 0.9) = -x^2$

25. $(x - 6.8)(x + 10.8) - 4x = 0$

26. $(x - 12.6)(x + 23.6) - 11x = 86.8$

EXERCISE 18–2B

Solve the following incomplete quadratic equation practical applications.

1. The formula $d = 16t^2$ can be used to determine the time (t) it takes an object to fall a given distance (d). How many seconds does it take an object to fall 920 feet? Give the answer to 2 significant digits.

2. A formula used in electricity is $I^2 = \dfrac{P}{R}$ where I is current in amperes (A), P is power in watts (W), and R is resistance in ohms (Ω). Compute the current in amperes present in a circuit if the resistance is 1.32 ohms and the power is 20.63 watts. Give the answer to 3 significant digits.

3. The stopping distance (d), in feet, of a car is given by the formula $d = \dfrac{1.1\,V^2}{r}$ where V is speed in miles per hour and r is the rate of retardation or the reduction of speed each second. A car stops in 225 feet with a rate of retardation of 26.2 feet per second. What was the speed of the car when the brake was first applied? Give the answer to 3 significant digits.

4. Figure 18–1 shows a vertical force (a) of 60.50 kilograms and a horizontal force (b) of 140.6 kilograms which are applied at the same time at point 0. The resultant force (R) is determined by the equation $R^2 = a^2 + b^2$. Solve for R. Give the answer to 3 significant digits.

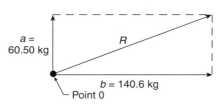

Figure 18–1

5. An object fired with an initial velocity (v) in feet per second reaching a height (h) in feet is given by the formula $h = \dfrac{v^2}{64}$ (neglecting air resistance). Compute the velocity needed to fire a projectile 3,780 feet high. Give the answer to the nearest foot per second.

6. The formula $t^2 = 4\pi^2\dfrac{l}{32}$ can be used to determine the time (t) in seconds for a complete swing of a pendulum of a given length (l) in feet. How many seconds does it take for a 1.75-foot-long pendulum to make a complete swing? Use $\pi = 3.1416$. Give the answer to 3 significant digits.

7. Determine the base diameter (d) in centimeters of a cone in Figure 18–2. Use the formula $V = 0.2618\,d^2h$ where volume (V) = 385 cubic centimeters and height (h) = 13.1 centimeters. Give the answer to 3 significant digits.

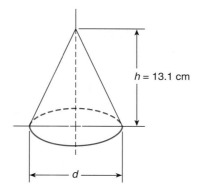

$h = 13.1$ cm

d

Figure 18–2

8. The formula $l = c\dfrac{d}{2}Av^2$ is used in aviation

   where l is the lift (weight of airplane) in pounds
            c is the coefficient of lift
            d is the air density
            A is the wing area in square feet
            v is the air speed in feet per second

   Compute the velocity needed to sustain a small airplane weighing (lift) 4,720 pounds with a wing area of 967 square feet. The air density is 0.0026 and the coefficient of lift is 0.480. Give the answer to 3 significant digits.

9. Kinetic energy (K) of a body equals one-half the product of the mass (m) and the square of the velocity (v).

$$K = \frac{1}{2}mv^2$$

   Compute the velocity of a body, in centimeters per second, when the kinetic energy is 20000 ergs and the mass is 100 grams.

10. Determine the area of metal, in square millimeters, of the square spacer with a circular hole shown in Figure 18–3.

    Area of square = s^2
    Area of circle = $3.1416r^2$

    Give the answer to the nearest square millimeter.

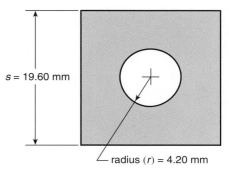

$s = 19.60$ mm

radius (r) = 4.20 mm

Figure 18–3

11. A proposed paved area in the shape of a half ring is shown in Figure 18–4 on page 408. Use the formula:

$$A = \frac{3.1416(R^2 - r^2)}{2}$$

   Compute the inner radius (r) which results in an area (A) of 6,220 square feet with an outer radius (R) of 81 feet. Give the answer to the nearest tenth foot.

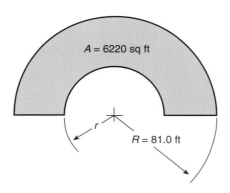

Figure 18–4

12. A concrete slab, which is to be constructed in the shape of a segment of a circle, is shown in Figure 18–5. Use the formula:

$$r = \frac{l^2 + 4h^2}{8h}$$

Compute the length (l) of the straight side of the slab. Give the answer to 3 significant digits.

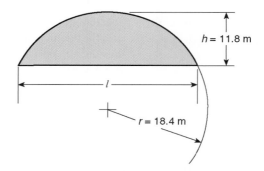

Figure 18–5

18–3   Complete Quadratic Equations

A quadratic equation that contains both the second power variable (x^2 term) and the first power variable (x term) is called a *complete quadratic equation.*

The following methods can be used in solving complete quadratic equations.

• **Factoring.** The factoring method has limited application. Only certain quadratic equations can be solved by factoring.

• **Completing the Square.** This method can be used for solving all quadratic equations in one variable. It can be a rather long and complicated procedure and is seldom used in practical applications.

• **Quadratic Formula.** This is the most useful method for solving complete quadratic equations. It can be used for solving all quadratic equations in one variable.

Because completing the square is so complicated and used only in a few places, it will not be studied it this book. Factoring is a very quick way to solve some quadratic equations. But, factoring takes a lot of practice to learn and, in spite of the fact that every quadratic equation can be factored, the factors of most equations are not easily seen.

The quadratic equation $x^2 - 3x - 10 = 0$ can be factored as $(x + 2)(x - 5) = 0$. The quadratic equation $x^2 + 4x - 1 = 0$ can also be factored. But no amount of factoring practice would let you see that the factored form is $\left(x + 2 - \sqrt{5}\right)\left(x + 2 + \sqrt{5}\right) = 0$. That is why we will just study the quadratic formula. The quadratic formula may not always be as fast as factoring, but it ***Always Works***!

Quadratic Formula

In this book, only the quadratic formula method is presented. The quadratic formula is generally the method used to solve complete quadratic equations. Recall that the general or standard form of a quadratic equation is:

$ax^2 + bx + c = 0$ (*a* cannot equal 0)   where *a* is the coefficient of the x^2 term
b is the coefficient of the *x* term
c is the constant term that does not contain *x* in any form.

If the general form of a quadratic equation is transformed and solved for *x*, the resulting equation is called the *quadratic formula.*

Quadratic Formula

$$x = \frac{-b \pm \sqrt{b^2 - 4ac}}{2a}$$

The $\pm$ sign indicates that one solution is obtained using $+\sqrt{b^2 - 4ac}$ and the second solution using $-\sqrt{b^2 - 4ac}$.

Procedure for Solving Complete Quadratic Equations

- Compare the given equation to the general form of a quadratic equation. It may be necessary to rearrange the given equation. All the terms must be on one side of the equation and zero (0) on the other side.
- Identify coefficients *a* and *b* and the constant term *c*. A complete quadratic equation may or may not have the term *c*.
- Substitute the numerical values for *a*, *b*, and *c* in the quadratic formula and solve for *x*. Remember the $\pm$ sign in the formula indicates two solutions.
- Check. Substitute each of the solutions in the original equation.

EXAMPLES

1. $5x^2 - 3x - 2 = 0$. Solve for *x*.

Identify *a*, *b*, and *c*.

$a = 5, b = -3, c = -2$

Substitute the numerical values of *a*, *b*, and *c* in the quadratic formula.

$$x = \frac{-(-3) \pm \sqrt{(-3)^2 - 4(5)(-2)}}{2(5)}$$

Solve for *x*.

$$x = \frac{3 \pm \sqrt{9 + 40}}{10}$$

$$x = \frac{3 \pm \sqrt{49}}{10}$$

$$x = \frac{3 + 7}{10} \text{ or } x = \frac{3 - 7}{10}$$

$$x = 1 \text{ or } x = -0.4 \text{ } Ans$$

Check. Substitute the *x* values in the original equation and simplify.

For $x = 1$
$$5x^2 - 3x - 2 = 0$$
$$5(1^2) - 3(1) - 2 = 0$$
$$5 - 3 - 2 = 0$$
$$0 = 0$$

For $x = -0.4$
$$5x^2 - 3x - 2 = 0$$
$$5(-0.4)^2 - 3(-0.4) - 2 = 0$$
$$0.8 + 1.2 - 2 = 0$$
$$0 = 0$$

2. $7.6x^2 = -5.4x$. Solve for x.

Rearrange the equation with all terms on one side of the equation and 0 on the other.

$$7.6x^2 + 5.4x = 0$$

Identify a, b, and c.

$a = 7.6$, $b = -5.4$. There is no c value; $c = 0$.

Substitute numerical values for a, b, and c in the quadratic formula and solve for x.

$$x = \frac{-5.4 \pm \sqrt{5.4^2 - 4(7.6)(0)}}{2(7.6)}$$

$$x = \frac{-5.4 \pm \sqrt{29.16 - 0}}{15.2}$$

$$x = \frac{-5.4 + \sqrt{29.16}}{15.2} \text{ or } x = \frac{-5.4 - \sqrt{29.16}}{15.2}$$

$$x = \frac{-5.4 + 5.4}{15.2} \text{ or } x = \frac{-5.4 - 5.4}{15.2}$$

$$x = 0 \text{ or } x \approx -0.71 \text{ (rounded to 2 decimal places) } Ans$$

Check. Substitute the x values in the original equation.

For $x = 0$
$$7.6(0)^2 = -5.4(0)$$
$$0 = 0$$

For $x \approx -0.71$
$$7.6(-0.71)^2 \approx -5.4(-0.71)$$
$$3.831 \approx 3.834$$

Observe the small difference between both sides of the equation because of rounding.

NOTE: When solving a complete quadratic equation, both roots must be computed. A common error that must be avoided is to divide both sides of the equation by x. Dividing both sides of the equation $7.6x^2 = -5.4x$ by x results in the new equation $7.6x = -5.4$, $x = \dfrac{-5.4}{7.6}$, $x \approx -0.71$. The root $x = 0$ is lost.

3. $x^2 - 10x + 25 = 0$. Solve for x.

Identify a, b, and c.

$a = 1$, $b = -10$, $c = 25$

Substitute numerical values for a, b, and c in the quadratic formula and solve for x.

$$x = \frac{-(-10) \pm \sqrt{(-10)^2 - 4(1)(25)}}{2(1)}$$

$$x = \frac{10 \pm \sqrt{100 - 100}}{2}$$

$$x = \frac{10 + 0}{2} \text{ or } x = \frac{10 - 0}{2}$$

$$x = 5 \text{ or } x = 5 \text{ } Ans$$

The two roots are equal. It may appear that the equation has only one root. All quadratic equations have two roots. This equation has two equal roots, which are sometimes known as double roots.

4. Solve $5x^2 - 3x + 9 = 3x^2 + 29$ for x.

Solution. First, rearrange the equation so all terms are on one side of the equal sign. Remember, you can only add and subtract like terms.

Subtract $3x^2$ from both sides.

$$\begin{array}{rcrr} 5x^2 - 3x + & 9 = & 3x^2 + 29 \\ -3x^2 & & -3x^2 \\ \hline 2x^2 - 3x + & 9 = & 29 \end{array}$$

Subtract 29 from both sides of the equation.

$$\begin{array}{rr} & -\ 29 \quad\quad -\ 29 \\ \hline 2x^2 - 3x - 20 = 0 \end{array}$$

Now, solve the equation using the quadratic formula.

Identify a, b, and c. $a = 2$, $b = -3$, and $c = -20$.

Substitute numerical values for a, b, and c in the quadratic formula and solve for x.

$$x = \frac{-(-3) \pm \sqrt{(-3)^2 - 4(2)(-20)}}{2(2)}$$

$$x = \frac{3 \pm \sqrt{9 + 160}}{4}$$

$$x = \frac{3 \pm \sqrt{169}}{4}$$

$$x = \frac{3 + 13}{4} \text{ or } \frac{3 - 13}{4}$$

$$x = 4 \text{ or } -2.5 \text{ Ans}$$

EXERCISE 18–3

Solve the following complete quadratic equations. Round the answers to 2 decimal places where necessary.

1. $x^2 - 10x = 0$
2. $2x^2 - 5x = 0$
3. $9x^2 + 6x = 0$
4. $x^2 = -9x$
5. $3x^2 = 15x$
6. $x^2 - 6x = -5$
7. $8x^2 - 24 - 26x = 0$
8. $x^2 + 4x - 32 = 0$
9. $x^2 - 5x = 24$
10. $x^2 - 8x = -12$
11. $x^2 - 10x - 75 = 0$
12. $x^2 + 9x - 36 = 0$
13. $4x^2 - 12 = 13x$
14. $x^2 + 1.5x = 1$
15. $16x^2 - 16x + 3 = 0$
16. $x^2 - 1.25 = 2x$
17. $x^2 - 0.25x = 0.75$

18. $x^2 + 12x + 35 = 0$
19. $6x^2 + 12 = 17x$
20. $x^2 + 8x = -12$
21. $6x^2 - 13x = 5$
22. $4x^2 + 2x - 3 = 0$
23. $\dfrac{x^2}{2} + \dfrac{3x}{4} = 11$
24. $\dfrac{6}{x + 3} = \dfrac{x + 2}{5}$
25. $\dfrac{x}{2x - 1} = \dfrac{2x + 3}{15}$
26. $x^2 - 0.35x + 0.015 = 0$
27. $0.2x^2 - 1.75x + 1.2 = 0$
28. $2x^2 - 3x + 6 = x^2 + 2x$
29. $x^2 - 3(x + 7) = x$
30. $(x + 2)^2 = 2(5x - 2)$

18–4   Practical Applications of Complete Quadratic Equations. Equations Given.

Quadratic equations have many applications in both science and technology. In the following examples, the equations are directly given and the quadratic formula is applied in their solutions.

EXAMPLES

1. A commercial building is designed to have an entrance opening with an arched top as shown in Figure 18–6 on page 412. An architectural drafter uses the following equation to determine the height (h) of the arched portion of the entrance opening.

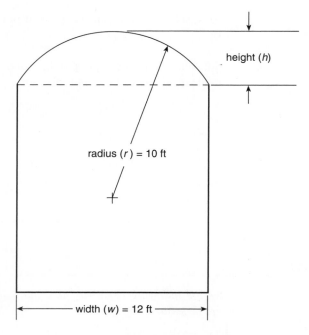

radius (r) = 10 ft

height (h)

width (w) = 12 ft

Figure 18–6

$w^2 = 8rh - 4h^2$, where w is the width of the opening and r is the radius of the arch. Determine the height (h).

Substitute the numerical values for w and r, simplify, and rearrange the equation.

$$w^2 = 8rh - 4h^2$$
$$12^2 = 8(10)h - 4h^2$$
$$144 = 80h - 4h^2$$
$$4h^2 - 80h + 144 = 0$$

Identify a, b, and c and substitute the values in the quadratic formula. Solve for h.

$$a = 4, b = -80, c = 144$$
$$h = \frac{-(-80) \pm \sqrt{(-80)^2 - 4(4)(144)}}{2(4)}$$
$$h = \frac{80 \pm \sqrt{6400 - 2304}}{8}$$
$$h = \frac{80 \pm \sqrt{4096}}{8}$$
$$h = \frac{80 \pm 64}{8}$$
$$h = \frac{80 + 64}{8} \text{ or } h = \frac{80 - 64}{8}$$
$$h = 18 \text{ ft or } h = 2 \text{ ft}$$

Since the height (h) of the arch cannot be greater than the radius (10 ft), $h = 18$ ft makes no sense and is eliminated.

The height (h) of the arch is 2 feet. *Ans*

Check. Substitute the numerical values for w, r, and h in the original equation and simplify.

$$w^2 = 8rh - 4h^2$$
$$(12 \text{ ft})^2 = 8(10 \text{ ft})(2 \text{ ft}) - 4(2 \text{ ft})^2$$
$$144 \text{ ft}^2 = 160 \text{ ft}^2 - 16 \text{ ft}^2$$
$$144 \text{ ft}^2 = 144 \text{ ft}^2$$

2. From physics, the formula $h = vt - 16t^2$ can be used to directly compute the height (h) in feet of an object at the end of t seconds when the object is thrown upward with a velocity (v) in feet per second. Using this formula:

  a. Determine the time (t) in seconds it will take an object to reach a height (h) of 115.0 feet when it is thrown upward at a velocity (v) of 92.0 feet per second.

  b. As the object falls, when is it again at a height of 115.0 feet?

Substitute numerical values:
$h = 115.0$ and $v = 92.0$.

$$h = vt - 16t^2$$
$$115.0 = 92.0t - 16t^2$$

Simplify and rearrange the equation.

$$16t^2 - 92.0t + 115.0 = 0$$

Identify a, b, and c and substitute the values in the quadratic equation. Solve for t.

$$a = 16, b = -92.0, c = 115.0$$

$$t = \frac{-(-92.0) \pm \sqrt{(-92.0)^2 - 4(16)(115.0)}}{2(16)}$$

$$t = \frac{92.0 \pm \sqrt{8464 - 7360}}{32}$$

$$t = \frac{92.0 \pm \sqrt{1104}}{32}$$

$$t \approx \frac{92.0 \pm 33.2265}{32}$$

$$t \approx \frac{92.0 + 33.2265}{32} \quad \text{or } t \approx \frac{92.0 - 33.2265}{32}$$

$$t \approx 3.9133 \text{ or } t \approx 1.8367$$

  a. When thrown upward, it takes 1.84 seconds to reach 115.0 feet. *Ans* (rounded to 3 significant digits)

  b. When falling, it takes 3.91 seconds from the time it is thrown to again be at a height of 115.0 feet. *Ans* (rounded to 3 significant digits)

Check. Substitute the numerical values in the original equation and simplify.

  a. For $t = 1.8367$

  $$115.0 = 92.0(1.8367) - 16(1.8367)^2$$

  Calculator Application

  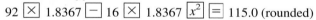
  92 $\boxed{\times}$ 1.8367 $\boxed{-}$ 16 $\boxed{\times}$ 1.8367 $\boxed{x^2}$ $\boxed{=}$ 115.0 (rounded)

  b. For $t = 3.9133$

  $$115.0 = 92.0(3.9133) - 16(3.9133)^2$$

  Calculator Application

  92 $\boxed{\times}$ 3.9133 $\boxed{-}$ 16 $\boxed{\times}$ 3.9133 $\boxed{x^2}$ $\boxed{=}$ 115.0 (rounded)

EXERCISE 18–4

Solve and check the following complete quadratic equation practical applications. The equations used in solving the problems are given.

  1. The formula $d^2 = h(h + 8000)$ can be used to determine distance (d) and height (h) relative to the horizon. What must the altitude or height of an airplane be, in miles, for a passenger to be able to see the horizon 65.5 miles away? Give the answer to three significant digits.

  2. The equation $e = t^2 - 7t + 9$ gives a variable voltage (e) with respect to time (t) in seconds. When e is 6.5 volts, what are the two values of t in seconds? Give the answers to 2 significant digits.

3. The formula $\dfrac{w}{l} = \dfrac{l}{w + l}$ is used to determine dimensions of length and width that result in a well-proportioned rectangle. Compute the length (l) of the rectangle in meters if the width (w) is 3.85 meters. Give the answer to 3 significant digits.

4. Determine the time (t) in seconds that it takes an object thrown vertically upward at a velocity (v) of 97 feet per second to reach a height (h) of 122 feet. Use the formula $h = vt - 16t^2$. Give the answer to 2 significant digits.

5. A segment of a circle is shown in Figure 18–7. Radius (r) = 73.80 centimeters and length (c) = 130.85 centimeters. Compute the height (h) of the segment using the formula

$$r = \frac{c^2 + 4h^2}{8h}$$

Give the answer to 2 decimal places.

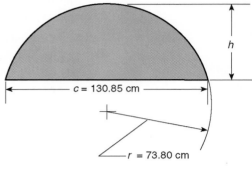

Figure 18–7

6. The cross-section of a steel structural member is shown in Figure 18–8. The cross-sectional area (A) is 42.78 square inches. Determine the thickness (t) using the formula $A = t[b + 2(a - t)]$. Give the answer to 2 decimal places.

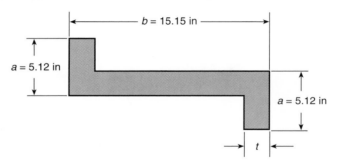

Figure 18–8

7. The front view and left side view of a solid tapered support base are shown in Figure 18–9. Given height (h), slant height (s), and radius (r) of the small circle, compute the radius (R) of the large circle. Use the formula

$$s = \sqrt{(R - r)^2 + h^2}$$

Give the answer to 3 decimal places.

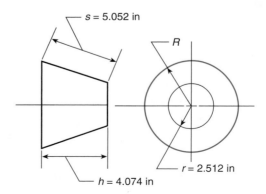

Figure 18–9

8. A gutter is to be made by folding up the edges of a strip of metal as shown in Figure 18–10. If the metal is 12 in wide and the cross-sectional area of the gutter is to be $16\frac{7}{8}$ in², what are the width and depth of the gutter?

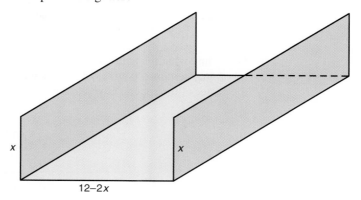

x x

12–2x

Figure 18–10

9. The angle iron in Figure 18–11 has a cross-sectional area of 70.56 cm². If the width of the angle iron is w, then the cross-sectional area is

$$12.6w + (18.65 - w)w$$

What is the width of this angle iron?

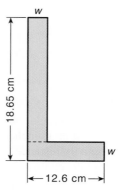

w

18.65 cm

w

←—12.6 cm—→

Figure 18–11

10. The center of gravity of a body is that point at which, if the body were suspended, it would be perfectly balanced in all positions. The center of gravity of a cylindrical surface or shell with one closed end and one open end, shown in Figure 18–12, is located using the formula

$$a = \frac{2h^2}{4h + d}$$

Determine what the height (h) must be in inches with a cylindrical surface diameter (d) of 8.35 inches to result in the center of gravity distance (a) of 5.06 inches from the closed end. Give the answer to 3 significant digits.

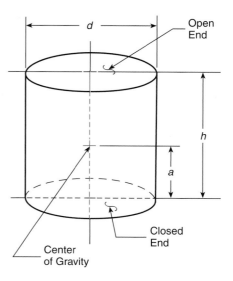

d Open End

h

a

Center of Gravity

Closed End

Figure 18–12

11. A solid segment of a sphere is shown in Figure 18–13. The center of gravity is located using the formula

$$a = \frac{3(2r - h)^2}{4(3r - h)}$$

Determine what the solid segment height (h) must be in centimeters with a sphere radius (r) of 46.8 centimeters to result in a center of gravity distance (a) of 32.8 centimeters from the center of the sphere. Give the answer to 3 significant digits.

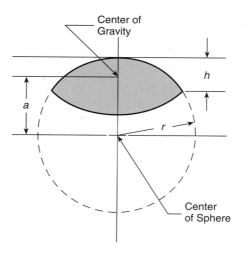

Figure 18–13

12. Use the formula and Figure 18–13 given in problem 11. Determine what the sphere radius (r) must be in inches with a solid segment height (h) of 30.40 inches to result in a center of gravity distance (a) of 34.85 inches from the center of the sphere. Give the answer to 4 significant digits.

13. A belt and pulley drive is shown in Figure 18–14. The diameter of the large pulley is 12.25 inches, the distance across centers (C) is 14.50 inches, and the length of the belt (L) is 61.13 inches. Compute the required diameter of the small pulley (d). Use the formula:

$$L = 2C + 1.57(D + d) + \frac{1}{4C}(D - d)^2$$

Give the answer to 2 decimal places.

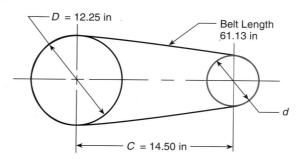

Figure 18–14

18–5   Word Problems Involving Complete Quadratic Equations. Equations Not Given.

In Unit 14, problems were solved that required expressing given word statements as linear equations. In linear equations, the variable is raised only to the first power. The same basic procedure is used in expressing word statements as quadratic equations as is used in solving linear word problems. The linear problem-solving procedure is given on page 337. A problem should be fully understood, broken down into simple parts, the unknown identified, and the equation written.

After the equation is written, it is usually necessary to rearrange the equation in general quadratic equation form. As with previous examples, solve the general form of a quadratic equation using the quadratic formula.

The primary purpose of the following examples is to show how to set up equations. Only the procedure for writing the equation is given, followed by rearranging the equation in the form of the general quadratic equation. The step-by-step procedure used in solving for the variable is left to the student. Answers are given.

EXAMPLES OF NUMBER PROBLEMS

1. The sum of two numbers is 13 and their product is 40. What are the numbers?

   Let x = one of the numbers. Since the sum of the two numbers is 13, the other number is $13 - x$. The product is 40. Write the equation.

$$x(13 - x) = 40$$

   Simplify. $13x - x^2 = 40$

   Rearrange. $-x^2 + 13x - 40 = 0$

   Solve and check. Use the quadratic formula, solve for x, and check using the same procedures as in the previous examples and exercises.

   5 and 8 *Ans*

2. If five times the square of a number is decreased by 10 and equals 25 more than 30 times the number, what is the number?

   Let x = the number. Write the equation and simplify.

$$5x^2 - 10 = 30x + 25$$

   Rearrange. $5x^2 - 30x - 35 = 0$

   Solve for x using the quadratic formula. Check.

   7 and -1 *Ans*

EXAMPLES OF PRACTICAL APPLICATIONS

3. A rectangular flower bed is 45 feet long and 38 feet wide. A landscaper proposes to increase the area of the bed by 1400 square feet. Both the length and width of the present bed are to be increased by the same number of feet. Determine the number of feet that are to be added to the length and width of the present bed.

   The area of the present flower bed is 1710 square feet.

   Area = length × width; Area = 45 feet × 38 feet; Area = 1710 square feet.

   The area of the proposed flower bed is 1710 square feet + 1400 square feet or 3110 square feet.

   Let x = the number of feet by which both the length and width of the present bed are increased.

   Length of proposed bed = $45 + x$.

   Width of proposed bed = $38 + x$.

The present and proposed flower beds are shown in Figure 18–15.

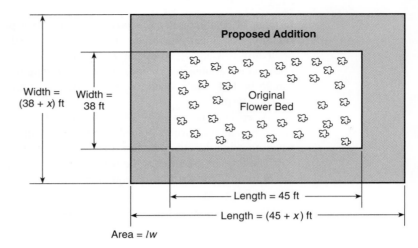

Figure 18–15

$$(45 + x)(38 + x) = 3110$$

Multiply. $(45)(38) + 45x + 38x + x^2 = 3110$

Simplify. $1710 + 83x + x^2 = 3110$

Rearrange. $x^2 + 83x - 1400 = 0$

Solve for x using the quadratic formula. Check.

$$x \approx 14.377 \text{ ft and } x \approx -97.377 \text{ ft}$$

Since -97.377 feet is ridiculous, it is eliminated.

The number of feet added to the length and width is 14 feet. *Ans* (rounded to 2 significant digits)

4. In building a condominium complex, a contractor plans to set aside a plot of land to be used as a recreation area. The plot is to be a rectangle with an area of 75,300 square feet and a length 50.0 feet greater than the width. A fence is going to completely enclose the recreation area. Determine the length of fencing required.

Let x = width, and $x + 50.0$ = length.

$$\text{Area} = lw$$
$$(x + 50.0)x = 75,300$$

Multiply. $x^2 + 50.0x = 75,300$

Rearrange. $x^2 + 50.0x - 75,300 = 0$

Solve for x using the quadratic formula.

$$x = 250.545 \text{ ft and } x = -300.545 \text{ ft}$$

Since, -300.545 ft is ridiculous, it is eliminated.

Width $= 250.545$ ft

Length $= 250.545$ ft $+ 50.0$ ft $= 300.545$ ft.

Compute the number of feet of   Perimeter $(P) = 2(l + w)$
fencing required. $P = 2(300.545 \text{ ft} + 250.545 \text{ ft})$
    $P = 2(551.09 \text{ ft})$
    $P = 1102.18 \text{ ft}$

1103 feet of fencing are required *Ans*

Normally, an answer like the one in the previous example would be rounded to the nearest foot. But rounding the answer *down* to 1102 feet of fencing would have been 0.18 ft ≈ 2.16 in. too little and would have left a 2.16 inch gap in the fence. Thus, the answer was round *up* to 1103 feet.

EXERCISE 18–5A

Solve and check the following number word problems, which involve complete quadratic equations. The equations used in solving the problems are not given.

1. The difference between two numbers is 5, and their product is 104. What are the numbers?

2. The sum of two numbers is 11, and twice their product is 56. What are the numbers?

3. One number is 7 more than a second number. If the product of the two numbers is 260, what are the numbers?

4. If the square of a number is increased by four times the number, the sum is 96. What is the number?

5. Twice the difference between two numbers is 8, and their product is 45. What are the numbers?

6. The sum of two numbers is 16, and twice their product minus one of the two numbers is 54. What are the numbers?

7. One-half the difference of two numbers is 4, and twice their product is 418. What are the numbers?

8. Five times the square of a number increased by 16.80 equals 24 times the number increased by 6.05. What is the number?

9. The sum of two numbers is 9.9, and the sum of the squares of the numbers is 53.21. What are the numbers?

10. Eight times the square of a number decreased by 27 equals the number squared plus twice the number increased by 11.75. Give the answer to 1 decimal place.

EXERCISE 18–5B

Solve and check the following practical application word problems, which involve complete quadratic equations. The equations used in solving these problems are not given. Round the answers to these problems to 3 significant digits where necessary unless otherwise specified.

NOTE: The following formulas are used in solving many of the problems. Refer to these formulas.

Area of rectangle = length × width: $A = lw$.

Perimeter of a rectangle = 2 × length + 2 × width: $P = 2l + 2w$ or $P = 2(l + w)$.

1. A rectangular piece is to be cut from a large flat piece of sheet metal. The length of the cut piece is to be 7.20 inches longer than the width, and the piece is to contain 215 square inches. Determine the length and width of the cut piece.

2. A rectangular box is designed to be 11.25 centimeters high and to have a volume of 3580 cubic centimeters. The perimeter of the base of the box is 81.3 centimeters. Determine the dimensions of the length and width of the box.

   NOTE: Volume of a rectangular box = height × area of the base. *Hint:* Perimeter = 2 × length + 2 × width.

3. Refer to Figure 18–16. A person walks 616 feet from point A directly to point B. A second person walks from point A to point C, then turns 90° and walks to point B. This distance from point A to point C is 306 feet greater than the distance from point C to point B. How many more feet did the second person walk than the first person?

   NOTE: $(AB)^2 = (AC)^2 + (BC)^2$.

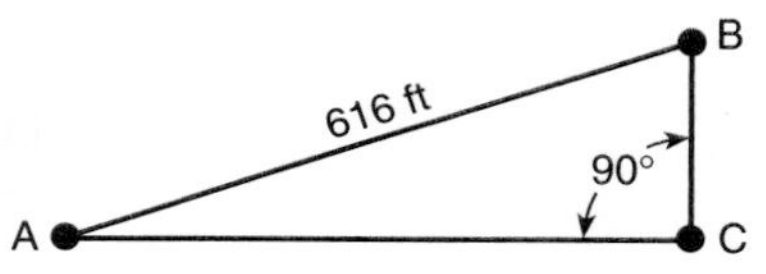

Figure 18–16

4. The original plans for a building called for a square foundation. Changes are made in the plans that call for a rectangular foundation whose length is 22 feet longer and whose width is 8 feet shorter than a side of the original square foundation. The area enclosed by the new foundation is 2176 square feet. What was the length of each side of the original square foundation?

    NOTE: Area of a square = side squared: $A = s^2$.

5. The selling price per unit of a certain product is expressed as x. The number of units of a product sold is expressed as $85 - 0.15x$. The total cost in dollars of producing and merchandizing the number of units sold is expressed as $0.15x^2 + 1.2x + 2$. If the total profit made is $1,090, how many units were sold? Round the answer to the nearest whole unit.

    NOTE: Total Profit = Unit Selling Price × Number of Units Sold − Total Cost.

6. A rectangular sheet metal box is made by cutting away squares of equal size from each of the four corners of a flat rectangular sheet which is 14.2 inches wide and 22.6 inches long. After the squares are cut away, the edges are folded up along the broken lines as shown in Figure 18–17. The area of the base of the box is 145 square inches. Compute the total number of square inches of scrap (the area of the four cut-away corners).

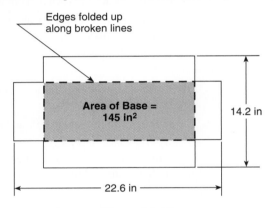

Figure 18–17

7. The requirements for fabricating a storage unit with rectangular sides and base call for a volume of 139.5 cubic meters, a base perimeter of 27.4 meters, and a height of 3.00 meters. Calculate the length and width of the unit.

    NOTE: Volume = height × base area; Perimeter = 2 × length + 2 × width.

8. A rectangular piece of sheet metal is rolled into a cylinder as shown in Figure 18–18.

    The height (h) of the open cylinder is 2.50 inches greater than its diameter (d). What is the diameter of the cylinder? Circumference (distance around) of the circle = π × diameter; $C = \pi d$; $\pi \approx 3.1416$.

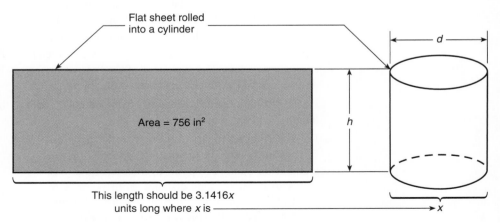

Figure 18–18

9. A parcel of land in the shape of a right triangle is shown in Figure 18–19. The parcel is laid out so that side b is 16 feet longer than twice the length of side a. The area of the parcel is 6,130 square feet. Determine the length of side c to the nearest foot.

Area of triangle $= \frac{1}{2}ab$; $c^2 = a^2 + b^2$.

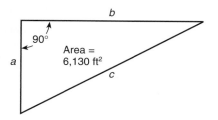

Figure 18–19

UNIT EXERCISE AND PROBLEM REVIEW

INCOMPLETE QUADRATIC EQUATIONS

Solve and check the following incomplete quadratic equations. Round the answers to 2 decimal places where necessary.

1. $x^2 = 64$

2. $x^2 = 75.68$

3. $4x^2 = 13.7$

4. $7x^2 = 42$

5. $x^2 + 3x^2 = 105$

6. $0.75x^2 + 0.03x^2 = 2.05$

7. $15.3x^2 - 12.9 = -9.77x^2 + 1.92$

8. $\dfrac{x}{9} = \dfrac{15}{x}$

9. $\dfrac{x + 3}{2} = \dfrac{8}{x - 3}$

10. $\dfrac{16.5 - x}{4.6} = \dfrac{12.5}{16.5 + x}$

11. $(x + 0.8)(x - 0.8) = 23.6$

12. $(x - 3.2)(x + 3.2) + 4.6x^2 = 14.1$

COMPLETE QUADRATIC EQUATIONS

Solve and check the following complete quadratic equations. Round the answers to 2 decimal places where necessary.

13. $8x^2 - 4x = 0$

14. $25x^2 - 12x = 13$

15. $10x = 24 - x^2$

16. $x^2 + 12x = -27$

17. $x^2 - 26x + 48 = 0$

18. $6x^2 + x - 12 = 0$

19. $x^2 - 0.75 = x$

20. $x^2 + 2x = -0.75$

21. $x^2 - 3.5x = -1.5$

22. $x^2 - 6x + 8 = 0$

23. $(x + 3)^2 = 9(x + 1)$

24. $x(x - 10) + 24 = 0$

25. $2x^2 + 7x - 13 = 0$

26. $3x^2 - 6x = -1$

27. $\dfrac{2}{x - 1} = \dfrac{x}{10}$

28. $x^2 = 1.7x + 0.6$

29. $x^2 + 0.2x - 0.8 = 0$

30. $1.6x - 12 = -0.16x^2$

COMPLETE QUADRATIC EQUATION NUMBER WORD PROBLEMS

Solve and check the following number word problems, which involve complete quadratic equations. The equations used in solving the problems are not given. Round the answers to 2 decimal places where necessary.

31. The sum of two numbers is 18, and three times their product is 231. What are the numbers?

32. One number is 12 less than a second number. If twice the product of the numbers increased by 7 is 35, what are the numbers?

33. The sum of two numbers is 23, and one-half of their product is 65. What are the numbers?

34. Four times the square of a number decreased by 3.20 equals three times the number increased by 1.84. What is the number?

35. The difference between two numbers is 6.2 and the sum of the squares of the numbers is 170.6. What are the numbers?

36. Three times the square of a number increased by 4.12 equals six times the number squared increased by 1.36 times the number. What is the number?

PRACTICAL APPLICATION PROBLEMS INVOLVING BOTH COMPLETE AND INCOMPLETE QUADRATIC EQUATIONS

Solve and check the following applied problems, which involve both complete and incomplete quadratic equations. Round the answers to 3 significant digits.

37. Compute the base diameter (d) of a cone, in inches. Use the formula $V = 0.2618d^2h$ where volume (V) = 97.6 cubic inches and height (h) = 5.30 inches.

38. Determine the time (t) in seconds that it takes an object thrown vertically upward at a velocity (v) of 86.5 feet per second to reach a height of 106 feet. Use the formula $h = vt - 16\,t^2$.

39. A rectangular piece is to be cut from a sheet of plywood. The length of the piece is to be 10.5 inches longer than the width, and the piece is to have an area of 11.8 square feet. Determine the length and width of the piece.

40. A rectangular parking lot is 272 feet long and 218 feet wide. The area of the parking lot is to be increased by 40%. Both the length and the width are to be increased by the same number of feet. Determine the number of feet that are to be added to both the length and width.

41. The center of gravity of a solid segment of a sphere is located using the formula $a = \dfrac{3(2r - h)^2}{4(3r - h)}$. The location of the center of gravity (a) from the center of the sphere is 35.3 centimeters. The sphere radius (r) is 51.5 centimeters. Determine the height (h) of the solid segment.

Fundamentals of Plane Geometry

UNIT 19 ▦ Introduction to Plane Geometry

O B J E C T I V E S

After studying this unit you should be able to

• identify axioms and postulates that apply to geometric statements.

• write geometric statements in symbol form.

• illustrate geometric statements.

Geometry is the branch of mathematics in which the properties of points, lines, surfaces, and solids are studied. Many geometric principles were first recognized by the Babylonians and Egyptians more than 5,000 years ago. The Egyptians used geometry for land surveying. This is the earliest known use of geometry. Throughout the centuries, geometry has been used in many ways that have greatly influenced modern living.

Much of the environment has been affected by the use of geometric principles. Practically everything in modern living depends on geometry. Geometric applications are used in building houses, apartments, offices, and shops where people live and work. Roads, bridges, and airports could not be constructed without the use of geometry. Automobiles, airplanes, and ships could not be designed and produced without the application of geometric principles. The manufacture of clothes and the processing and distribution of food depend on geometric applications.

Many occupations require a knowledge of geometry and the ability to apply this knowledge to practical on-the-job uses. Carpentry, plumbing, machining, drafting, and auto body repair are but a few of the occupations in which geometry is used regularly.

In addition to occupational uses, a knowledge of geometry is also of value in daily living. It is used, for example, to estimate the amount of paint or wallpaper required for a room, to determine the number of bags of fertilizer needed for a lawn, and to compute the number of feet of lumber needed for a home project.

19–1   Plane Geometry

Plane geometry deals with points, lines, and various figures that are made of combinations of points and line segments. The figures lie on a flat surface or *plane.* Examples of common plane figures that are discussed in this book are shown in Figure 19–1.

Since geometry is used in many occupational and nonoccupational applications, it is essential that the definitions and terms of geometry be understood. It is even more important to be able to apply geometric principles in problem solving. The methods and procedures used in problem solving are the same as those required in actual occupational situations.

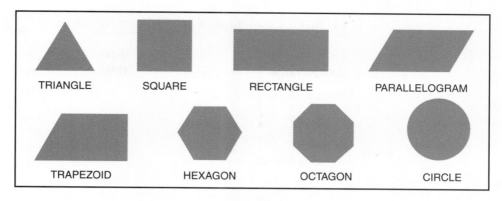

Figure 19–1

Procedure for Solving Geometry Problems

- Study the figure.
- Relate the figure to the principle or principles that are needed for the solution.
- Base all conclusions on given information and geometric principles.
- *Do not* assume that something is true because of its appearance or the way it is drawn.

19–2 Axioms and Postulates

In the study of plane geometry certain basic statements called axioms or postulates are accepted as true without requiring proof. Axioms or postulates may be compared to the rules of a game. Some axioms and postulates are listed. Others will be given as they are required for problem solving.

Quantities equal to the same quantities or to equal quantities are equal to each other.

EXAMPLE •————————————————————————————————

Refer to Figure 19–2.

Given: $a = 15$ cm

$d = 15$ cm

Conclusion: $a = d$

Figure 19–2

A quantity may be substituted for an equal quantity.

EXAMPLE •————————————————————————————————

Refer to Figure 19–2.

Given: $a = 15$ cm   Conclusion: $c = 15$ cm $+ 5$ cm

$b = 5$ cm $c = 20$ cm

$c = a + b$

If equals are added to equals, the sums are equal.

EXAMPLE •————————————————————————————————

Refer to Figure 19–2.

Given: $a = d$
            $b = e$

Conclusion: $a + b = d + e$
                  $c = f$

If equals are subtracted from equals, the remainders are equal.

EXAMPLE •————————————————————————————————

Refer to Figure 19–2.

Given: $c = f$
            $a = d$

Conclusion: $c - a = f - d$
                  $b = e$

If equals are multiplied by equals, the products are equal.

EXAMPLE •————————————————————————————————

Refer to Figure 19–3.

Given: $a = c$

Conclusion: $2a = 2c$
                  $b = d$

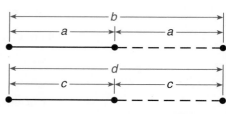

Figure 19–3

If equals are divided by equals, the quotients are equal.

EXAMPLE •————————————————————————————————

Refer to Figure 19–3.

Given: $2a = 2c$

Conclusion: $2a \div 2 = 2c \div 2$
                  $a = c$

The whole is equal to the sum of all its parts.

EXAMPLE

Refer to Figure 19–4.

$e = a + b + c + d$

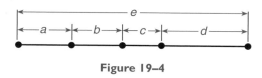

Figure 19–4

The whole is greater than any of its parts.

EXAMPLE

Refer to Figure 19–4.

The whole, e, is greater than a or b or c or d.

 One and only one straight line can be drawn between two given points.

EXAMPLE

In Figure 19–5, only one straight line can be drawn between point A and point B.

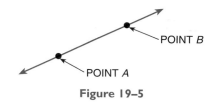

POINT B

POINT A

Figure 19–5

 Through a given point, one and only one line can be drawn parallel to a given straight line.

EXAMPLE

In Figure 19–6, lines a and b will never touch or cross.

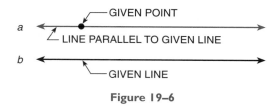

GIVEN POINT

a LINE PARALLEL TO GIVEN LINE

b GIVEN LINE

Figure 19–6

Two straight lines can intersect at only one point.

EXAMPLE •——————————————————————————————

In Figure 19–7, the lines cross at only one point.

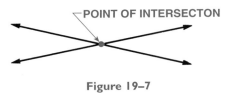

Figure 19–7

19–3   Points and Lines

Figure 19–8

Figure 19–9

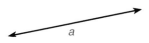

Figure 19–10

A *point* is shown as a dot. It is usually named by a capital letter, as shown in Figure 19–8. A point has no size or form; it has location only. For example, points on a map locate places; they do not show size or shape.

As used in this book, a *line* always means a straight line. A line other than a straight line, such as a curved line, is identified. A line extends without end in two directions. A line has no width; it is an infinite number of points. Arrowheads are used in drawing a line to show that there are no end points. A line is usually named by two points on the line. A double-headed arrow is placed over the letters that name the line, Figure 19–9. A line can also be named by a single lowercase letter, Figure 19–10.

A *curved line* is a line no part of which is straight, Figure 19–11.

A *line segment* is that part of a line that lies between two definite points, Figure 19–12. When we speak of a definite distance on a line we are dealing with a line segment. A line segment is usually referred to as simply a segment. Line segments are often named by placing a bar over the end point letters. Segment AB may be shown as $\overline{AB}$. In this book, segments are shown without a bar. Segment AB is shown as AB.

Parallel lines do not meet regardless of how far they are extended. They are the same distance apart (equidistant) at all points. The symbol $\parallel$ means parallel. In Figure 19–13 line AB is parallel to line CD ($\overleftrightarrow{AB} \parallel \overleftrightarrow{CD}$). Therefore, $\overleftrightarrow{AB}$ and $\overleftrightarrow{CD}$ are equidistant (distance x) at all points.

Figure 19–11   Figure 19–12   Figure 19–13

Perpendicular lines meet or intersect at a right or 90° angle. The symbol $\perp$ means perpendicular. Figure 19–14 shows examples of perpendicular lines. $\overleftrightarrow{AB} \perp \overleftrightarrow{CD}$ and $EF \perp EG$.

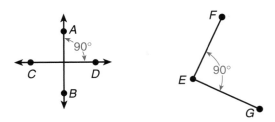

Figure 19–14

Oblique lines are neither parallel nor perpendicular. They meet or intersect at an angle other than 90°, Figure 19–15.

Figure 19–15

☷ UNIT EXERCISE AND PROBLEM REVIEW

1. Define geometry.
2. Name the kind of surface used in plane geometry.
3. Identify the postulate that applies to each of these statements.
   a. If $a = 5$ and $b = 5$, then $a = b$.
   b. If $EF = GH$, then $EF - KL = GH - KL$.
   c. Refer to Figure 19–16.
      $x = AB + BC + CD$

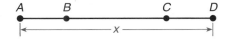

Figure 19–16

   d. If $m = p$, then $m + 8 = p + 8$.
   e. If $BC = DE$, then $15BC = 15DE$.
   f. If $e = AB + BC + CD$, then $e - g = AB + BC + CD - g$.
   g. If $HK - 4DE = 25$, and $2LM + ST = 25$, then $HK - 4DE = 2LM + ST$.
4. Write each statement using symbols.
   a. Segment BC is parallel to segment DE.
   b. Line FG is perpendicular to line HK.
   c. Segment AB is parallel to line CD.
5. Sketch each of the following statements.
   a. Line MP is parallel to line RS, and the distance between the two lines is represented by x.
   b. Segment AB is perpendicular to segment CD, and point C lies on segment AB.
   c. Oblique lines EF and GH intersect at point R.
   d. $\overleftrightarrow{AB} \perp CD$ and $\overleftrightarrow{AB}$ intersects CD at point M.
   e. $\overleftrightarrow{EF} \parallel \overleftrightarrow{GH}$ and $\overleftrightarrow{EF} \perp \overleftrightarrow{LM}$ at point P.

UNIT 20 ⦙⦙⦙ Angular Measure

After studying this unit you should be able to

- express degrees, minutes, and seconds as decimal degrees.
- express decimal degrees as degrees, minutes, and seconds.
- add, subtract, multiply, and divide angles given in degrees, minutes, and seconds.
- solve problems that require combinations of two or more arithmetic operations on angles.
- measure angles with a simple protractor.
- compute complements and supplements of angles.

The ability to compute and measure angles is required in a wide range of occupations. A plumber computes pipe lengths by making pipe diagrams using fitting angles of various degrees. A cabinetmaker determines angles and adjusts table saw miter gauges to ensure proper angular fits of stock. A land surveyor measures angles and distances to points from transit stations. A sheet metal technician computes and measures angles required in bending material.

Angular measure is required when working on many of our home projects and hobbies as well as for occupational uses.

An *angle* is a figure made by two lines that intersect. An *angle* is also described as the union of two rays having a common end point. The two rays are called the *sides* of the angle, and their common end point is called a *vertex*. A *ray* starts with an end point and continues indefinitely as shown in Figure 20–1. An example of an angle is shown in Figure 20–2. The size of an angle is determined by the number of degrees one side is rotated from the other. The size of an angle does *not* depend on the length of its sides.

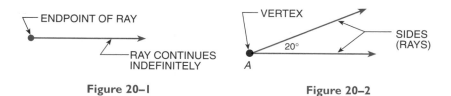

Figure 20–1   **Figure 20–2**

20–1   Units of Angular Measure

In order to measure the size of angles, the early Babylonians divided a circle into 360 parts. Each part was called one *degree*. The degree is still the basic unit of angular measure.

Angles are measured in degrees. The symbol for degree is °; the symbol for angle is ∠. The measure of an angle is often written with an "m" directly preceding the angle. For example, in Figure 20–2, the measure of angle A can be written as m∠A = 20°. In this book, the measure of an angle is shown without the m and with the degree symbol: ∠A = 20°.

A circle may be thought of as a ray with a fixed end point; the ray is rotated. One rotation makes a complete circle of 360° as shown in Figure 20–3.

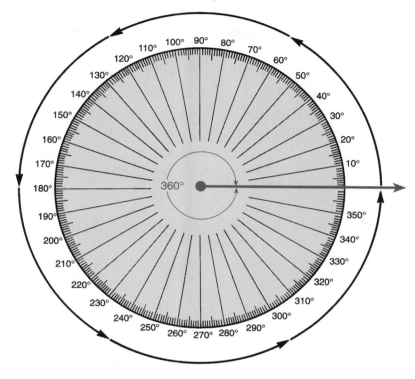

Figure 20–3

The degree of precision required in computing and measuring angles depends on how the angle is used. In woodworking, measuring to the nearest whole degree is often close enough. An automobile mechanic is required to make caster and camber adjustments to within $\frac{1}{2}$° or $\frac{1}{4}$°. Some manufactured parts are designed and processed to a very high degree of precision.

In metric calculations, the decimal degree is generally the preferred unit of measurement. In the customary system, angular measure is expressed in the following three ways.

1. As decimal degrees, such as 6.5 degrees and 108.274 degrees.

2. As fractional degrees, such as $12\frac{1}{4}$ degrees and $53\frac{1}{10}$ degrees.

3. As degrees, minutes, and seconds, such as 37 degrees, 18 minutes and 123 degrees, 46 minutes, 53 seconds.

Decimal and fractional degrees are added, subtracted, multiplied, and divided the same as any other numbers.

20–2 Units of Angular Measure in Degrees, Minutes, and Seconds

A degree is divided into 60 equal parts called *minutes.* The symbols for minute is ′. A minute is divided into 60 equal parts called *seconds.* The symbol for second is ″. The relationship among degrees, minutes, and seconds is shown in Figure 20–4.

1 Circle = 360 Degrees (°)	1 Degree (°) = $\frac{1}{360}$ of a Circle
1 Degree (°) = 60 Minutes (′)	1 Minute (′) = $\frac{1}{60}$ Degree (°)
1 Minute (′) = 60 Seconds (″)	1 Second (″) = $\frac{1}{60}$ Minute (′)

Figure 20–4

20–3   Expressing Degrees, Minutes, and Seconds as Decimal Degrees

The measure of an angle in the form of degrees, minutes, and seconds, such as $18°39'46''$, is sometimes expressed as decimal degrees. This is often the case when computations involve metric system units of measure.

Procedure for Expressing Degrees, Minutes, and Seconds as Decimal Degrees

- Divide the seconds by 60 to obtain the decimal minute.
- Add the decimal minute to the given number of minutes.
- Divide the sum of minutes by 60 to obtain the decimal degree.
- Add the decimal degree to the given number of degrees.

EXAMPLES ●

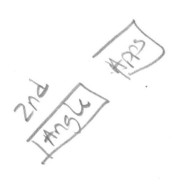

1. Express $78°43'$ as decimal degrees.

   Divide 43 minutes by 60 to obtain the decimal degree.

   $$43' \div 60 \approx 0.7167° \text{ (rounded)}$$

   Add the decimal degree ($0.7167°$) to the given degrees ($78°$).

   $$78° + 0.7167° \approx 78.7167° \text{ Ans}$$

2. Express $135°7'39''$ as decimal degrees.

   Divide 39 seconds by 60 to obtain the decimal minute.

   $$39'' \div 60 = 0.65'$$

   Add the decimal minute ($0.65'$) to the given minutes ($7'$).

   $$7' + 0.65' = 7.65'$$

   Divide the sum of minutes ($7.65'$) by 60 to obtain the decimal degree.

   $$7.65' \div 60 = 0.1275°$$

   Add the decimal degree to the given degrees.

   $$135° + 0.1275° = 135.1275° \text{ Ans}$$

20–4   Expressing Decimal Degrees as Degrees, Minutes, and Seconds

The measure of an angle given in the form of decimal degrees, such as $53.2763°$, must often be expressed as degrees, minutes, and seconds.

Procedure for Expressing Decimal Degrees as Degrees, Minutes, and Seconds

- Multiply the decimal part of the degrees by 60 minutes to obtain minutes.
- If the number of minutes obtained is not a whole number, multiply the decimal part of the minutes by 60 seconds to obtain seconds.
- Combine degrees, minutes, and seconds.

EXAMPLES ●

1. Express $53.45°$ as degrees and minutes.

   Multiply the decimal part of the degrees by 60 minutes to obtain minutes.

   $$0.45 \times 60' = 27'$$

   Combine degrees and minutes.

   $$53.45° = 53°27' \text{ Ans}$$

2. Express 28.2763° as degrees, minutes, and seconds.

Multiply the decimal part of the degrees by 60 minutes to obtain minutes.

$$0.2763 \times 60' = 16.578'$$

Multiply the decimal part of the minutes by 60 seconds to obtain seconds.

$0.578 \times 60'' = 34.68'' = 35''$ (rounded to the nearest whole second).

Combine degrees, minutes, and seconds.

$$28.2763° \approx 28°16'35''\ Ans$$

Calculator Applications

There are two basic formats used in degrees, minutes, seconds, and decimal degrees conversions. Depending on the make and model of your calculator, one of the two formats should apply.

1. Calculators with a $\boxed{°\,'\,''}$ key (degrees, minutes, seconds)

To convert degrees, minutes, seconds to decimal degrees:

Enter degrees, press $\boxed{°\,'\,''}$, enter minutes, press $\boxed{°\,'\,''}$, enter seconds, press $\boxed{°\,'\,''}$. The angle is directly displayed as decimal degrees.

EXAMPLE •────────────────────────────────────

Convert 53°47'25'' to decimal degrees.

53 $\boxed{°\,'\,''}$ 47 $\boxed{°\,'\,''}$ 25 $\boxed{°\,'\,''}$ → 53.79027778° *Ans*

On certain calculators, the execute key $\boxed{\text{EXE}}$ must be pressed last to convert display to decimal degrees.

────────────────────────────────────

To convert decimal degrees to degrees, minutes, seconds:

Enter decimal degrees, press $\boxed{\text{SHIFT}}$, press $\boxed{°\,'\,''}$.

NOTE: ← is the second function of the primary function key $\boxed{°\,'\,''}$.

The angle is directly displayed as degrees, minutes, seconds.

EXAMPLE •────────────────────────────────────

Convert 53.79027778° to degrees, minutes, seconds.

53.79027778 $\boxed{\text{SHIFT}}$ $\boxed{°\,'\,''}$ → 53°47'25'' *Ans*

or the execute key $\boxed{\text{EXE}}$ must be pressed directly after the angle is entered.

NOTE: Your calculator may display 50°47°25, which is interpreted as 50°47'25''

EXAMPLE •────────────────────────────────────

Convert 53.79027778° to degrees, minutes, seconds.

53.79027778 $\boxed{\text{EXE}}$ $\boxed{\text{SHIFT}}$ $\boxed{°\,'\,''}$ → 53°47'25'' *Ans*

────────────────────────────────────

2. Calculators with ▶ DD (decimal degrees) and ▶ DMS (degrees, minutes, seconds) functions with $\boxed{\text{2nd}}$ and $\boxed{\text{3rd}}$ function keys

NOTES: ▶ DD is the second function of the $\boxed{\blacktriangleright\underline{\text{DD}}}$ key and
   ▶ DMS is the third function of the $\boxed{\blacktriangleright\underline{\text{DD}}}$ key.

With certain model calculators without the $\boxed{\text{3rd}}$ function key, the procedures will be somewhat different than the procedures given. If so, refer to your user's guide or manual.

On some calculators, such as graphing calculators, to get the ▶ DMS command, first press the [2nd] key and then either [∠] or [ANGLE].

To convert degrees, minutes, seconds to decimal degrees:

Enter degrees, press [·], enter minutes, enter seconds, press [2nd], press [▶DD].

EXAMPLE ●

Convert 53°47′25″ to decimal degrees.

53 [·] 4725 [2nd] [▶DD] → 53.79027778° *Ans*

NOTE: When minutes and seconds have only one digit, enter zeros to position the digits in their proper place positions.

EXAMPLE ●

Convert 7°8′4″ to decimal degrees.

7 [·] 0804 [2nd] [▶DD] → 7.134444444 *Ans*

To convert decimal degrees to degrees, minutes, seconds:

Enter decimal degrees, press [3rd], press [▶DD].

[▶DMS] degrees, minutes, seconds (3rd function)

EXAMPLE ●

Convert 53.79027778° to degrees, minutes, seconds.

53.79027778 [3rd] [▶DD] → 53°47′25″0, 53°47′25″ *Ans*

EXERCISE 20–4

Express the following degrees and minutes as decimal degrees. Round the answers to 2 decimal places where necessary.

1. 15°30′	**4.** 67°23′	**7.** 2°59′
2. 78°45′	**5.** 105°47′	**8.** 0°15′
3. 59°12′	**6.** 96°1′	**9.** 256°19′

Express the following degrees, minutes, and seconds as decimal degrees. Round the answers to 4 decimal places where necessary.

10. 12°15′45″	**13.** 107°18′24″	**16.** 218°27′16″
11. 7°8′30″	**14.** 66°59′17″	**17.** 312°59′57″
12. 96°0′15″	**15.** 1°1′1″	**18.** 0°41′6″

Express the following decimal degrees as degrees and minutes. Round the answer to the nearest minute where necessary.

19. 7.75°	**22.** 281.92°	**25.** 44.616°
20. 96.25°	**23.** 113.285°	**26.** 307.031°
21. 10.06°	**24.** 0.913°	**27.** 26.009°

Express the following decimal degrees as degrees, minutes, and seconds. Round the answers to the nearest second where necessary.

28. 48.8610°

29. 123.0635°

30. 7.6678°

31. 0.5383°

32. 216.6079°

33. 44.0866° *4*

34. 7.36081°

35. 406.93058°

36. 0.00074°

20–5   Arithmetic Operations on Angular Measure in Degrees, Minutes, and Seconds

The division of minutes and seconds permits very precise computations and measurements. For example, precise measurements are made in land surveying. In machining operations, dimensions at times are computed to seconds in order to ensure the proper functioning of parts.

When computing with degrees, minutes, and seconds, it is sometimes necessary to exchange units. To exchange units, keep in mind that 1 degree equals 60 minutes and 1 minute equals 60 seconds. The following examples illustrate adding, subtracting, multiplying, and dividing angles in degrees, minutes, and seconds.

Performing arithmetic operations with degrees, minutes, and seconds computed with a calculator requires entering an angle as degrees, minutes, and seconds, converting to decimal degrees, and converting back to degrees, minutes, and seconds. Calculator applications for each of the arithmetic operations are shown.

Adding Angles

EXAMPLES •————————————————————————————————

1. Refer to Figure 20–5 and determine ∠1.

  ∠1 = 16°35′ + 57°16′

                 16°35′
              + 57°16′
              ──────────
                 73°51′ *Ans*

TI 30

✱ Change to DD each time, add, the change back DMS

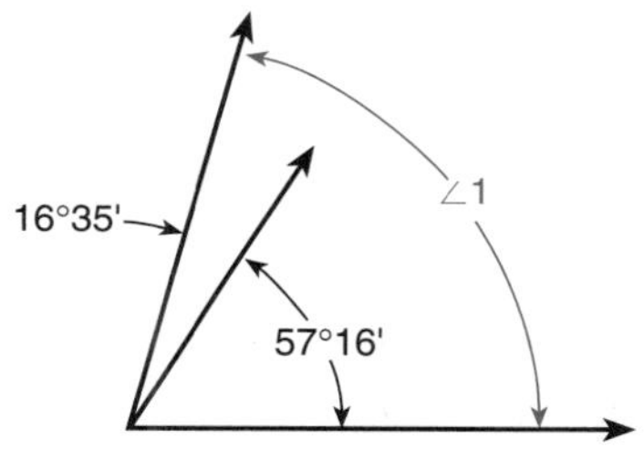

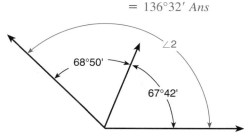

Figure 20–5

2. Refer to Figure 20–6 and determine ∠2.

  ∠2 = 68°50′ + 67°42′

  Express 92′ as degrees and minutes.
  92′ = 60′ + 32′ = 1°32′

  Add.
  135° + 1°32′

                 68°50′
              + 67°42′
              ──────────
                135°92′

Add the extra to degrees

                                      = 136°32′ *Ans*

Figure 20–6

3. Refer to Figure 20–7 and determine ∠3.

∠3 = 59°43′35″ + 77°31′48″

Express 83″ as minutes and seconds.
83″ = 60″ + 23″ = 1′23″

$$\begin{array}{r} 59°43′35″ \\ +\ 77°31′48″ \\ \hline 136°74′83″ \end{array}$$

Add.
136°74′ + 1′23″   = 136°75′23″

Express 75′ as degrees and minutes.
75′ = 60′ + 15′ = 1°15′

Add.
136°0′23″ + 1°15′   = 137°15′23″ *Ans*

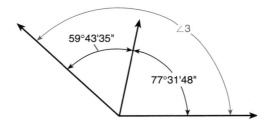

Figure 20–7

Calculator Application

∠3 = 59°43′35″ + 77°31′48″

or EXE

59 [°′″] 43 [°′″] 35 [°′″] [+] 77 [°′″] 31 [°′″] 48 [°′″] [=] [SHIFT] [°′″] → 137°15′23″ *Ans*

or

59.4335 [2nd] [▶DD] [+] 77.3148 [2nd] [▶DD] [▶DD] [3rd] [▶DD] → 137°15′23″ *Ans*

Subtracting Angles

EXAMPLES •

1. Refer to Figure 20–8 and determine ∠1.

∠1 = 134°53′46″ − 84°22′19″

$$\begin{array}{r} 134°53′46″ \\ -\ 84°22′19″ \\ \hline 50°31′27″\ Ans \end{array}$$

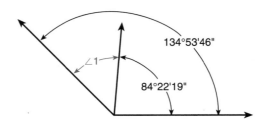

Figure 20–8

2. Refer to Figure 20–9 and determine ∠2.

∠2 = 79°15′ − 31°46′

Since 46′ cannot be subtracted from
15′, 1° is exchanged for 60 minutes.

$$\begin{array}{r} 79°15′ \\ -\ 31°46′ \end{array}$$

$$\begin{array}{r} 79°15′ = 78°75′ \\ -\ 31°46′ = 31°46′ \\ \hline 47°29′\ Ans \end{array}$$

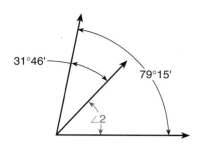

Figure 20-9

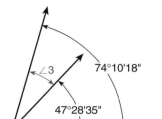

Figure 20-10

3. Refer to Figure 20–10 and determine ∠3.

$$\angle 3 = 74°10'18'' - 47°28'35''$$

Since 28′ cannot be subtracted from 10′, and 35″ cannot be subtracted from 18″, units are exchanged.

$$
\begin{array}{r}
74°10'18'' \\
-\ 47°28'35''
\end{array}
$$

Borrow to help

$$
\begin{array}{rcccl}
74°10'18'' & = & 73°70'18'' & = & 73°69'78'' \\
-\ 47°28'35'' & = & 47°28'35'' & = & 47°28'35'' \\
\hline
 & & & & 26°41'43''\ Ans
\end{array}
$$

Calculator Application

$$\angle 3 = 74°10'18'' - 47°28'35''$$

74 $\boxed{°'''}$ 10 $\boxed{°'''}$ 18 $\boxed{°'''}$ $\boxed{-}$ 47 $\boxed{°'''}$ 28 $\boxed{°'''}$ 35 $\boxed{°'''}$ $\boxed{=}$ ⌐ or $\boxed{EXE}$ $\boxed{SHIFT}$ $\boxed{°'''}$ → 26°41'43″ *Ans*

or

74.1018 $\boxed{2nd}$ $\boxed{▶DD}$ $\boxed{-}$ 47.2835 $\boxed{2nd}$ $\boxed{▶DD}$ $\boxed{▶DD}$ $\boxed{3rd}$ $\boxed{▶DD}$ → 26°41'43″ *Ans*

EXERCISE 20–5A

These exercises require the addition of angles. Determine the value of each.

1–35 odd

1. $23°$
 $+59°$

2. $128°$
 $+\ 43°$

3. $17°18'$
 $+\ 9°31'$

4. $27°44'$
 $+88°15'$

5. $68°34'$
 $+17°39'$

6. $21°50'$
 $+\ 7°17'$

7. $42°12'14''$
 $+33°17'15''$

8. $70°19'55''$
 $+42°\ 0'\ 2''$

9. $5°36'11''$
 $+18°51'16''$

10. $41°13'20''$
 $+19°58'12''$

11. $108°48'28''$
 $+\ 24°\ 0'47''$

12. $67°14'29''$
 $+13°56'44''$

13. $90°54'33'' + 11°17'33''$ s

14. $23°20'14'' + 31°19'22'' + 14°36'49''$

15. $51°19'28'' + 0°43'27'' + 12°9' + 33°0'14''$

These exercises require the subtraction of angles. Determine the value of each.

16.   67°
           −28°

17.   116°
           − 99°

18.   −74°27′
           −67°16′

19.   48°17′
           −32° 9′

20.   102°16′
           −100°54′

21.   61°41′
           − 7°47′

22.   88°
           −22°31′

23.   44°14′54″
           −41°27′13″

24.   31° 0′12″
           −27°28′ 4″

25.   70° 1′3″
           −66°59′2″

26.   120°17′44″
           −112°48′53″

27.   6°16′
           −4°18′19″

28. 87°0′4″ − 73°11′23″

29. 55°30′29″ − 23°32′50″

30. 68° − 67°59′59″

These problems require the addition or subtraction of angles. Determine the value of each.

31. Refer to Figure 20–11 and determine
∠1.

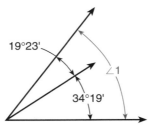

Figure 20–11

32. Refer to Figure 20–12 and determine
∠2.

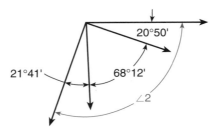

Figure 20–12

33. Refer to Figure 20–13 and determine
∠3.

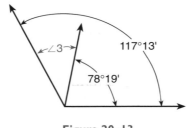

Figure 20–13

34. Refer to Figure 20–14 and determine
∠4.

Figure 20–14

35. Refer to Figure 20–15 and determine the
value of ∠5 − ∠6.

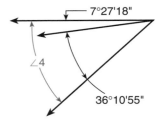

Figure 20–15

36. Refer to Figure 20–16 and determine the
value of ∠1.

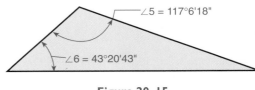

Figure 20–16

Multiplying Angles

EXAMPLES

1. Five holes are drilled on a circle as shown in Figure 20–17. The angular measure between two consecutive holes is 32°18′. Determine the angular measure, ∠1, between hole 1 and hole 5.

∠1 = 4 × (32°18′)   32°18′

Express 72′ as degrees and minutes.   ×   4
72′ = 60′ + 12′ = 1°12′   128°72′

Add.
128° + 1°12′   = 129°12′ *Ans*

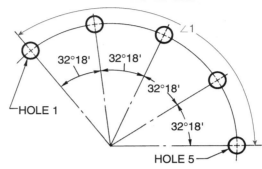

Figure 20–17

2. Refer to Figure 20–18 and determine ∠2 when ∠a = 43°28′45″.

∠2 = 5 × (43°28′45″)   43°28′45″

Express 225″ as minutes and seconds.   ×   5
225″ = 3 × 60″ + 45″ = 3′45″   215°140′225″

Add.
215°140′ + 3′45″   = 215°143′45″

Express 143′ as degrees and minutes.
143′ = 2 × 60′ + 23′ = 2°23′

Add.
215°0′45″ + 2°23′   = 217°23′45″ *Ans*

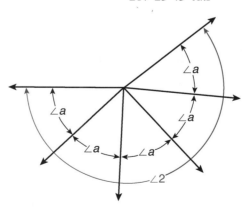

Figure 20–18

Calculator Application

∠2 = 5 × (43°28′45″)

or EXE

5 × 43 °′″ 28 °′″ 45 °′″ = SHIFT °′″ → 217°23′45″ *Ans*

or

5 × 43.2845 2nd ▶DD ▶DD 3rd ▶DD → 217°23′45″ *Ans*

Dividing Angles

EXAMPLES •

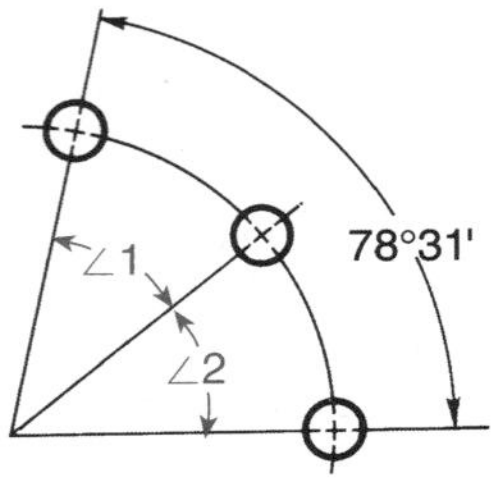

Figure 20–19

1. In Figure 20–19, $\angle 1$ equals $\angle 2$. Determine the value of $\angle 1$.

$\angle 1 = (78°31') \div 2$

Express 15.5 minutes as minutes and seconds.

$0.5 \times 60'' = 30''$

Add.

$39°15'' + 30''$

$= 39°15'30''$ *Ans*

$$\begin{array}{r} 39°15.5' \\ 2\overline{)78°31'} \end{array}$$

2. Refer to Figure 20–20. If $\angle 1$, $\angle 2$, and $\angle 3$ are equal, determine the value of each of these angles.

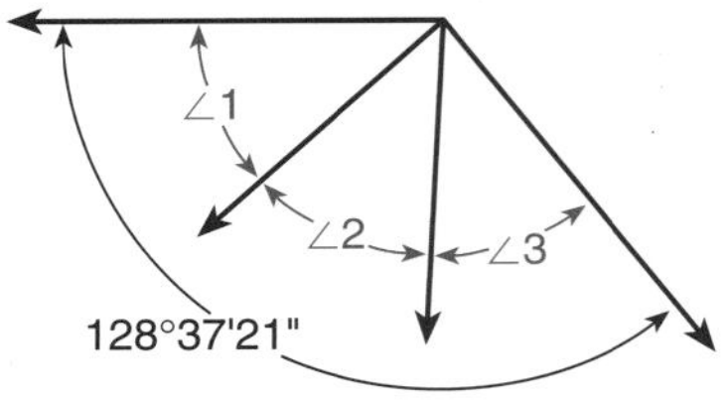

(figure shows angles ∠1, ∠2, ∠3 with 128°37'21")

Figure 20–20

$\angle 1 = \angle 2 = \angle 3 = (128°37'21'') \div 3$ $3\overline{)128°37'21''}$

Divide 128° by 3.

$$\begin{array}{r} 42° \\ 3\overline{)128°} \\ \underline{126°} \\ 2° \end{array}$$

Add the remainder of 2° to the 37'.
$2° = 120'$
$120' + 37' = 157'$

Divide 157' by 3.

$$\begin{array}{r} 52' \\ 3\overline{)157'} \\ \underline{156'} \\ 1' \end{array}$$

Add the remainder of 1' to the 21''.
$1' = 60''$

$60'' + 21'' = 81''$

Divide 81'' by 3.

Combine.

$$\begin{array}{r} 27'' \\ 3\overline{)81''} \end{array}$$

$42°52'27''$ *Ans*

Calculator Application

$\angle 1 = \angle 2 = \angle 3 = (128°37'21'') \div 3$

                                        or $\boxed{EXE}$

$128 \boxed{°'''} 37 \boxed{°'''} 21 \boxed{°'''} \div 3 \boxed{=} \boxed{SHIFT} \boxed{°'''} \rightarrow 42°52'27''$ *Ans*

or

$128.3721 \boxed{2nd} \boxed{\blacktriangleright DD} \div 3 \boxed{\blacktriangleright DD} \boxed{3rd} \boxed{\blacktriangleright DD} \rightarrow 42°52'27''$ *Ans*

(handwritten annotations:) * Change to DD 1st then divide by 3 then change back DMS

(handwritten:) 128.3721 2nd ⊞ ÷ 3   2nd ⊟

EXERCISE 20–5B

These exercises require multiplication of angles. Determine the value of each.

1. $6 \times 18°$ **3.** $2 \times (15°19')$ **5.** $2 \times (36°54')$

2. $14 \times 9°$ **4.** $5 \times (21°7')$ **6.** $4 \times (42°23')$

(handwritten:) # 7-15 All   #21-29

7. 9 × (6°14′) **10.** 5 × (15°11′8″) **13.** 4 × (10°21′12″)

8. 3 × (20°14′10″) **11.** 3 × (36°14′27″) **14.** 8 × (9°23′15″)

9. 12 × (7°2′4″) **12.** 7 × (2°8′12″) **15.** 6 × (45°52′49″)

These exercises require division of angles. Determine the value of each.

16. 56° ÷ 2 **21.** (46°12′) ÷ 4 **26.** (84°32′26″) ÷ 9

17. (27°18′) ÷ 3 **22.** (123°43′) ÷ 5 **27.** (278°5′57″) ÷ 3

18. (51°30′) ÷ 5 **23.** (56°42′21″) ÷ 7 **28.** (0°59′42″) ÷ 6

19. (73°8′) ÷ 4 **24.** (132°12′48″) ÷ 12 **29.** (333°5′30″) ÷ 15

20. (19°3′) ÷ 6 **25.** (97°30′50″) ÷ 2 **30.** (116°49′48″) ÷ 9

These problems require multiplication or division of angles. Some problems may also require addition or subtraction of angles. Determine the value of each.

31. Refer to Figure 20–21 and determine ∠1 when ∠A = 32°43′.

33. Refer to Figure 20–23 and determine ∠C.

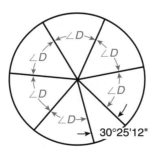

Figure 20–21

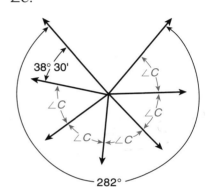

Figure 20–23

32. Refer to Figure 20–22 and determine ∠D.

34. Refer to Figure 20–24 and determine ∠B.

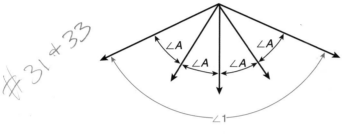

Figure 20–22

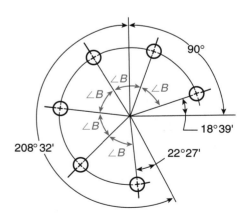

Figure 20–24

20–6 Simple Semicircular Protractor

Protractors are used for measuring, drawing, and laying out angles. Various types of protractors are available, such as the simple semicircular protractor, swinging blade protractor, and bevel protractor. The type of protractor used depends on its application and the degree of precision required. Protractors have wide occupational use, particularly in the metal and woodworking trades.

A simple semicircular protractor has two scales, each graduated from 0° to 180° so that it can be read from either the left or right side. The vertex of the angle to be measured or drawn is located at the center of the base of the protractor. A simple semicircular protractor is shown in Figure 20–25.

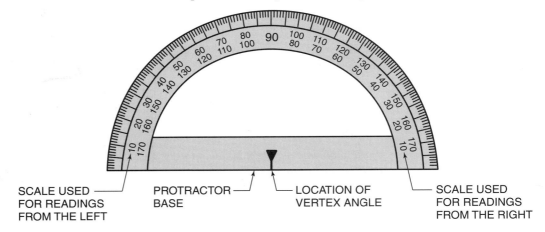

SCALE USED
FOR READINGS
FROM THE LEFT

PROTRACTOR
BASE

LOCATION OF
VERTEX ANGLE

SCALE USED
FOR READINGS
FROM THE RIGHT

Figure 20–25

Procedure for Measuring an Angle

• Place the protractor base on one side of the angle with the protractor center on the angle vertex.

• If the angle rotates from the right, choose the scale that has the zero-degree reading on the right side of the protractor. If the angle rotates from the left, choose the scale that has the zero-degree reading on the left side of the protractor. Read the measurement where the side crosses the protractor scale.

EXAMPLE •——

Measure ∠1 in Figure 20–26.

Extend the sides *OA* and *OB* of ∠1 as shown.

Place the protractor base on side *OB* with the protractor center on the angle vertex, point *O*.

Angle 1 is rotated from the right. The angle measurement is read from the inside scale, since the inside scale has a zero-degree (0°) reading on the right side of the protractor base. Read the measurement where the extension of side *OA* crosses the protractor scale. Angle 1 = 40° *Ans*

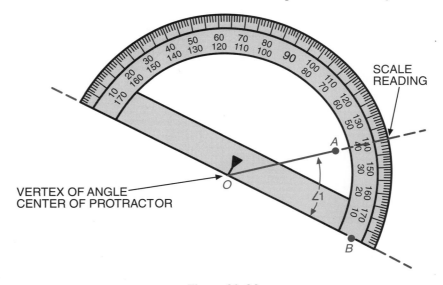

SCALE
READING

VERTEX OF ANGLE
CENTER OF PROTRACTOR

Figure 20–26

EXERCISE 20–6

1. Write the values of angles *A* through *J* shown in Figure 20–27.

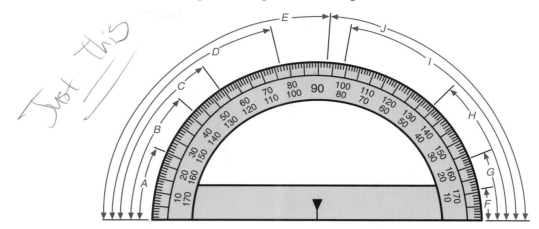

Just this

Figure 20–27

Use a protractor to measure the angles in Exercises 2 through 7.

2.

Figure 20–28

3.

Figure 20–29

4.

Figure 20–30

5.

Figure 20–31

6.

Figure 20–32

7.

Figure 20–33

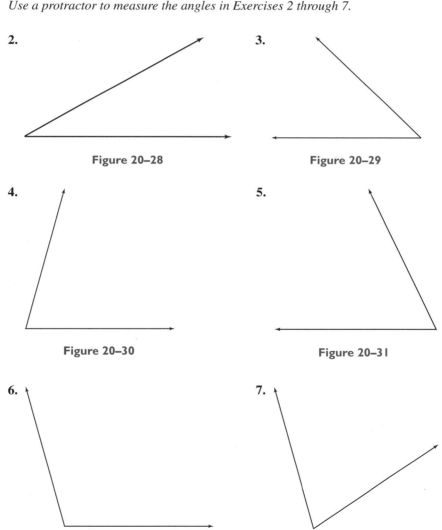

Class
No
Homework
#8-12
All

Without using a protractor, estimate the correct size of each angle in Exercises 8 through 13.

8. The size of the angle in Figure 20–34 is

   a. 30°

   b. 45°

   c. 60°

   d. 120°

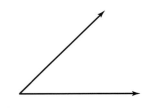

Figure 20–34

9. The size of the angle in Figure 20–35 is

   a. 30°

   b. 45°

   c. 60°

   d. 120°

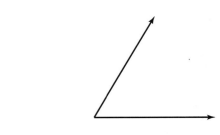

Figure 20–35

10. The size of the angle in Figure 20–36 is

   a. 90°

   b. 135°

   c. 150°

   d. 180°

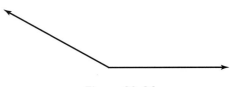

Figure 20–36

11. The size of the angle in Figure 20–37 is

   a. 90°

   b. 135°

   c. 150°

   d. 180°

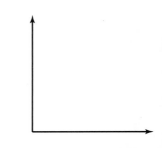

Figure 20–37

12. The size of the angle in Figure 20–38 is about:

Figure 20–38

13. The size of the angle in Figure 20–39 is about:

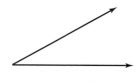

Figure 20–39

20–7 Complements and Supplements of Scale Readings

When measuring with a bevel protractor, the user must determine whether the desired angle of the object measured is the actual reading on the protractor or the complement or supplement of the protractor reading. Particular caution must be taken when measuring angles close to 45° and 90°.

Two angles are *complementary* when their sum is 90°. For example, in Figure 20–40, 42° + 48° = 90°. Therefore, 42° is the complement of 48°, and 48° is the complement of 42°.

Two angles are *supplementary* when their sum is 180°. For example, in Figure 20–41, 87° + 93° = 180°. Therefore, 87° is the supplement of 93°, and 93° is the supplement of 87°.

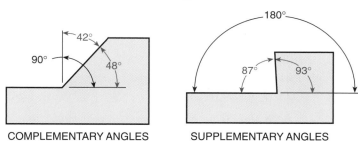

COMPLEMENTARY ANGLES

Figure 20–40

SUPPLEMENTARY ANGLES

Figure 20–41

EXERCISE 20–7

Write the complements of these angles.

1. 59°	**5.** 56°18′	**9.** 49°0′58″	
2. 18°	**6.** 44°59′	**10.** 89°59′59″	
3. 46°	**7.** 73°18′27″		
4. 1°	**8.** 0°43′19″		

Write the supplements of these angles.

11. 84°	**15.** 65°36′	**19.** 90°1′2″
12. 19°	**16.** 179°59′	**20.** 0°3′12″
13. 97°	**17.** 3°37′18″	
14. 129°	**18.** 89°57′50″	

☷ UNIT EXERCISE AND PROBLEM REVIEW

ADDING AND SUBTRACTING ANGLES

Add or subtract each of these exercises as indicated.

1. 34°
     + 97°

2. 14°27′
     + 66°15′

3.   43°38'
      + 16°51'

4.   50°19'20"
      + 28°30'14"

5. 5°43'18" + 13°40'26"

6. 17°27'53" + 92°56'27"

7. 20°34'19" + 44°0'27" + 16°54'49"

8.   94°
       − 67°

9.   68°26'
       − 54°17'

10.   74°19'
        − 68°34'

11.   53°41'18"
        − 17°22' 7"

12. 127°17'51" − 96°41'12"

13. 55°9'18" − 51°57'30"

14. 64°0'8" − 9°41'17"

15. 90° − 59°58'57"

MULTIPLYING AND DIVIDING ANGLES

Multiply or divide each of these exercises as indicated.

16. 12 × 23°

17. 6 × (18°7')

18. 3 × (27°19')

19. 3 × (53°19'12")

20. 5 × (34°15'6")

21. 9 × (12°5'23")

22. 7 × (28°16'25")

23. 107° ÷ 4

24. (35°42') ÷ 7

25. (53°28') ÷ 5

26. (192°56'40") ÷ 8

27. (91°42'36") ÷ 3

28. (103°44'18") ÷ 6

29. (47°18'27") ÷ 9

Each of these problems requires two or more arithmetic operations on angles for its solution. Determine the value of each.

30. Refer to Figure 20–42. Determine the value of ∠4 if ∠1 + ∠2 + ∠3 + ∠4 = 360°.

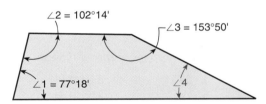

Figure 20–42

31. Refer to Figure 20–43. Determine ∠C when ∠B = 51°17'.

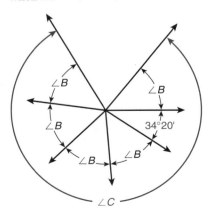

Figure 20–43

32. Refer to Figure 20–44. Determine ∠A and ∠B if ∠A = 3∠B.

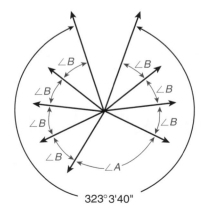

Figure 20–44

33. Refer to Figure 20–45. The sum of all angles = 540°. ∠1 = ∠5 = 119°53'. ∠2 = ∠4 = 29°16'. Determine ∠3.

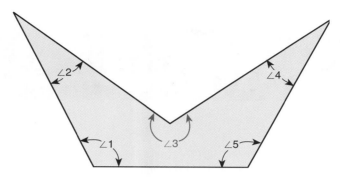

Figure 20–45

34. Refer to Figure 20–46. The sum of all angles = 720°. ∠1 = ∠3 = ∠4 = ∠6. ∠2 = ∠5 = 280°24'10". Determine ∠1.

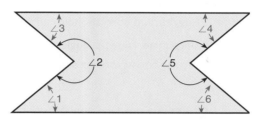

Figure 20–46

CONVERTING BETWEEN DECIMAL DEGREES AND DEGREES, MINUTES, AND SECONDS

Express the following degrees, minutes, and seconds as decimal degrees. Where necessary, round the answers to 4 decimal places.

35. 18°30'0"   **38.** 3°54'0"   **41.** 214°7'26"

36. 57°18'0"   **39.** 67°54'43"   **42.** 99°59'59"

37. 109°48'0"   **40.** 2°0'17"   **43.** 49°23'48"

Express the following decimal degrees as degrees, minutes, and seconds. Where necessary, round the answers to the nearest second.

44. 66.25°   **47.** 80.923°   **50.** 0.0531°

45. 207.75°   **48.** 117.063°   **51.** 196.5104°

46. 4.125°   **49.** 19.9775°   **52.** 84.0004°

ANGLE COMPLEMENTS AND SUPPLEMENTS

Write the complements of these angles.

53. 63°   **54.** 27°19'   **55.** 86°19'48"   **56.** 45°2'7"   **57.** 44°3'4"

Write the supplements of these angles.

58. 78°   **59.** 109°56'   **60.** 89°13'32"   **61.** 178°9'21"   **62.** 2°0'59"

UNIT 21 ⁝ Angular Geometric Principles

OBJECTIVES

After studying this unit you should be able to

- identify different types of angles.

- identify pairs of adjacent, alternate interior, corresponding, and vertical angles.

- determine values of angles in geometric figures, applying theorems of opposite, alternate interior, and corresponding angles.

- determine values of angles in geometric figures, applying theorems of parallel and perpendicular corresponding sides.

Solving a practical application may require working with a number of different angles. To avoid confusion, angles must be properly named and their types identified. Determination of required unknown angular and linear dimensions is often based on the knowledge and understanding of angular geometric principles and their practical applications.

21–1   Naming Angles

Angles are named by a number, a letter, or three letters. When an angle is named with three letters, the vertex must be the middle letter. For example, the angle shown in Figure 21–1 can be called $\angle 1$, $\angle C$, $\angle ACB$, or $\angle BCA$.

In cases where a point is the vertex of more than one angle, a single letter cannot be used to name an angle. In Figure 21–2, the single letter D cannot be used in naming an angle, since point D is the vertex of three different angles. The three angles are named $\angle 1$, $\angle FDG$, or $\angle GDF$; $\angle 2$, $\angle FDE$, or $\angle EDF$; and $\angle 3$, $\angle GDE$, or $\angle EDG$.

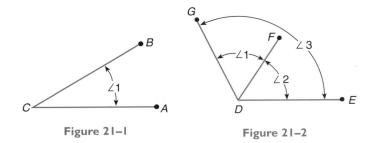

Figure 21–1   Figure 21–2

21–2   Types of Angles

The following five terms are used to describe the relative size of an angle.

An *acute angle* is an angle that is less than 90°. Angle 1 in Figure 21–3 is acute.

A *right angle* is an angle of 90°. Angle A in Figure 21–4 is a right angle.

An *obtuse angle* is an angle greater than 90° and less than 180°. Angle ABC in Figure 21–5 is an obtuse angle.

A *straight angle* is an angle of 180°. A straight line is a straight angle. Line segment *DBC* in Figure 21–5 is a straight angle.

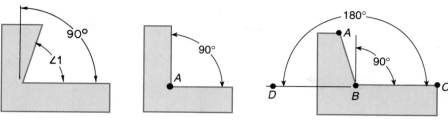

Figure 21–3   Figure 21–4   Figure 21–5

A *reflex angle* is an angle greater than 180° and less than 360°. For example, in Figure 21–6, the clock shows a time change from 10:00 to 10:40. The minute hand moves from the 12 to the 8. Since the minute hand moves through 30° (360° ÷ 12 = 30°) for each 5 minutes of time, it moves through 8 × 30° or 240°. The 240° moved by the minute hand is a reflex angle.

Two angles are *adjacent* if they have a common vertex and a common side. Angle 1 and angle 2 in Figure 21–7 are adjacent since they both contain the common vertex *B* and the common side *BC*.

Found out the minutes then multiply by 6

Each minute = 6°
Every 5 minute = 30°

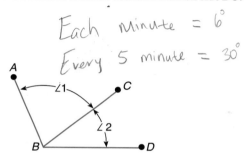

Figure 21–6   Figure 21–7

21–3   Angles Formed by a Transversal

A *transversal* is a line that intersects (cuts) two or more lines. Line *EF* in Figures 21–8 and 21–9 is a transversal since it cuts lines *AB* and *CD*.

Alternate interior angles are pairs of interior angles on opposite sides of a transversal. The angles have different vertices. In Figure 21–8, angles 1 and 4 and angles 2 and 3 are pairs of alternate interior angles.

Corresponding angles are pairs of angles, one interior and one exterior. Both angles are on the same side of the transversal with different vertices. In Figure 21–9, angles 1 and 5, 2 and 6, 3 and 7, and 4 and 8 are pairs of corresponding angles.

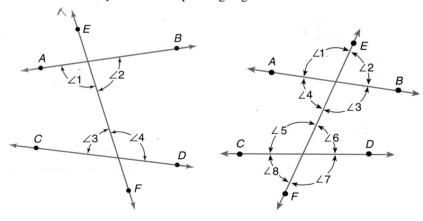

Figure 21–8   Figure 21–9

EXERCISE 21–3

Name each of the angles in Figure 21–10 in three additional ways.

1. ∠1
2. ∠2
3. ∠D
4. ∠DEF
5. ∠F

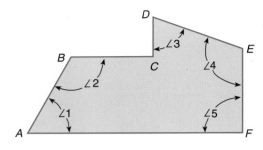

Figure 21–10

Name each of the angles in Figure 21–11 in two additional ways.

6. ∠1
7. ∠CBF
8. ∠3
9. ∠ECB
10. ∠5
11. ∠BCD

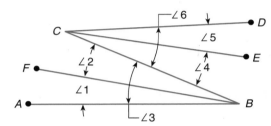

Figure 21–11

Identify each of the angles in Figure 21–12 as acute, right, obtuse, straight, or reflex.

12. ∠B
13. ∠ACB
14. ∠CDE
15. ∠BAF
16. ∠ABC
17. ∠BCD
18. ∠1
19. ∠EFG
20. ∠EFA
21. ∠EDF

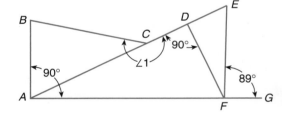

Figure 21–12

For each of the time changes in problems 22 through 30:

   a. Determine the number of degrees through which the minute hand moves.

   b. Write the type of angle (acute, right, obtuse, straight, or reflex) made by the movement of the minute hand.

22. from 8:00 to 8:25
23. from 11:00 to 11:55
24. from 6:30 to 6:45
25. from 5:50 to 6:20
26. from 3:15 to 4:05

27. from 4:00 to 4:12
28. from 2:22 to 2:54
29. from 7:18 to 7:48
30. from 1:39 to 2:26

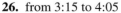

31. Name all pairs of adjacent angles shown in Figure 21–13.

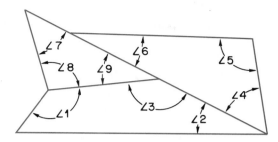

Figure 21–13

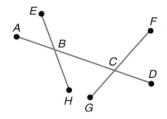

Figure 21–14

Refer to Figure 21–14 for problems 32 through 33.

32. Name all pairs of alternate interior angles.

33. Name all pairs of corresponding angles.

21–4   Theorems and Corollaries

Based on postulates, other statements about points, lines, and planes can be proved. A statement in geometry that can be proved is called a *theorem.* Theorems that are proved can then be used with postulates and definitions in proving other theorems.

A *corollary* is a statement based on a theorem. A corollary is often a special case of the theorem that it follows, and can be proved by applying the theorem.

In this book, theorems and corollaries are not proved. They are used as the geometric rules for problem solving. The theorems and corollaries in this unit appear in bold print in colored boxes.

NOTE: Angles that are referred to or shown as equal are angles of equal measure. For example, $\angle A = \angle B$ means m$\angle A$ = m$\angle B$. Line segments that are referred to or shown as equal are segments of equal length. For example, $AB = CD$ means length AB = length CD.

 If two lines intersect, the opposite, or vertical, angles are equal.

EXAMPLE •

Refer to Figure 21–15.

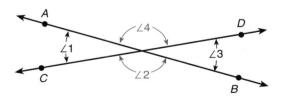

Figure 21–15

Given: Lines $\overleftrightarrow{AB}$ and $\overleftrightarrow{CD}$ intersect, and angles 1 and 3 and angles 2 and 4 are pairs of opposite or vertical angles.

Conclusion: $\angle 1 = \angle 3$ and $\angle 2 = \angle 4$

If two parallel lines are intersected by a transversal, the alternate interior angles are equal.

EXAMPLE •————————————————————————————————

Refer to Figure 21–16.

   Given: $\overleftrightarrow{AB} \parallel \overleftrightarrow{CD}$

   Conclusion: $\angle 1 = \angle 4$ and $\angle 2 = \angle 3$

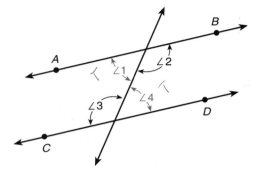

Figure 21–16

If two lines are intersected by a transversal and a pair of alternate interior angles are equal, the lines are parallel.

EXAMPLE •————————————————————————————————

Refer to Figure 21–17.

   Given: $\angle 1 = \angle 2$
   Conclusion: $\overleftrightarrow{AB} \parallel \overleftrightarrow{CD}$

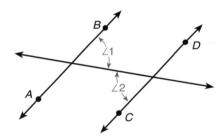

Figure 21–17

If two parallel lines are intersected by a transversal, the corresponding angles are equal.

EXAMPLE •————————————————————————————————

Refer to Figure 21–18.

   Given: $\overleftrightarrow{AB} \parallel \overleftrightarrow{CD}$

   Conclusion: $\angle 1 = \angle 5$; $\angle 2 = \angle 6$;
               $\angle 3 = \angle 7$; and $\angle 4 = \angle 8$

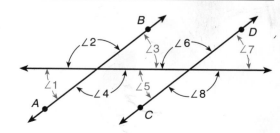

Figure 21–18

If two lines are intersected by a transversal and a pair of corresponding angles are equal, the lines are parallel.

EXAMPLE •————————————————————————————————

Refer to Figure 21–19.

    Given: $\angle 1 = \angle 2$
    Conclusion: $\overleftrightarrow{AB} \parallel \overleftrightarrow{CD}$

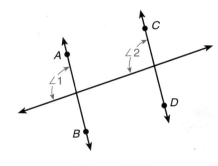

Figure 21–19

Two angles are either equal or supplementary if their corresponding sides are parallel.

EXAMPLE •————————————————————————————————

Refer to Figure 21–20.

    Given: Side $AB \parallel$ side FG, and
           side $BC \parallel$ side DE

    Conclusion: $\angle 1 = \angle 3$
                  $\angle 1 + \angle 2 = 180°$

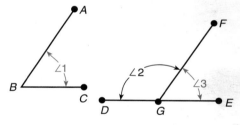

Figure 21–20

Two angles are either equal or supplementary if their corresponding sides are perpendicular.

EXAMPLE •————————————————————————————————

Refer to Figure 21–21.

    Given: Side $AB \perp$ side DH,
           and side $BC \perp$ side EF

    Conclusion: $\angle 1 = \angle 2$
                  $\angle 1 + \angle 3 = 180°$

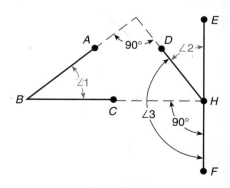

Figure 21–21

This example illustrates methods of solution for angular measure problems.

EXAMPLE •——————————————————————————————————

Refer to Figure 21–22.

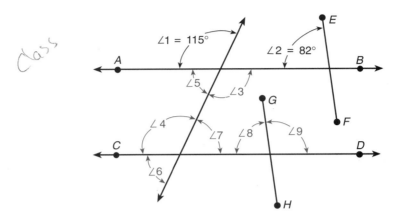

Figure 21–22

Given: $\overleftrightarrow{AB} \parallel \overleftrightarrow{CD}$, $EF \parallel GH$,

$\angle 1 = 115°$, and $\angle 2 = 82°$

Determine the values of $\angle 3$ through $\angle 9$.

Statement	*Reason*
a. (1) $\angle 3 = \angle 1$	**a. (1)** If two lines intersect, the opposite or vertical angles are equal.
(2) $\angle 3 = 115°$ *Ans*	**(2)** A quantity may be substituted for an equal quantity.
b. (1) $\angle 4 = \angle 3$	**b. (1)** If two parallel lines are intersected by a transversal, the alternate interior angles are equal.
(2) $\angle 4 = 115°$ *Ans*	**(2)** A quantity may be substituted for an equal quantity.
c. (1) $\angle 5 + \angle 1 = 180°$	**c. (1)** A straight angle is an angle of 180°.
(2) $\angle 5 + 115° = 180°$	**(2)** A quantity may be substituted for an equal quantity.
(3) $\angle 5 = 180° - 115° = 65°$ *Ans*	**(3)** If equals are subtracted from equals, the differences are equal.
d. (1) $\angle 6 = \angle 5$	**d. (1)** If two parallel lines are intersected by a transversal, the corresponding angles are equal.
(2) $\angle 6 = 65°$ *Ans*	**(2)** A quantity may be substituted for an equal quantity.
e. (1) $\angle 7 = \angle 6$	**e. (1)** If two lines intersect, the opposite or vertical angles are equal.
(2) $\angle 7 = 65°$ *Ans*	**(2)** A quantity may be substituted for an equal quantity.

f. (1) ∠8 = ∠2

   (2) ∠8 = 82° *Ans*

g. (1) ∠8 + ∠9 = 180°
   (2) ∠9 = 180° − ∠8

   (3) ∠9 = 180° − 82° = 98° *Ans*

f. (1) Two angles are either equal or supplementary if their corresponding sides are parallel. ∠8 and ∠2 are both acute angles.

   (2) A quantity may be substituted for an equal quantity.

g. (1) A straight angle equals 180°.

   (2) If equals are subtracted from equals, the differences are equal.

   (3) A quantity may be substituted for an equal quantity.

EXERCISE 21–4

Solve these problems.

1. Refer to Figure 21–23 and determine the values of angles 1 through 5.

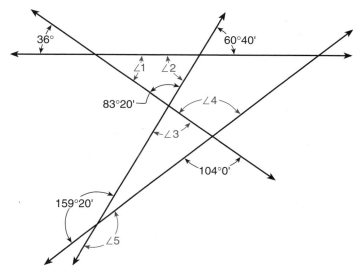

Figure 21–23

2. Refer to Figure 21–24. Determine the values of angles 2, 3, and 4 for these values of angle 1.
   a. ∠1 = 30.07°
   b. ∠1 = 29°53′

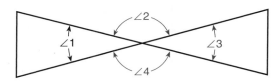

Figure 21–24

3. Refer to Figure 21–25. $\overleftrightarrow{AB} \parallel \overleftrightarrow{CD}$. Determine the values of angles 2 through 8 for these values of angle 1.

   a. ∠1 = 59°

   b. ∠1 = 63°18′

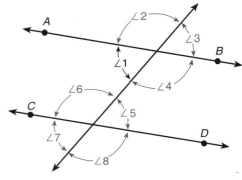

Figure 21–25

4. Refer to Figure 21–26. Hole centerline segments $EF \parallel GH$ and segments $MP \parallel KL$. Determine the values of angles 1 through 15 for these values of angle 16.

   a. ∠16 = 77°

   b. ∠16 = 81°13′

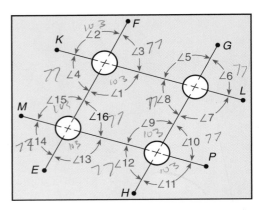

Figure 21–26

5. Refer to Figure 21–27. Hole centerline segments $AB \parallel CD$ and segments $EF \parallel GH$. Determine the values of angles 1 through 22 when ∠23 = 95°, ∠24 = 32°, and ∠25 = 104°.

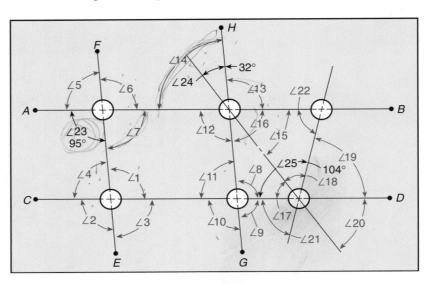

Figure 21–27

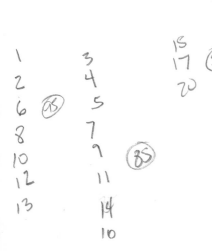

6. Refer to Figure 21–28. Segments *AB* ∥ *CD* and segments *AC* ∥ *ED*. Determine the values of angles *C* and *D* for these values of angle *A*.

    a. $\angle A = 64.875°$

    b. $\angle A = 70°27'$

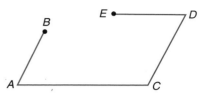

Figure 21–28

7. Refer to Figure 21–29. Segments *FH* ∥ *GS* ∥ *KM* and segments *FG* ∥ *HK*. Determine the values of angles *F*, *G*, and *H* for these values of angle *K*.

    a. $\angle K = 85.05°$

    b. $\angle K = 78°17'$

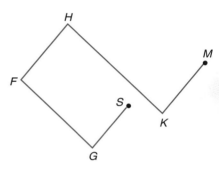

Figure 21–29

8. Refer to Figure 21–30. Segments *AE* ⊥ *FD* and segments *AD* ⊥ *CE*. Determine the values of ∠*D* and ∠*DBE* for these values of ∠*E* and ∠*CBD*.

    a. $\angle E = 52.634°$ and $\angle CBD = 37.366°$

    b. $\angle E = 50°19'$ and $\angle CBD = 39°41'$

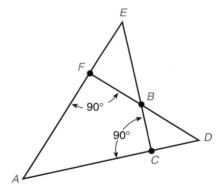

Figure 21–30

▦ UNIT EXERCISE AND PROBLEM REVIEW

NAMING ANGLES

Name each of the angles shown in Figure 21–31 in two additional ways.

1. ∠CBA
2. ∠4
3. ∠HFD
4. ∠EDF
5. ∠1
6. ∠5

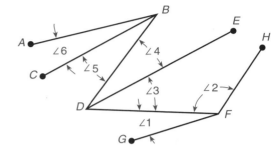

Figure 21–31

TYPES OF ANGLES

Identify each of the angles shown in Figure 21–32 as acute, right, obtuse, straight, or reflex.

7. ∠2
8. ∠DCA
9. ∠ACB
10. ∠DCB
11. ∠3
12. ∠1
13. ∠D
14. ∠DAC
15. ∠AFB
16. ∠AEF

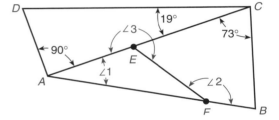

Figure 21–32

ADJACENT ANGLES

17. Name all pairs of adjacent angles shown in Figure 21–33.

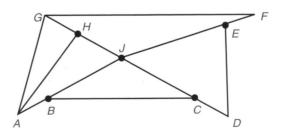

Figure 21–33

ANGLES FORMED BY A TRANSVERSAL, VERTICAL ANGLES

Refer to Figure 21–34 for problems 18 through 20.

18. Name all pairs of alternate interior angles.

19. Name all pairs of corresponding angles.

20. Name all pairs of opposite or vertical angles.

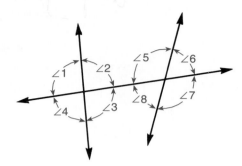

Figure 21–34

ANGULAR PROBLEMS

Solve these problems.

21. Refer to Figure 21–35 and determine the values of angles 1 through 10.

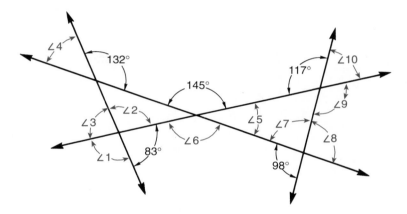

Figure 21–35

22. Refer to Figure 21–36. $\overleftrightarrow{AB} \parallel \overleftrightarrow{CD}$. Determine the values of angles 1 through 5.

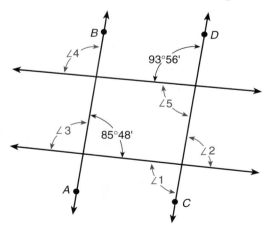

Figure 21–36

23. Refer to Figure 21–37. Hole centerline segments *AB* ∥ *CD* and *EF* ∥ *GH*. Determine the values of angles 1 through 10.

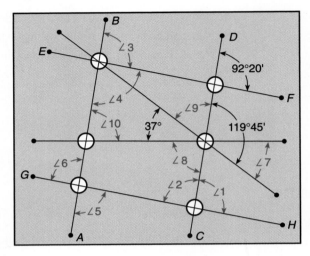

Figure 21–37

UNIT 22 ⠿ Triangles

OBJECTIVES

After studying this unit you should be able to

- make figures rigid by applying the fact that a triangle is a rigid figure.
- identify types of triangles by sides or angles.
- apply triangle theorems in determining unknown angles of a triangle.
- apply isosceles and equilateral triangle theorems in computing triangle sides and angles.
- solve practical applied problems.
- compute unknown sides of right triangles by applying the Pythagorean theorem.

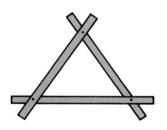

Figure 22–1 A rigid triangular frame.

A closed plane figure formed by three or more line segments is called a *polygon.*

A *triangle* is a three-sided polygon. It is the simplest kind of polygon. The symbol $\triangle$ means triangle. Triangles are used in art, manufacturing, surveying, navigation, astronomy, architecture, and engineering. The triangle is the basic figure in many designs and structures.

A triangle is a rigid figure. The rigidity of a triangle is illustrated by this simple example. Three strips of wood are fastened to form a triangular frame with one nail at each vertex, Figure 22–1. The size and shape of the frame cannot be changed without bending or breaking the strips. The triangular frame is said to be rigid.

A figure with more than three sides is not rigid. The frame made of four strips of wood fastened with one nail at each vertex is not rigid, Figure 22–2. The sides can be readily moved without bending or breaking to change its shape, Figure 22–3.

Figure 22–2

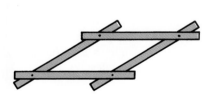

Figure 22–3

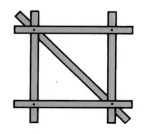

Figure 22–4

The four-sided frame can be made rigid by adding a cross brace, Figure 22–4. Observe that the addition of the brace divides the figure into two triangles.

The design and construction of many structures such as buildings and bridges are based on a system of triangles. Cross bridging is used between joists to reinforce floors, Figure 22–5. A roof truss gives support and rigidity to a roof, Figure 22–6. Notice the truss is made up of a combination of triangles.

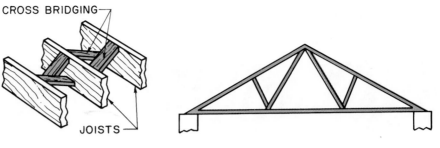

Figure 22–5   **Figure 22–6**   Roof truss.

Many objects used in the home are made rigid by means of triangles. The stepladder and shelf brackets show common applications of triangles, Figures 22–7 and 22–8.

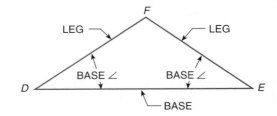

Figure 22–7   **Figure 22–8**

The designer, drafter, sheet metal technician, welder, and structural steel worker are some of the many craftspeople who require a knowledge of triangles in laying out work. Plumbers and pipefitters compute pipe lengths in diagonal pipe assemblies using triangular relationships. Auto body technicians apply principles of triangles when measuring and repairing automobile frames. Carpenters make use of triangular relationships when checking the squareness of wall corners and computing rafter lengths.

22–1   Types of Triangles

An *equilateral triangle* has three equal sides. It also has three equal angles. In the equilateral triangle *ABC* in Figure 22–9, sides *AB* = *AC* = *BC* and ∠*A* = ∠*B* = ∠*C*.

An *isosceles triangle* has two equal sides. The equal sides are called *legs.* The third side is called the *base.* The base angles of an isosceles triangle are equal. *Base angles* are the angles that are opposite the legs. In the isosceles triangle *DEF* shown in Figure 22–10, leg *DF* = leg *EF*. Since ∠*D* is opposite leg *EF* and ∠*E* is opposite leg *DF*, ∠*D* and ∠*E* are base angles.

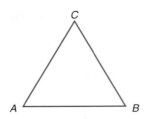

Figure 22–9   Equilateral triangle.

Figure 22–10   Isosceles triangle.

② A *scalene triangle* has three unequal sides. It has three unequal angles. Triangle *ABC* in Figure 22–11 is scalene. Sides *AB*, *AC*, and *BC* are unequal, and angles *A*, *B*, and *C* are unequal.

③ A *right triangle* has a right or 90° angle. The symbol for a right angle is a small square placed at the vertex of the angle. The side opposite the right angle is called the *hypotenuse*. The other two sides are called *legs*. Figure 22–12 shows a right triangle with the right angle at *D* and the hypotenuse *EF*.

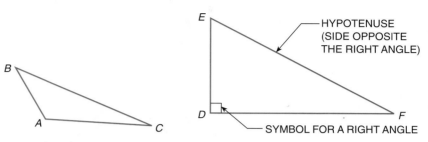

Figure 22–11   Scalene triangle.   Figure 22–12   Right triangle.

An *acute triangle* has three acute angles. An acute triangle is shown in Figure 22–13.

An *obtuse triangle* has one obtuse angle and two acute angles. An obtuse triangle is shown in Figure 22–14.

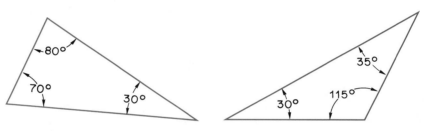

Figure 22–13   Acute triangle.   Figure 22–14   Obtuse triangle.

EXERCISE 22–1

1. Sketch each shape in Figure 22–15. Show how each shape can be made rigid by using a minimum number of braces.

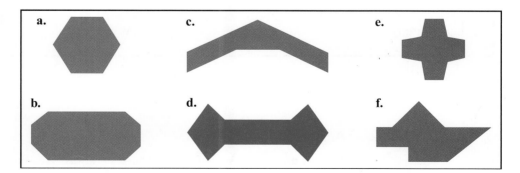

Figure 22–15

2. List at least five examples of practical uses of triangles as rigid figures found in and around your home.

3. Identify each of the triangles in Figure 22–16 as scalene, isosceles, or equilateral.

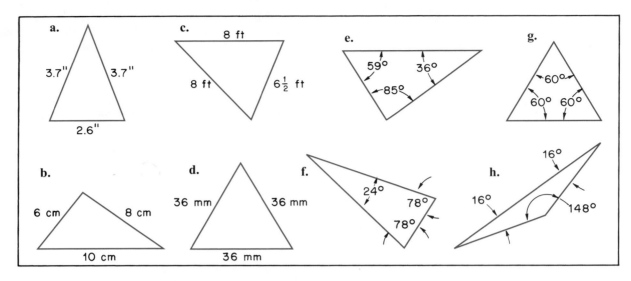

Figure 22–16

4. Identify each of the triangles in Figure 22–17 as right, acute, or obtuse.

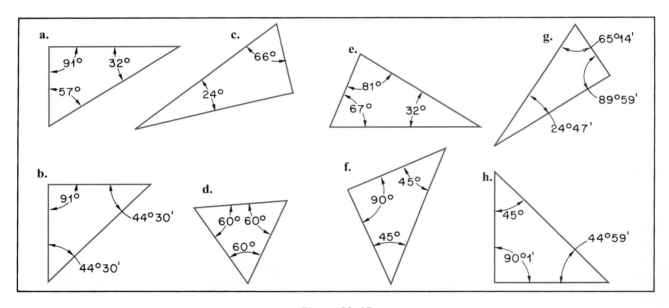

Figure 22–17

22–2   Angles of a Triangle

In every triangle the sum of the three angles is always the same.

> **The sum of the angles of any triangle is equal to 180°.**

The sum of the angles in a triangle is used in many practical applications.

EXAMPLES

1. In Figure 22–18, angles *A*, *B*, and *C* are hole centerline angles. Angle *A* equals 48°35′52″ and angle *C* = 87°55′27″. Find angle *B*.

$\angle B = 180° - (48°35′52″ + 87°55′27″)$

43°28′41″ Ans

Calculator Application

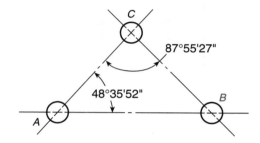

180 ⎯ (| 48 °′″ 35 °′″ 52 °′″ + 87 °′″ 55 °′″ 27 °′″ |) = SHIFT °′″ →
43°28′41″ Ans

or

180 ⎯ (| 48.3552 2nd ▶DD + 87.5527 2nd ▶DD |) ▶DD 3rd ▶DD →
43°28′41″ Ans

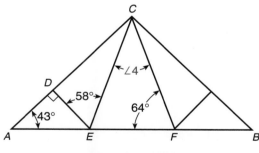

Figure 22–18

2. A roof truss is shown in Figure 22–19. Determine ∠4.

The sum of the angles in a triangle is equal to 180°.

$\angle AED = 180° - (43° + 90°)$
$\angle AED = 47°$

A straight angle equals 180°.

$\angle CEF = 180° - (47° + 58°)$
$\angle CEF = 75°$

The sum of the angles in a triangle is equal to 180°.

$\angle 4 = 180° - (64° + 75°)$
$\angle 4 = 41°$ *Ans*

Figure 22–19

EXERCISE 22–2

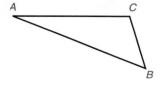

Figure 22–20

Solve these problems.

1. Refer to Figure 22–20. Determine the value of $\angle A + \angle B + \angle C$.

2. Refer to Figure 22–21.

    a. Find ∠3 when ∠1 = 68° and ∠2 = 85°.

    b. Find ∠1 when ∠2 = 81° and ∠3 = 33°.

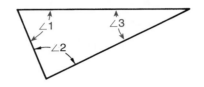

Figure 22–21

3. Refer to Figure 22–22.

    a. Find ∠6 when ∠4 = 31°23′ and ∠5 = 122°16′.

    b. Find ∠4 when ∠5 = 124°57′0″ and ∠6 = 27°15′28″.

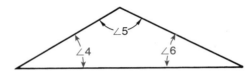

Figure 22–22

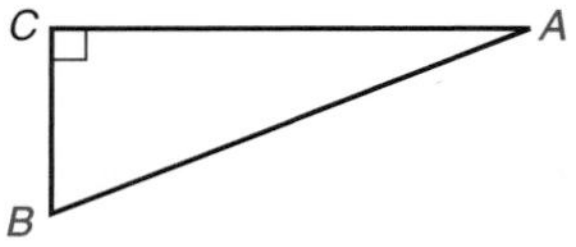

4. Refer to Figure 22–23.

    a. Find ∠B when ∠A = 21°36′.

    b. Find ∠A when ∠B = 64°19′27″.

Figure 22–23

5. Refer to Figure 22–24.

    a. Find ∠G when ∠E = 83°.

    b. Find ∠F when ∠G = 85°22′.

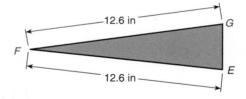

Figure 22–24

6. Refer to Figure 22–25.

    a. Find ∠1 when ∠3 = 18°.

    b. Find ∠2 when ∠3 = 23.65°.

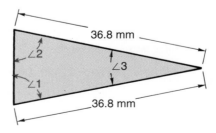

Figure 22–25

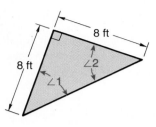

Figure 22–26

7. Refer to Figure 22–26.
    a. Find ∠1.
    b. Find ∠2.

8. Refer to Figure 22–27.
    a. Find ∠2 when ∠1 = 25.728° and ∠3 = 49.489°.
    b. Find ∠3 when ∠1 = 28°30′ and ∠2 = 16°22′.

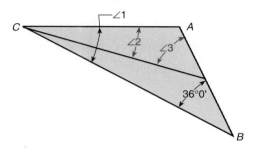

Figure 22–27

9. Refer to Figure 22–28. *AB* ‖ *DE*, and *BC* is an extension of *AB*.
    a. Find ∠A when ∠E = 64°21′.
    b. Find ∠E when ∠A = 20°59′47″.

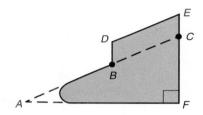

Figure 22–28

10. Refer to Figure 22–29. Hole centerline *AB* ‖ *CD*.
    a. Find ∠2 when ∠1 = 87°51′.
    b. Find ∠1 when ∠2 = 68°29′.

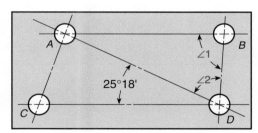

Figure 22–29

11. Refer to the roof truss in Figure 22–30.
    a. Determine the value of ∠1.
    b. Determine the value of ∠2.

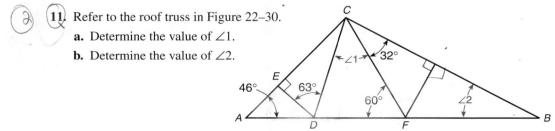

Figure 22–30

22–3   Isosceles and Equilateral Triangles

The *vertex angle* is the angle opposite the base. To *bisect* means to divide into two equal parts. An *altitude* is a line segment from a vertex perpendicular to the side opposite the vertex or perpendicular to that side extended. Every triangle has three altitudes.

In an isosceles triangle, an altitude to the base bisects the base and the vertex angle.

EXAMPLE •————————————————————————————

Refer to isosceles △*ABC* in Figure 22–31.

    Given: *AC* = *BC*

          CD is an altitude to base *AB*.

    Conclusion: *AD* = *BD*

              ∠1 = ∠2

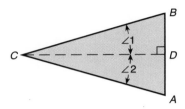

Figure 22–31

22–4   Isosceles Triangle Practical Application

Sides *DE* and *EF* of the triangular frame *DEF* in Figure 22–32 are equal. Piece *EG* is to be fastened to the frame for additional support. Determine ∠1, ∠2, and distance *DG*.

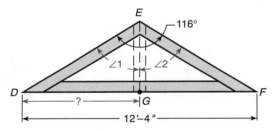

Figure 22–32

Solution. Since *EG* ⊥ *DF*, *EG* is an altitude to the base. *EG* bisects ∠*DEF* (116°) and base *DE* (12′4″).

    ∠1 = ∠2 = 116° ÷ 2 = 58° *Ans*

    DG = 12′4″ ÷ 2 = 6′2″ *Ans*

In an equilateral triangle, an altitude to any side bisects the side and the vertex angle.

EXAMPLE •————————————————————————————

Refer to equilateral △*ABC* in Figure 22–33.

    Given: *BD* is an altitude to side *AC*.

    Conclusion: *AD* = *DC*

              ∠1 = ∠2

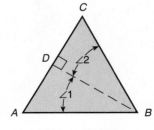

Figure 22–33

22–5 Equilateral Triangle Practical Application

Three holes are equally spaced on a circle as shown in Figure 22–34. The centerlines between holes form equilateral △MPT. Determine ∠1 and distance MW.

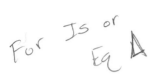

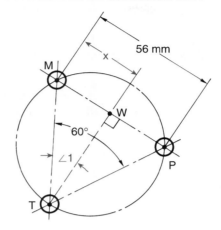

Figure 22–34

Solution. Since TW ⊥ MP, TW is an altitude to side MP. TW bisects MP (56 mm) and ∠MTP (60°),

MW = 56 mm ÷ 2 = 28 mm *Ans*

∠1 = 60° ÷ 2 = 30° *Ans*

EXERCISE 22–5

Solve problems 1 through 6 in Figure 22–35.

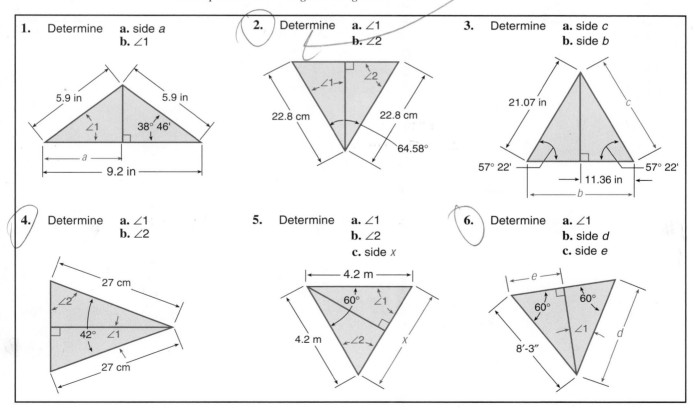

1. Determine   **a.** side *a*   **b.** ∠1

2. Determine   **a.** ∠1   **b.** ∠2

3. Determine   **a.** side *c*   **b.** side *b*

4. Determine   **a.** ∠1   **b.** ∠2

5. Determine   **a.** ∠1   **b.** ∠2   **c.** side *x*

6. Determine   **a.** ∠1   **b.** side *d*   **c.** side *e*

Figure 22–35

22–6   The Pythagorean Theorem

The Pythagorean theorem deals with the relationship of the sides of a right triangle. The application of the Pythagorean theorem permits accurate indirect measure. If the lengths of the sides of a triangle are known, it can be determined if the triangle contains a right angle.

If two sides of a right triangle are known, the third side can be calculated. The theorem is the basis of many formulas used in diverse occupational fields, such as construction, metal fabrication, woodworking, electricity, and electronics.

> **In a right triangle, the square of the length of the hypotenuse is equal to the sum of the squares of the lengths of the legs.**

EXAMPLE •————————————————————

Refer to right triangle *ABC* shown in Figure 22–36. Side *c* is the hypotenuse.

$$c^2 = a^2 + b^2$$

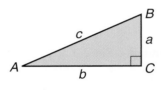

Figure 22–36

22–7   Pythagorean Theorem Practical Applications

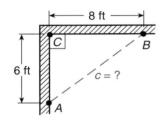

Figure 22–37

1. Carpenters and masons often apply the Pythagorean theorem in squaring-up corners of buildings. Squaring-up means making corner walls at right angles. A 6-foot length is marked off on one wall at point *A*, Figure 22–37. An 8-foot length is marked off on the other wall at point *B*. For the corner to be square, what should the measurement be from point *A* to point *B* (length *c*)?

   Solution.   The 6-foot and 8-foot legs are given, and the hypotenuse is determined.
   $$c^2 = (6 \text{ ft})^2 + (8 \text{ ft})^2$$
   $$c^2 = 36 \text{ sq ft} + 64 \text{ sq ft}$$
   $$c^2 = 100 \text{ sq ft}$$
   $$c = 10 \text{ ft } Ans$$

Calculator Application
$$c = \sqrt{(6 \text{ ft})^2 + (8 \text{ ft})^2}$$
$$\boxed{(}\; 6 \;\boxed{x^2}\; \boxed{+}\; 8 \;\boxed{x^2}\; \boxed{)}\; \boxed{\sqrt{x}} \rightarrow 10$$
$$c = 10 \text{ ft } Ans$$

2. Rafters 18 feet long are used on an equally pitched roof with a 24-foot span, Figure 22–38 on page 471. Determine the rise, *FG*, of the roof.

   Solution.   Since $\triangle DEF$ is an isosceles triangle, the rise *FG* is an altitude to the base (24-foot span).
   $$DG = 24 \text{ ft} \div 2$$
   $$DG = 12 \text{ ft}$$
   $$DF = 18 \text{ ft} - 2 \text{ ft} - 3 \text{ in}$$
   $$DF = 15.75 \text{ ft}$$

$$(15.75 \text{ ft})^2 = (FG)^2 + (12 \text{ ft})^2$$
$$248.0625 \text{ sq ft} = (FG)^2 + 144 \text{ sq ft}$$
$$(FG)^2 = 104.0625 \text{ sq ft}$$
$$FG \approx 10.201 \text{ ft}$$
$$10.201 \text{ ft} \approx 10 \text{ ft } 2.4 \text{ in } Ans$$

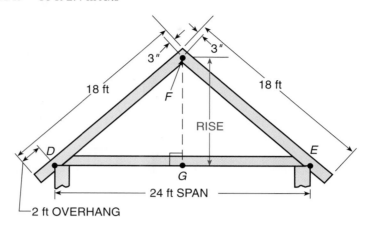

Figure 22–38

Calculator Application

$$(15.75)^2 = (FG)^2 + (12 \text{ ft})^2$$

Rearrange the equation in terms of *FG*.

$$\sqrt{(15.75 \text{ ft})^2 - (12 \text{ ft})^2} = FG$$

$\boxed{(}\ 15.75 \boxed{x^2} \boxed{-} 12 \boxed{x^2} \boxed{)} \boxed{\sqrt{x}} \rightarrow 10.201$ (rounded)

$FG = 10.201$ ft (rounded) 0.201 ft $\approx .201$ $\boxed{\times}$ 12 $\boxed{=}$ 2.4 in (rounded)

$$10.201 \text{ ft} \approx 10 \text{ ft } 2.4 \text{ in } Ans$$

EXERCISE 22–7

Apply the Pythagorean theorem in solving problems 1 through 6, Figure 22–39. Refer to the right triangle shown in Figure 22–40. Where necessary, express the answers to two decimal places.

	SIDE *a*	SIDE *b*	SIDE *c*
1.	15 in	20 in	?
2.	6.00 in	7.50 in	?
3.	18.0 mm	?	30.0 mm
4.	?	12.5 ft	16.0 ft
5.	45 mm	?	75 mm
6.	?	10.20 in	14.60 in

Figure 22–39

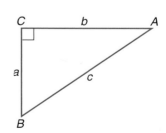

Figure 22–40

Solve these practical problems. In addition to the Pythagorean theorem, certain problems require the application of other triangle theorems.

7. A machined piece is shown in Figure 22–41 on page 472.

   a. Determine dimension *x*.

   b. Determine ∠1.

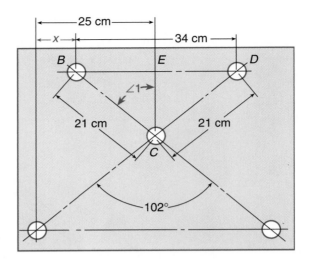

Figure 22–41

8. A person travels 15.00 miles directly north and then travels 7.00 miles directly west. How far is the person from the starting point? Express the answer to the nearest tenth mile.

9. A pole 16.20 meters high is supported by a cable attached 2.20 meters from the top of the pole. What is the length of cable required if the cable is fastened to the ground 6.80 meters from the foot of the pole? Assume the ground is level. Express the answer to 3 significant digits.

10. Three holes are drilled in a plate as shown in Figure 22–42. Determine dimensions A and B to one decimal place.

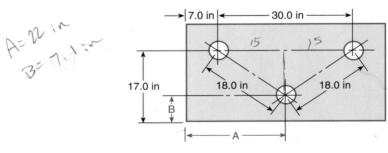

Figure 22–42

11. In remodeling a house, a carpenter checks the squareness of a wall and floor. A $4\frac{1}{2}$-foot height is marked off on the wall (point A), and a 6-foot length is marked off on the floor (point B), as shown in Figure 22–43. The measurement from point A to point B is 7'9''. Are the wall and floor square? Show your computations.

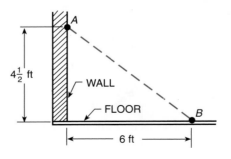

Figure 22–43

12. Can the piece shown in Figure 22–44 be shared (cut) from the sheet shown in Figure 22–45? Show your computations.

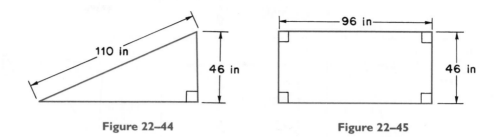

Figure 22–44   Figure 22–45

13. Two cars start from the same point at the same time. One car travels south at 60.0 kilometers per hour. The other car travels east at 50.0 kilometers per hour. How far apart, to the nearest kilometer, are the cars after 1.50 hours?

14. An irregularly-shaped plot of land is shown in Figure 22–46. Determine length *x* and length *y* to the nearest whole meter.

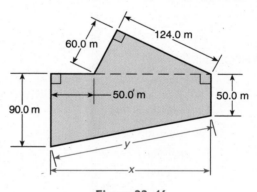

Figure 22–46

UNIT EXERCISE AND PROBLEM REVIEW

MAKING FIGURES RIGID BY BRACING

1. Sketch each shape in Figure 22–47. Show how each shape can be made rigid by using a minimum number of braces.

Figure 22–47

TYPES OF TRIANGLES

2. Identify each of the triangles in Figure 22–48 in these two ways:

- as scalene, isosceles, or equilateral.
- as right, acute, or obtuse.

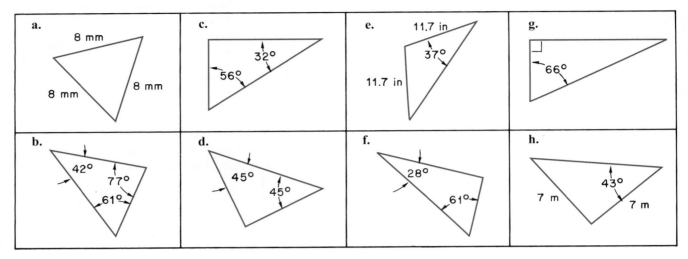

Figure 22–48

APPLICATIONS OF THEOREMS

Solve these problems by applying the theorem "the sum of the angles of any triangle is equal to 180°." In addition, certain problems require the application of other angular and triangular facts and theorems.

3. Refer to Figure 22–49.
    a. Find ∠3 when ∠1 = 19°37′.
    b. Find ∠2 when ∠3 = 142°36′.

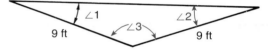

Figure 22–49

4. Refer to Figure 22–50. Line *AB* ∥ line *CD*.
    a. Find ∠3 when ∠1 = 110°.
    b. Find ∠4 when ∠2 = 146°.

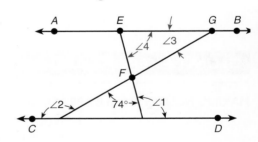

Figure 22–50

5. Refer to Figure 22–51.
    a. Find ∠2 when
       ∠1 = 18° and ∠3 = 41°.
    b. Find ∠3 when
       ∠1 = 31°17′ and ∠2 = 19°52′.

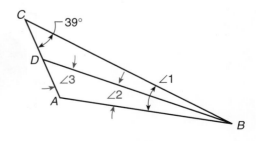

Figure 22–51

6. Refer to Figure 22–52. Segment AC = segment BC.

    a. Find $\angle 1$ when $\angle 2 = 66.87°$.

    b. Find $\angle 2$ when $\angle 1 = 27°29'15''$.

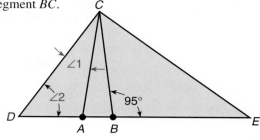

Figure 22–52

ISOSCELES AND EQUILATERAL TRIANGLES

Solve problems 7 through 9, using Figure 22–53.

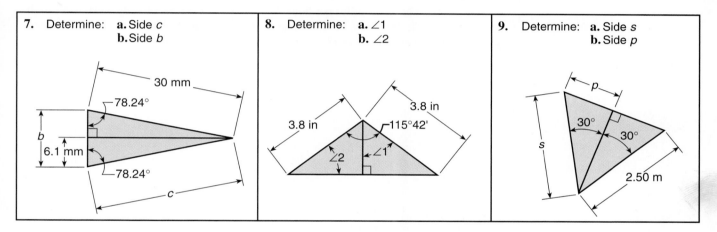

7. Determine: **a.** Side c **b.** Side b

8. Determine: **a.** $\angle 1$ **b.** $\angle 2$

9. Determine: **a.** Side s **b.** Side p

Figure 22–53

PYTHAGOREAN THEOREM

10. In a right triangle, a and b represent the legs and c represents the hypotenuse. Determine the unknown side. Express the answer to one decimal place where necessary.

    a. $a = 12$ ft and $b = 16$ ft; find c.

    b. $a = 1.5$ m and $c = 2.5$ m; find b.

    c. $b = 8.0$ in and $c = 11.0$ in; find a.

    d. $a = 10.5$ cm and $b = 14.2$ cm; find c.

11. A mason checks the squareness of a building foundation. From a corner, a 9-foot length is marked off on one wall and a 12-foot length is marked off on the adjacent wall. A measurement of 15 feet is made between the two markings. Is the foundation wall square? Show your computations.

12. Town B is 8.3 miles directly north of town A. Town C is directly east of town B. If the distance between town A and town C is 14.6 miles, what is the distance from town B to town C? Round the answer to one decimal place.

13. A beam is set on the tops of two vertical support columns that are 30.0 feet apart. One column is 10.0 feet high. The other column is 18.0 feet high. If the beam extends an additional 2.0 feet beyond each column, what length beam, to the nearest tenth foot, is required?

14. Determine distance x of the parcel of land shown in Figure 22–54. Express the answer to the nearest whole meter.

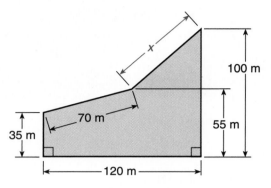

Figure 22–54

15. Two trucks start from the same place at the same time. One travels directly south at an average rate of speed of 52.25 miles per hour. The other truck travels directly west. After each truck travels for 7.500 hours, the trucks are 523.5 miles apart. What is the average rate of speed of the truck traveling west? Round the answer to 4 significant digits.

UNIT 23 ⁛ Congruent and Similar Figures

OBJECTIVES

After studying this unit you should be able to

- identify corresponding sides and corresponding angles of congruent triangles.
- identify similar pairs of polygons.
- compute lengths of sides and perimeters of similar polygons.
- determine whether given pairs of triangles are similar.
- informally prove that pairs of triangles are similar and determine unknown sides and angles.
- solve practical applied problems using theorems discussed in this unit.

Examples of congruency and similarity are continuously experienced in our daily activities. In manufacturing, mass production is based on the ability to produce congruent products. A road map is similar to the territory that it represents and a photograph is similar to the object that is photographed. In plane geometry, proving or verifying that certain angles are equal and certain sides of two or more figures are equal or proportional is based on congruence or similarity.

23–1   Congruent Figures

   Congruent figures have exactly the same size and shape. If congruent plane figures are placed on top of each other, the figures coincide or fit exactly. The symbol ≅ means congruent.

> **Corresponding parts of congruent triangles are equal.**

The corresponding parts of triangles are corresponding sides and corresponding angles. If congruent triangles are placed one on the other (superimposed), the parts that fit exactly are corresponding.

Corresponding Sides of Congruent Triangles

The sides that lie opposite equal angles are *corresponding sides.*

EXAMPLE •————————————————————————————————

Refer to Figure 23–1. Triangles *ABC* and *DEF* are congruent triangles. Determine the corresponding equal sides.

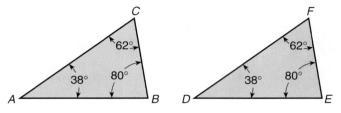

Figure 23–I

AB and *DE* both lie opposite 62° angles. *AB = DE Ans*

BC and *EF* both lie opposite 38° angles. *BC = EF Ans*

AC and *DF* both lie opposite 80° angles. *AC = DF Ans*

Corresponding Angles of Congruent Triangles

The angles that lie opposite equal sides are *corresponding angles.*

EXAMPLE •————————————————————————————————

Refer to Figure 23–2. Triangles *MNP* and *RST* are congruent triangles. Determine the corresponding equal angles.

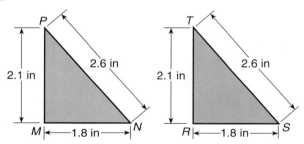

Figure 23–2

∠*M* and ∠*R* both lie opposite 2.6-inch sides. ∠*M = ∠R Ans*

∠*N* and ∠*S* both lie opposite 2.1-inch sides. ∠*N = ∠S Ans*

∠*P* and ∠*T* both lie opposite 1.8-inch sides. ∠*P = ∠T Ans*

It is important to remember that corresponding parts of congruent triangles are *not* determined by the positions of the triangles. For example, with the congruent triangles shown in Figure 23–3, sides *AC* and *DF* are *not* corresponding. They do *not* lie opposite equal angles. Sides *AC* and *ED* are corresponding and equal since they both lie opposite 41° angles.

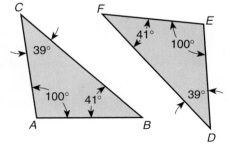

Figure 23–3

Congruent triangles are often marked, as shown in Figure 23–4, to indicate corresponding sides and angles. When side *AB* is given one mark, corresponding side *DE* is also given one mark. Angle *C*, which is opposite *AB*, and angle *F*, which is opposite *DE*, are also given one mark. The same procedure is used to mark the other corresponding sides and angles.

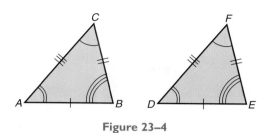

Figure 23–4

EXERCISE 23–1

1. Three pairs of congruent triangles are shown in Figure 23–5. Identify the pairs of corresponding sides in problems a, b, and c.

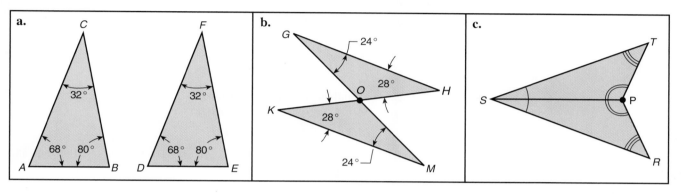

Figure 23–5

2. Three pairs of congruent triangles are shown in Figure 23–6. Identify the pairs of corresponding angles in problems a, b, and c.

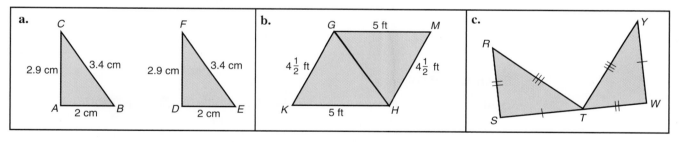

Figure 23–6

23–2   Similar Figures

Stated in a general way, *similar figures* mean figures that are alike in shape but may be different in size. Congruent figures are special types of similar figures that have the same size and the same shape.

Scale drawings, which are commonly used in a great variety of occupations, are examples of similar figures. Often, scale drawings are in the form of similar polygons or combinations of similar polygons. *Similar polygons* have the same number of sides, equal corresponding angles, and proportional corresponding sides. The symbol ~ means similar.

Corresponding angles of similar polygons are equal.

Corresponding sides of similar polygons are proportional.

EXAMPLES •————————————————————

1. Refer to similar polygons in Figure 23–7.

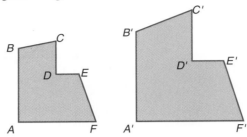

Figure 23–7

Given: $ABCDEF \sim A'B'C'D'E'F'$

Conclusion: The corresponding angles are equal.

$$\angle A = \angle A' \qquad \angle D = \angle D'$$
$$\angle B = \angle B' \qquad \angle E = \angle E'$$
$$\angle C = \angle C' \qquad \angle F = \angle F'$$

The corresponding sides are proportional.

$$\frac{AB}{A'B'} = \frac{BC}{B'C'} = \frac{CD}{C'D'} = \frac{DE}{D'E'} = \frac{EF}{E'F'} = \frac{FA}{F'A'}$$

2. Refer to the similar polygons in Figure 23–7. If $AB = 9.1$ cm, $A'B' = 10.5$ cm, and $EF = 5.2$ cm, what is the length of $E'F'$?

We have $\dfrac{AB}{A'B'} = \dfrac{EF}{E'F'}$. Substitute the given data

$$\frac{9.1 \text{ cm}}{10.5 \text{ cm}} = \frac{5.2 \text{ cm}}{E'F'}$$

$$E'F' = \frac{(5.2 \text{ cm})(10.5 \text{ cm})}{9.1 \text{ cm}}$$

$$= 6.0 \text{ cm } Ans$$

———•

Perimeters of Polygons

The distance around a polygon, or the sum of all sides, is called the *perimeter* (*P*). In Figure 23–7:

$$P = AB + BC + CD + DE + EF + FA$$
$$P' = A'B' + B'C' + C'D' + D'E' + E'F' + F'A'$$

The perimeters of two similar polygons have the same ratio as any two corresponding sides. For example, in Figure 23–7:

$$\frac{P}{P'} = \frac{AB}{A'B'}$$

Dissimilar Polygons

Except in the case of triangles, if two polygons have equal corresponding angles, it does *not* necessarily follow that the corresponding sides are proportional. This fact is illustrated in Figure 23–8.

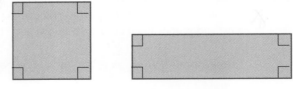

Figure 23–8 Corresponding angles are equal. Corresponding sides are *not* proportional. Polygons are not similar.

Except in the case of triangles, if two polygons have proportional corresponding sides, it does *not* necessarily follow that the corresponding angles are equal. This fact is illustrated in Figure 23–9.

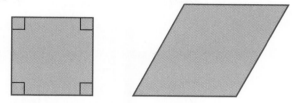

Figure 23–9   Corresponding sides are proportional. Corresponding angles are *not* equal. Polygons are not similar.

EXERCISE 23–2

1. Identify which of the 6 pairs of polygons, a–f, in Figure 23–10 are similar. For each pair of polygons, state the reason why the polygons are similar or why the polygons are not similar.

Classwork

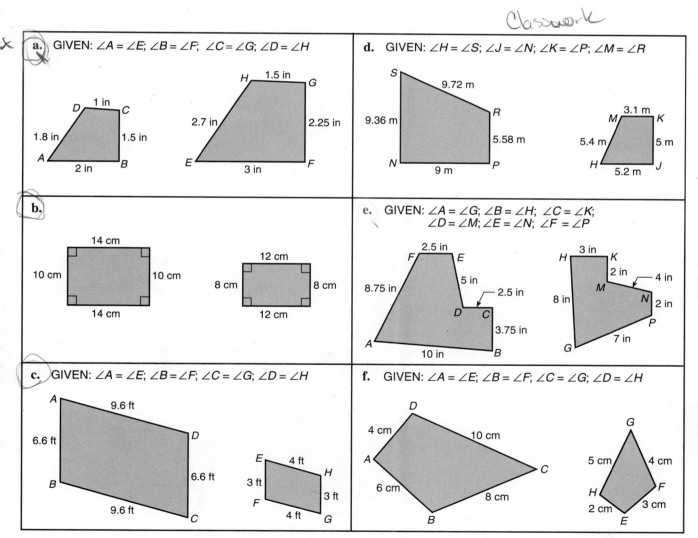

Figure 23–10

2. Refer to Figure 23–11. The two polygons are similar. $\angle A = \angle A'$; $\angle B = \angle B'$; $\angle C = \angle C'$; $\angle D = \angle D'$. Determine these values.

   a. side $A'B'$ c. side $C'D'$

   b. side $B'C'$ d. the perimeter of polygon $A'B'C'D'$

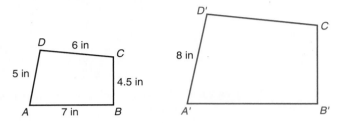

Figure 23–11

3. Refer to Figure 23–12. The two polygons are similar. $\angle A = \angle A'$; $\angle B = \angle B'$; $\angle C = \angle C'$; $\angle D = \angle D'$; $\angle E = \angle E'$. Determine these values.

   a. side BC

   b. side CD

   c. side DE

   d. side AE

   e. the perimeter of polygon $ABCDE$

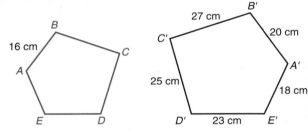

Figure 23–12

23–3   Practical Applications of Similar Triangles

Similar triangles have many practical applications in industry, construction, navigation, and land surveying. As with congruent triangles, the solutions to many similar triangle problems are based on a knowledge of corresponding parts.

Corresponding sides of similar triangles lie opposite equal corresponding angles.

EXAMPLE •————————————————————————————

Refer to similar triangles ABC and DEF shown in Figure 23–13.

   AB and EF are corresponding sides. Both lie opposite 55° angles.

   BC and DE are corresponding sides. Both lie opposite 50° angles.

   AC and DF are corresponding sides. Both lie opposite 75° angles.

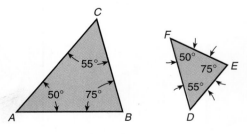

Figure 23–13

Draw a Picture 'or
tell me 3 ways
similar
A cam be
They are
S

It is often required to first informally prove triangles are similar, then to determine unknown angles or sides based on the relationship of corresponding parts.

> **If two angles of a triangle are equal to two angles of another triangle, the triangles are similar.**

Practical Application

Five holes are located on a piece as shown in Figure 23–14. Determine distance *AB*.

Solution. $\angle A = \angle C$

$\angle AEB = \angle CED$ (Vertical angles are equal.)

Therefore, $\triangle ABE \sim \triangle CDE$

Determine *AB*.

$$\frac{7 \text{ in}}{10.5 \text{ in}} = \frac{AB}{9 \text{ in}}$$

$AB = 6$ in *Ans*

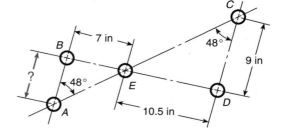

Figure 23–14

cw

> **If the corresponding sides of two triangles are proportional, the triangles are similar.**

EXAMPLE

Determine $\angle 1$ shown in Figure 23–15.

Solution.   First determine whether the corresponding sides are proportional.

$$\frac{6 \text{ cm}}{15 \text{ cm}} = 0.4$$

$$\frac{3 \text{ cm}}{7.5 \text{ cm}} = 0.4$$

$$\frac{5.2 \text{ cm}}{13 \text{ cm}} = 0.4$$

The sides are proportional.   Therefore, $\triangle FGH \sim \triangle PMH$
Determine $\angle 1$. $\angle 1 = \angle FHG$
    $\angle 1 = 30°$ *Ans*

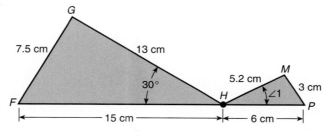

Figure 23–15

If two sides of a triangle are proportional to two sides of another triangle and if the angles included between these sides are equal, the triangles are similar.

Practical Application

An obstruction lies between points *A* and *B* as shown in Figure 23–16. Distance *AB* can be determined by making sides *AE* and *BE* proportional to sides *DE* and *CE*.

Solution.   Distances *AE*, *ED*, *BE*, and *EC* are measured and marked off as shown. Prove triangles *AEB* and *CED* similar.

$$\angle AEB = \angle CED \text{ (Vertical angles are equal.)}$$

$$\frac{90 \text{ ft}}{30 \text{ ft}} = 3 \text{ (The sides are proportional.)}$$

$$\frac{75 \text{ ft}}{25 \text{ ft}} = 3$$

Therefore, $\triangle AEB \sim \triangle CED$

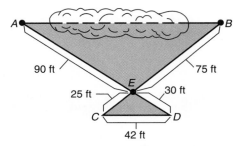

Figure 23–16

Distance *CD* is measured; *CD* measures 42 feet.
Determine distance *AB*.

$$\frac{AB}{CD} = \frac{AE}{ED}$$

$$\frac{AB}{42 \text{ ft}} = \frac{90 \text{ ft}}{30 \text{ ft}}$$

$$AB = 126 \text{ ft } Ans$$

Within a triangle, if a line is parallel to one side and intersects the other two sides, the triangle formed and the given triangle are similar.

Consider $\triangle ABC$ in Figure 23–17. If *DE* is parallel to *AC*, then $\triangle DBE \sim \triangle ABC$.

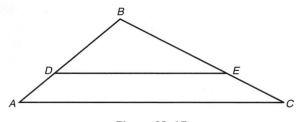

Figure 23–17

Practical Application

A design change is made in a building. A roof span is changed from 36 feet to 30 feet as shown in Figure 23–18. The roof pitch and rafter overhang are to remain unchanged. The changed rafter, *GE*, is parallel to the original rafter, *FC*. Compute the changed rafter length, *GE*.

Solution. $\triangle ACB \sim \triangle DEB$

$$\frac{DE}{AC} = \frac{DB}{AB}$$

$$\frac{DE}{18 \text{ ft}} = \frac{30 \text{ ft}}{36 \text{ ft}}$$

$$DE = 15 \text{ ft}$$

$$GE = 15 \text{ ft} + 1 \text{ ft} = 16 \text{ ft } Ans$$

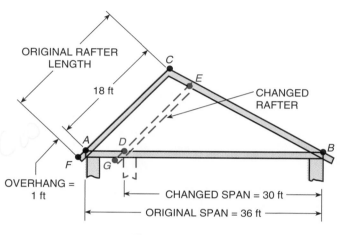

Figure 23–18

If the altitude is drawn to the hypotenuse of a right triangle, the two triangles formed are similar to each other and to the given triangle.

In Figure 23–19, $\triangle ABC$ is a right triangle with hypotenuse *AB*. An altitude, *CD*, is drawn from *C* to *D*. Then, $\triangle ADC \sim \triangle CDB \sim \triangle ACB$.

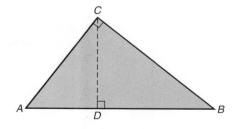

Figure 23–19

Practical Application

A plot of land, right $\triangle MSP$, shown in Figure 23–20, is to be divided into two building lots. Lot 1 is shown as $\triangle MTS$ and lot 2 is shown as $\triangle STP$. Determine the frontage, *MT* and *PT*, of each lot to the nearest tenth foot.

Solution.   Since *ST* is ⊥ to the hypotenuse *MP* of △*MSP*, *ST* is the altitude to the hypotenuse.
△*MTS* ~ △*STP* ~ △*MSP*. Determine the length of *MT*.

$$\frac{MT}{MS} = \frac{MS}{MP}$$

$$\frac{MT}{180.0 \text{ ft}} = \frac{180.0 \text{ ft}}{234.0 \text{ ft}}$$

$$MT \approx 138.5 \text{ ft } \textit{Ans}$$

$$PT \approx 234.0 \text{ ft} - 138.5 \text{ ft} \approx 95.5 \text{ ft } \textit{Ans}$$

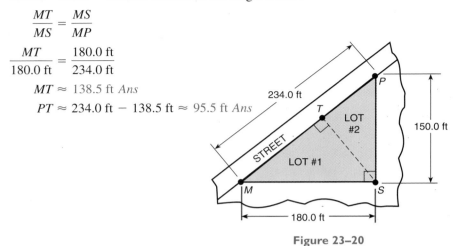

Figure 23–20

EXERCISE 23–3

For each of the problems, 1 through 6, in Figure 23–21, do the following:

a. *Informally prove that pairs of triangles are similar. Each pair must meet the requirements of one of the similar triangle theorems. Certain problems also require the application of angle theorems.*

b. *Determine the values of the required sides and angles.*

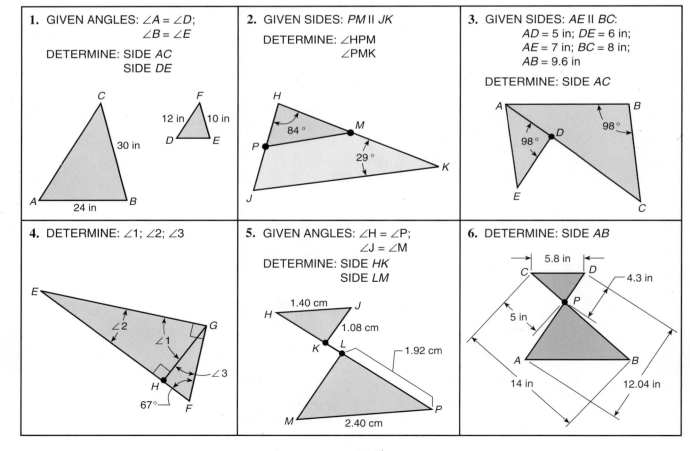

1. GIVEN ANGLES: ∠A = ∠D;
∠B = ∠E

DETERMINE: SIDE *AC*
SIDE *DE*

2. GIVEN SIDES: *PM* II *JK*

DETERMINE: ∠HPM
∠PMK

3. GIVEN SIDES: *AE* II *BC*:
AD = 5 in; *DE* = 6 in;
AE = 7 in; *BC* = 8 in;
AB = 9.6 in

DETERMINE: SIDE *AC*

4. DETERMINE: ∠1; ∠2; ∠3

5. GIVEN ANGLES: ∠H = ∠P;
∠J = ∠M

DETERMINE: SIDE *HK*
SIDE *LM*

6. DETERMINE: SIDE *AB*

Figure 23–21

7. A cluster of 5 holes is drilled in a casting as shown in Figure 23–22. Determine missing dimensions *x* and *y* to 2 decimal places.

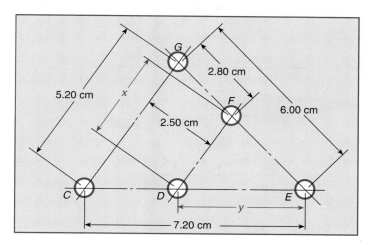

Figure 23–22

8. A highway ramp with a 12% grade is shown in Figure 23–23. A 12% grade means that for 100 feet of horizontal distance there is a vertical rise of 12 feet.

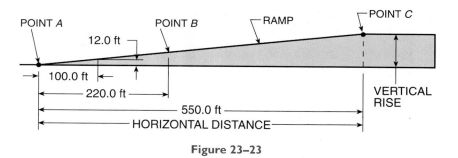

Figure 23–23

a. How much higher is point *C* than point *B*? Round the answer to the nearest tenth foot.

b. A car travels 221.6 feet from the foot of the ramp (point *A*) to point *B*. How far does the car travel from point *B* to the end of the ramp (point *C*)? Round the answer to the nearest tenth foot.

9. The side view of a tapered shaft is shown in Figure 23–24. Since the taper is uniform, △*ABC* and △*DEF* are congruent. Compute the shaft diameter at point *R*.

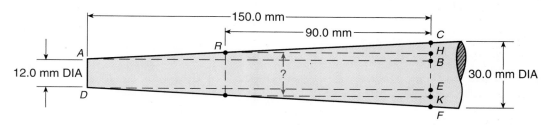

Figure 23–24

▦ UNIT EXERCISE AND PROBLEM REVIEW

CONGRUENT TRIANGLES: IDENTIFYING CORRESPONDING PARTS

Four pairs of congruent triangles are shown in Figures 23–25 through 23–28. Identify all pairs of corresponding angles and sides in each figure.

1.

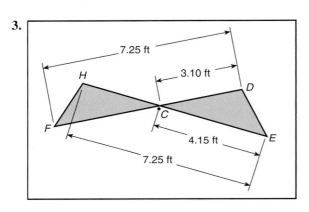

Figure 23–25

2.

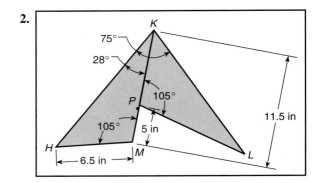

Figure 23–26

3.

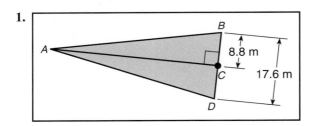

Figure 23–27

4.

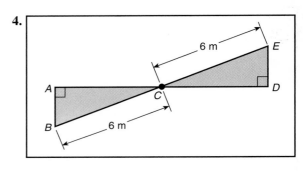

Figure 23–28

SIMILAR POLYGONS

5. Identify which of the 2 pairs of polygons, a and b, in Figure 23–29 are similar. For each pair of polygons state the reason why the polygons are similar or why the polygons are not similar.

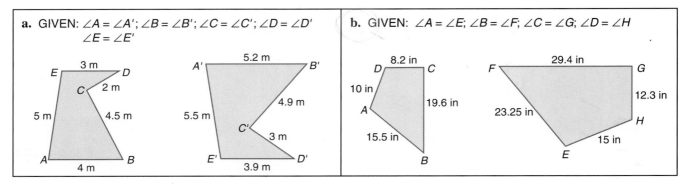

a. GIVEN: ∠A = ∠A′; ∠B = ∠B′; ∠C = ∠C′; ∠D = ∠D′
∠E = ∠E′

b. GIVEN: ∠A = ∠E; ∠B = ∠F; ∠C = ∠G; ∠D = ∠H

Figure 23–29

6. Refer to Figure 23–30. The two polygons are similar. $\angle A = \angle F$; $\angle B = \angle G$; $\angle C = \angle H$; $\angle D = \angle J$; $\angle E = \angle K$. Determine the following lengths. Round answers to 1 decimal place.

    a. side *FG*

    b. side *GH*

    c. side *JK*

    d. side *FK*

    e. the perimeter of polygon *FGHJK*

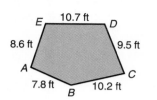
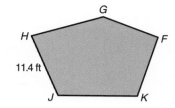

Figure 23–30

SIMILAR TRIANGLES

For each of the problems, 7 and 8, shown in Figure 23–31, do the following:

    a. *Informally prove that pairs of triangles are similar. Each pair must meet the requirements of one of the similar triangle theorems. Certain problems also require the application of angle theorems.*

    b. *Determine the values of the required sides or angles.*

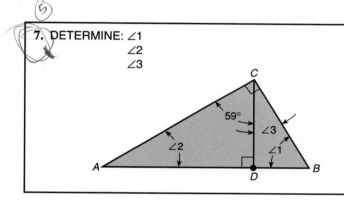

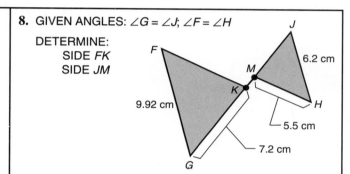

Figure 23–31

9. An obstruction lies between points *A* and *B* as shown in Figure 23–32. Distances *AE*, *ED*, *BE*, and *EC* are measured and marked off as shown. Distance *CD* is measured. Determine distance *AB*.

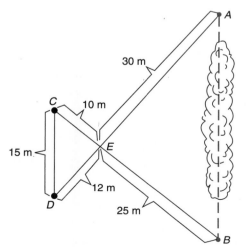

Figure 23–32

10. Vertical braces A and B are added to the frame shown in Figure 23–33. Determine dimensions *E*, *F*, *G*, and *H* in feet and inches.

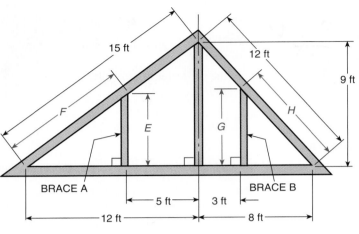

Figure 23–33

UNIT 24 ::: Polygons

OBJECTIVES

After studying this unit you should be able to

• identify various types of polygons.

• determine unknown sides and angles of quadrilaterals.

• compute interior and exterior angles of polygons by applying polygon angle theorems.

• apply parallelogram and trapezoid theorems in computing lengths.

• solve practical applied problems, using theorems discussed in this unit.

As previously stated, a *polygon* is a closed plane figure formed by three or more straight-line segments. Examples of polygons are shown in Figure 24–1.

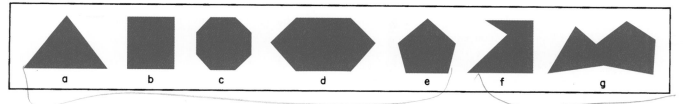

Figure 24–1

Concave

24–1   Types of Polygons

A *convex polygon* is a polygon in which no side, if extended, cuts inside the polygon. In Figure 24–1, polygons *a* through *e* are convex. In this book, unless otherwise stated, the word *polygon* means convex polygon.

A *concave polygon* is a polygon in which two or more sides, if extended, cut inside the polygon. In Figure 24–1, polygons *f* and *g* are concave.

An *equilateral polygon* is a polygon with all sides equal. In Figure 24–2, polygons *a*, *c*, and *e* are equilateral.

An *equiangular polygon* is a polygon with all angles equal. In Figure 24–2, polygon *b* is equiangular.

A *regular polygon* is a polygon that is both equilateral and equiangular; all sides and all angles are equal. In Figure 24–2, polygon *c* is regular.

Except in the case of triangles, a polygon can be equilateral without being equiangular and equiangular without being equilateral. This fact is shown by polygons *a* and *b* in Figure 24–2.

491

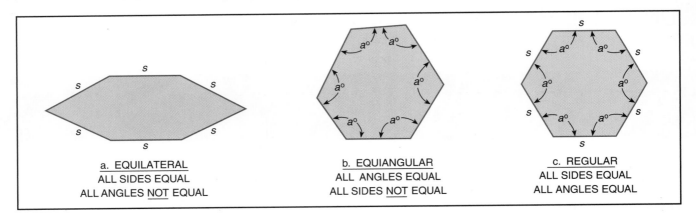

Figure 24–2

Polygons are often classified according to their number of sides. In addition to the triangle, the following polygons are the most common:

A *quadrilateral* is a polygon with four sides.

A *pentagon* is a polygon with five sides.

A *hexagon* is a polygon with six sides.

An *octagon* is a polygon with eight sides.

EXERCISE 24–1

1. Of the figures, a–j, in Figure 24–3, which are polygons? Identify each of the polygons as convex or concave.

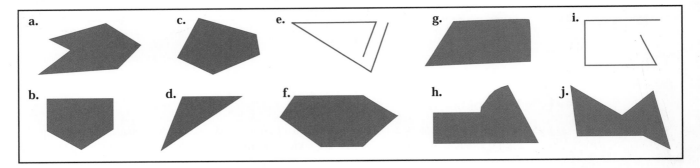

Figure 24–3

2. Identify each of the polygons, a–g, in Figure 24–4 as:

    (1) equilateral, equiangular, or regular.

    (2) a quadrilateral, a pentagon, a hexagon, or an octagon.

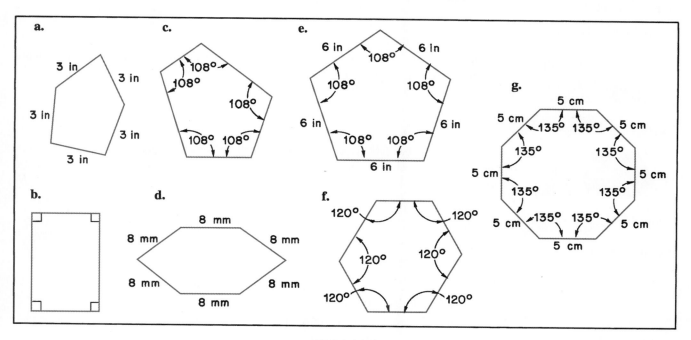

Figure 24–4

24–2   Types of Quadrilaterals

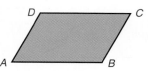

A *parallelogram* is a quadrilateral whose opposite sides are parallel and equal. Opposite angles of a parallelogram are equal. In the parallelogram in Figure 24–5, $AB \parallel DC$; $AB = DC$; $AD \parallel BC$; $AD = BC$; $\angle A = \angle C$; $\angle B = \angle D$.

Figure 24–5   Parallelogram.

A *rectangle* is a special type of parallelogram whose angles each equal 90°. In the rectangle in Figure 24–6, $EF \parallel GH$; $EF = GH$; $EG \parallel FH$; $EG = FH$; $\angle E = \angle F = \angle G = \angle H = 90°$.

Figure 24–6   Rectangle.

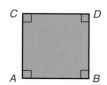

A *square* is a special type of rectangle whose sides are all equal. In the square in Figure 24–7, $AB \parallel CD$; $AC \parallel BD$; $AB = BD = CD = AC$; $\angle A = \angle B = \angle C = \angle D = 90°$.

Figure 24–7   Square.

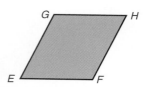

Figure 24–8   Rhombus.

A *rhombus* is a special type of parallelogram whose sides are all equal, but whose angles are not all equal. In the rhombus in Figure 24–8, *EF* ‖ *GH*; *EG* ‖ *FH*; *EF* = *FH* = *GH* = *EG*; ∠E = ∠H; ∠G = ∠F.

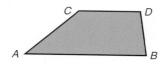

Figure 24–9   Trapezoid.

A *trapezoid* is a quadrilateral that has only two sides parallel. In the trapezoid in Figure 24–9, *AB* ‖ *CD*.

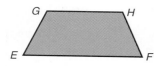

Figure 24–10   Isosceles trapezoid.

An *isosceles trapezoid* is a trapezoid in which the legs are equal. The legs are the sides that are *not* parallel. The base angles of an isosceles trapezoid are equal. The parallel sides of a trapezoid are called *bases*. In the isosceles trapezoid in Figure 24–10, *EF* ‖ *GH*; *EG* = *FH*; ∠E = ∠F; ∠G = ∠H.

EXERCISE 24–2

Answer each of these questions.

   1. Are all parallelograms quadrilaterals? Explain.
   2. Are all quadrilaterals parallelograms? Explain.
   3. Are all rectangles parallelograms? Explain.
   4. Are all parallelograms rectangles? Explain.
   5. Are all squares parallelograms? Explain.
   6. Are all rhombuses rectangles? Explain.
   7. Are all rectangles rhombuses? Explain.
   8. Are all trapezoids rhombuses? Explain.

Determine the unknown sides or angles of the quadrilaterals, 9 through 12, in Figure 24–11.

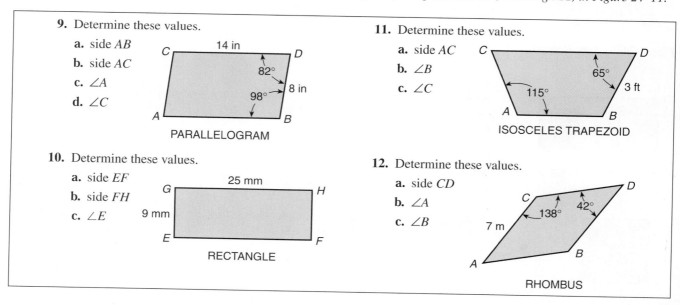

   9. Determine these values.
      a. side *AB*
      b. side *AC*
      c. ∠A
      d. ∠C

   PARALLELOGRAM

  11. Determine these values.
      a. side *AC*
      b. ∠B
      c. ∠C

   ISOSCELES TRAPEZOID

  10. Determine these values.
      a. side *EF*
      b. side *FH*
      c. ∠E

   RECTANGLE

  12. Determine these values.
      a. side *CD*
      b. ∠A
      c. ∠B

   RHOMBUS

Figure 24–11

24–3 Polygon Interior and Exterior Angles

The fields of navigation, land surveying, and manufacturing use the sum of the interior and exterior angles of polygons in many practical applications.

A *diagonal* is a line segment that connects two nonadjacent vertices. If diagonals are drawn in a polygon from any one vertex, there are two fewer triangles $(n - 2)$ formed than the number of sides (n) of the polygon.

EXAMPLE •————————————————————————————————

A quadrilateral is shown in Figure 24–12. Find the sum of the interior angles.

A diagonal is drawn from vertex D to vertex B. Two triangles, ABD and DBC, are formed. Since the sum of the angles of a triangle equals 180°, the sum of the angles in quadrilateral $ABCD = 2(180°) = 360°$ *Ans*

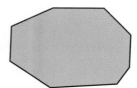

Figure 24–12

 The sum of the interior angles of a polygon of n sides is equal to $(n - 2)$ times 180°.

EXAMPLE •————————————————————————————————

An octagon is shown in Figure 24–13. Find the sum of the interior angles.

The sum of angles in an octagon $= (n - 2)180°$
$$= (8 - 2)180°$$
$$= 1{,}080° \; Ans$$

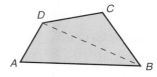

Figure 24–13

24–4 Practical Applications of Polygon Interior and Exterior Angles

1. A survey map of a piece of property is shown in Figure 24–14 on page 496. The surveyor's transit stations are shown by the small triangles at each vertex. The interior angular measurements made at each station are recorded on the survey map. Check the angular measurements using the sum of the interior angles of a polygon.

Solution. The sum of the interior angles $= (n - 2)180°$
$$= (6 - 2)180°$$
$$= 720°$$

Add the angles recorded at each station on the map.

117°10′0″ + 130°46′12″ + 123°18′27″ + 140°33′49″ + 97°53′40″ + 110°17′28″
= 719°59′36″

Subtract. 720° − 719°59′36″ = 0°0′24″

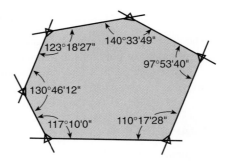

Figure 24–14

There is an angular error of $0°0'24''$ and a correction of $0°0'24''$ must be made. Since there are 6 stations (vertices), $24''$ is divided by 6 ($24'' \div 6 = 4''$). Four seconds are added to each angle at each station.

Calculator Application

117 $\boxed{°\,'\,''}$ 10 $\boxed{°\,'\,''}$ $\boxed{+}$ 130 $\boxed{°\,'\,''}$ 46 $\boxed{°\,'\,''}$ 12 $\boxed{°\,'\,''}$ $\boxed{+}$ 123 $\boxed{°\,'\,''}$ 18 $\boxed{°\,'\,''}$ 27 $\boxed{°\,'\,''}$ $\boxed{+}$ 140 $\boxed{°\,'\,''}$ $\boxed{33}$ $\boxed{°\,'\,''}$ $\boxed{49}$ $\boxed{°\,'\,''}$ $\boxed{+}$ 97 $\boxed{°\,'\,''}$ 53 $\boxed{°\,'\,''}$ 40 $\boxed{°\,'\,''}$ $\boxed{+}$ 110 $\boxed{°\,'\,''}$ 17 $\boxed{°\,'\,''}$ 28 $\boxed{°\,'\,''}$ $\boxed{=}$ $\boxed{\text{SHIFT}}$ $\boxed{°\,'\,''}$ $\rightarrow$
719°59'36''   ⌐ or $\boxed{\text{EXE}}$

or

117.1000 $\boxed{\text{2nd}}$ $\boxed{\blacktriangleright\underline{DD}}$ $\boxed{+}$ 130.4612 $\boxed{\text{2nd}}$ $\boxed{\blacktriangleright\underline{DD}}$ $\boxed{+}$ 123.1827 $\boxed{\text{2nd}}$ $\boxed{\blacktriangleright\underline{DD}}$ $\boxed{+}$ 140.3349 $\boxed{\text{2nd}}$ $\boxed{\blacktriangleright\underline{DD}}$ $\boxed{+}$ 97.5340 $\boxed{\text{2nd}}$ $\boxed{\blacktriangleright\underline{DD}}$ $\boxed{+}$ 110.1728 $\boxed{\text{2nd}}$ $\boxed{\blacktriangleright\underline{DD}}$ $\boxed{\blacktriangleright\underline{DD}}$ $\boxed{\text{3rd}}$ $\boxed{\blacktriangleright\underline{DD}}$ $\rightarrow$ 719°59'36''

2. In laying out the piece shown in Figure 24–15, a sheet metal technician marks off and scribes angles and lengths at each vertex. The angular value at vertex A is not given on the drawing. Determine the angle at vertex A.

Solution.   The sum of the interior angles $= (n - 2)180°$
$$= (5 - 2)180°$$
$$= 540°$$

Add the 4 angles given on the drawing.
$93° + 98° + 105° + 118° = 414°$

The angle at vertex $A = 540° - 414° = 126°$ *Ans*

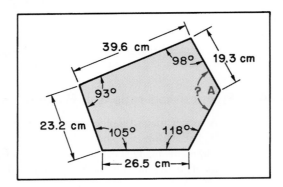

Figure 24–15

Exterior Angles

If a side of a polygon is extended, the angle between the extended side and the adjacent side of the polygon is an *exterior angle.* For example, in Figure 24–16, angles 1, 2, 3, 4, and 5 are exterior angles.

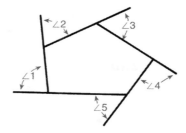

Figure 24–16

EXAMPLE •————————————————————————————————————

A polygon is shown in Figure 24–17. Find the sum of the exterior angles.

From a point inside the polygon, lines are drawn parallel to the sides of the polygon. The respective angles are equal to the exterior angles of the polygon and equal to a complete revolution or a full circle. The sum of the angles equals 360°. *Ans*

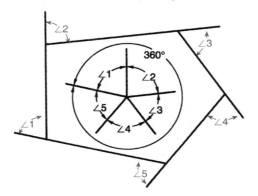

Figure 24–17

The sum of the exterior angles of any polygon, formed as each side is extended in succession, is equal to 360°.

EXAMPLE •————————————————————————————————————

Determine ∠A of the polygon in Figure 24–18.

Compute exterior ∠6.

$62° + 80° + 26° + 58° + 84° + ∠6 = 360°$

$62° + 80° + 26° + 58° + 84° = 310°$

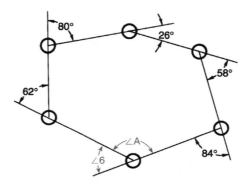

Figure 24–18

$$\angle6 = 360° - 310°$$
$$\angle6 = 50°$$
∠6 and ∠A are supplementary.
$$\angle A = 180° - \angle6 = 130°\ Ans$$

EXERCISE 24–4

1. Find the number of degrees in the sum of the interior angles of each figure listed.
   a. quadrilateral   b. pentagon   c. hexagon   d. octagon

2. Find the number of degrees in the unknown interior angles in each of the polygons in Figure 24–19.

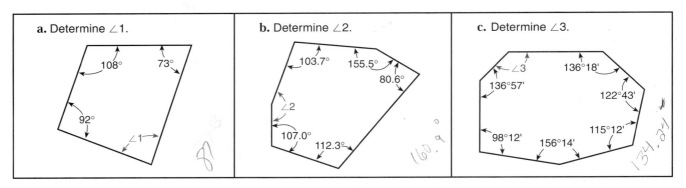

a. Determine ∠1.
108° 73°
92° ∠1

b. Determine ∠2.
103.7° 155.5°
80.6°
∠2
107.0°
112.3°

c. Determine ∠3.
∠3 136°18'
136°57'
122°43'
98°12' 156°14' 115°12'

Figure 24–19

3. Find the number of degrees in each interior angle of the regular figures listed.
   a. pentagon   b. hexagon   c. octagon   d. polygon of 12 sides

4. Find the number of degrees in the unknown exterior angles in each of the polygons in Figure 24–20.

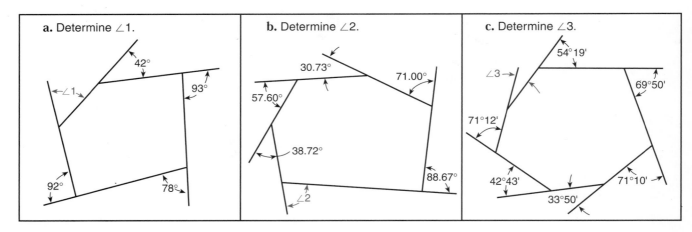

a. Determine ∠1.
42°
∠1 93°
92° 78°

b. Determine ∠2.
30.73° 71.00°
57.60°
38.72°
88.67°
∠2

c. Determine ∠3.
54°19'
∠3 69°50'
71°12'
42°43' 71°10'
33°50'

Figure 24–20

5. Find the number of degrees in each exterior angle of the regular figures listed.
   a. pentagon   b. hexagon   c. octagon   d. polygon of 16 sides

6. Find the number of sides of each regular polygon having an exterior angle of:
    a. 60°   **b.** 24°   **c.** 45°   **d.** 7°30′

7. Each interior angle of a regular polygon of n sides equals $\dfrac{(n-2)180°}{n}$. Find the number of sides of each regular polygon having an interior angle of:
    a. 120°   **b.** 135°   **c.** 108°   **d.** 140°

8. Refer to the template in Figure 24–21.
    a. Determine ∠1 when ∠2 = 83°15′.
    b. Determine ∠2 when ∠1 = 114°30′.

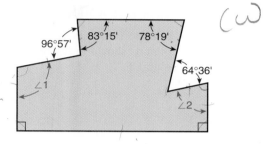

Figure 24–21

9. Refer to the drill jig in Figure 24–22. Determine ∠1.

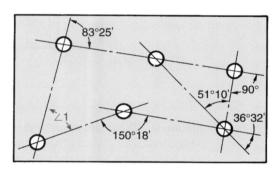

Figure 24–22

10. A survey map of a piece of property is shown in Figure 24–23. Determine ∠1.

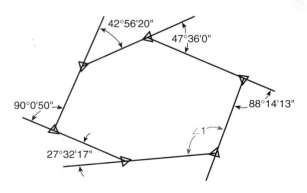

Figure 24–23

Median of a Trapezoid

A *median* of a trapezoid is a line that joins the midpoints of the nonparallel sides (legs).

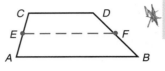

Figure 24–24

> **The median of a trapezoid is parallel to the bases and equal to one-half the sum of the bases.**

In the trapezoid in Figure 24–24, $CD \parallel AB$ and EF is the median. Therefore, $EF \parallel CD \parallel AB$ and $EF = \frac{1}{2}(AB + CD)$.

24–5   Practical Applications of Trapezoid Median

1. A support column (column B) is located midway between the 7'6" and 10'6" high columns shown in the rear view of a carport in Figure 24–25. Determine the height of column B.

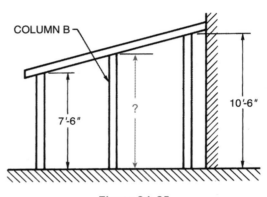

Figure 24–25

Solution.   The figure made by the 7'6" column, the 10'6" column, the roof, and the floor is a trapezoid. The 7'6" and 10'6" columns are the bases of the trapezoid, and column B is midway between the bases. The required height of column B = $\frac{1}{2}$(7'6" + 10'6") or 9' *Ans*

Calculator Application

   B = 0.5(7.5' + 10.5')
   .5 ☒ 〖(7.5 ⊞ 10.5 〗) ⊟ 9
   B = 9' *Ans*

2. The plot of land in Figure 24–26 is in the shape of a trapezoid with $AB \parallel CD$. Stakes have been driven at corners A, B, C, and D of the plot. Measurements are made between stakes. Distances AB, BD, and AC are measured and recorded as shown. The section of land between stakes C and D is heavily wooded, and a direct measurement CD cannot be made. Find distance CD.

   Solution.   One-half of AC (115') and one half of BD (154') are measured, and stakes are driven at points E and F. Distance EF is measured as 258'.

$$EF = \tfrac{1}{2}(AB + CD)$$
$$258' = \tfrac{1}{2}(326' + CD)$$
$$CD = 190' \; Ans$$

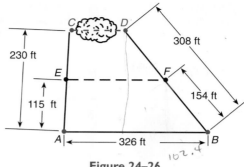

Figure 24–26

Calculator Application

$258' = \frac{1}{2}(326' + CD)$

2 ⊠ 258 ⊟ 326 ⊟ 190

$CD = 190'$ *Ans*

EXERCISE 24–5

1. The sheet stock in the shape of a trapezoid in Figure 24–27 is sheared into 4 pieces of equal length as shown by the broken lines. Determine widths *A*, *B*, and *C*.

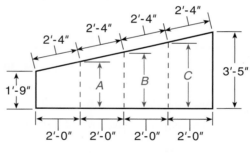

Figure 24–27

2. The frame *ABCD* in Figure 24–28 is in the form of a trapezoid. Cross member *EF* is located midway on side *AD* and side *BC*. If vertical member *EG* is added to the frame, how far should it be located horizontally from vertex *A* (distance *AG*)?

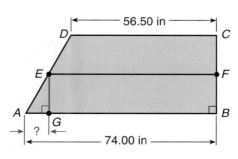

Figure 24–28

▦ UNIT EXERCISE AND PROBLEM REVIEW

IDENTIFYING POLYGONS

Of the shapes in Figure 24–29, which are polygons? Identify the polygons as convex or concave.

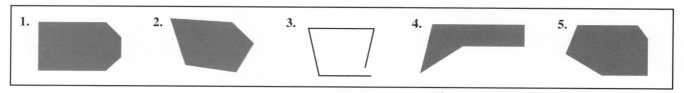

Figure 24–29

Identify each of the polygons in Figure 24–30 as:

(1) *equilateral, equiangular, or regular.*

(2) *a quadrilateral, a pentagon, a hexagon, or an octagon.*

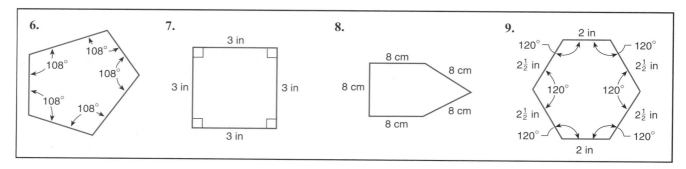

Figure 24–30

DETERMINING SIDES AND ANGLES OF QUADRILATERALS

Determine the unknown sides or angles of the quadrilaterals shown in Figure 24–31.

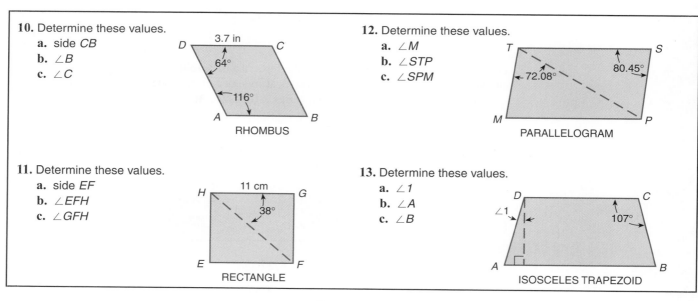

10. Determine these values.
   a. side *CB*
   b. ∠*B*
   c. ∠*C*

RHOMBUS

11. Determine these values.
   a. side *EF*
   b. ∠*EFH*
   c. ∠*GFH*

RECTANGLE

12. Determine these values.
   a. ∠*M*
   b. ∠*STP*
   c. ∠*SPM*

PARALLELOGRAM

13. Determine these values.
   a. ∠1
   b. ∠*A*
   c. ∠*B*

ISOSCELES TRAPEZOID

Figure 24–31

POLYGON INTERIOR AND EXTERIOR ANGLES

14. Find the number of degrees in the sum of the interior angles of each polygon.

   a. hexagon   **b.** 9-sided polygon   **c.** 12-sided polygon

15. Find the number of degrees in each interior angle of these regular polygons.

   a. quadrilateral   **b.** 10-sided polygon   **c.** 15-sided polygon

16. Determine the value of each unknown angle.

   a. Four interior angles of a pentagon are 97°, 142°, 76°, and 103°. Find the fifth angle.

   b. Seven interior angles of an octagon are 94°, 157°, 132°, 119°, 163°, 170°, and 127°. Find the eighth angle.

17. Find the number of degrees in each exterior angle of these regular polygons.

   a. quadrilateral   **b.** 9-sided polygon   **c.** 25-sided polygon

18. Find the number of sides of each regular polygon having an exterior angle of:

   a. 12°   **b.** 22.50°   **c.** 3°45′

19. Find the number of sides of each regular polygon having an interior angle of:

   a. 144°   **b.** 150°   **c.** 162°

20. Refer to the pattern in Figure 24–32.

   a. Determine ∠1 when ∠2 = 66°18′.

   b. Determine ∠2 when ∠1 = 319°51′.

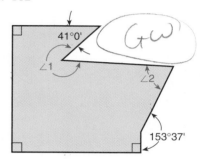

Figure 24–32

21. Refer to the truss in Figure 24–33.

   a. Determine ∠1.

   b. Determine ∠2.

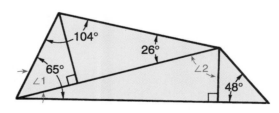

Figure 24–33

TRAPEZOID PROBLEMS

22. Determine the length of the median of these trapezoids.

   a. Bases of 21.6 cm and 36.2 cm

   b. Bases of 14′9″ and 20′7″

23. Determine the length of the other base of each of these trapezoids.

   a. Median of $14\frac{3}{16}$ in and one base of $21\frac{1}{16}$ in

   b. Median of 8.36 m and one base of 0.78 m

24. The frame shown in Figure 24–34 is in the form of an isosceles trapezoid *ABCD*. Cross member *EF* is located midway on sides *AB* and *CD*.

 a. Determine distance *AG* and *HD*.

 b. Determine lengths *AB* and *CD* to the nearest inch.

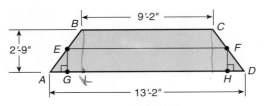

Figure 24–34

UNIT 25 ⦂⦂⦂ Circles

OBJECTIVES

After studying this unit you should be able to

• identify lines and angles used in describing the properties of circles.

• apply circumference and arc length formulas in computations.

• express radians as degrees and degrees as radians.

• apply chord, tangent, arc, and central angle theorems in computations.

• apply tangent and secant theorems in computing arcs and angles formed inside, on, and outside a circle.

• apply internally and externally tangent circles theorems in computations.

• solve practical applied problems, using theorems discussed in this unit.

A circle is a closed curve; every point on the curve is the same distance from a fixed point called the *center*. A *circle* is also defined as the set of all points in a plane that are at a given distance from a given point in the plane.

The circle is the simplest of all closed curves. Circles are easily drawn with a compass and their basic properties are readily understood. Students become acquainted with circles early in their education in making simple designs and constructions.

The uses of circles in everyday living and in occupations are almost unlimited. Circles are important in art, architecture, construction, and manufacturing. Circular designs are often used to create artistic effects. Machines operate by the use of combinations of gears and pulleys. Circular forms are widely found in nature. The earth and most plants have a circular cross section.

The relation of lines to circles is presented in this unit. Radii, diameters, chords, secants, and tangents have wide practical application.

25–1 Definitions

The following terms are commonly used to describe the properties of circles. It is necessary to know and understand these definitions.

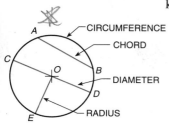

Figure 25–1

The *circumference* is the length of the curved line that forms the circle. See Figure 25–1.

A *chord* is a straight line segment that joins two points on the circle. In Figure 25–1, *AB* is a chord.

A *diameter* is a chord that passes through the center of a circle. In Figure 25–1, *CD* is a diameter.

505

A *radius* (plural radii) is a straight line segment that connects the center of a circle with a point on the circle. A radius is equal to one-half the diameter of a circle. In Figure 25–1, *OE* is a radius. *OC* and *OD* are also radii of the circle in Figure 25–1.

An *arc* is that part of a circle between any two points on the circle. In Figure 25–2, $\overset{\frown}{AB}$ is an arc. The symbol written above the letters means arc.

A *tangent* is a straight line that touches the circle at only one point. The point on the circle touched by the tangent is called the point of tangency. In Figure 25–2, $\overleftrightarrow{CD}$ is a tangent and point *P* is the *point of tangency*.

A *secant* is a straight line passing through a circle and intersecting the circle at two points. In Figure 25–2, $\overleftrightarrow{EF}$ is a secant.

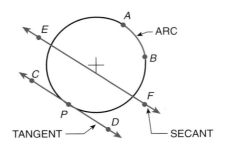

Figure 25–2

A *segment* is a figure formed by an arc and the chord joining the end points of the arc. In Figure 25–3, the shaded figure *ABC* is a segment.

A *sector* is a figure formed by two radii and the arc intercepted by the radii. In Figure 25–3, the shaded figure *EOF* is a sector.

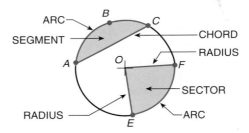

Figure 25–3

A *central angle* is an angle whose vertex is at the center of the circle and whose sides are radii. In Figure 25–4, ∠*MON* is a central angle.

An *inscribed angle* is an angle in a circle whose vertex is on the circle and whose sides are chords. In Figure 25–4, ∠*SRT* is an inscribed angle.

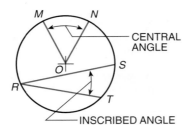

Figure 25–4

EXERCISE 25–1

These problems require the identification of terms used to describe the properties of circles.

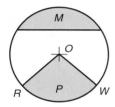

Figure 25–5

1. Refer to Figure 25–5. Write the word that identifies each of the following.

    a. *AB*   **c.** *EO*

    b. *CD*   **d.** Point *O*

2. Refer to Figure 25–6. Write the word that identifies each of the following.

    a. $\overleftrightarrow{HK}$ **d.** $\overrightarrow{LM}$

    b. $\widehat{GF}$ **e.** Point *P*

    c. *GF*

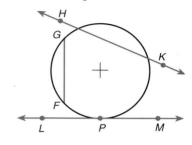

Figure 25–6

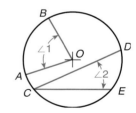

Figure 25–7

3. Refer to Figure 25–7. Write the word that identifies each of the following.

    a. *M*

    b. *P*

    c. $\widehat{RW}$

4. Refer to Figure 25–8. Write the word that identifies each of the following.

    a. ∠1   **d.** *CE*

    b. ∠2   **e.** *CD*

    c. $\widehat{CE}$

Figure 25–8

25–2   **Circumference Formula**

A polygon is *inscribed* in a circle when each vertex of the polygon is a point on the circle. In Figure 25–9, regular polygons are inscribed in circles. As the number of sides increases, the perimeter increases and approaches the circumference.

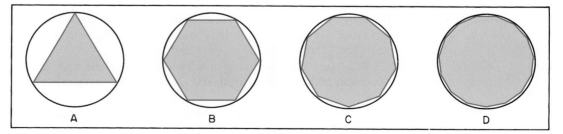

Figure 25–9

An important relationship exists between the circumference and the diameter of a circle. If the circumference of any circle is divided by the length of its diameter, the quotient is always the same number. The number is called *pi* and has a value of 3.1416 to four decimal places. The symbol for pi is π. The value of π cannot be expressed exactly with digits. Pi is called an irrational number.

For computations made without using a calculator, the value of π used depends on the type of problem to be solved and the degree of precision required. For example, a welder working to the nearest tenth inch would compute by using a value of π of 3.14. However, a tool and die maker, when working to the nearest ten-thousandth inch, would use a more precise value of π, such as 3.14159. The most commonly used approximations for π are $3\frac{1}{7}$, 3.14, and 3.1416. As is discussed on the next page, calculators have a π key, $\boxed{\pi}$.

> **The circumference of a circle is equal to pi times the diameter or two pi times the radius.**

Expressed as a formula where C is the circumference, d is the diameter, and r is the radius:

$$C = \pi d \qquad \text{or} \qquad C = 2\pi r$$

EXAMPLES

1. What is the circumference of a circle if the diameter equals 14.5 inches?

   Solution. Since the diameter is given to only 1 decimal place, use 3.14 for the value of π. Substitute values.

   $$C = \pi d$$
   $$C \approx 3.14(14.5 \text{ in})$$
   $$C \approx 45.5 \text{ in } \textit{Ans}$$

2. What is the circumference of a circle if the radius equals 23.764 centimeters?

   Solution. Since the radius is given to 3 decimal places, use 3.1416 for the value of π. Substitute values.

   $$C = 2\pi r$$
   $$C \approx 2(3.1416)\,(23.764 \text{ cm})$$
   $$C \approx 149.314 \text{ cm } \textit{Ans}$$

3. Determine the radius of a circle that has a circumference of 18.4 meters.

   Solution. Substitute values.

   $$C = 2\pi r$$
   $$18.4 \text{ m} \approx 2(3.14)r$$
   $$18.4 \text{ m} \approx (6.28)r$$
   $$r \approx 2.9 \text{ m } \textit{Ans}$$

25–3 Arc Length Formula

There are the same number of degrees in the arc of a central angle as there are in the central angle itself. In Figure 25–10, if central $\angle 1 = 62°$, then $\overset{\frown}{AB} = 62°$ and if $\overset{\frown}{CD} = 150°$, then $\angle 2 = 150°$.

When computing lengths of arcs, consider a complete circle as an arc of 360°. The ratio of the number of degrees of an arc to 360° gives the fractional part of the circumference for the arc.

> **The length of an arc equals the ratio of the number of degrees of the arc to 360° times the circumference.**

Figure 25–10

Expressed as a formula:

$$\text{Length of arc} = \frac{\text{Arc degrees}}{360°}(2\pi r)$$

or

$$\text{Length of arc} = \frac{\text{Central angle}}{360°}(2\pi r)$$

EXAMPLES

1. Determine the length of a 65° arc on a circle with a 4.2-inch radius.

Solution. Substitute values in the formula.

$$\text{Length of arc} = \frac{\text{Arc degrees}}{360°}(2\pi r)$$

$$\approx \frac{65°}{360°}[2(3.14)(4.2 \text{ in})]$$

$$\approx 4.8 \text{ in } Ans$$

Calculator Application

Calculators have the pi key, $\boxed{\pi}$. Depressing the pi key, $\boxed{\pi}$, enters the value of pi to 11 digits (3.1415926536) on most calculators. The display shows the value rounded to 8 digits (3.1415927) on 8-digit display calculators. On many calculators, π is the second or third function. Depending on the calculator used, press the $\boxed{\text{SHIFT}}$, $\boxed{\text{2nd}}$, or $\boxed{\text{3rd}}$ key; then press the appropriate key for which π is the second or third function.

$$\text{Length of arc} = \frac{65°}{360°}[2\pi(4.2 \text{ in})]$$

The following application shows the procedure used with a calculator with the pi key, $\boxed{\pi}$.
65 $\boxed{\div}$ 360 $\boxed{\times}$ 2 $\boxed{\times}$ $\boxed{\pi}$ $\boxed{\times}$ 4.2 $\boxed{=}$ 4.764748858

$$\text{Length of arc} \approx 4.8 \text{ in } Ans$$

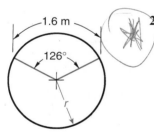

Figure 25–11

2. Determine the radius of the circle in Figure 25–11.

Solution. Substitute values in the formula.

$$\text{Length of arc} = \frac{\text{Central angle}}{360°}(2\pi r)$$

$$1.6 \text{ m} \approx \frac{126°}{360°}(2)(3.14)r$$

$$1.6 \text{ m} \approx 2.20r$$

$$r \approx 0.7 \text{ m } Ans$$

3. Determine the central angle that cuts off an arc length of 2.80 inches on a circle with a 5.00-inch radius.

Solution. Substitute values in the formula.

$$\text{Length of arc} = \frac{\text{Central angle}}{360°}(2\pi r)$$

$$2.80 \text{ in} \approx \frac{\text{Central angle}}{360°}[2(3.14)(5.00 \text{ in})]$$

$$(2.80 \text{ in})(360°) \approx \text{Central angle }(31.4 \text{ in})$$

$$\text{Central angle} \approx 32.1° \text{ } Ans$$

Calculator Application

$$\frac{2.80 \text{ in}(360°)}{2\pi(5.00 \text{ in})} = \text{Central angle}$$

2.8 ☒ 360 ÷ (2 ☒ π ☒ 5) = 32.08563653
Central angle ≈ 32.1° *Ans*

25–4   Radian Measure

In addition to degrees, another unit of angular measure is the *radian*. Radians are used for certain science and engineering applications, such as angular velocity and acceleration.

> **A radian is a central angle that cuts off (intercepts) an arc that is equal to the length of the radius of the circle.**

In Figure 25–12, $\overset{\frown}{AB} = OB = OA$; $\angle AOB = 1$ radian.

Since a full circle can be considered as an arc of 360°, then 360° = 2π (1 radian). Expressing the equation in terms of 1 radian, we write

$$2\pi \text{ (1 radian)} = 360°$$

$$1 \text{ radian} = \frac{\overset{180°}{\cancel{360°}}}{\underset{1}{\cancel{2\pi}}}$$

$$1 \text{ radian} = \frac{180°}{\pi} \approx 57.29578° \text{ (rounded to 5 decimal places)}$$

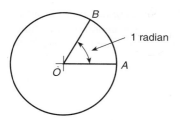

Figure 25–12

It is sometimes necessary to express radians as degrees or degrees as radians. The proportion $\frac{D}{R} = \frac{180°}{\pi}$ can be used to convert between degrees and radians. Here D stands for degrees and R for radians.

EXAMPLES

1. Express 2.7500 radians as degrees, using a calculator.

   First set up and simplify the proportion. Here $R = 2.7500$ and we want to find D.

   $$\frac{D}{R} = \frac{180°}{\pi}$$

   $$\frac{D}{2.7500} = \frac{180°}{\pi}$$

   $$D = \frac{2.7500 \times 180°}{\pi}$$

2.75 ☒ 180 ÷ π = 157.56339, 157.56° (rounded to 2 decimal places) *Ans*

2. Express 35.65° as radians, using a calculator.

   Set up and simplify the proportion with $D = 35.65°$; find R.

   $$\frac{D}{R} = \frac{180°}{\pi}$$

   $$\frac{35.65°}{R} = \frac{180°}{\pi}$$

$$35.65°\pi = 180°R$$

$$R = \frac{35.65°\pi}{180°}$$

35.65 ☒ π ☒ 180 ☒ 0.6222099, 0.62 radian (rounded to 2 decimal places) *Ans*

EXERCISE 25–4

Express the answers to 1 decimal place unless otherwise specified.

1. Determine the circumference of each circle.

   a. $d = 28.0$ in   **c.** $d = 7.50$ ft   **e.** $r = 17.70$ in

   b. $d = 32.80$ cm   **d.** $r = 3.10$ m   **f.** $r = 35.60$ mm

2. The circle size and the arc degrees are given. Determine the arc length.

   a. 6.0-in radius; 45° arc   **d.** 5.0-m diameter; 108° arc

   b. 14.6-cm radius; 62.0° arc   **e.** 40.0-cm diameter; 15° arc

   c. 2.0-ft radius; 130° arc

3. Express each of the following radians as degrees.

   a. 1.62 radians   **b.** 2.080 radians   **c.** 0.876 radian   **d.** 4.094 radians

4. Express each of the following degrees as radians. Round the answers to 4 significant digits.

   a. 75.00°   **b.** 107.63°   **c.** 319.15°   **d.** 16.42°

5. Find the radius of a circle that has a circumference of 38 inches.

6. Determine the central angle that cuts off an arc length of 1.20 meters on a circle with a 3.00-meter radius.

7. Find the radius of a circle in which a 62.8° central angle cuts off an arc length of 18.4 centimeters.

8. Determine the length of belt required to connect the two pulleys in Figure 25–13.

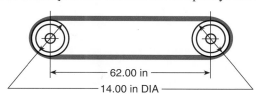

Figure 25–13

9. A circular walk is 1.50 meters wide, as shown in Figure 25–14. If the outer circumference is 72.0 meters, what is the inside diameter of the walk?

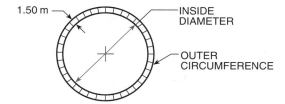

Figure 25–14

10. Determine the length of wire in feet in a coil of 65 turns if the average diameter of the coil is 27.20 inches.

11. An automobile wheel has an outside tire diameter of 27 inches. In going 1.0 mile, how many revolutions does the wheel make? Round the answer to 2 significant digits.

12. A spur gear is shown in Figure 25–15. Pitch circles of spur gears are the imaginary circles of meshing gears that make contact with each other. A pitch diameter is the diameter of a pitch circle. Circular pitch is the length of the arc measured on the pitch circle between the centers of two adjacent teeth. Determine the circular pitch of a spur gear that has 26 teeth and a pitch diameter of 4.1250 inches. Express the answer to 4 decimal places.

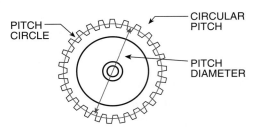

PITCH CIRCLE

CIRCULAR PITCH

PITCH DIAMETER

Figure 25–15

13. The flywheel of a machine has a 0.60-meter diameter and revolves 260 times per minute. How many meters does a point on the outside of the flywheel rim travel in 5 minutes? Round the answer to 2 significant digits.

14. A 1 000.00-meter track is shown in Figure 25–16. The track consists of two semicircles and two equal and parallel straightaways, *AB* and *CD*.

   a. Find the length of each straightaway.

   b. Find distance *L*.

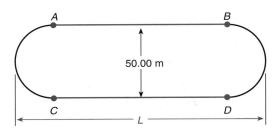

50.00 m

L

Figure 25–16

15. Determine the total distance around (perimeter) the patio in Figure 25–17. Round the answer to the nearest foot.

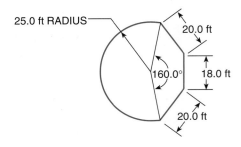

25.0 ft RADIUS

20.0 ft

160.0°

18.0 ft

20.0 ft

Figure 25–17

25–5   Circle Postulates

• All radii of the same circle, or of congruent circles, are equal.

EXAMPLE •————————————————————————————————

Refer to Figure 25–18.

  Given: Circle $A \cong$ Circle E.

  Conclusion: $AB = AC = AD = EF$.

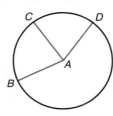
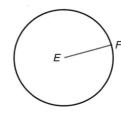

Figure 25–18

• A diameter of a circle bisects the circle and the surface enclosed by the circle; if a line bisects a circle, it is a diameter.

EXAMPLE •————————————————————————————————

Refer to Figure 25–19.

  Given: Circle O with diameter AB.

  Conclusion: $\overarc{ACB} = \overarc{ADB}$

The shaded surface inside the circle is half the total surface inside the circle.

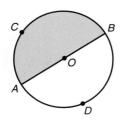

Figure 25–19

• The diameter of a circle is longer than any other chord of that circle.

EXAMPLE •————————————————————————————————

Refer to Figure 25–20.

  Given: Circle O, diameter AB, and chord CD.

  Conclusion: AB is longer than CD.

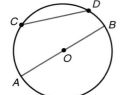

Figure 25–20

• A straight line passing through a point within a circle intersects the circle at two and only two points.

25–6   Chords, Arcs, and Central Angles

Arcs that are referred to or shown as equal are arcs of equal length.

 In the same circle or in congruent circles, equal chords subtend (cut off) equal arcs.

EXAMPLE •————————————————————————————————

Refer to the circles in Figure 25–21.

  Given: Circle A $\cong$ Circle B

      $CD = EF = GH = MS$

  Conclusion: $\overarc{CD} = \overarc{EF} = \overarc{GH} = \overarc{MS}$

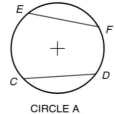

CIRCLE A

CIRCLE B

Equal chords

Figure 25–21

In the same circle or in congruent circles, equal central angles subtend (cut off) equal arcs.

EXAMPLE

Refer to the circles in Figure 25–22.

Given: Circle D ≅ Circle E

∠1 = ∠2 = ∠3 = ∠4

Conclusion: $\widehat{AB} = \widehat{GF} = \widehat{HK} = \widehat{MP}$

arcs are equal

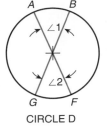

CIRCLE D

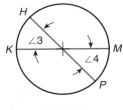

CIRCLE E

Figure 25–22

In the same circle or in congruent circles, two central angles have the same ratio as the arcs that are subtended (cut off) by the angles.

EXAMPLE

Refer to the circles in Figure 25–23.

a. Find $\widehat{EF}$.

b. Find ∠GOH.

Given: Circle A ≅ Circle B

∠COD = 75°

∠EOF = 42°

$\widehat{CD}$ = 18″

$\widehat{GH}$ = 24″

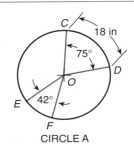

CIRCLE A

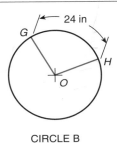

CIRCLE B

Figure 25–23

Solution

a. Set up a proportion with $\widehat{CD}$, $\widehat{EF}$, and their respective central angles.

$$\frac{75°}{42°} = \frac{18″}{\widehat{EF}}$$

$\widehat{EF} = 10.08″$ *Ans*

b. Set up a proportion with $\widehat{CD}$, $\widehat{GH}$, and their respective central angles.

$$\frac{18″}{24″} = \frac{75°}{∠GOH}$$

∠GOH = 100° *Ans*

A diameter perpendicular to a chord bisects the chord and the arcs subtended (cut off) by the chord; the perpendicular bisector of a chord passes through the center of the circle.

EXAMPLE

Refer to the circle in Figure 25–24.

Given: Diameter $DE \perp AB$ at C.

Conclusion: $AC = CB$

$\widehat{AD} = \widehat{DB}$

$\widehat{AE} = \widehat{EB}$

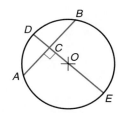

Figure 25–24

25–7 Practical Applications of Circle Chord Bisector

Holes A, B, and C are to be drilled in the plate shown in Figure 25–25. The centers of holes A and C lie on a circle with a diameter of 28.0 cm. The center of hole B lies on the intersection of chord AC and segment OB, which is perpendicular to AC. In order to locate the holes from the left and bottom edges of the plate, working dimensions F, G, and H must be computed.

Solution
Compute F.

$AB = 25.0$ cm $\div$ 2 $= 12.5$ cm

F $= 20.0$ cm $- 12.5$ cm $= 7.5$ cm *Ans*

Compute G.

$BC = AB = 12.5$ cm

G $= 20.0$ cm $+ 12.5$ cm $= 32.5$ cm *Ans*

Compute H. In right $\triangle ABO$, $AB = 12.5$ cm and $AO = 14.0$ cm. Using the Pythagorean theorem, determine OB.

$(14.0 \text{ cm})^2 = (OB)^2 + (12.5 \text{ cm})^2$

$(OB)^2 = 39.75 \text{ cm}^2$

$OB \approx 6.3$ cm (rounded)

H ≈ 18.0 cm $+ 6.3$ cm ≈ 24.3 cm *Ans*

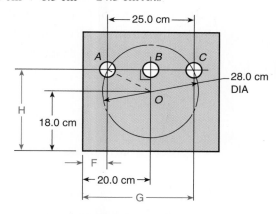

Figure 25–25

Calculator Application

H = 18 $\boxed{+}$ $\boxed{(}$ 14 $\boxed{x^2}$ $\boxed{-}$ 12.5 $\boxed{x^2}$ $\boxed{)}$ $\boxed{\sqrt{x}}$ $\boxed{=}$ 24.30476011, 24.3 cm (rounded) *Ans*

or

H = 18 $\boxed{+}$ $\boxed{\sqrt{}}$ $\boxed{(}$ 14 $\boxed{x^2}$ $\boxed{-}$ 12.5 $\boxed{x^2}$ $\boxed{)}$ $\boxed{\text{EXE}}$ 24.30476011, 24.3 cm (rounded) *Ans*

EXERCISE 25–7

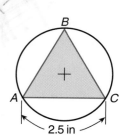

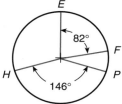

Figure 25–26

Solve these problems. Express the answers to one decimal place unless otherwise specified. Certain problems require the application of two or more theorems in the solutions.

1. Refer to Figure 25–26. △ABC is equilateral. Determine the length of these arcs.
   a. $\widehat{AB}$
   b. $\widehat{BC}$

2. Refer to Figure 25–27. AC and DB are diameters. Determine the length of these arcs.
   a. $\widehat{AB}$
   b. $\widehat{BC}$

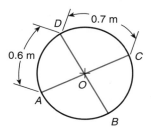

Figure 25–27

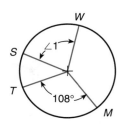

Figure 25–28

3. Refer to Figure 25–28. Determine the length of these arcs.
   a. $\widehat{HP}$ when $\widehat{EF} = 2.8''$
   b. $\widehat{EF}$ when $\widehat{HP} = 5.9''$

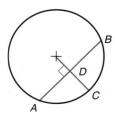

Figure 25–29

4. Refer to Figure 25–29. Determine these values.
   a. $\angle 1$ when $\widehat{SW} = 7.5$ cm and $\widehat{TM} = 10.0$ cm
   b. $\angle 1$ when $\widehat{TM} = 56.85$ mm and $\widehat{SW} = 29.06$ mm

5. Refer to Figure 25–30. Determine these lengths.
   a. DB and $\widehat{ACB}$ when AB = 0.8 m and $\widehat{AC} = 0.6$ m
   b. AB and $\widehat{CB}$ when DB = 1.20 m and $\widehat{ACB} = 2.70$ m

Figure 25–30

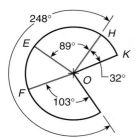

Figure 25–31

6. Refer to Figure 25–31. Determine the length of $\overset{\frown}{HK}$ when $\overset{\frown}{EF}$ = 8.4 in.

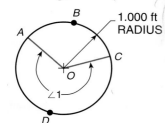

Figure 25–32

7. Refer to Figure 25–32. Determine these values.
  a. The length of $\overset{\frown}{ABC}$ when $\angle 1 = 236°$
  b. $\angle 1$ when $\overset{\frown}{ABC}$ = 2.500 ft

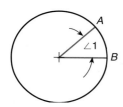

Figure 25–33

8. Refer to Figure 25–33. The circumference of the circle is 160 mm. Determine these values.
  a. The length of $\overset{\frown}{AB}$ when $\angle 1 = 40.05°$
  b. $\angle 1$ when $\overset{\frown}{AB}$ = 35.62 mm

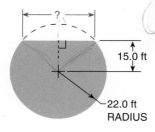

Figure 25–34

9. A circular-shaped concrete slab with one straight edge is to be constructed to the dimensions shown in Figure 25–34. Determine the length of the straight edge.

10. Determine arc length x between the centers of two holes shown in Figure 25–35. Compute the answer to the nearest hundredth millimeter.

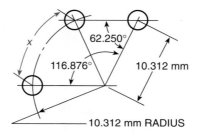

Figure 25–35

11. A circle with a radius OE of 5.6 meters has a chord EF 8.4 meters long. How far is chord EF from the center O of the circle?

  NOTE: It is helpful to sketch and label this problem.

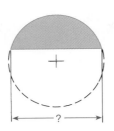

Figure 25–36

12. The only portion of an old tabletop that remains is a segment as shown in Figure 25–36. The segment is less than a semicircle. In order to make a new top the same size as the original, a furniture restorer must determine the diameter of the original top. Describe how the required diameter is determined.

25–8 Circle Tangents and Chord Segments

Tangents and chord segments are used to compute unknown lengths and angles and are often applied in design, layout, and problem solving in the construction and manufacturing occupations.

> **A line perpendicular to a radius at its extremity is tangent to the circle; a tangent to a circle is perpendicular to the radius at the tangent point.**

EXAMPLES •—————————————————————————————————

1. Refer to the circle in Figure 25–37.

   Given: $AB \perp$ radius OC at point C.

   Conclusion: AB is tangent to the circle at point C.

2. Refer to the circle in Figure 25–37.

   Given: DE tangent to the circle at point F.

   Conclusion: $DE \perp$ radius OF.

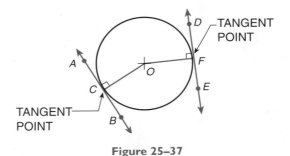

Figure 25–37

25–9 Practical Application of Circle Tangent

In Figure 25–38, AC is tangent to circle O at point C. Determine distance AO.

Solution

    $AC \perp$ radius CO at point C

    $\triangle AOC$ is a right triangle

      $(AO)^2 = (9.25 \text{ m})^2 + (7.35 \text{ m})^2$

      $(AO)^2 = 139.585 \text{ m}^2$

        $AO \approx 11.8 \text{ m } Ans$

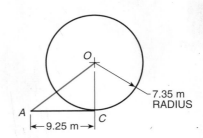

Figure 25–38

> **Two tangents drawn to a circle from a point outside the circle are equal and make equal angles with the line joining the point to the center.**

EXAMPLE

Refer to the circle in Figure 25–39.

> Given: Tangents *AP* and *BP* are drawn to circle *O* from point *P*. Line *PO* is drawn from point *P* to center point *O*.
>
> Conclusion: $AP = BP$
>
> $\angle APO = \angle BPO$

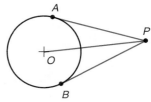

Figure 25–39

25–10 Practical Applications of Tangents from a Common Point

A drawing of a proposed park flower garden is shown in Figure 25–40. *PA* and *PB* are tangent to circle *O*. A walk is to be made along the complete lengths of *PA* and *PB*. Determine the total length of walk required to the nearest inch.

Solution

> $PA \perp$ radius *OA*
>
> $\triangle PAO$ is a right triangle
>
> $52'9'' = 52.75'$ and $34'6'' = 34.5'$
>
> $(52.75 \text{ ft})^2 = (PA)^2 + (34.50 \text{ ft})^2$
>
> $(PA)^2 = 1592.3125 \text{ ft}$
>
> $PA = 39.904 \text{ ft}$
>
> $PB = PA$
>
> $PB = 39.904 \text{ ft}$
>
> Length $= 39.904 \text{ ft} + 39.904 \text{ ft} = 79.81 \text{ ft}, 79'10''$ (rounded) *Ans*

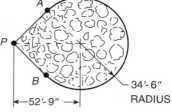

Figure 25–40

Calculator Application

Length $= \sqrt{(PA)^2} \times 2 = \sqrt{(52.75 \text{ ft})^2 - (34.50 \text{ ft})^2} \times 2$

Length $=$ (52.75 x^2 − 34.5 x^2) $\sqrt{x}$ × 2 = 79.80758109, 79.81 ft

or $\sqrt{}$ (52.75 x^2 − 34.5 x^2) × 2 EXE 79.80758109, 79.81 ft

.81 × 12 = 9.72, 9.72 in ≈ 10 in

$79.81' \approx 79'10''$ *Ans*

> **If two chords intersect inside a circle, the product of the two segments of one chord is equal to the product of the two segments of the other chord.**

EXAMPLE •————————————————————————————————————

Refer to the circle in Figure 25–41.

    Given: *AB* and *CD* intersect at point *E*.

    Conclusion: *AE*(*EB*) = *CE*(*ED*)

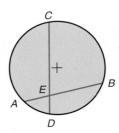

Figure 25–41

If two chords intersect and both segments of one chord are known and one segment of the other chord is known, the second segment can be determined.

EXAMPLE •————————————————————————————————————

In the circle in Figure 25–41, *CE* = 15 cm, *ED* = 3.4 cm, and *AE* = 4.5 cm. Find *EB*.

$$(4.5 \text{ cm})(EB) = (15 \text{ cm})(3.4 \text{ cm})$$
$$EB \approx 11.3 \text{ cm } Ans$$

EXERCISE 25–10 ————————————————————————————————————

Solve these problems. Express the answers to two decimal places unless otherwise specified. Certain problems require the application of two or more theorems in their solutions.

1. Refer to Figure 25–42. Point *P* is a tangent point and $\angle 1 = 109°26'$. Determine these values.
    a. $\angle E$ and $\angle F$ when $\angle 2 = 44°18'$
    b. $\angle 2$ when $\angle E = 46°20'$

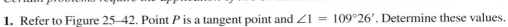

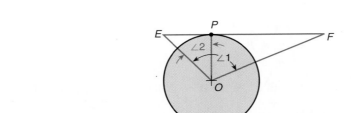

Figure 25–42

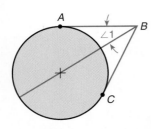

Figure 25–43

2. Refer to Figure 25–43. *AB* and *BC* are tangents. Determine these values.
    a. $\angle 1$ and *BC* when *AB* = 2.78 in and $\angle ABC = 65°0'$
    b. $\angle ABC$ and *AB* when *BC* = 3.93 in and $\angle 1 = 36°47'$

3. Refer to Figure 25–44. Points *E*, *G*, and *F* are tangent points. Determine these values.
    a. $\angle 2$
    b. $\angle 3$ when $\angle 1 = 125°31'$

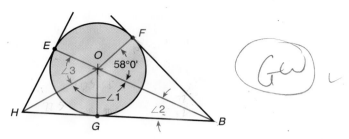

Figure 25–44

4. Refer to Figure 25–45. Determine these values.
   a. *GK* when *EK* = 7.03 m
   b. *EK* when *GK* = 4.98 m

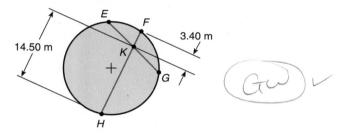

Figure 25–45

5. In the layout shown in Figure 25–46, determine dimension *x* to 3 decimal places.

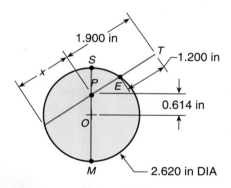

Figure 25–46

6. In the layout shown in Figure 25–47, points *E*, *F*, and *G* are tangent points. Determine lengths *OA*, *OB*, and *OC*.

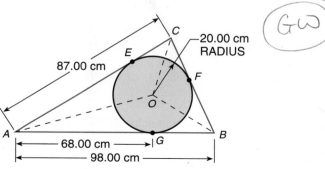

Figure 25–47

7. A circular railing is installed around a portion of the platform shown in Figure 25–48. The railing extends around the circumference from point A to point B ($\overparen{ACB}$). Determine the length of railing required. Round the answer to the nearest foot.

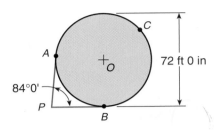

Figure 25–48

8. The front view of a cylindrical shaft setting in the V-groove of a block is shown in Figure 25–49. The center of circular cross section (point O) is directly above vertex B. Determine dimension x to 3 decimal places.

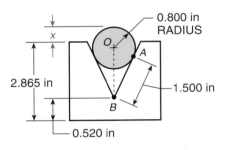

Figure 25–49

25–11   Angles Formed Inside and on a Circle

Arcs and angles inside and on circles have wide practical use, particularly in design and layout in the architectural, mechanical, and manufacturing fields.

Angles Inside a Circle

 An angle formed by two chords that intersect within a circle is measured by one-half the sum of its two intercepted arcs.

EXAMPLES •——————————————————————————————————————

1. Refer to the circle in Figure 25–50.

Given: Chords CD and EF intersect within circle O at point P.

Conclusion: $\angle EPD = \dfrac{1}{2}(\overparen{CF} + \overparen{DE})$

2. In Figure 25–50, $\overparen{CF} = 104°37'$ and $\overparen{DE} = 38°21'$.

Determine $\angle EPD$.

$\angle EPD = \dfrac{1}{2}(104°37' + 38°21')$

$\angle EPD = 71°29'$ *Ans*

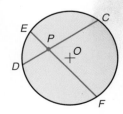

Figure 25–50

3. In Figure 25–50, $\angle EPD = 68°$ and $\overparen{CF} = 96°$.
Determine the number of degrees in $\overparen{DE}$.

$$68° = \frac{1}{2}(96° + \overparen{DE})$$

$$\overparen{DE} = 40° \; Ans$$

Angles on a Circle

An inscribed angle is measured by one-half its intercepted arc.

EXAMPLE

In Figure 25–51, vertex B is a point on circle O and $\overparen{AC} = 103°$.

$$\angle ABC = \frac{1}{2}(103°)$$

$$\angle ABC = 51.5° = 51°30' \; Ans$$

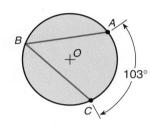

Figures 25–51

25–12 Practical Applications of Inscribed Angles

The center of the disk in Figure 25–52 is located as follows:

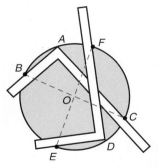

Figure 25–52

- Place the vertex of the carpenter's square anywhere on the circle, such as point A. The outside edges of the square intersect the circle at points B and C. Draw a line, BC, connecting these points. Since $\angle A$ of the carpenter's square is 90°, $\overparen{BC}$ is twice 90° or 180°. Chord BC is a diameter since it cuts off an arc of 180° or a semicircle.

- Place the vertex of the carpenter's square at another location on the circle, such as point D. Connect points E and F. Chord EF is also a diameter.

- The intersection of diameters BC and EF locates the center of the circle, point O.

An angle formed by a tangent and a chord at the tangent point is measured by one-half its intercepted arc.

EXAMPLE

Refer to the circle in Figure 25–53 on page 524.

Given: Tangent CD meets chord AB at tangent point A.
$\overparen{AEB} = 110°36'$

Conclusion: $\angle CAB = \dfrac{1}{2}\overparen{AEB} = \dfrac{1}{2}(110°36') = 55°18'$

$\overparen{AFB} = 360° - 110°36' = 249°24'$

$\angle DAB = \dfrac{1}{2}\overparen{AFB} = \dfrac{1}{2}(249°24') = 124°42'$

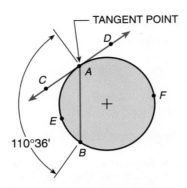

Figure 25–53

25–13   Practical Applications of Tangent and Chord

Refer to Figure 25–54. The centers of 3 holes lie on line *AC*. Line *AC* is tangent to circle *O* at hole center point *B*. The center *D* of a fourth hole lies on the circle. Determine ∠*ABD*.

Solution. $\overset{\frown}{DEB}$ = central ∠*DOB* = 132°54′48″

$$\angle ABD = \frac{1}{2}\overset{\frown}{DEB} = \frac{1}{2}(132°54′48″) = 66°27′24″\ Ans$$

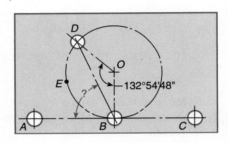

Figure 25–54

Calculator Application

∠*ABD* = 132 $\boxed{°\,'\,''}$ 54 $\boxed{°\,'\,''}$ 48 $\boxed{°\,'\,''}$ ÷ 2 $\boxed{=}$ $\boxed{\text{SHIFT}}$ $\boxed{°\,'\,''}$ → 66°27′24″ *Ans*

or

∠*ABD* = 132.5448 $\boxed{\text{2nd}}$ $\boxed{\blacktriangleright\underline{\text{DD}}}$ ÷ 2 $\boxed{\blacktriangleright\underline{\text{DD}}}$ $\boxed{\text{3rd}}$ $\boxed{\blacktriangleright\underline{\text{DD}}}$ → 66°27′24″ *Ans*

EXERCISE 25–13

Figure 25–55

Solve these problems. Express the answers to the nearest minute. Certain problems require the application of two or more theorems in the solution.

1. Refer to Figure 25–55. Determine these values.
   a. ∠1
   b. ∠2

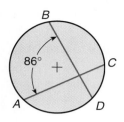

Figure 25–56

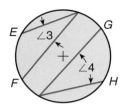

Figure 25–57

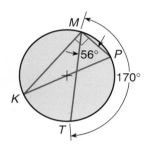

Figure 25–58

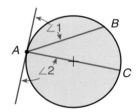

Figure 25–59

2. Refer to Figure 25–56. Determine the number of degrees in these arcs.
  a. $\overarc{AB}$ when $\overarc{DC} = 30°$
  b. $\overarc{DC}$ when $\overarc{AB} = 135°$

3. Refer to Figure 25–57. Determine the number of degrees for each arc or angle.
  a. $\overarc{EF}$ and $\angle 4$ when $\angle 3 = 49°$ and $\overarc{GH} = 84°$.
  b. $\overarc{GH}$ and $\angle 3$ when $\angle 4 = 18°50'$ and $\overarc{EF} = 105°$

4. Refer to Figure 25–58. Determine the number of degrees in these arcs.
  a. $\overarc{PT}$
  b. $\overarc{KT}$
  c. $\overarc{MP}$

5. Refer to Figure 25–59. When $\overarc{AB} = 122°24'$, determine the number of degrees for each arc or angle.
  a. $\angle 1$
  b. $\angle 2$
  c. $\overarc{BC}$

6. Refer to Figure 25–60. Determine the number of degrees for each arc or angle.
  a. $\angle 1$
  b. $\overarc{ADC}$
  c. $\angle 2$

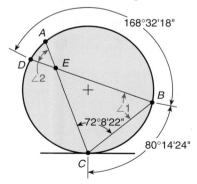

Figure 25–60

7. A triangle is inscribed in a circle. Two angles of the triangle cut off arcs of $62°26'$ and $106°58'$, respectively. Determine the third angle of the triangle.

8. In a circle, an inscribed angle and a central angle cut off the same arc. How do these two angles compare in size?

9. If the end points of two diameters of a circle are connected, what kind of a quadrilateral is formed? Explain your answer.

10. Describe how the carpenter's square is used to check the accuracy of the semicircular cutout in the trim piece in Figure 25–61.

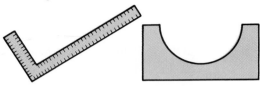

Figure 25–61

11. In Figure 25–62, $\angle CAD = 38°$, $\angle BEC = 40°$, $\widehat{ABC} = 130°$, and $\widehat{CDE} = 134°$. Determine angles 1 through 10.

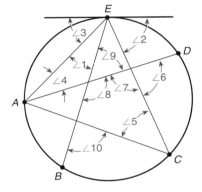

Figure 25–62

25–14   Angles Outside a Circle

An angle formed outside a circle by two secants, two tangents, or a secant and a tangent is measured by one-half the difference of the intercepted arcs.

Two Secants

EXAMPLES

1. Refer to the circle in Figure 25–63.

   Given: Secants ABP and DCP meet at point P and intercept $\widehat{BC}$ and $\widehat{AD}$.

   Conclusion: $\angle P = \dfrac{1}{2}(\widehat{AD} - \widehat{BC})$

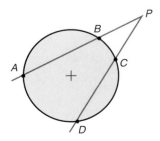

Figure 25–63

2. In Figure 25–63, $\widehat{AD} = 109°$ and $\widehat{BC} = 43°$.

Determine $\angle P$.

$$\angle P = \frac{1}{2}(109° - 43°) = 33° \; Ans$$

3. In Figure 25–63, $\angle P = 28°$ and $\widehat{BC} = 40°$.

Determine $\widehat{AD}$.

$$28° = \frac{1}{2}(\widehat{AD} - 40°)$$

$$\widehat{AD} = 96° \; Ans$$

Two Tangents

EXAMPLES •

1. Refer to Figure 25–64.

Given: Tangents DP and EP meet at point P and intercept $\widehat{DE}$ and $\widehat{DCE}$.

Conclusion: $\angle P = \frac{1}{2}(\widehat{DCE} - \widehat{DE})$

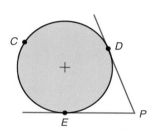

Figure 25–64

2. In Figure 25–64, $\widehat{DCE} = 247°$. Determine $\angle P$.

$$\widehat{DE} = 360° - 247° = 113°$$

$$\angle P = \frac{1}{2}(247° - 113°) = 67° \; Ans$$

A Tangent and a Secant

EXAMPLES •

1. Refer to Figure 25–65.

Given: Tangent AP and secant CBP meet at point P and intercept $\widehat{AC}$ and $\widehat{AB}$.

Conclusion: $\angle P = \frac{1}{2}(\widehat{AC} - \widehat{AB})$

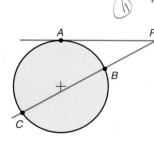

Figure 25–65

2. In Figure 25–65, $\widehat{AC} = 135°$ and $\widehat{AB} = 74°$. Determine $\angle P$.

$$\angle P = \frac{1}{2}(135° - 74°) = 30.5° = 30°30' \; Ans$$

3. In Figure 25–65, $\widehat{AC} = 126°38'$ and $\angle P = 28°50'$.

Determine the number of degrees in $\widehat{AB}$.

$$28°50' = \frac{1}{2}(126°38' - \widehat{AB})$$

$$\widehat{AB} = 126°38' - 2(28°50')$$

$$\widehat{AB} = 68°58' \; Ans$$

Calculator Application

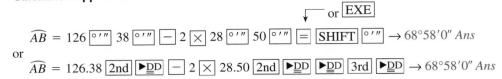

$\overparen{AB}$ = 126 $\boxed{° ′ ″}$ 38 $\boxed{° ′ ″}$ $\boxed{-}$ 2 $\boxed{\times}$ 28 $\boxed{° ′ ″}$ 50 $\boxed{° ′ ″}$ $\boxed{=}$ $\boxed{\text{SHIFT}}$ $\boxed{° ′ ″}$ → 68°58′0″ *Ans*

or

$\overparen{AB}$ = 126.38 $\boxed{\text{2nd}}$ $\boxed{\blacktriangleright\underline{\text{DD}}}$ $\boxed{-}$ 2 $\boxed{\times}$ 28.50 $\boxed{\text{2nd}}$ $\boxed{\blacktriangleright\underline{\text{DD}}}$ $\boxed{\blacktriangleright\underline{\text{DD}}}$ $\boxed{\text{3rd}}$ $\boxed{\blacktriangleright\underline{\text{DD}}}$ → 68°58′0″ *Ans*

25–15 Internally and Externally Tangent Circles

Two circles that are tangent to the same line at the same point are tangent to each other. Circles may be tangent internally or externally.

Two circles are *internally tangent* if both are on the same side of the common tangent line as shown in Figure 25–66.

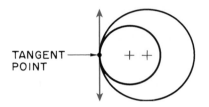

Figure 25–66   Internally tangent circles.

Two circles are *externally tangent* if the circles are on opposite sides of the common tangent line as shown in Figure 25–67.

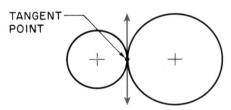

Figure 25–67   External tangent circles.

> **If two circles are either internally or externally tangent, a line connecting the centers of the circles passes through the point of tangency and is perpendicular to the tangent line.**

Internally Tangent Circles

EXAMPLE

Refer to Figure 25–68.

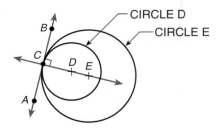

Figure 25–68

Given: Circle D and circle E are internally tangent at point *C*. *D* is the center of circle D. *E* is the center of circle E, and line *AB* is tangent to both circles at point *C*.

Conclusion: Line *DE*, which connects centers *D* and *E*, passes through tangent point *C*, and line *CDE* is ⊥ to tangent line *AB*.

Computing dimensions of objects on which two or more radii blend to give a smooth curved surface is illustrated by this practical application.

25–16 Practical Applications of Internally Tangent Circles

A sheet metal section is to be fabricated as shown in Figure 25–69. The proper location of the two radii results in a smooth curve from point *A* to point *B*. Note that the curve from *A* to *B* is *not* an arc of one circle. It is made up of arcs of two different size circles. In order to lay out the section, the location to the center of the 12.00-inch radius (dimension *x*) must be determined.

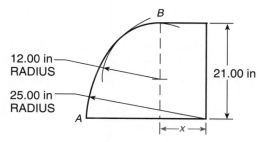

Figure 25–69

Solution.   Refer to Figure 25–70.

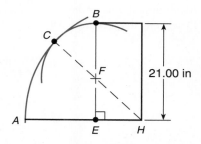

Figure 25–70

- The 12.00″ radius arc and the 25.00″ radius arc are internally tangent. A line connecting centers *F* and *H* of the arcs passes through the point of tangency, point *C*.

- Since tangent point *C* is the end point of the 25.00″ radius, *CFH* = 25.00″. Also, since point *C* is the end point of the 12.00″ radius, *CF* = 12.00″.

  FH = 25.00″ − 12.00″ = 13.00″

- Since *BFE* is vertical and *AEH* is horizontal, ∠*FEH* is a right angle. △*FEH* is a right triangle.

- In right △*FEH*,

  FH = 13.00″

  FE = 21.00″ − 12.00″ = 9.00″

  Compute *EH*, using the Pythagorean theorem.

  $(13.00″)^2 = (EH)^2 + (9.00″)^2$

  $EH \approx 9.38″$

  $EH = x \approx 9.38″$ *Ans*

Externally Tangent Circles

EXAMPLE

Refer to Figure 25–71.

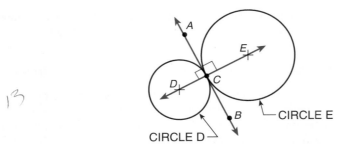

Figure 25–71

Given: Circle D and circle E are externally tangent at point C. D is the center of circle D. E is the center of circle E, and line AB is tangent to both circles at point C.

Conclusion: Line DE, which connects centers D and E, passes through tangent point C, and line DCE is ⊥ to tangent line AB at point C.

25–17   Practical Applications of Externally Tangent Circles

Three holes are to be bored in a metal plate as shown in Figure 25–72. The 42.00-mm and 61.40-mm diameter holes are tangent at point D, and CD is the common tangent line. Determine the distances between hole centers AB, AC, and BC.

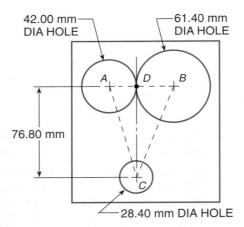

Figure 25–72

Solution

• Determine AB: Since AB connects the centers of two tangent circles, AB passes through tangent point D.

$AD = 42.00 \text{ mm} \div 2 = 21.00 \text{ mm}$

$DB = 61.40 \text{ mm} \div 2 = 30.70 \text{ mm}$

$AB = 21.00 \text{ mm} + 30.70 \text{ mm} = 51.70 \text{ mm}$ *Ans*

• Determine *AC* and *BC*: Since *AB* connects the centers of two tangent circles, *AB* is ⊥ to tangent line *DC*. Therefore, ∠*ADC* and ∠*BDC* are both right angles.

$$(AC)^2 = (21.00 \text{ mm})^2 + (76.80 \text{ mm})^2$$
$$(AC)^2 = 6339.24 \text{ mm}^2$$
$$AC \approx 79.62 \text{ mm } \textit{Ans}$$
$$(BC)^2 = (30.70 \text{ mm})^2 + (76.80 \text{ mm})^2$$
$$BC \approx 82.71 \text{ mm } \textit{Ans}$$

EXERCISE 25–17

Solve these problems. Where necessary, express angular value answers to the nearest minute or hundredth degree and length answers to two decimal places. Certain problems require the application of two or more theorems in their solutions.

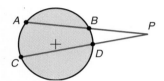

Figure 25–73

1. Refer to Figure 25–73.
   a. Determine ∠*P* when $\overset{\frown}{AC} = 58°$ and $\overset{\frown}{BD} = 32°$.
   b. Determine ∠*P* when $\overset{\frown}{AC} = 63.30°$ and $\overset{\frown}{BD} = 28.70°$.
   c. Determine the arc degrees for $\overset{\frown}{BD}$ when ∠*P* = 16° and $\overset{\frown}{AC} = 64°$.
   d. Determine the arc degrees for $\overset{\frown}{AC}$ when ∠*P* = 37° and $\overset{\frown}{BD} = 30°$.
   e. Determine ∠*P* when $\overset{\frown}{CAB} = 178°19'$, $\overset{\frown}{CD} = 156°47'$, and $\overset{\frown}{AC} = 52°0'$.
   f. Determine the arc degrees for $\overset{\frown}{AC}$ when $\overset{\frown}{CD} = 140°$, $\overset{\frown}{CAB} = 193°$, and ∠*P* = 39°.

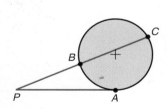

Figure 25–74

2. Refer to Figure 25–74. *A* and *B* are tangent points.
   a. Determine ∠*P* when $\overset{\frown}{AB} = 120°$.
   b. Determine ∠*P* when $\overset{\frown}{ACB} = 237.62°$.
   c. Determine ∠*P* when $\overset{\frown}{AC} = 160°$ and $\overset{\frown}{CB} = 88°$.
   d. Compare $\overset{\frown}{ACB}$ with $\overset{\frown}{AB}$ when ∠*P* is a very small value.

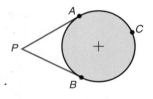

Figure 25–75

3. Refer to Figure 25–75. *A* is a tangent point.
   a. Determine ∠*P* when $\overset{\frown}{AC} = 120°$ and $\overset{\frown}{BA} = 70°$.
   b. Determine ∠*P* when $\overset{\frown}{AC} = 113.07°$ and $\overset{\frown}{BA} = 63.94°$
   c. Determine the arc degrees for $\overset{\frown}{AC}$ when ∠*P* = 25° and $\overset{\frown}{BA} = 58°$.
   d. Determine the arc degrees for $\overset{\frown}{BA}$ when ∠*P* = 37° and $\overset{\frown}{AC} = 85°$.
   e. Determine ∠*P* when $\overset{\frown}{BC} = 160°12'46''$ and $\overset{\frown}{AC} = 112°58'50''$.
   f. Determine ∠*P* when $\overset{\frown}{BC} = 150°$ and $\overset{\frown}{AC} = 2\overset{\frown}{BA}$.

4. Refer to Figure 25–76. *A* and *D* are tangent points. Determine ∠1, ∠2, and ∠3 when $\overset{\frown}{AB} = 72°$ and $\overset{\frown}{CD} = 50°$.

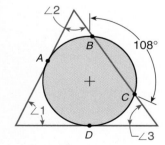

Figure 25–76

Figure 25–77

5. Refer to Figure 25–77. *E* is a tangent point.

   a. Find the arc degrees for $\widehat{DH}$ and $\widehat{EDH}$ when $\angle 1 = 26°$ and $\angle 2 = 63°$.

   b. Find the arc degrees for $\widehat{HK}$ when $\angle 2 = 49.08°$.

6. Refer to Figure 25–78. *A* is a tangent point.

   a. Find *x* when diameter $A = 41.25$ in and diameter $B = 23.65$ in.

   b. Find diameter *A* when $x = 9.83$ in and diameter $B = 11.58$ in.

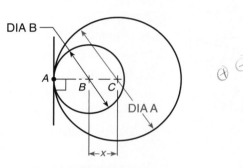

Figure 25–78

7. Refer to Figure 25–79.

   a. Find diameter *A* when $x = 9.82$ cm and $y = 11.94$ cm.

   b. Find *y* when diameter $A = 30.36$ cm and $x = 23.72$ cm.

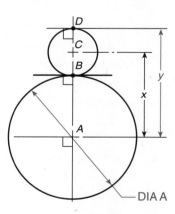

Figure 25–79

8. Refer to Figure 25–80.

   a. Find the arc degrees for $\widehat{AB}$ and $\widehat{DE}$ when $\angle 1 = 71°$ and $\angle 2 = 97°$.

   b. Find the arc degrees for $\widehat{AB}$ and $\widehat{DE}$ when $\angle 1 = 66°12'$ and $\angle 2 = 91°46'$.

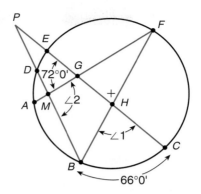

Figure 25–80

9. Refer to Figure 25–81. *T* and *M* are tangent points.
   a. Find *x* when diameter *A* = 10.00 m.
   b. Find diameter *A* when *x* = 18.00 m.

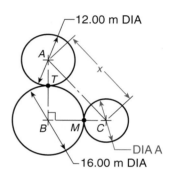

Figure 25–81

10. Determine the length of stock, dimension *L*, required to make the gauge shown in Figure 25–82.

   NOTE: The gauge is symmetrical (identical) on each side of the vertical centerline (℄).

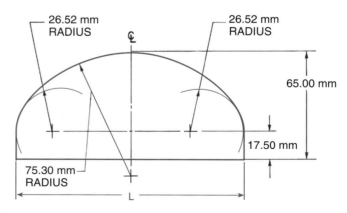

Figure 25–82

11. Three posts are mounted on the fixture shown in Figure 25–83 on page 534. Each post is tangent to the arc made by the 0.650-inch radius. Determine dimension *A* and dimension *B* to 3 decimal places.

   NOTE: The fixture is symmetrical (identical) on each side of the horizontal centerline (℄).

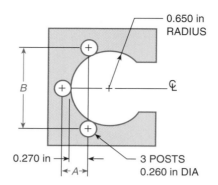

Figure 25–83

12. Three holes are to be located on the layout shown in Figure 25–84. The 7.24-cm- and 3.08-cm-diameter holes are tangent at point *T*, and *TA* is the common tangent line between the two holes. Determine dimension *C* and dimension *D*.

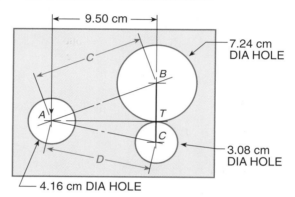

Figure 25–84

UNIT EXERCISE AND PROBLEM REVIEW

IDENTIFYING PARTS

These problems require the identification of terms.

1. Refer to Figure 25–85. Write the word that identifies each of the following.

   a. *AB*   **d.** $\overset{\frown}{AC}$

   b. *CD*   **e.** $\overleftrightarrow{EF}$

   c. *OD*   **f.** point *G*

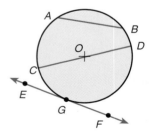

Figure 25–85

2. Refer to Figure 25–86. Write the word that identifies each of the following.

   a. $\overleftrightarrow{AB}$ **d.** *E*

   b. ∠1   **e.** *F*

   c. ∠2

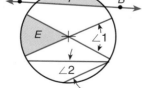

Figure 25–86

CIRCUMFERENCE AND ARC LENGTH

Express the answers to one decimal place unless otherwise specified.

3. Determine the circumference of each circle.

    a. $d = 32.00$ cm   **b.** $d = 5.200$ in   **c.** $r = 0.80$ m

4. The circle size and the arc degrees are given. Determine the arc length.

    a. 8.0-in radius; 60° arc

    b. 6.0-m diameter; 120° arc

    c. 3.0-ft diameter; 25° arc

5. Determine the radius of a circle that has a circumference of 15.27 meters.

6. Determine the central angle that cuts off an arc length of 7.500 feet on a circle with a 4.000-foot radius.

7. A pipe with a wall thickness of 0.20 inch has an outside diameter of 3.50 inches. Determine the inside circumference of the pipe.

8. The centers of two 15.00-centimeter diameter pulleys are 50.00 centimeters apart. Determine the length of belt required to connect the two pulleys.

9. A truck wheel has an outside tire diameter of 45.0 inches. In going one-half mile, how many revolutions does the wheel make? Round the answer to 3 significant digits.

RADIAN MEASURE

Express the answers to one decimal place.

10. Express each of the following radians as degrees.

    a. 1.50 radians   **b.** 5.080 radians   **c.** 0.860 radian

11. Express each of the following degrees as radians.

    a. 96.3°   **b.** 15.09°   **c.** 193.78°

CHORD, ARC, AND CENTRAL ANGLE APPLICATIONS

Solve these problems. Express answers to one decimal place unless otherwise specified. Certain problems require the application of two or more theorems in the solution.

12. Refer to Figure 25–87. *ABCDEF* is a regular hexagon. Determine the number of degrees in each arc or angle.

    a. $\overset{\frown}{BC}$ **b.** $\overset{\frown}{CDE}$ **c.** ∠1

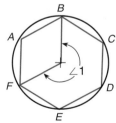

Figure 25–87

13. Refer to Figure 25–88.

    a. Find the length of $\overset{\frown}{MH}$ when $\overset{\frown}{EF} = 1.5$ m.

    b. Find the length of $\overset{\frown}{EF}$ when $\overset{\frown}{MH} = 3.2$ m.

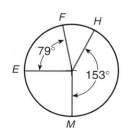

Figure 25–88

14. Refer to Figure 25–89.

   a. Find the length of $\overparen{ABC}$ when $\angle 1 = 115°$
   b. Find $\angle 1$ when $\overparen{ADC} = 32.00$ in.

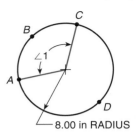

8.00 in RADIUS

Figure 25–89

15. Refer to Figure 25–90.

   a. Find the lengths of EH and $\overparen{EGF}$ when
       $EF = 23$ cm and $\overparen{EG} = 14$ cm.
   b. Find the lengths of $\overparen{EGF}$ and HF when
       $EF = 35.86$ cm and $\overparen{GF} = 20.52$ cm.

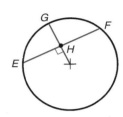

Figure 25–90

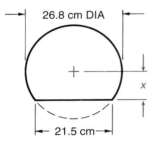

26.8 cm DIA

x

21.5 cm

Figure 25–91

16. Circle O has a diameter AB of 14.50 feet and a chord CD of 8.00 feet. Chord CD is perpendicular to diameter AB. How far is chord CD from the center O of the circle? Express the answer to 2 decimal places.

NOTE: It is helpful to sketch and label this problem.

17. A flat is cut on a circular piece as shown in Figure 25–91. Determine the distance from the center of the circle to the flat, dimension x.

TANGENT AND CHORD SEGMENT APPLICATIONS

Solve these problems. Express the answers to one decimal place unless otherwise specified. Certain problems require the application of two or more theorems in the solution.

18. Refer to Figure 25–92. AT and BT are tangents.

   a. Find $\angle 1$ and AT when $BT = 5'3''$ and $\angle BTA = 59°20'$.
   b. Find $\angle ATB$ and BT when $\angle 1 = 30.28°$ and $AT = 2.65$ m.

Figure 25–92

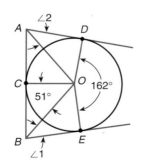

51°

162°

$\angle 2$

$\angle 1$

19. Refer to Figure 25–93. Points C, D, and E are tangent points.

   a. Find $\angle 1$.   **b.** Find $\angle 2$.

Figure 25–93

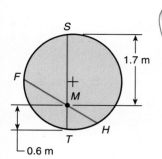

Figure 25–94

20. Refer to Figure 25–94.

   a. Find *FM* when *MH* = 0.8 m.

   b. Find *MH* when *FM* = 1.3 m.

21. Refer to Figure 25–95.

   Determine dimension *x*.

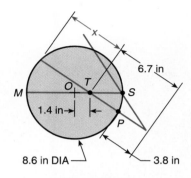

Figure 25–95

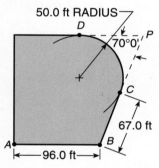

Figure 25–96

22. A sidewalk is constructed along distance *ABCD* of the parcel of land shown in Figure 25–96. Determine the total length of sidewalk required. Round the answer to 3 significant digits.

ANGLES FORMED INSIDE, ON, AND OUTSIDE A CIRCLE

Solve these problems. Certain problems require the application of two or more theorems in the solution. Express answers to the nearest minute or hundredth of a degree unless otherwise specified.

23. Refer to Figure 25–97. Find the number of degrees in each arc or angle.

   a. $\angle 1$ when $\widehat{BC} = 62°18'$ and $\widehat{AD} = 57°6'$

   b. $\widehat{AD}$ when $\angle 1 = 42°$ and $\widehat{BC} = 50°$

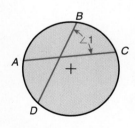

Figure 25–97

Figure 25–98

24. Refer to Figure 25–98. Find the number of degrees in each angle.

   a. $\angle 1$ when $\widehat{JH} = 86°$

   b. $\angle 2$ when $\widehat{FGHJK} = 197.08°$

25. Refer to Figure 25–99. Find the number of
degrees in each arc.

   a. $\overset{\frown}{BC}$

   b. $\overset{\frown}{DC}$

   c. $\overset{\frown}{AB}$

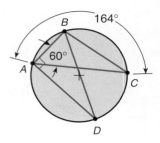

Figure 25–99

26. Refer to Figure 25–100. When $\overset{\frown}{BCA} = 255°18'$,
find the number of degrees in each angle.

   a. $\angle 1$

   b. $\angle 2$

   c. $\angle 3$

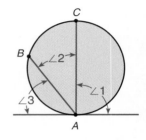

Figure 25–100

27. Refer to Figure 25–101. Round the answers to the nearest second.

   a. Determine $\angle P$ when $\overset{\frown}{AD} = 28°34'6''$ and $\overset{\frown}{BC} = 60°12'38''$.

   b. Determine the arc degrees for $\overset{\frown}{BC}$ when $\angle P = 34°0'0''$ and $\overset{\frown}{AD} = 20°56'14''$.

   c. Determine the arc degrees for $\overset{\frown}{AD}$ when $\overset{\frown}{AB} = 94°12'0''$, $\overset{\frown}{ABC} = 195°24'0''$ and
$\angle P = 35°8'20''$.

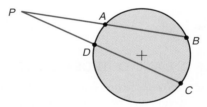

Figure 25–101

28. Refer to Figure 25–102. *E* and *F* are tangent points.

   a. Determine $\angle P$ when $\overset{\frown}{EF} = 156°$.

   b. Determine $\angle P$ when $\overset{\frown}{EGF} = 221°$.

   c. Determine $\angle P$ when $\overset{\frown}{EGF} = 3\overset{\frown}{EF}$.

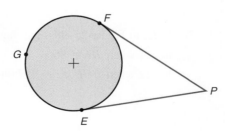

Figure 25–102

29. Refer to Figure 25–103. *C* is a tangent point.

    a. Determine $\angle P$ when $\widehat{BC} = 64.00°$ and $\widehat{AC} = 134.46°$.

    b. Determine the arc degrees for $\widehat{AC}$ when $\angle P = 59.00°$ and $\widehat{BC} = 80.00°$.

    c. Determine $\angle P$ when $\widehat{BA} = 119.32°$ and $\widehat{BC} = \frac{1}{2}\widehat{AC}$.

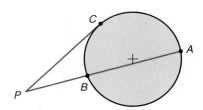

Figure 25–103

30. A quadrilateral is inscribed in a circle. Three vertex angles of the quadrilateral cut off arcs of 180°, 170°, and 230°, respectively. Determine the fourth vertex angle of the quadrilateral.

31. In Figure 25–104, points *A* and *C* are tangent points, *DC* is a diameter, $\widehat{AHC} = 116°$, $\widehat{EFC} = 140°$, $\widehat{EF} = 64°$, and $\widehat{CH} = 42°$. Determine angles 1 through 10.

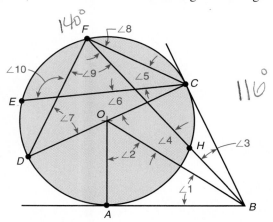

Figure 25–104

INTERNALLY AND EXTERNALLY TANGENT CIRCLES

Solve these problems. Certain problems require the application of two or more theorems in the solution. Express answers to two decimal places.

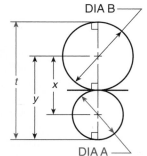

Figure 25–105

32. Refer to Figure 25–105.

    a. Find diameter A when $x = 3.60$ inches and $y = 5.10$ inches.

    b. Find *t* when diameter A $= 8.76$ inches and $x = 10.52$ inches.

33. Refer to Figure 25–106.

    a. Find diameter *M* when diameter H = 7.20 m, diameter T = 5.80 m, *e* = 4.70 m, and *d* = 4.10 m.

    b. Find dimension *f* when diameter T = 3.60 m, diameter H = 5.60 m, *e* = 7.80 m, and *d* = 6.20 m.

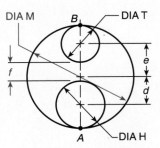

Figure 25–106

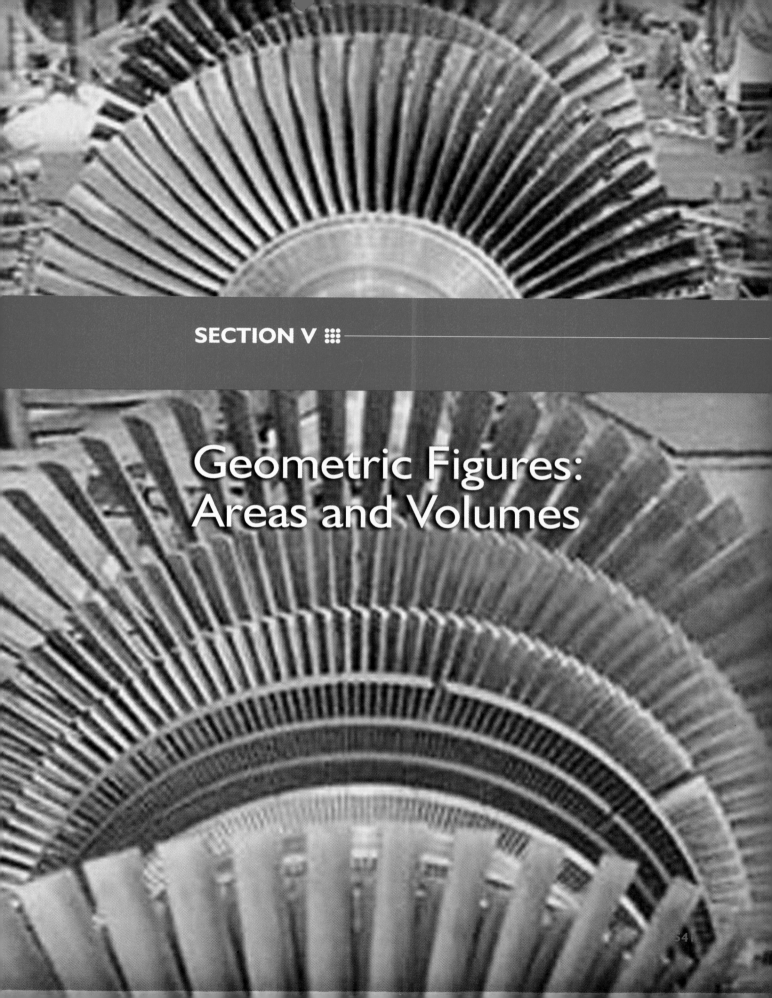

SECTION V ⁞⁞⁞

Geometric Figures: Areas and Volumes

UNIT 26 ⣿ Areas of Common Polygons

OBJECTIVES

After studying this unit you should be able to

- compute areas of common polygons, given bases and heights.
- compute heights of common polygons, given bases and areas.
- compute bases of common polygons, given heights and areas.
- compute areas of more complex figures that consist of two or more common polygons.
- solve applied problems by using principles discussed in this unit.

Many occupations require computations of common polygon areas in estimating job materials and costs. It is also sometimes necessary to find unknown lengths, widths, and heights of polygons when the areas are known.

Methods of computing areas, sides, and heights of rectangles, parallelograms, trapezoids, and triangles are presented in this unit. The areas of complex polygons are found by dividing the complex polygons into two or more of these simpler figures.

26–1 Areas of Rectangles

A *rectangle* is a four-sided polygon with opposite sides equal and parallel and with each angle equal to a right angle. The *area of a rectangle* is equal to the product of the length and width.

$$A = lw \qquad \text{where } A = \text{area}$$
$$l = \text{length}$$
$$w = \text{width}$$

EXAMPLES

1. A rectangular platform is 24 feet long and 14 feet wide. Find the area of the platform.

   $A = 24 \text{ ft} \times 14 \text{ ft}$

   $A = 336 \text{ sq ft } Ans$

2. Determine the area, in square meters, of a rectangular strip of sheet stock 24 centimeters wide and 3.65 meters long.

   To find the area in square meters, express both the length and width in meters.

   Length = 3.65 m

   Width = 24 cm = 0.24 m

   Find the area.

   $A = 3.65 \text{ m} \times 0.24 \text{ m}$

   $A \approx 0.88 \text{ m}^2 \text{ (rounded) } Ans$

3. A building floor plan is shown in Figure 26–1. Find the area of the floor.

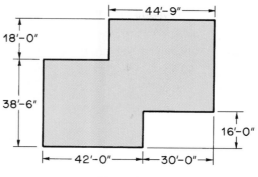

Figure 26–1

Divide the figure into rectangles. One way of dividing the figure is shown in Figure 26–2.

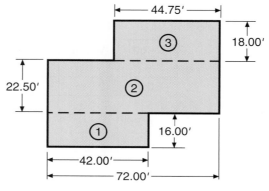

Figure 26–2

To find the area of the floor, compute the area of each rectangle and add the three areas.

• Area of rectangle ①.

  $A = 42.00 \text{ ft} \times 16.00 \text{ ft}$

  $A = 672.0 \text{ sq ft}$

• Area of rectangle ②.

  Length $= 42.00 \text{ ft} + 30.00 \text{ ft} = 72.00 \text{ ft}$

  Width $= 38.50 \text{ ft} - 16.00 \text{ ft} = 22.50 \text{ ft}$

  $A = 72.00 \text{ ft} \times 22.50 \text{ ft}$

  $A = 1,620 \text{ sq ft}$

• Area of rectangle ③.

  Length $= 44.75 \text{ ft}$

  Width $= 18.00 \text{ ft}$

  $A = 44.75 \text{ ft} \times 18.00 \text{ ft}$

  $A = 805.5 \text{ sq ft}$

• Total area $= 672.0 \text{ sq ft} + 1,620.0 \text{ sq ft} + 805.5 \text{ sq ft} \approx 3,098 \text{ sq ft}$ (rounded) *Ans*

Calculator Application

42 ⊠ 16 ⊞ 72 ⊠ 22.5 ⊞ 44.75 ⊠ 18 ⊟ 3097.5

area ①   area ②   area ③   3,098 sq ft (rounded) *Ans*

4. A rectangular patio is to have an area of 40.5 square meters. If the length is 9.0 meters, find the required width.

Substitute values in the formula and solve.

$$A = lw$$

$$40.5 \text{ m}^2 = 9.0 \text{ m}(w)$$

$$\frac{40.5 \text{ m}^2}{9.0 \text{ m}} = w$$

$$w = 4.5 \text{ m } Ans$$

EXERCISE 26–1

If necessary, use the tables found in appendix A for equivalent units of measure. Find the unknown area, length, or width for each of the rectangles, 1 through 16, in Figure 26–3. Round the answers to one decimal place.

	Length	Width	Area
1.	3.0 ft	7.0 ft	
2.	5.0 m	9.0 m	
3.	8.5 in	6.0 in	
4.	2.6 yd		11.7 sq yd
5.		23 cm	200.1 cm²
6.	0.4 km		0.2 km²
7.	12'6"	7'0"	
8.		0.086 mi	0.136 sq mi

	Length	Width	Area
9.	26.2 mm		366.8 mm²
10.	5.600 cm	18.80 cm	
11.		39.8 in	31.7 sq in
12.	64.2 m		3762 m²
13.	12'3.0"	16'9.0"	
14.	2.95 km	0.76 km	
15.	7.4 mi		6.7 sq mi
16.		125.0 ft	26,160 sq ft

Figure 26–3

Solve these problems. Round the answers to two decimal places unless otherwise specified.

17. A rectangular strip of steel is 9.00 inches wide and 6.50 feet long. Find the area of the strip in square feet.

18. A school shop 10.0 meters wide and 14.0 meters long is to be built. Allow 5.0 square meters for each workstation. How many workstations can be provided?

19. Carpet is installed in a room 12'0" wide and 24'0" long. The cost of the carpet is $18.75 per square yard. The installation cost is $75.00. Find the total cost of carpeting the room.

20. A square window contains 729 square inches of glass. Find the length of each of the sides of the window.

21. The cost of a rectangular plate of aluminum 3'0" wide and 4'0" long is $45. Find the cost of a rectangular plate 6'0" wide and 8'0" long, using this same stock.

22. The walls of a room 14'0" × 18'0" are wallpapered. The height of the walls is 7'6". The room has a doorway that is 3'0" × 7'0", and 4 windows each 3'0" × 5'0". Each roll of wallpaper has an area of 60.0 sq ft. An allowance of 20% is made for waste.

   a. Find the wall area to be wallpapered. Round the answer to the nearest ten square feet.

   b. How many rolls are required for this job? Round the answer to the nearest whole roll.

23. Find the area of the sheet metal piece in Figure 26–4.

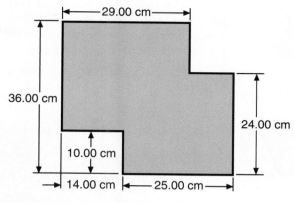

Figure 26–4

24. The bottom of a rectangular carton is to have an area of 2400 square centimeters. The length is to be one and one-half times the width. Compute the length and width dimensions.

25. The plot of land shown in Figure 26–5 has an area of 6350 square meters. Find distance x.

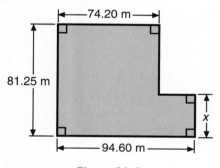

Figure 26–5

26. A structural supporting member is made in the shape shown in Figure 26–6. What is the cross-sectional area?

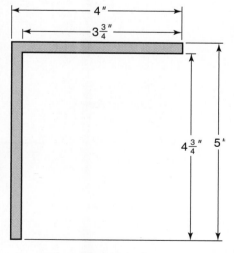

Figure 26–6

26–2 Areas of Parallelograms

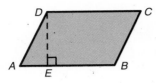

Figure 26–7

A *parallelogram* is a four-sided polygon with opposite sides parallel and equal. The *area of a parallelogram* is equal to the product of the base and height. An *altitude* is a segment perpendicular to the line containing the base drawn from a side opposite the base. The *height* is the length of the altitude.

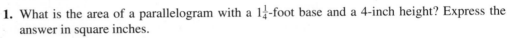

$$A = bh$$

where A = area
b = base
h = height or altitude

Figure 26–8

In Figure 26–7, AB is a base and DE is a height of parallelogram $ABCD$.

$$\text{Area of parallelogram } ABCD = AB(DE)$$

In Figure 26–8, BC is a base and DF is a height of parallelogram $ABCD$.

$$\text{Area of parallelogram } ABCD = BC(DF)$$

EXAMPLES

1. What is the area of a parallelogram with a $1\frac{1}{4}$-foot base and a 4-inch height? Express the answer in square inches.

   To find the area in square inches, express both the base and height in inches.

   $$\text{Base} = 1\frac{1}{4} \text{ ft} = 15 \text{ in}$$

   Altitude = 4 in

   $A = 15 \text{ in} \times 4 \text{ in}$

   $A = 60$ sq in *Ans*

2. The lot shown in Figure 26–9 is in the shape of a parallelogram. Find the area of the lot to the nearest ten meters.

   Find the base.

   Base = 46.5 m + 20.2 m = 66.7 m

   Find the height by the Pythagorean theorem.

   $$(40.4 \text{ m})^2 = (20.2 \text{ m})^2 + h^2$$
   $$1\,632.16 \text{ m}^2 = 408.04 \text{ m}^2 + h^2$$
   $$1\,224.12 \text{ m}^2 = h^2$$
   $$h = \sqrt{1\,224.12 \text{ m}^2}$$
   $$h \approx 34.987 \text{ m}$$

   Find the area.

   $$A \approx 66.7 \text{ m} \times 34.987 \text{ m}$$
   $$A \approx 2330 \text{ m}^2 \text{ Ans}$$

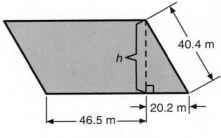

Figure 26–9

Calculator Application

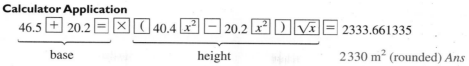

$$46.5 \boxed{+} 20.2 \boxed{=} \boxed{\times} \boxed{(}\ 40.4 \boxed{x^2}\ \boxed{-}\ 20.2 \boxed{x^2}\ \boxed{)}\ \boxed{\sqrt{x}}\ \boxed{=}\ 2333.661335$$

$$\underbrace{\hspace{4cm}}_{\text{base}} \quad \underbrace{\hspace{5cm}}_{\text{height}} \qquad 2330 \text{ m}^2 \text{ (rounded) } Ans$$

or

$$46.5 \boxed{+} 20.2 \boxed{\text{EXE}} \boxed{\times} \boxed{\sqrt{\ }} \boxed{(}\ 40.4 \boxed{x^2}\ -20.2 \boxed{x^2}\ \boxed{)}\ \boxed{\text{EXE}}\ 22333.661335$$

$$\underbrace{\hspace{4cm}}_{\text{base}} \quad \underbrace{\hspace{5cm}}_{\text{height}} \qquad 2330 \text{ m}^2 \text{ (rounded) } Ans$$

3. A drawing of a baseplate is shown in Figure 26–10. The plate is made of number 2 gage (thickness) aluminum, which weighs 3.4 pounds per square foot. Find the weight of the plate to 2 significant digits.

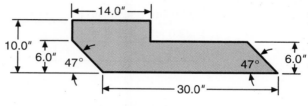

Figure 26–10

The area of the plate must be found. By studying the drawing, one method for finding the area is to divide the figure into a rectangle and a parallelogram as shown in Figure 26–11.

- Find the area of the rectangle.

  $A = 14.0 \text{ in} \times 4.0 \text{ in}$

  $A = 56 \text{ sq in}$

- Find the area of the parallelogram.

  $A = 30.0 \text{ in} \times 6.0 \text{ in}$

  $A = 180 \text{ sq in}$

- Find the total area of the plate.

  Total area $= 56 \text{ sq in} + 180 \text{ sq in} = 236 \text{ sq in}$

Compute the weight.

- Find the area in square feet.

$$\frac{236 \text{ sq in}}{1} \times \frac{1 \text{ sq ft}}{144 \text{ sq in}} \approx 1.64 \text{ sq ft}$$

- Weight of plate $\approx 1.64 \text{ sq ft} \times 3.4 \text{ lb/sq ft} \approx 5.6 \text{ lb } Ans$

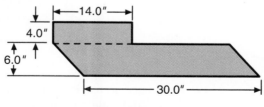

Figure 26–11

Calculator Application

$$\boxed{(}\ 14 \boxed{\times} 4 \boxed{+} 30 \boxed{\times} 6 \boxed{)} \boxed{\div} 144 \boxed{\times} 3.4 \boxed{=} 5.572222222$$

$$\underbrace{\hspace{5cm}}$$

Area of rectangle and
parallelogram in square inches

Weight of plate $\approx 5.6 \text{ lb } Ans$

EXERCISE 26–2

Use the tables in appendix A for equivalent units of measure. Find the unknown area, base, or height for each of the parallelograms, 1 through 16, in Figure 26–12. Where necessary, round the answers to one decimal place.

	Base	Height	Area
1.	20.00 cm	5.20 cm	
2.	6.00 yd	9.80 yd	
3.	26.0 in		486.2 sq in
4.		37.4 mm	2057 mm²
5.	0.07 mi		0.014 sq mi
6.	24.0 km	4.50 km	
7.	22.0 m	0.900 m	
8.	18′6″		312.5 sq ft

	Base	Height	Area
9.		0.60 km	5.1 km²
10.	56.00 mm	6.800 mm	
11.	17.00 mi	18.30 mi	
12.	58.1 cm		5.81 cm²
13.		38.0 in	1,887.2 sq ft
14.	20′9″		830 sq ft
15.	0.38 yd	0.266 yd	
16.	7.2 m	0.09 m	

Figure 26–12

Solve these problems. Round the answers to two decimal places.

17. The cross section of the piece of tool steel shown in Figure 26–13 is in the shape of a parallelogram. Find the cross-sectional area.

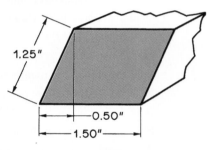

Figure 26–13

18. Two cutouts in the shape of parallelograms are stamped in a strip of metal as shown in Figure 26–14. Segment *AB* is parallel to segment *CD*, and dimension *E* equals dimension *F*. Compare the areas of the two cutouts.

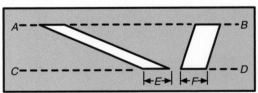

Figure 26–14

19. An oblique groove is cut in a block as shown in Figure 26–15. Before the groove was cut, the top of the block was in the shape of a rectangle. Determine the area of the top after the groove is cut.

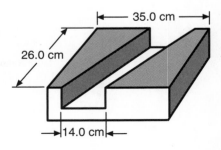

Figure 26–15

20. Two hundred meters of fencing are to be used to fence in a garden. The garden can be made in the shape of a square, a rectangle, or a parallelogram, such as those shown in Figure 26–16. Which of the three shapes permits the largest garden?

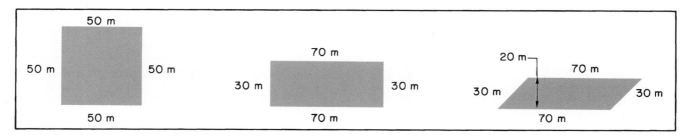

Figure 26–16

21. In Figure 26–17, building floor plans in the shape of a square, a rectangle, and a parallelogram are shown. Each floor plan contains 1,600.0 square feet. The walls of each building are 10.0 feet high.

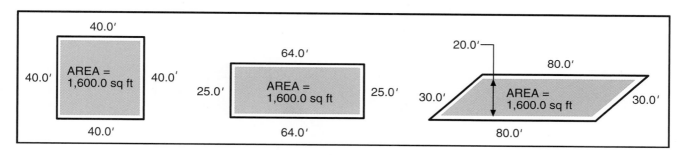

Figure 26–17

a. For each building, compute the wall area required to provide a floor area of 1,600.0 square feet.

b. Which shape provides the most area for the least amount of material and cost?

22. Find the area of the template in Figure 26–18.

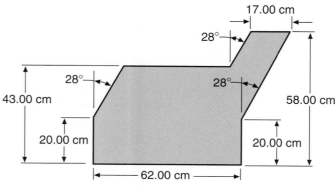

Figure 26–18

26–3   Areas of Trapezoids

A *trapezoid* is a four-sided polygon that has only two sides parallel. The parallel sides are called *bases*. The *area of a trapezoid* is equal to one-half the product of the height and the sum of the bases.

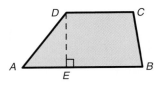

$$A = \frac{1}{2}h(b_1 + b_2)$$

where A = area
h = height
b_1 and b_2 = bases

In Figure 26–19, DE is the height, and AB and DC are the bases of trapezoid $ABCD$.

$$\text{Area of trapezoid } ABCD = \frac{1}{2}DE\,(AB + DC)$$

Figure 26–19

EXAMPLES •————————————————————————————————

1. Find the area of the stairway wall $ABCD$ in Figure 26–20. Round the answer to 2 significant digits.

$$A = \frac{1}{2}(4.2 \text{ m})(7.0 \text{ m} + 3.8 \text{ m})$$

$$A = \frac{1}{2}(4.2 \text{ m})(10.8 \text{ m})$$

$$A \approx 23 \text{ m}^2 \text{ (rounded) } Ans$$

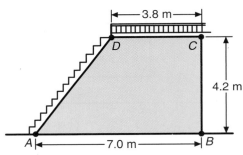

Figure 26–20

Calculator Application

.5 ×̄ 4.2 ×̄ (̄ 7 +̄ 3.8)̄ =̄ 22.68

$A \approx 23 \text{ m}^2$ (rounded) *Ans*

2. The area of a trapezoid is 376.58 square centimeters. The height is 16.25 centimeters, and one base is 35.56 centimeters. Find the other base. Round the answer to 4 significant digits.

Substitute values in the formula for the area of a trapezoid and solve.

$$376.58 \text{ cm}^2 = \frac{1}{2}(16.25 \text{ cm})(35.56 \text{ cm} + b_2)$$

$$376.58 \text{ cm}^2 = 8.125 \text{ cm}(35.56 \text{ cm} + b_2)$$
$$376.58 \text{ cm}^2 = 288.925 \text{ cm}^2 + 8.125 \text{ cm}(b_2)$$
$$87.655 \text{ cm}^2 = 8.125 \text{ cm}(b_2)$$
$$b_2 \approx 10.79 \text{ cm } Ans$$

3. The section of land shown in Figure 26–21 is to be graded and paved. The cost is $10.35 per square yard. What is the total cost of grading and paving the section? Express the answer to the nearest ten dollars.

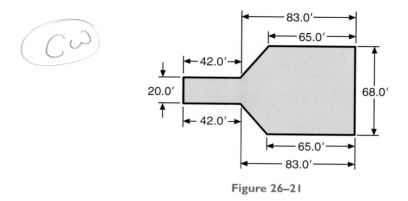

Figure 26–21

The area of the land must be found. Divide the section of land into two rectangles and a trapezoid as shown in Figure 26–22.

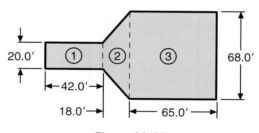

Figure 26–22

- Find area ①.

    $A = 42.0 \text{ ft} \times 20.0 \text{ ft}$

    $A = 840 \text{ sq ft}$

- Find area ②.

    Height $= 83.0 \text{ ft} - 65.0 \text{ ft} = 18.0 \text{ ft}$

    First base $= 68.0 \text{ ft}$

    Second base $= 20.0 \text{ ft}$

    $A = \frac{1}{2}(18.0 \text{ ft})(68.0 \text{ ft} + 20.0 \text{ ft})$

    $A = \frac{1}{2}(18.0 \text{ ft})(88.0 \text{ ft})$

    $A = 792 \text{ sq ft}$

- Find area ③.

    $A = 65.0 \text{ ft} \times 68.0 \text{ ft}$

    $A = 4,420 \text{ sq ft}$

- Find the total area of the land.

    Total area $= 840 \text{ sq ft} + 792 \text{ sq ft} + 4,420 \text{ sq ft} = 6,052 \text{ sq ft}$

Compute the cost.

- Find the area in square yards.

$$\frac{6,052 \text{ sq ft}}{1} \times \frac{1 \text{ sq yd}}{9 \text{ sq ft}} = 672.44 \text{ sq yd}$$

- Cost $= \$10.35/\text{sq yd} \times 672.44 \text{ sq yd} \approx \$6,960 \text{ } Ans$

Calculator Application

20 ☒ 42 ⊞ .5 ☒ 18 ☒ ⦅ 68 ⊞ 20 ⦆ ⊞ 65 ☒ 68 ⊟ ⊘ 9 ☒ 10.35 ⊟ 6959.8

sq ft ①   sq ft ②   sq ft ③ $6,960 (rounded) *Ans*

EXERCISE 26–3

If necessary, use the tables found in appendix A for equivalent units of measure. Find the unknown area, height, or base for each of the trapezoids, 1 through 16, in Figure 26–23. Where necessary, round the answers to one decimal place.

	Height (h)	Bases b₁	Bases b₂	Area (A)
1.	8.00 in	16.00 in	10.00 in	
2.	28.0 mm	47.0 mm	38.0 mm	
3.	0.60 m	6.50 m	2.40 m	
4.		8.00 ft	4.00 ft	64.0 sq ft
5.	1.2 yd		5.5 yd	7.7 sq yd
6.	0.6 km	0.8 km		0.4 km²
7.		56.00 cm	48.00 cm	738.4 cm²
8.	0.1 mi	1.2 mi	0.6 mi	

	Height (h)	Bases b₁	Bases b₂	Area (A)
9.	18.70 m	36.00 m	28.40 m	
10.	8'6.0"	14'4.0"	12'8.0"	
11.	3.8 km		8.7 km	62.1 km²
12.		66.37 in	43.86 in	2125 sq in
13.	0.3 yd	0.8 yd		0.2 sq yd
14.	14.00 cm	20.00 cm	3.200 cm	
15.		19'9"	13'3"	132 sq ft
16.	0.86 km	2.05 km	0.76 km	

Figure 26–23

Solve these problems.

17. A section of land in the shape of a trapezoid has a height of 530.0 feet and bases of 680.0 feet and 960.0 feet. How many acres are in the section of land? Round the answer to 2 decimal places.

18. A wooden ramp form is shown in Figure 26–24. The form has an open top and bottom.

    a. Find the number of square meters of lumber in the form.

    b. The lumber used to construct the form weighs 9.80 kilograms per square meter. Find the total weight of the form.

Round the answers for a and b to 3 significant digits.

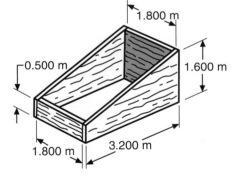

Figure 26–24

Figure 26–25

19. A cross section of a structural steel beam is shown in Figure 26–25. The ultimate strength of material is the unit stress that causes the material to break. The ultimate tensile (pulling) strength of the beam is 52,000 pounds per square inch.

    a. Find the cross-sectional area.

    b. What is the total ultimate tensile strength of the beam?

Round the answers for a and b to 2 significant digits.

20. A cross section of a concrete retaining wall in the shape of a trapezoid is shown in Figure 26–26.

   a. Find the cross-sectional area of this wall in square yards.

   b. Find the length of side *AB*. Round the answer to the nearest inch.

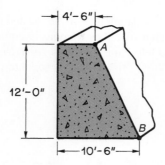

Figure 26–26

21. A common unit of measure used in carpentry and other woodworking occupations is the board foot. A board foot is equal to 1 square foot of lumber that is 1 inch thick or less. Oak flooring is installed on the floor in Figure 26–27. The cost of oak flooring is $965 per 1,000 board feet. An additional 25% must be purchased to allow for waste.

   a. Find the number of board feet of oak flooring purchased. Round the answer to the nearest ten board feet.

   b. What is the cost of the oak flooring for the building? Round the answer to the nearest ten dollars.

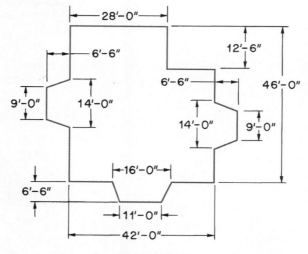

Figure 26–27

22. An industrial designer decided that the front plate of an appliance should be in the shape of an isosceles trapezoid with an area of 420 square centimeters. To give the desired appearance, the lower base dimension is to be equal to the height dimension, and the upper base dimension is to be equal to three-quarters of the lower base dimension. Compute the dimensions of the height and each base.

26–4   Areas of Triangles Given the Base and Height

In parallelogram *ABCD* shown in Figure 26–28, segment *DE* is the altitude to the base *AB*. Diagonal *DB* divides the parallelogram into two congruent triangles.

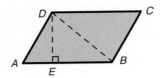

Figure 26–28

$$AB = DC \Big\}$$
$$AD = BC \Big\}$$ The opposite sides of a parallelogram are equal.
$$DB = DB$$

therefore, $\triangle ABD \cong \triangle CDB$ If three sides of one triangle are equal to three sides of another triangle, the triangles are congruent.

Parallelogram *ABCD* and triangles *ABD* and *CDE* have equal bases and equal heights. The area of either triangle is equal to one-half the area of the parallelogram. The area of parallelogram *ABCD* = *AB*(*DE*). Therefore, the area of $\triangle ABD$ or $\triangle CDB = \frac{1}{2}AB(DE)$. The *area of a triangle* is equal to one-half the product of the base and height.

$$A = \frac{1}{2}bh \qquad \text{where } A = \text{area}$$
$$b = \text{base}$$
$$h = \text{height}$$

EXAMPLES •———————————————————————

1. Find the area of the triangle shown in Figure 26–29.

$$A = \frac{1}{2}(22.0 \text{ cm})(19.0 \text{ cm})$$

$$A = 209 \text{ cm}^2 \; Ans$$

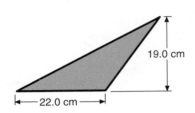

19.0 cm

22.0 cm

Figure 26–29

2. The triangular piece of land in Figure 26–30 is graded and seeded at a cost of $1,550. What is the grading and seeding cost per square foot?

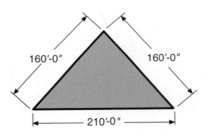

160'-0"   160'-0"

210'-0"

Figure 26–30

The area of the land must be found. To find the area, the base and height must be known. Since two sides of the triangle are equal, the triangle is isosceles. A line segment perpendicular to the base of an isosceles triangle from the vertex opposite the base bisects the base. In Figure 26–31, height *CE* bisects base *AB*.

• Find the height, using the Pythagorean theorem.

$$(160.0 \text{ ft})^2 = (105.0 \text{ ft})^2 + CE^2$$
$$25,600 \text{ sq ft} = 11,025 \text{ sq ft} + CE^2$$
$$14,575 \text{ sq ft} = CE^2$$
$$\sqrt{14,575 \text{ sq ft}} \approx CE$$
$$CE \approx 120.73 \text{ ft}$$

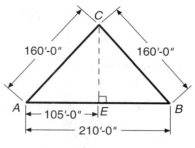

Figure 26–31

• Find the area of $\triangle ABC$.

$$A \approx \frac{1}{2}(210.0 \text{ ft})(120.73 \text{ ft})$$

$$A \approx 12{,}677 \text{ sq ft}$$

Compute the cost per square foot.

$$\frac{\$1{,}550}{12{,}677 \text{ sq ft}} \approx \$0.12/\text{sq ft (rounded) } Ans$$

26–5 Areas of Triangles Given Three Sides

Often three sides of a triangle are known, but a height is not known. A height can be determined by applying the Pythagorean theorem and a system of equations. However, with a calculator, it is quicker and easier to compute areas of triangles, given three sides, using a formula called *Hero's* or *Heron's formula.*

Hero's (Heron's) Formula

$$A = \sqrt{s(s-a)(s-b)(s-c)}$$

where A = area
a, b, and c = sides
$$s = \frac{1}{2}(a + b + c)$$

EXAMPLE •────────────────────────

Refer to the triangle in Figure 26–32.

    a. Find the area of the triangle.

    b. Find the altitude *JK*.

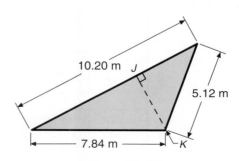

Figure 26–32

a. Compute the area by using Hero's formula.

$$s = \frac{1}{2}(7.84\text{ m} + 5.12\text{ m} + 10.20\text{ m})$$

$$s = 11.58\text{ m}$$

$$A = \sqrt{(11.58\text{ m})(11.58\text{ m} - 7.84\text{ m})(11.58\text{ m} - 5.12\text{ m})(11.58\text{ m} - 10.20\text{ m})}$$

$$A = \sqrt{(11.58\text{ m})(3.74\text{ m})(6.46\text{ m})(1.38\text{ m})}$$

$$A \approx \sqrt{386.093\text{ m}^4} \approx 19.6\text{ m}^2 \; Ans$$

b. Compute altitude JK from the formula $A = \frac{1}{2}bh$.

$$19.649\text{ m}^2 \approx \frac{1}{2}(10.20\text{ m})(JK)$$

$$19.649\text{ m}^2 \approx (5.10\text{ m})(JK)$$

$$JK \approx 3.85\text{ m} \; Ans$$

Calculator Application

a. Compute the area.

$s = .5 \boxed{\times} \boxed{(}\,7.84\,\boxed{+}\,5.12\,\boxed{+}\,10.2\,\boxed{)} = 11.58$

$A = \boxed{(}\,11.58\,\boxed{\times}\,\boxed{(}\,11.58\,\boxed{-}\,7.84\,\boxed{)}\,\boxed{\times}\,\boxed{(}\,11.58\,\boxed{-}\,5.12\,\boxed{)}\,\boxed{\times}\,\boxed{(}\,11.58\,\boxed{-}\,10.2\,\boxed{)}$

$\boxed{)}\,\boxed{\sqrt{x}} \rightarrow 19.64924569, \; A \approx 19.6\text{ m}^2 \; Ans$

or $\boxed{\sqrt{}}\,\boxed{(}\,11.58\,\boxed{\times}\,\boxed{(}\,11.58\,\boxed{-}\,7.84\,\boxed{)}\,\boxed{\times}\,\boxed{(}\,11.58\,\boxed{-}\,5.12\,\boxed{)}\,\boxed{\times}\,\boxed{(}\,11.58\,\boxed{-}\,10.2$

$\boxed{)}\,\boxed{)}\,\boxed{\text{EXE}}\; 19.64924569, \; A \approx 19.6\text{ m}^2 \; Ans$

b. Compute altitude JK.

$$JK = \frac{A}{\frac{1}{2}b}$$

$JK = 19.649\,\boxed{\div}\,\boxed{(}\,.5\,\boxed{\times}\,10.2\,\boxed{)}\,\boxed{=}\; 3.852745098$

$JK \approx 3.85\text{ m} \; Ans$

EXERCISE 26–5

Find the unknown area, base, or altitude for each of the triangles, 1 through 6, in Figure 26–33. Where necessary, round the answers to one decimal place.

	Base	Height	Area
1.	21.00 in	17.00 in	
2.	43.00 cm	29.00 cm	
3.	2.3 m	1.8 m	
4.		6.0 yd	78.2 sq yd
5.	0.2 mi		0.02 sq mi
6.		1.40 km	3.22 km²
7.	0.8 km	0.4 km	
8.	18'0"	7'3"	

	Base	Height	Area
9.	30'6"		427 sq ft
10.		38.0 mm	1919 mm²
11.	63.00 cm	0.900 cm	
12.	17.0 m	9.80 m	
13.		0.8 yd	16 sq yd
14.	45.412 in		249.7 sq in
15.	0.92 mi	0.71 mi	
16.		3.43 km	1.76 km²

Figure 26–33

Given 3 sides of these triangles, find the area of each triangle, 17 through 24, in Figure 26–34. Where necessary, round the answers to one decimal place unless otherwise specified.

	Side *a*	Side *b*	Side *c*
17.	4.00 m	6.00 m	8.00 m
18.	2.0 ft	5.0 ft	6.0 ft
19.	3'6"	4'0"	2'6"
20.	20.0 cm	15.0 cm	25.0 cm

	Side *a*	Side *b*	Side *c*
21.	3.2 yd	3.6 yd	0.8 yd
22.	9.100 in	30.86 in	28.57 in
23.	7.20 cm	10.00 cm	9.00 cm
24.	0.5 m	1.0 m	0.8 m

Figure 26–34

Solve these problems.

25. Find the cross-sectional area of metal in the triangular tubing in Figure 26–35. Round the answer to 3 significant digits.

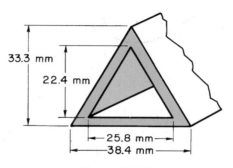

Figure 26–35

26. Two triangular pieces are cut from the sheet of plywood in Figure 26–36. After the triangular pieces are cut, the sheet is discarded. Find the number of square feet of plywood wasted. Round the answer to 1 decimal place.

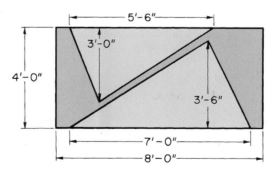

Figure 26–36

27. A hotel lobby, with a triangular-shaped floor, is remodeled. One side of the lobby is 11.60 meters long, and the altitude to the 11.60 meter side is 9.10 meters long. At a cost of $146.00 per square meter, what is the cost of remodeling the lobby? Round the answer to 3 significant digits.

28. The cross section of a concrete support column is in the shape of an equilateral triangle. Each side of the cross section measures 1'2". The compressive (crushing) strength of the concrete used for the column is 4,500 pounds per square inch. What is the total compressive strength of the column? Round the answer to 2 significant digits.

29. The area of the irregularly shaped sheet metal piece in Figure 26–37 is to be determined. The longest diagonal is drawn on the figure as shown in Figure 26–38. Perpendiculars are drawn to the diagonal from each of the other vertices. The perpendicular segments are measured as shown. From the measurements, the areas of each of the common polygons are computed. This is one method often used to compute areas of irregular figures. Compute the area of the sheet metal piece. Round the answer to 4 significant digits.

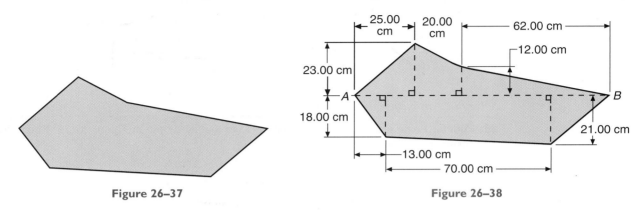

Figure 26–37   Figure 26–38

30. Determine the area of the building lot in Figure 26–39. Round the answer to 3 significant digits.

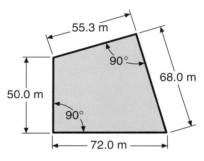

Figure 26–39

31. A hip roof is shown in Figure 26–40. The roof front and back are in the shape of congruent isosceles trapezoids. The two sides are in the shape of congruent isosceles triangles.

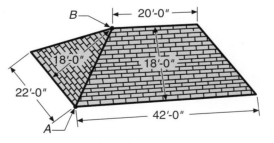

Figure 26–40

Round each answer to 3 significant digits.

a. Find the total roof area.

b. The shingles used on the roof cover 50.0 square feet per bundle. How many bundles are required to cover the roof allowing 15% for waste?

c. Find the length of the roof edge *AB*.

32. In Figure 26–41, 3 playgrounds shaped as equilateral, isosceles, and scalene triangles are shown. Each playground is enclosed by a concrete wall. The inside perimeter of each wall is 300.0 feet.

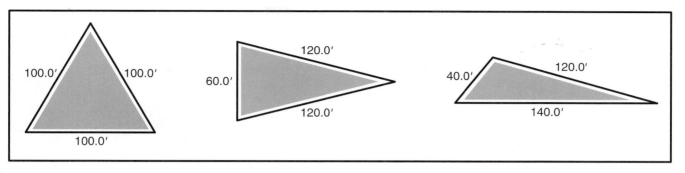

Figure 26–41

For a, b, and c, round each answer to the nearest square foot.

a. Find the area of the playground shaped as an equilateral triangle.

b. Find the area of the playground shaped as an isosceles triangle.

c. Find the area of the playground shaped as a scalene triangle.

d. What general conclusion do you reach as to the type of triangle that provides the most area for the least amount of enclosure cost?

UNIT EXERCISE AND PROBLEM REVIEW

RECTANGLES

Find the unknown area, length, or width for each of the rectangles, 1 through 8, in Figure 26–42. Where necessary, round the answers to one decimal place.

	Length	Width	Area
1.	17.500 in	9.000 in	
2.	26.00 cm	18.60 cm	
3.		6.3 m	88.2 m²
4.	15 ft		127.5 sq ft

	Length	Width	Area
5.		0.2 mi	0.12 sq mi
6.	1.2 km	0.90 km	
7.	43.0 mm		851.4 mm²
8.		6'9"	270 sq ft

Figure 26–42

PARALLELOGRAMS

Find the unknown area, base, or height for each of the parallelograms, 9 through 16, in Figure 26–43. Where necessary, round the answers to one decimal place.

	Base	Height	Area
9.	17.0 m	12.0 m	
10.	3.0 yd		13.8 sq yd
11.	14.60 in	9.500 in	
12.		3.2 mi	1.6 sq mi

	Base	Height	Area
13.	0.70 km		1.3 km²
14.	36.57 cm	19.16 cm	
15.		20'3"	486 sq ft
16.	$5\frac{1}{4}$ yd	$\frac{1}{2}$ yd	

Figure 26–43

TRAPEZOIDS

Find the unknown area, height, or base for each of the trapezoids, 17 through 24, in Figure 26–44. Where necessary, round the answers to one decimal place.

	Height	Bases		Area
		b_1	b_2	
17.	14.0 cm	23.0 cm	17.0 cm	
18.	0.6250 in	28.517 in	16.803 in	
19.		4 yd	5 yd	9 sq yd
20.	0.4 km		1 km	0.8 km²

	Height	Bases		Area
		b_1	b_2	
21.	6'0"	14'9"		69 sq ft
22.	0.3 mi	0.5 mi	0.4 mi	
23.		24.8 mm	21.7 mm	202 mm²
24.	17.6 m		11.2 m	255.8 m²

Figure 26–44

TRIANGLES

Find the unknown area, base, or height for each of the triangles, 25 through 32, given in Figure 26–45. Where necessary, round the answers to one decimal place.

	Base	Height	Area
25.	35.00 in	27.00 in	
26.	7.60 cm	5.50 cm	
27.	2.0 m		1.6 m²
28.		4.30 yd	32.7 sq yd

	Base	Height	Area
29.	0.85 km		0.21 m²
30.	0.9 mi	0.7 mi	
31.		22'9"	819 sq ft
32.	30.92 mm		2075 mm²

Figure 26–45

Given 3 sides of triangles, find the areas of each triangle, 33 through 36, in Figure 26–46. Where necessary, round the answers to one decimal place.

	Side *a*	Side *b*	Side *c*
33.	5.00 ft	7.00 ft	10.00 ft
34.	12.62 cm	8.04 cm	16.56 cm
35.	1.4 m	1.2 m	0.8 m
36.	24.00 in	30.00 in	14.00 in

Figure 26–46

APPLIED PROBLEMS

Solve these common polygon problems.

37. Find the area, in square feet, of a rectangular piece of lumber $3\frac{1}{4}$ inches wide and 8 feet $3\frac{1}{2}$ inches long. Round the answer to the nearest square foot.

38. Pieces in the shape of parallelograms are stamped from rectangular strips of stock as shown in Figure 26–47. If 24 pieces are stamped from a strip, how many square inches of strip are wasted? Round the answer to 3 significant digits.

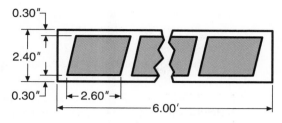

Figure 26–47

39. The cross section of a dovetail slide is shown in Figure 26–48. Before the dovetail was cut, the cross section was rectangular in shape. Find the cross-sectional area of the dovetail slide. Round the answer to 3 significant digits.

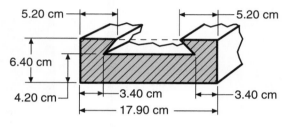

Figure 26–48

40. A bathroom 2.60 meters long and 2.00 meters wide is to have 4 walls covered with tile to a height of 1.50 meters. Deduct 2.30 square meters for door and window openings. How many square meters of wall are covered with tile? Round the answer to 3 significant digits.

41. Find the area of the side enclosed by the gambrel roof in Figure 26–49. Round the answer to 2 significant digits.

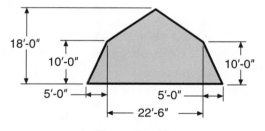

Figure 26–49

42. A baseplate in the shape of a parallelogram is shown in Figure 26–50.

   a. Find the plate area. Round the answer to 4 significant digits.

   b. Find the weight of the plate if 1.00 square meter weighs 68.4 kilograms. Round the answer to 3 significant digits.

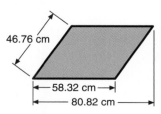

Figure 26–50

43. An A-frame building has triangular front and rear walls or sides and a roof reaching to the ground. Each of two sides of an A-frame building has a base of 24.0 feet and a height of 30.0 feet.

    a. Allowing 10% for waste, how many board feet of siding are required for the 2 sides? One board foot is equal to 1 square foot of lumber 1 inch or less in thickness.

    b. At a cost of $740.00 per 1,000 board feet, what is the total cost of siding for the two sides? Round the answer to the nearest dollar.

44. The floor of a building is covered with square vinyl tiles. The floor is in the shape of a trapezoid with an altitude of 32.0 feet and bases of 40.0 feet and 50.0 feet. One carton of tiles contains 80 tiles and has a coverage of 45 square feet. Allowing for 5% waste, how many cartons of tiles are purchased for the job?

45. A floor plan of a building is shown in Figure 26–51.

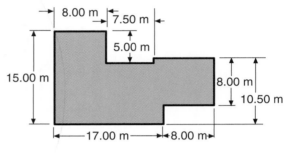

Figure 26–51

Round the answers for a, b, and c to 3 significant digits.

    a. Find the number of square meters of floor area.

    b. Find the number of square meters of exterior wall area of the building, disregarding wall thickness. The walls are 4.80 meters high. Make a 20% deduction for windows and doors.

    c. How many liters of paint are required to paint the exterior walls if 1 liter of paint covers 13.5 square meters of wall area?

46. The rectangular bottom of a shipping carton is designed to have an area of 17.5 square feet and a length of 60.0 inches. Can a thin object 6.00 feet long be packed in the carton? Show your computations in answering this question.

47. A concrete support column has a total ultimate compressive strength of 63,400 pounds. The concrete used for the column has an ultimate compressive strength of 4,350 pounds per square inch. If the cross section of the column is a square, what is the length of each side of the cross section? Round the answer to 3 significant digits.

48. A utility company purchased a power-line right-of-way from the owner of a piece of property. The piece is in the shape of the parallelogram shown in Figure 26–52. The right-of-way is the area of land between the diagonal broken lines.

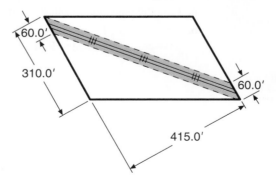

Figure 26–52

Round the answers to a and b to 3 significant digits.

a. The utility company paid the owner $0.85 per square foot for the land. How much was paid to the owner?

b. After the right-of-way was acquired, how many acres of land did the owner have left?

49. Compute the area of the pattern in Figure 26–53. Round the answer to 4 significant digits.

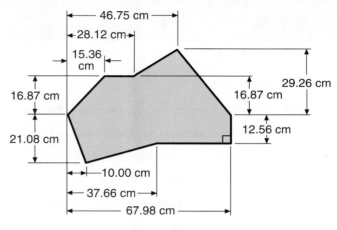

Figure 26–53

After studying this unit you should be able to

- compute areas, radii, and diameters of circles.
- compute areas, radii, and central angles of sectors.
- compute areas of segments.
- compute areas, major axes, and minor axes of ellipses.
- solve applied problems by using principles discussed in this unit.

Computations of areas of circular objects are often made on the job. Material quantities and weights are based on computed areas. Many industrial and construction material-strength computations are also based on circular cross-sectional areas of machined parts and structural members.

27–1 Areas of Circles

The *area of a circle* is equal to the product of π and the square of the radius.

$$A = \pi r^2 \qquad \text{where } A = \text{area}$$
$$r = \text{radius}$$
$$\pi \approx 3.1416$$

The formula for the area of a circle can be expressed in terms of the diameter. Since the radius is one-half the diameter, $\dfrac{d}{2}$ can be substituted in the formula for r.

$$A = \pi \left(\frac{d}{2}\right)^2$$
$$A = \frac{\pi d^2}{4}$$
$$A \approx \frac{3.1416 d^2}{4}$$
$$A \approx 0.7854 d^2$$

EXAMPLES

1. Find the area of a circle that has a radius of 6.500 inches.

    Substitute values in the formula and solve.

    $A = \pi r^2 \approx 3.1416(6.500 \text{ in})^2 \approx 3.1416(42.25 \text{ sq in}) \approx 132.7 \text{ sq in } Ans$

Calculator Application

$\boxed{\pi}$ $\boxed{\times}$ 6.5 $\boxed{x^2}$ $\boxed{=}$ 132.7322896

$A \approx 132.7$ sq in *Ans*

2. Determine the diameter of a circular table top which is to have an area of 1.65 square meters.

METHOD 1

$$A = \pi r^2$$

$$1.65 \text{ m}^2 \approx 3.1416 r^2$$

$$r^2 \approx \frac{1.65 \text{ m}^2}{3.1416}$$

$$r^2 \approx 0.52521 \text{ m}^2$$

$$r \approx \sqrt{0.52521 \text{ m}^2}$$

$$r \approx 0.725 \text{ m}$$

$$d \approx 2(0.725 \text{ m})$$

$$d \approx 1.45 \text{ m } Ans$$

METHOD 2

$$A \approx 0.7854 d^2$$

$$1.65 \text{ m}^2 \approx 0.7854 d^2$$

$$d^2 \approx \frac{1.65 \text{ m}^2}{0.7854}$$

$$d^2 \approx 2.10084 \text{ m}^2$$

$$d \approx \sqrt{2.10084 \text{ m}^2}$$

$$d \approx 1.45 \text{ m } Ans$$

3. A circular hole is cut in a square metal plate as shown in Figure 27–1. The plate weighs 8.3 pounds per square foot. What is the weight of the plate after the hole is cut?

Compute the area of the square.

$A = (10.30 \text{ in})^2 = 106.09$ sq in

Compute the area of the hole.

$A \approx 0.7854(7.00 \text{ in})^2 \approx 38.48$ sq in

Compute the area of the plate.

$A \approx 106.09$ sq in $- 38.48$ sq in ≈ 67.61 sq in

Compute the weight of the plate.

67.61 sq in $\div$ 144 sq in/sq ft ≈ 0.470 sq ft

Weight ≈ 0.470 sq ft $\times$ 8.3 lb/sq ft ≈ 3.9 lb *Ans*

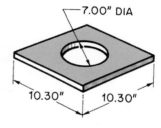

Figure 27–1

Calculator Application

$\boxed{(}$ 10.3 $\boxed{x^2}$ $\boxed{-}$ .7854 $\boxed{\times}$ 7 $\boxed{x^2}$ $\boxed{)}$ $\boxed{\div}$ 144 $\boxed{\times}$ 8.3 $\boxed{=}$ 3.896700139

Weight ≈ 3.9 lb *Ans*

27–2   Ratio of Two Circles

The areas of two circles have the same ratio as the squares of the radii or diameters.

$$\frac{A_1}{A_2} = \frac{r_1^2}{r_2^2} = \frac{d_1^2}{d_2^2}$$

EXAMPLE •

The radii of two circles are 5 feet and 2 feet. Compare the area of the 5-foot radius circle with the area of the 2-foot radius circle.

Form a ratio of the squares of the radii.

$$\frac{(5 \text{ ft})^2}{(2 \text{ ft})^2} = \frac{25 \text{ sq ft}}{4 \text{ sq ft}} = 6.25$$

The area of the 5-ft-radius circle is 6.25 times greater than the area of the 2-ft-radius circle. *Ans*

The radii or diameters of two circles have the same ratio as the square roots of the areas.

$$\frac{r_1}{r_2} = \frac{d_1}{d_2} = \frac{\sqrt{A_1}}{\sqrt{A_2}}$$

EXAMPLE

The areas of two circles are 36 cm² and 9 cm². Compare the diameter of the 36-cm² circle with the diameter of the 9-cm² circle.

Form a ratio of the square roots of the areas.

$$\frac{\sqrt{36 \text{ cm}^2}}{\sqrt{9 \text{ cm}^2}} = \frac{6 \text{ cm}}{3 \text{ cm}} = 2$$

The diameter of the circle with an area of 36 cm² is 2 times greater than the diameter of the circle with an area of 9 cm². *Ans*

EXERCISE 27–2

If necessary, use the tables in appendix A for equivalent units of measure. Find the unknown area, radius, or diameter for each of the circles, 1 through 12, in Figure 27–2. Round the answers to one decimal place.

	Radius	Diameter	Area
1.	7.000 in	–	
2.	10.80 cm	–	
3.	–	$15\frac{1}{2}$ ft	
4.	–	17.23 m	
5.		–	34 sq yd
6.	–		0.1 km²

	Radius	Diameter	Area
7.		–	380.0 mm²
8.	27.875 in	–	
9.	–	0.75 mi	
10.	–		102.6 cm²
11.		–	3.60 m²
12.	2'9.0"	–	

Figure 27–2

In these exercises, express the areas, radii, or diameters of the two circles as ratios; then solve. Where necessary, round the answers to two decimal places unless otherwise specified.

13. The radii of two circles are 6 inches and 2 inches. Compare the area of the larger circle with the area of the smaller circle.

14. The diameters of two circles are 15.0 cm and 10.0 cm. Compare the area of the larger circle with the area of the smaller circle.

15. The areas of two circles are 112 sq ft and 24.6 sq ft. Compare the radius of the larger circle with the radius of the smaller circle.

16. The areas of two circles are 6.20 m² and 3.60 m². Compare the diameter of the larger circle with the diameter of the smaller circle.

17. The radius of a circle is 6.3 times greater than the radius of a second circle. Compare the area of the larger circle with the area of the smaller circle.

18. The area of one circle is 12.60 times greater than the area of a second circle. Compare the diameter of the larger circle with the diameter of the smaller circle.

Solve these problems.

19. How many square yards are contained in a circular pavement with an 85.0-foot diameter? Round the answer to 3 significant digits.

20. A sheet metal reducer is shown in Figure 27–3. Find the difference in the cross-sectional areas of the two ends. Round the answer to 4 significant digits.

←26.80 cm DIA→

—10.60 cm
DIA

Figure 27–3

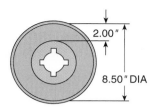

2.00″

8.50″ DIA

Figure 27–4

21. An electromagnetic brake uses friction disks as shown in Figure 27–4. The area shaded grey shows the brake lining surface of a disk. If 4 brake linings are used in the brake, what is the total brake lining area? Round the answer to the nearest square inch.

22. Hydraulic pressure of 705.0 pounds per square inch is exerted on a 3.150-inch diameter piston. Find the total force exerted on the piston. Round the answer to the nearest pound.

23. A circular walk has an outside diameter of 56′6″ and an inside diameter of 47′6″. Compute the cost of constructing the walk if labor and materials are estimated at a cost of $1.20 per square foot. Round the answer to the nearest dollar.

24. A force of 62,125 pounds pulls on a steel rod that has a diameter of 1.800 inches. Find the force pulling on 1 square inch of cross-sectional area. Round the answer to the nearest ten pounds per square inch.

25. A circular base is shown in Figure 27–5. The base is cut from a steel plate that weighs 34 kilograms per square meter of surface area. Find the weight of the circular base. Round the answer to the nearest tenth kilogram.

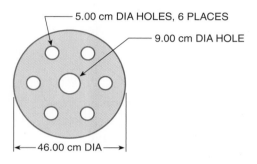

5.00 cm DIA HOLES, 6 PLACES

9.00 cm DIA HOLE

46.00 cm DIA

Figure 27–5

26. Three water pipes, 3.80 cm, 5.00 cm, and 7.50 cm in diameter are connected to a single pipe. The single pipe permits the same amount of water to flow as the total of the three pipes. Find the diameter of the single pipe. Round the answer to 3 significant digits.

27. Find the area of the template in Figure 27–6. Round the answer to 4 significant digits.

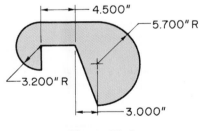

4.500″

5.700″ R

3.200″ R

3.000″

Figure 27–6

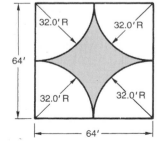

32.0′ R   32.0′ R

64′

32.0′ R   32.0′ R

64′

Figure 27–7

28. A rectangular sheet of plywood 2.00 meters long and 1.50 meters wide weighs 15.2 kilograms. Find the weight of the sheet after three 0.600-meter diameter holes are cut. Round the answer to the nearest tenth kilogram.

29. A support column is replaced with another column, which has 6.8 times the cross-sectional area of the original column. The diameter of the original column is 19.4 centimeters. Find the diameter of the new column. Round the answer to the nearest tenth centimeter.

30. A landscaper designs a garden area in the square piece of land shown in Figure 27–7. The shaded portion is planted with shrubs and ground cover. Find the number of square feet of area planted. Round the answer to the nearest square foot.

27–3 Areas of Sectors

A sector of a circle is a figure formed by two radii and the arc intercepted by the radii. To find the *area of a sector* of a circle, first find the fractional part of a circle represented by the central angle. Then multiply the fraction by the area of the circle.

$$A = \frac{\theta}{360°}(\pi r^2)$$

where A = area
θ = central angle
$\pi \approx 3.1416$
r = radius

EXAMPLES

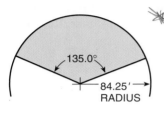

Figure 27–8

1. The parking area in the shape of a sector as shown in Figure 27–8 is graded and paved. At a cost of $0.40 per square foot, find, to the nearest dollar, the total cost of grading and paving the area.

Find the area of the sector:

$$A = \frac{135.0°}{360°}(3.1416)(84.25 \text{ ft})^2 = 8{,}362.23 \text{ sq ft}$$

Find the cost:

8,362.23 sq ft × $0.40/sq ft ≈ $3,345 (rounded) *Ans*

Calculator Application

135 ÷ 360 ⊠ π ⊠ 84.25 x^2 ⊠ .4 = 3344.883151
Cost ≈ $3,345 *Ans*

2. An apartment balcony is designed in the shape of a sector. The balcony has a central angle of 225° and contains 24.0 square meters of surface area. Find the radius.

Substitute values in the formula and solve.

$$24.0 \text{ m}^2 \approx \frac{225°}{360°}(3.1416)r^2$$

$$24.0 \text{ m}^2 \approx 1.9635r^2$$

$$r^2 \approx 12.223 \text{ m}^2$$

$$r \approx 3.50 \text{ m } Ans$$

EXERCISE 27–3

Use the tables in appendix A for equivalent units of measure. Find the unknown area, radius, or central angle for each of these sectors, 1 through 12, in Figure 27–9. Round the answers to one decimal place.

	Radius	Central Angle	Area
1.	10.00 cm	120.0°	
2.	3.5 ft	90.0°	
3.	45.00 in	40.00°	
4.		65.0°	300.0 m²
5.		180.5°	750.0 mm²
6.	9.570 yd		94.62 sq yd

	Radius	Central Angle	Area
7.	20′3.0″		1,028 sq ft
8.	0.2 km	220°	
9.	54.08 cm	26.30°	
10.		307.2°	79.4 sq in
11.	3.273 in		15.882 sq in
12.	150.78 m	15.286°	

Figure 27–9

Solve these problems.

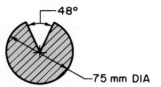

Figure 27–10

13. A cross section of a piece of round stock with a V-groove cut is shown in Figure 27–10. Find the cross-sectional area of the stock. Round the answer to 2 significant digits.

14. The entrance hall of an office building is constructed in the shape of a sector with a 22′6.0″ radius and a 118.50° central angle. Compute the cost of carpeting the hall if carpeting costs $46.50 per square yard. Allow an additional 20% for waste. Round the answer to the nearest dollar.

15. A kitchen countertop is shown in Figure 27–11.

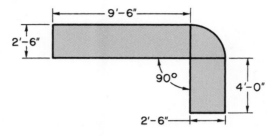

Figure 27–11

a. Find the surface area of the countertop. Round the answer to the nearest square foot.

b. The top is covered with laminated plastic that costs $1.35 per square foot. Allowing 15% for waste, find the cost of covering the top. Round the answer to the nearest dollar.

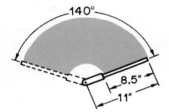

Figure 27–12

16. A vehicle windshield wiper swings an arc of 140° as shown in Figure 27–12. The wiper is 11 inches long with a wiper blade of 8.5 inches. Assuming the windshield is a plane surface, find the area swept by the wiper blade. Round the answer to 2 significant digits.

17. Three pieces, each in the shape of a sector, are cut from the rectangular sheet of steel shown in Figure 27–13. How many square meters are wasted after the 3 pieces are cut? Round the answer to the nearest tenth square meter.

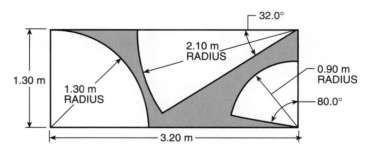

Figure 27–13

18. A walk around the arc of a sector as shown in Figure 27–14 is constructed at a cost of $19 per square meter. Find the total cost of constructing the walk. Round the answer to the nearest ten dollars.

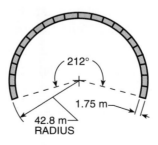

Figure 27–14

19. A patio is designed in the shape of an isosceles triangle as shown in Figure 27–15. The sector portion, *ABEC*, is constructed of concrete. *AB*, *AE*, and *AC* are radii. Portions *BDE* and *CFE* are planted with flowers.

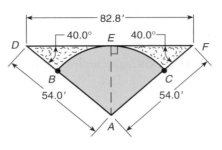

Figure 27–15

Round the answers to a and b to the nearest ten square feet.

a. Find the area of the concrete portion (sector *ABEC*).

b. Find the area of the flower portion (portions *BDE* and *CFE*).

27–4   Areas of Segments

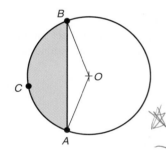

Figure 27–16

A *segment* of a circle is a figure formed by an arc and the chord joining the end points of the arc. In the circle shown in Figure 27–16, *the area of segment ACB* is found by subtracting the area of triangle *AOB* from the area of sector *OACB*.

EXAMPLE •

Segment *ACB* is cut from the circular plate shown in Figure 27–17. Find the area of the segment.

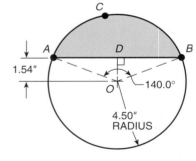

Figure 27–17

Find the area of the sector.

$$A \approx \frac{140°}{360°} (3.1416)(4.50 \text{ in})^2 \approx 24.740 \text{ sq in}$$

Find the area of isosceles triangle *AOB*.

$$c^2 = a^2 + b^2$$
$$(4.50 \text{ in})^2 = (1.54 \text{ in})^2 + AD^2$$
$$20.25 \text{ sq in} = 2.3716 \text{ sq in} + AD^2$$
$$AD^2 = 17.8784 \text{ sq in}$$
$$AD \approx 4.228 \text{ in}$$

The base *AB* of isosceles triangle *AOB* is bisected by altitude *DO*.

$$AB = 2(AD) \approx 2(4.228 \text{ in}) \approx 8.456 \text{ in}$$
$$A \approx 0.5(8.456 \text{ in})(1.54 \text{ in}) \approx 6.511 \text{ sq in}$$

Find the area of segment *ACB*.

$$A \approx 24.740 \text{ sq in} - 6.511 \text{ sq in} \approx 18.2 \text{ sq in } Ans$$

EXERCISE 27–4

If necessary, use the tables in appendix A for equivalent units of measure. Find the area of each of the segments ACB for exercises 1 through 5 in Figure 27–18. Refer to Figure 27–19.

	Area of Isosceles △AOB	Area of Sector OACB	Area of Segment ACB
1.	16.5 sq ft	24.8 sq ft	
2.	6.98 m²	9.35 m²	
3.	156 mm²	213.5 mm²	
4.	85.33 sq in	109.27 sq in	
5.	0.26 km²	0.39 km²	

Figure 27–18

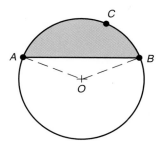

Figure 27–19

Solve these problems.

6. Find the area of the shaded segment in Figure 27–20. Round the answer to 3 significant digits.

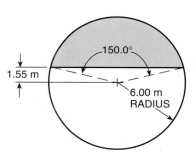

Figure 27–20

7. Paving cost is $10.30 per square yard. Find the total cost of paving the shaded area in Figure 27–21. Round the answer to the nearest ten dollars.

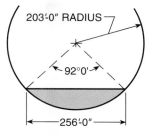

Figure 27–21

8. A pattern is shown in Figure 27–22.

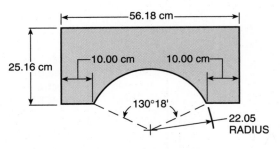

Figure 27–22

   a. Find the surface area of the pattern. Round the answer to the nearest square centimeter.

   b. The metal from which the pattern is made weighs 7.85 kilograms per square meter of area. Find the weight of the pattern. Round the answer to the nearest hundredth kilogram.

9. The shaded piece shown in Figure 27–23 is cut from a circular disk. Find the area of the piece. Round the answer to 3 significant digits.

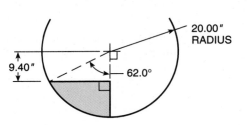

Figure 27–23

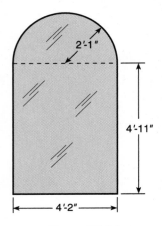

10. A window is made in the shape shown in Figure 27–24.

   a. Find the length of the molding used around the window.

   b. What is the area of the glass needed for this window?

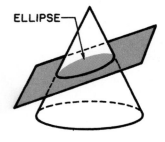

Figure 27–24

27–5   Areas of Ellipses

The curve of intersection of a plane that diagonally cuts through a cone is an ellipse, as shown in Figure 27–25. An *ellipse* is a closed oval-shaped curve that is symmetrical to two lines or axes that are perpendicular to each other. The curve shown in Figure 27–26 is an ellipse. It is symmetrical to axes *AB* and *CD*. The longer axis is called the *major axis,* and the shorter axis *CD* is called the *minor axis.* The *area of an ellipse* is equal to the product of π and one-half the major axis and one-half the minor axis.

ELLIPSE

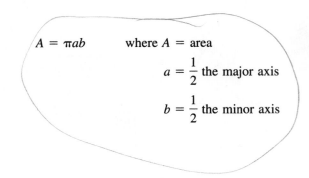

$$A = \pi ab \qquad \text{where } A = \text{area}$$
$$a = \frac{1}{2} \text{ the major axis}$$
$$b = \frac{1}{2} \text{ the minor axis}$$

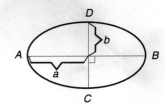

Figure 27–25

Figure 27–26

EXAMPLES

1. Find the area of an elliptical swimming pool that is 9.00 meters long (major axis) and 6.00 meters wide (minor axis).

$a = 0.5(9.00 \text{ m}) = 4.50 \text{ m}$

$b = 0.5(6.00 \text{ m}) = 3.00 \text{ m}$

$A \approx 3.1416(4.50 \text{ m})(3.00 \text{ m}) \approx 42.4 \text{ m}^2 \text{ Ans}$

Calculator Application

$\boxed{\pi}\ \boxed{\times}\ .5\ \boxed{\times}\ 9\ \boxed{\times}\ .5\ \boxed{\times}\ 6\ \boxed{=}\ 42.41150082$

$A \approx 42.4 \text{ m}^2 \text{ Ans}$

2. An elliptical platform is designed to have a surface area of 725 square feet. To give the desired form, the major axis must be $1\frac{1}{2}$ times as long as the minor axis. Determine the dimensions of the major and minor axes.

Let x = minor axis $b = 0.5x$

$1.5x$ = major axis $a = 0.75x$

$725 \text{ sq ft} \approx 3.1416(0.75x)(0.5x)$

$725 \text{ sq ft} \approx 1.1781x^2$

$x^2 \approx 615.3977 \text{ sq ft}$

$x \approx 24.807 \text{ ft}$

Minor axis ≈ 24.8 ft *Ans*

Major axis $\approx 1.5(24.807 \text{ ft}) \approx 37.2$ ft *Ans*

Calculator Application

$$x = \sqrt{\dfrac{725}{\pi(0.75)(0.5)}}$$

$x = \boxed{(}\ 725\ \boxed{\div}\ \boxed{(}\ \boxed{(}\ \boxed{\pi}\ \boxed{\times}\ .75\ \boxed{\times}\ .5\ \boxed{)}\ \boxed{)}\ \boxed{\sqrt{x}} \rightarrow 24.80723913$

or $x = \boxed{\sqrt{}}\ \boxed{(}\ 725\ \boxed{\div}\ \boxed{(}\ \boxed{(}\ \boxed{\pi}\ \boxed{\times}\ .75\ \boxed{\times}\ .5\ \boxed{)}\ \boxed{)}\ \boxed{\text{EXE}}\ 24.80723913$

Minor axis ≈ 24.8 ft *Ans*

Major axis $\approx 1.5\ \boxed{\times}\ 24.80723913\ \boxed{=}\ 37.2108587,\ 37.2$ ft *Ans*

EXERCISE 27–5

If necessary, use the tables in appendix A for equivalent units of measure. Find the unknown area, major axis, or minor axis of each of the ellipses, 1 through 10, in Figure 27–27. Round the answers to one decimal place.

	Major Axis	Minor Axis	Area
1.	10.00 in	7.00 in	
2.	23.20 cm	15.00 cm	
3.	8.60 m	4.20 m	
4.	43′0″ ft		1,215 sq ft
5.		4.6 yd	19.5 sq yd

	Major Axis	Minor Axis	Area
6.	86.0 mm		4730 mm²
7.	36.60 in	25.00 in	
8.	19′6.0″	15′3.0″	
9.		0.73 m	0.87 m²
10.	68.8 cm		2970 cm²

Figure 27–27

Solve these problems.

11. A concrete wall 3′6.0″ wide encloses an elliptical courtyard shown in Figure 27–28. Find the surface area of the top of the wall. Round the answer to the nearest square foot.

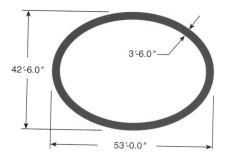

3′-6.0″

42′-6.0″

53′-0.0″

Figure 27–28

12. A platform was originally designed as a rectangle 10.00 meters long and 7.00 meters wide. The rectangular design is replaced with a platform in the shape of an ellipse. The elliptical platform will have the same area as the original rectangular platform, and the major axis is 14.00 meters. Find the minor axis. Round the answer to 3 significant digits.

13. A room divider panel is made with 8 identical elliptical cutouts as shown in Figure 27–29. The panel is made of plywood that weighs 1.4 pounds per square foot. Find the weight of the completed panel. Round the answer to the nearest pound.

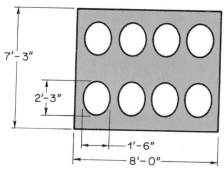

7′- 3″

2′-3″

1′-6″

8′-0″

Figure 27–29

14. An elliptical tabletop is designed to have a surface area of 2.30 square meters. To give the desired form, the minor axis must be $\frac{3}{4}$ as long as the major axis. Find the dimensions of the major and minor axes. Round the answer to 3 significant digits.

UNIT EXERCISE AND PROBLEM REVIEW

If necessary, use the tables in appendix A for equivalent units of measure.

CIRCLES

Find the unknown area, radius, or diameter for each of the circles, 1 through 5, in Figure 27–30. Round the answers to two decimal places.

		Radius	Diameter	Area
	1.	14.000 in	—	
	2.	—	28.600 cm	
	3.		—	45.0 sq yd
	4.	—		110.0 m²
	5.	—	12'-9.00"	

Figure 27–30

In these exercises, express the areas, radii, or diameters of two circles as ratios, and then solve. Where necessary, round the answers to two decimal places.

6. The radii of two circles are 12.0 inches and 3.0 inches. Compare the area of the larger circle with the area of the smaller circle.

7. The areas of two circles are 2.800 m² and 0.900 m². Compare the radius of the larger circle with the radius of the smaller circle.

8. The diameter of a circle is 6.24 times greater than the diameter of a second circle. Compare the area of the larger circle with the area of the smaller circle.

SECTORS

Determine the unknown area, radius, or central angle for each sector, 9 through 13, in Figure 27–31. Round the answers to one decimal place.

	Radius	Central Angle	Area
9.	5.50 in	120.0°	
10.	13.40 m	65°0'	
11.		230°25'	54.36 sq in
12.	0.200 km		0.0300 km²
13.	18'9.0"	78°20'	

Figure 27–31

SEGMENTS

Find the area of each of the segments ACB for exercises 14 through 16 in Figure 27–32. Refer to Figure 27–33. Round the answers to one decimal place.

	Area of Isosceles △AOB	Area of Sector OACB	Area of Segment ACB
14.	58.0 cm²	72.3 cm²	
15.	135.3 sq yd	207.8 sq yd	
16.	587.0 sq in	719.5 sq in	

Figure 27–32

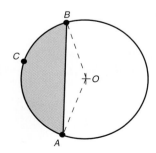

Figure 27–33

ELLIPSES

Find the unknown area, major axis, or minor axis of each of these ellipses, 17 through 21, in Figure 27–34. Round the answers to one decimal place.

	Major Axis	Minor Axis	Area
17.	27.00 yd	19.00 yd	
18.	9.60 m	7.00 m	
19.	54.6 cm		1720 cm²
20.		18.26 in	259.7 sq in
21.	14'0.0"	10'9.0"	

Figure 27–34

APPLIED PROBLEMS

Solve these problems. Round the answers to two decimal places unless otherwise specified.

22. Find the cross-sectional area of the spacer in Figure 27–35.

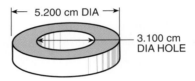

Figure 27–35

23. A circular wall is 0.500 meter thick and has an inside diameter of 10.20 meters. Find the cross-sectional area of the wall. Round the answer to the nearest tenth square meter.

24. A minipark is constructed in the shape of a sector with a 73'6" radius and a 135° central angle. The cost of grading and landscaping the park is $7.20 per square yard. Find the total cost of grading and landscaping the park. Round the answer to 3 significant digits.

25. A compressive force of 3,025 pounds per square inch is exerted on a concrete column. The column has a circular cross section with a 1.75-foot diameter. Find the total compressive force exerted on the column. Round the answer to 3 significant digits.

26. An elliptical shaped track is shown in Figure 27–36. The cost of resurfacing the track is $0.45 per square foot. Find the total cost of resurfacing the track. Round the answer to the nearest ten dollars.

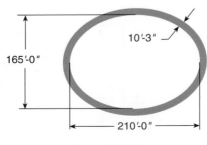

Figure 27–36

27. Find the area of the shaded segment shown in Figure 27–37.

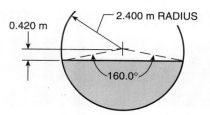

Figure 27–37

28. A circular plate with a 15.00-inch radius is cut from a rectangular aluminum piece that is 3′0.0″ wide and 4′0.0″ long. Before the circular plate was cut, the rectangular piece weighed 30.00 pounds. Find the weight of the circular plate. Round the answer to the nearest tenth pound.

29. Two sewer pipes 8.0 inches and 10.0 inches in diameter are joined to a single pipe. The single pipe is to carry the same amount of sewage as the combined amount carried by the 8.0-inch and 10.0-inch diameter pipes. Find the diameter of the single pipe. Round the answer to the nearest inch.

30. Compute the cross-sectional area of the grooved block in Figure 27–38.

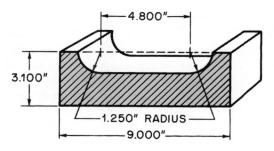

Figure 27–38

31. A pool designed in the shape of an ellipse has a surface area of 2,550.0 square feet. The major axis is 1.30 times as long as the minor axis. Find the major and minor axes. Round the answer to 3 significant digits.

32. Find the area of the pattern in Figure 27–39. Round the answer to 3 significant digits.

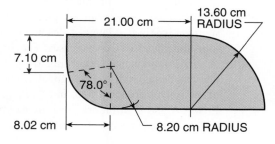

Figure 27–39

UNIT 28 ⫶⫶ Prisms and Cylinders: Volumes, Surface Areas, and Weights

OBJECTIVES

After studying this unit you should be able to

• compute volumes of prisms and cylinders.

• compute surface areas of prisms and cylinders.

• compute capacities and weights of prisms and cylinders.

• solve applied problems by using principles discussed in this unit.

The ability to compute volumes of prisms and cylinders is required in various occupations. Heating and air-conditioning technicians compute the volume of air in a building when determining heating and cooling system requirements. In the construction field, the type and size of structural supports used depend on the volume of the materials being supported. The displacement of an automobile engine is based on the volume of its cylinders. Nurses use volume measure when giving medications to patients.

Surface areas of the faces or sides of prisms and cylinders are often needed. To determine material requirements and costs, a welder computes the surface area of a cylindrical weldment. A packaging designer considers the surface area for the material of a carton when pricing a product.

28–1 Prisms

A *polyhedron* is a three-dimensional (solid) figure whose surfaces are polygons. In practical work, perhaps the most widely used solid is the prism. A *prism* is a polyhedron that has two identical (congruent) parallel polygon faces called *bases* and parallel lateral edges. The other sides or faces of a prism are parallelograms called *lateral faces.* A *lateral edge* is the line segment where two lateral faces meet. An *altitude* of a prism is a perpendicular segment that joins the planes of the two bases. The *height* of the prism is the length of an altitude.

Prisms are named according to the shape of their bases, such as triangular, rectangular, pentagonal, hexagonal, and octagonal. Some common prisms are shown in Figure 28–1. The parts of the prisms are identified. In a *right prism,* the lateral edges are perpendicular to the bases. Prisms A and B in Figure 28–1 are examples of right prisms. In an *oblique prism,* the lateral edges are *not* perpendicular to the bases. Prisms C and D are examples of oblique prisms.

28–2 Volumes of Prisms

The *volume of any prism* (right or oblique) is equal to the product of the base area and altitude.

$$V = A_B h \qquad \text{where } V = \text{volume}$$
$$A_B = \text{area of base}$$
$$h = \text{height}$$

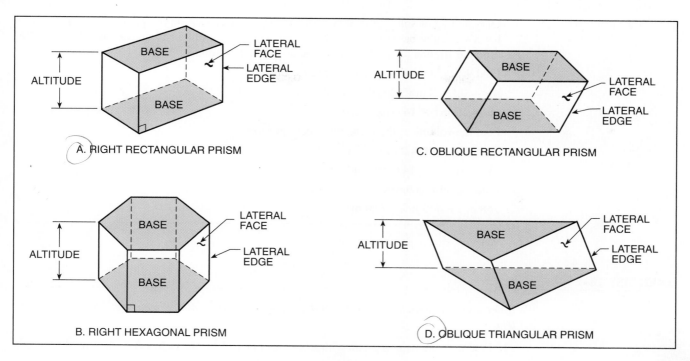

Figure 28–1

EXAMPLES

1. Compute the volume of a prism that has a base area of 15 square inches and a height of 6 inches.

   $V = 15 \text{ sq in} \times 6 \text{ in} = 90 \text{ cu in } Ans$

2. A concrete slab with a rectangular base is shown in Figure 28–2.

   a. Find the number of cubic yards of concrete required.

   b. One cubic yard of concrete weighs 3,700 pounds. Find the weight of the slab.

   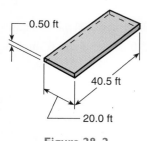

   Figure 28–2

   a. Find the area of the rectangular base.

   $A_B = 40.5 \text{ ft} \times 20.0 \text{ ft} = 810 \text{ sq ft}$

   Find the volume of the slab.

   $V = 810 \text{ sq ft} \times 0.50 \text{ ft} = 405 \text{ cu ft}$

   Find the volume in cubic yards.

   $405 \text{ cu ft} \div 27 \text{ cu ft/cu yd} = 15 \text{ cu yd } Ans$

   b. Find the weight of the slab.

   $15 \text{ cu yd} \times 3,700 \text{ lb/cu yd} \approx 56,000 \text{ lb } Ans$

3. Find the approximate volume of air contained in the building in Figure 28–3. Disregard the wall and ceiling thicknesses.

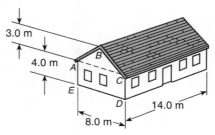

Figure 28–3

One method of solution is to consider the building as two simple prisms. The roof portion has a triangular base, *ABC*, and the remainder of the building has a rectangular base, *ACDE*. Find the volume of each portion separately and then add.

Find the volume of the triangular base portion.

$A_B = 0.5(8.0 \text{ m})(3.0 \text{ m}) = 12 \text{ m}^2$

$V = 12 \text{ m}^2 \times 14.0 \text{ m} = 168 \text{ m}^3$

Find the volume of the rectangular base portion.

$A_B = 8.0 \text{ m} \times 4.0 \text{ m} = 32 \text{ m}^2$

$V = 32 \text{ m}^2 \times 14.0 \text{ m} = 448 \text{ m}^3$

Find the total volume.

$168 \text{ m}^3 + 448 \text{ m}^3 \approx 620 \text{ m}^3$ *Ans*

EXERCISE 28–2

If necessary, use the tables in appendix A for equivalent units of measure. Solve these problems.

1. Find the volume of a prism with a base area of 125 square inches and a height of 8 inches.

2. Compute the volume of a prism with a height of 26.500 centimeters and a base of 610.00 square centimeters.

3. One cubic inch of cast iron weighs 0.26 pound. Find the weight of a block of cast iron with a base area of 48 square inches and an altitude of 5.3 inches. Round the answer to the nearest pound.

4. Find the capacity in gallons of a rectangular tank with a base area of 325.0 square feet and a height of 12.8 feet. Round the answer to 2 significant digits.

5. Compute the volume of a wall that is 0.800 meter thick, 7.00 meters long, and 2.00 meters high.

6. How many cubic feet of air are heated in a room 28'6" long, 22'0" wide, and 8'6" high? Round the answer to 2 significant digits.

7. A solid steel wedge is shown in Figure 28–4.

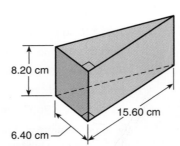

8.20 cm

15.60 cm

6.40 cm

Figure 28–4

   a. Find the volume of the wedge. Round the answer to the nearest cubic centimeter.

   b. The steel used for the wedge weighs 0.0080 kilogram per cubic centimeter. Find the weight of the wedge. Round the answer to the nearest tenth kilogram.

8. An excavation for a building foundation is 60.0 feet long, 35.0 feet wide, and 14.0 feet deep.

   a. How many cubic yards of soil are removed? Round the answer to 3 significant digits.

   b. How many truck loads are required to haul the soil from the building site if the average truck load is 3.5 cubic yards? Round the answer to 2 significant digits.

9. A length of angle iron is shown in Figure 28–5.

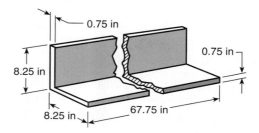

Figure 28–5

Round the answers for a and b to 2 significant digits.

a. Find the volume of the angle iron.

b. Find the weight of the angle iron if the material weighs 490 pounds per cubic foot.

10. A rectangular fuel oil storage tank is 10.50 feet long, 8.00 feet wide, and 6.30 feet high. Round the answers for a and b to 3 significant digits.

a. How many gallons of fuel oil are contained in the tank when it is 70% full?

b. Oil is pumped into the tank at a rate of 28.5 gallons per minute. Starting with a 70% full tank, how long will it take to fill the tank?

11. A concrete retaining wall is shown in Figure 28–6.

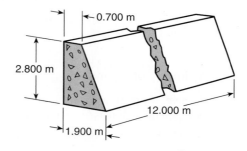

Figure 28–6

Round the answers for a and b to 3 significant digits.

a. Find the number of cubic meters of concrete required for the wall.

b. At a cost of $96.50 per cubic meter, what is the total cost of concrete in the wall?

12. The cross section of a road is shaped like a rectangle with a segment of a circle on top as shown in Figure 28–7. The road is two miles long.

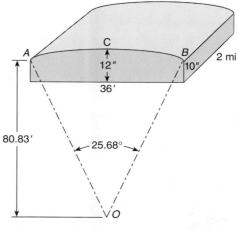

Figure 28–7

    a. Find the area in square yards of the cross section of the road.

    b. How many cubic yards of concrete will be needed for the road? Round the answer to the nearest hundred cubic yards.

28–3   Cylinders

Cylinders are used in many industrial and construction applications. Pipes, shafts, support columns, and tanks are a few of the practical uses made of cylinders.

    A *circular cylinder* is a solid that has identical (congruent) circular parallel bases. The surface between the bases is called *lateral surface.* The *altitude* of a circular cylinder is a perpendicular segment that joins the planes of the bases. The *height* of a cylinder is the length of an altitude. The *axis* of a circular cylinder is a line that connects the centers of the bases.

    In a *right circular cylinder* the axis is perpendicular to the bases. A right circular cylinder with its parts identified is shown in Figure 28–8. Only right circular cylinders are considered in this book.

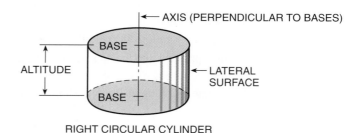

RIGHT CIRCULAR CYLINDER

Figure 28–8

28–4   Volumes of Cylinders

As with a prism, a right circular cylinder has a uniform cross-section area. The formula for computing volumes of right circular cylinders is the same as that of prisms. The *volume of a right circular cylinder* is equal to the product of the base area and height.

$$V = A_B h$$

    where V = volume

            A_B = area of base; $A_B = \pi r^2$ or $0.7854d^2$ (rounded)

            h = height

EXAMPLES •————————————————————————————————

1. Find the volume of a cylinder with a base area of 30.0 square centimeters and a height of 6.0 centimeters.

    $V = 30.0 \text{ cm}^2 \times 6.0 \text{ cm} = 180 \text{ cm}^3$ *Ans*

2. A cylindrical tank has a 6′3.0″ diameter and a height of 5′9.0″.

    a. Find the volume of the tank.

    b. Find the capacity in gallons.

        a. Find the area of the circular base.

            $A_B \approx 0.7854d^2$

            $A_B \approx (0.7854)(6.250 \text{ ft})^2 \approx 30.6797 \text{ sq ft}$

            Find the volume.

            $V \approx 30.6797 \text{ sq ft} \times 5.750 \text{ ft} \approx 176.4 \text{ cu ft}$ *Ans*

        b. Find the capacity in gallons.

            NOTE: There are 7.5 gal/cu ft.

            176.4 cu ft $\times$ 7.5 gal/cu ft $\approx$ 1,323 gal *Ans*

3. A length of pipe is shown in Figure 28–9. Find the volume of metal in the pipe.

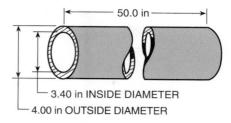

50.0 in

3.40 in INSIDE DIAMETER
4.00 in OUTSIDE DIAMETER

Figure 28–9

Find the area of the outside circle.

$A_B \approx 0.7854d^2$

$A_B \approx (0.7854)(4.00 \text{ in})^2 \approx 12.5664$ sq in

Find the area of the hole.

$A_B \approx (0.7854)(3.40 \text{ in})^2 \approx 9.079224$ sq in

Find the cross-sectional area.

12.5664 sq in $-$ 9.079224 sq in $\approx$ 3.487176 sq in

Find the volume.

$V \approx 3.487176$ sq in $\times$ 50.0 in $\approx$ 174 cu in *Ans*

Calculator Application

$(\boxed{} .7854 \boxed{\times} 4 \boxed{x^2} \boxed{-} .7854 \boxed{\times} 3.4 \boxed{x^2} \boxed{)} \boxed{\times} 50 \boxed{=} 174.3588$

$V \approx 174$ cu in *Ans*

EXERCISE 28–4

If necessary, use the tables in appendix A for equivalent units of measure. Solve these problems. Where necessary, round the answers to two decimal places unless otherwise specified.

1. Find the volume of a right circular cylinder with a base area of 76.00 square inches and a height of 8.600 inches.

2. Compute the volume of a right circular cylinder with an altitude of 0.40 meter and a base area of 0.30 square meter.

3. A cylindrical container has a base area of 154.0 square inches and a height of 16.00 inches. Find the capacity, in gallons, of the container.

4. Compute the weight of a cedar post that has a base area of 20.60 square inches and a length of 8.25 feet. Cedar weighs 23.0 pounds per cubic foot. Round the answer to 3 significant digits.

5. Find the volume of a steel shaft that is 59.00 centimeters long and has a diameter of 3.840 centimeters. Round the answer to 4 significant digits.

6. Each cylinder of a 6-cylinder engine has a 3.125-inch diameter and a piston stroke of 4.570 inches. Find the total piston displacement of the engine. Round the answer to 4 significant digits.

7. A cylindrical hotwater tank is 1.800 meters high and has a 0.500-meter diameter.

   a. Compute the volume of the tank.

   b. How many liters of water are held in this tank when full? Round the answer to 3 significant digits.

8. A 0.460-inch diameter brass rod is $8'5\frac{7}{32}''$ long. Round answers a and b to 3 significant digits.

   a. Find the volume, in cubic inches, of the rod.

   b. Compute the total weight of 40 rods. Brass weighs 0.300 pound per cubic inch.

9. A gasoline storage tank has a diameter of 12.00 meters and a height of 7.50 meters. Round answers a and b to 3 significant digits.

   a. Compute the volume of the tank.

   b. How many liters of gasoline are contained in the tank when it is one-third full?

10. A bronze bushing is shown in Figure 28–10.

    a. Compute the volume of bronze in the bushing.

    b. The bushing weighs 1.50 pounds. Find the weight of 1.00 cubic inch of bronze.

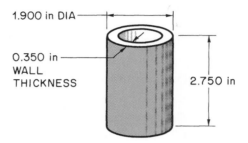

1.900 in DIA

0.350 in WALL THICKNESS

2.750 in

Figure 28–10

11. Each of two cylindrical vessels is 15.00 centimeters high. The larger vessel has a diameter of 7.800 centimeters. The smaller vessel has a diameter of 6.120 centimeters. How many more milliliters of liquid does the larger vessel hold? Round the answer to the nearest tenth milliliter.

12. A water pipe has a 5.00-inch inside diameter. Water flows through the pipe at a rate of 0.600 foot per second. How many gallons of water flow through the pipe in 10.0 minutes? Round the answer to the nearest gallon.

28–5 Computing Heights and Bases of Prisms and Cylinders

The height of a prism or a right circular cylinder can be determined if the base area and volume are known. Also, the base area can be found if the height and volume are known. Substitute the known values in the volume formula and solve for the unknown value.

EXAMPLES •————————————————————————

1. The volume of a steel shaft 20 inches long is 60 cubic inches. Compute the cross-sectional area.

   Substitute the values in the formula and solve.

   60 cu in $= A_B(20$ in$)$

   $A_B = 3$ sq in *Ans*

2. A concrete retaining wall is shown in Figure 28–11. Find the length of wall that can be constructed with 50.0 cubic yards of concrete.

   The cross section or base is in the shape of a trapezoid. Find the cross-sectional area.

   $A_B = 0.5(6.0$ ft$)(4$ ft $+ 2.0$ ft$) = 18$ sq ft

   Find the length.

   50.0 cu yd $\times$ 27 cu ft/cu yd $= 1,350$ cu ft

   1,350 cu ft $= 18$ sq ft (h)

   $h = 75$ ft *Ans*

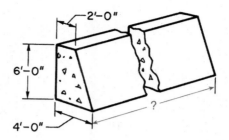

Figure 28–11

3. An engine piston has a height of 10.800 centimeters and a volume of 450.0 cubic centimeters. Find the piston diameter.

Find the base area.

$$450.0 \text{ cm}^3 = A_B(10.800 \text{ cm})$$

$$A_B \approx 41.6667 \text{ cm}^2$$

Find the piston diameter.

$$41.6667 \text{ cm}^2 \approx 0.7854d^2$$

$$d^2 \approx 53.051566 \text{ cm}^2$$

$$d \approx 7.284 \text{ cm } Ans$$

Calculator Application

$d = 450 \boxed{\div} 10.8 \boxed{\div} .7854 \boxed{=} \boxed{\sqrt{x}} \rightarrow 7.283647688$, 7.284 cm *Ans* (rounded)

or

$d = \boxed{\sqrt{}} \boxed{(} 450 \boxed{\div} 10.8 \boxed{\div} .7854 \boxed{)} \boxed{\text{EXE}}$ 7.283647688, 7.284 cm *Ans* (rounded)

EXERCISE 28–5

If necessary, use the tables in appendix A for equivalent units of measure. Solve these problems. Where necessary, round the answers to two decimal places unless otherwise specified.

1. Find the height of a prism that has a base area of 64 square inches and a volume of 448 cubic inches.

2. A solid right circular cylinder is 15.5 centimeters high and contains 110.0 cubic centimeters of material. Compute the cross-sectional area of the cylinder.

3. A room with a floor area of 440 square feet contains 3,740 cubic feet of air. Find the height of the room.

4. A solid steel post 27.6 inches long has a square base. The post has a volume of 104 cubic inches. Compute the length of a side of the base.

5. A cylindrical quart can has a 3.86-inch diameter. What is the height of the can?

6. A rectangular carton is designed to have a volume of 1.520 cubic meters. The carton is 1.18 meters high and 1.46 meters long. Compute the width of the carton.

7. A 22.50-foot-high fuel tank contains 70,550 gallons of fuel when full. Compute the tank diameter.

8. A triangular block of marble is shown in Figure 28–12 on page 586. The block weighs 5,544 pounds. Marble weighs 168 pounds per cubic foot.

  a. Compute the number of cubic feet of marble contained in the block.

  b. Compute the height of the block.

1-9
odd

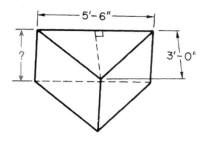

Figure 28–12

9. A rectangular aluminum plate required for a job is 4 feet 0 inches wide and 5 feet 0 inches long. The maximum allowable weight is 505.5 pounds. Aluminum weighs 168.5 pounds per cubic foot. What is the maximum thickness of the plate? Round the answer to the nearest tenth inch.

10. Water is pumped into a cylindrical storage tank at the rate of 325 liters per minute for a total of 4.75 hours. The tank has a base diameter of 6.46 meters. What is the depth of water in the tank if the tank was empty when the pumping started?

28–6   Lateral Areas and Surface Areas of Right Prisms and Cylinders

It is often necessary to determine surface areas of prisms and cylinders. A heating technician computes the number of square feet of stock for an air duct, a carpenter calculates the number of bundles of shingles for building siding, and a painting contractor computes the amount of paint for a cylindrical storage tank.

Lateral Areas

The *lateral area of a prism* is the sum of the areas of the lateral faces. The *lateral area of a cylinder* is the area of the curved or lateral surface. The lateral area of any prism can be determined by computing the area of each lateral face, then adding all of the areas of the faces. The *lateral area of a right prism* equals the product of the perimeter of the base and height.

$$LA = P_B h$$

where LA = lateral area
P_B = perimeter of base
h = height

To derive this formula, think of the lateral faces as being spread out on a flat surface or plane. For example, with the triangular base prism in Figure 28–13, imagine that a cut is made along edge E. Then the left face is unfolded back along edge F, and the right face is unfolded back along edge G. The three faces now lie on a flat surface or plane as shown in Figure 28–14. It can be seen that the total area of the three prism faces is equal to the area of the rectangle in Figure 28–14. Distance $A + B + C$ is the perimeter of the triangular prism base. Distance $A + B + C$ is also the length of the rectangle formed by unfolding the prism faces.

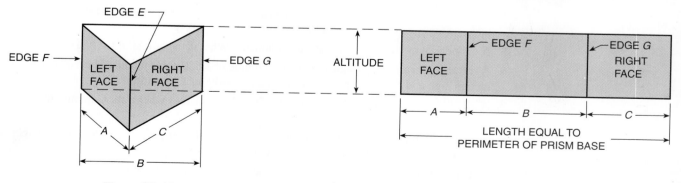

Figure 28–13

Figure 28–14

The *lateral area of a right circular cylinder* is equal to the product of the circumference of the base and height.

$$4 \quad LA = C_B h$$

where LA = lateral area
C_B = circumference of base
h = height

This formula is developed in much the same way as that of the prism. Imagine that a vertical cut is made along the lateral surface of the cylinder in Figure 28–15. The lateral surface is then unrolled and spread out flat as shown in Figure 28–16. The length of the rectangle in Figure 28–16 is equal to the circumference of the cylinder base in Figure 28–15.

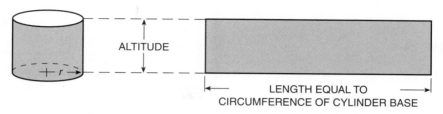

ALTITUDE

LENGTH EQUAL TO
CIRCUMFERENCE OF CYLINDER BASE

Figure 28–15   **Figure 28–16**

Surface Areas

The surface area of a prism or a cylinder must include the area of both bases as well as the lateral area. The *surface area of a prism or a cylinder* equals the sum of the lateral area and the two base areas.

$$5 \quad SA = LA + 2A_B$$

where SA = surface area
LA = lateral area
A_B = area of base

These examples illustrate the method of computing lateral areas and surface areas of right prisms and right circular cylinders.

EXAMPLES •————————————————————————————————

1. A shipping crate is shown in Figure 28–17.

   a. Compute the lateral area.

   b. Compute the total surface area.

   Find the perimeter of the base.

   P_B = 2(6.0 ft) + 2(4.5 ft) = 21 ft

   Find the lateral area.

   a. LA = 21 ft × 4.0 ft = 84 sq ft *Ans*

   Find the area of the base.

   A_B = 6.0 ft × 4.5 ft = 27 sq ft

   Find the surface area.

   b. SA = 84 sq ft + 2(27 sq ft) = 138 sq ft *Ans*

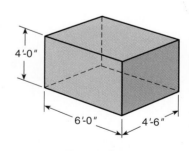

4'-0"

6'-0"   4'-6"

Figure 28–17

2. A cylinder hotwater tank has a diameter of 0.625 meter and a height of 1.85 meters.

   a. Find the lateral area.

   b. Find the surface area.

   Find the circumference of the circular base.

   $C_B \approx 3.1416(0.625 \text{ m}) \approx 1.9635 \text{ m}$

Find the lateral area.

   a. $LA \approx 1.9635$ m $\times$ 1.85 m ≈ 3.63 m^2 (rounded) *Ans*

   Find the area of the circular base.

   $A_B \approx 3.1416(0.3125$ m$)^2 \approx 0.3068$ m^2

Find the surface area.

   b. $SA \approx 3.63$ m^2 + 2(0.3068 m^2) ≈ 4.24 m^2 *Ans*

Calculator Application

   a. $LA = $ $\boxed{\pi}$ $\boxed{\times}$ .625 $\boxed{\times}$ 1.85 $\boxed{=}$ 3.632466506, 3.63 m^2 (rounded) *Ans*

   b. $SA = $ 3.63 $\boxed{+}$ 2 $\boxed{\times}$ $\boxed{\pi}$ $\boxed{\times}$ .3125 $\boxed{x^2}$ $\boxed{=}$ 4.243592315, 4.24 m^2 (rounded) *Ans*

EXERCISE 28–6

If necessary, use the tables in appendix A for equivalent units of measure. Solve these problems. Where necessary, round the answers to two decimal places unless otherwise specified.

1. A rectangular crate is 2 feet 6 inches wide, 6 feet 0 inches long, and 3 feet 0 inches high.

   a. Find the lateral area of the crate.

   b. Find the surface area of the crate.

2. A 0.460-meter diameter concrete pillar is 4.200 meters high. Compute the lateral area of the pillar.

3. How many square feet of insulation are required to cover a 5.75-inch outside diameter steam pipe that is 105 feet long? Round the answer to the nearest square foot.

4. An aboveground swimming pool is 2.75 meters high. The base is a regular octagon whose sides are each 2.20 meters long. Compute the lateral area of the pool.

5. A rectangular metal container with an open top is fabricated from 0.072-inch thick aluminum sheet. The container is 18.00 inches wide, 25.00 inches long, and 23.00 inches high. The aluminum sheet weighs 1.045 pounds per square foot. Find the total weight of the container.

6. A room 20 feet 0 inches wide, 25 feet 0 inches long, and 7 feet 6 inches high is wallpapered. One roll of wallpaper covers 55 square feet. How many rolls are required for the job? Deduct 15% for window and door openings. Round the answer to the nearest whole roll.

7. The sides and top of a cylindrical fuel storage tank are painted with 2 coats of paint. The tank is 75.0 feet in diameter and 60.0 feet high. One gallon of paint covers 550 square feet. How many whole gallons of paint are required?

8. A firm manufactures sheet metal pipe as shown in Figure 28–18. There is a 0.50-inch seam overlap. The material cost is $0.65 per square foot. What is the material cost for 1,200 pieces of pipe? Round the answer to the nearest ten dollars.

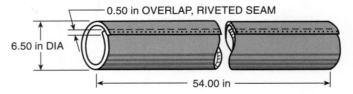

0.50 in OVERLAP, RIVETED SEAM

6.50 in DIA

54.00 in

Figure 28–18

9. A container manufacturer is presently producing 1.000-liter closed cylindrical containers that have a 12.80-centimeter diameter. The manufacturer will replace the present containers with 1.000-liter closed cylindrical containers having a 10.40-centimeter diameter.

   a. Find, in square centimeters, the material per container saved by replacing the 12.80-centimeter diameter containers with the 10.40-centimeter diameter containers.

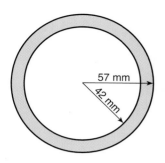

57 mm

42 mm

Figure 28–19

b. The cost of material used in making the containers is $2.10 per square meter. Based on an annual production of 5,800,000 containers, find the amount saved in material costs in 1 year by replacing the present containers. Round the answer to the nearest hundred dollars.

10. The cross section of a pipe is shown in Figure 28–19. If the pipe is 2 meters long, what is the volume, in cubic centimeters, of the material needed to make the pipe?

::: UNIT EXERCISE AND PROBLEM REVIEW

If necessary, use the tables in appendix A for equivalent units of measure. Solve these prism and cylinder problems. Where necessary, round the answers to two decimal places unless otherwise specified.

1. Compute the volume of a prism with a base area of 220.0 square centimeters and a height of 7.600 centimeters.

2. Find the volume of a right circular cylinder that has a height of 4.600 inches and a base area of 53.00 square inches.

3. Compute the height of a prism with a base area of 2.7 square feet and a volume of 4.86 cubic feet.

4. A solid right cylinder 0.90 meter high contains 0.80 cubic meter of material. Compute the cross-sectional area of the cylinder.

5. Find the lateral area of a rectangular box 7.30 inches wide, 9.60 inches long, and 5.40 inches high. Round the answer to 3 significant digits.

6. Each side of a square concrete platform is 8.70 meters long. The platform is 0.20 meter thick. Compute the number of cubic meters of concrete contained in the platform. Round the answer to the nearest cubic meter.

7. A circular $2\frac{1}{2}$-inch-thick pine tabletop has a 4-foot-0-inch diameter. The pine used for the top weighs 25 pounds per cubic foot. What is the weight of the top? Round the answer to the nearest pound.

8. Find the capacity, in gallons, of a cylindrical container with a base diameter of 21.50 inches and a height of 23.60 inches.

9. A carton with a square base is designed to contain 1.80 cubic meters. The carton is 1.15 meters high. What is the required length of each side of the base?

10. A cylindrical 1.000-liter vessel has a base diameter of 5.000 centimeters. How high is the vessel?

11. The walls and ceiling of a room are covered with sheetrock. The room is 25 feet 0 inches long, 16 feet 0 inches wide, and 8 feet 0 inches high. Deduct for 3 windows, each 3 feet 0 inches × 4 feet 6 inches, and 2 doorways, each 3 feet 6 inches × 7 feet 0 inches. How many square feet of sheetrock are required? Round the answer to the nearest ten square feet.

12. An automobile gasoline tank is shown in Figure 28–20 on page 590.

    a. Compute the capacity of the tank in liters.

    b. Compute the number of square meters of tank surface area.

    c. The sheet steel used to fabricate the tank weighs 11.08 kilograms per square meter. Compute the weight of the tank. Round the answer to the nearest tenth kilogram.

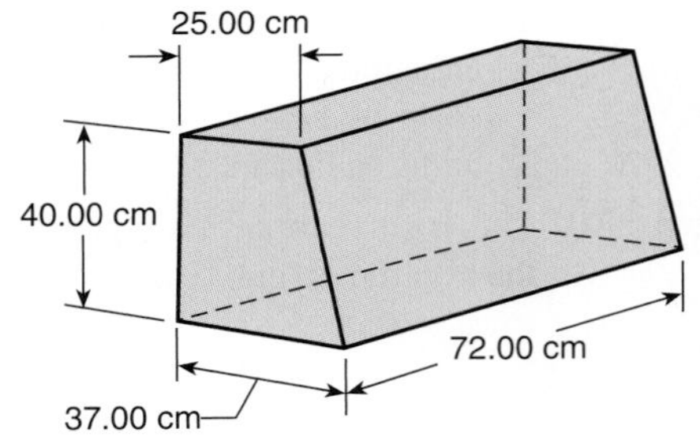

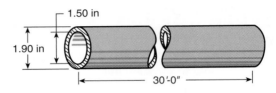

25.00 cm

40.00 cm

72.00 cm

37.00 cm

Figure 28–20

13. A length of brass pipe is shown in Figure 28–21.

1.50 in

1.90 in

30'-0"

Figure 28–21

Round the answers to a and b to 3 significant digits.

a. Find the number of cubic inches of brass contained in the pipe.

b. Brass weighs 526 pounds per cubic foot. What is the weight of the pipe?

14. The basement of a building is flooded with water to a depth of 18.0 inches. The basement is 52 feet 0 inches long and 38 feet 0 inches wide. Two pumps, each pumping at the rate of 34.0 gallons per minute, are used to drain the water. How many hours does it take to completely drain the water from the basement?

15. Topsoil is spread evenly over the entire surface of a rectangular lot that is 216 feet long and 187 feet wide. If 325 cubic yards of topsoil are spread over the lot, find the average thickness, in inches, of the topsoil. Round the answer to the nearest tenth inch.

UNIT 29 ⣿ Pyramids and Cones: Volumes, Surface Areas, and Weights

O B J E C T I V E S

After studying this unit you should be able to

* compute volumes of pyramids and cones.
* compute surface areas of pyramids and cones.
* compute capacities and weights of pyramids and cones.
* compute volumes of frustums of pyramids and cones.
* compute surface areas of frustums of pyramids and cones.
* compute capacities and weights of frustums of pyramids and cones.
* solve applied problems by using principles discussed in this unit.

Construction and mechanical applications of pyramids and cones are common. Certain types of building roofs and steeples are in the shape of pyramids. Tapered shafts, conical pulleys and clutches, conical compression springs, and roller bearings are some of the mechanical uses of portions (frustums) of cones. A tent is an example of a pyramid and food containers, buckets, lampshades, and funnels are examples of portions of cones.

29–1 Pyramids

A *pyramid* is a polyhedron whose base can be any polygon, and the other faces are triangles that meet at a common point called the *vertex* of the pyramid. The triangular faces that meet at the vertex are called *lateral faces.* A *lateral edge* is the line segment where two lateral faces meet. The *altitude* of a pyramid is the perpendicular segment from the vertex to the plane of the base. The *height* is the length of the altitude.

Pyramids are named according to the shape of the bases, such as triangular, quadrangular, pentagonal, hexagonal, and octagonal. In a *regular pyramid,* the base is a regular polygon, and the lateral edges are all equal in length. Only regular pyramids are considered in this book. Some common regular pyramids are shown in Figure 29–1. The parts of the pyramids are identified.

Draw Pictures w/ formulas on test. I will pro

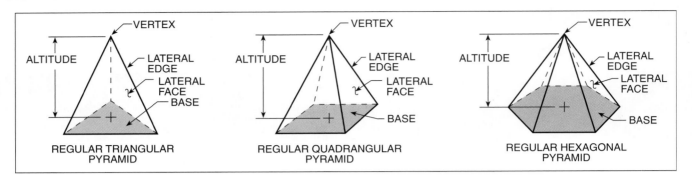

Figure 29–1

29–2   Cones

A *circular cone* is a solid figure with a circular base and a surface that tapers from the base to a point called the *vertex*. The surface lying between the base and the vertex is called the *lateral surface*. The *altitude* of a right circular cone is the perpendicular segment from the vertex to the center of the base. The *height* is the length of the altitude. The *axis* of a circular cone is a line that connects the vertex to the center of the circular base.

In a *right circular cone* the axis is perpendicular to the base. A right circular cone with the parts identified is shown in Figure 29–2. Only right circular cones are considered in this book.

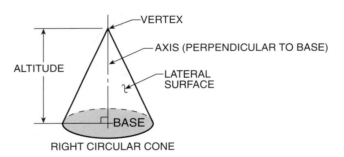

Figure 29–2

29–3   Volumes of Regular Pyramids and Right Circular Cones

Consider a prism and a pyramid that have identical base areas and altitudes. If the volumes of the prism and pyramid are measured, the volume of the pyramid will be one-third the volume of the prism. Also, if the volumes of a cylinder and a cone with identical bases and heights are measured, the volume of the cone will be one-third the volume of the cylinder. The formulas for computing volumes of prisms and right circular cylinders are the same. Therefore, the formulas for computing volumes of regular pyramids and right circular cones are the same. The *volume of a regular pyramid or a right circular cone* equals one-third the product of the area of the base and height.

$$V = \frac{1}{3}A_B h$$

where V = volume
A_B = area of the base
h = height

These examples illustrate the method of computing volumes of regular pyramids and right circular cones.

EXAMPLES •————————————————————————————————————

1. Compute the volume of a pyramid that has a base area of 24.0 square feet and a height of 6.0 feet.

$$V = \frac{24.0 \text{ sq ft} \times 6.0 \text{ ft}}{3} = 48 \text{ cu ft } Ans$$

2. A bronze casting in the shape of a right circular cone is 14.52 inches high and has a base diameter of 10.86 inches.

   a. Find the volume of bronze required for the casting.

   b. Find the weight of the casting to the nearest pound. Bronze weighs 547.9 pounds per cubic foot.

   a. $V \approx \dfrac{(3.1416)(5.430 \text{ in})^2(14.52 \text{ in})}{3} \approx 448.3 \text{ cu in } Ans$

   b. 448.3 cu in $\div$ $1{,}728$ cu in/cu ft ≈ 0.2594 cu ft

   0.2594 cu ft $\times 547.9$ lb/cu ft ≈ 142 lb Ans

Calculator Application

   a. $V = \boxed{\pi} \boxed{\times} 5.43 \boxed{x^2} \boxed{\times} 14.52 \boxed{\div} 3 \boxed{=} 448.3269989$, 448.3 cu in (rounded) Ans

   b. Weight $= 448.3 \boxed{\div} 1728 \boxed{\times} 547.9 \boxed{=} 142.1432697$, 142 lb (rounded) Ans

3. The roof of the building in Figure 29–3 is in the shape of a regular pyramid. Find the approximate number of cubic meters of attic space.

   $A_B = (12.00 \text{ m})^2 = 144.0 \text{ m}^2$

   $V = \dfrac{(144.0 \text{ m}^2)(5.00 \text{ m})}{3} = 240 \text{ m}^3 \ Ans$

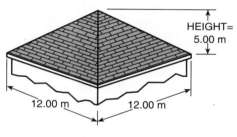

HEIGHT=
5.00 m

12.00 m   12.00 m

Figure 29–3

——•

EXERCISE 29–3 ——

If necessary, use the tables in appendix A for equivalent units of measure. Solve these problems. Where necessary, round the answers to two decimal places unless otherwise specified.

1. Compute the volume of a regular pyramid with a base area of 230 square feet and a height of 12 feet.

2. Find the volume of a right circular cone with a base area of 38.60 square centimeters and a height of 5.000 centimeters.

3. Find the volume of a regular pyramid with a height of 10.80 inches and a base area of 98.00 square inches.

4. A container is in the shape of a right circular cone. The base area is 4.8 square feet and the height is 1.5 feet. Compute the capacity of the container in gallons.

5. A solid granite monument is in the shape of a regular pyramid. The base area is 60.0 square feet, and the height is 10 feet 3 inches. Find the weight of the monument. Granite weighs 168 pounds per cubic foot. Round the answer to 3 significant digits.

6. A brass casting is in the shape of a right circular cone with a base diameter of 8.26 centimeters and a height of 18.36 centimeters. Find the volume. Round the answer to 3 significant digits.

7. Compute the number of cubic feet of attic space in a building that has a roof in the shape of a regular pyramid. Each of the 4 outside walls of the building is 42 feet 0 inches long, and the roof is 18 feet 0 inches high. Round the answer to the nearest hundred cubic feet.

8. Two solid pieces of aluminum in the shape of right circular cones with different base diameters are machined. The heights of both pieces are 6 inches. The base of the smaller piece is 2 inches in diameter. The base of the larger piece is twice as large, or 4 inches in diameter. How many times heavier is the larger piece than the smaller?

9. A vessel is in the shape of a right circular cone. This vessel contains liquid to a depth of 12.8 centimeters, as shown in Figure 29–4. How many liters of liquid must be added in order to fill the vessel?

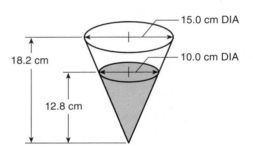

Figure 29–4

10. The steeple of a building is in the shape of a regular pyramid with a triangular base. Each of the 3 base sides is 4.6 meters long, and the steeple is 7.0 meters high. Compute the number of cubic meters of airspace contained in the steeple. Round the answer to the nearest cubic meter.

29–4   Computing Heights and Bases of Regular Pyramids and Right Circular Cones

As with prisms and cylinders, heights and base areas of regular pyramids and right circular cones are readily determined. Substitute known values in the volume formula and solve for the unknown value.

EXAMPLES

1. The volume of a regular pyramid is 270 cubic centimeters, and the height is 18 centimeters. Compute the base area.

Substitute the values in the formula and solve.

$$270 \text{ cm}^3 = \frac{A_B(18 \text{ cm})}{3}$$

$$A_B = 45 \text{ cm}^2 \; Ans$$

2. A disposable plastic drinking cup is designed in the shape of a right circular cone. The cup holds $\frac{1}{3}$ pint (9.63 cubic inches) of liquid when full. The rim (base) diameter is 3.60 inches. Find the cup depth (height).

Substitute the values in the formula and solve.

$$9.63 \text{ cu in} \approx \frac{3.1416(1.80 \text{ in})^2(h)}{3}$$

$$h \approx 2.84 \text{ in } Ans$$

Calculator Application

$h = 9.63 \text{ cu in } (3) \div \pi(1.80 \text{ in})^2$

9.63 $\boxed{\times}$ 3 $\boxed{\div}$ $\boxed{(}$ $\boxed{\pi}$ $\boxed{\times}$ 1.8 $\boxed{\times}$ $\boxed{)}$ $\boxed{=}$ 2.838256515

$h \approx 2.84 \text{ in } Ans$

EXERCISE 29–4

If necessary, use the tables in appendix A for equivalent units of measure. Solve these problems. Where necessary, round the answers to two decimal places unless otherwise specified.

1. Compute the height of a regular pyramid with a base area of 32.00 square feet and a volume of 152.0 cubic feet.

2. A right circular cone 1.2 meters high contains 0.80 cubic meter of material. Find the cone base area.

3. The base area of a wooden form in the shape of a regular pyramid is 28.0 square feet. The form contains 21.0 cubic feet of airspace. How high is the form?

4. A container with a capacity of 6.0 gallons is in the shape of a right circular cone. The container is 1.5 feet high. Find the container base area.

5. A tent in the shape of a regular pyramid is designed to contain 5.60 cubic meters of airspace. The base of the tent is square with each side 2.50 meters long. What is the height of the tent?

6. Find the base diameter of a right circular cone that has a volume of 922.4 cubic centimeters and a height of 14.85 centimeters.

7. A concrete monument in the shape of a regular pyramid with a square base weighs 13,400 pounds. The monument is 7.30 feet high. Compute the length of a base side. Concrete weighs 137 pounds per cubic foot.

8. The tank shown in Figure 29–5 is in the shape of a cone. It has a dipstick to measure the level of the water in the tank by measuring the vertical height of the water level. The cone holds 10000 liters and its height is 100 cm. If the water level is 50 cm, determine the radius of the water level. Round the answer to 3 decimal places.

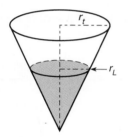

Figure 29–5

29–5 Lateral Areas and Surface Areas of Regular Pyramids and Right Circular Cones

To determine material requirements and weights of pyramids and conical-shaped objects, surface areas are computed. These types of computations have wide application in the industrial and construction fields.

Slant heights are used in determining lateral areas of pyramids and cones. The *slant height of a regular pyramid* is the length of the altitude of any of the lateral faces. The slant height of a regular pyramid is shown in Figure 29–6. The *slant height of a right circular cone* is the distance from the vertex to any point on the edge of the circular base. The slant height of a right circular cone is shown in Figure 29–7.

Figure 29–6   **Figure 29–7**

Lateral Areas

The lateral area of a pyramid is the sum of the areas of the lateral faces. The *lateral area of a regular pyramid* equals one-half the product of the perimeter of the base and the slant height.

$$LA = \frac{1}{2}P_B h_s$$ where LA = lateral area
    P_B = perimeter of base
    h_s = slant height

The lateral area of a circular cone is the area of the lateral surface. The *lateral area of a right circular cone* equals one-half the product of the circumference of the base and the slant height.

$$LA = \frac{1}{2}C_B h_s$$ where LA = lateral area
    C_B = circumference of base
    h_s = slant height

Surface Areas

The total surface area of a pyramid or cone includes the base as well as the lateral area. The *total surface area of a pyramid or cone* equals the sum of the lateral area and the base area.

$$SA = LA + A_B$$ where SA = total surface area
    LA = lateral area
    A_B = area of base

These examples illustrate the procedure for computing lateral areas and total surface areas of regular pyramids and right circular cones.

EXAMPLES •————————————————————————————

1. Refer to the right circular cone in Figure 29–8.

   a. Compute the lateral area

   b. Compute the total surface area.

   a. Find the lateral area.

   $LA \approx 0.5(3.1416)(14.00$ in$)(16.00$ in$) \approx 351.9$ sq in *Ans*

   b. Find the total surface area.

   $A_B \approx 3.1416(7.000$ in$)^2 \approx 153.9$ sq in

   $SA \approx 351.9$ sq in $+ 153.9$ sq in ≈ 505.8 sq in *Ans*

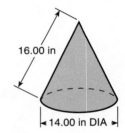

16.00 in

14.00 in DIA

Figure 29–8

2. The pyramid in Figure 29–9 has a square base with each base side 20.00 centimeters long. The pyramid altitude, 25.00 centimeters, is given. Find the lateral area of the pyramid.

   The slant height is not known and must be computed. In Figure 29–10, right triangle *ACB* is formed within the pyramid.

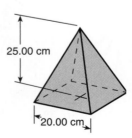

25.00 cm

20.00 cm

Figure 29–9

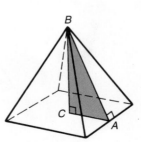

B

C

A

Figure 29–10

The triangle is formed by altitude CB, slant height AB, and triangle base CA. Slant height AB is computed by applying the Pythagorean theorem.

$$c^2 = a^2 + b^2$$
$$AB^2 = CA^2 + CB^2$$
$$AB^2 = (10.00 \text{ cm})^2 + (25.00 \text{ cm})^2$$
$$AB^2 = 100.0 \text{ cm}^2 + 625.0 \text{ cm}^2$$
$$AB^2 = 725.0 \text{ cm}^2$$
$$AB \approx 26.93 \text{ cm}$$

Find the lateral area.

$$P_B = 4(20.00 \text{ cm}) = 80.00 \text{ cm}$$
$$LA \approx 0.5(80.00 \text{ cm})(26.93 \text{ cm}) \approx 1{,}077 \text{ cm}^2 \text{ } Ans$$

Calculator Application

$AB = $ ⌈ $($ ⌉ 10 $\boxed{x^2}$ $\boxed{+}$ 25 $\boxed{x^2}$ $\boxed{)}$ $\boxed{\sqrt{x}}$ → 26.92582404

$LA = $ .5 $\boxed{\times}$ 4 $\boxed{\times}$ 20 $\boxed{\times}$ 26.93 $\boxed{=}$ 1077.2, 1,077 cm² (rounded) *Ans*

EXERCISE 29–5

If necessary, use the tables in appendix A for equivalent units of measure. Solve these problems. Where necessary, round the answers to two decimal places unless otherwise specified.

1. Find the lateral area of a regular pyramid that has a base perimeter of 92.0 inches and a slant height of 20.0 inches.

2. Find the lateral area of a right circular cone with a slant height of 1.80 meters and a base perimeter of 6.40 meters.

3. A regular pyramid has a base perimeter of 58.4 centimeters, a slant height of 17.8 centimeters, and a base area of 213.16 square centimeters.

   Round the answers to a and b to 3 significant digits.

   a. Compute the lateral area of the pyramid.

   b. Compute the total surface area of the pyramid.

4. A right circular cone has a slant height of 4.250 feet, a base circumference of 18.84 feet, and a base area of 28.26 square feet.

   a. Find the lateral area of the cone.

   b. Find the total surface area of the cone.

5. A regular pyramid with a square base has a slant height of 10.00 inches. Each side of the base is 8.00 inches long.

   a. Find the lateral area of the pyramid.

   b. Find the total surface area of the pyramid.

6. A right circular cone with a slant height of 14.05 centimeters has a base diameter of 9.72 centimeters.

   Round the answers to a and b to 3 significant digits.

   a. Compute the lateral area of the cone.

   b. Compute the total surface area of the cone.

7. The building roof in Figure 29–11 is in the shape of a pyramid with a square base.

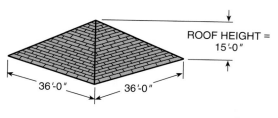

ROOF HEIGHT =
15'-0"

36'-0"   36'-0"

Figure 29–11

    a. Find the surface area of the roof. Round the answer to the nearest 10 square feet.

    b. Find the number of bundles of shingles required to cover the roof if 4 bundles of shingles are required for each 100 square feet of roof area. Allow 15% for waste. Round the answer to the nearest whole bundle.

8. A conical sheet copper cover with an open bottom is shown in Figure 29–12.

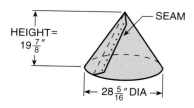

SEAM

HEIGHT=
$19\frac{7}{8}$"

$28\frac{5}{16}$" DIA

Figure 29–12

Round the answers to a and b to 3 significant digits.

    a. Find the number of square feet of copper contained in the cover. Allow 5% for the overlapping seam.

    b. Find the weight of the cover. The sheet copper used for the fabrication weighs 2.65 pounds per square foot.

9. A plywood form in the shape of a right pyramid with a regular hexagon base is constructed. Each side of the base is 1.20 meters long, and the form is 2.70 meters high. Allow 20% for waste. Find the number of square meters of plywood required for the lateral area of the form. Round the answer to the nearest tenth square meter.

10. The conical spire of a building has a 10.00-meter diameter. The altitude is 18.00 meters.

    a. Find the lateral area of the spire. Round the answer to the nearest tenth square meter.

    b. How many liters of paint are required to apply 2 coats of paint to the spire? One liter of paint covers 12.3 square meters of surface area. Round the answer to the nearest liter.

29–6   Frustums of Pyramids and Cones

When a pyramid or a cone is cut by a plane parallel to the base, the part that remains is called a *frustum.* Frustums of pyramids and cones are often found in architecture. As well as architectural applications, containers, tapered shafts, funnels, and lampshades are a few familiar examples of frustums of pyramids and cones. A frustum has two bases, upper and lower.

    The *larger base* is the base of the cone or pyramid. The *smaller base* is the circle or polygon formed by the parallel cutting plane. The smaller base of the pyramid has the same shape as the larger base. The two bases are similar. The *altitude* is the perpendicular segment that joins the planes of the bases. The *height* is the length of the altitude. A frustum of a pyramid and a frustum of a cone with their parts identified are shown in Figure 29–13.

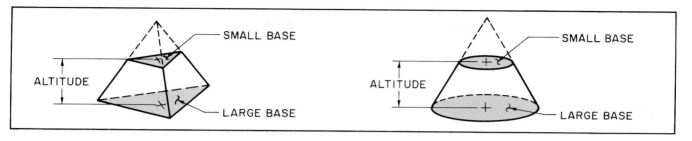

Figure 29–13

STOP & start

29–7 Volumes of Frustums of Regular Pyramids and Right Circular Cones

The *volume of the frustum of a pyramid or cone* is computed from the formula

$$V = \frac{1}{3}h\left(A_B + A_b + \sqrt{A_B A_b}\right)$$

where V = volume of the frustum
of a pyramid or cone
h = height
A_B = area of larger base
A_b = area of smaller base

The formula for the volume of a frustum of a right circular cone is expressed in this form.

$$V = \frac{1}{3}\pi h(R^2 + r^2 + Rr)$$

where V = volume of a right circular cone
h = height
R = radius of larger base
r = radius of smaller base

These examples illustrate the method of computing volumes of frustums of regular pyramids and right circular cones.

EXAMPLES •

1. A wastebasket is designed in the shape of a frustum of a pyramid with a square base as shown in Figure 29–14. Find the volume of the basket in cubic feet.

Find the larger base area.

$A_B = (14.0 \text{ in})^2 = 196 \text{ sq in}$

Find the smaller base area.

$A_b = (11.0 \text{ in})^2 = 121 \text{ sq in}$

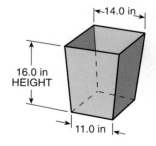

Figure 29–14

Find the volume.

$$V = \frac{(16.0 \text{ in})\left[196 \text{ sq in} + 121 \text{ sq in} + \sqrt{(196 \text{ sq in})(121 \text{ sq in})}\right]}{3}$$

$$V = \frac{(16.0 \text{ in})(196 \text{ sq in} + 121 \text{ sq in} + 154 \text{ sq in})}{3}$$

$V = 2,512$ cu in

Express the volume in cubic feet.

2,512 cu in ÷ 1,728 cu in/cu ft ≈ 1.45 cu ft *Ans*

Calculator Application

16 $\boxed{\times}$ $\boxed{(}$ 14 $\boxed{x^2}$ $\boxed{+}$ 11 $\boxed{x^2}$ $\boxed{+}$ $\boxed{(}$ 14 $\boxed{x^2}$ $\boxed{\times}$ 11 $\boxed{x^2}$ $\boxed{)}$ $\boxed{\sqrt{x}}$ $\boxed{)}$ $\boxed{\div}$ 3 $\boxed{\div}$ 1728 $\boxed{=}$
1.453703704

Volume ≈ 1.45 cu ft *Ans*

2. A tapered sheet shaft is shown in Figure 29–15.

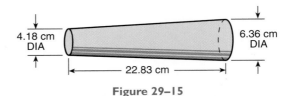

4.18 cm DIA   6.36 cm DIA   22.83 cm

Figure 29–15

a. Find the number of cubic centimeters of steel contained in the shaft.

b. Find the weight of the shaft. The steel in the shaft weighs 0.0078 kilogram per cubic centimeter.

a. Find the volume.

$$V \approx \frac{(3.1416)(22.83 \text{ cm})[(3.18 \text{ cm})^2 + (2.09 \text{ cm})^2 + (3.18 \text{ cm})(2.09 \text{ cm})]}{3}$$

≈ 505.09

$V \approx 505$ cm^3 *Ans*

b. Compute the weight.

505 cm^3 × 0.0078 kg/cm^3 ≈ 3.9 kg *Ans*

Calculator Application

a. $V =$ $\boxed{\pi}$ $\boxed{\times}$ 22.83 $\boxed{\times}$ $\boxed{(}$ 3.18 $\boxed{x^2}$ $\boxed{+}$ 2.09 $\boxed{x^2}$ $\boxed{+}$ 3.18 $\boxed{\times}$ 2.09 $\boxed{)}$ $\boxed{=}$ $\boxed{\div}$ 3 $\boxed{=}$
505.0870048

$V \approx 505$ cm^3 *Ans*

b. Weight = 505 $\boxed{\times}$ .0078 $\boxed{=}$ 3.939, 3.9 kg (rounded) *Ans*

EXERCISE 29–7

If necessary, use the tables in appendix A for equivalent units of measure. Solve these problems. Round the answers to 3 significant digits unless otherwise specified.

1. Compute the volume of the frustum of a regular pyramid with a height of 18.00 feet. The larger base area is 150.00 square feet, and the smaller base area is 90.00 square feet. Round the answer to 4 significant digits.

2. The frustum of a right circular cone has the larger base area equal to 32.00 square inches and the smaller base area equal to 15.00 square inches. The height is 16.00 inches. Compute the volume. Round the answer to 4 significant digits.

3. A pail is in the shape of a frustum of a right circular cone. The smaller base area is 426 square centimeters, and the larger base area is 875 square centimeters. The height is 29.5 centimeters. Compute the capacity of the pail in liters.

4. A solid oak trophy base is in the shape of a frustum of a regular pyramid. It has a larger base area of 175.0 square inches and a smaller base area of 120.0 square inches. The height is 8.50 inches. Compute the weight of the trophy base. Oak weighs 47.0 pounds per cubic foot.

5. The bottom of a drinking glass is 2.30 inches in diameter. The top is 2.80 inches in diameter. The height is 3.50 inches.

   a. Compute the volume of space contained in the glass.

   b. Compute the capacity of the glass in ounces.

6. The side view of a tapered steel shaft is shown in Figure 29–16. The length of the shaft is reduced from 18.40 inches to 13.60 inches. How many cubic inches of stock are removed?

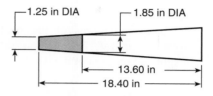

Figure 29–16

7. A sand storage bin in the shape of a frustum of a regular pyramid with square bases is shown in Figure 29–17. Round the answers to a and b to 2 significant digits.

   a. Find the maximum number of cubic yards of sand that can be stored in the bin.

   b. Find the weight of sand in the bin when it is full. Sand weighs 100.0 pounds per cubic foot.

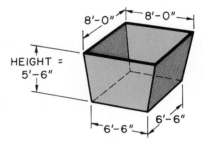

Figure 29–17

8. The side view of a rivet is shown in Figure 29–18.

   a. Find the volume of the rivet.

   b. Find the weight of the rivet. Steel weighs 7721 kg/m³.

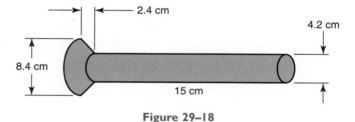

Figure 29–18

29–8 Lateral Areas and Surface Areas of Frustums of Regular Pyramids and Right Circular Cones

All lateral faces of frustums of pyramids are trapezoids. The *slant height* of the frustum of a regular pyramid is the length of the altitude of each trapezoidal lateral face. The slant height of the frustum of a regular pyramid is shown in Figure 29–19. The slant height of the frustum of a right circular cone is the shortest distance between the bases on the lateral surface. The slant height of the frustum of a right circular cone is shown in Figure 29–20.

Figure 29–19   **Figure 29–20**

Lateral Areas

The lateral area of the frustum of a pyramid is the sum of the areas of the lateral faces. The lateral area can be determined by computing the area of each trapezoidal face and adding the face areas. However, it is easier to compute lateral areas by using slant heights and base perimeters. The *lateral area of the frustum of a regular pyramid* equals one-half the product of the slant height and the sum of the two base perimeters.

$$LA = \frac{1}{2}h_s(P_B + P_b)$$

where LA = lateral area
h_s = slant height
P_B = perimeter of larger base
P_b = perimeter of smaller base

The lateral area of the frustum of a cone is the area of the lateral surface. The *lateral area of the frustum of a right circular cone* equals one-half the product of the slant height and the sum of the two base circumferences.

$$LA = \frac{1}{2}h_s(C_B + C_b)$$

where LA = lateral area
h_s = slant height
C_B = circumference of larger base
C_b = circumference of smaller base

The formula for the lateral area of the frustum of a right circular cone is simplified to this form.

$$LA = \pi h_s(R + r)$$

where LA = lateral area
h_s = slant height
R = radius of larger base
r = radius of smaller base

Surface Areas

The total surface area of the frustum of a pyramid or cone must include the area of both bases as well as the lateral area. The *total surface area of the frustum of a pyramid or cone* equals the sum of the lateral area, the larger base area, and the smaller base area.

$$SA = LA + A_B + A_b$$

where SA = total surface area
LA = lateral area
A_B = area of larger base
A_b = area of smaller base

These examples illustrate the procedure for computing lateral areas and total surface areas of frustums of regular pyramids and right circular cones.

EXAMPLES •————————————————————————————————————

1. A plywood pedestal has the shape of the frustum of a square-based regular pyramid. Each side of the larger base is 4 feet 6 inches long. Each side of the smaller base is 3 feet 3 inches long. The slant height is 3 feet 0 inches.

   a. Compute the lateral area of the pedestal.

   b. Compute the total surface area of the pedestal.

   a. Compute the lateral area.

   $P_B = 4 \times 4.5$ ft $= 18$ ft

   $P_b = 4 \times 3.25$ ft $= 13$ ft

   $LA = 0.5(3$ ft$)(18$ ft $+ 13$ ft$) = 46.5$ sq ft, 47 sq ft *Ans*

   b. Compute the total surface area.

   $A_B = (4.5$ ft$)^2 = 20.25$ sq ft

   $A_b = (3.25$ ft$)^2 = 10.56$ sq ft

   $SA = 46.5$ sq ft $+ 20.25$ sq ft $+ 10.56$ sq ft ≈ 77 sq ft *Ans*

Calculator Application

Compute total surface area (*SA*).

.5 ☒ 3 ☒ ⦅ 18 ⊞ 13 ⦆ ⊞ 4.5 x^2 ⊞ 3.25 x^2 ⊟ 77.3125

$SA \approx 77$ sq ft *Ans*

2. Compute the number of square centimeters of fabric contained in the lampshade in Figure 29–21.

The slant height must be found. In Figure 29–22, right triangle *ACB* is formed by altitude *CB*, slant height *AB*, and triangle base *AC*.

$$c^2 = a^2 + b^2$$
$$AB^2 = (36.0 \text{ cm})^2 + (9.0 \text{ cm})^2$$
$$AB^2 = 1,377 \text{ cm}^2$$
$$AB \approx 37.11 \text{ cm}$$

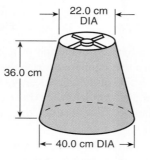

Figure 29–21

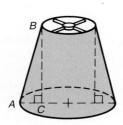

Figure 29–22

Find the lateral area.

$$LA \approx 3.1416(37.11 \text{ cm})(20.0 \text{ cm} + 11.0 \text{ cm}) \approx 3,600 \text{ cm}^2 \text{ } Ans$$

Calculator Application

$LA =$ π ☒ ⦅ 36 x^2 ⊞ 9 x^2 ⦆ $\sqrt{x}$ ☒ ⦅ 20 ⊞ 11 ⦆ ⊟ 3613.920018

$LA \approx 3,600 \text{ cm}^2 \text{ } Ans$

EXERCISE 29–8

If necessary, use the tables in appendix A for equivalent units of measure. Solve these problems. Where necessary, round the answers to 3 significant digits unless otherwise specified.

1. The frustum of a regular pyramid has a larger base perimeter of 60 feet 0 inches and a smaller base perimeter of 44 feet 0 inches. The slant height is 16 feet 0 inches. Compute the lateral area.

2. A frustum of a right circular cone has a slant height of 5.60 inches. The larger base circumference is 4.80 inches, and the smaller base circumference is 3.20 inches. Find the lateral area.

3. The frustum of a regular pyramid has square bases. The length of a side of the larger base is 3.60 meters, and the length of a side of the smaller base is 2.70 meters. The slant height is 4.20 meters.

   a. Find the lateral area of the pyramid.

   b. Find the surface area of the pyramid.

4. The frustum of a right circular cone has a smaller base diameter of 24.2 centimeters and a larger base diameter of 36.4 centimeters. The slant height is 29.3 centimeters.

   a. Compute the lateral area of the cone.

   b. Compute the surface area of the cone.

5. A redwood planter in the shape of a frustum of a regular pyramid is shown in Figure 29–23. The bases are regular hexagons. Find the number of square feet of redwood required for the lateral surface of the planter.

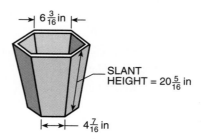

Figure 29–23

6. A wooden platform in the shape of a frustum of a regular pyramid is constructed. The platform bases are equilateral triangles. Each side of the bottom base is 4.80 meters long, and each side of the top base is 4.00 meters long. The slant height is 0.70 meter. The bottom of the platform is open. Allow 15% for waste. Find the number of square meters of lumber required to construct the platform. Round the answer to the nearest tenth square meter.

7. A support column in the shape of a frustum of a right circular cone has a slant height of 16.0 feet. The smaller base is 15.0 inches in diameter, and the larger base is 21.0 inches in diameter. Find, in square feet, the lateral area of the column.

8. An open-top pail is shown in Figure 29–24.

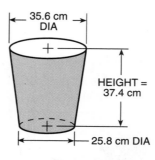

Figure 29–24

   a. Compute the lateral area of the pail.

   b. Compute the total number of square meters of metal contained in the pail.

   c. The pail is made of galvanized sheet steel, which weighs 11.20 kilograms per square meter. Compute the weight of the pail.

9. A granite monument in the shape of a frustum of a regular pyramid with regular octagonal bases is shown in Figure 29–25. The sides and the top of the monument are ground and polished. Find the number of square meters of surface area ground and polished.

10. A paper cup is made by first taking a circle and turning up (crimping) the outer 0.6 cm of the circle. This circle is glued inside the small base of the frustum of a cone as shown in Figure 29.26.

    a. What is the area of the paper needed for the lateral surface of the cup?

    b. What is the area of the paper needed to make the bottom of the cup?

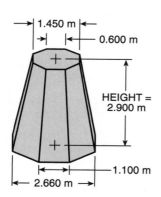

Figure 29–25

Figure 29–26

⠿ UNIT EXERCISE AND PROBLEM REVIEW

If necessary, use the tables in appendix A for equivalent units of measure. Solve these problems. Where necessary, round the answers to 3 significant digits unless otherwise specified.

1. Find the volume of a right circular cone with a base area of 0.800 square meter and a height of 1.240 meters.

2. Compute the volume of a regular pyramid that has a height of 2.60 feet and a base area of 2.80 square feet.

3. A regular pyramid with a base area of 54.6 square feet contains 210.5 cubic feet of material. Find the height of the pyramid.

4. Compute the base area of a right circular cone that is 15.8 centimeters high and has a volume of 1,070 cubic centimeters.

5. A regular pyramid has a base perimeter of 56.3 inches and a slant height of 14.9 inches. Find the lateral area of the pyramid.

6. A right circular cone has a slant height of 3.76 feet. The base circumference is 17.58 feet, and the base area is 24.62 square feet.

   a. Compute the lateral area of the cone.

   b. Compute the total surface area of the cone.

7. The frustum of a right circular cone has a larger base area of 40.0 square centimeters and a smaller base area of 19.0 square centimeters. The height is 22.0 centimeters. Find the volume.

8. The frustum of a regular pyramid has a smaller base perimeter of 18.0 meters and a larger base perimeter of 26.0 meters. The slant height is 5.60 meters. Find the lateral area.

9. A building has a roof in the shape of a regular pyramid. Each of the 4 outside walls of the building is 38 feet 0 inches long, and the roof is 16 feet 0 inches high. Compute the number of cubic feet of attic space in the building.

10. A solid brass casting in the shape of a right circular cone has a base diameter of 4.36 inches and a height of 3.94 inches. Find the weight of the casting. Brass weighs 0.302 pound per cubic inch.

11. A vessel in the shape of a right circular cone has a capacity of 0.690 liter. The base diameter is 12.3 centimeters. What is the height of the vessel?

12. A plywood form is constructed in the shape of a right pyramid with a square base. Each side of the base is 7 feet 6 inches long, and the form is 5 feet 3 inches high. Compute the number of square feet of plywood required for the lateral surface of the form. Allow 15% for waste. Round the answer to 2 significant digits.

13. A container in the shape of a frustum of a regular pyramid with square bases is shown in Figure 29–27. Compute the capacity of the container in liters. Round the answer to 2 significant digits.

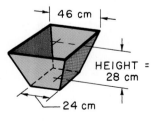

Figure 29–27

14. Compute the number of square inches of fabric in the lampshade in Figure 29–28.

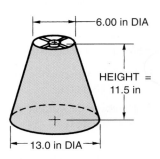

Figure 29–28

15. A decorative copper piece in the shape of a regular pyramid with an equilateral triangle base is shown in Figure 29–29. The lateral faces and base are covered with sheet copper, which weighs 1.87 pounds per square foot.

  a. Compute the total number of square feet of copper contained in the piece.

  b. Compute the weight of the piece.

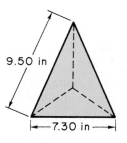

Figure 29–29

UNIT 30 ⠿ Spheres and Composite Figures: Volumes, Surface Areas, and Weights

OBJECTIVES

After studying this unit you should be able to

- compute surface areas and volumes of spheres.

- compute capacities and weights of spheres.

- solve applied sphere problems, using the principles discussed in this unit.

- solve applied composite solid figures, using principles discussed in section V.

One of the many practical applications of sphere formulas is in the design and construction of spherical holding tanks. Calculating volumes is required in determining capacities; surface areas must be computed to determine material sizes and weights.

Practical volume and surface area applications often require working with objects that are a combination of two or more simple solid shapes. These applications require working with related volume and surface area formulas.

30–1 Spheres

A *sphere* is a solid figure bounded by a curved surface such that every point on the surface is the same distance (equidistant) from a point called the *center.* A round ball, such as a baseball or basketball, is an example of a sphere.

The *radius* of a sphere is the length of any segment from the center to any point on the surface. A *diameter* is a segment through the center with its endpoints on the curved surface. The diameter of a sphere is twice the radius.

If a plane cuts through (intersects) a sphere and does *not* go through the center, the section is called a *small circle.* As intersecting planes move closer to the center, the circular sections get larger. A plane that cuts through (intersects) the center of a sphere is called a *great circle.* A great circle is the largest circle that can be cut by an intersecting plane. The sphere and the great circle have the same center. The circumference of a great circle is the circumference of the sphere. If a plane is passed through the center of a sphere, the sphere is cut in two equal parts. Each part is a half sphere, called a *hemisphere.* A sphere with its parts identified is shown in Figure 30–1.

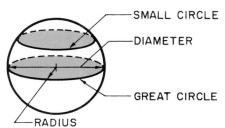

Figure 30–1

30–2   Surface Area of a Sphere

The *surface area of a sphere* equals four times the area of the great circle.

$$SA = 4\pi r^2$$ where SA = surface area of the sphere
r = radius of the sphere or great circle

EXAMPLE •

A spherical gas storage tank, which has a 96.4-foot diameter, is to be painted.

Compute the surface area of the tank to the nearest hundred square feet.

Compute, to the nearest gallon, the amount of paint required. One gallon of paint covers 530 square feet.

a. Find the surface area.
$$SA = 4(3.1416)(48.2 \text{ ft})^2 \approx 29{,}200 \text{ sq ft } Ans$$

b. Find the number of gallons of paint required.
$$29{,}200 \text{ sq ft} \div 530 \text{ sq ft/gal} \approx 55 \text{ gal } Ans$$

EXERCISE 30–2

If necessary, use the tables in appendix A for equivalent units of measure. Find the surface area for each sphere. Round the answers to 4 significant digits.

1. 3.000-inch diameter   **5.** 7.900-foot radius

2. 20′6.0″ diameter   **6.** 6.384-inch diameter

3. 1.400-meter radius   **7.** 4.873-meter diameter

4. 26.87-centimeter radius   **8.** 0.7856-centimeter radius

Solve these problems. Where necessary, round the answers to 3 significant digits unless otherwise specified.

9. A spherical storage tank has an 80.0-foot diameter. The storage tank will be repainted, and the cost of preparation, priming, and applying a finish coat of paint is estimated at $0.20 per square foot. Compute the total cost of repainting the tank.

10. A company manufactures plastic covers in the shape of hemispheres. The diameter of each cover is 38.6 centimeters. The material expense for the covers is based on a cost of $3.40 per square meter. What is the material cost for a production run of 55,800 covers?

11. A spherical copper float with a diameter of 1.40 inches weighs 0.80 ounce. Another spherical float made of the same material is twice the diameter, or 2.80 inches. What is the weight of the 2.80-inch diameter float? Round the answer to 2 significant digits.

12. Compute the surface area of a basketball that has a circumference (great circle) of $29\frac{3}{4}$ inches.

13. A spherical fuel storage tank has a diameter of 6.12 meters. A cylindrical fuel storage tank has the same diameter, 6.12 meters, and a height of 4.08 meters. Both tanks have the same fuel capacity of approximately 120,000 liters.

a. Which of the two tanks requires more surface area material?

b. How much more material is required by the larger tank? Round the answer to 2 significant digits.

30–3   Volume of a Sphere

The *volume of a sphere* equals the surface area multiplied by one-third the radius. Since the surface area of a sphere equals $4\pi r^2$, the volume equals $\frac{1}{3}r(4\pi r^2)$. The formula for the volume of a sphere is simplified to this form:

$$V = \frac{4}{3}\pi r^3 \qquad \text{where } V = \text{volume of the sphere}$$
$$r = \text{radius of the sphere}$$

EXAMPLE

A stainless steel ball bearing contains balls that are each 1.80 centimeters in diameter.

Find the volume of a ball.

Find the weight of a ball to the nearest gram. Stainless steel weighs 7.88 grams per cubic centimeter.

a. Find the volume.

$$V = \frac{4(3.1416)(0.900 \text{ cm})^3}{3} \approx 3.05 \text{ cm}^3 \text{ Ans}$$

b. Find the weight.

$$3.05 \text{ cm}^3 \times 7.88 \text{ g/cm}^3 \approx 24 \text{ g } Ans$$

Calculator Application

$4 \boxed{\times} \boxed{\pi} \boxed{\times} .9 \boxed{y^x} 3 \boxed{\div} 3 \boxed{=} 3.053628059 \boxed{\times} 7.88 \boxed{=} 24.06258911$
  └ Volume   └ Weight

a. $V \approx 3.05 \text{ cm}^3 \text{ Ans}$

b. Weight ≈ 24 g Ans

EXERCISE 30–3

If necessary, use the tables in appendix A for equivalent units of measure. Compute the volume of each sphere. Round the answers to 3 significant digits.

1. 2.00-meter radius
2. 28.0-centimeter diameter
3. 7.60-inch diameter
4. 6.00-foot radius

5. 4.78-inch radius
6. 0.075-meter diameter
7. 16.2-centimeter diameter
8. 25′6″ diameter

Solve these problems. Round the answers to 3 significant digits unless otherwise specified.

9. A thrust bearing contains 18 steel balls. The steel used weighs 0.283 pound per cubic inch. The diameter of each ball is 0.240 inch. Compute the total weight of the balls in the bearing.

10. A vat in the shape of a hemisphere with an 18.0-inch diameter contains liquid. What is the capacity of the vat in gallons?

11. An empty spherical tank that has a diameter of 8.6 meters is filled with water. Water is pumped into the tank at a rate of 870 liters per minute. How many hours of pumping are required to fill the tank? Round the answer to the nearest tenth hour.

12. Spheres are formed from molten bronze. The diameter of the mold in which the spheres are formed is 6.26 centimeters. When the bronze spheres solidify (turn solid), they shrink by 6% of the molten-state volume. Compute the volume of a sphere after the bronze solidifies.

13. A truck will deliver a shipment of solid concrete spheres. The concrete weighs 137 pounds per cubic foot. The diameter of the sphere is 22 inches. The maximum load weight limit of the truck is 6.0 short tons. What is the maximum number of spheres that can be carried by the truck?

14. A spherical shell of cast iron has an external diameter of 6.34 in and the thickness of the shell is 0.625 in. Find its weight if 1 in³ of cast iron weighs 0.2604 lb.

30–4   Volumes and Surface Areas of Composite Solid Figures

A shaft or a container may be a combination of a cylinder and the frustum of a cone. A round-head rivet is a combination of a cylinder and a hemisphere. Objects of this kind are called *composite solid figures* or *composite space figures.*

    To compute *volumes and surface areas of composite solid figures,* it is necessary to determine the volume or surface area of each simple solid figure separately. The individual volumes or areas are then added or subtracted.

EXAMPLES ●

1. An aluminum weldment is shown in Figure 30–2.

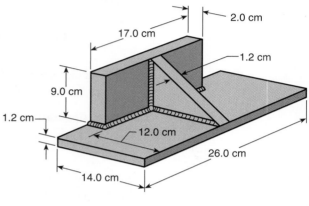

Figure 30–2

   a. Find the total volume of the weldment.

   b. Find the weight of the weldment. Aluminum weighs 0.0027 kilogram per cubic centimeter.

   a. Find the volume of the bottom plate. (Volume of a prism)

   $V_1 = A_b h$

   $V_1 = [(14.0 \text{ cm})(26.0 \text{ cm})](1.2 \text{ cm}) = 436.8 \text{ cm}^3$

   Find the volume of the top plate. (Volume of a prism)

   $V_2 = A_b h$

   $V_2 = [(9.0 \text{ cm})(17.0 \text{ cm})](2.0 \text{ cm}) = 306 \text{ cm}^3$

   Find the volume of the triangular plate. (Volume of a prism)

   $V_3 = A_b h$

   $V_3 = [0.5(12.0 \text{ cm})(9.0 \text{ cm})](1.2 \text{ cm}) = 64.8 \text{ cm}^3$

   Find the total volume.

   $V_T = 436.8 \text{ cm}^3 + 306 \text{ cm}^3 + 64.8 \text{ cm}^3 = 807.6 \text{ cm}^3$

   $V_T = 807.6 \text{ cm}^3 \; Ans$

   b. Find the weight.

   $0.0027 \text{ kg/cm}^3 \times 807.6 \text{ cm}^3 \approx 2.2 \text{ kg} \; Ans$

2. The side view of a flanged shaft is shown in Figure 30–3. Find the volume of metal in the shaft.

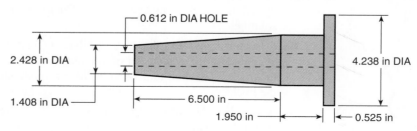

Figure 30–3

a. Find the volume of the 6.500-inch long frustum of a cone.

$$V = \frac{1}{3}\pi\, h(R^2 + r + Rr)$$

$R = 0.5(2.428 \text{ in}) = 1.214 \text{ in}$

$r = 0.5(1.408 \text{ in}) = 0.704 \text{ in}$

$$V_1 \approx \frac{(3.1416)(6.500 \text{ in})[(1.214 \text{ in})^2 + (0.704 \text{ in})^2 + (1.214 \text{ in})(0.704 \text{ in})]}{3}$$

$$V_1 \approx \frac{(3.1416)(6.500 \text{ in})(1.474 \text{ sq in} + 0.4956 \text{ sq in} + 0.8547 \text{ sq in})}{3}$$

$V_1 \approx 19.223 \text{ cu in}$

b. Find the volume of the 2.428-inch diameter cylinder.

$V = A_b h$

$r = 0.5(2.428 \text{ in}) = 1.214 \text{ in}$

$V_2 \approx [3.1416(1.214 \text{ in})^2](1.950 \text{ in}) \approx 9.029 \text{ cu in}$

c. Find the volume of the 4.238-inch diameter cylinder.

$V = A_b h$

$V_3 = [3.1416(2.119 \text{ in})^2](0.525 \text{ in}) \approx 7.406 \text{ cu in}$

d. Find the volume of the 0.612-inch diameter through hole.

$V = A_b h$

$r = 0.5(0.612 \text{ in}) = 0.306 \text{ in}$

$h = 6.500 \text{ in} + 1.950 \text{ in} + 0.525 \text{ in} = 8.975 \text{ in}$

$V_4 = [3.1416(0.306 \text{ in})^2](8.975 \text{ in}) \approx 2.640 \text{ cu in}$

Find the volume of the metal.

$V_T \approx 19.223 \text{ cu in} + 9.029 \text{ cu in} + 7.406 \text{ cu in} - 2.640 \text{ cu in} \approx 33.018 \text{ cu in}$

$V_T \approx 33.0 \text{ cu in } Ans$

Calculator Application

$V_1 \approx \boxed{\pi}\,\boxed{\times}\,6.5\,\boxed{\times}\,\boxed{(}\,1.214\,\boxed{x^2}\,\boxed{+}\,.704\,\boxed{x^2}\,\boxed{+}\,1.214\,\boxed{\times}\,.704\,\boxed{)}\,\boxed{=}\,\boxed{\div}\,3\,\boxed{=}\,19.22282111$

$V_2 \approx \boxed{\pi}\,\boxed{\times}\,1.214\,\boxed{x^2}\,\boxed{\times}\,1.95\,\boxed{=}\,9.028630039$

$V_3 \approx \boxed{\pi}\,\boxed{\times}\,2.119\,\boxed{x^2}\,\boxed{\times}\,.525\,\boxed{=}\,7.405784826$

$V_4 \approx \boxed{\pi}\,\boxed{\times}\,.306\,\boxed{x^2}\,\boxed{\times}\,\boxed{(}\,6.5\,\boxed{+}\,1.95\,\boxed{+}\,.525\,\boxed{)}\,\boxed{=}\,2.640141373$

Volume of metal: $V_T \approx 19.223\,\boxed{+}\,9.0286\,\boxed{+}\,7.4058\,\boxed{-}\,2.6401\,\boxed{=}\,33.0173$

$V_T \approx 33.0 \text{ cu in } Ans$

3. Compute the surface area of the sheet metal elbow in Figure 30–4.

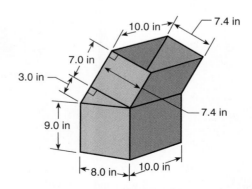

Figure 30–4

a. Find the lateral area of the rectangular base prism bottom section.

$LA = P_b h$

$LA = [2(8.0) + 2(10.0 \text{ in})] \times 9.0 \text{ in} = 324 \text{ sq in}$

b. Find the lateral area of the triangular base prism section.

The section consists of two triangular faces and one rectangular back face.

$A_1 = 0.5(3.0 \text{ in})(7.4 \text{ in}) = 11.1 \text{ sq in (triangular face)}$

$A_2 = 3.0 \text{ in} \times 10.0 \text{ in} = 30 \text{ sq in (rectangular back face)}$

$LA = 2(11.1 \text{ sq in}) + 30 \text{ sq in} = 52.2 \text{ sq in}$

c. Find the lateral area of the rectangular base prism top section.

$LA = P_b h$

$LA = [2(10.0 \text{ in}) + 2(7.4 \text{ in})] \times 7.0 \text{ in} = 243.6 \text{ sq in}$

Find the total surface area of the elbow.

$SA = 324 \text{ sq in} + 52.2 \text{ sq in} + 243.6 \text{ sq in} = 619.8 \text{ sq in}, 620 \text{ sq in (rounded)}$ *Ans*

EXERCISE 30–4

If necessary, use the tables in appendix A for equivalent units of measure. Solve these problems. Where necessary, round the answers to 3 significant digits unless otherwise specified.

1. Compute the number of cubic yards of concrete required to construct the steps in Figure 30–5.

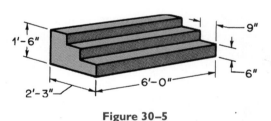

Figure 30–5

2. Find the weight of the steel baseplate in Figure 30–6. Steel weighs 490 pounds per cubic foot.

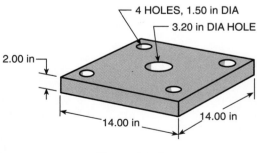

Figure 30–6

3. What is the total number of cubic meters of airspace in the building in Figure 30–7? Disregard wall and floor volumes.

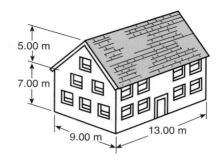

Figure 30–7

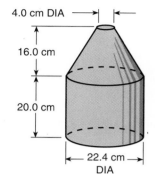

Figure 30–8

4. Compute the capacity, in liters, of the container in Figure 30–8. Round the answer to the nearest liter.

5. A sheet copper pipe and flange are shown in Figure 30–9. The pipe fits into a 5.00-inch diameter hole in the flange.

   a. Find the total surface area of the pipe and flange.

   b. Find the weight of the pipe and flange. The copper sheet weighs 2.355 pounds per square foot.

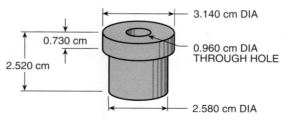

Figure 30–9

6. Find the number of cubic centimeters of material contained in the jig bushing in Figure 30–10.

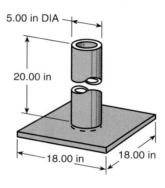

Figure 30–10

7. A sheet metal reducer is shown in Figure 30–11 on page 614. The top and the bottom sections are in the shape of rectangular prisms with square bases. The middle section is in the shape of a frustum of a regular pyramid with square bases. Allow 10% for waste and seam overlaps. Find the total number of square feet of material required for the lateral surface of the reducer. Round the answer to the nearest tenth square foot.

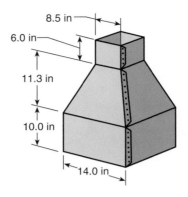

Figure 30–11

8. Compute the number of cubic centimetres of material in the locating saddle in Figure 30–12.

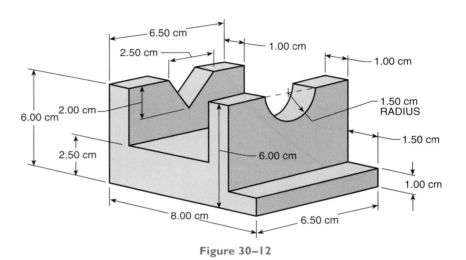

Figure 30–12

9. Find the number of cubic meters of topsoil required for the plot of land in Figure 30–13. The topsoil will be spread to an average thickness of 15 centimeters. Round the answer to 2 significant digits.

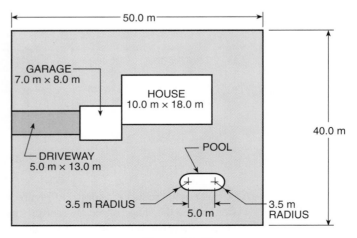

Figure 30–13

10. Compute the number of cubic yards of concrete required for the 3.0-inch thick concrete patio in Figure 30–14. Round the answer to 2 significant digits.

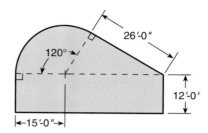

Figure 30–14

UNIT EXERCISE AND PROBLEM REVIEW

If necessary, use the tables in appendix A for equivalent units of measure. Round the answers to 4 significant digits.

For each sphere in exercises 1 through 4:

    a. Compute the surface area.

    b. Compute the volume.

1. 6.425-inch diameter

2. 26.80-centimeter diameter

3. 0.3006-meter radius

4. 7'6.00" radius

Solve these problems. Round the answers to 3 significant digits unless otherwise specified.

5. A firm manufactures lampshades in the shape of hemispheres with a 14.0-inch diameter. The cost of the shade material is $0.60 per square foot. Compute the total material expense of 2,500 lampshades.

6. The side view of a steel roundhead rivet is shown in Figure 30–15. What is the weight of the rivet? Steel weighs 0.283 pound per cubic inch.

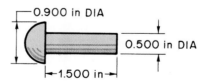

Figure 30–15

7. A hollow glass sphere has an outside circumference (great circle) of 23.80 centimeters. The wall thickness of the sphere is 0.50 centimeter. Compute the weight of the sphere. Glass weighs 1.5 grams per cubic centimeter. Round the answer to the nearest ten grams.

8. The material cost of a solid bronze sphere with a diameter of 3.80 centimeters is $1.05. Compute the material cost of a solid bronze sphere with a 5.70-centimeter diameter.

9. A wooden planter is shown in Figure 30–16. The top section is in the shape of a prism with a square base. The bottom section is in the shape of a frustum of a pyramid with square bases.

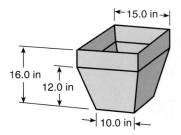

Figure 30–16

    a. Compute the number of cubic feet of soil that can be held by the planter when full. Disregard the thickness of the lumber.

    b. Compute the total number of square feet of lumber required in the construction. Disregard the thickness of the lumber. Allow 15% for waste.

10. A spherical tank has a diameter of 42 feet. The tank is $\frac{1}{4}$ full of water. The water will be drained from the tank at a rate of 225 gallons per minute. How many hours will it take to empty the tank? Round the answer to 2 significant digits.

11. Compute the number of cubic yards of asphalt required to pave the section of land in Figure 30–17. The average thickness of the asphalt is 3.0 inches. Round the answer to 2 significant digits.

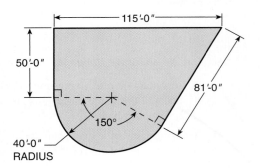

Figure 30–17

Basic Statistics

UNIT 31 ⠿ Graphs: Bar, Circle, and Line

OBJECTIVES

After studying this unit you should be able to

- read and interpret data from given vertical and horizontal bar graphs.

- draw and label vertical and horizontal bar graphs using given data.

- draw bar, circle, and broken-line graphs using a computer spreadsheet.

- read and interpret data from given circle graphs.

- read and interpret data from given broken-line, straight-line, and curved-line graphs.

- draw and label broken-line, straight-line, and curved-line graphs by directly using given data.

- draw and label straight-line and curved-line graphs by expressing given formulas as table data.

- identify given table data in terms of constant or variable rates of change and identify the type of graph that would be produced by plotting the data.

A graph shows the relationship between sets of quantities in picture form. Graphs are widely used in business, industry, government, and scientific and technical fields. Newspapers, magazines, books, and manuals often contain graphs. Since they are used in both occupations and everyday living, it is important to know how to interpret and construct basic types of graphs.

Statistical data are often time-consuming and difficult to interpret. Graphs present data in simple and concise picture form. Data, when graphed, often can be interpreted more quickly and are easier to understand.

The Cartesian coordinate system was studied in Unit 16. Now, three different types of graphs will be shown. Bar graphs, circle graphs or charts, and line graphs are three common ways of picturing statistical data. These three types of graphs can be seen on television, in magazines, books, and manuals, and, almost everyday, in newspapers.

31–1   Types and Structure of Graphs

Many kinds of graphs are designed for special-purpose applications. An understanding of basic graphs, such as bar graphs and line graphs, provides a background for the reading and construction of other more specialized graphs.

Circle graphs are constructed with the use of a protractor. Protractors were discussed in Unit 20, Angular Measure.

Graph paper, which is also called coordinate and cross-section paper, is used for graphing data. Cross-section paper is available in various line spacings. Paper with 5 or 10 equal spaces in a given length is generally used.

Bar graphs and line graphs contain two scales. The *horizontal scale* is usually called the *x-axis* and the *vertical scale* is normally called the *y-axis* as shown in Figure 31–1. The axes can be drawn at any convenient location, but are usually located on the bottom and left of the graph. The axes (scales) for all graphs in this unit are located at the bottom and left as shown.

A scale shows the values of the cross-sectional spaces. The scale values vary and depend on the data that are graphed.

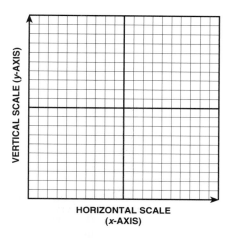

Figure 31–1

31–2   Reading Bar Graphs

On a bar graph, the lengths of the bars represent given data. The bars on a bar graph may be vertical, as in Figure 31–2, or horizontal, as in Figure 31–3. Some computer programs, such as Excel, refer to a vertical bar graph as a *column graph*. To read a bar graph, first determine the value of each space on the scale (axis). If the bars are horizontal, determine the space value on the horizontal scale. If the bars are vertical, determine the space value on the vertical scale.

Next, locate the end of each bar. If the bar is horizontal, project down (vertically) to the horizontal scale. If the bar is vertical, project across (horizontally) to the vertical scale. Read each value on the appropriate scale. If the end of a bar is not directly on a line, estimate its value.

EXAMPLE ●───

The bar graph in Figure 31–2 on page 620 shows the monthly production of a manufacturing firm over a five-month period.

a. How many units are produced each month?

b. How many units are produced during the entire five-month period?

c. How many units difference is there between the highest and lowest monthly production?

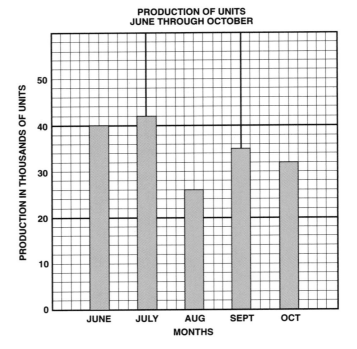

Figure 31–2

Find the vertical scale values. The major divisions represent 10,000 units. Each small space represents 10,000 units ÷ 5 = 2,000 units.

From the end of each bar, project over to the vertical scale and read each value.

June:		40,000 *Ans*
July:	40,000 + (1 × 2,000) =	42,000 *Ans*
Aug:	20,000 + (3 × 2,000) =	26,000 *Ans*
Sept:	30,000 + (2.5 × 2,000) =	35,000 *Ans*
Oct:	30,000 + (1 × 2,000) =	32,000 *Ans*

a. Add the monthly production from June to October.
   40,000 + 42,000 + 26,000 + 35,000 + 32,000 = 175,000 *Ans*

b. Subtract August's production from July's production.
   42,000 − 26,000 = 16,000 *Ans*

Bar graphs can be constructed so that each bar represents more than one quantity. Each bar is divided into two or more sections. To distinguish one section from another, sections are generally shaded, colored, or crosshatched.

EXAMPLE

The bar graph in Figure 31–3 shows United States employment in major occupational groups for a certain year. Each group is represented by a bar. Each bar is divided into the number of males and the number of females employed within the group. So, each bar represents the total number of workers. Because the variables in each group are stacked, this is often called a *stacked bar graph*.

EMPLOYMENT IN MAJOR OCCUPATIONAL GROUPS

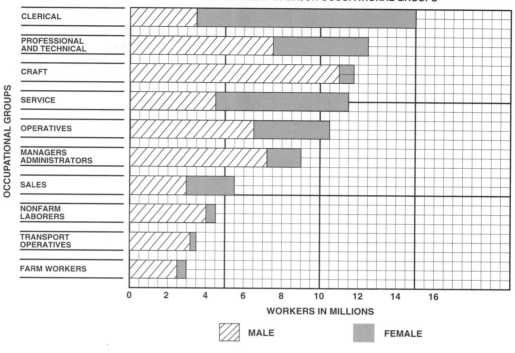

Figure 31–3

a. How many men are employed in service occupations?

b. How many women are employed in service occupations?

c. What percent, to the nearest whole percent, of the clerical group is made up of women?

Find the horizontal scale values. The numbered divisions each represent 2,000,000 workers. Each small space represents 2,000,000 ÷ 4 or 500,000 workers.

a. Find the bar that represents service occupations.

From the end of the male division of the bar, project down to the horizontal scale and read the value.

$$4,000,000 \text{ workers} + (1 \times 500,000 \text{ workers})$$
$$= 4,500,000 \text{ workers, or } 4.5 \text{ million workers}$$
in service occupations are men *Ans*

b. From the end of the service occupations bar, project down to the horizontal scale and read the value.

$$10,000,000 \text{ workers} + (3 \times 500,000 \text{ workers})$$
$$= 11,500,000 \text{ workers or } 11.5 \text{ million}$$

This is the total number of men and women workers.

Subtract the number of men from the total number of workers.

$$11,500,000 \text{ workers} - 4,500,000 \text{ workers}$$
$$= 7,000,000 \text{ workers, or } 7 \text{ million}$$
in service occupations are women *Ans*

c. From the end of the clerical bar, project down to the horizontal scale. The total number of workers is found to be 15 million. The number of men is $3\frac{1}{2}$ million. The number of women is found by subtracting the number of men from the total number of clerical workers. There are 11.5 million women in this group.

$$15,000,000 - 3,500,000 = 11,500,000 \text{ or } 11.5 \text{ million}$$

To find the percent of women, divide the number of women by the total number of workers in the group.

$$\frac{11.5 \text{ million}}{15 \text{ million}} \approx 0.7666 \approx 77\% \text{ (to the nearest whole number) of the clerical group is made up of women } Ans$$

EXERCISE 31–2

1. A firm's operating expenses for a certain year are shown on the horizontal bar graph in Figure 31–4.

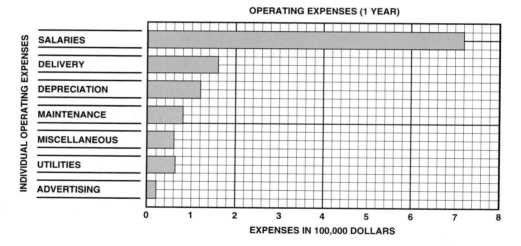

Figure 31–4

a. What is the amount of each of the seven operating expenses?

b. How many more dollars are spent for delivery than for utilities?

c. What percent of the total operating expenses is delivery? Express the answer to the nearest whole number.

d. The utilities expense shown is 20% greater than that of the previous year. How many dollars were spent on utilities during the previous year?

2. A vertical bar graph in Figure 31–5 shows United States production of aluminum for each of eight consecutive years. Notice the gap between the two jagged lines near the bottom of the graph. This gap is to show that the production numbers between 0.1 and 1.4 have been eliminated. This is often done to make it easier to read the graph.

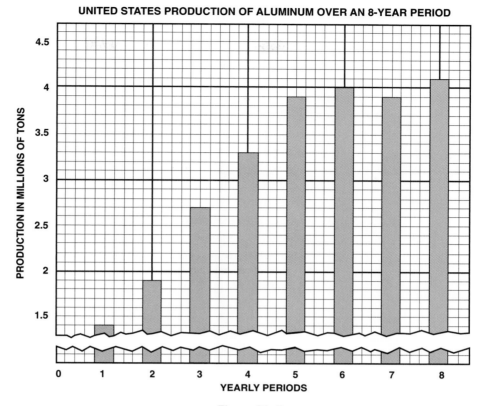

Figure 31–5

a. What is the number of tons of aluminum produced during each yearly period?

b. How many more tons of aluminum were produced during the last year than during the first year of the eight-year period?

c. What is the percent increase in production of the last year over the first year of the eight-year period? Express the answer to the nearest whole percent.

3. The stacked bar graph in Figure 31–6 shows the motor vehicle production for a certain year by the six leading national producers.

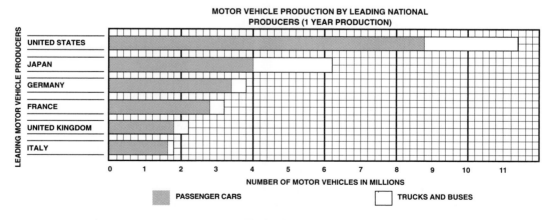

Figure 31–6

a. What is the total number of motor vehicles produced by the six nations?

b. What is the total number of trucks and buses produced by the six nations?

c. What percent of the total six-nation motor vehicle production was produced by the United States? Express the answer to the nearest whole percent.

31–3   Drawing Bar Graphs

The structure of bar graphs and how to read graph data have just been explained in this unit. Now it is possible to use what was learned to draw bar graphs from given data. Using graph paper saves time and increases the accuracy of measurements. These directions will be for drawing a general bar graph. In Unit 32–6, a specific type of bar graph, called a histogram, will be studied.

The following five steps are used to draw a bar graph.

1. Arrange the given data in a logical order. For example, group data from the smallest to the largest values or from the beginning to the end of a time period.

2. Decide which group of data is to be on the horizontal scale and which is to be on the vertical scale. Generally, the horizontal scale is on the bottom and the vertical scale is on the left of the graph.

3. Draw and label the horizontal and vertical scales. Although the starting point is usually zero, any convenient value can be assigned.

4. Assign values to the spaces on the scales that conveniently represent the data. The data should be clear and easy to read.

5. Draw each bar to the required length according to the given data.

EXAMPLE •——

A company's sales for each of eight years are listed in Figure 31–7. Indicate the yearly sales by a bar graph.

1992—$2,400,000	1993—$2,300,000	1989—$1,800,000
1990—$2,200,000	1991—$2,100,000	1994—$2,700,000
1995—$2,900,000	1988—$1,300,000	

Figure 31–7

Arrange the data in logical order. Rearrange the data in sequence from 1988 to 1995 as shown in the table in Figure 31–8.

Year	1988	1989	1990	1991	1992	1993	1994	1995
Yearly Sales	$1,300,000	$1,800,000	$2,200,000	$2,100,000	$2,400,000	$2,300,000	$2,700,000	$2,900,000

Figure 31–8

Decide which data are to be on the horizontal scale and which on the vertical scale. It is decided to make the vertical scale the sales scale. The bars will be vertical. The vertical scale will be on the left and the horizontal scale on the bottom of the graph. Draw and label the scales as shown in Figure 31–9. The starting point will be zero, as shown.

Assign values to scale spaces. Space years (centers of the bars) 5 small spaces apart. Assign each major division (10 small spaces) a value of $1,000,000. Label the major divisions 1, 2, and 3 to represent $1,000,000, $2,000,000, and $3,000,000, respectively. Each small space represents 0.1 million dollars or $100,000. For ease of reading, label each fifth small space in 0.5 million units. Refer to Figure 31–10.

Draw each bar starting with 1988. The sales for 1988 were $1,300,000 or 1.3 million dollars. From the 1988 location on the horizontal scale, project up one major division (1 million) + three small spaces (0.3 million) as shown in Figure 31–11. Locate the end points of the remaining seven bars the same way. Draw and shade each bar. The completed graph is shown in Figure 31–12.

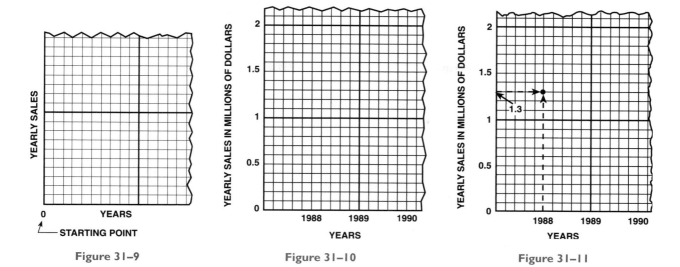

Figure 31–9   Figure 31–10   Figure 31–11

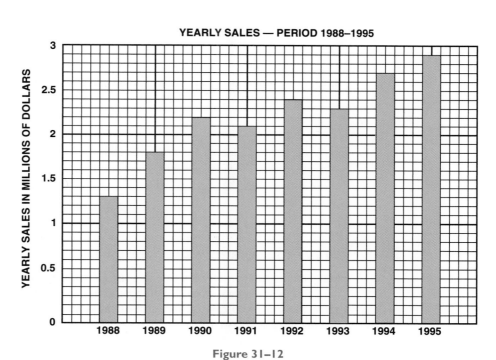

Figure 31–12

EXERCISE 31–3

Draw and label a bar graph for each of the following problems.

1. The table shown in Figure 31–13 lists five sources of electrical energy in the United States. The percent of each source of the total energy production for a certain year is given. Draw a horizontal bar graph showing the table data.

Energy Source	Coal	Gas	Oil	Hydropower	Nuclear
Percent of Total Energy	19%	30%	27%	18%	6%

Figure 31–13

2. Draw a vertical bar graph to show the dollar values of a chemical company's exports as listed in the table shown in Figure 31–14.

Month	Jan	Feb	Mar	April	May	June
Monthly Exports	$910,000	$930,000	$1,050,000	$1,000,000	$980,000	$1,100,000

Figure 31–14

3. The table shown in Figure 31–15 lists the six leading cattle producing states for a certain year. Draw a horizontal bar graph showing the table data.

State	Texas	Iowa	Kansas	Nebraska	Oklahoma	Missouri
Number of Head of Cattle	13,600,000	7,800,000	6,800,000	6,600,000	5,400,000	5,200,000

Figure 31–15

4. Draw a vertical bar graph showing the United States imports for eight consecutive years as shown in the table in Figure 31–16.

Year	1st	2nd	3rd	4th	5th	6th	7th	8th
Imports in Billions of Dollars	18	21	25	26	33	36	39	45

Figure 31–16

31–4   Drawing Bar Graphs with a Spreadsheet

Many employers consider it a valuable asset if you have ability to use a spreadsheet. The purpose of this text is not to teach you how to use a spreadsheet, but rather to introduce you to some ways that you can use spreadsheets in your job. A spreadsheet is not the only tool for making bar graphs. Statistical programs, such as MINITAB, SAS, and S-PLUS®, can be used to draw graphs, and some graphing calculators will draw bar graphs.

A spreadsheet can be used to draw a bar graph and can remove much of the drudgery of plotting graphs by hand. The examples and the instructions in this section are meant to supplement the user's guide for your computer spreadsheet, not to replace it. You should always consult the user's guide to get answers to "how to" questions.

To use a spreadsheet, first arrange the data in a logical order. Next, decide which items you want listed along the vertical axis and which values should be on the horizontal axis. Enter the data in two columns with the items for the vertical axis in Column A and those for the horizontal axis in Column B. Next, follow the program's directions. Examples in this section were run using Microsoft® Excel. However, you could use a different spreadsheet. While the directions may be a little different, the procedures are basically the same for all spreadsheets.

EXAMPLES •───

1. Figure 31–17 shows the expense at Junior's Auto Repair for the month of September. Use a computer spreadsheet to create a horizontal bar graph to indicate the expenses for the month.

Salaries	$8,895	Rent	$ 1,775
Parts	3,177	Office Supplies	106
Advertising	212	Insurance	215
Utilities	1,271	Taxes	1,690
Loan Payments	2,135	Legal	568

Figure 31–17

Solution. We will need two columns. We will use the first column, called Column A on the spreadsheet, for the types of expenses that will be listed on the vertical axis. Column B will contain the actual amount of the expense. Enter "Expense" in cell A1 and "Amount" in cell B1.

Next, enter the names of the expenses in Column A beginning with cell A2. Pressing the Return key after each name will move the cursor to the cell directly below. Finally, enter the amount of each expense in Column B starting with cell B2. The final result should look something like Figure 31–18.

	A	B
1	Expense	Amount
2	Salaries	8895
3	Parts	3177
4	Advertising	212
5	Utilities	1271
6	Loan Payment	2135
7	Rent	1775
8	Office Supplies	106
9	Insurance	215
10	Taxes	1690
11	Legal	568
12		

Figure 31–18

To graph the function, first highlight the table you just constructed. Left click on cell A2 and hold the left mouse key down as you move the cursor to cell B11. As you move the cursor, the cells should become highlighted.

Release the mouse key (leaving the cells highlighted) and move the cursor up to "Chart Wizard." (See Figure 31–19.) Chart Wizard should be indicated by a small icon on the top standard menu. Click on Chart Wizard to begin the process of constructing a graph.

Figure 31–19

The first step is to select the type of graph. The menu shows two general types of bar graphs. One has the bars going horizontally and is called "bar." The other, called "column," has the bars going vertically. We want the one with horizontal bars, so click on "bar."

When you click on "bar," six subtypes of bar graphs are shown. They are circled in Figure 31–20. None of them look like the type of graph we want, but the first one looks the closest, so click on the first drawing of a bar graph. Click on Next>. A Data Range is shown next. No changes are needed, so again click on Next>.

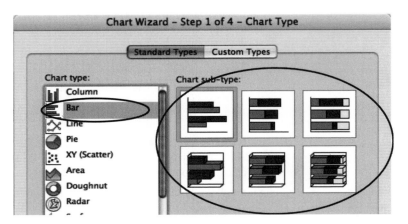

Figure 31–20

The next window allows you to title the graph and label the axes. A sample is shown in Figure 31–21. Not all of the title of the graph can be seen in the figure. After the titles have been filled in, click on $\boxed{\text{Next} >}$. To display the finished graph, click on $\boxed{\text{Finish}}$ with the result in Figure 31–22.

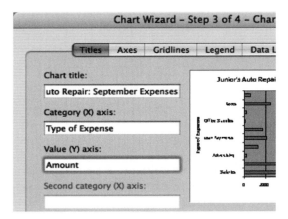

Figure 31–21

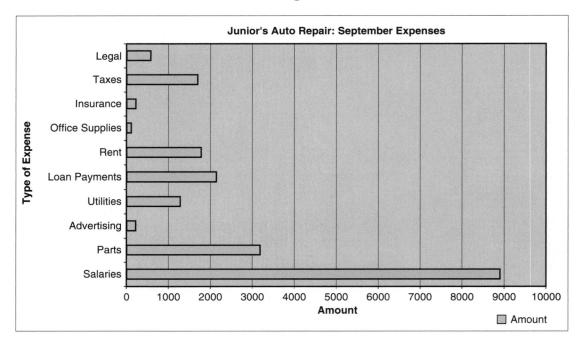

Figure 31–22

Once the computer has made the graph, you might want to change the background color or the color of the bars. Consult the help menu for the program to learn how to do these.

2. The table in Figure 31–23 below gives the energy consumption of electricity and natural gas for commercial building in four regions of the country. Use a computer spreadsheet to create a stacked vertical bar graph of this information.

Region	Consumption (trillion Btu)	
	Electricity	Natural Gas
Northeast	543	299
Midwest	662	709
South	1,247	618
West	645	396

Figure 31–23

Solution.   Enter the data in three columns as shown in Figure 31–24.

◇	A	B	C
1	Region	Electricity	Natural gas
2	East	543	299
3	Midwest	662	709
4	South	1247	618
5	West	645	396

Figure 31–24

To graph the function, highlight the table you just constructed. Left click on cell A2 and hold the left mouse key down as you move the cursor to cell C5. Release the mouse key (leaving the cells highlighted) and move the cursor up to "Chart Wizard." Click on Chart Wizard to begin the process of constructing a graph of the function.

The first step is to select the type of graph. The menu shows two general types of bar graphs. This time you want the type with vertical bars, so click on "column." It is circled in the left of Figure 31–25.

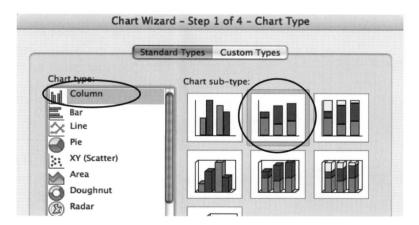

Figure 31–25

When you click on "column" seven subtypes of graphs are shown. They are shown on the right in Figure 31–25. The stacked type of column graph is the second one shown and is circled. Select it and press Next >.

Title the graph, label the axes, and display the finished graph. The finished graph should look something like the one in Figure 31–26. The background of the graph in Figure 31–26 has been changed from grey to white by double clicking on the background. This opens a menu, which allows you to change the border or the background (area) of the graph.

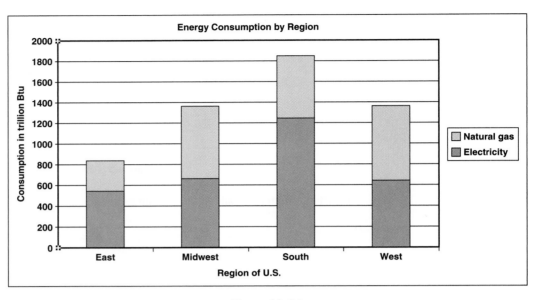

Figure 31–26

EXERCISE 31–4

1. Use a computer spreadsheet to draw a horizontal bar graph that shows the number of miles traveled by each type of vehicle listed in Figure 31–27.

Vehicle Type	Personal	Airplane	Bus	Train	Ship
Miles (×1,000,000)*	735,882	367,889	23,747	21,020	9,316

*735,882 represents 735,882 × 1,000,000 = 735,882,000,000.

Figure 31–27

2. The data in Figure 31–28 shows the amount in millions of dollars spent in different categories of health care. Use a computer spreadsheet to draw a vertical bar graph of this data.

Service	Hospital Care	Physician Services	Prescription Drugs	Nursing Homes	Public Administration	Public Health
Amount	316,445	114,814	36,239	66,054	33,319	51,159

Figure 31–28

3. The table shown in Figure 31–29 lists the eight leading wheat producing states for a certain year. Use a computer spreadsheet to draw a vertical bar graph showing the table data. Production figures are for millions of bushels.

State	Kansas	N. Dak.	Okla.	Wash.	Mont.	S. Dak.	Minn.	Texas
Production	480	317	179	139	138	116	105	97

Figure 31–29

4. The table in Figure 31–30 lists the daily crude oil production of selected countries for a certain year. Use a computer spreadsheet to draw a horizontal bar graph showing the table data. Production figures are for thousands of barrels per day.

Country	Canada	China	Iran	Mexico	Norway	Russia	Saudi Arabia	United States
Production	2,029	3,300	3,724	3,157	3,117	7,049	8,031	5,801

Figure 31–30

5. The table shown in Figure 31–31 lists the number of males and females working in selected occupations during a certain year. Use a computer spreadsheet to draw a stacked horizontal bar graph showing the table data. Figures are for thousands of people.

Occupation	Agriculture	Mining	Construction	Retail Trade	Education	Financial
Males	1695	452	9165	8318	6980	4318
Females	580	73	973	7932	21,310	5430

Figure 31–31

6. Use a computer spreadsheet to make a stacked vertical bar graph on the data in Figure 31–32 showing the number of retail drug prescriptions based on the type of sales outlet. The number of prescriptions are in millions.

	Year		
Unit	1995	1999	2003
Traditional Chain	914	1,246	1,494
Independent	666	680	731
Mass Merchant	238	319	345
Supermarkets	221	357	462
Mail Order	86	134	189

Figure 31–32

31–5 Circle Graphs

A common use of the protractor is in the construction of a circle graph or *pie chart*. A *circle graph* shows the comparison of parts to each other and to the whole. It compares quantities by means of angles constructed from the center of the circle. A circle graph is shown in Figure 31–33.

Procedure for Constructing a Circle Graph

• Add all of the items to be shown on the graph. The sum is equal to the whole, or 100%.

• Make a table showing:

The fractional part and percent of each item in relation to the whole.

The number of degrees representing each fractional part or percent. Degrees are obtained by multiplying each fractional part by 360 degrees. Round to the nearest whole degree.

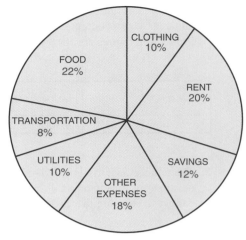

A FAMILY'S BUDGET

Figure 31–33

- Draw a circle of convenient size. With a protractor, construct angles using the number of degrees representing each part. The center of the circle is the vertex of each angle.
- Label each part with the item name and the percent that each item represents.
- Label the graph itself with a descriptive title.

This example illustrates the method of constructing a circle graph.

EXAMPLE •

During one year a certain city spent its total income as follows: educational services, $5,250,000; health, safety, and welfare, $4,200,000; public works, $1,800,000; interest on debt, $1,500,000; other services, $2,250,000. Construct a circle graph showing how the city income is spent.

Add all of the cost items: $5,250,000 + $4,200,000 + $1,800,000 + $1,500,000 + $2,250,000 = $15,000,000.

Make a table (Figure 31–34) showing the fractional part and percent that each cost item is of the whole, and the number of degrees represented by each.

	Educational Services	Health, Safety, and Welfare	Public Works	Interest on Debt	Other Services
a.	$\frac{\$5,250,000}{\$15,000,000}=$ 0.35 = 35%	$\frac{\$4,200,000}{\$15,000,000}=$ 0.28 = 28%	$\frac{\$1,800,000}{\$15,000,000}=$ 0.12 = 12%	$\frac{\$1,500,000}{\$15,000,000}=$ 0.10 = 10%	$\frac{\$2,250,000}{\$15,000,000}=$ 0.15 = 15%
b.	0.35 × 360° = 126°	0.28 × 360° = 100.8° ≈ 101°	0.12 × 360° = 43.2° ≈ 43°	0.10 × 360° = 36°	0.15 × 360° = 54°

Figure 31–34

Draw a circle and construct the respective angles. The center of the circle is the vertex of each angle. See Figure 31–35.

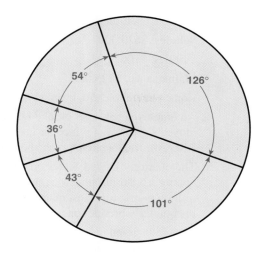

Figure 31–35

Label each part with the cost item name and the percent that each item represents. Identify the graph with a descriptive title. See Figure 31–36.

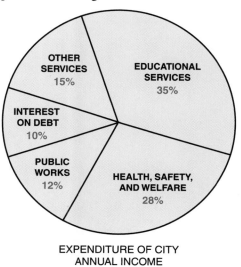

EXPENDITURE OF CITY
ANNUAL INCOME

Figure 31–36

EXERCISE 31–5

1. In producing a certain product, a small company had the following manufacturing costs: material costs, $136,800; labor costs, $167,200; overhead expenses, $76,000. Construct a circle graph showing these manufacturing costs.

2. World motor vehicle production for a certain year is shown in the table in Figure 31–37. Construct a circle graph using this data.

Country	United States	Japan	Germany	France	United Kingdom	Italy	Other
Number of Motor Vehicles Produced (in millions)	12.6	10.7	6.3	2.8	3.4	2.7	3.1

Figure 31–37

3. A firm's operating expenses for the first half year are as follows:

    Salaries, $620,000   Advertising, $55,000

    Utilities, $100,000   Delivery, $70,000

    Maintenance, $125,000   Depreciation, $140,000

    Construct a circle graph showing these operating expenses.

4. An employee has a net or take-home pay of $587. These deductions are made from the em-
    ployee's gross wage: income tax, $71; FICA, $53; retirement, $12; insurance, $15; other,
    $10. Construct a circle graph showing net pay and deductions.

5. The quarterly United States production of iron ore during a certain year is as follows: first
    quarter, 16.8 million metric tons; second quarter, 18.9 million metric tons; third quarter, 17.5
    million metric tons; last quarter, 15.4 million metric tons. Construct a circle graph showing
    the quarterly production.

6. The circle graph in Figure 31–38 shows the percent of new vehicle sales in the United States
    during a given year. If the total sales were 17,118,000 vehicles, use the circle graph to de-
    termine the following (round answers to the nearest thousand):

    a. The number of domestic new cars that were sold.

    b. The total number of imported light trucks that were sold.

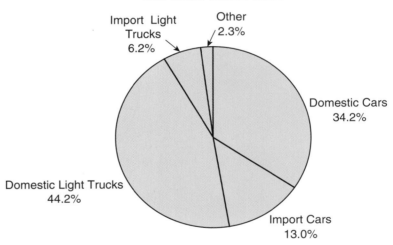

New Motor Vehicle Sales

Figure 31–38

7. The circle graph in Figure 31–39 shows the percent of the world population that lives on
    each continent. If the total population was 6,085,000,000, use the circle graph to answer the
    following (round answers to the nearest million):

    a. The number of people who live in North America.

    b. The number of people who live in Asia.

    c. The number of people who live in Oceania.

World Population by Continent, 2000

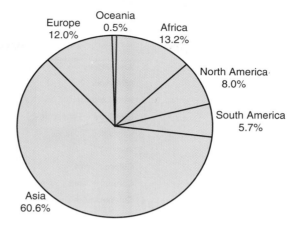

Figure 31–39

31–6   Drawing Circle Graphs with a Spreadsheet

Using a spreadsheet to draw a circle graph is done in much the same manner as drawing a bar graph. The main difference is selecting the graph type. The next example shows how to use a spreadsheet to draw the graph in the previous example.

EXAMPLE •——

During one year, a certain city spent its total income as shown in the table in Figure 31–40.

Educational Services	Health, Safety, and Welfare	Public Works	Interest on Debt	Other Services
$5,250,000	$4,200,000	$1,800,000	$1,500,000	$2,250,000

Figure 31–40

Solution.   Once the data is entered, use "Chart Wizard." This time select the "Pie" chart type and the first subtype, as indicated by the rings in Figure 31–41. If you want the graph to be in black and white, then press "Custom Types" and select "B&W Pie" as shown by the rings in Figure 31–42. For this example, we will use the B&W Pie format.

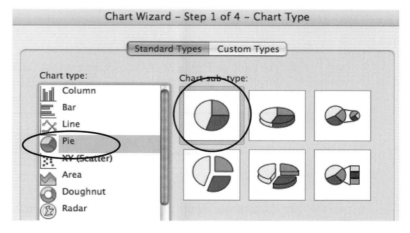

Figure 31–41

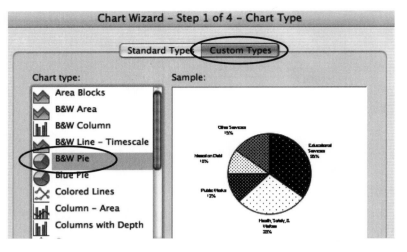

Figure 31–42

Proceed as before. The final result should look like Figure 31–43.

Expenditure of City Annual Income

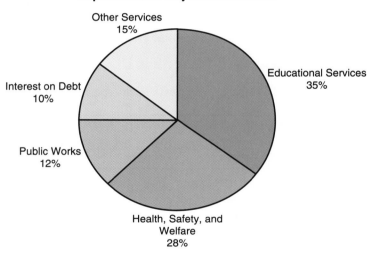

Figure 31–43

EXERCISE 31–6

1. The table in Figure 31–44 shows the expenses at Junior's Auto Repair during the month of July. Use a computer spreadsheet to draw a circle graph of this data.

Salaries and Wages	Parts	Utilities	Fixed Expenses	Taxes	Misc.
$43,629	$12,540	$2,357	$2,750	$9,525	$1,575

Figure 31–44

2. The table in Figure 31–45 shows the number, in thousands, of vehicles produced by various companies in the United States during 2003. Use a computer spreadsheet to draw a circle graph of this data.

Company	GM	Ford	Daimler Chrysler	Toyota	Honda	Nissan
Production (×1000)	3890.2	3118.5	1731.0	1122.5	845.3	522.3

Figure 31–45

3. The table in Figure 31–46 shows the United States military manpower in 2003 according to the four major branches of service. Use a computer spreadsheet to draw a circle graph of this data.

Branch	Air Force	Army	Navy	Marines
Manpower (×1000)	375	499	382	178

Figure 31–46

4. The data in Figure 31–47 contains the number of retail drug prescriptions in 2003 according to where each prescription was filled. Use a computer spreadsheet to draw a circle graph of this data.

Location	Mail Order	Supermarket	Mass Merchant	Independent	Traditional Chain
Number of Prescriptions (millions)	189	462	345	731	1,494

Figure 31–47

31–7   Line Graphs

Line graphs show changes and relationships between quantities. Line graphs are widely used to graph the following two general types of data: data where there is no causal relationship between quantities, and data where there is a causal relationship between quantities.

Data Where There Is *No* Causal Relationship between Quantities

When the data are graphed, the graph shows a changing condition usually identified by a broken line. This type of graph is called a *broken-line graph.* The time and temperature graph shown in Figure 31–48 is an example of a broken-line graph.

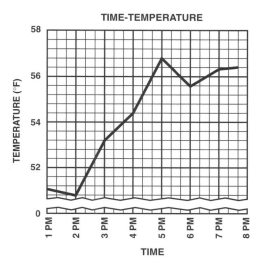

Figure 31–48   Broken-Line Graph

Data Where There *Is* a Causal Relationship between Quantities

The quantities are related to each other by a mathematical rule or formula. When the data are graphed, the line is usually a straight line or a smooth curve.

The graph shown in Figure 31–49 is an example of a *straight-line graph.* The quantities are the perimeters of squares in relation to the lengths of their sides. The perimeter of a square is the distance around the square. The formula for the perimeter of a square is Perimeter = 4 times the side length, $P = 4s$.

The graph shown in Figure 31–50 is an example of a *curved-line graph.* The quantities are the areas of squares in relation to the lengths of their sides. The formula for the area of a square is Area = the square of a side, $A = s^2$.

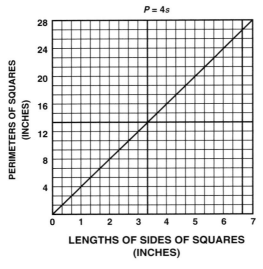

Figure 31–49   Straight-Line Graph

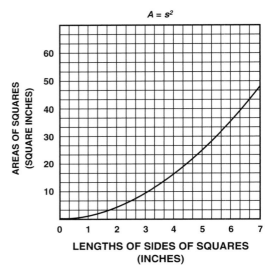

Figure 31–50   Curved-Line Graph

31–8   Reading Line Graphs

Information is read directly from a line graph by locating a value on one scale, projecting to a point on the graphed line, and projecting from the point to the other scale. More data can generally be obtained from a line graph than from a bar graph. Data between given scale values can be read. In most cases, the values read are close approximations of the true values.

To read a line graph, first determine the value of each space on the scales. Then locate the given value on the appropriate scale. The value may be on either the horizontal or vertical scale, depending on how the graph is organized. If the value does not lie directly on a line, estimate its location.

From the given value, project up from a horizontal scale or across from a vertical scale to a point on the graphed line. From the point, project across or down to the other scale. Read the scale value. If the value does not lie directly on the line, estimate the value.

EXAMPLES •————————————————————————————————

1. A quality control assistant constructs the broken-line graph shown in Figure 31–51. The percent of defective pieces of the total production for each of ten consecutive production days is given.

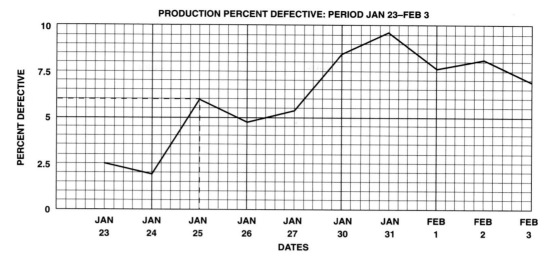

Figure 31–51

a. Find, to the nearest 0.5%, the percent of defective pieces for January 25.

Find the vertical scale values. There is 2.5% between each numbered division. Each small space represents 2.5% ÷ 5 = 0.5%.

Locate January 25 on the horizontal scale. Project up to a point on the graphed line. From this point, project across to the percent defective scale. Read the value to the nearest 0.5%   (Jan 25) 6% *Ans*

b. If the total production for January 25 is 1,550 pieces, how many pieces are defective?

Find 6% of 1,550 pieces.   0.06 × 1,550 = 93 *Ans*

2. The broken-line graph in Figure 31–52 is a continuation of Figure 31–51. It shows the percent of defective pieces of the total production for the next ten consecutive days of production. On what dates were there 6.5% defective pieces?

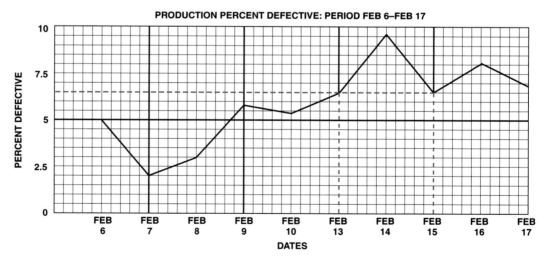

Figure 31–52

Locate 6.5% on the vertical scale. Project over to points on the graphed lines. Notice that there are there are two of these points. From each of these points, project down to the dates scale. Read the dates.

Feb 13 and Feb 15 *Ans*

31–9   Reading Combined-Data Line Graphs

Two or more sets of data are often combined on the same graph. Graphs of this type are useful in showing relationships and making comparisons between sets of data. Comparing information on two or more lines on a graph can be done more quickly and interpreted more easily than comparing listed data.

EXAMPLE •————————————————————————————————

Acceleration tests are made for two cars. One car is a manufacturer's standard production model. The other car is a high-performance competition model. Acceleration data obtained from tests are plotted on the graph shown in Figure 31–53. Observe that gear shift points are also indicated on the graph.

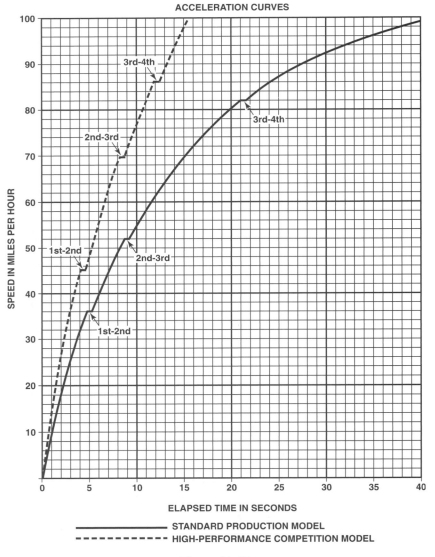

Figure 31–53

a. How many seconds are required for the high-performance model to accelerate from 0 to 60 miles per hour?

b. What is the speed of the production model after 22 seconds from a standing start?

c. How many seconds are required by the production model to accelerate from 40 to 70 miles per hour?

d. At the end of 12 seconds, how much greater is the speed of the high-performance model than of the production model?

e. At what speeds were the shifts through gears made on the production model?

Solutions

a. Locate 60 mi/h on the speed scale. Project across to a point on the graph for high-performance (broken line). Read the time to the nearest second.

7 seconds *Ans*

b. Locate 22 seconds on the time scale. Project up to a point on the graph for the production model (solid line). Read the speed.

82 mi/h *Ans*

c. Locate 40 mi/h on the speed scale. Project across to a point on the graph for the production model (solid line). The time is 6 seconds.

Locate 70 mi/h on the speed scale. Project across to a point on the graph for the production model (solid line). The time is 15 seconds.

Subtract.   15 seconds − 6 seconds = 9 seconds *Ans*

d. Locate 12 seconds on the time scale. Project up and across for each model.

Subtract.   86 mi/h − 61 mi/h = 25 mi/h *Ans*

e. Locate on the graph for the production model each place of gear change. Read the speed for each gear change.

36 mi/h *Ans*
52 mi/h *Ans*
82 mi/h *Ans*

EXERCISE 31–9

1. Temperatures in degrees Celsius for the different times of day are shown on the graph in Figure 31–54. Express the answers to the nearest 0.2 degree.

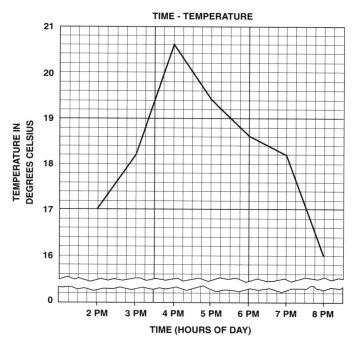

Figure 31–54

    a. What is the temperature for each of the hours shown on the graph?

    b. What is the average hourly temperature during the six-hour period?

    c. What is the temperature change from

        (1) 2 PM to 4 PM?

        (2) 4 PM to 5 PM?

        (3) 6 PM to 8 PM?

2. The surface or rim speed of a wheel is the number of feet that a point on the rim of the wheel travels in one minute. The surface speed depends on the size of the wheel diameter and on the number of revolutions per minute (r/min) that the wheel is turning. The graph in Figure 31–55 shows the surface speeds of different diameter wheels. All wheels are turning at 320 revolutions per minute. Express the answers for surface speeds to the nearest 10 feet per minute and for diameters to the nearest 0.2 inch.

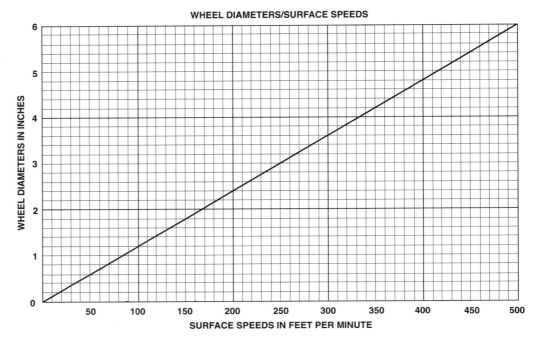

Figure 31–55

    a. What is the surface speed of each of the wheel diameters shown on the graph?

    b. What is the surface speed of a 3.6-inch diameter wheel?

    c. What is the surface speed of a 4.6-inch diameter wheel?

    d. What is the surface speed of a 5.4-inch diameter wheel?

    e. What diameter wheels are needed to give 270 feet per minute surface speed?

    f. What diameter wheels are needed to give 350 feet per minute surface speed?

    g. What diameter wheels are needed to give 420 feet per minute surface speed?

3. An electrical current/resistance graph is shown in Figure 31–56. A constant power of 150 watts is being consumed. Express the answers for resistance to the nearest ohm and for current to the nearest 0.2 ampere.

    a. What is the current for each of the resistances shown on the graph?

    b. What is the resistance for each of the currents shown on the graph? Disregard the 2-ampere current.

    c. What is the resistance for a current of 2.6 amperes?

d. What is the resistance for a current of 4.6 amperes?

e. What is the resistance for a current of 5.4 amperes?

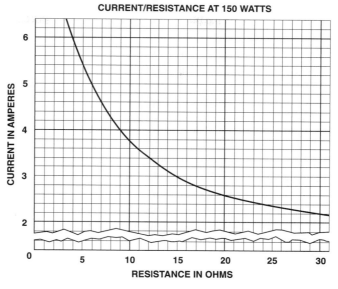

Figure 31–56

4. This graph in Figure 31–57 on page 644 shows the brake horsepower of two engines at various engine speeds. One engine is fitted with a medium-compression head, the other with a high-compression head. Express the answers to the nearest 5 brake horsepower or to the nearest 100 r/min.

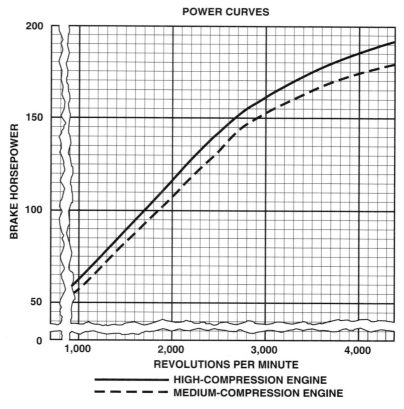

Figure 31–57

a. What is the brake horsepower for the medium-compression engine at 2,200 r/min?

b. What is the brake horsepower for the high-compression engine at 3,100 r/min?

c. What is the brake horsepower for the high-compression engine at 3,800 r/min?

d. How many revolutions per minute are required for the medium-compression engine when developing 140 brake horsepower?

e. How many revolutions per minute are required for the high-compression engine when developing 160 brake horsepower?

f. How many revolutions per minute are required for the high-compression engine when developing 185 brake horsepower?

g. What is the increase in brake horsepower of each engine when the engine speeds are increased from 2,200 r/min to 3,700 r/min?

h. How many brake horsepower greater is the high-compression engine than the medium-compression engine at 1,400 r/min?

i. How many brake horsepower greater is the high-compression engine than the medium-compression engine at 2,600 r/min?

j. How many brake horsepower greater is the high-compression engine than the medium-compression engine at 4,200 r/min?

31–10   Drawing Line Graphs

As with a bar graph, there are five steps to drawing a line graph. The first four steps of constructing a line graph are the same as for constructing a bar graph.

1. To draw a line graph, first arrange the given data in a logical order. For example, group data from the smallest to the largest values or from the beginning to the end of a time period. Sometimes data are not given directly, but must be computed from other given facts, such as formulas.

2. Decide which group of data is to be on the horizontal scale and which is to be on the vertical scale. Generally, the horizontal scale is on the bottom, and the vertical scale is on the left of the graph.

3. Draw and label the horizontal and vertical scales. Next, assign values to the spaces on the scales that conveniently represent the data. The data should be clear and easy to read.

4. Plot each pair of numbers (coordinates). Project up from the horizontal scale and across from the vertical scale. Place a dot on the graph where the two projections meet.

5. Connect the plotted points with a straightedge or curve. Depending on the given data, the line may be straight, broken, or curved.

31–11   Drawing Broken-Line Graphs

Quantities that are not related to each other by a mathematical rule or formula form a broken-line graph.

EXAMPLE •———————————————————————————————————————

A company's profit for each of six weeks is listed in the table shown in Figure 31–58. Draw a line graph showing this data.

Week	1	2	3	4	5	6
Weekly Profit	$1,350	$1,100	$1,600	$1,850	$1,750	$1,900

Figure 31–58

Arrange the data in logical order. Since the data are listed in order from week 1 to week 6, no rearrangement is necessary.

Decide which data are to be on the horizontal scale and which on the vertical scale. It is decided to make weeks the horizontal scale. Make the weeks scale on the bottom and the profit scale on the left of the graph.

Draw and label the scales. Refer to Figure 31–59.

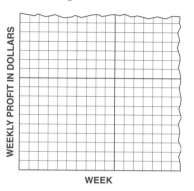

Figure 31–59

Assign values to scale spaces. Space each week 5 small spaces apart on the horizontal scale. On the vertical scale, start the numbering at $1,000. Observe there is no profit less than $1,000. Assign each major division (10 small spaces) a value of $500. Each small space represents $500 ÷ 10 or $50. Label each fifth space. Each fifth space represents 5 × $50 or $250. Refer to Figure 31–60.

Plot each pair of values. From the week 1 location project up, and from $1,350 project across. Place a small dot where the projections meet. Locate the remaining 5 points the same way. Connect the plotted points. Draw a straight line between each of 2 consecutive points. Refer to Figure 31–61.

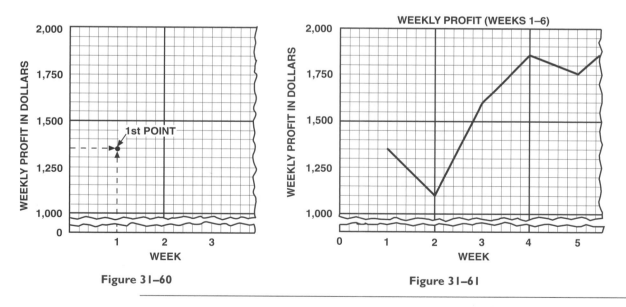

Figure 31–60   **Figure 31–61**

EXERCISE 31–11

1. Draw a broken-line graph to show the six-month production of iron ore in the United States as listed in the table in Figure 31–62 on page 646. Production is given in units of millions of metric tons.

Month	June	July	Aug	Sept	Oct	Nov
Monthly Production of Iron Ore in Millions of Metric Tons	6.2	6.4	5.9	5.7	6.0	6.1

Figure 31–62

2. The table shown in Figure 31–63 lists the percent of defective pieces of the total number of pieces produced daily by a manufacturer. The data are recorded for seven consecutive working days. Draw and label a broken-line graph showing the table data.

Date	March 27	March 28	March 29	March 30	March 31	April 3	April 4
Daily Percent Defective	3.8%	4.2%	5.6%	5.0%	6.2%	6.8%	5.4%

Figure 31–63

31–12   Drawing Broken-Line Graphs with a Spreadsheet

Using a spreadsheet to draw a broken-line graph is done in much the same manner as drawing a bar graph or a circle graph. The main difference is selecting the graph type. The next example shows how to use a spreadsheet to draw the graph in the previous example.

EXAMPLE •——

A company's profit for each of six weeks is listed in the table shown in Figure 31–64.

Week	1	2	3	4	5	6
Weekly Profit	$1,350	$1,100	$1,600	$1,859	$1,750	$1,900

Figure 31–64

Solution.   Once the data is entered, use "Chart Wizard." This time select the "XY (Scatter)" chart type and the fifth subtype, both are circled in Figure 31–65. You could also select the fourth type. It shows the data points and connects the points with lines.

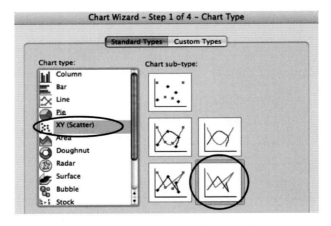

Figure 31–65

Proceed as before. The final result should look like Figure 31–66.

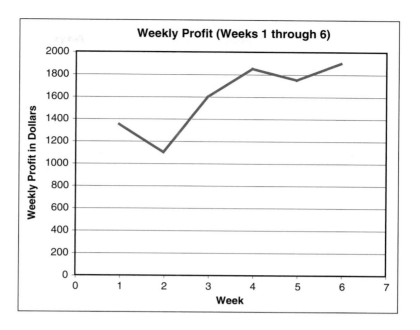

Figure 31–66

There is a lot of blank space at the bottom of this graph. This was eliminated in Figures 31–5 and 31–48 by putting a gap between the 0 at the bottom of the scale and the 1,000 level. A similar effect can be created with a spreadsheet. Double click on one of the numbers along the left side. A "Format Axis" window should open, part of which is shown in Figure 31–67. Change the minimum value to 1000 and click on $\boxed{\text{OK}}$. The result should look like the graph in Figure 31–68.

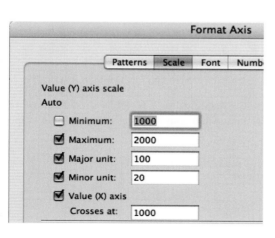

Figure 31–67

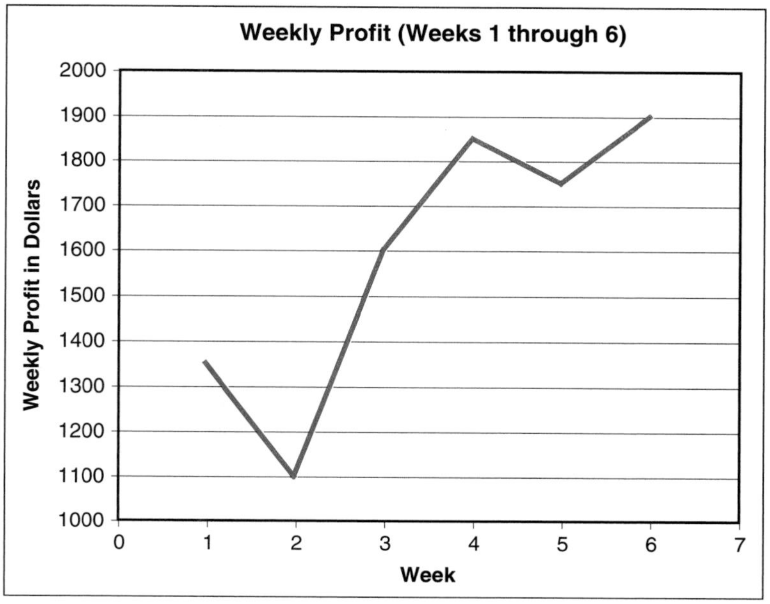

Figure 31–68

EXERCISE 31–12

1. The table in Figure 31–69 shows the net sales for Junior's Auto Repair for the last six months of the year. Use a computer spreadsheet to draw a broken-line graph of this data.

Month	July	August	September	October	November	December
Net Sales	$24,317	$22,179	$21,378	$23,964	$32,451	$17,642

Figure 31–69

2. The table in Figure 31–70 shows the value, in millions of dollars, of factory shipments of computers and industrial electronics from 1993–2001. Use a computer spreadsheet to draw a broken-line graph of this data.

Year	1993	1994	1995	1996	1997	1998	1999	2000	2001
Value	54,821	59,254	73,555	78,278	76,317	78,831	87,412	84,317	68,035

Figure 31–70

3. The readings in Figure 31–71 are test results from checking the volume of 100-μL variable volume pipettes. Samples were collected each half hour for six hours. Use a computer spreadsheet to draw a broken-line graph of this data.

Sample	1	2	3	4	5	6	7	8	9	10	11	12
Volume (μL)	100.37	100.17	98.13	100.33	101.92	100.57	98.23	99.25	99.86	98.64	101.27	98.40

Figure 31–71

4. Each day 100 integrated circuits are removed from production and checked to electrical specifications. The data in Figure 31–72 contains the results of testing the circuits for 15 days. Use a computer spreadsheet to draw a broken-line graph of this data.

Day	1	2	3	4	5	6	7	8	9	10	11	12	13	14	15
No. Defective	24	38	62	35	37	38	48	52	33	21	44	29	30	34	45

Figure 31–72

31–13   Drawing Straight-Line Graphs

Quantities that are related to each other by a mathematical rule or formula form a curved or straight line when graphed. If the two quantities change at a constant rate, a straight-line graph is formed.

EXAMPLE •────────────────────────────────────

Draw a line graph for the perimeters of squares.

$$P = 4s$$

Use side lengths of 1 centimeter through 8 centimeters.

The data are not given directly. The perimeters related to each of the sides must be computed. Substitute each side length in the formula to determine the corresponding perimeter value. The change in the perimeter is a constant 4 cm for each centimeter change in the length of the side.

$$4 \times 1 \text{ cm} = 4 \text{ cm} \qquad 4 \times 5 \text{ cm} = 20 \text{ cm}$$
$$4 \times 2 \text{ cm} = 8 \text{ cm} \qquad 4 \times 6 \text{ cm} = 24 \text{ cm}$$
$$4 \times 3 \text{ cm} = 12 \text{ cm} \qquad 4 \times 7 \text{ cm} = 28 \text{ cm}$$
$$4 \times 4 \text{ cm} = 16 \text{ cm} \qquad 4 \times 8 \text{ cm} = 32 \text{ cm}$$

Organize the data in table form as shown in Figure 31–73.

Lengths of Sides in Centimeters	1	2	3	4	5	6	7	8
Perimeters of Sides in Centimeters ($P = 4s$)	4	8	12	16	20	24	28	32

Figure 31–73

The completed graph is shown in Figure 31–74.

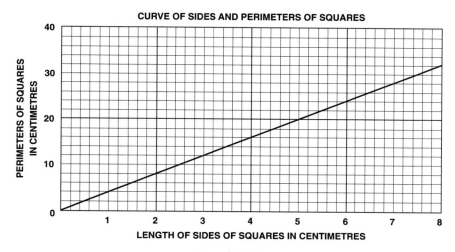

Figure 31–74

31–14   Drawing Curved-Line Graphs

The area of a square does not increase at a constant amount for each unit change in the length of the side. If the variation is not constant, a curved line is formed.

EXAMPLE

Draw a line graph for the areas of squares.

$$A = s^2$$

Use side lengths of 1 centimeter through 8 centimeters.

The data are not given directly. The areas related to each of the sides must be computed. Substitute each side length in the formula to determine the corresponding area value. The change in the area is not a constant amount for each centimeter change in the length of the side.

$$1 \text{ cm} \times 1 \text{ cm} = \ \ 1 \text{ cm}^2 \qquad 5 \text{ cm} \times 5 \text{ cm} = 25 \text{ cm}^2$$
$$2 \text{ cm} \times 2 \text{ cm} = \ \ 4 \text{ cm}^2 \qquad 6 \text{ cm} \times 6 \text{ cm} = 36 \text{ cm}^2$$
$$3 \text{ cm} \times 3 \text{ cm} = \ \ 9 \text{ cm}^2 \qquad 7 \text{ cm} \times 7 \text{ cm} = 49 \text{ cm}^2$$
$$4 \text{ cm} \times 4 \text{ cm} = 16 \text{ cm}^2 \qquad 8 \text{ cm} \times 8 \text{ cm} = 64 \text{ cm}^2$$

Organize the data in table form as shown in Figure 31–75.

Lengths of Sides in Centimeters	1	2	3	4	5	6	7	8
Areas in Square Centimeters ($A + s^2$)	1	4	9	16	25	36	49	64

Figure 31–75

The completed graph is shown in Figure 31–76.

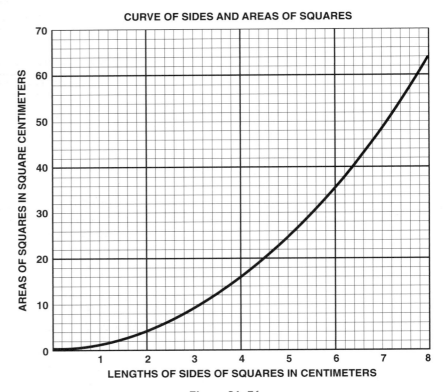

CURVE OF SIDES AND AREAS OF SQUARES

AREAS OF SQUARES IN SQUARE CENTIMETERS (vertical axis)

LENGTHS OF SIDES OF SQUARES IN CENTIMETERS (horizontal axis)

Figure 31–76

EXAMPLE •

Use a computer spreadsheet to graph the data in Figure 31–77.

Solution.   Once the data is entered, use "Chart Wizard." Select the "XY (Scatter)" chart type and the second or third subtype. (The third is indicated by the ring in Figure 31–77.) The second type shows the data points and connects the points with curves.

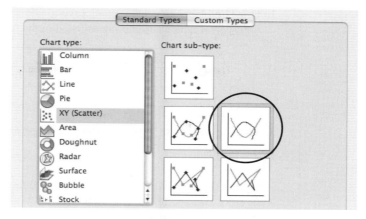

Figure 31–77

Proceed as before. The final result should look something like Figure 31–78.

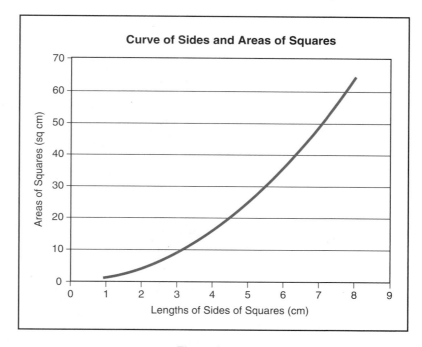

Figure 31–78

To get the curve between 0 and 1, you would have to enter a length of 0 and an area of 0 to the table.

EXERCISE 31–14

1. The data in the table in Figure 31–79 show electrical currents at various voltages. All listed currents are flowing through a constant resistance of 20 ohms. Draw and label a current-voltage graph using the table data.

Voltage (Volts)	10	20	30	40	50
Current (Amperes)	0.5	1.0	1.5	2.0	2.5

Figure 31–79

2. Draw and label an acceleration curve to show the acceleration data for a certain expensive, high-performance automobile as listed in the table in Figure 31–80.

Elapsed Time in Seconds	0	5	10	15	20	25	30
Speed in Miles per Hour	0	32	54	68	78	86	90

Figure 31–80

3. The rise and run of a roof are shown in Figure 31–81.

$$\text{Rise} = \text{pitch} \times 2 \times \text{run}$$

Draw a graph showing the relation of rise to run with a constant pitch value of $\frac{1}{3}$. Use run values of 8, 12, 16, 20, 24, and 28 feet.

NOTE: It is helpful to make a table of the values to be graphed.

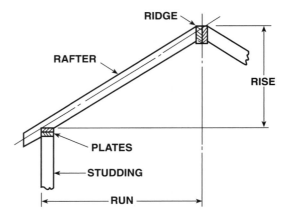

Figure 31–81

4. The area of a circle is approximately equal to 3.1416 times the square of the radius.

$$A \approx 3.1416 \times r^2$$

Draw a curved-line graph to show the relation of radii to areas. Use radii from 0.5 meter to 3.5 meters in 0.5-meter intervals. Label the radius scale in meters and the area scale in square meters.

NOTE: It is helpful to make a table of the values to be graphed.

5. The cutting speeds of different size (diameter) drills are shown in the table in Figure 31–82. All drills are turning at a constant 600 revolutions per minute. Copy the table and fill in the missing changes in cutting speed values. If the table data are plotted, is the graph a curved-line graph or a straight-line graph? Do not graph the table data.

Drill Diameters in Inches	0.250	0.500	0.750	1.000	1.250	1.500
Cutting Speeds in Feet per Minute	39.25	78.50	117.75	157.00	196.25	235.50
Changes in Cutting Speeds in Feet per Minute						

Figure 31–82

6. Lengths of sides of cubes with their respective volumes are given in the table in Figure 31–83. Copy the table and fill in the missing changes in volume values. If the table data are plotted, is the graph a curved-line graph or a straight-line graph? Do not graph the table data.

Lengths of Cube Sides in Meters	0.4	0.8	1.2	1.6	2.0	2.4
Volumes of Cubes in Cubic Meters	0.064	0.512	1.728	4.096	8.000	13.824
Changes in Volumes in Cubic Meters						

Figure 31–83

⠿ UNIT EXERCISE AND PROBLEM REVIEW

READING BAR GRAPHS

1. The bar graph in Figure 31–84 shows the quarterly production for a manufacturing firm over a period of a year.

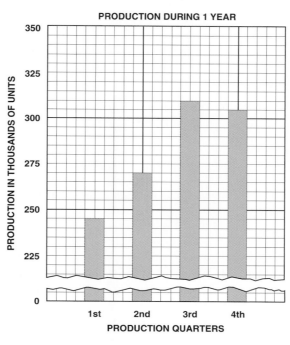

Figure 31–84

   a. How many units are produced during each quarter?

   b. What is the average quarterly production during the year?

   c. What percent of the total yearly production is manufactured during the first half of the year? Express the answer to the nearest whole percent.

2. The number of domestic and international passengers during one year using six major international airports is shown in Figure 31–85 on page 654.

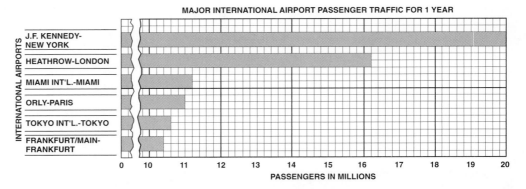

Figure 31–85

a. What is the total passenger traffic for the six airports?

b. What percent of the passenger traffic through J. F. Kennedy International Airport is the traffic through Heathrow International Airport? Express the answer to the nearest whole percent.

c. The number of passengers shown on the graph using Heathrow International Airport represents an 8% increase above the previous year. How many passengers used Heathrow the previous year? Round the answer to the nearest one hundred thousand passengers.

DRAWING BAR GRAPHS

3. The production of five major United States farm products for a certain year is shown in the table in Figure 31–86. Draw and label a horizontal bar graph showing the table data.

Crop	Corn, Grain	Wheat	Soybeans	Oats	Barley
Number of Bushels Produced in Billions of Bushels	5.4	1.5	1.3	0.7	0.5

Figure 31–86

4. Draw a vertical bar graph showing the United States imports for eight consecutive years as shown in the table in Figure 31–87.

Year	1st	2nd	3rd	4th	5th	6th	7th	8th
Imports in Billions of Dollars	18	21	25	26	33	36	39	45

Figure 31–87

READING LINE GRAPHS

5. A wholesale distributor's monthly profits for six consecutive months are shown on the graph in Figure 31–88.

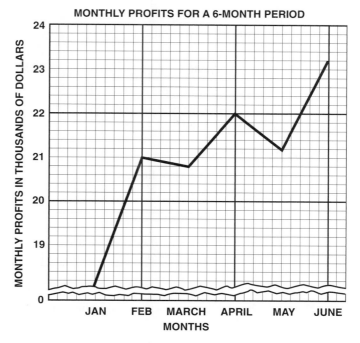

Figure 31–88

   a. What is the profit for each of the six months? Express the answers to the nearest $200.

   b. How much greater is the profit for June than the average monthly profit during the six-month period?

   c. What is the percent increase in profit in June over January? Express the answer to the nearest whole percent.

6. The carburetor of an internal combustion engine mixes air with gasoline. There is an ideal mixture of gasoline and air that gives maximum fuel economy. The mixture of gasoline and air is called an air-fuel ratio. The number of pounds of air in the mixture is compared with the number of pounds of gasoline. The graph in Figure 31–89 shows air-fuel ratios in relation to fuel consumption in miles per gallon of gasoline for a certain car. Express the answers to the nearest whole mile per gallon.

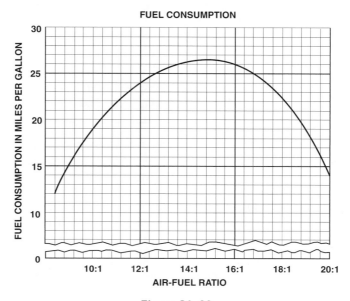

Figure 31–89

a. How many miles per gallon of gasoline are obtained at a 12 : 1 air-fuel ratio?

b. How many miles per gallon of gasoline are obtained at a 16 : 1 air-fuel ratio?

c. How many miles per gallon of gasoline are obtained at a 17.2 : 1 air-fuel ratio?

d. How many miles per gallon of gasoline are obtained at a 18.8 : 1 air-fuel ratio?

e. What air-fuel ratio gives the greatest number of miles per gallon?

f. How many more miles per gallon are obtained with a 14 : 1 air-fuel ratio than with a 20 : 1 ratio?

DRAWING LINE GRAPHS

7. Draw a broken-line graph to show the United States's percent of the total world trade for each year of an eight-year period as listed in the table in Figure 31–90.

Year	1st	2nd	3rd	4th	5th	6th	7th	8th
United States Percent of Total World Trade	14.6	14.8	14.6	14.3	13.8	15.4	13.9	13.4

Figure 31–90

8. This formula is used for finding current in an electrical circuit when power and resistance are known.

$$I = \sqrt{\frac{P}{R}}$$

where I = current in amperes
P = power in watts
R = resistance in ohms

Draw a graph using a constant power value of 250 watts. Use resistance values of 25, 50, 75, 100, 125, 150, and 175 ohms.

NOTE: It is helpful to make a table of the values to be graphed.

USING SPREADSHEETS

9. The table in Figure 31–91 shows the distance, in meters, of stopping distance for automobiles at various speeds. Use a computer spreadsheet to draw a broken-line graph of this data.

Speed (kph)	10	20	30	40	50	60	70	80	90	100
Distance (m)	2.5	5.6	8.6	11.7	14.8	17.9	20.9	24.0	27.1	30.2

Figure 31–91

10. Use a computer spreadsheet to draw a circle graph of the data in Figure 31–92. The table shows the passenger car production in major countries for a recent year.

Country	Canada	France	Germany	Italy	Japan	Korea	USA	Other
Production (×1000)	1211	3183	5145	1031	8478	2768	4510	16,096

Figure 31–92

11. Use a computer spreadsheet to draw a bar graph of the data in Figure 31–93. The table shows the temperature readings at different times of the day.

Time	1 PM	3 PM	5 PM	7 PM	9 PM	11 PM	1 AM	3 AM	5 AM	7 AM	9 AM	11 AM
Temp °F	51.0	46.0	41.0	41.0	35.0	57.1	49.0	55.0	53.1	54.0	54.0	52.0

Figure 31–93

UNIT 32 ⸬ Statistics

After completing this unit you should be able to:

- determine the probability of an event using the number of successes and trials

- determine the probability of two or more independent events

- determine the mean, median, mode, range, and standard deviation for a set of data and make decisions about which statistic is best to use for an application

- determine the quartile and a specific percentile for a set of data

- create a frequency chart and plot the frequency distribution

- create and interpret a statistical process control (SPC) chart.

Probability has many applications in technology and is of basic importance in statistics. Probability and statistics are needed in any problem dealing with large numbers of variables where it is impossible or impractical to have complete information. In many technical settings, information is needed about an operation. When it is not possible to gather information about the entire operation, information is gathered on a part, or sample, of the operation. This information is then analyzed and decisions are made as a result of this analysis. Statistics are the basis of this analysis.

32–1 Probability

Probability concerns the possible outcomes of experiments. An experiment can be something as simple as tossing a coin or something more complicated such as determining the number of bad computer chips at a production station. A result of an experiment is called an *outcome.* The group of all possible outcomes for an experiment is the *sample space.*

EXAMPLES •

1. If a coin is tossed, the sample space contains the two possible outcomes of heads (H) or tails (T) and can be written as $\{H, T\}$.

2. If a die is rolled, the sample space of the number of dots on the upper face is $\{1, 2, 3, 4, 5, 6\}$. Each trial will produce exactly one of these numbers.

3. If two coins are tossed, the sample space has the following four possible outcomes $\{HH, HT, TH, TT\}$.

The probability P of an event E occurring is

$$P(E) = \frac{n}{s}$$

where n is the number of times the event occurred and s is the size of the sample space.

The probability of an event may be written as a fraction, a decimal, or a percent. Any of these is acceptable, unless you are directed otherwise.

The *complement* of event E, written E', $\overline{E}$, or $\sim E$, are the outcomes in the sample space that are not in event E. We will use E' in this text.

If the probability P of an event E occuring is $\frac{s}{n}$, then the probability of E *not* occurring is

$$P(E') = \frac{s - n}{s}$$

where n is the number of times the event occurred and s is the size of the sample space.

EXAMPLES

1. Find the probability that a 5 will result when one die is rolled.

   Solution.   Here $n = 1$ and $s = 6$. So, $P(5) = \frac{1}{6}$.

2. Find the probability of at least one head when 2 coins are tossed.

   Solution.   The possible ways this event can occur are *HH, HT, TH*. Here $n = 3$ and $s = 4$. So, P(at least one head) $= \frac{3}{4}$.

3. A bag contains 3 red balls, 1 white ball, and 1 green ball. If 2 balls are drawn out at the same time, find the probability of that 2 red balls are drawn.

   Solution.   To help determine the sample space, label the 3 red balls R_1, R_2, and R_3. This gives a sample space of size 10: $\{R_1R_2, R_1R_3, R_2R_3, R_1W, R_1G, R_2W, R_2G, R_3W, R_3G, WG\}$. Here $n = 3$ and $s = 10$. So, $P(RR) = \frac{3}{10}$.

All probabilities are between 0 and 1, inclusive. The probability that an event *must* happen is 1, and the probability that an event *cannot* happen is 0. If the probability that something will happen is p, then the probability that it will not happen is $1 - p$. That is, if $P(E) = p$, then $P(E') = 1 - p$.

EXAMPLES

1. Find the probability that an 8 will result when one die is rolled.

   Solution.   The sample space when one die is rolled is $\{1, 2, 3, 4, 5, 6\}$. The probability that an 8 will happen is 0.

2. A bag contains 3 red balls, 1 white ball, and 1 green ball. If one ball is drawn, what is the probability that it is not red.

   Solution.   Since P(red) $= \frac{3}{5}$, then P(not red) $= P$(red$'$) $= 1 - \frac{3}{5} = \frac{2}{5}$.

EXERCISE 32–1

In Exercises 1–6, find each sample space.

1. The clubs from an ordinary deck of 52 playing cards.
2. Tossing three coins.
3. Rolling two dice.
4. The red face cards from a standard deck of cards.
5. Two marbles drawn at the same time from a bag of 3 red marbles, 2 white marbles, and 1 green marble.
6. Eight pieces of paper are numbered 1 through 8 and then placed in a hat. Two pieces of paper are drawn at the same time.
7. What is the probability that the K♣ will be drawn from the sample space in Exercise 1?
8. What is the probability of tossing exactly two tails from the sample space in Exercise 2?
9. What is the probability of rolling at least one 4 from the sample space in Exercise 3?

10. What is the probability that the Q♡ will be drawn from the sample space in Exercise 4?

11. What is the probability that the 2 ◇ will be drawn from the sample space in Exercise 4?

12. What is the probability that 2 red marbles will be drawn from the sample space in Exercise 5?

13. What is the probability that a 5 will be drawn from the sample space in Exercise 6?

14. What is the probability that the sum of the two numbers drawn from the sample space in Exercise 6 will be 8?

15. A machine produces 25 defective parts out of every 1,000. What is the probability of a defective part being produced?

16. Among 800 randomly selected drivers in the 20–24 age bracket, 316 were in a car accident during the last year. A driver of that age bracket is randomly selected.

   a. What is the probability that he or she will be in a car accident during the next year?

   b. What is the probability that he or she will not be in a car accident during the next year?

32–2   Independent Events

Sometimes we are interested in the probability of an event when we know that another event has already occurred. If the probability that the second event will occur is not affected by what happens to the first event, then we say that the events are *independent.*

EXAMPLES •——

1. A 3 is drawn from a deck of cards and then replaced. The cards are shuffled and a 3 is drawn again. Are these independent events?

   Solution.   Yes, the fact that the first card is replaced and the cards are shuffled makes the probability of each event $\frac{1}{13}$.

2. A die is rolled and a 4 is seen. The die is picked up and rolled again. This time a 5 is obtained. Are the two events independent?

   Solution.   Yes, in both cases the probability is $\frac{1}{6}$.

3. A bag contains 3 red marbles and 2 green marbles. A red marble is drawn and then replaced before a second red marble is drawn. Are these independent events?

   Solution.   Yes, in both cases the probability is $\frac{3}{5}$.

4. A bag contains 3 red marbles and 2 green marbles. A red marble is drawn and not replaced and then a second red marble is drawn. Are these independent events?

   Solution.   No, the probability of drawing the first red marble is $\frac{3}{5}$. Since this marble is not replaced, there are only 4 marbles in the bag when the second marble is drawn. Since there are now only 2 red marbles in the bag, the probability of drawing the second red marble is $\frac{2}{4} = \frac{1}{2}$.

——•

   If events A and B are independent, then the probability that both A and B will occur is the probability of A times the probability of B, or in symbols, $P(A \text{ and } B) = P(A) \cdot P(B)$.

EXAMPLES •——

1. A bag contains 3 red marbles and 2 green marbles. A marble is drawn and replaced and then a second marble is drawn. Find the probability that both marbles are red?

   Solution.   These are independent events. The probability of drawing a red marble is $\frac{3}{5}$. $P(\text{Red and Red}) = P(\text{Red}) \cdot P(\text{Red}) = \frac{3}{5} \cdot \frac{3}{5} = \frac{9}{25}$.

2. Two cards are to be drawn from a well-shuffled deck of cards. After the first card is drawn, it is replaced and the deck is shuffled again. What is the probability that 2 diamonds will be drawn?

   Solution.   There are 13 diamonds in the deck, so the probability of getting a diamond is $\frac{13}{52} = \frac{1}{4}$. Since the first card is replaced and the deck is shuffled, the probability of getting a diamond on the second draw is independent of what was drawn on the first card. The probability of getting a diamond on the second card is $\frac{1}{4}$, and the probability of getting 2 diamonds is $\frac{1}{4} \cdot \frac{1}{4} = \frac{1}{16}$.

3. A bag contains 3 red marbles and 2 green marbles. A marble is drawn and replaced and then a second marble is drawn. Find the probability that one marble is red and the other green?

   Solution. These are independent events, but there are two ways this can happen.

   a. First, a red marble can be drawn and then a green marble can be drawn or

   b. A green marble can be drawn first and then a red marble is drawn.

   The probability of drawing a red marble is $\frac{3}{5}$ and the probability of getting a green marble is $\frac{2}{5}$. $P(\text{R and then G}) = P(\text{R}) \cdot P(\text{G}) = \frac{3}{5} \cdot \frac{2}{5} = \frac{6}{25}$. $P(\text{G and then R}) = P(\text{G}) \cdot P(\text{R}) = \frac{2}{5} \cdot \frac{3}{5} = \frac{6}{25}$. So, the probability of getting one red and one green marble is $\frac{6}{25} + \frac{6}{25} = \frac{12}{25}$.

EXERCISE 32–2

1. A bag contains 3 red marbles and 2 green marbles. A marble is drawn and replaced and then a second marble is drawn. Find the probability that the first marble is red and the second is green?

2. A bag contains 5 red marbles, 3 green marbles, and 2 white marbles. A marble is drawn and replaced and then a second marble is drawn. Find the probability that the first marble is red and the second is green?

3. Two cards are to be drawn from a well-shuffled deck of cards. After the first card is drawn, it is replaced and the deck is shuffled again. What is the probability of first drawing one diamond and then drawing one club?

4. Two cards are to be drawn from a well-shuffled deck of cards. After the first card is drawn, it is replaced and the deck is shuffled again. What is the probability of drawing the ace of hearts and then drawing a 4?

5. Two cards are to be drawn from a well-shuffled deck of cards. After the first card is drawn, it is replaced and the deck is shuffled again. What is the probability of drawing a black card and then a face card?

6. Two cards are to be drawn from a well-shuffled deck of cards. After the first card is drawn, it is replaced and the deck is shuffled again. What is the probability of drawing a face card and then a non-face card?

7. A coin is tossed and a die is rolled. What is the probability of getting a head and a 4?

8. A card is drawn from a deck of cards. It is replaced, the deck is shuffled, and a card is drawn. This is repeated until four cards have been drawn. What is the probability of getting a jack, queen, king, and ace in that order?

9. Two cards are to be drawn from a well-shuffled deck of cards. After the first card is drawn, it is replaced and the deck is shuffled again. What is the probability of drawing one diamond and one club?

10. Two dice are rolled and the total of their sum is recorded. The dice are picked up, rolled again, and the second total is recorded.

    a. What is the probability of rolling a sum of 7 and then a sum of 11?

    b. What is the probability of rolling a sum of 7 and a sum of 11 in either order?

11. A certain medication is known to cure a specific illness for 75% of the people who have the illness. If two people with the illness are selected at random and take the medicine, what is the probability that

    a. both will be cured?

    b. neither will be cured?

    c. only one will be cured?

12. Blood groups for a certain sample of people are shown in the table in Figure 32–1.

Blood Group	Frequency
O	110
A	64
B	20
AB	6

Figure 32–1

If one person from this sample of people is randomly selected, what is the probability that he or she has type B blood?

32–3 Mean Measurement

Technology requires that people work with lots of information. This information is in the form of *data* or factual information, such as measurements, that are used as a basis for reasoning, discussion, or calculation. To use the data effectively, it has to be examined and its trends need to be summarized and analyzed. The summary is usually in the form of a statistical measurement.

The first statistical measurements that will be studied are averages. There are three types of averages: the mean, median, and mode. Together they are called the *measures of central tendency*. The mean is the one that is most used and is the one people usually intend when they say "average."

The *arithmetic mean*, or *mean*, is found by adding all the measurements and dividing this by the total number of measurements.

$$\text{mean} = \frac{\text{sum of measurements}}{\text{number of measurements}}$$

EXAMPLE •————————————————————————————————————

At an automobile engine plant, a quality control technician pulls crankshafts from the assembly line at regular intervals. The technician measures a critical dimension on each of these crankshafts. Even though the dimension is supposed to be 182.000 mm, some variation will occur during production. Here are the measurements for one morning's sample:

182.120   182.005   182.025   181.987   181.898   182.034

Find the mean of these measurements.

Solution.

$$\text{mean} = \frac{\text{sum of measurements}}{\text{number of measurements}}$$

$$= \frac{182.120 + 182.005 + 182.025 + 181.987 + 181.898 + 182.034}{6}$$

$$= \frac{1092.069}{6} = 182.0115, \; 182.012 \; Ans$$

————————————————————————————————————•

The mean measurement is written with the same precision as each of the measurements.

Sometimes the data are reported in terms of a frequency table. Here, to get the sum of the measurements, multiply each measurement by the number of times it occurs.

EXAMPLE •————————————————————————————————

A measuring instrument is placed along an interstate highway and every 15 minutes it measures the noise level in decibels. A frequency distribution is made of the readings for the first day. Decibels readings have been rounded to the nearest 10 decibels.

Decibels	50	60	70	80	90	100	110	120
Frequency	4	6	10	16	18	24	19	5

Find the mean of these measurements.

Solution.

$$\text{mean} = \frac{\text{sum of measurements}}{\text{number of measurements}}$$

$$= \frac{50 \cdot 4 + 60 \cdot 6 + 70 \cdot 10 + 80 \cdot 16 + 90 \cdot 18 + 100 \cdot 24 + 110 \cdot 19 + 120 \cdot 5}{4 + 6 + 10 + 16 + 18 + 24 + 19 + 5}$$

$$= \frac{9250}{102} = 90.686, \ 91 \text{ decibels } Ans$$

——— •

EXERCISE 32–3

Find the mean for each set of measurements in Exercises 1 through 4.

1. 4.2, 2.5, 6.4, 3.6, 7.4, 5.3, 6.9, 2.1, 8.3, 2.7

2. 50.1, 52.4, 52.6, 52.6, 54.8, 54.3, 54.2, 56.7, 58.3, 58.2

3. 80.0, 77.0, 82.0, 73.0, 92.0, 89.0, 100.0, 96.0, 96.0, 94.0, 74.0, 94.0, 94.0, 96.0, 83.0, 84.0, 96.0, 87.0, 84.0, 96.0

4. 100, 98, 96, 94, 93, 90, 89, 85, 82, 78, 76, 66, 64, 64, 78, 89, 93, 96, 98, 96, 93, 64, 96

5. A technician tested an electric circuit and found the following values in milliamperes on successive trials:

   5.24, 5.31, 5.42, 5.26, 5.31, 5.47, 5.32, 5.29, 5.35, 5.44,

   5.35, 5.31, 5.45, 5.46, 5.39, 5.34, 5.35, 5.46, 5.26, 5.32,

   5.47, 5.34, 5.28, 5.39, 5.34, 5.42, 5.43, 5.46, 5.34, 5.29

   Determine the mean for the given data.

6. An environmental officer measured the carbon monoxide emissions (in g/m) for several vehicles. The results are shown in the following table:

   5.02   12.36   13.46   6.92   7.44   8.52   12.82

   11.92   14.32   12.06   8.02   11.34   6.66   9.28

   Determine the mean for the given data.

7. A patrol officer using a laser gun recorded the following speeds for motorists driving on a highway:

   52   57   62   59   67   54

   55   64   65   59   63   72

   Determine the mean for the given data.

8. The blood alcohol content levels of 15 drivers involved in fatal accidents and then convicted with jail sentences are given below:

   0.14   0.16   0.21   0.10   0.13

   0.19   0.26   0.22   0.13   0.09

   0.11   0.18   0.12   0.24   0.32

   Determine the mean for the given data.

9. During a 24-hour time period, a World Wide Web site kept track of the number of times, or "hits," their home page received. The results are shown in the table in Figure 32–2. Here, hour 0 represents 12:00 midnight–1:00 AM, hour1 represents 1:00 AM–2:00 AM, etc.

Hour	0	1	2	3	4	5	6	7
Number of "Hits"	181	120	138	96	146	115	142	323

Hour	8	9	10	11	12	13	14	15
Number of "Hits"	776	697	836	886	922	838	892	947

Hour	16	17	18	19	20	21	22	23
Number of "Hits"	625	558	355	349	320	402	238	204

Figure 32–2

Determine the mean number of "hits" for the given data.

10. In a popcorn experiment 20 samples, each with 100 kernels, were heated in oil for three minutes. At the end of that time, the number of popped kernels were counted and recorded in the table in Figure 32–3.

23	77	20	12	19	54	15	44	41	15
73	31	41	31	79	70	80	69	79	83

Figure 32–3

Determine the mean, for the given data.

32–4   Other Average Measurements

As mentioned in Unit 32–3, the mean is just one average measure, or measure of central tendency. The other two are the median and the mode.

The *median* is the middle number of a group that is arranged in order of size. One-half of the values are larger than the median and one-half are smaller. If there is an even number of items, the median is the number halfway between the two middle items.

EXAMPLES •————————————————————————————————

1. Given the numbers 9, 8, 3, 2, 4, determine the median.

   Solution.   First arrange the numbers in increasing order: 2, 3, 4, 8, 9. There are five numbers, so the middle number is the third number. The third number is 4, so the median is 4.

2. Find the median of 11, 12, 15, 18, 20, 20.

   Solution.   These number are already in numerical order. There are six numbers. The median will be halfway between the third and fourth numbers. The third number is 15 and the fourth is 18. Midway between these is 16.5, thus the median is 16.5.

——•

The third, and final, measure of central tendency is the mode. The *mode* is the value that has the greatest frequency. A set of numbers can have more than one mode. If there are two modes, the data are said to be *bimodal*. Not every set of numbers has a mode.

1. What is the mode for the following data: 11, 12, 15, 18, 20, and 20?

   Solution.   The mode is 20, because that value occurs twice and all the other values occur once.

2. What is the mode for the data: 10, 12, 12, 17, 18, 19, 19, and 20?

   Solution.   There are two modes: 12 and 19, because each of these values occurs twice and all the other values occur once. This is a bimodal set of numbers.

──•

   Both the mean and the median are widely used when referring to an average. The median is a good choice when there is are some extreme values. The mode is not used as often.

A small company has six technicians. The president of the company has a salary of $112,500. The salaries of the technicians are $37,230, $37,950, $39,125, $42,375, $45,300, and $48,715. Determine the mean and the median of these seven salaries.

Solution.

   Mean:

$$\text{mean} = \frac{\text{sum of measurements}}{\text{number of measurements}}$$

$$= \frac{112{,}500 + 37{,}230 + 37{,}950 + 39{,}125 + 42{,}375 + 45{,}300 + 48{,}715}{7}$$

$$= \frac{363{,}195}{7} = 51{,}885, \ \$51{,}885 \ Ans$$

   Median: When the salaries are arranged in increasing order they are $37,230, $37,950, $39,125, $42,375, $45,300, $48,715, and $112,500. The fourth number is $42,375, so this is the median. $42,375 *Ans*

Notice that the mean is higher than six of the seven salaries. The one extreme value for the salary of the company president helps show why the median is often used as the average when there are extreme values.

──•

32–5   Quartiles and Percentiles

The median divides the items into two equally sized parts. In the same way, the *quartiles* Q_1, Q_2, and Q_3, divide the numbers into four equally sized parts when the numbers are arranged in increasing (or decreasing) order.

   There are four steps for finding quartiles.

1. Arrange the numbers in increasing order.

2. Q_2 is the median. It divides the numbers into a lower half and an upper half.

3. Q_1 is the median of the lower half of the numbers.

4. Q_3 is the median of the upper half of the numbers.

Determine the quartiles of 12, 15, 42, 37, 61, 14, 14, 9, 25, 32, 32, and 30.

Solution.   Start by arranging the numbers in order from lowest to highest.

9, 12, 14, 14, 15, 25, 27, 30, 32, 37, 42, and 61

There are 12 numbers, so the second quartile, Q_2 (or median), is midway between the sixth and seventh numbers. The sixth number is 25. The seventh number is 27.

$$Q_2 = \frac{25 + 27}{2} = 26$$

Q_1 is the median of the lower half, so it is the median of the smallest six numbers. Q_1 is midway between the third and fourth numbers. These are both 14, so $Q_1 = 14$.

Q_3 is the median of the upper half. The upper half of the items is 27, 30, 32, 37, 42, and 61. The median of these is midway between 32 and 37, so

$$Q_3 = \frac{32 + 37}{2} = 34.5$$

Percentiles are numbers that divide the data into 100 equal parts. The *n*th percentile is the number P_n, where *n* is the percent of data smaller than or equal to P_n when the data is ranked from smallest to largest.

EXAMPLE

Consider the ranked data in Figure 32–4.

a. Find the 40th percentile.

b. Find the 95th percentile.

15	18	19	20	22	26	32	28	29	34
35	37	42	46	48	51	57	62	63	65
70	71	72	72	77	82	85	86	88	90
92	93	93	93	96	98	101	103	105	106
109	110	115	117	121	122	124	128	129	129

Figure 32–4

Solution.

a. There are 50 numbers, so the 40th percentile, P_{40}, is 0.40 × 50 = 20 or the 20th number. The 20th number is 65, so $P_{40} = 65$. This means that 40% of the data is less than or equal to 65.

b. The 95th percentile is the 0.95 × 50 = 47.5 ranked number. Round this up to 48. The 95th percentile is the 48th number, or 128. Thus, 95% of the data is smaller than or equal to 128.

EXERCISE 32–5

*Find the **a.** median, **b.** mode, and **c.** quartiles for each set of measurements in Exercises 1 through 4.*

1. 4.2, 2.5, 6.4, 3.6, 7.4, 5.3, 6.9, 2.1, 8.3, 2.7

2. 50.1, 52.4, 52.6, 54.6, 54.8, 54.3, 54.2, 56.7, 58.3, 58.2

3. 80.0, 77.0, 82.0, 73.0, 92.0, 89.0, 100.0, 96.0, 96.0, 94.0, 74.0, 94.0, 94.0, 96.0, 83.0, 84.0, 96.0, 87.0, 84.0, 96.0

4. 100, 98, 96, 94, 93, 90, 89, 85, 82, 78, 76, 66, 64, 64, 78, 89, 93, 96, 98, 96, 93, 64, 96

5. A technician tested an electric circuit and found the following values in milliamperes on successive trials:

5.24, 5.31, 5.42, 5.26, 5.31, 5.47, 5.32, 5.29, 5.35, 5.44,

5.35, 5.31, 5.45, 5.46, 5.39, 5.34, 5.35, 5.46, 5.26, 5.32,

5.47, 5.34, 5.28, 5.39, 5.34, 5.42, 5.43, 5.46, 5.34, 5.29

a. Determine the median and mode for the given data.

b. Determine the quartiles for the given data.

c. Determine the 35th percentile for these data.

d. Determine the 90th percentile for these data.

6. An environmental officer measured the carbon monoxide emissions (in g/m) for several vehicles. The results are shown, ranked from smallest to largest, in the table in Figure 32–5.

5.02	5.78	5.81	6.53	6.66	6.87	6.92	6.92	6.94	7.34
7.42	7.44	7.69	8.02	8.07	8.20	8.21	8.34	8.45	8.52
8.52	8.63	8.74	9.10	9.28	9.34	9.36	9.53	9.57	9.62
9.73	9.95	10.32	10.35	10.63	11.08	11.21	11.34	11.54	11.54
11.54	11.92	12.06	12.34	12.36	12.62	12.82	13.46	13.58	14.32

Figure 32–5

a. Determine the median and mode for the given data.

b. Determine the quartiles for the given data.

c. Determine the 40th percentile for these data.

d. Determine the 90th percentile for these data.

e. Determine the 65th percentile for these data.

7. During a 24 hour time period, a World Wide Web site kept track of the number of times, or "hits," their home page received. The results are shown in the table in Figure 32–6. Here hour 0 represents 12:00 midnight–1:00 AM, hour 1 represents 1:00 AM–2:00 AM, etc.

Hour	0	1	2	3	4	5	6	7
Number of "Hits"	181	120	138	96	146	115	142	323

Hour	8	9	10	11	12	13	14	15
Number of "Hits"	776	697	836	886	922	838	892	947

Hour	16	17	18	19	20	21	22	23
Number of "Hits"	625	558	355	349	320	402	238	204

Figure 32–6

a. Determine the median and mode for the number of "hits."

b. Determine the quartiles for the given data.

8. A popcorn experiment used 40 samples each with 100 kernels of popcorn. Each sample was heated in oil for three minutes. At the end of that time, the number of popped kernels were counted and recorded in the table in Figure 32–7.

23	77	20	12	19	54	15	44	41	15
73	31	41	31	79	70	80	69	79	83
57	63	89	76	43	48	85	37	64	72
68	40	61	72	83	29	62	53	47	82

Figure 32–7

   a. Determine the median number of kernels popped.

   b. Determine the quartiles for the given data.

   c. Determine the 70th percentile for these data.

   d. Determine the 45th percentile for these data.

32–6   Grouped Data

Grouping data is one way to save time and reduce mistakes. Data that is grouped is often reported using a frequency table. In Unit 32–3 frequency tables were used to help compute the mean. The *range* is the difference between the highest value and the lowest value of the data.

Grouping data means to arrange the data in groups. This can be done by setting up intervals or classes. For example, the interval 2–6 would contain all numbers between 2 and 6 (and including 2 and 6).

For the interval *a–b*, the number *a* is called the *lower limit* and *b* is the *upper limit.* The *midpoint* of the interval is halfway between the lower and upper limits. The midpoint of the interval *a–b* is $\frac{a+b}{2}$. For the interval 2–6, the midpoint is $\frac{2+6}{2} = \frac{8}{2} = 4$.

The following general rules can be used for setting up intervals.

1. The number of intervals should be between six and 20.

2. The size of all intervals should be the same.

3. The upper limit in an interval will be less than the lower limit in the next interval. This will make it clear to which interval a measurement belongs.

4. The class limits should have the same number of decimal places as the original data.

5. The lower limit of the first interval should be less than or equal to the lowest measurement; the upper limit of the last interval should be greater than the highest measurement.

The grouped data is usually shown in a *frequency distribution.* In a frequency distribution, one line contains a list of possible values and a second line contains the number of times each value was observed in a particular time.

EXAMPLE •───────────────────────────────────

People who live near an interstate highway have complained about traffic noise. A measuring instrument is placed along the highway and every 15 minutes it measures the noise level in decibels. A frequency distribution of the readings for the first day is shown in the table in Figure 32–8.

Decibels	50–54	55–59	60–64	65–69	70–74	75–79	80–84
Frequency	2	2	4	4	4	6	8

Decibels	85–89	90–94	95–99	100–104	105–109	110–114	115–119
Frequency	8	10	12	14	10	8	4

Figure 32–8

───•

Tally marks can often be used to help make a frequency distribution, particularly when the data is not ranked. Once the intervals have been chosen, list each interval and its midpoint, and tally the number of measurements that lie in that interval.

EXAMPLE •───

Create a frequency distribution for the daily emission (in tons) of sulfur oxides at an industrial plant shown in Figure 32–9.

15.8	26.4	17.3	11.2	23.9	24.8	18.7	13.9	9.0	13.2
22.7	9.8	6.2	14.7	17.5	26.1	12.8	28.6	17.6	23.7
26.8	22.7	18.0	20.5	11.0	20.9	15.5	19.4	16.7	10.7
19.1	15.2	22.9	26.6	20.4	21.4	19.2	21.6	16.9	19.0
18.5	23.0	24.6	20.1	16.2	18.0	7.7	13.5	23.5	14.5
14.4	29.6	19.4	17.0	20.8	24.3	22.5	24.6	18.4	18.1
8.3	21.9	12.3	22.3	13.3	11.8	19.3	20.0	25.7	31.8
25.9	10.5	15.9	32.5	18.1	17.9	9.4	24.1	20.1	28.5

Figure 32–9

Solution.   The largest value is 31.8 and the smallest is 6.2. The range of values is $31.8 - 6.2 = 25.6$. Several choices could be made for the number of intervals. For example, six intervals could be chosen with the limits 5.0–9.9, 10.0–14.9, . . . , 30.0–34.9. However, seven intervals with the limits 5.0–8.9, 9.0–12.9, . . . , 29.0–32.9 could be chosen. Nine intervals, with the limits 5.0–7.9, 8.0–10.9, . . . , 29.0–31.9, could also be selected. Notice that in each case, the intervals do not overlap and are all of the same size. The number or size of the intervals is often up to the person making the frequency distribution.

Seven intervals are chosen for this example and the table in Figure 32–10 is created:

Sulfur Oxide Emissions (tons)	Midpoint x	Tally	Frequency f
5.0–8.9	6.95	///	3
9.0–12.9	10.95	LHT LHT	10
13.0–16.9	14.95	LHT LHT ////	14
17.0–20.9	18.95	LHT LHT LHT LHT LHT	25
21.0–24.9	22.95	LHT LHT LHT //	17
25.0–28.9	26.95	LHT ////	9
29.0–32.9	30.95	//	2
			80

Figure 32–10

To determine the mean from the frequency distribution,

a. Multiply the frequency f of each interval by the midpoint x of that interval to get the product fx

b. Add the products, and

c. Divide by the total frequency, the sum of the f-values.

$$\text{mean} = \frac{\text{sum of } fx}{\text{sum of } f}$$

EXAMPLE •———

Find the mean for the data in Figure 32–10.

Solution.   The frequency distribution table is modified by adding a column for the product fx. The result is shown in Figure 32–11.

Sulfur Oxide Emissions (tons)	Midpoint x	Tally	Frequency f	Product fx
5.0–8.9	6.95	///	3	20.85
9.0–12.9	10.95	⌿⌿⌿ ⌿⌿⌿	10	109.50
13.0–16.9	14.95	⌿⌿⌿ ⌿⌿⌿ ////	14	209.30
17.0–20.9	18.95	⌿⌿⌿ ⌿⌿⌿ ⌿⌿⌿ ⌿⌿⌿ ⌿⌿⌿	25	473.75
21.0–24.9	22.95	⌿⌿⌿ ⌿⌿⌿ ⌿⌿⌿ //	17	390.15
25.0–28.9	26.95	⌿⌿⌿ ////	9	242.55
29.0–32.9	30.95	//	2	61.90
			80	1508.00

Figure 32–11

The mean of the sulfur oxide emissions is

$$\text{mean} = \frac{\text{sum of } fx}{\text{sum of } f} = \frac{1508}{80} = 18.85 \text{ tons.}$$

•

It is often useful to graph the data in a frequency diagram. A *histogram,* like the one in Figure 32–12, is often used. A histogram is just a bar graph in which there is no space between the bars.

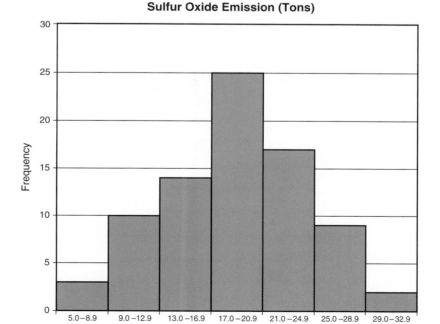

Figure 32–12

EXERCISE 32–6

1. The data in the table in Figure 32–13 are based on the energy consumption for one household's electric bills for 36 two-month periods.

   a. Construct a histogram that corresponds to this frequency table.

   b. Determine the mean of the data.

Energy (kWh)	700–719	720–739	740–759	760–779	780–799
Frequency	2	2	4	5	3

Energy (kWh)	800–819	820–839	840–859	860–879	880–899
Frequency	4	7	5	2	2

Figure 32–13

2. A sample of 100 batteries was selected from the day's production for a machine. The batteries were tested to see how long they would operate a flashlight with the results shown in the table in Figure 32–14.

   a. Construct a histogram that corresponds to this frequency table.

   b. Determine the mean of the data.

Hours	211–215	216–220	221–225	226–230
Frequency	4	9	19	23

Hours	231–235	236–240	241–245	246–250
Frequency	16	14	10	5

Figure 32–14

3. The data in the table in Figure 32–15 shows the number of registered nurses, in thousands, for a certain year grouped by age.

Age	Midpoint x	Frequency f	Product fx
20–24		66	
25–29		177	
30–34		248	
35–39		360	
40–44		464	
45–49		465	
50–54		342	
55–59		238	
60–64		156	
65–69		154	

Figure 32–15

Determine the mean of the data.

4. The average number of motor vehicle fatalities per 100,000 accidents is shown in the table in Figure 32–16.

Age	Midpoint x	Frequency f	Product fx
6–15		6.4	
16–25		35.2	
26–35		11.3	
36–45		8.4	
46–55		6.2	
56–65		9.7	
66–75		15.3	
76–86		26.9	

Figure 32–16

Determine the mean of the data.

5. The emission data in the table in Figure 32–9 were separated into seven intervals (see Figure 32–9).

    a. Create a frequency distribution for this data using six intervals with the limits 5.0–9.9, 10.0–14.9, . . . , 30.0–34.9.

    b. Determine the mean.

    c. Draw a histogram of this distribution.

6. The emission data in the table in Figure 32–9 was separated into seven intervals.

    a. Create a frequency distribution for this data using nine intervals with the limits 5.0–7.9, 8.0–10.9, . . . , 29.0–31.9.

    b. Determine the mean.

    c. Draw a histogram of this distribution.

    d. Since the same data was used in both Exercises 5 and 6, how do you explain that the means are not equal?

7. The data in the table in Figure 32–17 shows the ignition times when a certain upholstery material is exposed to a flame. Times are given to the nearest 0.01, sec.

2.58	2.51	4.04	6.43	1.68	6.42	2.25	4.10
5.79	6.20	1.54	1.36	5.87	4.64	6.12	5.15
4.50	5.92	6.54	3.46	5.90	2.47	3.21	3.22
7.65	8.54	7.80	4.07	7.42	3.62	2.46	5.87
5.62	7.68	1.74	2.68	9.54	11.78	2.14	1.87
6.73	8.79	4.83	8.65	5.19	4.13	7.63	5.40
11.36	4.65	5.63	9.46	7.41	7.86	10.64	3.64
3.87	3.57	5.33	3.01	6.34	1.90	3.42	1.67
4.75	3.79	6.87	5.62	9.80	5.11	4.62	1.57
4.21	1.67	8.76	1.38	6.88	2.94	7.46	11.68

Figure 32–17

    a. Create a frequency distribution for this data using a suitable number of intervals.

    b. Determine the mean.

    c. Draw a histogram of this distribution.

8. In a two-week study of the productivity of workers, the data in the table in Figure 32–18 were obtained on the number of acceptable pieces produced by 100 workers.

76	55	77	72	39	46	55	53	66	51
82	72	46	47	44	61	53	59	60	44
62	66	75	48	58	95	50	69	53	70
88	66	74	55	82	48	54	32	60	69
49	69	70	32	56	39	61	55	52	96
44	31	72	43	57	67	43	59	51	54
67	48	36	68	37	24	81	79	74	74
52	50	66	73	52	41	69	79	73	92
41	50	59	51	77	48	59	58	60	47
39	60	45	77	52	61	61	42	82	72

Figure 32–18

a. Create a frequency distribution for this data using eight intervals with the limits 20–29, 30–39, . . . , 90–99.

b. Determine the mean.

c. Draw a histogram of this distribution.

32–7   Variance and Standard Deviation

The mean and median give technicians two useful ways to measure the average of a sample. But neither one gives any information about how the actual data is distributed. Are the data close together or spread out? This section will give two ways to help give this information.

EXAMPLE •————————————————————————————————

Determine the range of 12, 15, 42, 37, 14, 9, 25, 27, 32, and 30.

Solution.   The lowest number is 9 and the highest value is 42. The range is $42 - 9 = 33$.

——•

While the range gives one indication how the information is spread out, the standard deviation is often more helpful. The *standard deviation* gives a measure of how much the numbers are spread out from the mean.

The standard deviation is defined in terms of the *variance*.

$$\text{variance} = \frac{\text{sum of (measurement} - \text{mean)}^2}{\text{number of measurements} - 1}$$

$$= \frac{\text{sum of } (x - \bar{x})^2}{n - 1}$$

where x is a measurement, $\bar{x}$ is the mean, and n is the total number of measurements.

However, using this formula requires a lot of work. An easier formula to use is

$$\text{variance} = \frac{n \cdot \text{sum of } (x^2) - (\text{sum of } x)^2}{n(n - 1)}$$

where x is a measurement and n is the total number of measurements.

The variance is not often used except to get the standard deviation. The standard deviation is the square root of the variance. One reason the variance is not used is because it does not have the same unit of measure as the original data (where the standard deviation does have the same unit of measurement as the original data).

$$\text{standard deviation} = \sqrt{\text{variance}}$$

EXAMPLE •

Determine the standard deviation of 12, 15, 42, 37, 14, 9, 25, 27, 32, and 30. This is the same data used in the previous example.

Solution.

Step 1: Find the sum of the measurements and their squares.

x	x^2
12	$12^2 = 144$
15	$15^2 = 225$
42	$42^2 = 1764$
37	$37^2 = 1369$
14	$14^2 = 196$
9	$9^2 = 81$
25	$25^2 = 625$
32	$32^2 = 1024$
27	$27^2 = 729$
30	$30^2 = 900$
243	7057

Step 2: Substitute these values in the formula for the variance.

$$\text{variance} = \frac{n(\text{sum of } x^2) - (\text{sum of } x)^2}{n(n-1)}$$

$$= \frac{10(7057) - 243^2}{10(10-1)}$$

$$= \frac{70,570 - 59,049}{90} = \frac{11,521}{90} \approx 128.01111$$

Step 3: Find the standard deviation by taking the square root of the answer in Step 2.

$$\text{standard deviation} = \sqrt{128.01111} \approx 11.3 \; Ans$$

Interval	Frequency f
15–24	26
25–34	33
35–44	41
45–54	36
55–64	25
65–74	15
75–84	12

Figure 32–19

The variance and standard deviation of grouped data is found in a way similar to the way the mean is found. A frequency table is used with one additional column for the product of frequency and the square of the midpoint of each interval, fx^2. The formula is

$$\text{variance of grouped data} = \frac{n[\text{sum of } (f \cdot x^2)] - (\text{sum of } fx)^2}{n(n-1)}$$

where x is the midpoint of a measurement interval, f is the frequency of that midpoint, and n is the total number of measurements.

EXAMPLE •

Find the mean, and standard deviation for the data in Figure 32–19.

Solution.   The frequency distribution table is modified by adding columns for the midpoint of each interval, x, the product of the frequency and the midpoint fx, and the product of the frequency and the square of the midpoint, fx^2. The result is shown in Figure 32–20.

Interval	Midpoint x	Frequency f	fx	fx^2
15–24	19.5	26	507.0	9,886.50
25–34	29.5	33	973.5	28,718.50
35–44	39.5	41	1,619.5	63,970.25
45–54	49.5	36	1,782.0	88,209.00
55–64	59.5	25	1,487.5	88,506.25
65–74	69.5	15	1,042.5	72,453.75
75–84	79.5	12	954.0	75,843.00
		188	8,366.0	432,587.25

Figure 32–20

We first use some of the totals from the table to determine the mean.

$$\text{mean} = \frac{\text{sum of } fx}{\text{sum of } f}$$

$$= \frac{8,366}{188} = 44.5 \ Ans$$

Substituting the values from the table in the formula for the variance of grouped data produces

$$\text{variance of grouped data} = \frac{n[\text{sum of }(f \cdot x^2)] - (\text{sum of } fx)^2}{n(n-1)}$$

$$= \frac{188[432,587.25] - 8,366^2}{188(188-1)}$$

$$= \frac{80,386,403 - 69,989,956}{188(187)} = \frac{10,396,447}{188(187)} \approx 295.7233$$

Now, find the standard deviation by taking the square root of the variance.

$$\text{standard deviation} = \sqrt{295.7233} \approx 17.20 \ Ans$$

Many scientific calculators have statistical functions. On these calculators the mean is usually denoted $\bar{x}$ and the standard deviation is denoted by Sx. Read the manual for your calculator to see how your calculator can be used to find the mean and standard deviation.

EXERCISE 32–7

Find the mean and standard deviation for each set of measurements.

Ungrouped data

1. 4.2, 2.5, 6.4, 3.6, 7.4, 5.3, 6.9, 2.1, 8.3, 2.7

2. 50.1, 52.4, 52.6, 52.6, 54.8, 54.3, 54.2, 56.7, 58.3, 58.2

3. A technician tested an electric circuit and found the following values in milliamperes on successive trials:

   5.24, 5.31, 5.42, 5.26, 5.31, 5.47, 5.32, 5.29, 5.35, 5.44,

   5.35, 5.31, 5.45, 5.46, 5.39, 5.34, 5.35, 5.46, 5.26, 5.32,

   5.47, 5.34, 5.28, 5.39, 5.34, 5.42, 5.43, 5.46, 5.34, 5.29

4. An environmental officer measured the carbon monoxide emissions (in g/m) for several vehicles. The results are shown in the following table:

5.02	12.36	13.46	6.92	7.44	8.52	12.82
11.92	14.32	12.06	8.02	11.34	6.66	9.28

5. During a 24-hour time period, a World Wide Web site kept track of the number of times, or "hits," their home page received. The results are shown in the table in Figure 32–21. Here hour 0 represents 12:00 midnight–1:00 AM, hour 1 represents 1:00 AM–2:00 AM, etc.

Hour	0	1	2	3	4	5	6	7
Number of "Hits"	181	120	138	96	146	115	142	323

Hour	8	9	10	11	12	13	14	15
Number of "Hits"	776	697	836	886	922	838	892	947

Hour	16	17	18	19	20	21	22	23
Number of "Hits"	625	558	355	349	320	402	238	204

Figure 32–21

6. In a popcorn experiment 100 kernels of different brands of popcorn were heated in oil for three minutes. At the end of that time, the number of popped kernels were counted and recorded in the table in Figure 32–22.

23	77	20	12	19	54	15	44	41	15
73	31	41	31	79	70	80	69	79	83

Figure 32–22

Grouped data

7. The data in the table in Figure 32–23 shows the number of registered nurses, in thousands, for a certain year grouped by age.

Age	Midpoint x	Frequency f	fx	fx^2
20–24		66		
25–29		177		
30–34		248		
35–39		360		
40–44		464		
45–49		465		
50–54		342		
55–59		238		
60–64		156		
65–69		154		

Figure 32–23

8. The average number of motor vehicle fatalities per 100,000 accidents is shown in the table in Figure 32–24.

Age	Midpoint x	Frequency f	fx	fx²
6–15		6.4		
16–25		35.2		
26–35		11.3		
36–45		8.4		
46–55		6.2		
56–65		9.7		
66–75		15.3		
76–85		26.9		

Figure 32–24

9. The data in the table in Figure 32–25 shows the ignition times when a certain upholstery material is exposed to a flame. Times are given to the nearest 0.01, sec.

Ignition Time (seconds)	Midpoint x	Frequency f	fx	fx²
1.00–1.99	1.495	10		
2.00–2.99	2.495	8		
3.00–3.99	3.495	10		
4.00–4.99	4.495	11		
5.00–5.99	5.495	13		
6.00–6.99	6.495	9		
7.00–7.99	7.495	8		
8.00–8.99	8.495	4		
9.00–9.99	9.495	3		
10.00–10.99	10.495	1		
11.00–11.99	11.495	3		

Figure 32–25

10. In a two-week study of the productivity of workers, the data in the table in Figure 32–26 were obtained on the number of acceptable pieces produced by 100 workers.

No. of Acceptable Pieces	Midpoint x	Frequency f	fx	fx²
20–29	24.5	2		
30–39	34.5	7		
40–49	44.5	17		
50–59	54.5	32		
60–69	64.5	20		
70–79	74.5	19		
80–89	84.5	5		
90–99	94.5	3		

Figure 32–26

32–8   Statistical Process Control: X-Bar Charts

One of the uses of statistics is in the reduction of defects in manufactured goods. This is accomplished by using statistics during production so that changes can be made early rather than waiting until a large number of defects have been produced. This method is called *Statistical Process Control (SPC)* and uses control charts.

A *control chart* gives a continuous series of snapshots from small inspections taken at regular intervals. At each inspection, samples are pulled from the production line and measured. These measurements are graphed to make it easier to notice trends or abnormalities in the production process.

There are two general types of control charts.

Variable chart: A dimension or characteristic is measured and the result is a number.

Attribute chart: A dimension or characteristic is not measured in numbers but is classified as either "good" or "bad."

We will study just one type of control chart—the variable chart known as the *X-Bar-R* chart. This is perhaps the most common variable chart and actually consists of two charts: the $\bar{x}$ (X-Bar) and the *R* chart. While these are separate charts, they are usually considered together.

The X-Bar chart indicates the changes that have occured in the central tendency of a process. These changes might be due to such factors as tool wear, or new and stronger materials. *R* chart values indicate that a gain or loss in dispersion has occured. Such a change might be due to worn bearings, a loose tool, an erratic flow of lubricants to a machine, or sloppiness on the part of the machine operator. The two types of charts go hand-in-hand when monitoring variables, because they measure the two critical parameters: central tendency and dispersion.

For each chart you need to compute, and then draw, the *central line* and the control limits. There are two control limits—the *Upper Control Limit (UCL)* and the *Lower Control Limit (LCL).* The LCL and UCL are used to determine if the process is in control or out of control.

For example, the X-Bar chart in Figure 32–27 shows a manufacturing process that is out of control because one point, indicated by the arrow, is outside the control limits. A process may also be out of control when there are long runs (8 or more consecutive points) either above or below the centerline.

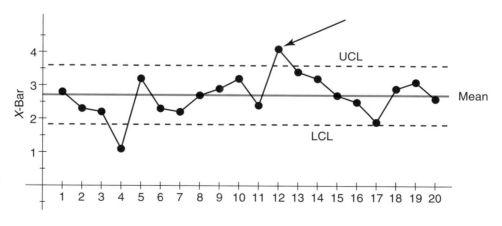

Figure 32–27

The following steps are used to prepare an $\overline{X}$-Bar chart.

1. Choose a sample size *n* and how often a sample is selected. Sample sizes usually contain 4 to 7 items.

2. Collect the data. Compute the mean and range of each sample and the average range $\overline{R}$ of all the samples.

3. Determine the central line by computing the average mean $\bar{\bar{x}}$.

4. Determine the control limits: $\bar{\bar{x}} \pm A_2 \cdot \overline{R}$ where A_2 is found in the table in Figure 32–28.

n	2	3	4	5	6	7	8	9
A_2	1.880	1.023	0.729	0.577	0.483	0.419	0.373	0.337

Figure 32–28

5. Plot the means of the samples.

6. Draw the central line and the upper and lower control limits.

EXAMPLE •

The data in the table in Figure 32–29 shows the mean and range of 20 sets of five measurements of the diameter, in cm, of an engine shaft. Construct an X-Bar control chart using this information.

Mean	2.000	2.001	2.001	2.000	2.001	2.003	1.998
Range	0.005	0.005	0.007	0.007	0.006	0.004	0.001

Mean	2.001	2.000	1.999	2.000	2.001	2.000	2.003
Range	0.002	0.008	0.006	0.004	0.007	0.004	0.005

Mean	1.998	1.997	2.002	2.000	1.997	2.003
Range	0.007	0.010	0.006	0.002	0.006	0.008

Figure 32–29

Solution.

Step 1: The mean of the means is $\bar{\bar{x}} = 2.0003$ and the mean of the ranges is $\bar{R} = 0.0055$.

Step 2: Compute the control limits. Look at the table in Figure 32–28. Since each set of data has five measurements, $n = 5$, and from the table we see that $A_2 = 0.557$.

$$UCL = \bar{\bar{x}} + A_2 \cdot \bar{R} = 2.0003 + 0.557(0.0055) = 2.0034$$
$$LCL = \bar{\bar{x}} + -A_2 \cdot \bar{R} = 2.0003 - 0.557(0.0055) = 1.9972$$

Step 3: Plot the means, the central line, and the control limits. The result is the graph in Figure 32–30.

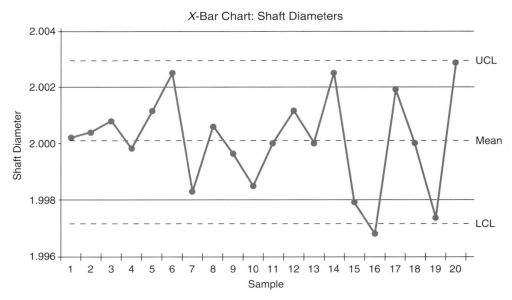

Figure 32–30

Notice that this process is out of control because the point at sample 16 is below the lower control limit. By this time, the technicians should have located the problem and fixed it.

EXERCISE 32–8

Use the table in Figure 32–28 for the value of A_2.

1. The data in the table in Figure 32–31 are the mean and range of dial indicator readings of a pin diameter (in mm). Six samples were collected each half-hour for an eight-hour period.

Sample	1	2	3	4	5	6	7	8
Mean, $\bar{x}$, in mm	2.50	2.50	2.49	2.50	2.52	2.52	2.50	2.65
Range (mm)	0.05	0.09	0.16	0.10	0.12	0.15	0.07	0.29

Sample	9	10	11	12	13	14	15	16
Mean, $\bar{x}$, in mm	2.73	2.52	2.51	2.56	2.51	2.50	2.49	2.50
Range (mm)	0.32	0.05	0.15	0.06	0.05	0.06	0.07	0.10

Figure 32–31

a. Determine the central line, upper control limit, and lower control limit for the mean ($\bar{x}$).

b. Construct an X-Bar control chart using this information.

c. Is the process out of control? If it is, at what sample does the X-Bar control chart signal lack of control?

2. The table in Figure 32–32 are test results from checking the volume of 100-μL variable volume pipettes. Samples were collected each half-hour on four pipettes.

Sample	1	2	3	4	5	6	7	8
$\bar{x}$ (μL)	99.75	99.54	99.95	99.32	100.20	101.73	100.87	100.59
Range	2.24	2.87	3.14	2.79	2.68	0.33	1.60	2.43

Sample	9	10	11	12	13	14	15	16
$\bar{x}$ (μL)	99.74	100.30	100.85	99.81	99.99	99.57	100.67	99.93
Range	3.35	2.77	2.40	3.52	3.69	2.28	0.07	3.18

Sample	17	18	19	20	21	22	23	24
$\bar{x}$ (μL)	99.84	98.89	99.61	100.15	100.10	100.18	99.58	99.91
Range	2.97	1.65	3.35	3.20	2.54	0.60	0.36	1.19

Figure 32–32

a. Determine the central line, upper control limit, and lower control limit for the mean ($\bar{x}$).

b. Construct an X-Bar control chart using this information.

c. Is the process out of control? If it is, at what sample does the X-Bar control chart signal lack of control?

3. The data in the table in Figure 32–33 are the mean and range of the length of a shaft in cm. Five samples were collected every 20 minutes for an seven-hour period.

Sample	1	2	3	4	5	6	7
$\bar{x}$ (cm)	11.9940	11.9980	11.9920	11.9940	12.0160	11.9920	12.0040
Range (cm)	0.080	0.070	0.040	0.060	0.030	0.030	0.050

Sample	8	9	10	11	12	13	14
$\bar{x}$ (cm)	12.0140	11.9920	12.0080	11.9720	11.9380	11.9520	11.9540
Range (cm)	0.040	0.060	0.080	0.050	0.040	0.050	0.060

Sample	15	16	17	18	19	20	21
$\bar{x}$ (cm)	11.9700	11.9800	11.9880	12.0060	12.0010	12.0060	11.9960
Range (cm)	0.070	0.040	0.050	0.030	0.060	0.030	0.040

Figure 32–33

   a. Determine the central line, upper control limit, and lower control limit for the mean ($\bar{x}$).

   b. Construct an X-Bar control chart using this information.

   c. Is the process out of control? If it is, at what sample does the X-Bar control chart signal lack of control?

4. Twelve samples of size $n = 5$ are taken of the diameter, in inches, of a bearing. The mean and range of these samples are shown in the table in Figure 32–34.

Sample	1	2	3	4	5	6
$\bar{x}$ (in.)	0.345	0.342	0.346	0.336	0.340	0.341
Range (in.)	0.003	0.004	0.002	0.009	0.005	0.006

Sample	7	8	9	10	11	12
$\bar{x}$ (in.)	0.346	0.347	0.348	0.345	0.349	0.356
Range (in.)	0.004	0.003	0.002	0.005	0.003	0.006

Figure 32–34

   a. Determine the central line, upper control limit, and lower control limit for the mean ($\bar{x}$).

   b. Construct an X-Bar control chart using this information.

   c. Is the process out of control? If it is, at what sample does the X-Bar control chart signal lack of control?

5. The data in the table in Figure 32–35 are the mean and range of the endplay in a motor shaft, measured in mm. Five samples were collected each half-hour for an ten-hour period.

Sample	1	2	3	4	5	6	7	8	9	10
$\bar{x}$ (mm)	37.8	33.6	43.6	43.4	34.2	41.0	41.2	43.4	45.4	42.0
Range (mm)	32	32	14	14	21	32	15	21	30	18

Sample	11	12	13	14	15	16	17	18	19	20
$\bar{x}$ (mm)	35.6	44.6	44.4	41.8	47.6	41.0	45.6	43.8	40.4	35.4
Range (mm)	13	6	20	17	16	22	19	21	20	15

Figure 32–35

a. Determine the central line, upper control limit, and lower control limit for the mean ($\bar{x}$).

b. Construct an X-Bar control chart using this information.

c. Is the process out of control? If it is, at what sample does the X-Bar control chart signal lack of control?

6. Sixteen samples of size $n = 4$ are taken of Amoxicillin capsules. Each capsule is supposed to contain 250 mg. The mean and range of the samples are shown in the table in Figure 32–36.

Sample	1	2	3	4	5	6	7	8
$\bar{x}$ (mg)	249.2	250.1	250.6	249.9	250.3	250.3	250.2	249.9
Range (mg)	1.6	1.5	0.7	0.3	2.1	1.7	1.2	1.4

Sample	9	10	11	12	13	14	15	16
$\bar{x}$ (mg)	250.4	249.8	249.2	250.6	250.7	250.7	251.0	250.8
Range (mg)	0.8	1.2	0.8	0.6	1.0	0.7	1.6	0.3

Figure 32–36

a. Determine the central line, upper control limit, and lower control limit for the mean ($\bar{x}$).

b. Construct an X-Bar control chart using this information.

c. Is the process out of control? If it is, at what sample does the X-Bar control chart signal lack of control?

32–9 Statistical Process Control: R-Charts

The following steps are used to prepare an R chart. These steps assume that you have already followed the steps for preparing an X-Bar chart listed on pages 677 and 678.

1. Determine the central line by computing the mean of the ranges, $\bar{R}$.

2. Determine the control limits: LCL $= D_3 \cdot \bar{R}$ and UCL $= D_4 \cdot \bar{R}$ where D_3 and D_4 are found in the table in Figure 32–37.

n	2	3	4	5	6	7	8	9
D_3	0	0	0	0	0	0.076	0.136	0.184
D_4	3.267	2.574	2.282	2.114	2.004	1.924	1.864	1.816

Figure 32–37

3. Draw the central line and the upper and lower control limits.

EXAMPLE

The data in the table in Figure 32–38 shows the mean and range of 20 sets of five measurements of the diameter, in cm, of an engine shaft. Construct an R chart using this information. Note that this is the same data that is in Figure 32–29 on page 678.

Mean	2.000	2.001	2.001	2.000	2.001	2.003	1.998
Range	0.005	0.005	0.007	0.007	0.006	0.004	0.001

Mean	2.001	2.000	1.999	2.000	2.001	2.000	2.003
Range	0.002	0.008	0.006	0.004	0.007	0.004	0.005

Mean	1.998	1.997	2.002	2.000	1.997	2.003
Range	0.007	0.010	0.006	0.002	0.006	0.008

Figure 32–38

Solution.

Step 1: The mean of the ranges is $\overline{R} = 0.0055$.

Step 2: Compute the control limits with $D_3 = 0$ and $D_4 = 2.114$.

$$UCL = D_3 \cdot \overline{R} = 0$$
$$LCL = D_4 \cdot \overline{R} = 2.114(0.0055) = 0.011632$$

Step 3: Plot the ranges, the central line, and the control limits. The result is the graph in Figure 32–39.

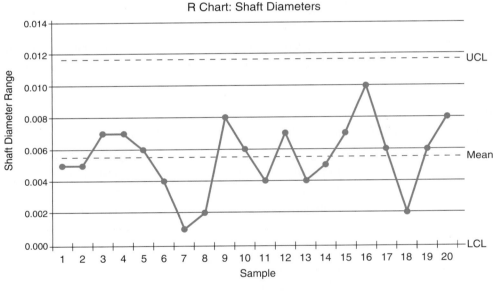

Figure 32–39

Notice that this process is in control because none of the points go above the upper control limit.

EXERCISE 32–9

The data for these exercises are the same as for those in Exercise Set 32–8. Use the table in Figure 32–27 for the values of D_3 and D_4.

1. The data in the table in Figure 32–40 are the mean and range of dial indicator readings of a pin diameter (in mm). Six samples were collected each half-hour for an eight-hour period.

Sample	1	2	3	4	5	6	7	8
Mean, $\bar{x}$, in mm	2.50	2.50	2.49	2.50	2.52	2.52	2.50	2.65
Range (mm)	0.05	0.09	0.16	0.10	0.12	0.15	0.07	0.29

Sample	9	10	11	12	13	14	15	16
Mean, $\bar{x}$, in mm	2.73	2.52	2.51	2.56	2.51	2.50	2.49	2.50
Range (mm)	0.32	0.05	0.15	0.06	0.05	0.06	0.07	0.10

Figure 32–40

    a. Determine the central line, upper control limit, and lower control limit for the range (R).

    b. Construct an R control chart using this information.

    c. Is the process out of control? If it is, at what sample does the R control chart signal lack of control?

2. The table in Figure 32–41 are test results from checking the volume of 100-μL variable volume pipettes. Samples were collected each half-hour on four pipettes.

Sample	1	2	3	4	5	6	7	8
$\bar{x}$ (μL)	99.75	99.54	99.95	99.32	100.20	101.73	100.87	100.59
Range	2.24	2.87	3.14	2.79	2.68	0.33	1.60	2.43

Sample	9	10	11	12	13	14	15	16
$\bar{x}$ (μL)	99.74	100.30	100.85	99.81	99.99	99.57	100.67	99.93
Range	3.35	2.77	2.40	3.52	3.69	2.28	0.07	3.18

Sample	17	18	19	20	21	22	23	24
$\bar{x}$ (μL)	99.84	98.89	99.61	100.15	100.10	100.18	99.58	99.91
Range	2.97	1.65	3.35	3.20	2.54	0.60	0.36	1.19

Figure 32–41

    a. Determine the central line, upper control limit, and lower control limit for the range (R).

    b. Construct an R control chart using this information.

    c. Is the process out of control? If it is, at what sample does the R control chart signal lack of control?

3. The data in the table in Figure 32–42 are the mean and range of the length of a shaft in cm. Five samples were collected every 20 minutes for an seven-hour period.

Sample	1	2	3	4	5	6	7
$\bar{x}$ (cm)	11.9940	11.9980	11.9920	11.9940	12.0160	11.9920	12.0040
Range (cm)	0.080	0.070	0.040	0.060	0.030	0.030	0.050

Sample	8	9	10	11	12	13	14
$\bar{x}$ (cm)	12.0140	11.9920	12.0080	11.9720	11.9380	11.9520	11.9540
Range (cm)	0.040	0.060	0.080	0.050	0.040	0.050	0.060

Sample	15	16	17	18	19	20	21
$\bar{x}$ (cm)	11.9700	11.9800	11.9880	12.0060	12.0010	12.0060	11.9960
Range (cm)	0.070	0.040	0.050	0.030	0.060	0.030	0.040

Figure 32–42

a. Determine the central line, upper control limit, and lower control limit for the range (R).

b. Construct an R control chart using this information.

c. Is the process out of control? If it is, at what sample does the R control chart signal lack of control?

4. Twelve samples of size n = 5 are taken of the diameter, in inches, of a bearing. The mean and range of these samples are shown in the table in Figure 32–43.

Sample	1	2	3	4	5	6
$\bar{x}$ (in.)	0.345	0.342	0.346	0.336	0.340	0.341
Range (in.)	0.003	0.004	0.002	0.009	0.005	0.006

Sample	7	8	9	10	11	12
$\bar{x}$ (in.)	0.346	0.347	0.348	0.345	0.349	0.356
Range (in.)	0.004	0.003	0.002	0.005	0.003	0.006

Figure 32–43

a. Determine the central line, upper control limit, and lower control limit for the range (R).

b. Construct an R control chart using this information.

c. Is the process out of control? If it is, at what sample does the R control chart signal lack of control?

5. The data in the table in Figure 32–44 are the mean and range of the endplay in a motor shaft, measured in mm. Five samples were collected each half-hour for an ten-hour period.

Sample	1	2	3	4	5	6	7	8	9	10
$\bar{x}$ (mm)	37.8	33.6	43.6	43.4	34.2	41.0	41.2	43.4	45.4	42.0
Range (mm)	32	32	14	14	21	32	15	21	30	18

Sample	11	12	13	14	15	16	17	18	19	20
$\bar{x}$ (mm)	35.6	44.6	44.4	41.8	47.6	41.0	45.6	43.8	40.4	35.4
Range (mm)	13	6	20	17	16	22	19	21	20	15

Figure 32–44

a. Determine the central line, upper control limit, and lower control limit for the range (R).

b. Construct an R control chart using this information.

c. Is the process out of control? If it is, at what sample does the R control chart signal lack of control?

6. Sixteen samples of size n = 4 are taken of Amoxicillin capsules. Each capsule is supposed to contain 250 mg. The mean and range of the samples are shown in the table in Figure 32–45.

Sample	1	2	3	4	5	6	7	8
$\bar{x}$ (mg)	249.2	250.1	250.6	249.9	250.3	250.3	250.2	249.9
Range (mg)	1.6	1.5	0.7	0.3	2.1	1.7	1.2	1.4

Sample	9	10	11	12	13	14	15	16
$\bar{x}$ (mg)	250.4	249.8	249.2	250.6	250.7	250.7	251.0	250.8
Range (mg)	0.8	1.2	0.8	0.6	1.0	0.7	1.6	0.3

Figure 32–45

a. Determine the central line, upper control limit, and lower control limit for the range (R).

b. Construct an R control chart using this information.

c. Is the process out of control? If it is, at what sample does the R control chart signal lack of control?

UNIT EXERCISE AND PROBLEM REVIEW

SAMPLE SPACE

1. What is the sample space when a die is rolled and a coin is tossed?

2. Two dice are rolled and the sum of the faces that are up is computed. What is the sample space?

PROBABILITY

Determine the probability in each of the following:

3. Getting a "tail" when a coin is tossed.

4. Drawing a green ball, blindfolded, from a bag containing 6 red and 10 green balls.

5. Not drawing a queen from a deck of 52 cards.

6. Rolling a 5 or higher with a single die.

7. Rolling a sum of 4 or less with a pair of die.

8. Getting a head and a tail when two coins are tossed.

INDEPENDENT EVENTS

9. A die is rolled and a coin is tossed. What is the probability of getting a 5 or higher on the die and a head on the coin?

10. The probability that someone will get a certain disease is 0.15. What is the probability that two people from different areas will not get the disease?

11. A spinner has the numbers 1–5 marked equally on it. If the spinner is spun twice, what is the probability of getting two 4s?

12. A spinner has the numbers 1–5 marked equally on it. If the spinner is spun twice, what is the probability of getting two numbers with a sum of 8?

MEAN, MEDIAN, MODE, AND STANDARD DEVIATION

An inspector at a disc-drive manufacturing company has collected 24 samples of disc drives and measured the number of revolutions per minute for each drive. The results are shown in the table in Figure 32–46.

3,600.2	3,600.1	3,600.4	3,599.9	3,599.2	3,598.6
3,599.7	3,598.9	3,601.2	3,600.8	3,598.7	3,600.4
3,599.4	3,599.6	3,600.4	3,598.2	3,600.1	3,600.2
3,600.5	3,599.6	3,600.4	3,599.2	3,601.6	3,601.4

Figure 32–46

Use the data in Figure 32–46 to answer questions 13 through 16.

13. Determine the mean of the data in Figure 32–46.

14. Determine the median of the data in Figure 32–46.

15. Determine the quartiles of the data in Figure 32–46.

16. Determine the mode of the data in Figure 32–46.

Use the data in Figure 32–47 to answer questions 17 and 18.

17. The data in the table in Figure 32–47 are the mean and range of the number of revolutions per minute of samples of four disk drives.

Sample	1	2	3	4	5	6	7	8
$\bar{x}$ (rpm)	3599.9	3599.9	3599.6	3600.2	3600.1	3599.7	3599.3	3600.7
Range	1.2	2.5	2.5	2.4	2.8	1.7	1.3	2.1

Sample	9	10	11	12	13	14	15	16
$\bar{x}$ (rpm)	3600.1	3599.2	3600.0	3600.7	3598.9	3598.8	3598.6	3599.7
Range	2.3	3.1	0.8	2.4	1.2	1.1	0.8	2.1

Figure 32–47

a. Determine the central line, upper control limit, and lower control limit for the mean ($\bar{x}$).

b. Construct an *X*-Bar control chart using this information.

c. Is the process out of control? If it is, at what sample does the *X*-Bar control chart signal lack of control?

18. a. Determine the central line, upper control limit, and lower control limit for the range (*R*).

b. Construct an *R* control chart using this information.

c. Is the process out of control? If it is, at what sample does the *XR* control chart signal lack of control?

Fundamentals of Trigonometry

UNIT 33 ⁙ Introduction to Trigonometric Functions

OBJECTIVES

After studying this unit you should be able to

- identify the sides of a right triangle with reference to any angle.
- state the ratios of the six trigonometric functions in relation to given triangles.
- find functions of angles given in decimal degrees and degrees, minutes, and seconds.
- find angles in decimal degrees and degrees, minutes, and seconds of given functions.

Trigonometry is the branch of mathematics that is used to compute unknown angles and sides of triangles. The word *trigonometry* is derived from the Greek words for triangle and measurement. Trigonometry is based on the principles of geometry. Many problems require the use of geometry and trigonometry.

As with geometry, much in our lives depends on trigonometry. The methods of trigonometry are used in constructing buildings, roads, and bridges. Trigonometry is used in the design of automobiles, trains, airplanes, and ships. The machines that produce the manufactured products we need could not be made without the use of trigonometry.

A knowledge of trigonometry and the ability to apply the knowledge in actual occupational uses is required in many skilled trades. Machinists, surveyors, drafters, electricians, and electronics technicians are a few of the many occupations in which trigonometry is a requirement.

Practical problems are often solved by using a combination of elements of algebra, geometry, and trigonometry. It is essential that you develop the ability to analyze a problem in order to determine the mathematical principles that are involved in the solution. The solution is done in orderly steps based on mathematical facts.

When solving a problem, it is important that you understand the trigonometric operations involved rather than mechanically plugging in values. To solve more complex problems, such as those found later in this section, an understanding of the principles involved is essential.

33–1 Ratio of Right Triangle Sides

In a right triangle, the ratio of two sides of the triangle determines the sizes of the angles, and the angles determine the ratio of the sides. For example, in Figure 33–1, the size of angle A is determined by the ratio of side a to side b. When side $a = 1$ inch and side $b = 2$ inches, the ratio of a to b is 1:2. If side a is increased to 2 inches and side b remains 2 inches, as shown in Figure 33–2, the ratio of a to b is 1:1. Figure 33–3 compares the two ratios and shows the change in angle A.

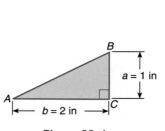

Figure 33–1

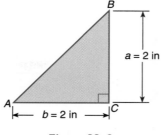

Figure 33–2

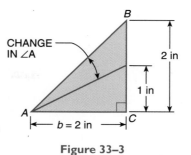

Figure 33–3

33–2   Identifying Right Triangle Sides by Name

The sides of a right triangle are named the opposite side, adjacent side, and hypotenuse. The hypotenuse is the longest side of a right triangle and is always the side opposite the right angle. The positions of the opposite and adjacent sides depend on the reference angle. The opposite side is opposite the reference angle. The adjacent side is next to the reference angle.

For example, in Figure 33–4, the hypotenuse (c) is opposite the right angle. In reference to angle A, b is the adjacent side and a is the opposite side. In Figure 33–5, the hypotenuse (c) is opposite the right angle. In reference to angle B, side b is the opposite side and side a is the adjacent side. It is important to be able to identify the opposite and adjacent sides of right triangles in reference to any angle regardless of the positions of the triangles.

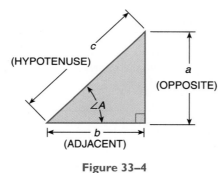

Figure 33–4

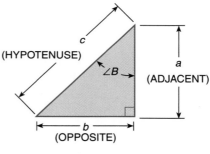

Figure 33–5

EXERCISE 33–2

With reference to ∠1, name the sides of each of these right triangles as opposite, adjacent, or hypotenuse.

1. Name sides r, x, and y.

3. Name sides a, b, and c.

5. Name sides a, b, and c.

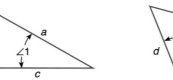

7. Name sides d, m, and p.

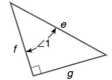

2. Name sides r, x, and y.

4. Name sides a, b, and c.

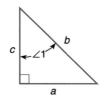

6. Name sides d, m, and p.

8. Name sides e, f, and g.

9. Name sides *h*, *k*, and *l*.

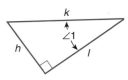

10. Name sides *h*, *k*, and *l*.

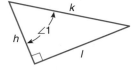

11. Name sides *m*, *p*, and *s*.

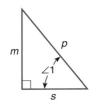

12. Name sides *m*, *p*, and *s*.

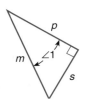

13. Name sides *m*, *r*, and *t*.

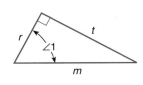

14. Name sides *m*, *r*, and *t*.

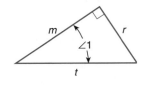

15. Name sides *f*, *g*, and *h*.

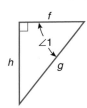

16. Name sides *f*, *g*, and *h*.

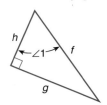

33–3   Trigonometric Functions: Ratio Method

There are two methods of defining trigonometric functions: the unity or unit circle method and the ratio method. Only the ratio method is presented in this book. Since a triangle has three sides and a ratio is the comparison of any two sides, there are six different ratios. The names of the ratios are the sine, cosine, tangent, cotangent, secant, and cosecant.

The six trigonometric functions are defined in Figure 33–6. They are defined in relation to the triangle in Figure 33–7, where the reference angle is *A*, the adjacent side is *b*, the opposite side is *a*, and the hypotenuse is *c*.

Function	Symbol	Definition of Function
sine of angle *A*	sin *A*	$\sin A = \dfrac{\text{opposite side}}{\text{hypotenuse}} = \dfrac{a}{c}$
cosine of angle *A*	cos *A*	$\cos A = \dfrac{\text{adjacent side}}{\text{hypotenuse}} = \dfrac{b}{c}$
tangent of angle *A*	tan *A*	$\tan A = \dfrac{\text{opposite side}}{\text{adjacent side}} = \dfrac{a}{b}$
cotangent of angle *A*	cot *A*	$\cot A = \dfrac{\text{adjacent side}}{\text{opposite side}} = \dfrac{b}{a}$
secant of angle *A*	sec *A*	$\sec A = \dfrac{\text{hypotenuse}}{\text{adjacent side}} = \dfrac{c}{b}$
cosecant of angle *A*	csc *A*	$\csc A = \dfrac{\text{hypotenuse}}{\text{opposite side}} = \dfrac{c}{a}$

Figure 33–6

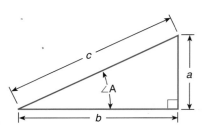

Figure 33–7

SO H
CA H
TO A

To properly use trigonometric functions, you must understand that the function of an angle depends on the ratio of the sides and *not the size* of the triangle. The functions of similar triangles are the same regardless of the sizes of the triangles, since the sides of similar triangles are proportional. For example, in Figure 33–8, the functions of angle A are the same for the three triangles. The equality of the tangent function is shown. Each of the other five functions has equal values for the three similar triangles.

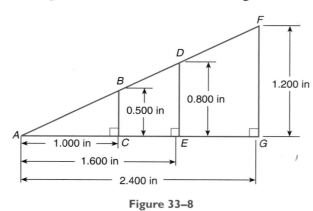

In $\triangle ACB$, $\tan A = \dfrac{0.500}{1.000} = 0.500$

In $\triangle AED$, $\tan A = \dfrac{0.800}{1.600} = 0.500$

In $\triangle AGF$, $\tan A = \dfrac{1.200}{2.400} = 0.500$

Figure 33–8

EXERCISE 33–3

#1-9

The sides of each triangle are labeled with different letters. State the ratio of each of the six functions in relation to $\angle 1$ for each of the triangles. For example, for the triangle in exercise 1,

$$sin\ \angle 1 = \frac{y}{r},\ cos\ \angle 1 = \frac{x}{r},\ tan\ \angle 1 = \frac{y}{x},\ cot\ \angle 1 = \frac{x}{y},\ sec\ \angle 1 = \frac{r}{x},\ and\ csc\ \angle 1 = \frac{r}{y}.$$

1.
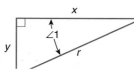

3.
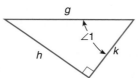

5.
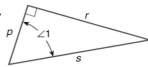

2.
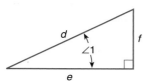

4.
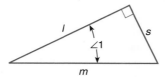

6.
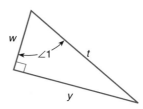

These exercises show three groups of triangles. Each group consists of four triangles. Within each group name the triangles, a, b, c, or d, in which angles A are equal.

7. a.
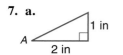

b.
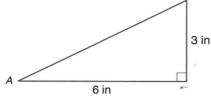

c.

d.

8. a.

b.
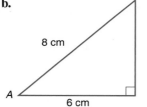

c.

d.

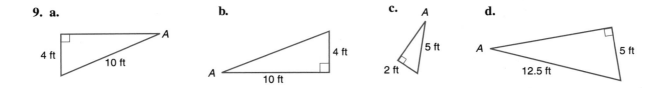

9. a.   **b.**   **c.**   **d.**

33–4   Customary and Metric Units of Angular Measure

As discussed in Unit 20, angular measure in the customary system is generally expressed in degrees and minutes or in degrees, minutes, and seconds for very precise measurements. In the metric system, the decimal degree is the preferred unit of measure. Unless otherwise specified, degrees and minutes or degrees, minutes, and seconds are to be used when solving customary system units of measure problems. Decimal degrees are to be used when solving metric system units of measure problems.

Calculator applications with degrees, minutes, and seconds and decimal degrees were presented on pages 433 and 434. You may want to review the material on these pages.

33–5   Determining Functions of Given Angles and Determining Angles of Given Functions

Calculator Applications

Determining functions of given angles or angles of given functions is readily accomplished using a calculator. As previously stated, calculator procedures vary among different makes of calculators. Also, different models of the same make calculator vary in some procedures. Generally, where procedures differ, there are basically two different procedures. Where relevant, both procedures are shown. However, because of the many makes and models of calculators, some procedures on your calculator may differ from the procedures shown. If so, it is essential that you refer to your user's guide or owner's manual.

The trigonometric keys, $\boxed{\text{sin}}$, $\boxed{\text{cos}}$, and $\boxed{\text{tan}}$, calculate the sine, cosine, and tangent of the angle in the display. An angle can be measured in degrees, radians, or gradients. *When calculating functions of angles measured in degrees, be certain that the calculator is in the degree mode.* Some calculators are in the degree mode when the abbreviation DEG or D appears in the display when the calculator is turned on. Some calcualtors require you to press a $\boxed{\text{MODE}}$ key to see if the calculator is in degree or radian mode.

Depending on the make and model of the calculator, for most calculators there are essentially three ways of putting the calculator in the degree mode.

1. Pressing the $\boxed{\text{DRG}}$ key changes the mode to radian, RAD, gradient, GRAD, or degree, DEG. Press $\boxed{\text{DRG}}$ until DEG is displayed.

    NOTE: On some calculators DRG is the second or third function.

2. Press $\boxed{\text{MODE}}$, press $\boxed{4}$, or press $\boxed{\text{MODE}}$ twice and press $\boxed{1}$. The calculator is in the degree mode and DEG or D is displayed.

3. Press $\boxed{\text{MODE}}$ $\boxed{\blacktriangledown}$ $\boxed{\blacktriangledown}$ $\boxed{\blacktriangleright}$ $\boxed{\text{ENTER}}$.

Procedure for Determining the Sine, Cosine, and Tangent Functions

The procedure for determining functions of angles varies with different calculators. Although there are exceptions, basically there are two different procedures.

1. The value of the angle is entered first, and then the appropriate function key, $\boxed{\text{sin}}$, $\boxed{\text{cos}}$, or $\boxed{\text{tan}}$ is pressed.

2. The appropriate function key, $\boxed{\text{sin}}$, $\boxed{\text{cos}}$, or $\boxed{\text{tan}}$ is pressed first, and then the value of the angle is entered.

    The following examples show both procedures.

NOTE: As previously stated, on certain calculators $\boxed{=}$ is used instead of $\boxed{\text{EXE}}$. Other calculators use an $\boxed{\text{ENTER}}$ key.

EXAMPLES

Round each answer to 5 decimal places.

1. Determine the sine of 43°.

43 $\boxed{\sin}$ → 0.68199836, 0.68200 *Ans*

or $\boxed{\sin}$ 43 $\boxed{\text{EXE}}$ 0.6819983601, 0.68200 *Ans*

2. Determine the cosine of 6.034°.

6.034 $\boxed{\cos}$ → 0.994459692, 0.99446 *Ans*

or $\boxed{\cos}$ 6.034 $\boxed{\text{EXE}}$ 0.9944596918, 0.99446 *Ans*

3. Determine the tangent of 51.9162°.

51.9162 $\boxed{\tan}$ → 1.276090171, 1.27609 *Ans*

or $\boxed{\tan}$ 51.9162 $\boxed{\text{EXE}}$ 1.276090171, 1.27609 *Ans*

4. Determine the sine of 61°49′.

61.49 $\boxed{\text{2nd}}$ $\boxed{\blacktriangleright \underline{\text{DD}}}$ $\boxed{\sin}$ → 0.881440874, 0.88144 *Ans*

or $\boxed{\sin}$ 61 $\boxed{° ′ ″}$ 49 $\boxed{° ′ ″}$ $\boxed{\text{EXE}}$ 0.8814408742, 0.88144 *Ans*

5. Determine the tangent of 32°7′23″.

32.0723 $\boxed{\text{2nd}}$ $\boxed{\blacktriangleright \underline{\text{DD}}}$ $\boxed{\tan}$ → 0.627859699, 0.62786 *Ans*

or $\boxed{\tan}$ 32 $\boxed{° ′ ″}$ 7 $\boxed{° ′ ″}$ 23 $\boxed{° ′ ″}$ $\boxed{\text{EXE}}$ 0.6278596985, 0.62786 *Ans*

2nd ⊕ when you have minutes or seconds

EXERCISE 33–5A

Determine the sine, cosine, or tangent functions of the following angles. Round the answers to 5 decimal places.

11 –29 odd

1. sin 36°	**11.** tan 73.86°	**21.** cos 19°42′
2. cos 53°	**12.** sin 50.05°	**22.** sin 71°59′
3. tan 47°	**13.** cos 16.77°	**23.** tan 42°36′
4. cos 18°	**14.** sin 0.86°	**24.** sin 20°28′
5. sin 79°	**15.** tan 61.07°	**25.** cos 6°16′
6. cos 4°	**16.** cos 60.605°	**26.** tan 37°26′12″
7. tan 65.18°	**17.** cos 77.144°	**27.** tan 9°4′50″
8. sin 27.06°	**18.** tan 10°18′	**28.** cos 86°30′38″
9. tan 12.92°	**19.** sin 26°29′	**29.** sin 53°46′19″
10. cos 5.98°	**20.** sin 6°53′	**30.** tan 70°51′44″

Procedure for Determining the Cosecant, Secant, and Cotangent Functions

The cosecant, secant, and cotangent functions are reciprocal functions.

The cosecant is the reciprocal of the sine.

$$\csc \angle A = \frac{1}{\sin \angle A}$$

The secant is the reciprocal of the cosine.

$$\sec \angle A = \frac{1}{\cos \angle A}$$

The cotangent is the reciprocal of the tangent.

$$\cot \angle A = \frac{1}{\tan \angle A}$$

Cosecants, secants, and cotangents are computed with the reciprocal key, $\boxed{1/x}$ or $\boxed{x^{-1}}$. On certain calculators, the reciprocal key is a second function. As with the sine, cosine, and tangent functions, basically there are two different procedures for determining reciprocal functions.

1. Enter the value of the angle, press the appropriate function key, $\boxed{\sin}$, $\boxed{\cos}$, or $\boxed{\tan}$; press $\boxed{1/x}$.
2. Press the appropriate function key; $\boxed{\sin}$, $\boxed{\cos}$, $\boxed{\tan}$; enter the value of the angle; press $\boxed{\text{EXE}}$; press $\boxed{x^{-1}}$; press $\boxed{\text{EXE}}$.
   └——— or press $\boxed{=}$, press $\boxed{1/x}$

The following examples show both procedures.

EXAMPLES •——————————————————————————————

Round each answer to 5 decimal places.

1. Determine the cosecant of 57.16°.

   57.16 $\boxed{\sin}$ $\boxed{1/x}$ → 1.190209506, 1.19021 *Ans*

   or $\boxed{\sin}$ 57.16 $\boxed{\text{EXE}}$ $\boxed{x^{-1}}$ $\boxed{\text{EXE}}$ 1.190209506, 1.19021 *Ans*

   └——— or $\boxed{=}$ $\boxed{1/x}$ → 1.190209506

   [handwritten: sin (57.61)⁻¹]

2. Determine the secant of 13.795°.

   13.795 $\boxed{\cos}$ $\boxed{1/x}$ → 1.029701649, 1.02970 *Ans*

   or $\boxed{\cos}$ 13.795 $\boxed{\text{EXE}}$ $\boxed{x^{-1}}$ $\boxed{\text{EXE}}$ 1.029701649, 1.02970 *Ans*

   └——— or $\boxed{=}$ $\boxed{1/x}$ → 1.029701649

3. Determine the cotangent of 78.63°.

   78.63 $\boxed{\tan}$ $\boxed{1/x}$ → 0.20109054, 0.20109 *Ans*

   or $\boxed{\tan}$ 78.63 $\boxed{\text{EXE}}$ $\boxed{x^{-1}}$ $\boxed{\text{EXE}}$ 0.2010905402, 0.20109 *Ans*

   └——— or $\boxed{=}$ $\boxed{1/x}$ → 0.2010905402

   [handwritten: tan (.20109)⁻¹]

4. Determine the cosecant of 24°51′.

   24.51 $\boxed{\text{2nd}}$ $\boxed{\blacktriangleright \text{DD}}$ $\boxed{\sin}$ $\boxed{1/x}$ → 2.379569353, 2.37957 *Ans*

   or $\boxed{\sin}$ 24 $\boxed{°\,′\,″}$ 51 $\boxed{°\,′\,″}$ $\boxed{\text{EXE}}$ $\boxed{x^{-1}}$ $\boxed{\text{EXE}}$ 2.379569353, 2.37957 *Ans*

   └——— or $\boxed{=}$ $\boxed{1/x}$ → 2.379569353

5. Determine the secant of 43°36′25″.

   43.3625 $\boxed{\text{2nd}}$ $\boxed{\blacktriangleright \text{DD}}$ $\boxed{\cos}$ $\boxed{1/x}$ → 1.381047089, 1.38105 *Ans*

   or $\boxed{\cos}$ 43 $\boxed{°\,′\,″}$ 36 $\boxed{°\,′\,″}$ 25 $\boxed{°\,′\,″}$ $\boxed{\text{EXE}}$ $\boxed{x^{-1}}$ $\boxed{\text{EXE}}$ 1.381047089, 1.38105 *Ans*

   └——— or $\boxed{=}$ $\boxed{1/x}$ → 1.381047089

EXERCISE 33–5B

Determine the cosecant, secant, or cotangent functions of the following angles. Round the answers to 5 decimal places.

[handwritten: # 9-21 odd]

1. csc 27°	8. sec 77.08°	15. cot 17°19′
2. sec 56°	9. sec 86.92°	16. sec 80°51′
3. cot 33°	10. csc 44.077°	17. sec 4°39′
4. sec 48°	11. csc 6.904°	18. csc 76°0′15″
5. csc 6.16°	12. cot 31.081°	19. cot 2°58′59″
6. cot 18.85°	13. sec 20°16′	20. sec 55°16′32″
7. cot 36.97°	14. csc 46°27′	21. csc 19°34′18″

Angles of Given Functions

Determining the angle of a given function is the inverse of determining the function of a given angle. When a certain function value is known, the angle can be found easily.

The term *arc* is often used as a prefix to any of the names of the trigonometric functions, such as arcsine, arctangent, etc. Such expressions are called inverse trigonometric functions and they mean angles. For example, sin 30°15′ ≈ 0.503774, then 30°15′ ≈ arcsin 0.503774 or 30°15′ is the angle whose sine ≈ 0.503774.

Arcsin is often written as $\sin^{-1}$, arccos is written as $\cos^{-1}$, and arctan is written as $\tan^{-1}$. So, if sin A = 0.625, then A = arcsin 0.625 or A = $\sin^{-1}$0.625.

On many calculators, the $\sin^{-1}$, $\cos^{-1}$, and $\tan^{-1}$ functions are located above the sin, cos, and tan keys, respectively. These inverse trignometric function keys are accessed by first pressing the 2nd key

Procedure for Determining Angles of Given Functions

The procedure for determining angles of given functions varies somewhat with the make and model of calculator. With most calculators, the inverse functions are shown as second functions [$\sin^{-1}$], [$\cos^{-1}$], and [$\tan^{-1}$] of function keys sin , cos , and tan .

With some calculators, the function value is entered before the function key is pressed. With other calculators, the function key is pressed before the function value is entered.

The following examples show the procedure for determining angles of given functions. All examples show the procedures where [$\sin^{-1}$], [$\cos^{-1}$], and [$\tan^{-1}$] are the second functions.

Remember, for certain calculators it is necessary to substitute = in place of EXE .

EXAMPLES •——————————————————————————

1. Find the angle whose tangent is 1.902. Round the answer to 2 decimal places.

1.902 2nd (or SHIFT) tan → 62.2662961, 62.27° *Ans*

or SHIFT tan 1.902 EXE 62.2662961, 62.27° *Ans* $\tan^{-1}(1.902)$

2. Find arcsin 0.21256. Round the answer to 2 decimal places.

Remember, arcsin 0.21256 means to find the angle that has a sine of 0.21256.

.21256 2nd (or SHIFT) sin → 12.27241712, 12.27° *Ans*

or SHIFT sin .21256 EXE 12.27241712, 12.27° *Ans* $\sin^{-1}(.21256)$

3. Determine $\cos^{-1}$0.732976. Give the answer in degrees, minutes, and seconds.

Here $\cos^{-1}$0.732976 means to find the angle that has a cosine of 0.732976.

.732976 2nd cos 3rd ▶DD → 42°51′48″7, 42°51′49″ *Ans*

or .732976 SHIFT cos SHIFT °′″ → 42°51′48.72″, 42°51′49″ *Ans*

or SHIFT cos .732976 EXE SHIFT °′″ → 42°51′48.72″, 42°51′49″ *Ans*

———•

Angles for the reciprocal functions—cosecant, secant, and cotangent—are calculated using the reciprocal key, $1/x$ or x^{-1} .

EXAMPLES •——————————————————————————

1. Find the angle whose secant is 1.2263. Round the answer to 2 decimal places.

1.2263 $1/x$ 2nd (or SHIFT) cos → 35.36701576, 35.37° *Ans*

or SHIFT cos 1.2263 x^{-1} (or $1/x$) EXE 35.36701576, 35.37° *Ans* 1.2263^{-1} then $\cos^{-1}(Ans)$

2. Find the angle whose cotangent is 0.4166. Give the answer in degrees and minutes.

.4166 $1/x$ 2nd tan 3rd ▶DD 67°23′00″2, 67°23′ *Ans*

or .4166 $1/x$ SHIFT tan SHIFT °′″ → 67°23′0.2″, 67°23′ *Ans*

or SHIFT tan 0.4166 x^{-1} (or $1/x$) EXE SHIFT °′″ → 67°23′0.2″, 67°23′ *Ans*

———•

EXERCISE 33–5C

Determine the value of angle A in decimal degrees for each of the given functions. Round the answers to the nearest hundredth of a degree.

1. $\sin A = 0.83692$
2. $\cos A = 0.23695$
3. $\tan A = 0.59334$
4. $\cos A = 0.97370$
5. $\tan A = 3.96324$
6. $\sin A = 0.77376$
7. $\sin A = 0.02539$
8. $\tan A = 1.56334$
9. $\tan A = 0.11884$
10. $\cos A = 0.20893$
11. $\cos A = 0.87736$
12. $\sin A = 0.10532$

13. $\cos A = 0.38591$
14. $\tan A = 0.67871$
15. $\sin A = 0.63634$
16. $\cos A = 0.00636$
17. $\sec A = 1.58732$
18. $\csc A = 2.08363$
19. $\cot A = 0.89538$
20. $\cot A = 6.06790$
21. $\csc A = 5.93632$
22. $\sec A = 1.02353$
23. $\csc A = 4.93317$
24. $\cot A = 2.89895$

25. $A = \arcsin 0.2953$
26. $A = \arccos 0.9163$
27. $A = \arctan 4.2156$
28. $A = \arctan 0.8176$
29. $A = \arccos 0.7156$
30. $A = \arcsin 0.0250$
31. $A = \cos^{-1} 0.7442$
32. $A = \tan^{-1} 1.500$
33. $A = \sin^{-1} 0.8240$
34. $A = \cos^{-1} 0.500$
35. $A = \tan^{-1} 0.4752$
36. $A = \sin^{-1} 0.1234$

EXERCISE 33–5D

Determine the value of angle A in degrees and minutes for each of the given functions. Round the answers to the nearest minute.

1. $\cos A = 0.23076$
2. $\tan A = 0.56731$
3. $\sin A = 0.92125$
4. $\tan A = 4.09652$
5. $\cos A = 0.03976$
6. $\sin A = 0.09741$
7. $\sin A = 0.70572$
8. $\tan A = 0.95300$
9. $\cos A = 0.00495$
10. $\cos A = 0.89994$
11. $\sin A = 0.30536$
12. $\tan A = 7.60385$

13. $\cos A = 0.69304$
14. $\tan A = 3.03030$
15. $\sin A = 0.70705$
16. $\cos A = 0.99063$
17. $\csc A = 1.38630$
18. $\sec A = 5.05377$
19. $\cot A = 0.27982$
20. $\csc A = 2.02103$
21. $\sec A = 9.90778$
22. $\cot A = 8.03012$
23. $\csc A = 3.03539$
24. $\sec A = 2.71177$

25. $A = \arcsin 0.9876$
26. $A = \arccos 0.0055$
27. $A = \arctan 7.258$
28. $A = \arctan 0.0025$
29. $A = \arccos 0.3572$
30. $A = \arcsin 0.8660$
31. $A = \cos^{-1} 0.4821$
32. $A = \tan^{-1} 1.000$
33. $A = \sin^{-1} 0.9513$
34. $A = \cos^{-1} 0.3579$
35. $A = \tan^{-1} 0.2684$
36. $A = \sin^{-1} 0.2486$

⠿ UNIT EXERCISE AND PROBLEM REVIEW

NAMING RIGHT TRIANGLE SIDES

With reference to ∠1, name the sides of each of the following right triangles as opposite, adjacent, or hypotenuse.

1. Name sides a, b, and c.

2. Name sides r, x, and y.

3. Name sides e, f, and g.

4. Name sides m, n, and p.

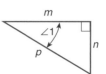

5. Name sides *h*, *j*, and *k*.

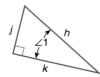

6. Name sides *s*, *t*, and *w*.

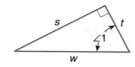

7. Name sides *a*, *d*, and *p*.

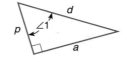

8. Name sides *b*, *f*, and *m*.

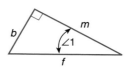

RATIOS OF RIGHT TRIANGLE SIDES

The sides of each of these triangles are labeled with different letters. State the ratio of each of the six functions in relation to ∠1 for each of the triangles.

9.

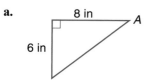

10.

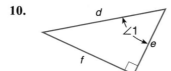

11.

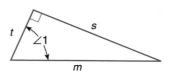

12. Of the five triangles shown, name the triangles, a, b, c, d, or e, in which angles *A* are equal.

a.

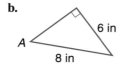

8 in
6 in

b.

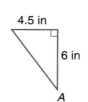

6 in
A
8 in

c.

4.5 in
6 in
A

d.

A
4 in
3 in

e.

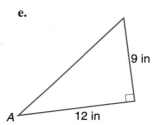

9 in
A   12 in

DETERMINING SINE, COSINE, AND TANGENT FUNCTIONS OF ANGLES

Determine the sine, cosine, or tangent functions of the following angles. Round the answers to 5 decimal places.

13. sin 54°

14. tan 23°

15. cos 37.98°

16. tan 76.05°

17. cos 40.495°

18. sin 7.861°

19. tan 78°19′

20. cos 5°53′

21. sin 83°17′

22. cos 20°32′10″

23. tan 89°12′59″

24. sin 53°13′41″

DETERMINING COSECANT, SECANT, AND COTANGENT FUNCTIONS OF ANGLES

Determine the cosecant, secant, or cotangent functions of the following angles. Round the answers to 5 decimal places.

25. csc 65°

26. sec 17°

27. cot 32.17°

28. csc 59.53°

29. sec 29.809°

30. cot 66.778°

31. csc 79°35′

32. sec 4°49′

33. cot 86°40′

34. sec 20°31′57″

35. csc 66°51′22″

36. cot 44°8′35″

DETERMINING ANGLES OF FUNCTIONS IN DECIMAL DEGREES

Determine the value of angle A in decimal degrees for each of the given functions. Round the answers to the nearest hundredth of a degree.

37. sin *A* = 0.79363

38. cos *A* = 0.31236

39. tan *A* = 0.89336

40. cos *A* = 0.90577

41. tan *A* = 4.97831

42. sin *A* = 0.08763

43. $\cos A = 0.98994$ **47.** $\cot A = 4.86731$ **51.** $A = \text{arccot } 4.2156$

44. $\tan A = 0.55314$ **48.** $\sec A = 1.93505$ **52.** $A = \tan^{-1} 0.4742$

45. $\sec A = 3.65306$ **49.** $A = \arcsin 0.7931$ **53.** $A = \sin^{-1} 0.5738$

46. $\csc A = 2.03953$ **50.** $A = \arctan 1.759$ **54.** $A = \sec^{-1} 5.8240$

DETERMINING ANGLES OF FUNCTIONS IN DEGREES AND MINUTES

Determine the value of angle A in degrees and minutes for each of the given functions. Round the answers to the nearest minute.

55. $\cos A = 0.37604$ **61.** $\tan A = 7.06072$ **67.** $A = \arccos 0.7931$

56. $\tan A = 0.63985$ **62.** $\cos A = 0.90519$ **68.** $A = \arcsin 0.9752$

57. $\sin A = 0.83036$ **63.** $\sec A = 4.06578$ **69.** $A = \text{arccsc } 4.3286$

58. $\tan A = 7.54982$ **64.** $\csc A = 2.93930$ **70.** $A = \cos^{-1} 0.6457$

59. $\cos A = 0.09561$ **65.** $\cot A = 0.17976$ **71.** $A = \tan^{-1} 2.6472$

60. $\sin A = 0.72344$ **66.** $\sec A = 3.86731$ **72.** $A = \cot^{-1} 3.7137$

UNIT 34 ⠿ Trigonometric Functions with Right Triangles

OBJECTIVES

After studying this unit you should be able to

- determine the variations of functions as angles change.
- compute cofunctions of complementary angles.
- compute unknown angles of right triangles when two sides are known.
- compute unknown sides of a right triangle when an angle and a side are known.

It is important to understand trigonometric function variation. This unit is designed to show the changes in triangle sides as the sizes of reference angles vary. An understanding of angle and function relationships reduces the possibility of error when solving applications that require a number of sequential mathematical steps.

34–1 Variation of Functions

As the size of an angle increases, the sine, tangent, and secant functions increase, but the cofunctions (cosine, cotangent, cosecant) decrease. As the reference angles approach 0° or 90°, the function variation can be shown. These examples illustrate variations of an increasing function and a decreasing function for a reference angle that is increasing in size. Use Figure 34–1 for these examples.

EXAMPLES •

1. Variation of an increasing function; the sine function. Refer to Figure 34–1.

   OP_1 and OP_2 are radii of the arc of a circle. $OP_1 = OP_2 = r$

   The sine of an angle $= \dfrac{\text{opposite side}}{\text{hypotenuse}}$

   $$\sin \angle 1 = \frac{A_1P_1}{r}$$

   $$\sin \angle 2 = \frac{A_2P_2}{r}$$

   A_2P_2 is greater than A_1P_1; therefore, $\sin \angle 2$ is greater than $\sin \angle 1$.

   Conclusion: As the angle is increased from $\angle 1$ to $\angle 2$, the sine of the angle increases. Observe that if $\angle 1$ decreases to 0°, side $A_1P_1 = 0$.

   $$\sin 0° = \frac{0}{r} = 0$$

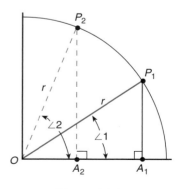

Figure 34–1

If $\angle 2$ increases to $90°$, since $A_2 P_2 = r$

$$\sin 90° = \frac{r}{r} = 1$$

2. Variation of a decreasing function; the cosine function. Refer to Figure 34–1.

$$\text{The cosine of an angle} = \frac{\text{adjacent side}}{\text{hypotenuse}}$$

$$\cos \angle 1 = \frac{OA_1}{r}$$

$$\cos \angle 2 = \frac{OA_2}{r}$$

OA_2 is less than OA_1; therefore, $\cos \angle 2$ is less than $\cos \angle 1$.

Conclusion: As the angle is increased from $\angle 1$ to $\angle 2$, the cosine of the angle decreases. Observe that if $\angle 1$ decreases to $0°$, side $OA_1 = r$.

$$\cos 0° = \frac{r}{r} = 1$$

If $\angle 2$ increases to $90°$, side $OA_2 = 0$.

$$\cos 90° = \frac{0}{r} = 0$$

It is helpful to sketch figures similar to Figure 34–1 for all functions in order to further develop an understanding of the relationship of angles and their functions. Particular attention should be given to functions of angles close to $0°$ and $90°$.

Variations of trigonometric functions are shown in Figure 34–2 for an angle increasing from $0°$ to $90°$.

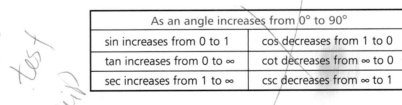

As an angle increases from $0°$ to $90°$	
sin increases from 0 to 1	cos decreases from 1 to 0
tan increases from 0 to ∞	cot decreases from ∞ to 0
sec increases from 1 to ∞	csc decreases from ∞ to 1

Figure 34–2

The contangent of $0°$, cosecant of $0°$, tangent of $90°$, and secant of $90°$ involve division by zero; since division by zero is not possible, these values are undefined. Although they are undefined, the values are often written as ∞.

NOTE: Depending on the make of the calculator, these undefined values are displayed as E, Error, or ERR:DOMAIN.

The symbol ∞ means infinity. Infinity is the quality of existing beyond, or being greater than, any countable value. It cannot be used for computations at this level of mathematics.

Rather than to attempt to treat ∞ as a value, think of the tangent and secant functions not at an angle of $90°$, but at angles very close to $90°$. Observe that as an angle approaches $90°$, the tangent and secant functions get very large. Think of the cotangent and cosecant functions not at an angle of $0°$, but as very small angles close to $0°$. Observe that as an angle approaches $0°$, the cotangent and cosecant functions get very large.

EXERCISE 34-1

Refer to Figure 34–3 to answer these questions. It may be helpful to sketch figures.

1. When ∠1 is almost 90°:
   a. how does side *y* compare to side *r*?
   b. how does side *x* compare to side *r*?
   c. how does side *x* compare to side *y*?

2. When ∠1 is 90°:
   a. what is the value of side *x*?
   b. how does side *y* compare to side *r*?

3. When ∠1 is slightly greater than 0°:
   a. how does side *y* compare to side *r*?
   b. how does side *x* compare to side *r*?
   c. how does side *x* compare to side *y*?

4. When ∠1 is 0°:
   a. what is the value of side *y*?
   b. how does side *x* compare to side *r*?

5. When side *x* = side *y*:
   a. what is the value of ∠1?
   b. what is the value of the tangent function?
   c. what is the value of the cotangent function?

6. When side *x* = side *r*:
   a. what is the value of the cosine function?
   b. what is the value of the secant function?
   c. what is the value of the sine function?
   d. what is the value of the tangent function?

7. When side *y* = side *r*:
   a. what is the value of the sine function?
   b. what is the value of the cosecant function?
   c. what is the value of the cosine function?
   d. what is the value of the cotangent function?

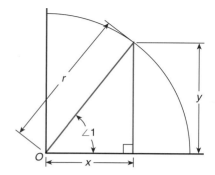

Figure 34–3

For each exercise, functions of two angles are given. Which of the functions of the two angles is greater? Do not use a calculator.

8. sin 38°; sin 43°
9. tan 17°; tan 18°
10. cos 53°; cos 61°
11. cot 40°; cot 36°
12. sec 5°; sec 8°
13. csc 22°; csc 25°
14. tan 19°20′; tan 16°40′
15. cos 81°19′; cos 81°20′
16. sin 0.42°; sin 0.37°
17. csc 39.30°; csc 39.25°
18. cot 27°23′; cot 87°0′
19. sec 55°; sec 54°50′

34–2   Functions of Complementary Angles

Two angles are complementary when their sum is 90°. For example, 20° is the complement of 70°, and 70° is the complement of 20°. In Figure 34–4, ∠A is the complement of ∠B, and ∠B is the complement of $\angle A$. The six functions of the angle and the cofunctions of the complementary angle are shown in the table in Figure 34–5.

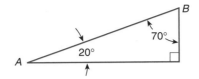

Figure 34–4

sin 20° = cos 70° ≈ 0.34202	cos 20° = sin 70° ≈ 0.93969
tan 20° = cot 70° ≈ 0.36397	cot 20° = tan 70° ≈ 2.7475
sec 20° = csc 70° ≈ 1.0642	csc 20° = sec 70° ≈ 2.9238

Figure 34–5

> **A function of an angle is equal to the cofunction of the complement of the angle.**

The complement of an angle equals 90° minus the angle. The relationships of the six functions of angles and the cofunctions of the complementary angles are shown in the table in Figure 34–6.

sin A = cos (90° − A)	cos A = sin (90° − A)
tan A = cot (90° − A)	cot A = tan (90° − A)
sec A = csc (90° − A)	csc A = sec (90° − A)

Figure 34–6

EXAMPLES

For each function of an angle, write the cofunction of the complement of the angle.

1. sin 30° = cos (90° − 30°) = cos 60° *Ans*

2. cot 10° = tan (90° − 10°) = tan 80° *Ans*

3. tan 72.53° = cot (90° − 72.53°) = cot 17.47° *Ans*

4. sec 40°20′ = csc (90° − 40°20′) = csc (89°60′ − 40°20′) = csc 49°40′ *Ans*

5. cos 90° = sin (90° − 90°) = sin 0° *Ans*

EXERCISE 34–2

For each function of an angle, write the cofunction of the complement of the angle.

1. tan 17°	**8.** sin 0°	**15.** cos 5.89°
2. sin 49°	**9.** tan 66.5°	**16.** cot 0°
3. cos 26°	**10.** cos 12.2°	**17.** tan 90°
4. sec 83°	**11.** cot 7°10′	**18.** sec 44°29′
5. cot 35°	**12.** sec 31°26′	**19.** cos 0.01°
6. csc 51°	**13.** csc 0°38′	**20.** sin 89°59′
7. cos 90°	**14.** sin 7.97°	

For each exercise, functions and cofunctions of two angles are given. Which of the functions or cofunctions of the two angles is greater? Do not use a calculator.

21. cos 55°; sin 20°   27. sin 12°; cos 80°

22. cos 55°; sin 40°   28. sin 12°; cos 75°

23. tan 21°; cot 56°   29. cot 89°10′; tan 1°20′

24. tan 30°; cot 45°   30. cot 89°10′; tan 0°40′

25. sec 43°; csc 56°   31. sec 0.2°; csc 89.9°

26. sec 43°; csc 58°   32. sec 0.2°; csc 89.0°

34–3   Determining an Unknown Angle When Two Sides of a Right Triangle Are Known

In order to solve for an unknown angle of a right triangle where neither acute angle is known, at least two sides must be known. The following procedure outlines the steps required in computing an angle:

Procedure for Determining an Unknown Angle When Two Sides Are Given

- In relation to the desired angle, identify two given sides as adjacent, opposite, or hypotenuse.
- Determine the functions that are ratios of the sides identified in relation to the desired angle.

NOTE: Two of the six trigonometric functions are ratios of the two known sides. Either of the two functions can be used. Both produce the same value for the unknown.

- Choose one of the two functions, substitute the given sides in the ratio.
- Determine the angle that corresponds to the quotient of the ratio.

EXAMPLES •————————————————————————

1. Determine ∠A of the right triangle in Figure 34–7 to the nearest minute.

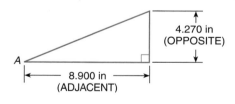

4.270 in
(OPPOSITE)

A

8.900 in
(ADJACENT)

Figure 34–7

Solution.   In relation to ∠A, the 8.900-inch side is the adjacent side, and the 4.270-inch side is the opposite side.

Determine the two functions whose ratios consist of the adjacent and opposite sides. Then, tan ∠A = opposite side/adjacent side, and cot ∠A = adjacent side/opposite side. Either the tangent or cotangent function can be used.

Choosing the tangent function: tan ∠A = 4.270 in/8.900 in.

Determine the angle whose tangent function is the quotient of 4.270/8.900.

Calculator Application

or

$\angle A$ = 4.27 $\boxed{\div}$ 8.9 $\boxed{\blacktriangleright\underline{\text{DD}}}$ $\boxed{\text{2nd}}$ $\boxed{\tan}$ $\boxed{\text{3rd}}$ $\boxed{\blacktriangleright\underline{\text{DD}}}$ 25°37′49.95″, 25°38′ *Ans*

$\angle A$ = $\boxed{\text{SHIFT}}$ $\boxed{\tan}$ $\boxed{(}$ 4.27 $\boxed{\div}$ 8.9 $\boxed{)}$ $\boxed{\text{EXE}}$ $\boxed{\text{SHIFT}}$ $\boxed{°\,′\,″}$ → 25°37′49.95″, 25°38′ *Ans*

2. Determine $\angle B$ of the right triangle in Figure 34–8 to the nearest hundredth degree.

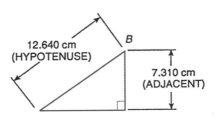

Figure 34–8

Solution.   In relation to $\angle B$, the 12.640-centimeter side is the hypotenuse, and the 7.310-centimeter side is the adjacent side.

Determine the two functions whose ratios consist of the adjacent side and the hypotenuse. Then, $\cos \angle B$ = adjacent side/hypotenuse; and $\sec \angle B$ = hypotenuse/adjacent side. Either the cosine or secant function can be used.

Choosing the cosine function: $\cos \angle B$ = 7.310 cm/12.640 cm.

Determine the angle whose cosine function is the quotient of 7.310/12.640.

Calculator Application

or

$\angle B$ = 7.31 $\boxed{\div}$ 12.64 $\boxed{\blacktriangleright\underline{\text{DD}}}$ $\boxed{\text{2nd}}$ $\boxed{\cos}$ → 54.66733748, 54.67° *Ans*

$\angle B$ = $\boxed{\text{SHIFT}}$ $\boxed{\cos}$ $\boxed{(}$ $\boxed{7.31}$ $\boxed{\div}$ 12.64 $\boxed{)}$ $\boxed{\text{EXE}}$ 54.66733748, 54.67° *Ans*

3. Determine $\angle 1$ and $\angle 2$ of the triangle in Figure 34–9 to the nearest minute.

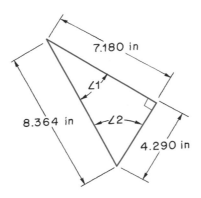

Figure 34–9

Solution.   Compute either $\angle 1$ or $\angle 2$. Choose any two of the three given sides for a ratio. In relation to $\angle 1$, the 4.290-inch side is the opposite side, and the 8.364-inch side is the hypotenuse.

Determine the two functions whose ratios consist of the opposite side and the hypotenuse. Then, $\sin \angle 1$ = opposite side/hypotenuse, and $\csc \angle 1$ = hypotenuse/opposite. Either the sine or cosecant can be used.

Choosing the sine function: $\sin \angle 1$ = 4.290 in/8.364 in.

Determine the angle whose sine function is the quotient of 4.290/8.364.

Calculator Application

∠1 = 4.29 ÷ 8.364 ▶DD 2nd sin 3rd ▶DD 30°51′28″8, 30°51′ *Ans*

or

∠1 = SHIFT sin (4.29 ÷ 8.364) EXE SHIFT °′″ → 30°51′28.89″, 30°51′ *Ans*

Since ∠1 + ∠2 = 90°, ∠2 ≈ 90° − 30°51′, ∠2 ≈ 59°9′ *Ans*

EXERCISE 34–3

Determine the unknown angles of these right triangles. Compute angles to the nearest minute in triangles with customary unit sides. Compute angles to the nearest hundredth degree in triangles with metric unit sides.

1. Determine ∠A.

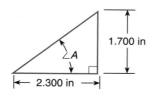

2. Determine ∠B.

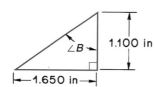

3. Determine ∠1.

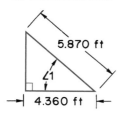

4. Determine ∠x.

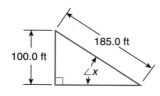

5. Determine ∠1.

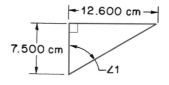

6. Determine ∠A.

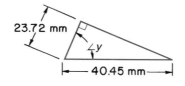

7. Determine ∠y.

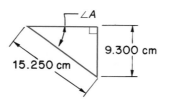

8. Determine ∠B.

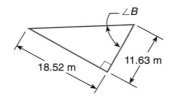

9. Determine ∠1 and ∠2.

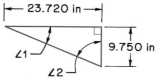

10. Determine ∠A and ∠B.

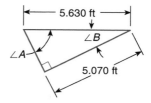

11. Determine ∠x and ∠y.

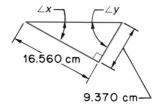

12. Determine ∠C. and ∠D.

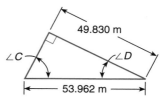

34–4   Determining an Unknown Side When an Acute Angle and One Side of a Right Triangle Are Known

In order to solve for an unknown side of a right triangle, at least an acute angle and one side must be known. The following procedure outlines the steps required to compute the unknown side.

Procedure for Determining an Unknown Side When an Angle and a Side Are Given

- In relation to the given angle, identify the given side and the unknown side as adjacent, opposite, or hypotenuse.
- Determine the trigonometric functions that are ratios of the sides identified in relation to the given angle.

NOTE: Two of the six functions will be found as ratios of the two identified sides. Either of the two functions can be used. Both produce the same values for the unknown. If the unknown side is made the numerator of the ratio, the problem is solved by multiplication. If the unknown side is made the denominator of the ratio, the problem is solved by division.

- Choose one of the two functions and substitute the given side and given angle.
- Solve as a proportion for the unknown side.

EXAMPLES •———————————————————————————————

1. Determine side x of the right triangle in Figure 34–10. Round the answer to 2 decimal places.

    Solution.   In relation to the 61°50′ angle, the 5.410-inch side is the adjacent side, and side x is the opposite side.

    Determine the two functions whose ratios consist of the adjacent and opposite sides. Tan 61°50′ = opposite side/adjacent side, and cot 61°50′ = adjacent side/opposite side. Either the tangent or cotangent function can be used.

    Choosing the tangent function: tan 61°50′ = x/5.410 in.

    Solve as a proportion.

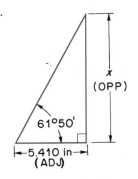

$$\frac{\tan 61°50'}{1} = \frac{x}{5.410 \text{ in}}$$
$$x = \tan 61°50' \ (5.410 \text{ in})$$
$$x \approx 1.86760 \ (5.410 \text{ in})$$
$$x \approx 10.10 \text{ in } Ans$$

Figure 34–10

Calculator Application

$x \approx$ 61.5 [2nd] [▶DD] [tan] [×] 5.41 [▶DD] 10.10371739, 10.10 in *Ans*

or

$x \approx$ [tan] 61 [°′″] 50 [°′″] [×] 5.41 [EXE] 10.10371739, 10.10 in *Ans*

2. Determine side r of the right triangle in Figure 34–11. Round the answer to 3 decimal places.

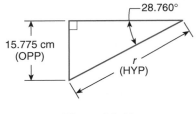

Figure 34–11

    Solution.   In relation to the 28.760° angle, the 15.775-centimeter side is the opposite side and side r is the hypotenuse.

Determine the two functions whose ratios consist of the opposite side and the hypotenuse. Sin 28.760° = opposite side/hypotenuse, and csc 28.760° = hypotenuse/opposite side. Either the sine or cosecant function can be used.

Choosing the sine function: sin 28.760° = 15.775 cm/r.

Solve as a proportion:

$$\frac{\sin 28.760°}{1} = \frac{15.775 \text{ cm}}{r}$$

$$r = \frac{15.775 \text{ cm}}{\sin 28.760°}$$

Calculator Application

$r \approx$ 15.775 $\boxed{\div}$ 28.76 $\boxed{\text{sin}}$ $\boxed{\blacktriangleright\text{DD}}$ 32.78659364, 32.787 cm *Ans*

or

$r \approx$ 15.775 $\boxed{\div}$ $\boxed{\text{sin}}$ 28.76 $\boxed{\text{EXE}}$ 32.78659364, 32.787 cm *Ans*

3. Determine side x, side y, and $\angle 1$ of the right triangle in Figure 34–12. Round the answer to 3 decimal places.

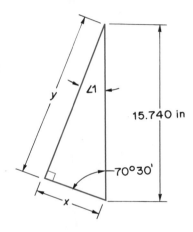

Figure 34–12

Solution.   Compute either side x or side y. Choosing side x, in relation to the 70°30' angle, side x is the adjacent side. The 15.740-inch side is the hypotenuse.

Determine the two functions whose ratios consist of the adjacent side and the hypotenuse. Either the cosine or secant function can be used.

Choosing the cosine function: cos 70°30' = x/15.740.

Solve as a proportion.

$$\frac{\cos 70°30'}{1} = \frac{x}{15.740 \text{ in}}$$

$$x = \cos 70°30' \ (15.740 \text{ in})$$

Calculator Application

$x \approx$ 70.30 $\boxed{\text{2nd}}$ $\boxed{\blacktriangleright\text{DD}}$ $\boxed{\text{cos}}$ $\boxed{\times}$ 15.74 $\boxed{\blacktriangleright\text{DD}}$ 5.254119964, 5.254 in *Ans*

or

$x \approx$ $\boxed{\text{cos}}$ 70 $\boxed{°'''}$ 30 $\boxed{°'''}$ $\boxed{\times}$ 15.74 $\boxed{\text{EXE}}$ 5.254119964, 5.254 in *Ans*

Solve for side y by using either a trigonometric function or the Pythagorean theorem. If the Pythagorean theorem is used to determine y, then $y^2 = (15.740)^2 - (5.254)^2$ and $y = \sqrt{(15.740)^2 - (5.254)^2}$. In cases like this, it is generally more convenient to solve for the side by using a trigonometric function. In relation to the 70°30′ angle, side y is the opposite side. The 15.740-inch side is the hypotenuse.

Determine the two functions whose ratios consist of the opposite side and the hypotenuse. Either the sine or cosecant function can be used.

Choosing the sine function: sin 70°30′ = y/15.740 in.

NOTE: Since side x has been calculated, it can be used with the 70°30′ angle to determine side y. However, it is better to use the given 15.740-inch hypotenuse rather than the calculated side x. Whenever possible, use given values rather than calculated values when solving problems. The calculated values could have been incorrectly computed or improperly rounded off resulting in an incorrect answer.

Solve as a proportion.

$$\frac{\sin 70°30′}{1} = \frac{y}{15.740 \text{ in}}$$

Calculator Application

$y \approx$ 70.30 [2nd] [▶DD] [sin] [×] 15.74 [▶DD] 14.83717707, 14.837 in *Ans*

or

$y \approx$ [sin] 70 [°′″] 30 [°′″] [×] 15.74 [EXE] 14.83717707, 14.837 in *Ans*

Determine ∠1: ∠1 = 90° − 70°30′ = 19°30′ *Ans*

EXERCISE 34–4

Determine the unknown sides in these right triangles. Compute sides to three decimal places.

1. Determine side b.

38°0'
6.800 in
b

2. Determine side c.

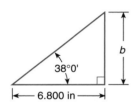

27°0'
8.950 in
c

3. Determine side x.

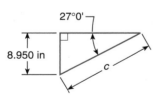

17°20'
15.750 ft
x

4. Determine side d.

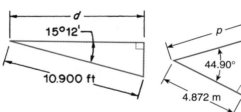

d
15°12'
10.900 ft

5. Determine side y.

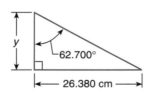

y
62.700°
26.380 cm

6. Determine side f.

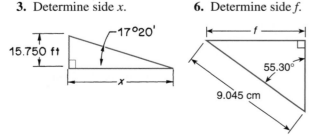

f
55.30°
9.045 cm

7. Determine side p.

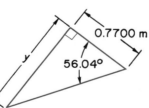

p
44.90°
4.872 m

8. Determine side y.

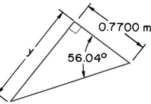

0.7700 m
y
56.04°

9. Determine sides d and e.

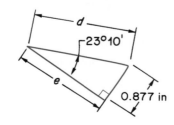

d
23°10'
e
0.877 in

10. Determine sides s and t.

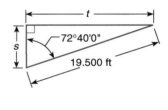

t
72°40'0"
s
19.500 ft

11. Determine sides x and y.

6.850 m
y
19.90°
x

12. Determine sides p and n.

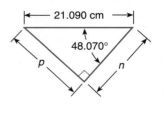

21.090 cm
48.070°
p
n

UNIT EXERCISE AND PROBLEM REVIEW

VARIATION OF FUNCTIONS

Refer to Figure 34–13 to answer these questions. It may be helpful to sketch figures.

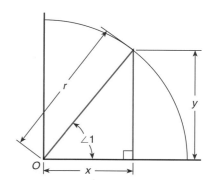

Figure 34–13

1. When $\angle 1 = 0°$:
   a. what is the value of side y?
   b. how does side x compare to side r?

2. When $\angle 1 = 90°$:
   a. what is the value of side x?
   b. how does side y compare to side r?

3. When side x = side y:
   a. what is the value of $\angle 1$?
   b. what is the value of the tangent function?
   c. what is the value of the cotangent function?

4. When side y = side r:
   a. what is the value of the cosine function?
   b. what is the value of the sine function?

5. When side x = side r:
   a. what is the value of the tangent function?
   b. what is the value of the secant function?

For each exercise, functions of two angles are given. Which of the functions of the two angles is greater? Do not use a calculator.

6. tan 28°; tan 31°
7. cot 43°; cot 48°
8. cos 86°; cos 37°

9. sin 19°30′; sin 12°18′
10. sec 47.85°; sec 40.36°
11. csc 81.66°; csc 79.12°

FUNCTIONS OF COMPLEMENTARY ANGLES

For each function of an angle, write the cofunction of the complement of the angle.

12. sin 39°
13. cos 81°
14. tan 77°

15. sec 25°
16. cot 11°19′
17. csc 0°

18. tan 90°
19. sin 51.88°
20. cos 89°59′

For each exercise, functions and cofunctions of two angles are given. Which of the functions or cofunctions of the two angles is greater? Do not use a calculator.

21. tan 40°; cot 60°
22. tan 42°; cot 47°

23. cos 44°; sin 41°
24. sin 39°; cos 63°

25. csc 54°; sec 45°
26. csc 68°; sec 50°

DETERMINING ANGLES AND SIDES OF RIGHT TRIANGLES

Determine the unknown angles or sides of these right triangles. Compute angles to the nearest minute in triangles with customary unit sides. Compute angles to the nearest hundredth degree in triangles with metric unit sides. Compute sides to three decimal places.

27. Determine $\angle A$.

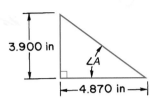

28. Determine $\angle B$.

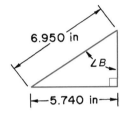

29. Determine side x.

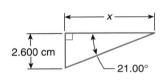

30. Determine side b.

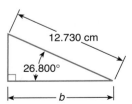

31. Determine $\angle B$.

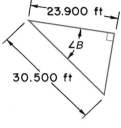

32. Determine side c.

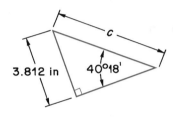

33. Determine $\angle 1$.

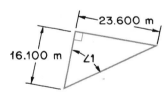

34. Determine side y.

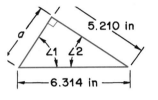

35. Determine $\angle B$, side x, and side y.

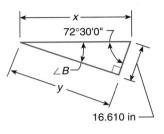

36. Determine $\angle 1$, $\angle 2$, and side a.

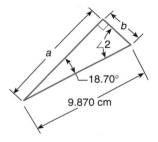

37. Determine side a, side b, and $\angle 2$.

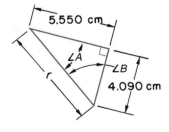

38. Determine $\angle A$, $\angle B$, and side r.

UNIT 35 ⁘ Practical Applications with Right Triangles

OBJECTIVES

After studying this unit you should be able to

- solve applied problems stated in word form.

- solve simple applied problems that require the projection of auxiliary lines and the application of geometric principles.

- solve complex applied problems that require forming two or more right triangles by the projection of auxiliary lines.

In the previous unit, you solved for unknown angles and sides of right triangles. Emphasis was placed on developing an understanding and the ability to apply proper procedures in solving for angles and sides. No attempt was made to show the many practical applications of right-angle trigonometry.

In this unit, practical applications from various occupational fields are presented. A great advantage of trigonometry is that it provides a method of computing angles and distances without actually having to physically measure them. Often problems are not given directly in the form of right triangles. They may be given in word form, which may require expressing word statements as pictures by sketching right triangles. Also, often when a problem is given in picture form, a right triangle does not appear. In these types of problems, right triangles must be developed within the given picture.

35–1 Solving Problems Stated in Word Form

When solving a problem stated in word form:

- Sketch a right triangle based on the given information.
- Label the known parts of the triangle with the given values. Label the angle or side to be found.
- Follow the procedure for determining an unknown angle or side of a right triangle.

EXAMPLES •————————————————————————————

1. A brace 15.0 feet long is to support a wall. One end of the brace is fastened to the floor at an angle of 40° with the floor. At what height from the floor will the brace be fastened to the wall? Round the answer to 1 decimal place.

   Solution. Sketch and label a right triangle as in Figure 35–1.

   Let h represent the unknown height.

   Compute h:

   $$\sin 4\overline{0}° = \frac{h}{15.0 \text{ ft}}$$

   $$h = \sin 4\overline{0}° \,(15.0 \text{ ft})$$

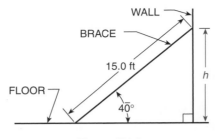

Figure 35–1

Calculator Application

$$h \approx 40 \boxed{\sin} \boxed{\times} 15 \boxed{\blacktriangleright \underline{\text{DD}}} \; 9.641814145, \; 9.6 \text{ ft } Ans$$

or

$$h \approx \boxed{\sin} 40 \boxed{\times} 15 \boxed{\text{EXE}} \; 9.641814145, \; 9.6 \text{ ft } Ans$$

2. The sides of a sheet metal piece in the shape of a right triangle measure 25.50 cm, 12.00 cm, and 28.18 cm. What are the measures of the two acute angles of the piece? Round the answers to the nearest hundredth degree.

Solution. Sketch and label a right triangle as in Figure 35–2.
Let ∠1 and ∠2 represent the unknown angles.

Compute ∠1. Choose any two sides for a ratio. Choose the 12.00-cm side and the 25.50-cm side.

$$\tan \angle 1 = \frac{12.00 \text{ cm}}{25.50 \text{ cm}}$$

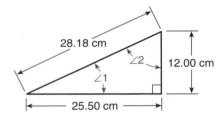

Figure 35–2

Calculator Application

$$\angle 1 \approx 12 \boxed{\div} 25.5 \boxed{\blacktriangleright \underline{\text{DD}}} \boxed{\text{2nd}} \boxed{\tan} \rightarrow 25.20112365, \; 25.20° \; Ans$$

or

$$\angle 1 \approx \boxed{\text{SHIFT}} \boxed{\tan} \boxed{(} 12 \boxed{\div} 25.5 \boxed{)} \; 25.20112365, \; 25.20° \; Ans$$

Compute ∠2.

$$\angle 2 \approx 90° - 25.20° \approx 64.80° \; Ans$$

Surveying and navigation computations are based on right-angle trigonometry. A surveyor uses a transit to measure angles between locations. By a combination of angle and distance measurements, distances that cannot be measured directly can be computed.

When a surveyor sights a point that is either above or below the horizontal, the measured angle is read on the transit vertical protractor. When a point above eye level is sighted, the transit telescope is pointed upward. The angle formed by the line of sight and the horizontal is called the *angle of elevation.* An angle of elevation is shown in Figure 35–3. When a point below eye level is sighted, the transit telescope is pointed downward. The angle formed by the line of sight and the horizontal is called the *angle of depression.* An angle of depression is shown in Figure 35–4.

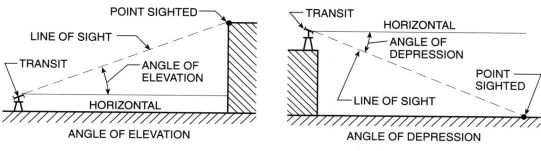

Figure 35–3   Figure 35–4

When angles of elevation or depression are measured, computed vertical distances must be corrected by adding or subtracting the height of the transit from the ground to the telescope. This type of problem is illustrated by the following example.

EXAMPLE

A surveyor is to determine the height of a tower. The transit is positioned at a horizontal distance of 35.00 meters from the foot of the tower. An angle of elevation of 58.00° is read in sighting the top of the tower. The height from the ground to the transit telescope is 1.70 meters. Determine the height of the tower. Round the answer to the nearest hundredth meter.

Solution

Sketch and label a right triangle as shown in Figure 35–5. Let x represent the side of the right triangle opposite the 58.00° angle of elevation. Let h represent the height of the tower: $h = x +$ transit height.

Compute x.

$$\tan 58.00° = \frac{x}{35.00 \text{ m}}$$

$$x = \tan 58.00° \,(35.00 \text{ m})$$

Calculator Application

$x \approx 58$ [tan] [×] 35 [▶DD] 56.01170852

or

$x \approx$ [tan] 58 [×] 35 [EXE] 56.01170852

Compute h.

$h = x +$ transit height

$h \approx 56.01 \text{ m} + 1.70 \text{ m} \approx 57.71 \text{ m } Ans$

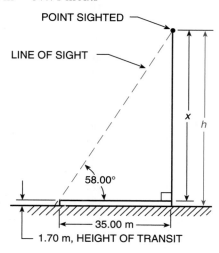

Figure 35–5

When a section of land is surveyed, horizontal distances that cannot be measured directly are computed by determining angles between horizontal points. From a horizontal line of sight, the surveyor turns the transit telescope to the left or to the right in sighting a point. The ángle between lines of sight is read on the transit horizontal protractor. This type of problem is illustrated by this example.

EXAMPLE •——————————————————————————————————

A surveyor wishes to measure the distance between two horizontal points. The two points, A and B, are separated by a river and cannot be directly measured. The surveyor does the following:

From point A, point B is sighted. Then the transit telescope is turned 90°0'. Along the 90°0' sighting, a distance of 80.00 feet is measured, and a stake is driven at the 80.0-foot distance (point C). From point C, the surveyor points the transit telescope back to point A. Then the transit telescope is turned to point B across the river. An angle of 70°20' is read on the transit. What is the distance between points A and B? Round the answer to the nearest tenth foot.

Solution

Sketch and label a right triangle as in Figure 35–6.

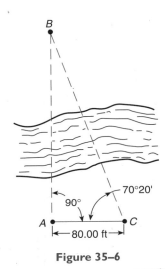

Compute distance AB.

$$\tan 70°20' = \frac{AB}{AC}$$

$$\tan 70°20' = \frac{AB}{80.00 \text{ ft}}$$

$$AB = \tan 70°20' \ (80.00 \text{ ft})$$

Calculator Application

$AB \approx 70.2$ [2nd] [▶DD] [tan] [×] 80 [▶DD]
223.8415835, 223.8 ft *Ans*

or

$AB \approx$ [tan] 70 [°'"] 20 [°'"] [×] 80 [EXE]
223.8415835, 223.8 ft *Ans*

Figure 35–6

——

EXERCISE 35–1

——

Sketch and label each of the following problems. Compute the unknown linear values to two decimal places unless otherwise noted; compute customary angular values to the nearest minute and metric angular values to the nearest hundredth of a degree.

1. The sides of a pattern, which is in the shape of a right triangle, measure 10.600 inches, 23.500 inches, and 25.780 inches. What are the measures of the two acute angles of the pattern?

2. A highway entrance ramp rises 37.25 feet in a horizontal distance of 180.00 feet. What is the measure of the angle that the ramp makes with the horizontal?

3. The roof of a building slopes at an angle of 32.0° from the horizontal. The horizontal distance (run) of the roof is 6.00 meters. Compute the rafter length of the roof.

4. The centers of two diagonal holes in a locating plate are 22.600 centimeters apart. The angles between the centerline of the 2 holes and a vertical line is 53.60°. Compute distances to 3 decimal places.

   a. What is the vertical distance between the centers of the 2 holes?

   b. What is the horizontal distance between the centers of the 2 holes?

5. A wall brace is to be positioned so that it makes an angle of 40° with the floor. It is to be fastened to the floor at a distance of 12.0 feet from the foot of the wall. Compute the length of the brace. Round the answer to 1 decimal place.

6. A surveyor wishes to determine the height of a building. The transit is positioned on level land at a distance of 160.00 feet from the foot of the building. An angle of 41°30′ is read in sighting the top of the building. The height from the ground to the transit telescope is 5.50 feet. What is the height of the building? Round the answer to 1 decimal place.

7. A 4 foot 0 inches × 8 foot 0 inches rectangular sheet of plywood is to be cut into 2 pieces of equal size by a diagonal cut made from the lower left corner to the upper right corner of the sheet. What is the measure of the angle that the diagonal cut makes with the 4 feet 0 inches side? Round the answer to the nearest degree.

8. From a center-punched starting point on a sheet of steel, a horizontal line segment 32.40 centimeters long is scribed, and the end point is center-punched. From this center-punched point, a vertical line segment 27.80 centimeters is scribed, and the end point is center-punched. A line segment is scribed between the starting point and the last center-punched point. What are the measures of the 2 acute angles of the scribed triangle?

9. A surveyor determines the horizontal distance between two locations. The transit is positioned at the first location, which is 18.00 meters higher in elevation than the second location. The second location is sighted, and a 34.000 angle of depression is read. The height from the ground to the transit telescope is 1.700 meters. Determine the horizontal distance between the two locations.

10. A hole is drilled through the entire thickness of a rectangular metal block at an angle of 46°20′ with the horizontal top surface. The block is 2.750 inches thick. What is the length of the drilled hole? Compute the answer to 3 decimal places.

11. A drain pipe is to be laid between 2 points. One point is 10.0 feet higher in elevation than the other point. The pipe is to slope at an angle of 12.0° with the horizontal. Compute the length of the drain pipe. Round the answer to 1 decimal place.

12. The rectangular bottom of a carton is designed so the length is $1\frac{1}{2}$ times as long as the width. What is the measure of the angle made by a diagonal across corners and the length of the bottom?

13. A surveyor determines the distance between two horizontal points on a piece of land. The two points, A and B, are separated by an obstruction and cannot be directly measured. The surveyor does the following:

From point A, point B is sighted. Then the transit telescope is turned 90.00°. Along the 90.00° sighting, a distance of 30.00 meters is measured, and a stake is driven at the 30 meter distance (point C). From point C, the surveyor points the transit telescope back to point A. The telescope is then turned to point B, and an angle of 66.00° is read on the horizontal protractor. Compute the distance between point A and point B.

14. An airplane flies in a direction 28° north of east at an average speed of 380.0 miles per hour. At the end of $2\frac{1}{2}$ hours of flying, how far due east is the airplane from its starting point? Round the answer to the nearest mile.

15. A cable used to brace a utility pole is attached 4.5 ft from the top of the 42 ft pole. If the cable makes a 47°44′ angle with the ground, how long, to the nearest inch, is the cable from the ground to where it is attached on the pole?

16 A guy wire for a power pole is anchored in the ground at a point 18.75 ft from the base of the pole. The wire will make an angle of 63° with the level ground.

   a. How high up the pole is the wire attached?

   b. What length of wire is needed if 21″ must be added to allow for the wire to be attached to both the pole and the ground?

35–2 Solving Problems Given in Picture Form that Require Auxiliary Lines

The following examples are practical applications of right-angle trigonometry, although they do not appear in the form of right triangles. To solve the problems, it is necessary to project auxiliary lines to produce right triangles. The unknown value and the given or computed values are parts of the produced right triangle.

The auxiliary lines may be projected between given points or from given points. Also, they may be projected parallel or perpendicular to centerlines, tangents, or other reference lines.

A knowledge of both geometric and trigonometric principles and the ability to apply the principles to specific situations are required to solve these problems. It is important to carefully study the procedures and use of auxiliary lines as they are applied in the solutions of these examples.

EXAMPLES •————————————————————————————————

1. Compute ∠1 in the pattern in Figure 35–7. Round the answer to the nearest tenth degree.

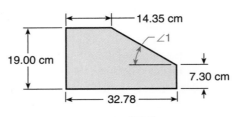

Figure 35–7

Solution. Angle 1 must be computed by forming a right triangle that contains ∠1. Refer to Figure 35–8.

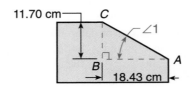

Figure 35–8

Project line segment *AB* parallel to the base of the pattern. Project vertical segment *CB*. Right △*ABC* is formed.

Compute sides *AB* and *CB*.

AB = 32.78 cm − 14.35 cm = 18.43 cm

CB = 19.00 cm − 7.30 cm = 11.70 cm

Solve for ∠1.

$$\tan \angle 1 = \frac{CB}{AB} = \frac{11.70 \text{ cm}}{18.43 \text{ cm}}$$

Calculator Application

∠1 ≈ 11.7 ÷ 18.43 ▶DD 2nd tan → 32.40878959, 32.4° *Ans*

or

∠1 ≈ SHIFT tan (11.7 ÷ 18.43) EXE 32.40878958, 32.4° *Ans*

2. Determine the included taper angle. $\angle T$, of the shaft in Figure 35–9. Round the answer to the nearest minute.

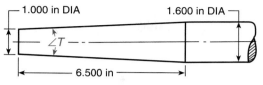

How much it opens up?

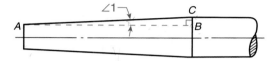

Figure 35–9

Solution. Refer to Figure 35–10. Project line segment AB parallel to the shaft centerline. Right $\triangle ABC$ is formed, in which $\angle 1$ is equal to the one-half the included taper angle, $\angle T$.

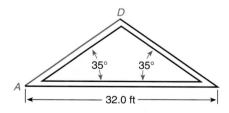

Figure 35–10

Determine sides AB and BC.

$AB = 6.500$ in

$BC = (1.600 \text{ in} - 1.000 \text{ in}) \div 2 = 0.300$ in

Solve for $\angle 1$.

$$\tan \angle 1 = \frac{BC}{AB} = \frac{0.300 \text{ in}}{6.500 \text{ in}}$$

Calculator Application

$\angle 1 = .3 \boxed{\div} 6.5 \boxed{\blacktriangleright \underline{DD}} \boxed{\text{2nd}} \boxed{\text{tan}} \boxed{\text{3rd}} \boxed{\blacktriangleright \underline{DD}} \; 2°38'33''16$

or

$\angle 1 = \boxed{\text{SHIFT}} \boxed{\text{tan}} \boxed{(} .3 \boxed{\div} 6.5 \boxed{)} \boxed{\text{EXE}} \boxed{\text{SHIFT}} \boxed{°'''} \; 2°38'33.16''$

Compute $\angle T$.

$\angle T \approx 2(\angle 1) \approx 2(2°38'33'') \approx 5°17'$ *Ans*

The solutions to many practical trigonometry problems are based on recognizing figures as isosceles triangles. A perpendicular projected from the vertex to the base of an isosceles triangle bisects the vertex angle and the base. This fact is illustrated by the following two problems.

EXAMPLES •

1. Determine the rafter length, AD, of the roof section in Figure 35–11. Round the answer to 1 decimal place.

Figure 35–11

Solution. Since both base angles equal 35°, the roof section is in the form of an isosceles triangle.

Refer to Figure 35–12. Project line segment DB perpendicular to base AC. Base AC is bisected by DB.

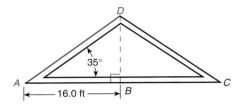

Figure 35–12

$AB = AC \div 2 = 32.0 \text{ ft} \div 2 = 16.0 \text{ ft}$

In right $\triangle ABD$, $AB = 16.0$ ft, $\angle C = 35°$.

Solve for side AD.

$$\cos 35° = \frac{AB}{AD} = \frac{16.0 \text{ ft}}{AD}$$

$AD = 16.0 \text{ ft} \div \cos 35°$

Calculator Application

$AD \approx 16 \boxed{\div} 35 \boxed{\cos} \boxed{\blacktriangleright \underline{\text{DD}}}$ 19.53239342, 19.5 ft *Ans*

or

$AD \approx 16 \boxed{\div} \boxed{\cos} 35 \boxed{\text{EXE}}$ 19.53239342, 19.5 ft *Ans*

2. In Figure 35–13, 5 holes are equally spaced on a 14.680-centimeter diameter circle.

Determine the straight-line distance between the centers of any two consecutive holes. Round the answer to 3 decimal places.

Solution. Refer to Figure 35–14.

Choosing any two consecutive holes, such as A and B, project radii from center O to hole centers A and B. Project a line segment from A to B. Since $OA = OB$, $\triangle AOB$ is isosceles.

Compute central $\angle AOB$.

$\angle AOB = 360° \div 5 = 72°$

Project line segment OC perpendicular to AB from point O. Line segment OC bisects $\angle AOB$ and side AB.

In right $\triangle AOC$,

$\angle AOC = 72° \div 2 = 36°$

$AO = 14.680 \text{ cm} \div 2 = 7.340 \text{ cm}$

Compute side AC.

$$\sin 36° = \frac{AC}{AO} = \frac{AC}{7.340 \text{ cm}}$$

$AC = \sin 36° \, (7.340 \text{ cm})$

14.680 cm DIA

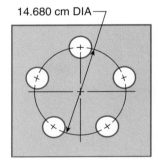

Figure 35–13

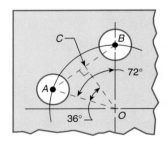

Figure 35–14

Calculator Application

$AC \approx 36 \boxed{\sin} \boxed{\times} 7.34 \boxed{\blacktriangleright \underline{\text{DD}}}$ 4.314343752

or

$AC \approx \boxed{\sin} 36 \boxed{\times} 7.34 \boxed{\text{EXE}}$ 4.314343752

Compute side AB.

$AB = 2(AC) \approx 2(4.3143 \text{ cm}) \approx 8.629 \text{ cm } Ans$

The solutions to the following two examples are based on the geometric theorem that a tangent is perpendicular to the radius of a circle at the tangent point. The solutions to applied trigonometry problems in many fields, such as construction and manufacturing, are based on this principle.

EXAMPLES •———

1. A park is shown in Figure 35–15. A fence is to be built from point T to point R. Line segment TR is tangent to the circle at point T. Compute the required length of fencing. Round the answer to the nearest meter.

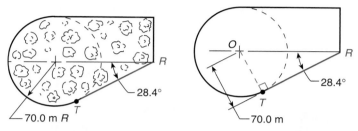

| Figure 35–15 | Figure 35–16 |

Solution.   Refer to Figure 35–16.

Connect a line segment from the center of the circle O to tangent point T. Line segment OT is a radius. (A tangent is perpendicular to a radius at its tangent point.)

$$OT = 70.0 \text{ m}$$

$$\angle OTR = 90°$$

In right $\triangle OTR$, $OT = 70.0$ m and $\angle R = 28.4°$.

Solve for side TR.

$$\tan 28.4° = \frac{OT}{TR}, \ \tan 28.4° = \frac{70.0 \text{ m}}{TR}$$

$$TR = 70.0 \text{ m} \div \tan 28.4°$$

Calculator Application

$TR \approx 70 \boxed{\div} 28.4 \boxed{\tan} \boxed{\blacktriangleright DD}$ 129.4622917, 129 m *Ans*

or

$TR \approx 70 \boxed{\div} \boxed{\tan} 28.4 \boxed{EXE}$ 129.4622917, 129 m *Ans*

2. The front view of the internal half of a dovetail slide is shown in Figure 35–17. Two pins or balls are used to check the dovetail slide for both location and accuracy. Compute check dimension x. Round the answer to 3 decimal places.

Solution.   Refer to Figure 35–18. Only the left side of the dovetail slide is shown. The left and right sides are congruent.

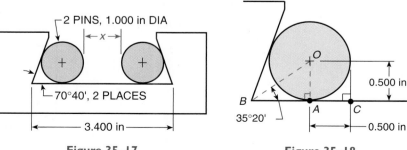

| Figure 35–17 | Figure 35–18 |

Project vertical line segment *OA* from pin center *O* to tangent point *A*. The angle at *A* equals 90°. (A tangent is perpendicular to a radius at its tangent point.)

Project line segment *OB* from pin center *O* to point *B*. Segment *OB* bisects the 70°40′ angle. (The angle formed by two tangents meeting at a point outside a circle is bisected by a segment drawn from the point to the center of the circle.)

In right △*ABO*,

∠*B* = 70°40′ ÷ 2 = 35°20′

OA = 1.000 in ÷ 2 = 0.500 in

Compute side *AB*.

$$\tan 35°20' = \frac{OA}{AB}, \tan 35°20' = \frac{0.500 \text{ in}}{AB}$$

AB = 0.500 in ÷ tan 35°20′

Calculator Application

$AB \approx .5 \boxed{\div} 35.20 \boxed{\text{2nd}} \boxed{\blacktriangleright\underline{\text{DD}}} \boxed{\tan} \boxed{\blacktriangleright\underline{\text{DD}}} 0.705304906$

or

$AB \approx .5 \boxed{\div} \boxed{\tan} 35 \boxed{\circ\,'\,''} 20 \boxed{\circ\,'\,''} \boxed{\text{EXE}} 0.705304906$

AC = pin radius = 0.500 in

BC = *AB* + *AC* ≈ 0.7053 in + 0.500 in ≈ 1.2053 in

Check dimension *x*.

x ≈ 3.400 in − 2(1.2053) ≈ 0.989 in *Ans*

EXERCISE 35–2

Compute the unknown values in each of these problems. Compute linear values to two decimal places, unless otherwise noted; compute customary angular values to the nearest minute and metric angular values to the nearest hundredth of a degree.

1. Compute ∠*A* in the template in Figure 35–19.

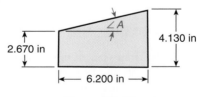

Figure 35–19

2. A plot of land is shown in Figure 35–20.

    a. Compute distance *AB*.

    b. Compute distance *BC*.

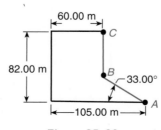

Figure 35–20

3. Compute the included taper angle, $\angle T$, of the shaft in Figure 35–21.

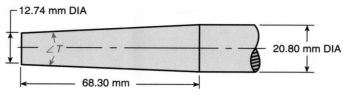

Figure 35–21

4. Compute diameter B of the tapered support column in Figure 35–22.
5. Compute the rafter length of the roof section in Figure 35–23.

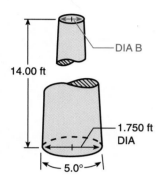

Figure 35–22

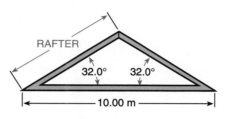

Figure 35–23

6. Compute the distance across the centers, dimension x, of two consecutive holes in the baseplate in Figure 35–24. Compute the answer to 3 decimal places.

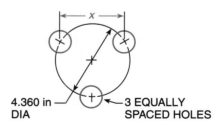

Figure 35–24

7. Compute the depth of cut y, in the machined block in Figure 35–25. Distance $AC = BC$.

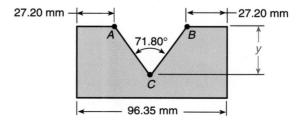

Figure 35–25

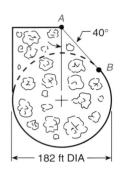

A

40°

B

←—— 182 ft DIA ——→

Figure 35–26

8. A sidewalk is constructed from point *A* to point *B* in the minipark in Figure 35–26. Point *B* is a tangent point. What is required length of sidewalk? Round the answer to the nearest foot.

9. Two sections of a brick wall, *AB* and *AC*, meet at point *A* as shown in Figure 35–27. A circular patio is to be constructed so that it is tangent to the wall at points *D* and *E*.

   a. What is the required diameter of the patio?

   b. At what distance must the center, *O*, of the patio be located from point *A*?

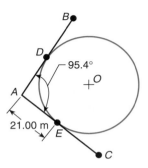

Figure 35–27

10. Compute check dimension *x* of the internal half of the dovetail slide in Figure 35–28.

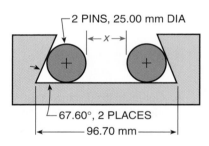

Figure 35–28

11. A gambrel roof is shown in Figure 35–29. Round the answers to the nearest degree.

   a. Compute $\angle 1$.

   b. Compute $\angle 2$.

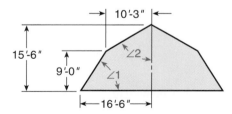

Figure 35–29

12. A plumber is to install a water pipe assembly as shown in Figure 35–30 on page 723. Round the answer to the nearest centimeter.

   a. Compute dimension *A*.

   b. Compute dimension *B*.

   c. Compute dimension *C*.

   d. Compute dimension *D*.

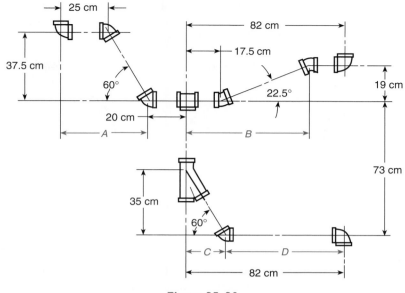

Figure 35–30

13. The top view of a platform is shown in Figure 35–31. Compute ∠A. Round the answer to the nearest tenth degree.

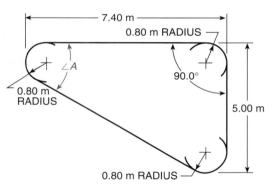

Figure 35–31

14. Determine gauge dimension *y* of the V-block in Figure 35–32. Compute the answer to 3 decimal places. Dimension *EF* = *GF*.

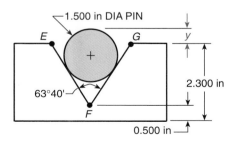

Figure 35–32

15. Eight circular columns located in a circular pattern, as shown in Figure 35–33, are proposed to support the roof of a structure. Compute the inside distance *x* between two adjacent columns.

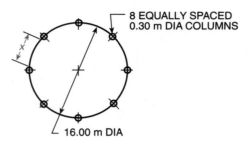

Figure 35–33

16. The side view of a sheet metal pipe and flange is shown in Figure 35–34. Round the answers to 1 decimal place.

   a. Compute dimension *A*.

   b. Compute dimension *B*.

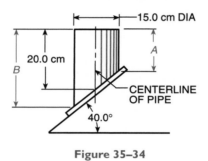

Figure 35–34

17. A portion of the framework for a building is shown in Figure 35–35. Round the answers to the nearest inch.

   a. Compute distance *AB*.

   b. Compute distance *BC*.

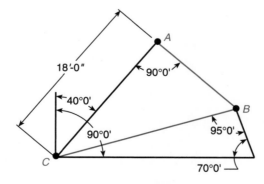

Figure 35–35

18. A plot of land was surveyed as shown in Figure 35–36. The distance between points *A* and *B* and the distance between points *C* and *B* could not be measured directly because of obstructions.

  a. Compute distance *AB*.

  b. Compute distance *BC*.

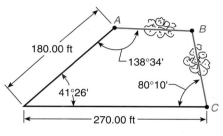

Figure 35–36

35–3 Solving Complex Problems that Require Auxiliary Lines

The following examples and problems are more challenging than those previously presented. These problems are also practical applications that require a combination of principles from geometry and trigonometry in their solutions. Two or more right triangles must be formed with auxiliary lines for the solution of each problem.

Typical examples from various occupational fields are discussed. It is essential that you study and, if necessary, restudy the procedures that are given in detail for solving the examples. There is a common tendency to begin writing computations before the complete solution to a problem has been thought through. This tendency must be avoided.

As problems become more complex, a greater proportion of time and effort is required to analyze the problems. After a problem has been completely analyzed, the written computations must be developed in clear and orderly steps.

Apply the following procedures when solving complex problems.

Method of Solving Complex Problems

Analyze the problem before writing computations.

• Relate given dimensions to the unknown, and determine whether other dimensions in addition to the given dimensions are required in the solution.

• Determine the auxiliary lines required to form right triangles that contain dimensions needed for the solution.

• Determine whether sufficient dimensions are known to obtain required values within the right triangles. If enough information is not available for solving a triangle, continue the analysis until enough information is obtained.

• Check each step in the analysis to verify that there are no gaps or false assumptions.

Write the computations.

EXAMPLES •——————————————————————————————————————

1. Determine the length of x of the template in Figure 35–37. Round the answer to 3 decimal places.

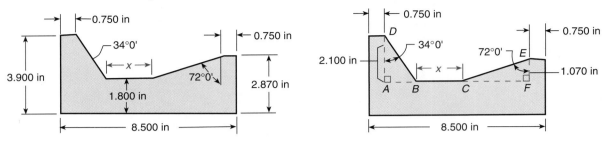

Figure 35–37   Figure 35–38

Analyze the problem.

Refer to Figure 35–38 where auxiliary line segments have been drawn to form right △ABD and right △CEF. If distances AB and CF can be determined, length x can be computed. Length $x = 8.500$ in $-$ (0.750 in $+ AB + CF + 0.750$ in)

Determine whether enough information is given to solve for AB. In right △ABD, ∠$D = 34°0'$ and $AD = 2.100$ inches. There is enough information to determine AB.

Determine whether enough information is given to solve for CF. In right △CEF, ∠$E = 72°0'$ and $EF = 1.070$ inches. There is enough information to determine CF.

Write the computations.

Solve for AB.

$$\tan \angle D = \frac{AB}{AD}$$

$$\tan 34°0' = \frac{AB}{2.100 \text{ in}}$$

$$AB = \tan 34°0' \, (2.100 \text{ in})$$

Calculator Application

$AB \approx 34$ $\boxed{\tan}$ $\boxed{\times}$ 2.1 $\boxed{\blacktriangleright\underline{DD}}$ 1.416467885

or

$AB \approx$ $\boxed{\tan}$ 34 $\boxed{\times}$ 2.1 $\boxed{EXE}$ 1.416467885

Solve for CF.

$$\tan \angle E = \frac{CF}{EF}$$

$$\tan 72°0' = \frac{CF}{1.070 \text{ in}}$$

$$CF = \tan 72°0' \, (1.070 \text{ in})$$

Calculator Application

$CF \approx 72$ $\boxed{\tan}$ $\boxed{\times}$ 1.07 $\boxed{\blacktriangleright\underline{DD}}$ 3.293121385

or

$CF \approx$ $\boxed{\tan}$ 72 $\boxed{\times}$ 1.07 $\boxed{EXE}$ 3.293121385

Solve for x.

$$x = 8.500 \text{ in} - (0.750 \text{ in} + AB + CF + 0.750 \text{ in})$$

$$x \approx 8.500 \text{ in} - (0.750 \text{ in} + 1.4165 \text{ in} + 3.2931 \text{ in} + 0.750 \text{ in})$$

Calculator Application

$x \approx 8.5 \boxed{-} \boxed{(} .75 \boxed{+} 1.4165 \boxed{+} 3.2931 \boxed{+} .75 \boxed{)} \boxed{=} 2.2904$

$x \approx 2.290$ in *Ans*

2. A plaza is to be constructed in a city redevelopment area. The shaded area in Figure 35–39 represents the proposed plaza. Determine $\angle y$. Round the answer to the nearest hundredth degree.

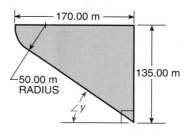

Figure 35–39

Analyze the problem.

Generally, when solving problems that involve an arc that is tangent to one or more lines, it is necessary to project the radius of the arc to the tangent point and to project a line from the vertex of the unknown angle to the center of the arc.

Refer to Figure 35–40. Project auxiliary line segments between points A and O and from point O to tangent point B. Right $\triangle ACO$ and right $\triangle ABO$ are formed.

If $\angle 1$ and $\angle 2$ can be determined, $\angle y$ can be computed.

Determine whether enough information is given to solve for $\angle 1$. In right $\triangle ACO$, $AC = 135.00$ m and $OC = 120.00$ m. There is enough information to determine $\angle 1$.

Determine whether enough information is given to solve for $\angle 2$. In right $\triangle ABO$, $OB = 50.00$ m. Side OA is also a side of right $\triangle ACO$ and can be computed by the Pythagorean theorem or after $\angle 1$ is computed. There is enough information given to determine $\angle 2$.

Write the computations.

Solve for $\angle 1$.

$$\tan \angle 1 = \frac{OC}{AC} = \frac{120.00 \text{ m}}{135.00 \text{ m}}$$

Calculator Application

$\angle 1 \approx 120 \boxed{\div} 135 \boxed{\blacktriangleright\text{DD}} \boxed{\text{2nd}} \boxed{\tan} \rightarrow 41.63353934$

or

$\angle 1 \approx \boxed{\text{SHIFT}} \boxed{\tan} \boxed{(} 120 \boxed{\div} 135 \boxed{)} \boxed{\text{EXE}} \ 41.63353934$

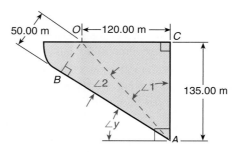

Figure 35–40

Solve for side *OA*.

$$OA^2 = OC^2 + AC^2$$
$$OA^2 = (120.00 \text{ m})^2 + (135.00 \text{ m})^2$$
$$OA = \sqrt{(120.00 \text{ m})^2 + (135.00 \text{ m})^2}$$

Calculator Application

$OA \approx$ (120 x^2 + 135 x^2) $\sqrt{x}$ → 180.6239187

or

$OA \approx$ $\sqrt{}$ (120 x^2 + 135 x^2) EXE 180.6239187

Solve for ∠2.

$$\sin \angle 2 = \frac{OB}{OA} \approx \frac{50.00 \text{ m}}{180.624 \text{ m}}$$

Calculator Application

∠2 ≈ 50 ÷ 180.624 ▶DD 2nd sin → 16.07039288

or

∠2 ≈ SHIFT sin (50 ÷ 180.624) EXE 16.07039288

Solve for ∠*y*.

$$\angle y = 90° - (\angle 1 + \angle 2)$$
$$\angle y \approx 90° - (41.63° + 16.07°)$$

Calculator Application

∠*y* ≈ 90 − (41.63 + 16.07) = 32.30, 32.30° *Ans*

3. The front view of a metal piece with a V-groove cut is shown in Figure 35–41. A 1.200-inch diameter pin is used to check the groove for depth and angular accuracy. Compute check dimension *y*. Round the answer to 3 decimal places.

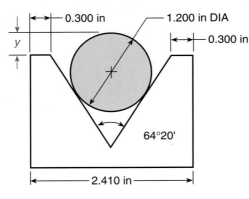

Figure 35–41

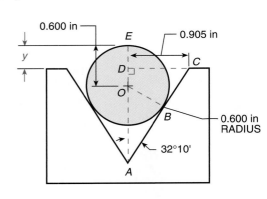

Figure 35–42

Analyze the problem.

Check dimension *y* is determined by the pin diameter, the points of tangency where the pin touches the groove, the angle of the V-groove, and the depth of the groove. Therefore, these dimensions and locations must be included in the calculations for *y*.

Refer to Figure 35–42. Project auxiliary line segments from point *A* through the center *O* of the pin, from point *O* to the tangent point *B*, and from point *C* horizontally intersect vertical segment *AE* at point *D*. Right △*ABO* and right △*ADC* are formed.

If *AO* and *AD* can be determined, check dimension *y* can be computed:

$$y = (AO + OE) - AD$$

Determine whether enough information is given to solve for *AO*. In right △*ABO*, *OB* = 0.600 inches and ∠*A* = 32°10′. (The angle formed by two tangents at a point outside a circle is bisected by a line drawn from the point to the center of the circle.) There is enough information to determine *AO*.

Determine whether enough information is given to solve for *AD*. In right △*ADC*, *DC* = 0.905 inches and ∠*A* = 32°10′. There is enough information to determine *AD*.

Write the computations.

Solve for *AO*.

$$\sin \angle A = \frac{OB}{AO}$$

$$\sin 32°10′ = \frac{0.600 \text{ in}}{AO}$$

$$AO = 0.600 \text{ in} \div \sin 32°10′$$

Calculator Application

$AO \approx .6 \boxed{\div} 32.10 \boxed{\text{2nd}} \boxed{\blacktriangleright\underline{DD}} \boxed{\sin} \boxed{\blacktriangleright\underline{DD}} \ 1.127006302$

or

$AO \approx .6 \boxed{\div} \boxed{\sin} 32 \boxed{°′″} 10 \boxed{°′″} \boxed{\text{EXE}} \ 1.127006302$

Solve for *AD*.

$$\tan \angle A = \frac{DC}{AD}$$

$$\tan 32°10′ = \frac{0.905 \text{ in}}{AD}$$

$$AD = 0.905 \text{ in} \div \tan 32°10′$$

Calculator Application

$AD \approx .905 \boxed{\div} 32.10 \boxed{\text{2nd}} \boxed{\blacktriangleright\underline{DD}} \boxed{\tan} \boxed{\blacktriangleright\underline{DD}} \ 1.438971506$

or

$AD \approx .905 \boxed{\div} \boxed{\tan} 32 \boxed{°′″} 10 \boxed{°′″} \boxed{\text{EXE}} \ 1.438971506$

Solve for check dimension *y*.

$y = (AO + OE) - AD \approx (1.1270 \text{ in} + 0.600 \text{ in}) - 1.4390 \text{ in} \approx 0.288 \text{ in } Ans$

4. Two proposed circular landscaped sections of a park are shown in Figure 35–43. A fence is to be constructed between points *A* and *B*. In laying out the plot, a drafter is required to compute ∠*x*. What is the value of ∠*x*? Round the answer to the nearest hundredth degree.

Analyze the problem:

Refer to Figure 35–44. Project an auxiliary line segment between centers *D* and *E* of the two circles. From center point *E*, project a horizontal auxiliary line segment that meets the vertical centerline at point *C*. Right △*CDE* is formed.

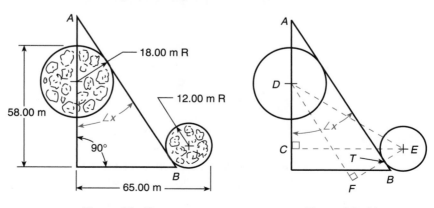

Figure 35–43   Figure 35–44

Project an auxiliary line segment from point D parallel to segment AB. Project an auxiliary line segment from point E through tangent point T. The two line segments meet at right angles at point F. Right $\triangle FDE$ is formed.

If $\angle CDE$ and $\angle FDE$ can be determined, $\angle x$ can be computed. $\angle CDF = \angle CDE - \angle FDE$. Angle $x = \angle CDF$. (If two parallel lines are intersected by a transversal, the corresponding angles are equal.)

Determine whether enough information is given to solve for $\angle CDE$. In right $\triangle CDE$, $CD = 58.00$ m $- (12.00$ m $+ 18.00$ m$) = 28.00$ m, and $CE = 65.00$ m $- 12.00$ m $= 53.00$ m. There is enough information to determine $\angle CDE$.

Determine whether enough information is given to solve for $\angle FDE$. In right $\triangle FDE$, $FE = FT + TE = 18.00$ m $+ 12.00$ m $= 30.00$ m. From right $\triangle CDE$, DE can be computed by the Pythagorean theorem. $DE^2 = CD^2 + CE^2$. There is enough information to determine $\angle FDE$.

Write the computations.

Solve for $\angle CDE$.

$$\tan \angle CDE = \frac{CE}{CD} = \frac{53.00 \text{ m}}{28.00 \text{ m}}$$

Calculator Application

$\angle CDE \approx 53 \boxed{\div} 28 \boxed{\blacktriangleright \underline{DD}} \boxed{\text{2nd}} \boxed{\tan} \rightarrow 62.15242174$

or

$\angle CDE \approx \boxed{\text{SHIFT}} \boxed{\tan} \boxed{(} 53 \boxed{\div} 28 \boxed{)} \boxed{\text{EXE}} \ 62.15242174$

$\angle CDE \approx 62.15°$

Solve for $\angle FDE$.

$$DE^2 = CD^2 + CE^2$$
$$DE = \sqrt{(28.00 \text{ m})^2 + (53.00 \text{ m})^2}$$

Calculator Application

$DE \approx \boxed{(} 28 \boxed{x^2} \boxed{+} 53 \boxed{x^2} \boxed{)} \boxed{\sqrt{x}} \rightarrow 59.94163828$

$DE \approx 59.952$ m

$$\sin \angle FDE = \frac{FE}{DE} \approx \frac{30.00 \text{ m}}{59.942 \text{ m}}$$

Calculator Application

$\angle FDE \approx 30 \boxed{\div} 59.942 \boxed{\blacktriangleright \underline{DD}} \boxed{\text{2nd}} \boxed{\sin} \rightarrow 30.03201318$

or

$\angle FDE \approx \boxed{\text{SHIFT}} \boxed{\sin} \boxed{(} 30 \boxed{\div} 59.942 \boxed{)} \boxed{\text{EXE}} \ 30.03201318$

$\angle FDE \approx 30.03°$

Solve for $\angle x$.

$\angle CDF = \angle CDE - \angle FDE \approx 62.15° - 30.03° \approx 32.12°$

Angle $x = \angle CDF \approx 32.12°$ *Ans*

EXERCISE 35–3

Compute the unknown values in each of these problems. Compute linear values to two decimal places unless otherwise noted, customary angular values to the nearest minute, and metric angular values to the nearest hundredth of a degree.

1. Compute length x of the pin in Figure 33–45. Compute the answer to 3 decimal places.

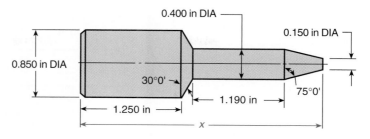

Figure 35–45

2. A plot of land is shown in Figure 35–46.
   a. Compute $\angle A$.
   b. Compute distance AB. Round the answer to the nearest tenth foot.

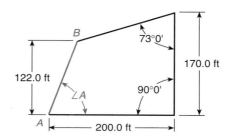

Figure 35–46

3. A roof truss is shown in Figure 35–47.
   a. Compute the length of cross member DE.
   b. Compute $\angle E$.

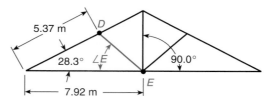

Figure 35–47

4. Compute $\angle x$ of the pattern in Figure 35–48. Round the answer to the nearest tenth degree.

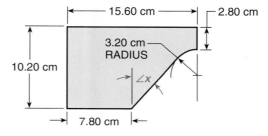

Figure 35–48

5. Compute ∠y of the gauge in Figure 35–49.

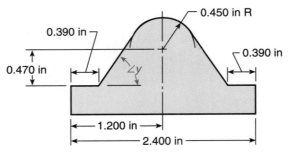

Figure 35–49

6. Compute check dimension y of the V-groove cut in Figure 35–50.

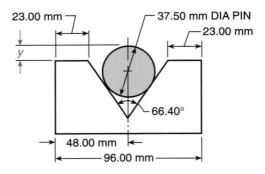

Figure 35–50

7. A sidewalk is constructed between points A and B of a mall as shown in Figure 35–51. Compute the length of the sidewalk. Round the answer to the nearest foot.

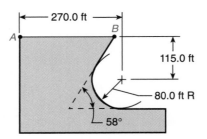

Figure 35–51

8. Compute the angular hole location ∠x, in the guide plate in Figure 35–52.

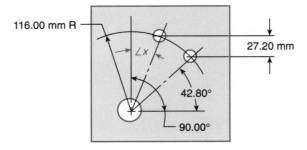

Figure 35–52

9. A traffic rotary is designed as shown in Figure 35–53. Compute distance *d*.

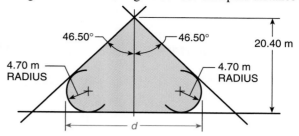

46.50°   46.50°

20.40 m

4.70 m
RADIUS

4.70 m
RADIUS

d

Figure 35–53

10. In surveying a piece of land, a surveyor made the measurements shown in Figure 35–54. Compute ∠1.

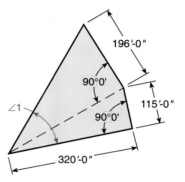

196′-0″

90°0′

∠1

115′-0″

90°0′

320′-0″

Figure 35–54

11. Compute dimension *x* of the template in Figure 35–55. Compute the answer to 3 decimal places.

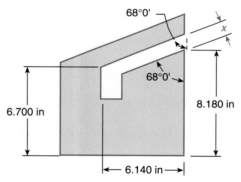

68°0′

x

68°0′

6.700 in

8.180 in

6.140 in

Figure 35–55

12. Determine ∠*y* made by the vertical and the tangent line of the two pins in Figure 35–56.

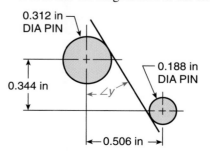

0.312 in
DIA PIN

0.188 in
DIA PIN

0.344 in

∠*y*

0.506 in

Figure 35–56

13. A section of a park in the shape shown in Figure 35–57 is designed as a botanical garden.

    a. Determine ∠1.

    b. Determine ∠2.

    c. Determine distance *AB*.

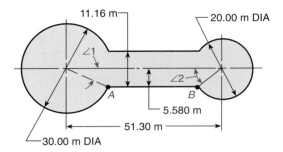

Figure 35–57

14. A curved driveway is shown in Figure 35–58. Compute ∠*y*.

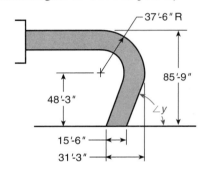

Figure 35–58

15. Compute hole location check dimension *x* in the piece in Figure 35–59.

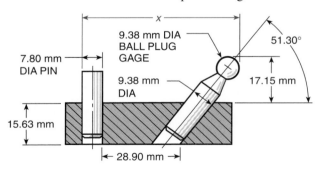

Figure 35–59

UNIT EXERCISE AND PROBLEM REVIEW

PROBLEMS STATED IN WORD FORM

Sketch and label each of these problems. Compute unknown linear values to two decimal places unless otherwise noted, customary angular values to the nearest minute, and metric angular values to the nearest hundredth of a degree.

1. A piece of sheet metal is sheared in the shape of a right triangle. The hypotenuse measures 17.48 inches, and one of the acute angles measures 37°30′. What is the length of the side opposite the 37°30′ angle?

2. A road rises uniformly along a horizontal distance of 450.00 meters. The rise at the end of the 450.00 meters is 95.00 meters.

   a. What is the measure of the angle that the road makes with the horizontal?

   b. What is the length of the road? Compute the answer to the nearest whole meter.

3. A surveyor wishes to determine the height of a tower. The transit is positioned at a distance of 200.00 feet from the foot of the tower. An angle of elevation of 46°50′ is read in sighting the top of the tower. The height from the ground to the transit telescope is 5′6″. What is the height of the tower?

4. A machinist drills 3 holes in a plate as follows: The first hole is drilled 20.00 millimeters from the left edge of the plate. The second hole is located and drilled 58.00 millimeters from the left edge of the plate, directly to the right of the first hole. The third hole is located and drilled 75.00 millimeters from the second hole, directly above the second hole. Compute the 2 acute angles of the triangle made by line segments connecting the centers of the 3 holes.

5. The horizontal distance between 2 points that are located at different elevations is to be determined. A surveyor positions the transit at a point that is 24.50 meters lower than the second point. The height from the ground to the transit telescope is 1.80 meters. The second point is sighted, and 42.60° angle of elevation is read. What is the horizontal distance between the 2 points?

6. A surveyor wishes to determine the distance between 2 horizontal points on a flat piece of land. The two points, A and B, are separated by an obstruction and cannot be directly measured. The surveyor does the following:

   From point A, point B is sighted. Then the transit telescope is turned 90°0′. Along the 90°0′ sighting, a distance of 150.00 feet is measured, and a stake is driven at the 150.00 foot distance (point C). From point C, the surveyor points the transit telescope back to point A. The telescope is then turned to point B, and an angle of 57°0′ is read on the horizontal protractor.

   Compute the distance between point A and point B.

PROBLEMS THAT REQUIRE AUXILIARY LINES

Each of these problems requires forming a right triangle by projecting auxiliary lines. Compute linear values to two decimal places unless otherwise noted, customary angular values to the nearest minute, and metric angular values to the nearest hundredth of a degree.

7. Compute ∠x and the length of edge AB of the retaining wall in Figure 35–60. Round the answer to the nearest degree.

8. Compute ∠T of the sheet metal reducer in Figure 35–61.

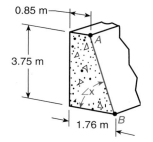

Figure 35–60

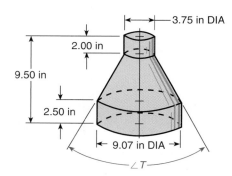

Figure 35–61

9. Compute the distance across centers, dimension D, of the holes in the locating plate in Figure 35–62.

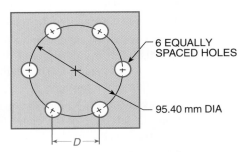

6 EQUALLY
SPACED HOLES

95.40 mm DIA

Figure 35–62

10. What is the diameter of the largest circular piece that can be cut from the triangular sheet of plywood in Figure 35–63?

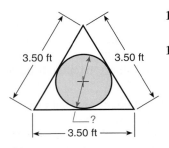

3.50 ft   3.50 ft

?

3.50 ft

Figure 35–63

11. Compute check dimension x of the external half of a dovetail slide shown in Figure 35–64. Compute the answer to 3 decimal places.

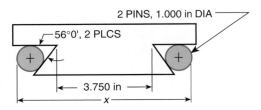

2 PINS, 1.000 in DIA

56°0', 2 PLCS

3.750 in

x

Figure 35–64

12. A patio is to be constructed as shown in Figure 35–65. Compute the straight-line distance between point A and point B.

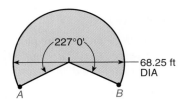

227°0'

68.25 ft
DIA

A B

Figure 35–65

13. A platform is laid out as shown in Figure 35–66. Compute $\angle x$. Round the answer to the nearest tenth degree.

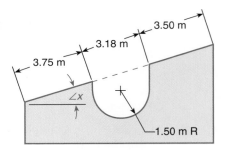

3.50 m

3.18 m

3.75 m

$\angle x$

1.50 m R

Figure 35–66

14. Pin locations on a positioning fixture are shown in Figure 35–67. Compute distance *y*.

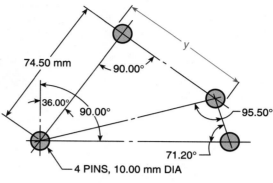

Figure 35–67

MORE COMPLEX PROBLEMS THAT REQUIRE AUXILIARY LINES

In the solution, each of these problems requires forming two or more right triangles by projecting auxiliary lines. Compute linear values to two decimal places unless otherwise noted, customary angular values to the nearest minute, and metric angular values to the nearest hundredth of a degree.

15. Compute the length of piece *AB* of the roof truss in Figure 35–68.

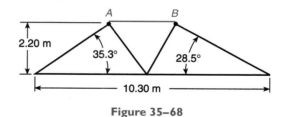

Figure 35–68

16. Determine ∠*x* of the pattern in Figure 35–69.

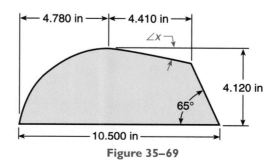

Figure 35–69

17. A section of a road is laid out as shown in Figure 35–70. Compute $\angle y$.

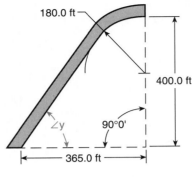

180.0 ft

400.0 ft

$\angle y$

90°0'

365.0 ft

Figure 35–70

18. Compute check dimension x of the angle cut in the piece in Figure 35–71.

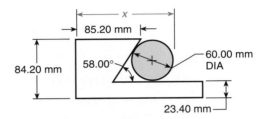

x

85.20 mm

60.00 mm DIA

84.20 mm

58.00°

23.40 mm

Figure 35–71

19. A wall is to be constructed along distance d in the courtyard in Figure 35–72. Compute the length of the wall.

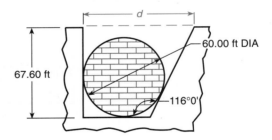

d

60.00 ft DIA

67.60 ft

116°0'

Figure 35–72

20. Compute $\angle x$ of the gauge in Figure 35–73.

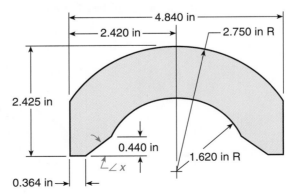

4.840 in

2.420 in

2.750 in R

2.425 in

0.440 in

1.620 in R

$\angle x$

0.364 in

Figure 35–73

UNIT 36 ⣿ Functions of Any Angle, Oblique Triangles

OBJECTIVES

After studying this unit you should be able to

- determine functions of angles in any quadrant.

- determine functions of angles greater than 360°.

- compute unknown angles and sides of oblique triangles by using the Law of Sines.

- compute unknown angles and sides of oblique triangles by using the Law of Cosines.

- solve applied problems by using principles of right and oblique triangles.

In a triangle that is not a right triangle, one of the angles can be greater than 90°. It is sometimes necessary to determine functions of angles greater than 90°. Computations using functions of angles greater than 90° are often required in solving obtuse triangle problems. In the fields of electricity and electronics, functions of angles greater than 90° are used when solving certain problems in alternating current.

36–1 Cartesian (Rectangular) Coordinate System

A function of any angle is described by the Cartesian coordinate system shown in Figure 36–1. The Cartesian coordinate system was presented in Unit 16. Following is a brief review of the system. A fixed point (*O*) called the *origin* is located at the intersection of a vertical axis and a horizontal axis. The horizontal axis is the *x-axis,* and the vertical axis is the *y-axis.* The *x-* and

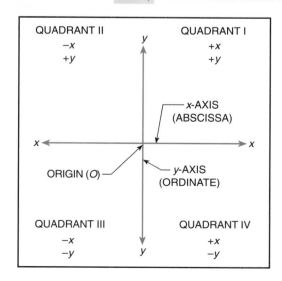

Figure 36–1

739

y-axes divide a plane into four parts, which are called *quadrants.* Quadrant I is the upper right section. Quadrants II, III, and IV are located going in a counterclockwise direction from quadrant I.

All points located to the right of the *y*-axis have positive (+) *x*-values; all points to the left of the *y*-axis have negative (−) *x*-values. All points above the *x*-axis have positive (+) *y*-values; all points below the *x*-axis have negative (−) *y*-values. The *x*-value is called the *abscissa,* and the *y*-value is called the *ordinate.*

The *x*- and *y*-values for each quadrant are listed in the table in Figure 36–2.

Quadrant I	Quadrant II	Quadrant III	Quadrant IV
+x	−x	−x	+x
+y	+y	−y	−y

Figure 36–2

36–2   Determining Functions of Angles in Any Quadrant

As a ray is rotated through any of the four quadrants, functions of an angle are determined as follows:

• The ray is rotated in a counterclockwise direction with its vertex at the origin (*O*). Zero degrees is on the *x*-axis.

• From a point on the rotated ray, a line segment is projected perpendicular to the *x*-axis. A right triangle is formed, of which the rotated side (ray) is the hypotenuse, the projected line segment is the opposite side, and the side on the *x*-axis is the adjacent side. The *reference angle* is the acute angle of the triangle that has the vertex at the origin (*O*).

• The signs of the functions of a reference angle are determined by noting the signs (+ or −) of the opposite and adjacent sides of the right triangle. The hypotenuse (*r*) is always positive in all four quadrants.

These examples illustrate the method of determining functions of angles greater than 90° in the various quadrants.

EXAMPLES •————————————————————————————

1. Determine the sine and cosine functions of 120°. Refer to Figure 36–3.

With the end point of the ray (*r*) at the origin (*O*), the ray is rotated 120° in a counterclowise direction.

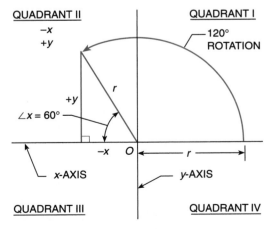

Figure 36–3

From a point on *r*, side *y* is projected perpendicular to the *x*-axis. In the right triangle formed, in relation to the reference angle (∠*x*), *r* is the hypotenuse, *y* is the opposite side, and *x* is the adjacent side.

$$\angle x = 180° - 120° = 60°$$

sin ∠*x* = opposite side/hypotenuse. In quadrant II, *y* is positive and *r* is always positive. Therefore, sin ∠*x* = +*y*/+*r*. In quadrant II, the sine is a positive (+) function.

$$\sin 120° = \sin (180° - 120°) = \sin 60°$$

Calculator Application

Using a calculator, functions of angles greater than 90° are computed using the same procedure as used in computing functions of acute angles.

$$\sin 120° \approx 120 \boxed{\sin} \rightarrow 0.866025404 \; Ans$$

or

$$\sin 120° \approx \boxed{\sin} \; 120 \; \boxed{EXE} \; 0.8660254038 \; Ans$$

cos ∠*x* = adjacent side/hypotenuse. Side *x* is negative (−); therefore, cos ∠*x* = −*x*/+*r*. Since the quotient of a negative value divided by a positive value is negative, in quadrant II, the cosine is a negative (−) function.

$$\cos 120° = -\cos (180° - 120°) = -\cos 60°$$

Calculator Application

$$\cos 120° = 120 \boxed{\cos} \rightarrow -0.5 \; Ans$$

or

$$\cos 120° = \boxed{\cos} \; 120 \; \boxed{EXE} \; -0.5 \; Ans$$

NOTE: A negative function of an angle does *not* mean that the angle is negative; it is a negative function of a positive angle. For example, −cos 70° does *not* mean cos (−70°).

2. Determine the tangent and secant functions of 220°. Refer to Figure 36–4.

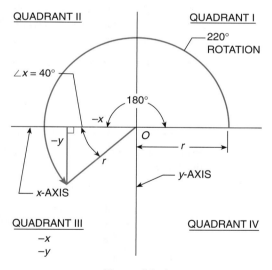

Figure 36–4

Rotate *r* 220° in a counterclockwise direction.

From a point on *r*, project side *y* perpendicular to the *x*-axis.

$$\angle x = 220° - 180° = 40°$$

tan ∠*x* = opposite side/adjacent side. In quadrant III, *y* is negative and *x* is negative. Therefore, tan ∠*x* = −*y*/−*x*. Since the quotient of a negative value divided by a negative value is positive, in quadrant III, the tangent is a positive (+) function.

$$\tan 220° = \tan (220° - 180°) = \tan 40°$$

Calculator Application

$\tan 220° \approx 220$ $\boxed{\tan}$ $\rightarrow 0.839099631$ *Ans*

or

$\tan 220° \approx$ $\boxed{\tan}$ 220 $\boxed{\text{EXE}}$ 0.8390996312 *Ans*

sec $\angle x$ = hypotenuse/adjacent side. In quadrant III, x is negative and r is always positive. Therefore, sec $\angle x = +r/-x$. Since the quotient of a positive value divided by a negative value is negative, in quadrant III, the secant is a negative ($-$) function.

$\sec 220° = -\sec (220° - 180°) = -\sec 40°$

Calculator Application

$\sec 220° \approx 220$ $\boxed{\cos}$ $\boxed{^1/_x}$ $\rightarrow -1.305407289$ *Ans*

or

$\sec 220° \approx$ $\boxed{\cos}$ 220 $\boxed{\text{EXE}}$ $\boxed{x^{-1}}$ $\boxed{\text{EXE}}$ -1.305407289 *Ans*

  ⌐— or $\boxed{^1/_x}$

3. Determine the cotangent and cosecant functions of 305°. Refer to Figure 36–5.

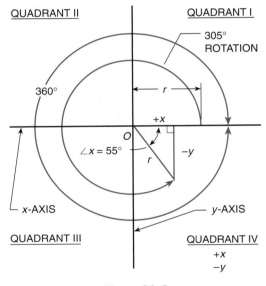

Figure 36–5

Rotate r 305° in a counterclockwise direction. From a point on r, project side y perpendicular to the x-axis.

$\angle x = 360° - 305° = 55°$

cot $\angle x$ = adjacent side/opposite side. In quadrant IV, y is negative and x is positive. Therefore, cot $\angle x = +x/-y$. Since the quotient of a positive value divided by a negative value is negative, in quadrant IV, the cotangent is a negative ($-$) function.

$\cot 305° = -\cot (360° - 305°) = -\cot 55°$

Calculator Application

$\cot 305° \approx 305$ $\boxed{\tan}$ $\boxed{^1/_x}$ $\rightarrow -0.700207538$ *Ans*

or

$\cot 305° \approx$ $\boxed{\tan}$ 305 $\boxed{\text{EXE}}$ $\boxed{x^{-1}}$ $\boxed{\text{EXE}}$ -0.700207538 *Ans*

  ⌐— or $\boxed{^1/_x}$

csc $\angle x$ = hypotenuse/opposite side. In quadrant IV, y is negative and r is always positive. Therefore, csc $\angle x = +r/-y$. The cosecant is a negative ($-$) function.

$\csc 305° = -\csc (360° - 305°) = -\csc 55°$

Calculator Application

$\csc 305° \approx 305$ $\boxed{\sin}$ $\boxed{1/x}$ $\rightarrow -1.220774589$ *Ans*

or

$\csc 305° \approx$ $\boxed{\sin}$ 305 $\boxed{\text{EXE}}$ $\boxed{x^{-1}}$ $\boxed{\text{EXE}}$ -1.220774589 *Ans*

— or $\boxed{1/x}$

36–3   Alternating Current Applications

An electric current that flows back and forth at regular intervals in a circuit is called an *alternating current.* The current and voltage each rise from zero to a maximum value and return to zero; then current and voltage increase to a maximum in the opposite direction and return to zero. This process is called a *cycle.*

The cycle is divided into 360 degrees. Current and electromotive force (voltage) are continuously changing during the cycle; therefore, their values must be stated at a given instant. A curve, such as that in Figure 36–6, is called a *sine curve.* This curve generally approximates the curves of electromotive force (emf) values of most alternating current generators. Since the current continues to flow back and forth, there are many cycles. Figure 36–6 shows just a little more than one cycle. Most electricity in the world is generated at either 50 or 60 cycles per second. One cycle per second is called a hertz (Hz), so most electricity is 50 Hz or 60 Hz.

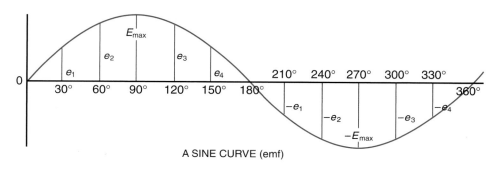

A SINE CURVE (emf)

Figure 36–6

NOTE: e_1 through $-e_4$ represent instantaneous emf (voltages) at various phases in a cycle. E_{max} represents the maximum emf (voltage) of a cycle. Following is the formula for computing the instantaneous value of the voltage of an alternating circuit current when the voltage follows the sine curve or wave.

$$e = E_{max} \sin \theta$$

where e = instantaneous voltage
E_{max} = maximum voltage
θ = angle in degrees

NOTE: θ is the Greek letter theta.

Since the cycle is divided into 360°, θ may be any angle from 0° to 360°. Therefore, the sine function of θ may be positive (+) or negative (−), depending on which of the four quadrants θ lies in when the voltage is determined at a certain instant.

Refer to the Cartesian coordinate system in Figure 36–7 on page 742. Observe that the sine function is positive (+) for the range of angles greater than 0° and less than 180°. The sine function is negative (−) for the range of angles greater than 180° and less than 360°. These positive and negative sine functions compare directly to the positive and negative signs of the instantaneous voltages (+e and −e) from 0° to 360° in the sine curve in Figure 36–6.

In the formula $e = e_{max} \sin \theta$, $\sin \theta$ is computed by the procedure for determining functions of angles in any of the four quadrants.

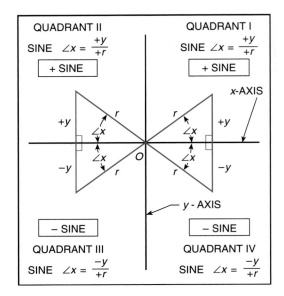

Figure 36–7

EXAMPLES •————————————————————————————————

1. What is the instantaneous voltage (e) of an alternating emf when it has reached 160° of the cycle? The maximum voltage (E_{max}) is 600.0 volts. Refer to Figure 36–8. Round the answer to the nearest tenth volt.

Compute $\sin \theta$.

$\sin 160° = \sin(180° - 160°) = \sin 20°$

Compute e.

$e = E_{max} \sin \theta = (600.0 \text{ V})(\sin 160°)$

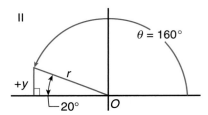

Figure 36–8

Calculator Application

$e = 600$ $\boxed{\times}$ 160 $\boxed{\sin}$ $\boxed{\blacktriangleright \text{DD}}$ 205.212086, 205.2 volts *Ans*

or

$e = 600$ $\boxed{\times}$ $\boxed{\sin}$ 160 $\boxed{\text{EXE}}$ 205.212086, 205.2 volts *Ans*

2. What is the instantaneous voltage (e) of an alternating emf when it has reached 320° of the cycle? The maximum voltage (E_{max}) is 550.0 volts. Refer to Figure 36–9. Round the answer to the nearest tenth volt.

Compute $\sin \theta$.

$\sin 320° = -\sin(360° - 320°) = -\sin 40°$

Compute e.

$e = E_{max} \sin \theta = (550.0 \text{ V})(\sin 320°)$

Calculator Application

$e = 550 \boxed{\times} 320 \boxed{\sin} \boxed{\blacktriangleright\underline{DD}}$ -353.5331853, -353.5 volts *Ans*

or

$e = 550 \boxed{\times} \boxed{\sin} 320 \boxed{EXE}$ -353.5331853, -353.5 volts *Ans*

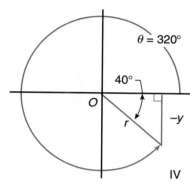

Figure 36–9

EXERCISE 36–3

Determine the sine, cosine, tangent, cotangent, secant, and cosecant for each of these angles. For each angle, sketch a right triangle similar to those in Figures 36–3, 36–4, and 36–5. Label the sides of the triangles + or −. Determine the reference angles and functions of the angles. Round the answers to 5 significant digits.

1. 118°	**5.** 300°	**9.** 139°16′
2. 207°	**6.** 350°	**10.** 202.6°
3. 260°	**7.** 216°20′	**11.** 313.2°
4. 168°	**8.** 96°50′	**12.** 179.9°

Compute the instantaneous voltage (e), in volts, of an alternating electromotive force (emf) for each of these problems. Compute the answer to 1 decimal place.

	Number of Degrees Reached in Cycle (θ)	Maximum Voltage (E_{max})	Instantaneous Voltage (e)
13.	120°	600.0 volts	
14.	165°	320.0 volts	
15.	210°	240.0 volts	
16.	255°	550.0 volts	
17.	300°	120.0 volts	
18.	330°	800.0 volts	
19.	90°	300.0 volts	
20.	180°	240.0 volts	
21.	270°	550.0 volts	
22.	360°	600.0 volts	

36–4   Determining Functions of Angles Greater Than 360°

Functions of any angle of a ray that is rotated more than 360° or one revolution are easily deter-mined. Functions of an angle greater than 360° are computed just as functions of angles from 0° to 360° are computed after 360° or a multiple of 360° is subtracted from the given angle. These two examples illustrate the method of computing functions of angles greater than 360°.

EXAMPLES •────────────────────────────────────

1. Determine the tangent of 472°.

   Subtract one complete revolution.

   tan 472° = tan (472° − 360°) = tan 112°

   Ray *r* is rotated one complete revolution plus an additional 112°. Reference ∠*x* lies in quadrant II, as shown in Figure 36–10.

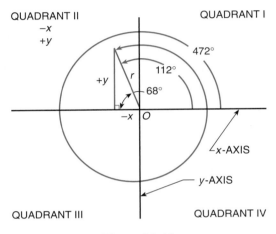

Figure 36–10

   In quadrant II, tan ∠*x* = +*y*/−*x*; therefore, the tangent function is negative (−).

   tan 112° = −tan (180° − 112°) = −tan 68°

   Calculator Application

   tan 472° ≈ 472 [tan] → −2.475086853 *Ans*

   or

   tan 472° ≈ [tan] 472 [EXE] −2.475086853 *Ans*

2. Determine the cosine of 1,055°.

   Divide by 360° to find the number of complete rotations.

   1,055° ÷ 360° = 2 complete revolutions plus 335°

   cos 1,055° = cos[1,055° − 2(360°)] = cos 335°

   Ray *r* is rotated two complete revolutions plus an additional 335°. Reference ∠*x* lies in quadrant IV, as shown in Figure 36–11. In quadrant IV, cos ∠*x* = +*x*/+*r*, therefore, the cosine function is positive (+).

   cos 335° = cos (360° − 335°) = cos 25°

   Calculator Application

   cos 1,055° ≈ 1055 [cos] → 0.906307787 *Ans*

   or

   cos 1,055° ≈ [cos] 1055 [EXE] 0.906307787 *Ans*

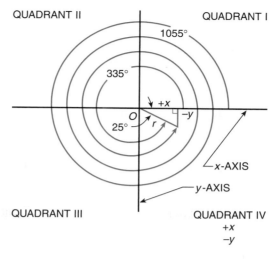

QUADRANT II

QUADRANT I

1055°

335°

25°

+x

−y

x-AXIS

y-AXIS

QUADRANT III

QUADRANT IV

+x

−y

Figure 36–11

36–5 Instantaneous Voltage Related to Time Application

The frequency of current is the number of times a cycle is repeated in 1 second of time. The standard frequency of 60 cycles per second means that the current makes 60 complete cycles in 1 second. Since one cycle is divided into 360°, the current goes through 60 × 360° or 21,600° in 1 second. Stated in terms of an alternating current generator, the angular velocity of the generator is 21,600° per second. The angle in degrees (θ) equals 21,600° at exactly one second in time. Therefore, the value of θ at any instant equals 21,600° times the number of seconds at that instant. After θ is determined, $e = E_{max} \sin \theta$ may again be used to compute an instantaneous voltage (e).

EXAMPLE •—————————————————————————————————————

In a 60-cycle alternating emf, the angular velocity is 21,600° per second. What is the instantaneous voltage (e) at the end of exactly 0.03 second? The maximum voltage (E_{max}) is 120.0 volts. Refer to Figure 36–12. Round the answer to 1 decimal place.

Compute $\sin \theta$.

$\theta = (21{,}600°/\text{s})(0.03 \text{ s}) = 648°$

Subtract one complete revolution from 648°.

$\sin 648° = \sin (648° - 360°) = \sin 288°$

$\sin 288° = -\sin (360° - 288°) = -\sin 72°$

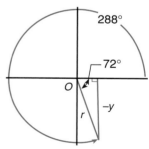

288°

72°

O

−y

r

Figure 36–12

Calculator Application

$\sin \theta \approx 216000 \;\boxed{\times}\; .03 \;\boxed{\blacktriangleright\underline{\text{DD}}}\; \boxed{\sin} \rightarrow -0.951056516$

or

$\sin \theta \approx \boxed{\sin}\; \boxed{(}\; 21600 \;\boxed{\times}\; .03 \;\boxed{)}\; \boxed{\text{EXE}}\; -0.951056516$

Compute e.

$e = E_{max} \sin \theta$

$e \approx (120 \text{ volts})(-0.951056516) \approx -114.1267819$

$e \approx -114.1 \text{ volts } Ans$

EXERCISE 36–5

Determine the sine, cosine, tangent, cotangent, secant, and cosecant for each of these angles that are greater than 360°. For each angle, sketch a right triangle similar to those in Figures 36–10 and 36–11. Label the sides of the triangles + or −. Determine the reference angles and the functions of the angles. Round the answers to 5 significant digits.

1. 510°		**6.** 743°
2. 405°		**7.** 937°
3. 555°		**8.** 1036°30′
4. 680°		**9.** 1248.4°
5. 531°		**10.** 1440°

Each of these problems has a 60-cycle alternating emf; the angular velocity is 21,600° per second. Compute the instantaneous voltage (e), in volts, for each problem. Compute the answer to one decimal place.

	Time	Maximum Voltage (E_{max})	Instantaneous Voltage (e)
11.	0.02 second	240.0 volts	
12.	0.016 second	550.0 volts	
13.	0.03 second	320.0 volts	
14.	0.027 second	600.0 volts	
15.	0.035 second	120.0 volts	
16.	0.01 second	300.0 volts	
17.	0.022 second	800.0 volts	
18.	0.08 second	240.0 volts	
19.	0.04 second	600.0 volts	
20.	0.071 second	450.0 volts	

36–6   Solving Oblique Triangles

An *oblique triangle* is a triangle that does not have a right angle. An oblique triangle may be either acute or obtuse. In an acute triangle, each of the three angles is acute or less than 90°. In an obtuse triangle, one of the angles is obtuse or greater than 90°.

Angles and sides must be computed in practical problems that involve oblique triangles. These problems can be solved as a series of right triangles, but the process is time-consuming.

Two formulas, called the Law of Sines and the Law of Cosines, are used to simplify oblique triangle computations. In order to use either formula, three parts of an oblique triangle must be known and at least one part must be a side.

36–7   Law of Sines

> **In any triangle the sides are proportional to the sines of their opposite angles.**

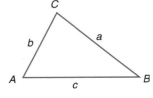

Figure 36–13

In reference to the triangle shown in Figure 36–13,

$$\frac{a}{\sin A} = \frac{b}{\sin B} = \frac{c}{\sin C}$$

The Law of Sines is used to solve the following two kinds of oblique triangle problems:

- Problems where any two angles and any side of an oblique triangle are known
- Problems where any two sides and an angle opposite one of the given sides of an oblique triangle are known

NOTE: Since an angle of an oblique triangle may be greater than 90°, you must often determine the sine of an angle greater than 90° and less than 180°. Recall that the angle lies in quadrant II of the Cartesian coordinate system. The sine of an angle between 90° and 180° equals the sine of the supplement of the angle. For example, the sine of 120°40′ = sin (180° − 120°40′) = sin 59°20′.

36–8 Solving Problems Given Two Angles and a Side, Using the Law of Sines

EXAMPLES

1. Given two angles and a side, determine side x of the oblique triangle in Figure 36–14. Round the answer to 3 decimal places.

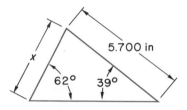

Figure 36–14

Since side x is opposite the 39° angle and the 5.700-inch side is opposite the 62° angle, the proportion is set up as

$$\frac{x}{\sin 39°} = \frac{5.700 \text{ in}}{\sin 62°}$$

$$x = \frac{\sin 39° (5.700 \text{ in})}{\sin 62°}$$

Calculator Application

$x \approx 39 \boxed{\sin} \boxed{\times} 5.7 \boxed{\div} 62 \boxed{\sin} \boxed{\blacktriangleright DD} \; 4.062671735$

or

$x \approx \boxed{\sin} 39 \boxed{\times} 5.7 \boxed{\div} \boxed{\sin} 62 \boxed{\text{EXE}} \; 4.062671735$

$x \approx 4.063$ in *Ans*

2. Given two angles and a side, determine $\angle A$, side a, and side b of the oblique triangle in Figure 36–15. Round the answer to 2 decimal places.

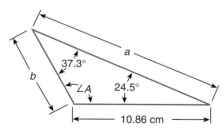

Figure 36–15

Solve for ∠A. The sum of the three angles of a triangle equals 180°.

∠A = 180° − (37.3° + 24.5°) = 118.2° *Ans*

Solve for side *a*.

Sin 118.2° = sin (180° − 118.2°) = sin 61.8°

$$\frac{a}{\sin 118.2°} = \frac{10.86 \text{ cm}}{\sin 37.3°}$$

$$a = \frac{\sin 118.2° (10.86 \text{ cm})}{\sin 37.3°}$$

Calculator Application

$a ≈ 118.2$ $\boxed{\sin}$ $\boxed{\times}$ 10.86 $\boxed{\div}$ 37.3 $\boxed{\sin}$ $\boxed{\blacktriangleright\underline{DD}}$ 15.79395824

or

$a ≈$ $\boxed{\sin}$ 118.2 $\boxed{\times}$ 10.86 $\boxed{\div}$ $\boxed{\sin}$ 37.3 $\boxed{EXE}$ 15.79395824

$a ≈ 15.79$ cm *Ans*

Solve for side *b*.

$$\frac{b}{\sin 24.5°} = \frac{10.86 \text{ cm}}{\sin 37.3°}$$

$$b = \frac{\sin 24.5° (10.86 \text{ cm})}{\sin 37.3°}$$

Calculator Application

$b ≈ 24.5$ $\boxed{\sin}$ $\boxed{\times}$ 10.86 $\boxed{\div}$ 37.3 $\boxed{\sin}$ $\boxed{\blacktriangleright\underline{DD}}$ 7.431773633

or

$b ≈$ $\boxed{\sin}$ 24.5 $\boxed{\times}$ 10.86 $\boxed{\div}$ $\boxed{\sin}$ 37.3 $\boxed{EXE}$ 7.431773633

$b ≈ 7.43$ cm *Ans*

EXERCISE 36–8

In each of these problems, two angles and a side are given. Determine the required values. Compute side lengths to three decimal places, customary angular values to the nearest minute, and metric angular values to the nearest hundredth of a degree.

1. Determine side *x*.

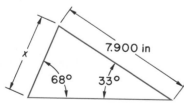

2. Determine side *a*.

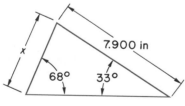

3. Determine side *d*.

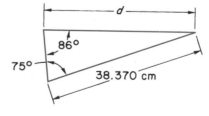

4. Determine side *y*.

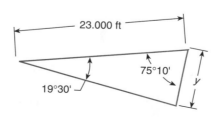

5. Determine side *b*.

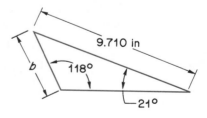

6. Determine side *c*.

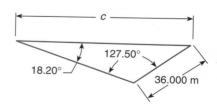

7. a. Determine ∠A.
  b. Determine side a.

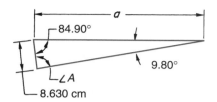

9. a. Determine ∠A.
  b. Determine side a.

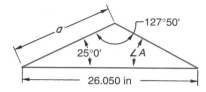

11. a. Determine ∠A.
  b. Determine side a.
  c. Determine side b.

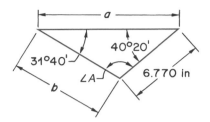

8. a. Determine ∠C.
  b. Determine side c.

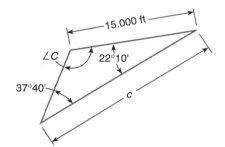

10. a. Determine ∠D.
  b. Determine side d.
  c. Determine side e.

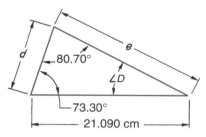

12. a. Determine ∠E.
  b. Determine side f.
  c. Determine side g.

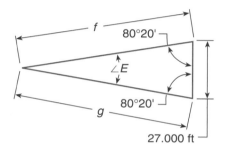

36–9 Solving Problems Given Two Sides and an Angle Opposite One of the Given Sides, Using the Law of Sines

A special condition exists when certain problems are solved in which two sides and an angle opposite one of the sides are given. If triangle data are given in word form or if a triangle is inaccurately sketched, there may be two solutions to a problem.

It is possible to have two different triangles with the same two sides and the same angle opposite one of the given sides. A situation of this kind is called an ambiguous case. The following example illustrates the ambiguous case or a problem with two solutions.

EXAMPLE (The Ambiguous Case or 2 Solutions) •————————————

A triangle has a 1.5-inch side, a 2.5-inch side, and an angle of 32° opposite the 1.5-inch side. From the given data, Figure 36–16 is accurately drawn. But two triangles can be constructed from the same data. Triangle BCA and triangle DCA both have a 1.5-inch side, a 2.5-inch side, and a 32° angle opposite the 1.5-inch side. The triangles are shown separately in Figure 36–17 on page 750.

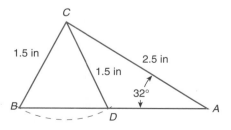

Figure 36–16

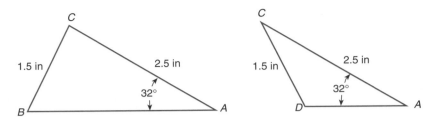

Figure 36–17

The only condition under which a problem can have two solutions is when the given angle is acute and the given side opposite the given angle is smaller than the other given side. For example, in the problem illustrated by Figures 36–16 and 36–17, the 32° angle is acute, and the 1.5-inch side opposite the 32° angle is smaller than the 2.5-inch side.

In most problems you do not get involved with two solutions. Even under the condition in which there can be two solutions, if the problem is shown in picture form as an accurately drawn triangle, it can readily be observed that there is only one solution.

EXAMPLES •

1. Given two sides and an angle opposite one of the given sides, determine ∠A, ∠C, and side c of the oblique triangle in Figure 36–18. Round ∠A and ∠C to the nearest minute, and round side c to 3 decimal places.

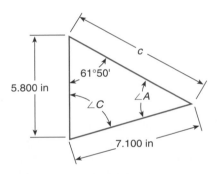

Figure 36–18

The 7.100-inch side opposite the 61°50′ angle is larger than the 5.800-inch side, therefore there is only one solution.

Solve for ∠A.

$$\frac{5.800 \text{ in}}{\sin \angle A} = \frac{7.100 \text{ in}}{\sin 61°50'}$$

$$\sin \angle A = \frac{5.800 \text{ in} (\sin 61°50')}{7.100 \text{ in}}$$

Calculator Application

sin ∠A ≈ 5.8 ⊠ 61.50 [2nd] [▶DD] [sin] ÷ 7.1 [▶DD] 0.720162491

or

sin ∠A ≈ 5.8 ⊠ [sin] 61 [°′″] 50 [°′″] ÷ 7.1 [EXE] 0.720162491

∠A ≈ .720162491 [2nd] [sin] [3rd] [▶DD] 46°04′04″4

or

∠A ≈ [SHIFT] [sin] .720162491 [EXE] [SHIFT] [←] → 46°4′4.43″

∠A ≈ 46°4′ *Ans*

Solve for $\angle C$.

$\angle C \approx 180° - (61°50' + 46°4') \approx 72°6'$ *Ans*

Solve for side c.

$$\frac{c}{\sin 72°6'} = \frac{7.100 \text{ in}}{\sin 61°50'}$$

$$c \approx \frac{\sin 72°6' \ (7.100 \text{ in})}{\sin 61°50'}$$

Calculator Application

$c \approx$ 72.06 [2nd] [▶DD] [sin] [×] 7.1 [÷] 61.50 [2nd] [▶DD] [sin] [▶DD] 7.663891984

or

$c \approx$ [sin] 72 [°'"] 6 [°'"] [×] 7.1 [÷] [sin] 61 [°'"] 50 [°'"] [EXE] 7.663891984

$c \approx 7.664$ in *Ans*

2. Given two sides and an angle opposite one of the given sides as shown in Figure 36–19, determine $\angle D$. Figure 36–19 is drawn accurately to scale. Observe that $\angle D$ is greater than 90°.

Set up the proportion and solve.

$$\frac{6.870 \text{ cm}}{\sin \angle D} = \frac{3.500 \text{ cm}}{\sin 27.6°}$$

$$\sin \angle D = \frac{6.870 \text{ cm} \ (\sin 27.6°)}{3.500 \text{ cm}}$$

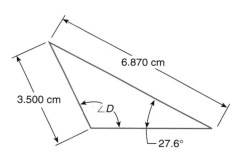

Figure 36–19

Calculator Application

$\sin \angle D \approx$ 6.87 [×] 27.6 [sin] [÷] 3.5 [▶DD] 0.909383932

or

$\sin \angle D \approx$ 6.87 [×] [sin] 27.6 [÷] 3.5 [EXE] 0.9093839318

.909383932 [2nd] [sin] → 65.42035385

or

[SHIFT] [sin] .909383932 [EXE] 65.42035385

The angle that corresponds to the sine function 0.909383932 is 65.42° (rounded). Since $\angle D$ is greater than 90°, $\angle D$ is the supplement of 65.42° (rounded)

$\angle D \approx 180° - 65.42° \approx 114.58°$ *Ans*

EXERCISE 36–9

Two sides and an angle opposite one of the sides are given in these triangle problems. Identify each problem as to whether it has one or two solutions. Do not *solve the problems for angles and sides.*

1. A 3″ side, a 5″ side, a 37° angle opposite the 3″ side.
2. A 9.5-cm side, a 9.8-cm side, a 75° angle opposite the 9.5-cm side.
3. A 21-m side, a 29-m side, a 41° angle opposite the 29-m side.
4. A 0.943″ side, a 1.612″ side, and an 82°15′ angle opposite the 1.612″ side.
5. A 210-ft side, a 305-ft side, a 29°30′ angle opposite the 305-ft side.
6. A 16.35-cm side, a 23.86-cm side, a 115° angle opposite the 23.86-cm side.
7. An 87.6-m side, a 124.8-m side, a 12.9° angle opposite the 87.6-m side.
8. A 33.86″ side, a 34.09″ side, a 46°18′ angle opposite the 33.86″ side.

In each of these problems, two sides and an angle opposite one of the sides are given. Compute side lengths to three decimal places, customary angular values to the nearest minute, and metric angular values to the nearest hundredth of a degree. The triangles are drawn accurately to scale.

9. Determine ∠A.

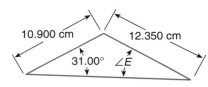

10. Determine ∠E.

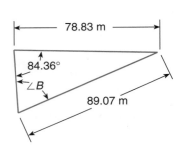

11. Determine ∠B.

12. Determine ∠C.

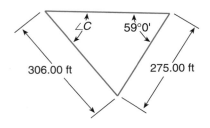

13. Determine ∠A.

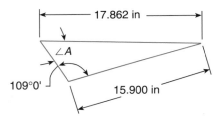

14. Determine ∠D.

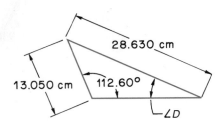

15. a. Determine ∠A.
    b. Determine ∠B.

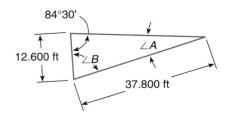

16. a. Determine ∠D.
    b. Determine ∠E.

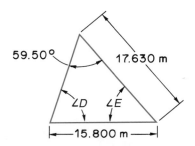

17. a. Determine ∠A.
    b. Determine ∠B.

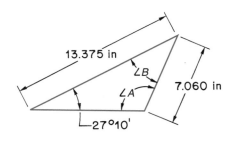

18. a. Determine ∠C.
   b. Determine ∠D.
   c. Determine side *d*.

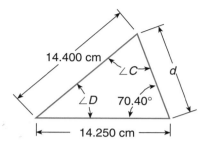

19. a. Determine ∠A.
   b. Determine ∠B.
   c. Determine side *b*.

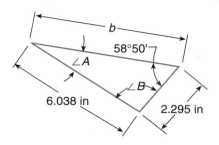

20. a. Determine ∠E.
   b. Determine ∠F.
   c. Determine side *f*.

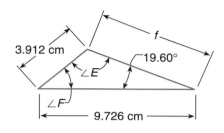

36–10   Law of Cosines (Given Two Sides and the Included Angle)

> **In any triangle, the square of any side is equal to the sum of the squares of the other two sides minus twice the product of these two sides multiplied by the cosine of their included angle.**

In reference to the triangle shown in Figure 36–20,

$$a^2 = b^2 + c^2 - 2bc(\cos A)$$
$$b^2 = a^2 + c^2 - 2ac(\cos B)$$
$$c^2 = a^2 + b^2 - 2ab(\cos C)$$

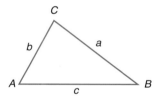

Figure 36–20

The Law of Cosines in the form shown is used to solve the following kind of oblique triangle problems:

• Problems where two sides and the included angle of an oblique triangle are known

NOTE: An angle of an oblique triangle may be greater than 90°. Therefore, you must often determine the cosine of an angle greater than 90° and less than 180°. These angles lie in quadrant II of the Cartesian coordinate system. Recall that the cosine of an angle between 90° and 180° equals the negative (−) cosine of the supplement of the angle. For example, the cosine of 118°10′ = −cos (180° − 118°10′) = −cos 61°50′.

36–11   Solving Problems Given Two Sides and the Included Angle, Using the Law of Cosines

EXAMPLES •

1. Given two sides and the included angle, determine side *x* of the oblique triangle in Figure 36–21 on page 756. Observe that 34.7° is included between the 8.700-cm and 9.100-cm sides. Round the answer to 3 decimal places.

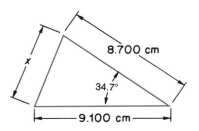

Figure 36–21

Substitute the given values in their appropriate places in the formula and solve for side *x*.

$$x^2 = (8.700 \text{ cm})^2 + (9.100 \text{ cm})^2 - 2(8.700 \text{ cm})(9.100 \text{ cm})(\cos 34.7°)$$

$$x = \sqrt{(8.700 \text{ cm})^2 + (9.100 \text{ cm})^2 - 2(8.700 \text{ cm})(9.100 \text{ cm})(\cos 34.7°)}$$

Calculator Application

$x \approx$ [(] 8.7 [x^2] [+] 9.1 [x^2] [−] 2 [×] 8.7 [×] 9.1 [×] 34.7 [cos] [)] [$\sqrt{x}$] → 5.321814779

or

$x \approx$ [$\sqrt{}$] [(] 8.7 [x^2] [+] 9.1 [x^2] [−] 2 [×] 8.7 [×] 9.1 [×] [cos] 34.7 [)] [EXE]
5.321814779

$x \approx 5.322$ cm *Ans*

2. Given two sides and the included angle, determine side *a*, ∠*B*, and ∠*C* of the oblique triangle in Figure 36–22. Round side *a* to 3 decimal places, and round ∠*B* and ∠*C* to the nearest minute.

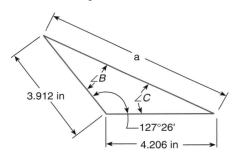

Figure 36–22

Solve for *a*, using the Law of Cosines.

$$a^2 = (3.912 \text{ in})^2 + (4.206 \text{ in})^2 - 2(3.912 \text{ in})(4.206 \text{ in})(\cos 127°26')$$

$$a = \sqrt{(3.912 \text{ in})^2 + (4.206 \text{ in})^2 - 2(3.912 \text{ in})(4.206 \text{ in})(\cos 127°26')}$$

Calculator Application

$a \approx$ [(] 3.912 [x^2] [+] 4.206 [x^2] [−] 2 [×] 3.912 [×] 4.206 [×] 127.26 [2nd] [►DD] [cos]
[►DD] [)] [$\sqrt{x}$] → 7.27988697

or

$a \approx$ [$\sqrt{}$] [(] 3.912 [x^2] [+] 4.206 [x^2] [−] 2 [×] 3.912 [×]4.206 [×] [cos] 127 [°'"] 26 [°'"]
[)] [EXE] 7.27988697

$a \approx 7.280$ in *Ans*

Solve for ∠*B*, using the Law of Sines.

$$\frac{4.206 \text{ in}}{\sin \angle B} = \frac{7.280 \text{ in}}{\sin 127°26'}$$

$$\sin \angle B = \frac{4.206 \text{ in } (\sin 127°26')}{7.280 \text{ in}}$$

Calculator Application

or

$\sin \angle B \approx 4.206$ ⨯ 127.26 2nd ▶DD sin ÷ 7.280 ▶DD 0.458766636

$\sin \angle B \approx 4.206$ ⨯ sin 127 °′″ 26 °′″ ÷ 7.280 EXE 0.4587666359

$\angle B \approx 0.458766636$ 2nd sin 3rd ▶DD $27°18'27''1$

or

$\angle B \approx$ SHIFT sin $.458766636$ EXE SHIFT °′″ $\rightarrow 27°18'27.18''$

$\angle B \approx 27°18'$ *Ans*

Solve for $\angle C$.

$\angle C \approx 180° - (127°26' + 27°18') \approx 25°16'$ *Ans*

EXERCISE 36–11

In each of the following problems, two sides and the included angle of a triangle are given. Compute side lengths to three decimal places, customary angular values to the nearest minute, and metric angular values to the nearest hundredth of a degree.

1. Determine side *x*.

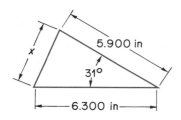

3. Determine side *b*.

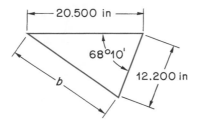

5. Determine side *x*.

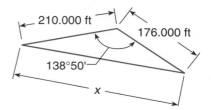

2. Determine side *a*.

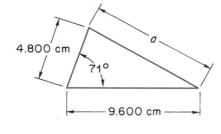

4. Determine side *y*.

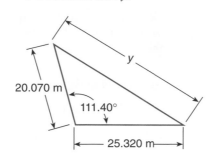

6. Determine side *c*.

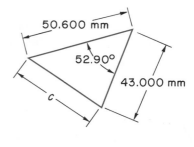

7. a. Determine side *c*.
 b. Determine ∠*A*.
 c. Determine ∠*B*.

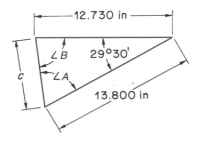

9. a. Determine side *a*.
 b. Determine ∠*B*.
 c. Determine ∠*C*.

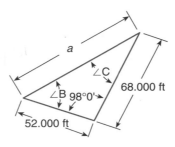

11. a. Determine side *a*.
 b. Determine ∠*B*.
 c. Determine ∠*C*.

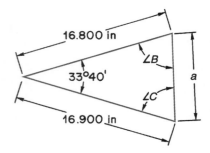

8. a. Determine side *e*.
 b. Determine ∠*F*.
 c. Determine ∠*G*.

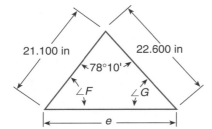

10. a. Determine side *n*.
 b. Determine ∠*M*.
 c. Determine ∠*P*.

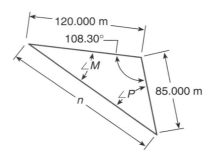

12. a. Determine side *c*.
 b. Determine ∠*D*.
 c. Determine ∠*E*.

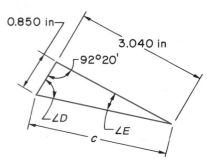

36–12   Law of Cosines (Given Three Sides)

> **In any triangle, the cosine of an angle is equal to the sum of the squares of the two adjacent sides minus the square of the opposite side, divided by twice the product of the two adjacent sides.**

In reference to the triangle shown in Figure 36–23.

$$\cos A = \frac{b^2 + c^2 - a^2}{2bc}$$

$$\cos B = \frac{a^2 + c^2 - b^2}{2ac}$$

$$\cos C = \frac{a^2 + b^2 - c^2}{2ab}$$

Figure 36–23

NOTE: These formulas, which are stated in terms of the cosines of angles, are rearrangements of the formulas on page 755, which are stated in terms of the squares of the sides.

The Law of Cosines in the form shown is used to solve the following kind of oblique triangle problems:

• Problems where three sides of an oblique triangle are known

NOTE: When an unknown angle is determined, its cosine function may be negative. A negative cosine function means that the angle being computed is greater than 90°. The angle lies in quadrant II of the Cartesian coordinate system. Recall that the cosine of an angle between 90° and 180° equals the negative cosine of the supplement of the angle. For example, the cosine of $147°40' = -\cos(180° - 147°40') = -\cos 32°20'$.

36–13   Solving Problems Given Three Sides, Using the Law of Cosines

EXAMPLES ●

1. Given three sides, determine ∠A of the oblique triangle in Figure 36–24. Round the answer to the nearest minute.

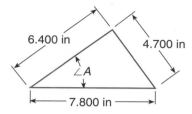

Figure 36–24

$$\cos \angle A = \frac{(6.400 \text{ in})^2 + (7.800 \text{ in})^2 - (4.700 \text{ in})^2}{2(6.400 \text{ in})(7.800 \text{ in})}$$

Calculator Application

— or [EXE]

$\cos \angle A \approx 6.4 \boxed{x^2} \boxed{+} 7.8 \boxed{x^2} \boxed{-} 4.7 \boxed{x^2} \boxed{\blacktriangleright \underline{DD}} \boxed{\div} \boxed{(} 2 \boxed{\times} 6.4 \boxed{\times} 7.8 \boxed{)} \boxed{\blacktriangleright \underline{DD}} 0.798377404$

— or [EXE]

$\angle A \approx .798377404 \boxed{2nd} \boxed{\cos} \boxed{3rd} \boxed{\blacktriangleright \underline{DD}} 37°01'28''4$

or

$\angle A \approx \boxed{SHIFT} \boxed{\cos} .798377404 \boxed{EXE} \boxed{SHIFT} \boxed{°\,'\,''} 37°1'28.44''$

$\angle A \approx 37°01'$ *Ans*

2. Given three sides, determine ∠P of the oblique triangle in Figure 36–25. Round the answer to the nearest hundredth degree.

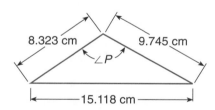

Figure 36–25

$$\cos \angle P = \frac{(8.323 \text{ cm})^2 + (9.745 \text{ cm})^2 - (15.118 \text{ cm})^2}{2(8.323 \text{ cm})(9.745 \text{ cm})}$$

Calculator Application

— or [EXE]

$\cos \angle P \approx 8.323 \boxed{x^2} \boxed{+} 9.745 \boxed{x^2} \boxed{-} 15.118 \boxed{x^2} \boxed{\blacktriangleright \underline{DD}} \boxed{\div} \boxed{(} 2 \boxed{\times} 8.323 \boxed{\times} 9.745 \boxed{)}$
$\boxed{\blacktriangleright \underline{DD}} -0.396488999$

— or [EXE]

$\angle P \approx .39648899 \boxed{+/-} \boxed{2nd} \boxed{\cos} \rightarrow 113.3588715$

or

$\angle P \approx \boxed{SHIFT} \boxed{\cos} \boxed{-} .396488999 \boxed{EXE} 113.3588715$

$\angle P \approx 113.36°$ *Ans*

3. Given three sides, determine $\angle A$, $\angle B$, and $\angle C$ of the oblique triangle in Figure 36–26. Round the answers to the nearest minute.

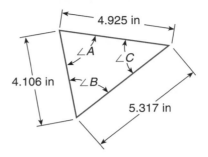

Figure 36–26

Solve for $\angle A$.

$$\cos \angle A = \frac{(4.106 \text{ in})^2 + (4.925 \text{ in})^2 - (5.317 \text{ in})^2}{2(4.106 \text{ in})(4.925 \text{ in})}$$

Calculator Application

or $\boxed{\text{EXE}}$

$\cos \angle A \approx 4.106 \boxed{x^2} \boxed{+} 4.925 \boxed{x^2} \boxed{-} 5.317 \boxed{x^2} \boxed{\blacktriangleright\text{DD}} \boxed{\div} \boxed{(} 2 \boxed{\times} 4.106 \boxed{\times} 4.925 \boxed{)}$

$\boxed{\blacktriangleright\text{DD}}$ 0.317583331

or $\boxed{\text{EXE}}$

$\angle A \approx .317583331 \boxed{\text{2nd}} \boxed{\cos} \boxed{\text{3rd}} \boxed{\blacktriangleright\text{DD}}$ 71°28'59"3

or

$\angle A \approx \boxed{\text{SHIFT}} \boxed{\cos} .317583331 \boxed{\text{EXE}} \boxed{\text{SHIFT}} \boxed{°'''} \rightarrow 71°28'59.38"$

$\angle A \approx 71°29'$ *Ans*

Solve for $\angle B$. Angle B may also be computed from the Law of Cosines formula, but it is simpler to use the Law of Sines formula.

$$\frac{4.925 \text{ in}}{\sin \angle B} = \frac{5.317 \text{ in}}{\sin \angle A}, \frac{4.925 \text{ in}}{\sin \angle B} \approx \frac{5.317 \text{ in}}{\sin 71°29'}$$

$$\sin \angle B \approx \frac{4.925 \text{ in} (\sin 71°29')}{5.317 \text{ in}}$$

Calculator Application

$\sin \angle B \approx 4.925 \boxed{\times} 71.29 \boxed{\text{2nd}} \boxed{\blacktriangleright\text{DD}} \boxed{\sin} \boxed{\div} 5.317 \boxed{\blacktriangleright\text{DD}}$ 0.878322217

or

$\sin \angle B \approx 4.925 \boxed{\times} \boxed{\sin} 71 \boxed{°'''} 29 \boxed{°'''} \boxed{\div} 5.317 \boxed{\text{EXE}}$ 0.8783222166

$\angle B \approx .878322217 \boxed{\text{2nd}} \boxed{\sin} \boxed{\text{3rd}} \boxed{\blacktriangleright\text{DD}}$ 61°26'26"2

or

$\angle B \approx \boxed{\text{SHIFT}} \boxed{\sin} .878322217 \boxed{\text{EXE}} \boxed{\text{SHIFT}} \boxed{°'''} \rightarrow 61°26'26.27"$

$\angle B \approx 61°26'$ *Ans*

Solve for $\angle C$.

$\angle C \approx 180° - (71°29' + 61°26') \approx 47°5'$ *Ans*

#1–11 add

EXERCISE 36–13

In each of these problems, three sides of a triangle are given. Compute customary angular values to the nearest minute and metric angular values to the nearest hundredth of a degree.

1. Determine ∠A.

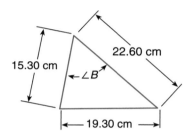

2. Determine ∠B.

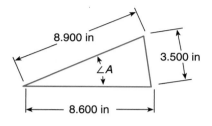

3. Determine ∠C.

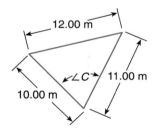

4. Determine ∠F.

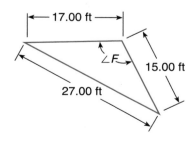

5. Determine ∠A.

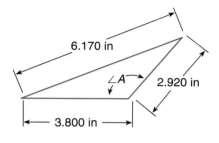

6. Determine ∠M.

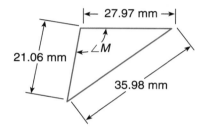

7. a. Determine ∠B.
   b. Determine ∠C.
   c. Determine ∠D.

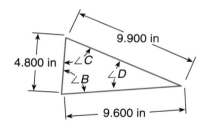

8. a. Determine ∠F.
   b. Determine ∠G.
   c. Determine ∠H.

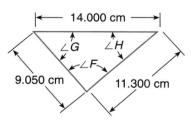

9. a. Determine ∠A.
   b. Determine ∠B.
   c. Determine ∠C.

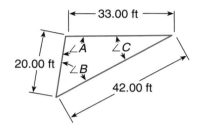

10. a. Determine ∠A.
   b. Determine ∠B.
   c. Determine ∠C.

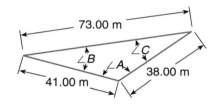

11. a. Determine ∠M.
   b. Determine ∠R.
   c. Determine ∠T.

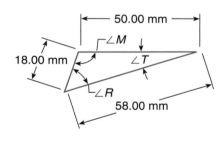

12. a. Determine ∠D.
   b. Determine ∠E.
   c. Determine ∠F.

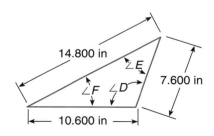

36–14   Practical Applications of Oblique Triangles

The oblique triangle examples and problems that have been presented were not given as practical applied problems. They were intended to develop skills in applying proper procedures in solving angles and sides of triangles, using the Law of Sines and the Law of Cosines.

   The practical applications of oblique triangles are now presented. Often problems are not given directly in the form of oblique triangles. As with right triangle problems, oblique triangle problems may be given in word form or in picture form where an oblique triangle does not appear.

   When solving an oblique triangle problem stated in word form, sketch and label a triangle using the given values. When solving an oblique triangle problem where an oblique triangle is not directly given, you may have to project auxiliary lines to form triangles. In addition, oblique triangle problems sometimes require a combination of right triangles and oblique triangles in the solution.

   These examples illustrate the methods of solving practical word-type and picture-type problems.

EXAMPLES •———————————————————————————————

1. A metal frame in the shape of an oblique triangle is to be fabricated. One side of the frame is 2.40 meters long. One end of the second side, which is 1.80 meters long, is to be fastened to an end of the 2.40-meter side at an angle of 58.00°. Compute the required length of the third side of the frame. Round the answer to 2 decimal places.

   Solution.   Sketch and label an oblique triangle as shown in Figure 36–27. Let c represent the third side.

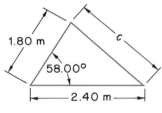

Figure 36–27

Compute side c.

Two sides and the included angle are known. Apply the Law of Cosines.

$$c^2 = (1.80 \text{ m})^2 + (2.40 \text{ m})^2 - 2(1.80 \text{ m})(2.40 \text{ m})(\cos 58.00°)$$
$$c = \sqrt{(1.80 \text{ m})^2 + (2.40 \text{ m})^2 - 2(1.80 \text{ m})(2.40 \text{ m})(\cos 58.00°)}$$

Calculator Application

$c \approx$ $($ 1.8 $\boxed{x^2}$ $\boxed{+}$ 2.4 $\boxed{x^2}$ $\boxed{-}$ 2 $\boxed{\times}$ 1.8 $\boxed{\times}$ 2.4 $\boxed{\times}$ 58 $\boxed{\cos}$ $\boxed{)}$ $\boxed{\sqrt{x}}$ → 2.102735732

or

$c \approx$ $\boxed{\sqrt{}}$ $($ 1.8 $\boxed{x^2}$ $\boxed{+}$ 2.4 $\boxed{x^2}$ $\boxed{-}$ 2 $\boxed{\times}$ 1.8 $\boxed{\times}$ 2.4 $\boxed{\times}$ $\boxed{\cos}$ 58 $)$ $\boxed{\text{EXE}}$ 2.102735732

$c \approx 2.10$ m *Ans*

2. A surveyor wishes to measure the distance between two horizontal points. The two points, A and B, are separated by a pond and the distance cannot be directly measured. The surveyor does the following:

On the same side of the pond as point A, a third point (point C) is measured at a distance of 150.00 feet toward the pond from point A. From point A, point C is sighted. Then the transit telescope is turned to point B across the pond. An angle of 29°30′ is read on the transit. The surveyor moves to point C. From point C, point A is sighted. Then the transit telescope is turned to point B across the pond. An angle of 138°0′ is recorded. Determine the distance between point A and point B. Round the answer to the nearest tenth foot.

Solution. Sketch and label an oblique triangle as shown in Figure 36–28. Compute *AB*. Two angles and a side are known. Apply the Law of Sines. Since ∠*B* lies opposite the known side, 150.00 feet, ∠*B* must first be determined.

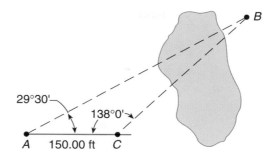

Figure 36–28

$$\angle B = 180° - (29°30' + 138°0') = 12°30'$$

Applying the Law of Sines, $\dfrac{AB}{\sin \angle C} = \dfrac{AC}{\sin \angle B}$

$$\frac{AB}{\sin 138°0'} = \frac{150.00 \text{ ft}}{\sin 12°30'}$$

$$AB = \frac{\sin 138°0'(150.00 \text{ ft})}{\sin 12°30'}$$

Calculator Application

$AB \approx 138$ ⎡sin⎤ ⎡×⎤ 150 ⎡÷⎤ 12.30 ⎡2nd⎤ ⎡▶DD⎤ ⎡sin⎤ ⎡▶DD⎤ 463.7302254

or

$AB \approx$ ⎡sin⎤ 138 ⎡×⎤ 150 ⎡÷⎤ ⎡sin⎤ 12 ⎡°′″⎤ 30 ⎡°′″⎤ ⎡EXE⎤ 463.7302254

$AB \approx 463.7$ ft *Ans*

3. A piece of land is measured off as shown in Figure 36–29. Sides *AB* and *DC* are parallel. Compute ∠*A*.

Solution. Angle *A* is computed by forming an oblique triangle that contains ∠*A*. Refer to Figure 36–30. Round the answer to the nearest hundredth degree.

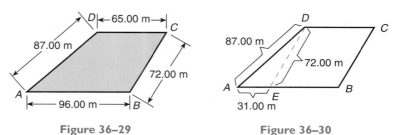

Figure 36–29   **Figure 36–30**

Project line segment *DE* parallel to side *BC*. Oblique triangle *AED* is formed.

In triangle *AED*, *AD* = 87.00 m, *ED* = *BC* = 72.00 m (*EDCB* is a parallelogram).

EB = *DC* = 65.00 m

AE = 96.00 m − 65.00 m = 31.00 m

Solve for ∠*A*. Three sides of triangle *AED* are known. Apply the Law of Cosines.

$$\cos \angle A = \frac{(87.00 \text{ m})^2 + (31.00 \text{ m})^2 - (72.00 \text{ m})^2}{2(87.00 \text{ m})(31.00 \text{ m})}$$

Calculator Application

or $\boxed{\text{EXE}}$ or $\boxed{\text{EXE}}$

$\cos \angle A \approx 87 \boxed{x^2} \boxed{+} 31 \boxed{x^2} \boxed{-} 72 \boxed{x^2} \boxed{\blacktriangleright\underline{\text{DD}}} \boxed{\div} \boxed{(} \boxed{(} 2 \boxed{\times} 87 \boxed{\times} 31 \boxed{)} \boxed{\blacktriangleright\underline{\text{DD}}} 0.620318873$

$\angle A \approx .620318873 \boxed{\text{2nd}} \boxed{\cos} \rightarrow 51.66057599$

or

$\angle A \approx \boxed{\text{SHIFT}} \boxed{\cos} .620318873 \boxed{\text{EXE}} 51.66057599$

$\angle A \approx 51.66° \, Ans$

4. A metal plate is to be machined to the dimensions shown in Figure 36–31. Compute dimension x. Round the answer to 3 decimal places.

Analyze the problem.

Refer to Figure 36–32. Project an auxiliary line segment from point A to point C. Two oblique triangles are formed, $\triangle ABC$ and $\triangle ACD$.

If AC can be determined, side x can be computed from the Law of Sines, since AD, AC, and $\angle D$ would be known in $\triangle ACD$.

Determine whether enough information is given to solve for AC. In oblique $\triangle ABC$, $AB = 12.300$ in, $BC = 8.900$ in, and $\angle B = 72°0'$. Side AC can be computed using the Law of Cosines.

Write the computations.

In $\triangle ABC$, solve for AC. Use the Law of Cosines.

$$AC^2 = (12.300 \text{ in})^2 + (8.900 \text{ in})^2 - 2(12.300 \text{ in})(8.900 \text{ in})(\cos 72°0')$$

$$AC^2 = \sqrt{(12.300 \text{ in})^2 + (8.900 \text{ in})^2 - 2(12.300 \text{ in})(8.900 \text{ in})(\cos 72°0')}$$

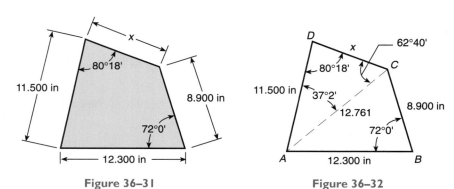

Figure 36–31   Figure 36–32

Calculator Application

$AC \approx \boxed{(} 12.3 \boxed{x^2} \boxed{+} 8.9 \boxed{x^2} \boxed{-} 2 \boxed{\times} 12.3 \boxed{\times} 8.9 \boxed{\times} 72 \boxed{\cos} \boxed{)} \boxed{\sqrt{x}} \rightarrow 12.76102736$

or

$AC \approx \boxed{\sqrt{}} \boxed{(} 12.3 \boxed{x^2} \boxed{+} 8.9 \boxed{x^2} \boxed{-} 2 \boxed{\times} 12.3 \boxed{\times} 8.9 \boxed{\times} \boxed{\cos} 72 \boxed{)} \boxed{\text{EXE}}$
12.76102736

$AC \approx 12.761$ in

In $\triangle ACD$ solve for dimension x. Angle ACD must first be computed using the Law of Sines.

$$\frac{AD}{\sin \angle ACD} = \frac{AC}{\sin 80°18'}$$

$$\frac{11.500 \text{ in}}{\sin \angle ACD} = \frac{12.761 \text{ in}}{\sin 80°18'}$$

$$\sin \angle ACD = \frac{11.500 \text{ in} (\sin 80°18')}{12.761 \text{ in}}$$

Calculator Application

sin ∠ACD ≈ 11.5 $\boxed{\times}$ 80.18 $\boxed{\text{2nd}}$ $\boxed{\blacktriangleright\underline{DD}}$ $\boxed{\sin}$ $\boxed{\div}$ 12.761 $\boxed{\blacktriangleright\underline{DD}}$ 0.888299498

or

sin ∠ACD ≈ 11.5 $\boxed{\times}$ $\boxed{\sin}$ 80 $\boxed{\circ\,'\,''}$ 18 $\boxed{\circ\,'\,''}$ $\boxed{\div}$ 12.761 $\boxed{\text{EXE}}$ 0.888299498

∠ACD ≈ .888299498 $\boxed{\text{2nd}}$ $\boxed{\sin}$ $\boxed{\text{3rd}}$ $\boxed{\blacktriangleright\underline{DD}}$ 62°39′37″2

or

∠ACD ≈ $\boxed{\text{SHIFT}}$ $\boxed{\sin}$ .888299498 $\boxed{\text{EXE}}$ $\boxed{\text{SHIFT}}$ $\boxed{\circ\,'\,''}$ → 62°39′37.2″

Compute ∠DAC.

∠DAC ≈ 180° − (80°18′ + 62°40′) = 37°2′

Compute dimension x. Use the Law of Sines.

$$\frac{x}{\sin 37°2'} \approx \frac{12.761 \text{ in}}{\sin 80°18'}$$

$$x \approx \frac{\sin 37°2' \,(12.761 \text{ in})}{\sin 80°18'}$$

Calculator Application

x ≈ 37.02 $\boxed{\text{2nd}}$ $\boxed{\blacktriangleright\underline{DD}}$ $\boxed{\sin}$ $\boxed{\times}$ 12.761 $\boxed{\div}$ 80.18 $\boxed{\text{2nd}}$ $\boxed{\blacktriangleright\underline{DD}}$ $\boxed{\sin}$ $\boxed{\blacktriangleright\underline{DD}}$ 7.797161682

or

x ≈ $\boxed{\sin}$ 37 $\boxed{\circ\,'\,''}$ 2 $\boxed{\circ\,'\,''}$ $\boxed{\times}$ 12.761 $\boxed{\div}$ $\boxed{\sin}$ 80 $\boxed{\circ\,'\,''}$ 18 $\boxed{\circ\,'\,''}$ $\boxed{\text{EXE}}$ 7.797161682

x ≈ 7.797 in *Ans*

EXERCISE 36–14

Solve these problems. Compute unknown linear values to two decimal places unless otherwise stated, customary angular values to the nearest minute, and metric angular values to the nearest hundredth of a degree.

1. Two sides of a triangular-shaped template are 9.300 inches and 8.600 inches. An angle of 57°0′ lies opposite the 9.300-inch side. Compute the angle that lies opposite the 8.600-inch side.

2. A triangular-shaped piece of land is to be fenced in. Two sides of the property are 120.0 feet and 160.0 feet. The included angle between these two sides is 62°0′. What is the length of fencing required for the third side? Round the answer to 1 decimal place.

3. Center locations of 3 holes are laid out on a piece of sheet metal. The centerline distance between hole 1 and hole 2 is 5.60 cm, between hole 1 and hole 3 it is 6.50 cm, and between hole 2 and hole 3 it is 6.10 cm.

   a. Compute the angle made by the centerlines at hole 1.

   b. Compute the angle made by the centerlines at hole 2.

   c. Compute the angle made by the centerlines at hole 3.

4. Sides AB and CD of the baseplate in Figure 36–33 are parallel. Compute ∠A.

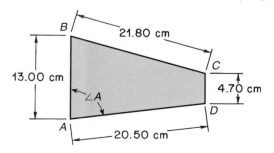

Figure 36–33

5. Three circles are to be cut out of the plywood panel in Figure 36–34. The 8.0-inch diameter and 11.0-inch diameter circles are each tangent to the 15.0-inch diameter circle. Determine the distance from the center of the 8.0-inch diameter circle to the center of the 11.0-inch diameter circle. Round the answer to 1 decimal place.

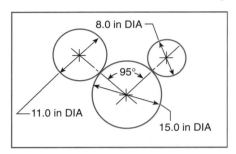

8.0 in DIA

95°

11.0 in DIA

15.0 in DIA

Figure 36–34

6. A cabinetmaker is to make a countertop to the dimensions given in Figure 36–35. Compute the length of side *AB*.

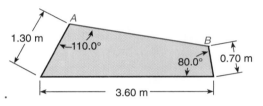

A

1.30 m

110.0°

B

80.0°   0.70 m

3.60 m

Figure 36–35

7. In laying out an acute oblique triangle, a drafter draws a line segment 5.0 inches long. From one end point of the 5.0-inch segment, a line segment is drawn at an angle of 48°. From the other end point of the 5.0-inch segment, another line segment is drawn at an angle of 59°.

   a. Find the length of the side opposite the 48° angle.

   b. Find the length of the side opposite the 59° angle.

   Round the answers to 1 decimal place.

8. An unequally pitched roof has a front rafter 4.500 meters long and a rear rafter 7.000 meters long. The horizontal distance (span) between the front and rear end points of the rafters is 8.500 meters.

   a. Compute the angle made by the front rafter and the horizontal (span).

   b. Compute the angle made by the rear rafter and the horizontal (span).

9. Compute dimension *b* of the drill jig in Figure 36–36. Express the answer to 3 decimal places.

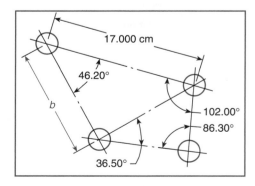

17.000 cm

46.20°

b

102.00°

86.30°

36.50°

Figure 36–36

10. A wall is built from point *A* to point *B* in the section of land shown in Figure 36–37. Sides *AC* and *BD* are parallel. Determine the length of *AB*. Round the answer to the nearest foot.

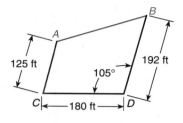

Figure 36–37

11. Compute ∠*B* of the template in Figure 36–38.

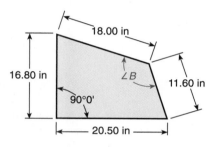

Figure 36–38

12. A frame in the shape of a parallelogram is to be strengthened by the addition of a diagonal brace fastened on opposite corners of the frame. The frame has two sides each 2.30 meters long and two sides each 3.70 meters long. The brace is to lie opposite an included angle of 115°, which is made by two sides of the frame. Compute the required length of the brace.

13. A structural section in the shape of an oblique triangle with 30 feet 0 inches, 22 feet 0 inches, and 16 feet 0 inches sides is to be fabricated.

   a. Compute the angle opposite the 22 feet 0 inches side.

   b. Compute the angle opposite the 16 feet 0 inches side.

   c. A brace is to be made that extends from the midpoint of the 30 feet 0 inches side to the vertex where the 16 feet 0 inches and 22 feet 0 inches sides meet. What is the required length of the brace? Round the answer to the nearest inch.

14. In a proposed housing development plan, 4 streets meet as shown in Figure 36–39. Compute ∠*A*.

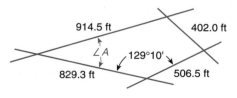

Figure 36–39

15. Determine ∠x for the pattern in Figure 36–40.

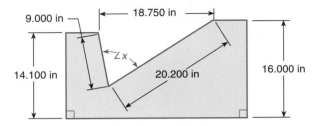

Figure 36–40

16. A concrete platform is shown in Figure 36–41. Compute ∠A.

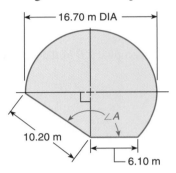

Figure 36–41

17. A surveyor wishes to measure the distance between two horizontal points. The two points, A and B, are separated by a river and cannot be directly measured. The surveyor does the following:

On the same side of the river as point A, a third point (point C) is measured at a distance of 120.00 feet toward the river from point A. From point A, point C is sighted. Then the transit telescope is turned to point B across the river. An angle of 33°20′ is read on the transit. The surveyor moves to point C. From point C, point A is sighted. Then the transit telescope is turned to point B across the river. An angle of 132°0′ is recorded.

Determine the distance between point A and point B.

18. A length of sidewalk is to be constructed in a square lot. Each side of the lot is 130.0 feet long. The sidewalk is laid out as follows:

Stakes are driven at two opposite corners of the lot, and a string is stretched between the two stakes. From one of the staked corners, a distance of 60.0 feet is measured along the string, and a third stake is driven. From this stake, a measurement is made to one of the corners of the lot where a stake has not been driven.

What is the length of this measurement?

19. Compute ∠x of the fixture hole location in Figure 36–42.

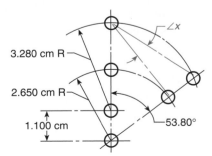

Figure 36–42

20. In determining the height of a tower, a surveyor made the distance and angular measurements shown in Figure 36–43. What is the height of the tower? Round the answer to 4 significant digits.

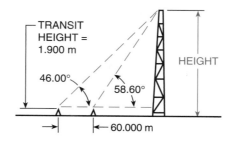

Figure 36–43

21. Compute $\angle D$ of the template in Figure 36–44.

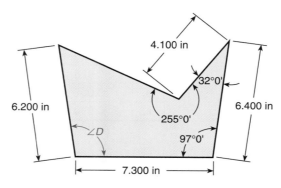

Figure 36–44

UNIT EXERCISE AND PROBLEM REVIEW

DETERMINING FUNCTIONS OF ANGLES IN ANY QUADRANT

Determine the sine, cosine, tangent, cotagent, secant, and cosecant of these angles. For each angle, sketch a right triangle similar to those in Figures 36–3, 36–4, and 36–5. Label the sides of the triangles + or −. Determine the reference angles and functions of the angles. Round the answers to 5 significant digits.

1. 115°

2. 220°

3. 290°20′

4. 95.6°

5. 580°

6. 490°

7. 739.7°

8. 1,060°50′

DETERMINING INSTANTANEOUS VOLTAGES

9. Compute the instantaneous voltage (*e*) to the nearest tenth volt of an alternating electromotive force (emf) for each of these problems. Use the formula

$$e = E_{max} \sin \theta$$

where e = instantaneous voltage
E_{max} = maximum voltage
θ = angle in degrees

	Number of Degrees Reached in Cycle (θ)	Maximum Voltage (E_{max})	Instantaneous Voltage (e)
a.	140°	240.0 volts	
b.	235°	600.0 volts	
c.	310°	120.0 volts	
d.	180°	800.0 volts	
e.	340°	300.0 volts	

10. Each of these problems has a 60-cycle alternating emf. The angular velocity is 21,600° per second. Compute the instantaneous voltage (e) to the nearest tenth volt for each problem. Compute the angle in degrees (θ); then apply the formula $e = E_{max} \sin \theta$.

	Time	Maximum Voltage (E_{max})	Instantaneous Voltage (e)
a.	0.015 second	550.0 volts	
b.	0.020 second	120.0 volts	
c.	0.026 second	600.0 volts	
d.	0.007 second	240.0 volts	
e.	0.030 second	780.0 volts	

SOLVING OBLIQUE TRIANGLE PROBLEMS USING THE LAW OF SINES

Solve these problems by using the Law of Sines. Compute side lengths to three decimal places, customary angular values to the nearest minute, and metric angular values to the nearest hundredth of a degree.

11. Determine side *a*.

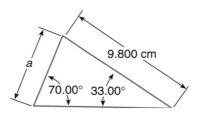

13. Determine side *x*.

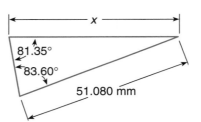

15. Determine side *c*.

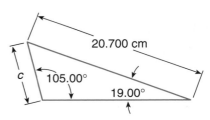

12. Determine ∠*B*.

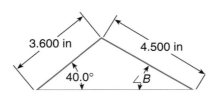

14. Determine ∠*A*.

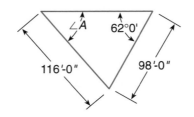

16. Determine ∠*M*.

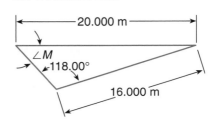

17. a. Determine $\angle B$.
    b. Determine side b.

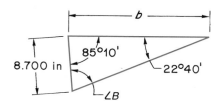

19. a. Determine $\angle A$.
    b. Determine side a.

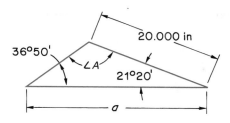

21. a. Determine $\angle E$.
    b. Determine side e.
    c. Determine side f.

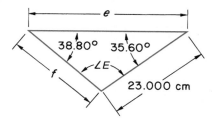

18. a. Determine $\angle F$.
    b. Determine $\angle G$.

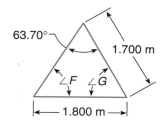

20. a. Determine $\angle A$.
    b. Determine $\angle B$.
    c. Determine side b.

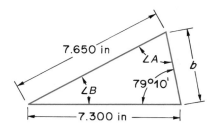

22. a. Determine $\angle N$.
    b. Determine $\angle P$.
    c. Determine side p.

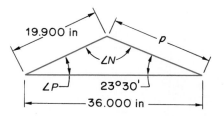

SOLVING OBLIQUE TRIANGLE PROBLEMS USING THE LAW OF COSINES

Solve these problems by using the Law of Cosines. Also, apply the Law of Sines in the solution of certain problems. Compute side lengths to three decimal places, customary angular values to the nearest minute, and metric angular values to the nearest hundredth of a degree.

23. a. Determine side a.

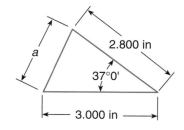

25. Determine side m.

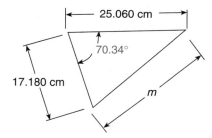

27. Determine side a.

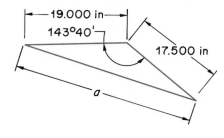

24. Determine $\angle B$.

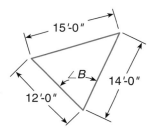

26. Determine $\angle C$.

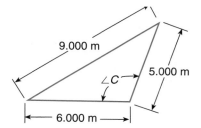

28. Determine $\angle B$.

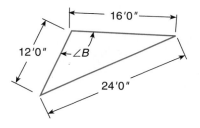

29. a. Determine side *d*.
   b. Determine ∠*E*.
   c. Determine ∠*F*.

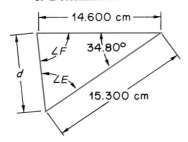

31. a. Determine side *d*.
   b. Determine ∠*E*.
   c. Determine ∠*F*.

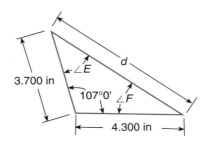

33. a. Determine side *a*.
   b. Determine ∠*B*.
   c. Determine ∠*C*.

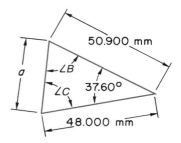

30. a. Determine ∠*A*.
   b. Determine ∠*B*.
   c. Determine ∠*C*.

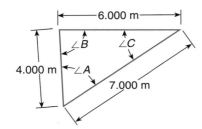

32. a. Determine ∠*B*.
   b. Determine ∠*C*.
   c. Determine ∠*D*.

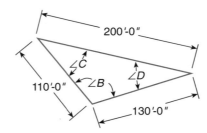

34. a. Determine ∠*M*.
   b. Determine ∠*R*
   c. Determine ∠*T*.

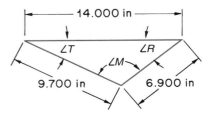

SOLVING APPLIED OBLIQUE TRIANGLE PROBLEMS

Solve these problems. Compute unknown linear values to two decimal places unless otherwise noted, customary angular values to the nearest minute, and metric angular values to the nearest hundredth of a degree.

35. A triangular-shaped panel is to be machined. Two sides are given as 14.200 inches and 11.700 inches with an included angle of 70°20′. Determine the length of the third side. Round the answer to 3 decimal places.

36. A triangular roof truss has sides of 9.00 meters, 7.00 meters, and 6.00 meters.

   a. Compute the angle opposite the 6.00-meter side.

   b. Compute the angles opposite the 7.00-meter side.

   c. Compute the angle opposite the 9.00-meter side.

37. Determine ∠*A* of the structural frame in Figure 36–45.

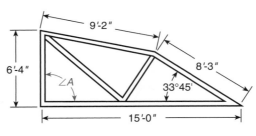

Figure 36–45

38. Determine side x of the assembly cover plate in Figure 36–46.

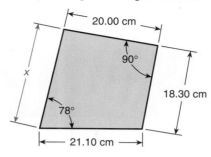

Figure 36–46

39. Three externally tangent holes are to be bored in a casting. The hole diameters are 3.400 inches, 2.800 inches, and 2.200 inches. Compute the 3 angles of a triangle made by connecting line segments between the centers of the 3 holes.

40. A building lot in the shape of a parallelogram has 2 sides each 65.0 meters long and 2 sides each 40.0 meters long. A measurement of 90.0 meters across opposite acute angle corners of the lot is made. What are the values of the 2 pairs of opposite angles of the lot? Round the answers to the nearest degree.

41. Compute dimension x of the template in Figure 36–47.

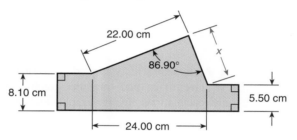

Figure 36–47

42. A piece of stock is to be machined as shown in Figure 36–48. Determine dimension b. Compute the answer to 3 decimal places.

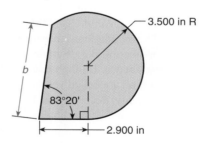

Figure 36–48

43. A surveyor wishes to measure the distance between two horizontal points. The two points, A and B, are separated by an obstruction and cannot be directly measured. The surveyor does the following:

On the same side of the obstruction as point A, a third point (point C) is measured at a distance of 50.000 meters toward the obstruction from point A. From point A, point C is sighted. Then the transit telescope is turned to point B on the other side of the obstruction. An angle of 40.50° is read on the transit. The surveyor moves to point C. From point C, point A is sighted. Then the transit telescope is turned to point B on the other side of the obstruction. An angle of 121.30° is recorded.

Determine the distance between point A and point B.

UNIT 37 ::: Vectors

OBJECTIVES

After studying this unit you should be able to

- identify vector and scalar quantities.
- identify vectors in standard position.
- determine vector sums (resultant vectors) graphically.
- determine vector sums (resultant vectors) using trigonometry.
- determine component vectors using trigonometry.
- solve applied problems using methods of determining resultant vectors.

Vectors are widely applied in physical science, navigation, and engineering fields. Electrical, electronics, and computer graphics, as well as mechanical and construction technologies and occupations, involve graphic and trigonometric applications of vectors.

37–1 Scalar and Vector Quantities

Most quantities are completely described by stating the magnitude of the quantity. *Magnitude* is the size or amount of a quantity; quantities of this kind are called *scalar* quantities. Weight, length, area, volume, and time are some examples of scalar quantities.

To be completely described, some quantities require direction to be specified as well as size or amount. For example, a distance may be described as 25 miles due south, or a force may be described as 1500 pounds exerted at 35 degrees to the right above the horizontal. Quantities of this kind are called vectors. A *vector* is a quantity that has both magnitude and direction.

Velocity, acceleration, force, and displacement (position changes) are common examples of vectors in the two-dimensional (x,y) coordinate system. Only vectors in the two-dimensional coordinate system are presented in this book. Vectors in the three-dimensional (x,y,z) coordinate system are not discussed.

37–2 Description and Naming of Vectors

Vectors are shown as directed line segments. The length of the segment represents the magnitude, and the arrowhead represents the direction of the quantity. Vectors have an *initial point* and a *terminal point.* An arrowhead represents the terminal point.

A vector is named by its two end points or by a single lowercase letter. Arrows are placed above the line segments. Three vectors are shown in Figure 37–1; $\overrightarrow{AB}$, $\overrightarrow{CD}$, and $\overrightarrow{e}$ are vectors. In $\overrightarrow{AB}$ the initial point is point A, and the terminal point is point B. In $\overrightarrow{CD}$ the initial point is point C, and the terminal point is point D. In $\overrightarrow{e}$ the terminal point is the end point indicated by the arrowhead.

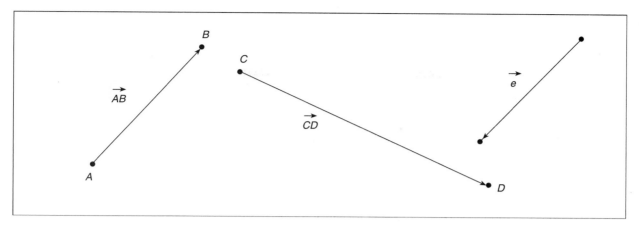

Figure 37–1

The naming of vectors is not entirely uniform. In this book, vectors are always named with an arrow. The lengths (magnitudes) are shown as line segments, which is consistent with the way lengths have been shown.

$\overrightarrow{AB}$ is a vector; AB is the length (magnitude) of $\overrightarrow{AB}$.

$\vec{e}$ is a vector; e is the length (magnitude) of $\vec{e}$.

Equal vectors have identical magnitudes and directions. A vector can be repositioned, provided its magnitude and direction remain the same. Equal vectors are shown in Figure 37–2; unequal vectors are shown in Figure 37–3.

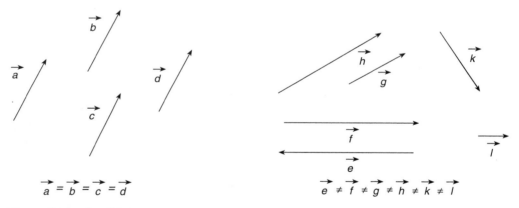

Figure 37–2   Equal Vectors

Figure 37–3   Unequal Vectors

Vectors are usually shown on the rectangular coordinate (x,y) system. A vector is in *standard position* when the initial point is at the origin of the rectangular coordinate (x,y) system and its angle is measured counterclockwise from the positive x-axis.

37–3   Vector Ordered Pair Notation

A vector can be represented on the rectangular coordinate system using *ordered pair notation* (x,y). A vector can be represented by the x and y coordinates of its terminal point. The initial point is at the origin $(0,0)$. The horizontal change from the initial point to the terminal point is represented by x. The vertical change from the initial point to the terminal point is represented by y. Vectors $\overrightarrow{OA}$ and $\overrightarrow{OB}$ are shown in Figure 37–4; $\overrightarrow{OA} = (5,3)$ and $\overrightarrow{OB} = (-2,6)$.

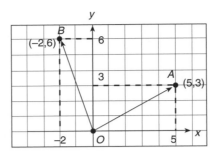

Figure 37–4

In most practical applications, such as force, velocity, and displacement, the ordered pair notation vector representation is seldom used.

37–4   Vector Length and Angle Notation

Instead of ordered pair notation, vectors are shown and solved as lengths and angles, usually in standard position. The angle is measured as a counterclockwise rotation from the x-axis. Three vectors, $\vec{a}$, $\vec{b}$, and $\vec{c}$ are shown in standard position in Figure 37–5. Vector angles are often represented as θ (theta).

$a = 20$ cm, $\theta_1 = 30°$, $\vec{a} = 20$ cm at $30°$
$b = 28$ cm, $\theta_2 = 110°$, $\vec{b} = 28$ cm at $110°$
$c = 35$ cm, $\theta_3 = 205°$, $\vec{c} = 35$ cm at $205°$

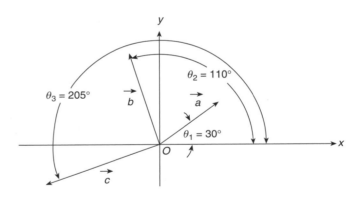

Figure 37–5

37–5   Adding Vectors

A *resultant vector* is the sum of two or more vectors. *Component vectors* are the vectors that are added to produce the resultant vector.

> **The resultant vector produces the same effect as the combination of two or more component vectors. The resultant vector can be substituted for the component vectors or the component vectors can be substituted for the resultant vector.**

Adding vector quantities is *not* the same process as adding scalar quantities; vector addition is not the same as arithmetic or algebraic addition. For example, in arithmetic addition,

5 feet + 3 feet = 8 feet. In vector addition, the sum of a vector with length of 5 feet and a vector with a length of 3 feet could perhaps result in a vector with length of 7 feet.

$$\text{If } \vec{a} + \vec{b} = \vec{c}, \text{ then } a + b > \text{ or } = c$$

where a, b, and c are the lengths of $\vec{a}$, $\vec{b}$, and $\vec{c}$, respectively.

Usually $a + b > c$. The only way $a + b = c$ can occur is when a and b have exactly the same direction.

There are two basic methods of adding vectors: the graphic method and the trigonometric method. The graphic method, using a ruler and a protractor, can be used in applications where a relatively low degree of accuracy is sufficient. The trigonometric method of vector addition permits the sum of the vectors to be computed to virtually any degree of accuracy. Both magnitude and direction can be computed to a great number of significant digits. Depending on a given situation, either right triangle or oblique triangle computations are used in adding vectors.

EXERCISE 37–5

1. Which of the following quantities are vectors?
   a. 65 miles per hour
   b. 20.3 centimeters to the right along the y-axis
   c. 698 pounds per square foot
   d. 150 kilograms vertically down
   e. 17.06 meters per second
   f. 514 feet due west
   g. 28.5 inches from the starting point

2. Which of the following are vectors?

   a.   b.   c.   d.   e.   f.

3. Which of the following name a vector?
   a. $\overleftrightarrow{AB}$ b. $\overrightarrow{AB}$ c. AB d. $\overline{AB}$ e. v f. $\vec{v}$ g. $\bar{v}$

4. Which pairs of vectors are equal?

   a.   b.   c.   d.

   e.   f.

5. Which of the following represents a vector in standard position on the rectangular coordinate system?

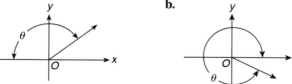

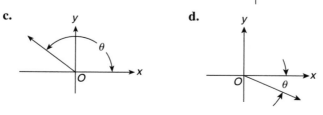

6. Which of the following represents a vector on the rectangular coordinate system using ordered pair notation?

a.

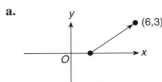

b.

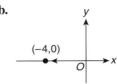

c.

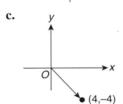

d.

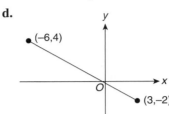

7. Given: $\vec{a} + \vec{b} = \vec{c}$; a, b, and c are the vector lengths. If $c = 15$ km,

    a. Can $a + b = 17$ km?

    b. Can $a + b = 14$ km?

    c. Can $a + b = 0.10$ km?

    d. Can $a + b = 20.00$ km?

37–6   Graphic Addition of Vectors

Vectors are drawn and measured to an appropriate scale using a ruler and a protractor. The larger the vectors are drawn, the greater the accuracy. The resultant is in standard position. Two methods of graphic addition are used. Both the parallelogram method and triangle or head-to-tail method are presented.

Parallelogram Method of Adding Vectors Graphically

Vectors $\vec{a}$ and $\vec{b}$ are shown in Figure 37–6. To determine the resultant ($\vec{R}$) as shown in Figure 37–7, do the following:

- Draw $\vec{a}$ and $\vec{b}$ in standard position.
- From the terminal point of $\vec{a}$ draw a vector equal to $\vec{a}$, and from the terminal point of $\vec{b}$ draw a vector equal to $\vec{b}$ to form a parallelogram.
- Draw a diagonal from the origin (O) to the opposite vertex of the parallelogram.
- The diagonal is the sum or resultant ($\vec{R}$) of $\vec{a}$ and $\vec{b}$.
- Measure the length of $\vec{R}$ and measure the angle (θ) that $\vec{R}$ makes with the horizontal (x-axis). The angle (θ) is measured as a counterclockwise rotation from the x-axis.

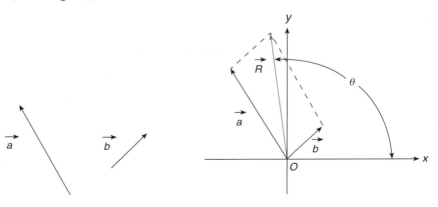

Figure 37–6   Figure 37–7

Triangle (Head-to-Tail) Method of Adding Vectors Graphically

Vectors $\vec{a}$ and $\vec{b}$ are shown in Figure 37–8. Observe that Figures 37–6 and 37–8 are identical. To determine the resultant ($\vec{R}$) as shown in Figure 37–9, do the following:

- Draw either $\vec{a}$ or $\vec{b}$ in standard position. In this case $\vec{a}$ is in standard position.
- Draw $\vec{b}$ with its initial point on the terminal point of $\vec{a}$.
- Draw a vector from the initial point of $\vec{a}$ to the terminal point of $\vec{b}$. This vector is the sum or resultant ($\vec{R}$) of $\vec{a}$ and $\vec{b}$.
- Measure the length of $\vec{R}$ and measure the angle θ that $\vec{R}$ makes with the horizontal (x-axis). The angle (θ) is measured as a counterclockwise rotation from the x-axis.

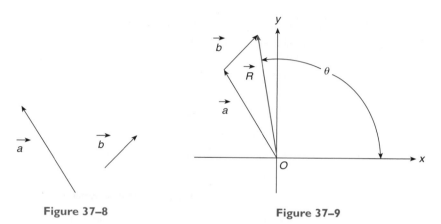

Figure 37–8   Figure 37–9

It is more efficient to use the triangle method of vector addition than the parallelogram method when determining the sum (resultant) of three or more vectors. An example with six vectors, $\vec{a}$, $\vec{b}$, $\vec{c}$, $\vec{d}$, $\vec{e}$, and $\vec{f}$ is shown in Figure 37–10. The triangle method of vector addition with the resultant ($\vec{R}$) in standard position is shown in Figure 37–11.

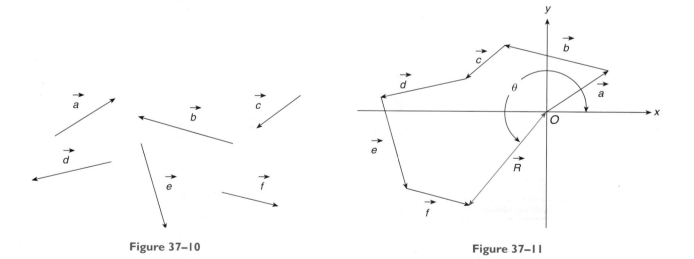

Figure 37–10   Figure 37–11

Graphic Addition Force Application

A *force* is a vector quantity that can change the state of rest or motion of a body; it is a push or pull.

A *contact force* is a force that is in contact with a body. Forces are widely used in construction and mechanical technology design computations.

EXAMPLE

A force of 800 pounds pulls upward and to the right on a body at an angle of 35° with the horizontal. A second force of 650 pounds pulls vertically up on the body. Find the vector sum (resultant vector) of the two forces in standard position.

Determine a practical and convenient scale. In this case make 1 inch = 500 pounds and calculate vector lengths. The first force is represented by $\vec{a}$ and the second force by $\vec{b}$.

$$\text{For } \vec{a}, a = \text{length of } \vec{a}: \frac{a}{1 \text{ in}} = \frac{800 \text{ lb}}{500 \text{ lb}}, a = 1.6 \text{ in}$$

$$\text{For } \vec{b}, b = \text{length of } \vec{b}: \frac{b}{1 \text{ in}} = \frac{650 \text{ lb}}{500 \text{ lb}}, b = 1.3 \text{ in}$$

Both methods of vector addition are shown. Figure 37–12 shows the parallelogram method and Figure 37–13 shows a triangle or head-to-tail method. Use a ruler and a protractor. Draw vectors $\vec{a}$ and $\vec{b}$ in standard position using the procedures given on pages 778 and 779.

Make $a \approx 1.6$ in and $b \approx 1.3$ in.

Measure an angle of 35° for $\vec{a}$ and 90° for $\vec{b}$.

Draw the resultant vector ($\vec{R}$) using the procedures previously given. Measure the length of $\vec{R}$; $R \approx 2.6$ in.

Calculate the force in pounds of R.

$$\frac{R}{500 \text{ lb}} \approx \frac{2.6 \text{ in}}{1 \text{ in}}, R \approx 1300 \text{ lb}$$

With a protractor, measure the angle θ of $\vec{R}$, $\theta \approx 59°$

$\vec{R} \approx 1300$ lb at 59° *Ans*

A single force of 1300 lb at 59° exerts the same force as a force of 800 lb at 35° and a force of 650 lb at 90° combined.

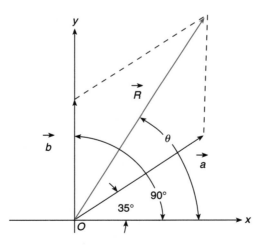

Figure 37–12   Parallelogram Method

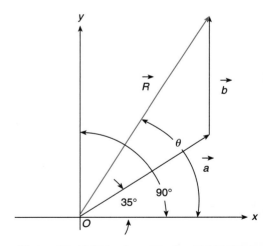

Figure 37–13   Triangle or Head-to-Tail Method

Graphic Addition Displacement Application

Displacement is a vector quantity that is a change of position of a body. It is the distance and the angle between the starting (initial) point and the end (terminal) point of a body.

EXAMPLE •————————————————————————————————————

The flight pattern of a projectile is as follows:

> The projectile first travels 140 miles at an angle of 20° west of north, then travels 120 miles at an angle of 50° east of north, then travels 200 miles at an angle of 75° east of north.

Determine the change in position (displacement) in standard position of the body. Let $\vec{a}$, $\vec{b}$, and $\vec{c}$ be the component vectors and $\vec{R}$ the resultant vector.

$$\vec{a} = 140 \text{ mi at } 110° \ (90° + 20° = 110°)$$
$$\vec{b} = 120 \text{ mi at } 40° \ (90° - 50° = 40°)$$
$$\vec{c} = 200 \text{ mi at } 15° \ (90° - 75° = 15°)$$

A scale of 1 in = 100 mi is used. The lengths to scale and the angles are drawn as shown in Figure 37–14. The resultant vector ($\vec{R}$) is drawn and measured:

$$\vec{R} \approx 350 \text{ mi and } \theta \approx 48°.$$

The displacement ($\vec{R}$) ≈ 350 mi at 48° *Ans*

A displacement of 350 mi at 48° is the same displacement as the combined displacements of 140 mi at 110°, 120 mi at 40°, and 200 mi at 15°.

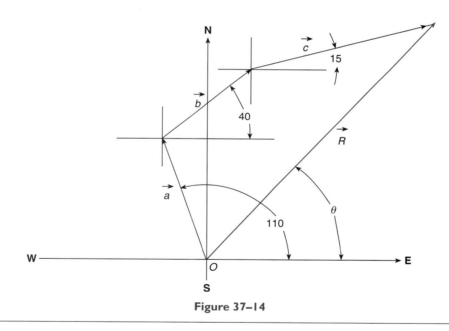

Figure 37–14

EXERCISE 37–6 ——

Make a scale drawing and determine the vector sum (resultant vector) in standard position of each exercise. Express magnitudes to two significant digits and angles to the nearest degree. Use either the parallelogram or triangle (head-to-tail) method for exercises 1–15.

1. Forces of 820 lb horizontally to the left and 460 lb vertically up.
2. Velocities of 85 mi/h due south and 55 mi/h due east.
3. Displacements of 23 km due north and 18 km due east.
4. Forces of 1500 lb vertically down and 2100 lb horizontally to the right.
5. Displacements of $9\overline{0}$ mi due west and 130 mi due south.

Exercises 6–19 are given in standard position.

6. $\vec{a} = 5\overline{0}0$ lb at 0°, $\vec{b} = 8\overline{0}0$ lb at 90°

7. $\vec{c} = 250$ mi at 60°, $\vec{d} = 7\overline{0}$ mi at 35°

8. $\vec{e} = 75$ km/h at 115°, $\vec{f} = 6\overline{0}$ km/h at 87°

9. $\vec{v} = 160$ ft/s at 152°, $\vec{w} = 190$ ft/s at 190°

10. $\vec{a} = 1600$ lb at 212°, $\vec{b} = 590$ lb at 262°

11. $\vec{c} = 710$ m at 235°, $\vec{d} = 940$ m at 340°

12. $\vec{e} = 66$ mi/h at 88°, $\vec{f} = 53$ mi/h at 209°

13. $\vec{g} = 32{,}000$ lb at 307°, $\vec{h} = 19{,}000$ lb at 14°

14. $\vec{v} = 27$ km at 77°, $\vec{w} = 39$ km at 183°

15. $\vec{b} = 52$ mi at 293°, $\vec{c} = 49$ mi at 94°

For exercises 16–19, draw the vectors using the triangle (head-to-tail) method.

16. $\vec{a} = 680$ lb at 173°
    $\vec{b} = 270$ lb at 108°
    $\vec{c} = 410$ lb at 77°

17. $\vec{d} = 71$ mi at 190°
    $\vec{e} = 1\overline{0}0$ mi at 213°
    $\vec{f} = 44$ mi at 280°

18. $\vec{u} = 120$ km/h at 345°
    $\vec{v} = 87$ km/h at 27°
    $\vec{w} = 52$ km/h at 111°
    $\vec{z} = 75$ km/h at 180°

19. $\vec{a} = 53$ m/s at 0°
    $\vec{b} = 76$ m/s at 90°
    $\vec{c} = 93$ m/s at 126°
    $\vec{d} = 7\overline{0}$ m/s at 270°

20. A force of 1,200 pounds pulls upward and to the left on a body at an angle of 34° with the horizontal. A second force of 1,700 pounds pulls horizontally to the right. Find the vector sum (resultant vector) in standard position of the two forces.

21. A projectile travels 210 miles at an angle of 32° east of south and then travels 330 miles at an angle of 38° west of south. Determine the change in position (displacement) of the projectile.

22. From point 0, a vehicle travels to point *A* at 72 kilometers per hour for 2.5 hours in a direction of 40° west of north. From point *A*, the vehicle travels due west to point *B* at 65 kilometers per hour for 1.4 hours. Determine the resultant vector in standard position from point 0 to point *B*.

37–7 Addition of Vectors Using Trigonometry

The sum of vectors can be computed to great accuracy using trigonometry. Both magnitude and direction are determined to virtually any desired number of significant digits. As with the graphic method, both the parallelogram and triangle methods are used in computing vector quantities. In addition, the general component method of trigonometric vector addition is usually used when three or more component vectors are involved.

Recall that the *resultant vector* is the sum of two or more vectors, and *component vectors* are the vectors that are added to produce the resultant vector. In certain applications, the component vectors are given and the resultant vector is computed. In other applications, the resultant vector is given and one or more component vectors are determined.

Resultant vector and component vector computations are shown in the following examples.

Right Triangle Force Application: Determining Resultant Vector

Although either of two procedures can be used to compute $\vec{R}$, generally a procedure using the Pythagorean theorem is applied.

EXAMPLE •————————————————————————————————

Two forces, $\vec{V}_x$ and $\vec{V}_y$, pull on a body as shown in Figure 37–15.

Given: $\vec{V}_x$ = 786.0 lb horizontally to the right and $\vec{V}_y$ = 462.0 lb vertically up.

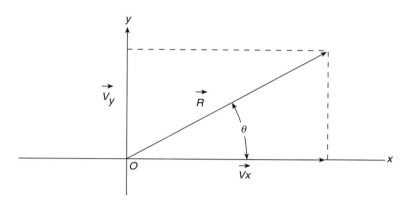

Figure 37–15

Compute the resultant force $\vec{R}$ in standard position.

Since two legs of a right triangle are known, the Pythagorean theorem is used to compute R.

$$R = \sqrt{V_x^2 + V_y^2}$$
$$R = \sqrt{(786.0 \text{ lb})^2 + (462.0 \text{ lb})^2}$$
$$R \approx 911.7 \text{ lb}$$

Solve for θ:

$$\tan \theta = \frac{V_y}{V_x} = \frac{462.0 \text{ lb}}{786.0 \text{ lb}} \approx 0.587786$$

$$\theta \approx 30.45°$$

$$\vec{R} \approx 911.7 \text{ lb at } 30.45° \; Ans$$

——•

Right Triangle Electronics Application: Determining Resultant Vector

In electrical and electronics technology, a vector is used to indicate the magnitude and direction of a quantity such as a voltage or a current. A vector can also represent a resistance, a reactance, or an impedance. Some rotating vectors are called *phasors.* A phasor is a voltage, current, or impedance expressed as a vector.

Many electronics vector applications require at least a basic understanding of electronics theory. Therefore, in this book, a relatively simple alternating current circuit in series application is given.

Figure 37–16 shows the phasors (vectors) for resistance ($\vec{R}$) and capacitive reactance ($\vec{X}_C$). The addition of these vectors results in the alternating current series circuit impedance ($\vec{Z}$). Resistance ($\vec{R}$) is the opposition, expressed in ohms (Ω), that a material has to current flow. Capacitive reactance ($\vec{X}_C$) is the opposition, expressed in ohms (Ω), to an alternating current due to capacitance. Impedance ($\vec{Z}$) is the total opposition, expressed in ohms (Ω), to an alternating current.

NOTE: In actual applications in the electronics field, vector quantities are shown without the arrow above the letter. Vector quantities are shown as Z, R, and X_C rather than $\vec{Z}$, $\vec{R}$, and $\vec{X}_C$. However, for purposes of consistency in this book vector arrows are shown. Direction is often indicated with the symbol, $\angle$. For example, $\vec{R} = 12.6\ \Omega$ at $0°$ is written $\vec{R} = 12.6\ \Omega\ \underline{/0°}$.

EXAMPLE •————————————————————————————————————

Determine the impedance ($\vec{Z}$), given resistance ($\vec{R}$), and capacitive reactance ($\vec{X}_C$) as shown in Figure 37–16.

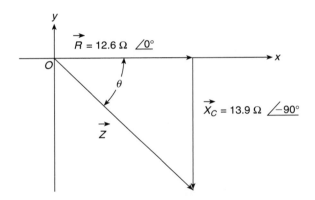

Figure 37–16

$$Z = \sqrt{R^2 + X_C^2}$$
$$Z = \sqrt{(12.6\ \Omega)^2 + (-13.9\ \Omega)^2} \approx 18.8\ \Omega$$
$$\tan\theta = \frac{-13.9\ \Omega}{12.6\ \Omega} \approx -1.103175$$

$$\theta \approx -47.8°$$

$$\vec{Z} \approx 18.8\ \Omega \text{ at } -47.8° \text{ or } 18.8\ \Omega\ \underline{/-47.8°} \quad Ans$$

Right Triangle Velocity Application: Determining Component Vectors

Speed and velocity are often used interchangeably, but they are not the same. Speed indicates how fast an object is moving. Velocity includes direction as well as speed; therefore, velocity is a vector quantity.

EXAMPLE •──────────────────

An airplane flies at 835.6 kilometers per hour at 36.08° west of south as shown in Figure 37–17. Compute the airplane's velocity due west (horizontal component) and velocity due south (vertical component)

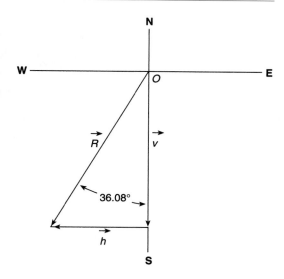

Figure 37–17

Horizontal Component Vector ($\vec{h}$)

$$\sin 36.08° = \frac{h}{835.6 \text{ km/h}}$$

   $h = (\sin 36.08°)(835.6 \text{ km/h})$
   $h \approx (0.588914)(835.6 \text{ km/h})$
   $h \approx 492.1 \text{ km/h}$
   $\vec{h} \approx 492.1 \text{ km/h due west}$
or $\vec{h} \approx 492.1 \text{ km/h at } 180°$
   (standard position) *Ans*

Vertical Component Vector ($\vec{v}$)

$$\cos 36.08° = \frac{v}{835.6 \text{ km/h}}$$

   $v = (\cos 36.08°)(835.6 \text{ km/h})$
   $v \approx (0.808195)(835.6 \text{ km/h})$
   $v \approx 675.3 \text{ km/h}$
   $\vec{v} \approx 675.3 \text{ km/h due south}$
or $\vec{v} \approx 675.3 \text{ km at } 270°$
   (standard position) *Ans*

Oblique Triangle Force Application: Determining Resultant Vector

EXAMPLE •──────────────────

Two forces act on a body as shown in Figure 37–18. Vector $\overrightarrow{OC}$ is $257\overline{0}$ pounds at 143.6° and vector $\overrightarrow{OA}$ is 2045 pounds at 202.8°. Determine the resultant force, $\overrightarrow{OB}$ in standard position.

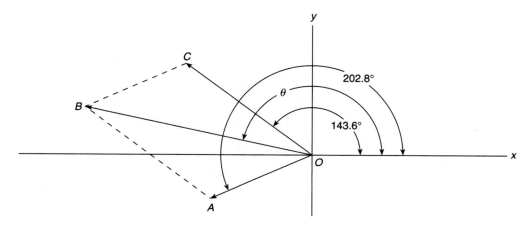

Figure 37–18

In $\triangle OAB$, $OA = 2045$ lb and $AB = OC = 257\overline{0}$ lb. Compute $\angle OAB$, then compute OB using the Law of Cosines. The sum of the interior angles of a parallelogram equals 360°.

Angle $AOC = 202.8° - 143.6° = 59.2°$. Angle $ABC = \angle AOC = 59.2°$ (opposite angles are equal). Angle $OAB = (360° - 2(59.2°)) \div 2 = 120.8°$.

$OB^2 = OA^2 + AB^2 - 2(OA)(AB)(\cos \angle OAB)$

$OB = \sqrt{(2045 \text{ lb})^2 + (257\overline{0} \text{ lb})^2 - 2(2045 \text{ lb})(257\overline{0} \text{ lb})(\cos 120.8°)}$

$OB \approx 4021$ lb

Compute $\angle AOB$ using the Law of Sines.

$$\frac{AB}{\angle AOB} = \frac{\angle OB}{\angle OAB}$$

$$\frac{257\overline{0} \text{ lb}}{\sin \angle AOB} = \frac{4021 \text{ lb}}{\sin 120.8°}$$

$$\sin \angle AOB \approx \frac{257\overline{0} \text{ lb}(\sin 120.8°)}{4021 \text{ lb}}$$

$$\sin \angle AOB \approx 0.5489995$$

$$\angle AOB \approx 33.3°$$

Compute θ.

$\theta \approx 202.8° - 33.3° \approx 169.5°$

Resultant vector $\overrightarrow{OB} \approx 4021$ lb at 169.5° *Ans*

EXERCISE 37–7

Determine the vector sum (resultant vector) in standard position for each of exercises 1–20. Express magnitudes to three significant digits and angles to the nearest tenth degree for all the exercises. It is often helpful to make a rough sketch in solving vector problems using trigonometry.

1. Displacements of 16.4 mi due south and 29.9 mi due west.
2. Forces of 2760 lb vertically up and 3890 lb horizontally to the left.
3. Velocities of 212 km/h due north and 307 km/h due west.
4. Displacements of 343 km due south and 516 km due east.
5. Forces of 12,700 lb horizontally to the right and 8330 lb vertically down.

Exercises 6–20 are given in standard position.

6. $\overrightarrow{v} = 612$ lb at 90.0°, $\overrightarrow{h} = 817$ lb at 0°
7. $\overrightarrow{d} = 14.7$ km at 18$\overline{0}$°, $\overrightarrow{e} = 10.8$ km at 27$\overline{0}$°
8. $\overrightarrow{g} = 70.8$ mi/h at 0°, $\overrightarrow{h} = 65.9$ mi/h at 46.0°
9. $\overrightarrow{a} = 5780$ lb at 35.6°, $\overrightarrow{b} = 3070$ lb at 88.3°
10. $\overrightarrow{k} = 93.6$ ft/s at 12.8°, $\overrightarrow{l} = 106$ ft/s at 92.0°
11. $\overrightarrow{c} = 59.7$ km at 107.0°, $\overrightarrow{d} = 70.6$ km at 168.0°
12. $\overrightarrow{f} = 202$ mi/h at 193.0°, $\overrightarrow{g} = 155.0$ mi/h at 250.7°
13. $\overrightarrow{b} = 998$ lb at 202.0°, $\overrightarrow{c} = 719$ lb at 292.8°
14. $\overrightarrow{d} = 195$ m/s at 343.0°, $\overrightarrow{e} = 216$ m/s at 13.6°
15. $\overrightarrow{f} = 303$ km at 19.6°, $\overrightarrow{g} = 414$ km at 243.0°
16. $\overrightarrow{m} = 43.3$ mi/h at 102.0°, $\overrightarrow{h} = 50.6$ mi/h at 250.9°
17. $\overrightarrow{d} = 1050$ lb at 88.6°, $\overrightarrow{e} = 2160$ lb at 167.0°

18. $\vec{a}$ = 93.3 km at 260.6°, $\vec{b}$ = 103 km at 333.0°

19. $\vec{c}$ = 108 ft/s at 12.9°, $\vec{d}$ = 143 ft/s at 13.5°

20. $\vec{g}$ = 1020 lb at 187.0°, $\vec{h}$ = 1570 lb at 286.0°

Given a single vector, determine the horizontal and vertical component vectors. Exercises 21–28 are given in standard position.

21. $\vec{R}$ = 527 lb at 46.8°   **25.** $\vec{R}$ = 703 mi at 208.0°

22. $\vec{R}$ = 316 mi/h at 196.6°   **26.** $\vec{R}$ = 88.8 km/h at 179.9°

23. $\vec{R}$ = 17.6 km at 103.0°   **27.** $\vec{R}$ = 1040 lb at 0.6°

24. $\vec{R}$ = 87.9 ft/s at 295.5°   **28.** $\vec{R}$ = 476 m at 90.9°

Solve the following vector problems:

29. A force of 2160 pounds pushes down and to the right of a body at an angle of 70.3° with the horizontal. Another force of 3060 pounds pushes down and to the left at an angle of 21.0° with the horizontal. Compute the resultant force in standard position.

30. From the start, a vehicle travels 52.5 miles in a direction 34.6° east of north and then travels 63.0 miles due north. Determine the change in position (displacement) from start to finish in standard position.

31. Two forces act on a point. One force of 318 pounds acts at an angle of 38.7° with the horizontal. The other force of 206 pounds acts at an angle of 54.9° with the horizontal. Both angles are inclined to the right. Determine the total force that tends to move the body horizontally.

32. Determine the impedance ($\vec{Z}$) of an alternating current series circuit given resistance ($\vec{R}$) and capacitive reactance ($\vec{X}_C$) as shown in Figure 37–19.

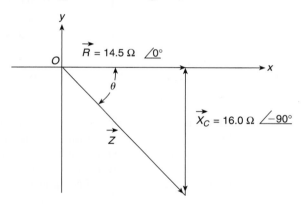

Figure 37–19

33. Determine the impedance ($\vec{Z}$) of an alternating current series circuit given resistance ($\vec{R}$) and inductive reactance (X_L) as shown in Figure 37–20.

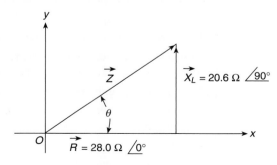

Figure 37–20

34. Determine the resistance ($\vec{R}$) and the inductive reactance ($\vec{X}_L$) of an alternating current series circuit when impedance $\vec{Z} = 40.8\ \Omega\ \underline{/37.0°}$.

35. Two forces, $\overrightarrow{OA}$ and $\overrightarrow{OC}$ are acting on a body as shown in Figure 37–21. Magnitudes $OA = 4850$ lb and $OC = 7070$ lb. Determine the resultant force $\overrightarrow{OB}$ in standard position.

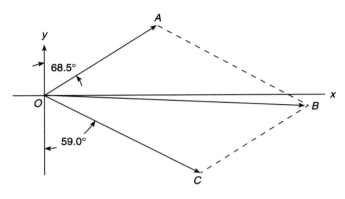

Figure 37–21

36. An airplane flies at 792 kilometers per hour at 25.0° west of north for 1.35 hours and then flies at 849 kilometers per hours at 43.8° west of south for 0.790 hour. Determine the displacement in standard position for the total time traveled.

37. A roof is inclined at an angle of 32.0° with the horizontal. Wind blows from due east horizontally against the roof with a force of 2900 pounds.

    a. Find the force perpendicular to the roof.

    b. Pressure is force acting upon a unit area of surface. Pressure = force ÷ area. The roof has an area of 1240 square feet. What is the horizontal pressure in pounds per square foot?

37–8   General (Component Vector) Procedure for Adding Vectors Using Trigonometry

Component vectors that are not vertical or horizontal can be added (resultants determined) by applying the Law of Cosines and the Law of Sines as was shown on pages 785 and 786. However, the following procedure is a more efficient way of adding vectors when three or more component vectors are involved. This general procedure is applied to any vector situation regardless of the number of component vectors and their directions.

General Procedure for Vector Addition

With known component vectors, the resultant vector $\vec{R}$ is computed as follows:

- Compute the horizontal component, x, of each vector. Add the x-components algebraically. The sum is the x-component of the resultant vector, $\vec{R}$.
- Compute the vertical component, y, of each vector. Add the y-components algebraically. The sum is the y-component of the resultant vector, $\vec{R}$.
- Compute the magnitude, R, or the resultant vector, $\vec{R}$, using the Pythagorean theorem.
- Compute the reference angle if the resultant vector lies in quadrants II, III, or IV. Use the tangent function.
- Compute θ.
- Show $\vec{R}$ as magnitude R and direction θ in standard position.

Multiple Vector Component Displacement Application: Determining
Resultant Vector

EXAMPLE •—————————————————————————————————

A plot of land is laid out as shown in Figure 37–22. Angles and distances are measured
beginning at point O, from points O to A, A to B, B to C, and C to D. An obstruction lies
between points O and D. Vector directions (angles) are given in the figure. The lengths of the
vectors are:

$\overrightarrow{OA}: OA = 88.7$ m
$\overrightarrow{AB}: AB = 75.3$ m
$\overrightarrow{BC}: BC = 178.5$ m
$\overrightarrow{CD}: CD = 215.9$ m

Determine the resultant vector $\vec{R}$ in standard position.

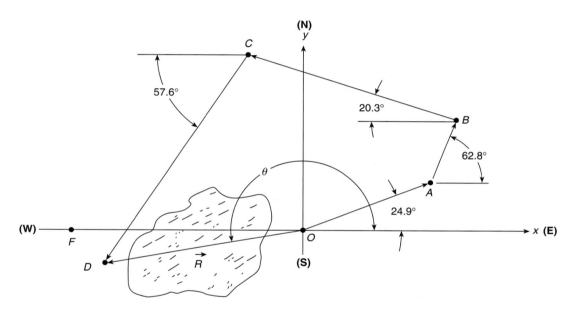

Figure 37–22

Compute the x and y components of each vector. Consider each of the points, O, A, B, and
C as an origin of a rectangular coordinate system. Be aware of the quadrant a vector lies in.

x-components   **y-components**

$$\overrightarrow{OA} : \text{Quadrant I} \begin{pmatrix} +x \\ +y \end{pmatrix}$$

$\cos 24.9° = \dfrac{\overrightarrow{OA}_x}{OA}$ $\sin 24.9° = \dfrac{\overrightarrow{OA}_y}{OA}$

$\overrightarrow{OA}_x = (88.7 \text{ m})(\cos 24.9°)$ $\overrightarrow{OA}_y = (88.7 \text{ m})(\sin 24.9°)$

$\overrightarrow{OA}_x \approx 80.5$ m $\overrightarrow{OA}_y \approx 37.3$ m

| **x-components** | **y-components** |

$$\overrightarrow{AB}: \text{Quadrant I} \begin{pmatrix} +x \\ +y \end{pmatrix}$$

$$\cos 62.8° = \frac{\overrightarrow{AB}_x}{AB}$$ $\qquad\qquad$ $$\sin 62.8° = \frac{\overrightarrow{AB}_y}{AB}$$

$$\overrightarrow{AB}_x = (75.3 \text{ m})(\cos 62.8°)$$ $\qquad$ $$\overrightarrow{AB}_y = (75.3 \text{ m})(\sin 62.8°)$$

$$\overrightarrow{AB}_x \approx 34.4 \text{ m}$$ $\qquad\qquad$ $$\overrightarrow{AB}_y \approx 67.0 \text{ m}$$

$$\overrightarrow{BC}: \text{Quadrant II} \begin{pmatrix} -x \\ +y \end{pmatrix}$$

$$-\cos 20.3° = \frac{\overrightarrow{BC}_x}{BC}$$ $\qquad\qquad$ $$\sin 20.3° = \frac{\overrightarrow{BC}_y}{BC}$$

$$\overrightarrow{BC}_x = (178.5 \text{ m})(-\cos 20.3°)$$ $\qquad$ $$\overrightarrow{BC}_y = (178.5 \text{ m})(\sin 20.3°)$$

$$\overrightarrow{BC}_x \approx -167.4 \text{ m}$$ $\qquad\qquad$ $$\overrightarrow{BC}_y \approx 61.9 \text{ m}$$

$$\overrightarrow{CD}: \text{Quadrant III} \begin{pmatrix} -x \\ -y \end{pmatrix}$$

$$-\cos 57.6° = \frac{\overrightarrow{CD}_x}{CD}$$ $\qquad\qquad$ $$-\sin 57.6° = \frac{\overrightarrow{CD}_y}{CD}$$

$$\overrightarrow{CD}_x = (215.9 \text{ m})(-\cos 57.6°)$$ $\qquad$ $$\overrightarrow{CD}_y = (215.9 \text{ m})(-\sin 57.6°)$$

$$\overrightarrow{CD}_x \approx -115.7 \text{ m}$$ $\qquad\qquad$ $$\overrightarrow{CD}_y \approx -182.3 \text{ m}$$

Compute $\overrightarrow{R}_x$ the sum of the x components. $\qquad\qquad$ Compute $\overrightarrow{R}_y$ the sum of the y components.

$$\overrightarrow{R}_x \approx 80.5 \text{ m} + 34.4 \text{ m} +$$
$$(-167.4 \text{ m}) + (-115.7 \text{ m})$$
$$\overrightarrow{R}_x \approx -168.2 \text{ m}$$

$$\overrightarrow{R}_y \approx 37.3 \text{ m} + 67.0 \text{ m} +$$
$$61.9 \text{ m} + (-182.3 \text{ m})$$
$$\overrightarrow{R}_y \approx -16.1 \text{ m}$$

Compute the length R of resultant vector R.

$$R = \sqrt{\overrightarrow{R}_x^2 + \overrightarrow{R}_y^2}$$

$$R \approx \sqrt{(-168.2 \text{ m})^2 + (-16.1 \text{ m})^2}$$

$$R \approx 169.0 \text{ m}$$

Compute reference angle, *FOD*.

$$\tan \angle FOD = \frac{\overrightarrow{R}_y}{\overrightarrow{R}_x}$$

$$\tan \angle FOD \approx \frac{-16.1 \text{ m}}{-168.2 \text{ m}} \approx 0.095719$$

$$\angle FOD \approx 5.5°$$

Compute θ.

$$\theta \approx 180° + 5.5°$$
$$\theta \approx 185.5°$$
$$\overrightarrow{R} \approx 169.0 \text{ m at } 185.5° \; Ans$$

EXERCISE 37–8

Use the component vector procedure for adding vectors. For all exercises and problems, compute magnitudes to three significant digits and angles to the nearest tenth of a degree.

Compute the vector sum (resultant vector) in standard position for each set of vectors. All angles are given in standard position.

1. $\vec{c}$ = 895 lb at 38.4°
   $\vec{d}$ = 605 lb at 113.0°
   $\vec{e}$ = 333 lb at 176.0°

2. $\vec{a}$ = 118 km/h at 323.0°
   $\vec{b}$ = 95.0 km/h at 39.6°
   $\vec{c}$ = 77.7 km/h at 108.8°

3. $\vec{k}$ = 43.6 mi at 206.0°
   $\vec{l}$ = 10.7 mi at 166.6°
   $\vec{m}$ = 35.0 mi at 107.3°

4. $\vec{u}$ = 870 ft at 343.0°
   $\vec{v}$ = 107 ft at 39.6°
   $\vec{w}$ = 521 ft at 102.0°

5. $\overrightarrow{OB}$ = 316 km 180.0°
   $\overrightarrow{OC}$ = 244 km at 126.0°
   $\overrightarrow{OD}$ = 417 km at 88.8°

6. $\overrightarrow{AB}$ = 389 mi/h at 0°
   $\overrightarrow{BC}$ = 404 mi/h at 90.0°
   $\overrightarrow{CD}$ = 517 mi/h at 107.0°

7. $\vec{a}$ = 87.6 ft at 216.3°
   $\vec{b}$ = 51.1 ft at 270.8°
   $\vec{c}$ = 97.6 ft at 316.0°

8. $\vec{e}$ = 2320 m at 180.0°
   $\vec{f}$ = 1790 m at 217.0°
   $\vec{g}$ = 3040 m at 17.9°

9. $\vec{k}$ = 793 km/h at 38.6°
   $\vec{l}$ = 685 km/h at 106.3°
   $\vec{m}$ = 611 km/h at 166.7°
   $\vec{n}$ = 479 km/h at 180.0°

10. $\vec{p}$ = 1070 lb at 329.0°
    $\vec{r}$ = 2190 lb at 213.0°
    $\vec{s}$ = 1880 lb at 90.0°
    $\vec{t}$ = 1550 lb at 298.0°

11. $\vec{m}$ = 2980 ft at 106.0°
    $\vec{n}$ = 1760 ft at 26.6°
    $\vec{p}$ = 3170 ft at 318.0°
    $\vec{r}$ = 2010 ft at 77.9°

12. $\vec{c}$ = 99.2 ft/s at 180.0°
    $\vec{d}$ = 70.8 ft/s at 178.0°
    $\vec{e}$ = 55.5 ft/s at 219.7°
    $\vec{f}$ = 68.8 ft/s at 280.0°

13. $\vec{b}$ = 972 lb at 282.0°
    $\vec{c}$ = 1080 lb at 267.0°
    $\vec{d}$ = 843 lb at 243.3°
    $\vec{e}$ = 912 lb at 206.5°

14. $\vec{l}$ = 5.26 mi at 14.8°
    $\vec{m}$ = 7.98 mi at 167.0°
    $\vec{n}$ = 6.13 mi at 213.5°
    $\vec{p}$ = 4.05 mi at 310.0°

Solve the following problems:

15. Three forces act on a point as shown in Figure 37–23. Determine the resultant force in standard position.

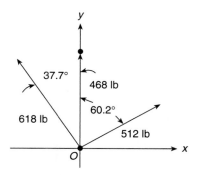

Figure 37–23

16. Determine the resultant force in standard position of the four forces acting on the point shown in Figure 37–24.

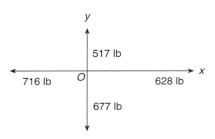

Figure 37–24

17. The flight pattern of an airplane is as follows: The airplane first travels 278 km at an angle of 18.6° east of north, then travels 222 km at an angle of 52.8° west of north, then travels 406 km at an angle of 72.0° west of north. Determine the change in position (displacement) in standard position from the beginning to the end of the flight.

18. A projectile is launched from point A and flies the distances at the angles of elevation and depression shown in Figure 37–25. Determine the horizontal distance AC and the vertical distance BC from launching point A.

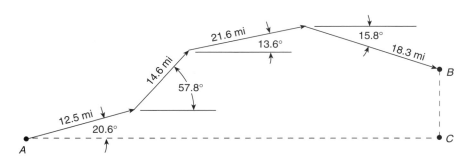

Figure 37–25

19. Three forces act on a body. The first force equals 1250 lb at 46.6°, the second force equals 1480 lb at 324.0°, and the resultant force equals 1540 lb 303.0°. All angles are in standard position. Determine the third force.

20. In laying out a plot of land, the following vectors are measured:

From point O, $\overrightarrow{OA}$ = 317 m at 32.8° north of west.

From point A, $\overrightarrow{AB}$ = 264 m at 39.0° north of west.

From point B, $\overrightarrow{BC}$ = 408 m at 27.6° north of east.

From point C, $\overrightarrow{CD}$ = 518 m at 40.9° south of east.

Determine the resultant vector in standard position.

21. Forces of 50.8 lb at 97.6°, 79.0 lb at 143.0°, and 93.5 lb at 170.0° are exerted on a body. The forces are in standard position. How many pounds of force are exerted on the body in a horizontal direction to the left?

22. From the start (point A), an airplane travels at 715 km/h for 0.650 hr at 323.0°, then travels at 608 km/h for 1.08 h at 285.0°, then travels at 685 km/h for 2.17 h at 256.5° to the completion of the flight (point B). Angles are given in standard position. How many miles less than the route traveled would be a direct flight from point A to point B?

UNIT EXERCISE AND PROBLEM REVIEW

VECTOR IDENTIFICATION AND REPRESENTATION

1. Which of the following quantities are vectors?
   a. 27.6 meters per second
   b. 130.0 pounds per square foot
   c. 0.063 mile due east
   d. 17.6 kilometers from the origin

2. Which of the following show or name a vector?

   a. 
   b. 
   c.

   d. 
   e. $\overline{AB}$
   f. $\vec{b}$

   g. $\overrightarrow{CD}$
   h. $\overleftrightarrow{C}$

3. Which of the following represents a vector in standard position on the rectangular coordinate system?

   a. 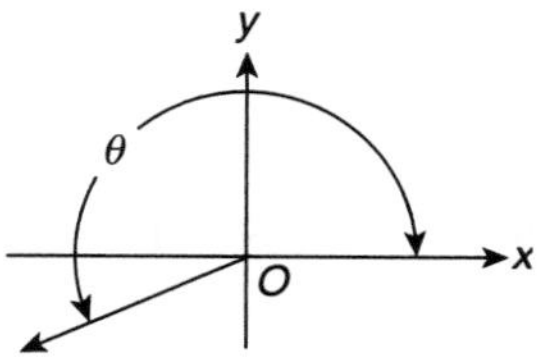
   b. 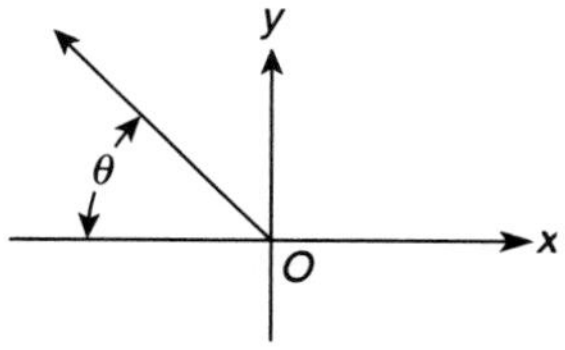
   c.

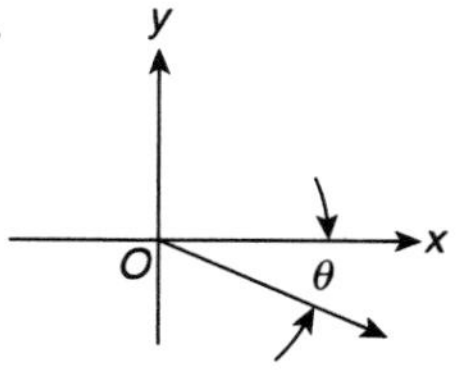

GRAPHIC ADDITIONS OF VECTORS

The following sets of vectors are given in standard position. Make a scale drawing and determine the vector sum (resultant vector) in standard position of each set. Use the triangle method for exercises 6 and 7. Express magnitudes to two significant digits and angles to the nearest degree.

4. $\vec{v}$ = 23 km at 90°, $\vec{w}$ = 19 km at 180°

5. $\vec{a}$ = 1900 lb at 188°, $\vec{b}$ = 1200 lb at 258°

6. $\vec{d}$ = 73 mi at 187°
   $\vec{e}$ = 110 mi at 222°
   $\vec{f}$ = 98 mi at 295°

7. $\vec{f}$ = 46 ft/s at 13°
   $\vec{g}$ = 66 ft/s at 110°
   $\vec{h}$ = 88 ft/s at 176°
   $\vec{k}$ = 12 ft/s at 214°

ADDITION OF VECTORS USING TRIGONOMETRY

For the following exercises, 8–19 express magnitudes to three significant digits and angles to the nearest tenth of a degree.

Determine the vector sums (resultant vectors) in standard position of exercises 8–15. Vectors are given in standard position.

8. $\vec{g}$ = 68.6 mi/h at 0°, $\vec{h}$ = 72.4 mi/h at 38.4°
9. $\vec{m}$ = 3130 lb at 108.0°, $\vec{h}$ = 2860 lb at 168.0°
10. $\vec{a}$ = 103 ft/s at 19.6°, $\vec{b}$ = 190 ft/s at 106.0°
11. $\vec{c}$ = 73.0 km at 118.0°, $\vec{d}$ = 68.6 km at 177.0°
12. $\vec{f}$ = 212 m/s at 303.0°, $\vec{g}$ = 198 m/s at 166.6°, $\vec{h}$ = 213 m/s at 107.0°
13. $\vec{k}$ = 1070 lb at 98.6°, $\vec{l}$ = 2050 lb at 173.0°, $\vec{m}$ = 1690 lb at 342.5°
14. $\vec{d}$ = 56.0 mi/h at 0°, $\vec{e}$ = 73.8 mi/h at 303.0°, $\vec{f}$ = 67.4 m/h at 278.2°
15. $\vec{t}$ = 216 km at 21.6°, $\vec{v}$ = 455 km at 278.3°, $\vec{w}$ = 198 km at 303.0°

Given a single vector, determine the horizontal and vertical component vectors. Exercises 16–19 are given in standard position.

16. $\vec{R}$ = 813 mi at 78.3°   18. $\vec{R}$ = 21.1 km at 142.0°
17. $\vec{R}$ = 608 lb at 203.0°   19. $\vec{R}$ = 93.7 ft/s at 316.0°

Solve the following vector problems. Express magnitudes to three significant digits and angles to the nearest tenth of a degree.

20. From the start, a vehicle travels 61.6 mi in a direction 29.7° west of north and then travels 58.2 mi due north. Determine the change in position (displacement) in standard position from start to finish.

21. Determine the impedance ($\vec{Z}$) of an alternating current series circuit given resistance ($\vec{R}$) and capacitive reactance ($\vec{X}_C$) as shown in Figure 37–26.

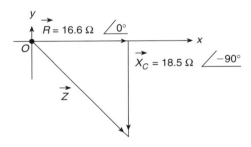

Figure 37–26

22. A projectile's velocity is 1470 m/s at angle of elevation of 38.0°. How many kilometers has the projectile traveled horizontally in 2.15 minutes?

23. Two forces act on a point. One force of 1120 lb acts at an angle of 29.2° with the horizontal. The second force of 1780 lb acts at an angle of 60.8° with the horizontal. Both angles are inclined up and to the left. Determine the total force that tends to move the body horizontally.

24. Determine the resistance ($\vec{R}$) and the inductive reactance ($\vec{X}_L$) of an alternating current series circuit when impedance $\vec{Z}$ = 38.6 Ω /39.5°

25. Determine the displacement (resultant vector) in standard position of the following flight. From the start, an airplane flies at 686 km/h for 1.14 h at 174.0°, then travels at 708 k/h for 2.05 hr at 187.0°, then travels at 754 k/h for 0.930 hr at 224.0° to the completion of the flight. Angles are given in standard position.

26. Four forces act on a point (0) as shown in Figure 37–27. Determine the resultant force in standard position.

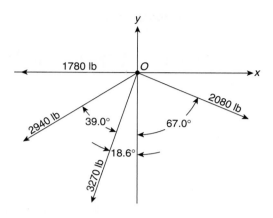

Figure 37–27

27. Fencing is to be installed to partially enclose a parcel of land. Starting at point *A*, distances and angles are measured and stakes driven at points *A*, *B*, *C*, *D*, and *E* as follows:

$\overrightarrow{AB}$ = 205 ft at 29.8° east of south

$\overrightarrow{BC}$ = 334 ft at 34.7° south of west

$\overrightarrow{CD}$ = 186 ft at 38.0° north of west

$\overrightarrow{DE}$ = 397 ft at 12.6° east of north

Because an obstruction lies between points *A* and *E*, the segment *AE* will not be fenced. Compute $\overrightarrow{AE}$ in standard position.

United States Customary and Metric Units of Measure

CUSTOMARY UNITS

CUSTOMARY UNITS OF LINEAR MEASURE

1 foot (ft)	= 12 inches (in)
1 yard (yd)	= 3 feet (ft)
1 yard (yd)	= 36 inches (in)
1 rod (rd)	= 16.5 feet (ft)
1 rod (rd)	= 5.5 yards (yd)
1 furlong	= 220 yards (yd)
1 mile (mi)	= 5,280 feet (ft)
1 mile (mi)	= 1,760 yards (yd)
1 mile (mi)	= 320 rods (rd)
1 mile (mi)	= 8 furlongs

CUSTOMARY UNITS OF AREA MEASURE

1 square foot (sq ft)	= 144 square inches (sq in)
1 square yard (sq yd)	= 9 square feet (sq ft)
1 square rod (sq rd)	= 30.25 square yards (sq yd)
1 acre (A)	= 160 square rods (sq rd)
1 acre (A)	= 43,560 square feet (sq ft)
1 square mile (sq mi)	= 640 acres (A)

CUSTOMARY UNITS OF VOLUME MEASURE

1 cubic foot (cu ft)	= 1.728 cubic inches (cu in)
1 cubic yard (cu yd)	= 27 cubic feet (cu ft)

METRIC UNITS

METRIC UNITS OF LINEAR MEASURE

1 millimeter (mm)	= 0.001 meter (m)	1000 millimeters (mm)	= 1 meter (m)
1 centimeter (cm)	= 0.01 meter (m)	100 centimeters (cm)	= 1 meter (m)
1 decimeter (dm)	= 0.1 meter (m)	10 decimeters (dm)	= 1 meter (m)
1 meter (m)	= 1 meter (m)	1 meter (m)	= 1 meter (m)
1 dekameter (dam)	= 10 meters (m)	0.1 dekameter (dam)	= 1 meter (m)
1 hectometer (hm)	= 100 meters (m)	0.01 hectometer (hm)	= 1 meter (m)
1 kilometer (km)	= 1000 meters (m)	0.001 kilometer (km)	= 1 meter (m)

METRIC UNITS OF AREA MEASURE

1 square millimeter (mm^2)	= 0.000 001 square meter (m^2)
1 square centimeter (cm^2)	= 0.0001 square meter (m^2)
1 square decimeter (dm^2)	= 0.01 square meter (m^2)
1 square meter (m^2)	= 1 square meter (m^2)
1 square dekameter (dam^2)	= 100 square meters (m^2)
1 square hectometer (hm^2)	= 10 000 square meters (m^2)
1 square kilometer (km^2)	= 1 000 000 square meters (m^2)

1 000 000 square millimeters (mm^2)	= 1 square meter (m^2)
10 000 square centimeters (cm^2)	= 1 square meter (m^2)
100 square decimeters (dm^2)	= 1 square meter (m^2)
1 square meter (m^2)	= 1 square meter (m^2)
0.01 square dekameter (dam^2)	= 1 square meter (m^2)
0.0001 square hectometer (hm^2)	= 1 square meter (m^2)
0.000 001 square kilometer (km^2)	= 1 square meter (m^2)

METRIC UNITS OF VOLUME MEASURE

1 cubic millimeter (mm^3)	= 0.000 000 001 cubic meter (m^3)
1 cubic centimeter (cm^3)	= 0.000 001 cubic meter (m^3)
1 cubic decimeter (dm^3)	= 0.001 cubic meter (m^3)
1 cubic meter (m^3)	= 1 cubic meter (m^3)

1 000 000 000 cubic millimeters (mm^3)	= 1 cubic meter (m^3)
1 000 000 cubic centimeters (cm^3)	= 1 cubic meter (m^3)
1000 cubic decimeters (dm^3)	= 1 cubic meter (m^3)
1 cubic meter (m^3)	= 1 cubic meter (m^3)

CUSTOMARY UNITS

CUSTOMARY UNITS OF CAPACITY MEASURE
16 ounces (oz) = 1 pint (pt)
2 pints (pt)   = 1 quart (qt)
4 quarts (qt)   = 1 gallon (gal)
COMMONLY USED CAPACITY-CUBIC MEASURE EQUIVALENTS
1 gallon (gal)   = 231 cubic inches (cu in)
7.5 gallons (gal) = 1 cubic foot (cu ft)

CUSTOMARY UNITS OF WEIGHT MEASURE
16 ounces (oz)   = 1 pound (lb)
2,000 pounds (lb) = 1 net or short ton
2,240 pounds (lb) = 1 gross or long ton

METRIC UNITS

METRIC UNITS OF CAPACITY MEASURE
1000 milliliters (mL) = 1 liter (L)
COMMONLY USED CAPACITY-CUBIC MEASURE EQUIVALENTS
1 milliliter (mL) = 1 cubic centimeter (cm^3)
1 liter (L)   = 1 cubic decimeter (dm^3)
1 liter (L)   = 1000 cubic centimeters (cm^3)
1000 liters (L)   = 1 cubic meter (m^3)

METRIC UNITS OF WEIGHT (MASS) MEASURE
1000 milligrams (mg) = 1 gram (g)
1000 grams (g)   = 1 kilogram (kg)
1000 kilograms (kg)   = 1 metric ton (t)

METRIC-CUSTOMARY

METRIC-CUSTOMARY LENGTH CONVERSIONS
1 in = 2.54 cm
1 ft = 30.48 cm
1 yd = 0.9144 m
1 mi $\approx$ 1.6093 km

METRIC-CUSTOMARY AREA CONVERSIONS
1 square inch (sq in or in^2) = 6.4516 cm^2
1 square foot (sq ft or ft^2) $\approx$ 0.0929 m^2
1 square yard (sq yd or yd^2) $\approx$ 0.8361 m^2
1 acre $\approx$ 0.4047 ha
1 square mile (sq mi or mi^2) $\approx$ 2.59 km^2

METRIC-CUSTOMARY VOLUME CONVERSIONS
1 cubic inch (cu in or in^3) = 16.387 cm^3
1 fluid ounce (fl oz) $\approx$ 29.574 cm^3
1 teaspoon (tsp) $\approx$ 4.929 mL
1 tablespoon (tbsp) $\approx$ 14.787 mL
1 cup $\approx$ 236.6 mL
1 quart (qt) $\approx$ 0.9464 L
1 gallon (gal) $\approx$ 3.785 L

METRIC-CUSTOMARY WEIGHT CONVERSIONS
1 ounce (oz) = 28.35 g
1 pound (lb) $\approx$ 0.4536 kg
1 ton $\approx$ 907.2 kg

Formulas for Areas (A) of Plane Figures

NOTE: The page where the formula and its application can be found is noted after each formula.

Rectangle $\quad A = lw$ (p. 542): $l =$ length, $w =$ width

Parallelogram $A = bh$ (p. 546): $b =$ base, $h =$ height

Trapezoid $\quad A = \frac{1}{2}h\,(b_1 + b_2)$ (p. 550): $h =$ height, b_1 and $b_2 =$ bases

Triangle $\quad A = \frac{1}{2}bh$ (p. 554): $b =$ base, $h =$ height

$\quad\quad\quad\quad\;\; A = \sqrt{s(s - a)(s - b)(s - c)}$ (p. 555): $a, b,$ and $c =$ sides, and

$\quad\quad\quad\quad\;\; s = \frac{1}{2}(a + b + c)$

Circle $\quad\quad A = \pi r^2$ (p. 564): $r =$ radius

$\quad\quad\quad\quad\;\; A = 0.7854d^2$ (p. 564): $d =$ diameter

Sector $\quad\quad A = \dfrac{\theta}{360°}(\pi r^2)$ (p. 568): $\theta =$ central angle, $r =$ radius

Ellipse $\quad\quad A = \pi ab$ (p. 572): $a = \frac{1}{2}$ the major axis, $b = \frac{1}{2}$ the minor axis

Formulas for Volumes and Areas of Solid Figures

NOTE: The page where the formula and its application can be found is noted after each formula.

FIGURE	VOLUME (V)	LATERAL AREA (LA) AND SURFACE AREA (SA)
Prism	$V = A_B h$ (p. 578) A_B = area of base h = altitude	$LA = P_B h$ (p. 586) P_B = perimeter of base h = altitude
Right Circular Cylinder	$V = A_B h$ (p. 582) A_B = area of base h = altitude	$LA = C_B h$ (p. 587) C_B = circumference of base h = altitude
Regular Pyramid	$V = \frac{1}{3} A_B h$ (p. 592) A_B = area of base h = altitude	$LA = \frac{1}{2} P_B h_s$ (p. 596) P_B = perimeter of base h_s = slant height
Right Circular Cone	$V = \frac{1}{3} A_B h$ (p. 592) A_B = area of base h = altitude	$LA = \frac{1}{2} C_B h_s$ (p. 596) C_B = circumference of base h_s = slant height
Frustum of a Regular Pyramid	$V = \frac{1}{3} h (A_B + A_b + \sqrt{A_B A_b})$ (p. 599) h = altitude A_B = area of larger base A_b = area of smaller base	$LA = \frac{1}{2} h_s (P_B + P_b)$ (p. 602) h_s = slant height P_B = perimeter of larger base P_b = perimeter of smaller base
Frustum of a Right Circular Cone	$V = \frac{1}{3} \pi h (R^2 + r^2 + Rr)$ (p. 599) h = altitude R = radius of larger base r = radius of smaller base	$LA = \frac{1}{2} h_s (C_B + C_b)$ (p. 602) h_s = slant height C_B = circumference of larger base C_b = circumference of smaller base $LA = \pi h_s (R + r)$ (p. 602) h_s = slant height R = radius of larger base r = radius of smaller base
Sphere	$V = \frac{4}{3} \pi r^3$ (p. 609) r = radius of sphere	$SA = 4 \pi r^2$ (p. 608) r = radius of sphere

Answers to Odd-Numbered Exercises

SECTION I: Fundamentals of General Mathematics

UNIT 1: WHOLE NUMBERS

Exercise 1–2

1. Hundreds **5.** Ten thousands **9.** Ten thousands **13.** Ten thousands **15.** Units
3. Thousands **7.** Millions **11.** Tens
17. $(8 \times 100) + (5 \times 10) + (7 \times 1)$ **27.** $(3 \times 100) + (3 \times 10) + (3 \times 1)$
19. $(9 \times 100) + (4 \times 10) + (2 \times 1)$ **29.** $(6 \times 100) + (0 \times 10) + (0 \times 1)$
21. $(1 \times 10) + (0 \times 1)$ **31.** $(9 \times 10,000) + (0 \times 1,000) + (5 \times 100) + (0 \times 10) + (7 \times 1)$
23. $(3 \times 100) + (7 \times 10) + (9 \times 1)$ **33.** $(9 \times 10,000) + (7 \times 1,000) + (5 \times 100) + (6 \times 10) + (0 \times 1)$
25. $(5 \times 100) + (0 \times 10) + (4 \times 1)$
35. $(2 \times 100,000) + (3 \times 10,000) + (4 \times 1,000) + (1 \times 100) + (2 \times 10) + (3 \times 1)$
37. $(4 \times 100,000,000) + (2 \times 10,000,000) + (8 \times 1,000,000) + (0 \times 100,000) + (0 \times 10,000) + (0 \times 1,000) + (9 \times 100) + (7 \times 10) + (5 \times 1)$

Exercise 1–3A

1. 60 **3.** 800 **5.** 8,000 **7.** 47,000 **9.** 470,000

Exercise 1–3B

1. 780 **3.** 1,400 **5.** 32,000 **7.** 26,500

Exercise 1–4

1. 121 **3.** 1,016 **5.** 1,985 **7.** 18,436 **9.** 24,317 **11.** 188,447

Exercise 1–5

1. 17 **5.** 115 **9.** 7,899 **13.** 1,756 lb **17.** 33 **21.** 6,353
3. 9 **7.** 1,898 **11.** 219 in **15.** 2,949 L **19.** 2,048

Exercise 1–7

1. 265 cu yd **3.** $88 **5. a.** 2 gal **7.** $621 **9.** A = 10 mm   C = 90 mm
b. 5 rolls   B = 40 mm   D = 50 mm

Exercise 1–8

1. 600 **7.** 60,995 **13.** 1,006,223 **19.** 3,199,065 **25.** 911,586
3. 16,893 **9.** 2,545,872 **15.** 15,537,447 **21.** 45,427,018 **27.** 3,203,600
5. 3,138 **11.** 4,617 **17.** 89,817,200 **23.** 79,724,666 **29.** 1,344

Exercise 1–9

1. 87 **7.** 68,571 **13.** 18 **19.** 49 R411 **25.** 73 R2,520
3. 68 **9.** 10,608 R3 **15.** 241 **21.** 127 **27.** 2,067 R3,518
5. 2,547 **11.** 137,935 R2 **17.** 538 R66 **23.** 81 R200

Exercise 1–10

1. 28,700 impressions **5. a.** 30 mm **7.** 161,664 lb **11.** 22 pieces **15.** 21 cans
3. $935 **b.** 20 mm **9.** 13,200 gal **13.** 99,314 Btu

Exercise 1–11

1. 9	**9.** 34	**17.** 101	**23.** 44	**29.** 64
3. 61	**11.** 14	**19.** 13	**25.** 3	**31.** 1
5. 30	**13.** 108	**21.** 79	**27.** 128	**33.** 9
7. 49	**15.** 42			

Exercise 1–12

1. a. 11 kW
   b. 23 kW
   c. 33 kW
   d. 23 kW
   e. 11 kW

3. a. 1 A
   b. 3 A
   c. 3 A

5. a. °C = 75
   b. °F = 131
   c. °F = 68
   d. °C = 5

Unit Exercise and Problem Review

1. Tens   **3.** Thousands   **5.** Thousands
7. $(4 \times 10) + (8 \times 1)$
9. $(1 \times 10,000) + (3 \times 1,000) + (6 \times 100) + (9 \times 10) + (2 \times 1)$
11. $(5 \times 1000) + (1 \times 100) + (0 \times 10) + (3 \times 1)$

13. 97	**31.** 313	**49.** 279,510	**65.** 37	**81.** 8
15. 219	**33.** 8	**51.** 9,955,200	**67.** 17	**83.** 5
17. 8,851	**35.** 3,524	**53.** 1,002	**69.** 108	**85.** 240 circuits
19. 184,481	**37.** 823,797	**55.** 52,262 R4	**71.** 17	**87. a.** 150 mm
21. 1,652	**39.** 3,708	**57.** 303	**73.** 147	**b.** 130 mm
23. 170,628	**41.** 7,273,665	**59.** 17	**75.** 5	**c.** 170 mm
25. 33	**43.** 72,957,640	**61.** 535 R107	**77.** 109	**89.** 4:00 PM
27. 53	**45.** 5,600,000,000	**63.** 78 R2,388	**79.** 54	**91.** 57,168 sq ft
29. 175	**47.** 1,320			

UNIT 2: COMMON FRACTIONS

Exercise 2–2

1. $A = \dfrac{2}{10}$ $B = \dfrac{4}{10}$ $C = \dfrac{5}{10}$ $D = \dfrac{7}{10}$ $E = \dfrac{8}{10}$ $F = \dfrac{9}{10}$

Exercise 2–4

1. $\dfrac{8}{16}$ **3.** $\dfrac{24}{32}$ **5.** $\dfrac{28}{32}$ **7.** $\dfrac{44}{64}$ **9.** $\dfrac{5}{20}$ **11.** $\dfrac{114}{72}$

Exercise 2–5

1. $\dfrac{1}{4}$ **3.** $\dfrac{3}{8}$ **5.** $\dfrac{9}{8}$ **7.** $\dfrac{1}{4}$ **9.** $\dfrac{3}{4}$ **11.** $\dfrac{9}{4}$

Exercise 2–7

1. $\dfrac{5}{2}$ **5.** $\dfrac{19}{16}$ **9.** $\dfrac{191}{32}$ **13.** $2\dfrac{4}{5}$ **17.** $4\dfrac{15}{16}$ **21.** $9\dfrac{15}{16}$ **25.** $\dfrac{12}{8}$ **29.** $\dfrac{2604}{32}$

3. $\dfrac{13}{8}$ **7.** $\dfrac{137}{32}$ **11.** $\dfrac{219}{5}$ **15.** $15\dfrac{3}{4}$ **19.** $10\dfrac{7}{8}$ **23.** $7\dfrac{3}{64}$ **27.** $\dfrac{244}{32}$

Exercise 2–8

1. $6\dfrac{1}{7}$ **3.** $13\dfrac{4}{11}$ **5.** $17\dfrac{1}{5}$ **7.** $53\dfrac{11}{12}$

Exercise 2–9

1. a. $\dfrac{29}{122}$ **d.** $\dfrac{31}{122}$
   b. $\dfrac{29}{122}$ **e.** $\dfrac{58}{122}$ or $\dfrac{29}{61}$
   c. $\dfrac{33}{122}$ **f.** $\dfrac{93}{122}$

3. Piece #1: $\dfrac{1}{4}$
Piece #2: $\dfrac{3}{16}$
Piece #3: $\dfrac{1}{2}$
Piece #4: $\dfrac{1}{16}$

5. Fractional part 1st day: $\dfrac{961}{3,007}$
Fractional part 2nd day: $\dfrac{930}{3,007}$
Fractional part 3rd day: $\dfrac{1,116}{3,007}$

Exercise 2–10A

1. 4   9. 24   17. 168   21. $\frac{5}{12}, \frac{3}{4}, \frac{7}{8}$

3. 10   11. 120   19. $\frac{3}{8}, \frac{1}{2}, \frac{9}{16}$

5. 12   13. 120   23. $\frac{5}{16}, \frac{1}{2}, \frac{9}{12}, \frac{7}{8}$

7. 48   15. 210

Exercise 2–10B

1. $\frac{1}{2}$ 3. $1\frac{21}{64}$ 5. $1\frac{3}{4}$ 7. $1\frac{15}{16}$ 9. $\frac{31}{32}$ 11. $2\frac{5}{6}$ 13. $1\frac{37}{72}$

Exercise 2–10C

1. $2\frac{3}{4}$ 3. $8\frac{3}{8}$ 5. $31\frac{7}{10}$ 7. $36\frac{51}{64}$ 9. $30\frac{1}{6}$ 11. $62\frac{3}{4}$ 13. $84\frac{1}{4}$ reams

Exercise 2–11A

1. $\frac{3}{5}$ 3. $\frac{2}{25}$ 5. $\frac{11}{32}$ 7. $\frac{3}{20}$ 9. $\frac{1}{8}$ 11. $\frac{1}{20}$ 13. $\frac{29}{64}$ 15. $\frac{37}{48}$

Exercise 2–11B

1. $6\frac{1}{2}$ 3. $17\frac{9}{16}$ 5. $2\frac{1}{8}$ 7. $6\frac{2}{5}$ 9. $1\frac{21}{32}$ 11. $\frac{3}{64}$

Exercise 2–11C

1. $7\frac{1}{4}$ 3. $6\frac{19}{24}$ 5. $\frac{14}{15}$ 7. $5\frac{2}{5}$ 9. $10\frac{5}{8}$ 11. $2\frac{1}{4}$ 13. $26\frac{3}{25}$ 15. $70\frac{55}{56}$

Exercise 2–11D

1. $1\frac{13}{16}$ in   3. $104\frac{19}{24}$ gal

Exercise 2–12

1. $12\frac{13}{32}''$ C: $4\frac{15}{64}''$ F: $1\frac{21}{64}''$ 7. A: $\frac{15}{32}''$ D: $\frac{15}{32}''$ G: $1\frac{7}{32}''$ 9. $13\frac{5}{32}$ in

3. A: $2\frac{9}{32}''$ D: $2\frac{19}{32}''$ G: $4\frac{45}{64}''$ B: $\frac{21}{32}''$ E: $\frac{9}{32}''$ H: $\frac{31}{64}''$ 11. $6\frac{11}{12}$ h

B: $3\frac{55}{64}''$ E: $3\frac{3}{8}''$ 5. $64\frac{1}{3}$ lb   C: $\frac{7}{16}''$ F: $\frac{11}{16}''$ I: $\frac{39}{64}''$

13a. Sub 16: $10\frac{7}{8}$ reams   13b. Sub 16: $4\frac{3}{8}$ reams

Sub 20: $14\frac{1}{3}$ reams   Sub 20: $5\frac{1}{3}$ reams

Sub 24: $5\frac{1}{24}$ reams   Sub 24: $4\frac{7}{24}$ reams

Sub 28: 11 reams   Sub 28: $\frac{5}{8}$ ream

Exercise 2–13A

1. $a = 5, b = 2$ 3. $a = 3, b = 5$ 5. $a = 64, b = 127$

Exercise 2–13B

1. $\frac{1}{8}$ 3. $\frac{2}{9}$ 5. $\frac{25}{64}$ 7. $\frac{57}{100}$ 9. $\frac{9}{20}$ 11. $\frac{11}{96}$

Exercise 2–13C

1. $2\frac{1}{2}$ 3. $9\frac{5}{8}$ 5. $9\frac{11}{16}$ 7. $\frac{6}{35}$ 9. $\frac{2}{9}$ 11. $5\frac{3}{16}$ 13. $29\frac{7}{32}$ 15. $14\frac{5}{8}$

Exercise 2–14

1. a. $A = \dfrac{5}{128}''$

   $B = \dfrac{85}{384}''$

b. $A = \dfrac{3}{64}''$

   $B = \dfrac{17}{64}''$

c. $A = \dfrac{15}{512}''$

   $B = \dfrac{85}{512}''$

d. $A = \dfrac{1}{16}''$

   $B = \dfrac{17}{48}''$

e. $A = \dfrac{3}{32}''$

   $B = \dfrac{17}{32}''$

f. $C = \dfrac{1}{32}''$

g. $C = \dfrac{1}{48}''$

h. $C = \dfrac{1}{160}''$

i. $C = \dfrac{1}{224}''$

j. $C = \dfrac{1}{256}''$

3. a. $58\dfrac{1}{4}'$

b. $22\dfrac{1}{3}'$

c. $80\dfrac{7}{12}'$

5. $30\dfrac{11}{32}''$

7. $\dfrac{1}{8}$ gr

9. $191\dfrac{5}{8}$ in

Exercise 2–15A

1. 2   **3.** 5.   **5.** $\dfrac{1}{2}$ **7.** $1\dfrac{1}{10}$ **9.** $2\dfrac{2}{3}$ **11.** $2\dfrac{1}{22}$

Exercise 2–15B

1. 24   **3.** $\dfrac{1}{27}$ **5.** $\dfrac{9}{2} = 4\dfrac{1}{2}$ **7.** $\dfrac{7}{36}$ **9.** $19\dfrac{1}{11}$ **11.** $3\dfrac{1}{6}$ **13.** $\dfrac{8}{13}$ **15.** $1\dfrac{22}{23}$

Exercise 2–16

1. a. $2\dfrac{3}{16}''$

b. $1\dfrac{3}{64}''$

c. $1\dfrac{7}{8}''$

d. $1\dfrac{5}{32}''$

e. $1\dfrac{31}{32}''$

3. #1: $6\dfrac{1}{4}''$

   #2: $7\dfrac{3}{16}''$

   #3: $10\dfrac{7}{8}''$

   #4: $13\dfrac{9}{16}''$

5. a. Full size (city): $17\dfrac{1}{2}$ mi/gal

   Intermediate (city): 20 mi/gal

   Compact (city): $26\dfrac{2}{3}$ mi/gal

   Full size (highway): 21 mi/gal

   Intermediate (highway): 24 mi/gal

   Compact (highway): $32\dfrac{1}{2}$ mi/gal

b. Compact: $100\dfrac{5}{8}$ mi

7. a. 72 mAs

b. 48 mAs

c. 32 mAs

9. $8\dfrac{28}{33}$

Exercise 2–17

1. $1\dfrac{1}{8}$ **5.** $2\dfrac{3}{4}$ **9.** $\dfrac{67}{200}$ **13.** $11\dfrac{57}{64}$ **17.** $1\dfrac{5}{38}$ **21.** $76\dfrac{7}{15}$ **25.** $8\dfrac{172}{277}$

3. $1\dfrac{1}{6}$ **7.** $18\dfrac{43}{48}$ **11.** $11\dfrac{1}{6}$ **15.** $21\dfrac{1}{3}$ **19.** $1\dfrac{125}{147}$ **23.** $12\dfrac{1}{6}$

Exercise 2–18

1. a. hp $= 1\dfrac{1}{2}$

b. hp $= 2$

c. hp $= 2\dfrac{3}{4}$

d. hp $= 3\dfrac{1}{3}$

3. a. $34\dfrac{1}{4}''$

b. $38\dfrac{5}{8}''$

c. $28\dfrac{1}{4}''$

d. $37\dfrac{1}{2}''$

5. $1{,}885\dfrac{115}{128}$ lb

Unit Exercise and Problem Review

1. $\dfrac{4}{8}$

3. $\dfrac{40}{35}$

5. $\dfrac{48}{256}$

7. $\dfrac{2}{5}$

9. $\dfrac{1}{7}$

11. $\dfrac{1}{6}$

13. $\dfrac{3}{2}$

15. $\dfrac{39}{4}$

17. $\dfrac{255}{64}$

19. $6\dfrac{1}{2}$

21. $21\dfrac{1}{4}$

23. $1\dfrac{39}{128}$

25. 8

27. 90

29. 1

31. $2\dfrac{14}{15}$

33. $7\dfrac{2}{3}$

35. $76\dfrac{9}{50}$

37. $\dfrac{3}{4}$

39. $\dfrac{1}{4}$

41. $\dfrac{19}{64}$

43. $7\dfrac{3}{8}$

45. $4\dfrac{5}{64}$

47. $77\dfrac{17}{25}$

49. $5\dfrac{29}{32}$

51. $\dfrac{1}{12}$

53. $\dfrac{3}{64}$

55. 6

57. $16\dfrac{1}{2}$

59. $\dfrac{1}{2}$

61. $1\dfrac{7}{12}$

63. $\dfrac{2}{5}$

65. 27

67. 320

69. $4\dfrac{4}{19}$

71. $\dfrac{7}{16}$

73. $2\dfrac{5}{9}$

75. $5\dfrac{4}{9}$

77. $11\dfrac{37}{40}$

79. $4\dfrac{53}{140}$

81. $10\dfrac{3}{8}$ lb

83. $3{,}493\dfrac{1}{2}$ ft

85. a. Item A: $21\dfrac{2}{3}$

b. Item B: $1\dfrac{1}{2}$

c. Item C: $10\dfrac{1}{2}$

d. Total: $33\dfrac{2}{3}$

87. $21\dfrac{7}{8}$ ft

UNIT 3:   DECIMAL FRACTIONS

Exercise 3–4

1. three tenths
3. one hundred seventy-five thousandths
5. ninety-eight ten-thousandths
7. fifteen and eight hundred seventy-six thousandths
9. twenty-seven and twenty-seven ten-thousandths
11. two hundred ninety-nine and nine ten-thousandths
13. 0.9   15. 0.0002   17. 0.00301   19. 12.001   21. 0.19   23. 0.0287   25. 0.999   27. 0.8111

Exercise 3–5

1. 0.84   3. 0.007   5. 0.889   7. 22.196   9. 89.899   11. 722.010

Exercise 3–7

1. 0.25   7. 0.7344   13. $\dfrac{3}{10}$ 17. $\dfrac{1}{200}$ 21. $\dfrac{999}{1,000}$
3. 0.4063   9. 0.0625
5. 0.8333   11. 0.95   15. $\dfrac{13}{40}$ 19. $\dfrac{7}{250}$ 23. $\dfrac{13}{16}$

Exercise 3–8

1. 0.4   3. a. 0.363   b. 0.093   c. 0.302   d. 0.242   5. a. 0.28   b. 0.3   c. 0.42

Exercise 3–10

1. 0.632   7. 636.993   13. 44.8   19. 0.0009   25. 19.9951   31. 115.947 lb
3. 38.533   9. 100.508   15. 123.1508   21. 0.051   27. 63.8288   33. 6.912 in
5. 0.1359   11. 1.083   17. 4.825   23. 0.164   29. 4.0063

Exercise 3–11

1. $36.13   c. 1.607″   5. a. 92.3 cu ft/min   d. 6.6 cu ft/min   b. 53.81 kWh
3. a. 0.550″   d. 1.463″   b. 48.5 cu ft/min   e. 45.8 cu ft/min   c. 35.42 kWh
   b. 1.280″   e. 0.636″   c. 22.0 cu ft/min   7. a. 8,053.73 kWh

Exercise 3–12A

1. 0.54   7. 14.1825   13. 0.005   19. 0.000729   25. 0.1629
3. 1.23   9. 394.18743   15. 8,087.58   21. 75.625   27. 0.19898
5. 0.0212   11. 0.000152   17. 0.00268   23. 1.185921

Exercise 3–12B

1. 7.2   3. 0.5   5. 31,288,000   7. 0.0008   9. 0.0843   11. 3.3   13. 9,350   15. 0.0707

Exercise 3–13

1. 460.75 lb   5. a. 413 cu ft/min   c. 1,250 cu ft/min   e. 88 cu ft/min   g. 550 cu ft/min
3. $3.42   b. 1,006 cu ft/min   d. 304 cu ft/min   f. 7,590 cu ft/min   h. 7,523 cu ft/min

Exercise 3–14A

1. 2   5. 1.47   9. 0.2   13. 0.125   17. 40   21. 0.672   25. 0.564   29. 0.01
3. 18   7. 4.003   11. 40   15. 3.23   19. 6.36   23. 0.0657   27. 0.0027

Exercise 3–14B

1. 0.037   5. 520   9. 0.080658   13. 0.000087
3. 0.732   7. 0.0639   11. 9,080

Exercise 3–15

1. Mon: 17.0 cu yd   3. a. 5,420 tiles   7. Circuit #1: 5.1 A
   Tues: 15.1 cu yd   b. 3,780 tiles   Circuit #2: 5.5 A
   Wed: 24.1 cu yd   5. 6 holes
   Thurs: 23.1 cu yd
   Fri: 14.4 cu yd

Exercise 3–16A

1. 0.64
3. 1
5. 0.000064
7. 99,225

9. 0.013824
11. 0.002197
13. 243
15. 0.000000729

17. 0.125 or $\dfrac{1}{8}$
19. 81
21. 0.140625 or $\dfrac{9}{64}$

23. 243
25. 6.25
27. 0.4225 or $\dfrac{169}{400}$

29. 1
31. 49
33. 0.0016

Exercise 3–16B

1. 5
3. 7
5. 3

7. 12
9. 4
11. 2

13. 4
15. 7
17. 10

19. 4
21. 2
23. 5

Exercise 3–17

1. a. 1.56 sq ft
b. 0.11 cm^2
c. 5.52 sq yd
d. 0.44 km^2
e. 0.01 sq ft

3 a. 11.18 ft
b. 2.96 km
c. 7.60 in
d. 0.94 m
e. 49.79 ft

5 a. 11.4 A
b. 19.8 A
c. 17.8 A
d. 18.9 A

7. $15,010
9. 210 m
11. $2,610

Exercise 3–18

1. 0.9375
3. 0.625
5. $\dfrac{9}{32}$
7. $\dfrac{5}{64}$

9. $\dfrac{13}{64}$
11. $\dfrac{11}{32}$
13. $\dfrac{3}{32}$

15. $\dfrac{63}{64}$
17. $\dfrac{9}{64}$

19. A $= \dfrac{33''}{64}$
   B $= \dfrac{15''}{64}$
   C $= \dfrac{57''}{64}$

   D $= \dfrac{5\,''}{16}$
   E $= \dfrac{3''}{8}$
   F $= \dfrac{31''}{32}$

   G $= \dfrac{11''}{32}$
   H $= \dfrac{9\,''}{32}$

Exercise 3–19

1. 16.81
3. 3.8

5. 16.93
7. 55.10

9. 1.35
11. 3.99

13. 275.62
15. 1.36

17. 21.03
19. 3.16

21. 0.37
23. 22.74

25. 353.20
27. 215.20

29. 316.59
31. 52.82

Exercise 3–20

1. $AB = 189.8$ ft
3. a. 106,000 Btu/min
b. 35,000 Btu/min

c. 3,000 Btu/min
d. 10,000 Btu/min
e. 18,000 Btu/min

5. a. 1.9 Ω
b. 1.2 Ω
7. 3.4 m^3

Unit Exercise and Problem Review

1. 0.94
3. 0.010
5. 17.0
7. 306.3001
9. 0.875
11. 0.167
13. 0.516
15. 0.9
17. $\dfrac{3}{5}$
19. $\dfrac{1}{16}$
21. $\dfrac{5}{32}$
23. $\dfrac{499}{500}$
25. 0.446
27. 0.0333

29. 403.81392
31. 88.192
33. 0.105
35. 0.0011
37. 0.2002
39. 0.5927
41. 0.56
43. 2.5382
45. 0.00144
47. 79.79148
49. 3.159
51. 195.738
53. 0.4452
55. 8.1
57. 16,300
59. 0.175
61. 4
63. 0.03

65. 2.1
67. 339.6
69. 0.007
71. 0.0861
73. 0.035872
75. 0.2
77. 36
79. 29.791
81. 10.648
83. 32
85. 0.00000064
87. 0.0625
89. 6,561
91. 1
93. 15
95. 5
97. 3
99. 3

101. 6
103. 9
105. 3
107. 15.72
109. 2.127
111. 15.5
113. 0.09375
115. 0.265625
117. $\dfrac{19}{64}$
119. $\dfrac{1}{16}$
121. $\dfrac{33}{64}$
123. $\dfrac{61}{64}$
125. 6.84
127. 8.97

129. 13.87
131. 13.74
133. 70.40
135. 0.86
137. 9.89
139. 10.42
141. 186 brake hp
143. a. $252.51
b. $197.35
145. a. 7.54 ft
b. 3.06 m
c. 13.15 cm
d. 5.16 ft
147. $2,400
149. a. 2.3 A
b. 2.3 A
c. 3.2 A
151. 1940 m^2

UNIT 4: RATIO AND PROPORTION

Exercise 4–2

1. $\dfrac{3}{7}$ 11. $\dfrac{13}{4}$ 21. $\dfrac{1}{2}$ 29. a. $\dfrac{9}{1}$ b. $\dfrac{6}{11}$ g. $\dfrac{11}{5}$

3. $\dfrac{1}{2}$ 13. $\dfrac{4}{3}$ 23. $\dfrac{2}{5}$ b. $\dfrac{8}{1}$ c. $\dfrac{5}{11}$ h. $\dfrac{17}{11}$

5. $\dfrac{4}{15}$ 15. $\dfrac{32}{3}$ 25. $\dfrac{3}{4}$ c. $\dfrac{11}{1}$ d. $\dfrac{5}{17}$

7. $\dfrac{6}{23}$ 17. $\dfrac{1}{12}$ 27. $\dfrac{1}{3}$ d. $\dfrac{7}{1}$ e. $\dfrac{11}{17}$

9. $\dfrac{9}{22}$ 19. $\dfrac{30}{23}$ 31. a. $\dfrac{6}{5}$ f. $\dfrac{17}{6}$

Exercise 4–3

1. 1   9. 2.2   15. $\dfrac{5}{6}$ 19. 4.0 m

3. 35   11. 8.2   21. 11.1 cm^3

5. 36   13. 4   17. $-31\dfrac{1}{2}$ 23. $32.74

7. 17.5

Exercise 4–5

1. 8 gal   5. $3,600

3. a. 288 r/min   7. a. 112 ft

   b. 157.5 r/min   b. $7\dfrac{1}{2}$ lb

   c. 28 teeth   c. 4.625 ft

   d. 25 teeth   d. 35 lb

   e. 154.3 r/min   9. 1.2 lb

Unit Exercise and Problem Review

1. $\dfrac{15}{32}$

3. $\dfrac{6}{23}$

5. $\dfrac{7}{11}$

7. $\dfrac{1}{2}$

9. $\dfrac{5}{1}$

11. $\dfrac{3}{5}$

13. $\dfrac{3}{4}$

15. a. Cost to selling price: $\dfrac{5}{8}$

   Cost to profit: $\dfrac{5}{3}$

   b. Cost to selling price: $\dfrac{7}{12}$

   Cost to profit: $\dfrac{7}{5}$

   c. Cost to selling price: $\dfrac{6}{11}$

   Cost to profit: $\dfrac{6}{5}$

   d. Cost to selling price: $\dfrac{51}{110}$

   Cost to profit: $\dfrac{51}{59}$

17. 2

19. 24

21. −4.5

23. 56

25. a. 67.5 lb/sq in

   b. 90 lb/sq in

   c. 1 cu ft

   d. 0.6 cu ft

27. $687.23

29. $3\dfrac{2}{3}$ h

31. a. Gear B: 320.0 r/min
   Gear C: 320.0 r/min
   Gear D: 800.0 r/min

   b. Gear B: 20 teeth
   Gear D: 30 teeth
   Gear C: 300.0 r/min

   c. Gear C: 168.0 r/min
   Gear B: 168.0 r/min
   Gear A: 28 teeth

   d. Gear C: 30 teeth
   Gear B: 175.0 r/min
   Gear A: 79.6 r/min

UNIT 5: PERCENTS

Exercise 5–3

1. 44%   **7.** 4%   **13.** 0.02%   **17.** 15%   **21.** 159%
3. 25%   **9.** 0.8%   **15.** 25%   **19.** 53.125% **23.** 1462.5%
5. 35%   **11.** 207.6%

Exercise 5–5

1. 0.82   **9.** 0.0473   **17.** $\dfrac{1}{2}$ **21.** $\dfrac{4}{25}$ **25.** $\dfrac{37}{1000}$
3. 0.03   **11.** 0.0075
5. 0.2776   **13.** 0.02375   **19.** $\dfrac{5}{8}$ **23.** $1\dfrac{9}{10}$ **27.** $\dfrac{9}{1000}$
7. 2.249   **15.** 0.436

Exercise 5–6

1. 16   **5.** 129.98   **9.** 37.47   **13.** 7.14   **17.** 1.40
3. 120   **7.** 101.4   **11.** 392   **15.** 0.13   **19.** 5.38

Exercise 5–7A

1. 2.4 gal   **3.** 6.8 h   **5.** 29.9 V   **7.** 1,400 bu

Exercise 5–7B

1. 50%   **5.** 118.95%   **9.** 40%   **13.** 42.86%
3. 37%   **7.** 155.46%   **11.** 30.77%   **15.** 125.47%

Exercise 5–8A

1. 7.14%   **3.** 1.05%   **5.** 12%   **7.** 25%   **9.** 5.24%

Exercise 5–8B

1. 150   **5.** 170   **9.** 2.34   **13.** 42.93
3. 320   **7.** 240   **11.** 270.57   **15.** 0.5

Exercise 5–9

1. 2,154 units   **3.** $650,000   **5.** $26,723   **7.** 500 lb   **9.** 31.1 Ωh

Exercise 5–10

1. 14%   **5.** 282 reams   **9.** 151 pieces   **13.** 190 g   **17.** 9.12 in
3. 159.9 mL   **7.** 17%   **11.** 30 Ω   **15.** 74 lb

Unit Exercise and Problem Review

1. 72%   **17.** $\dfrac{3}{10}$ **25.** 9   **41.** 144.24%   **57.** 75%   **73. a.** 0.38 kg   **83.** $344.12
3. 203.7%   **27.** 87.36   **43.** 24.49%   **59.** 15.60   **b.** 0.63 kg   **85.** 128%
5. 4%   **19.** $1\dfrac{2}{5}$ **29.** 5.68   **45.** 50%   **61.** 19.05%   **75. a.** 19.5 m^3 **87.** 142 ft
7. 0.28%   **31.** 275.6   **47.** 33.33   **63.** 153.99   **b.** 5.25 m^3 **89.** 119 lb
9. 0.19   **21.** $\dfrac{1}{8}$ **33.** 4   **49.** 16.47   **65.** 57.99   **c.** 0.25 m^3 **91.** 24.3%
11. 0.1809   **35.** 9.31   **51.** 74.38   **67.** 3.38   **77.** 32.4 A   **93.** 1.7%
13. 0.007   **23.** $\dfrac{49}{5000}$ **37.** 20%   **53.** 41.18   **69.** 125.14   **79.** 30%
15. 0.0075   **39.** 24.69% **55.** 0.61   **71.** 30.89   **81.** 0.26%

UNIT 6:   SIGNED NUMBERS

Exercise 6–1

1. −8 mi/h   **3.** −$18   **5.** −$167,000   **7.** +9 volts   **9.** −$280

Exercise 6–2

1. +6
3. +2
5. +10
7. −11
9. +6
11. −20
13. −11

15. −4
17. +6
19. +11
21. $+5\frac{1}{2}$
23. −3.5
25. −7.25

27. $-2\frac{3}{4}$
29. −3.75
31. −4.25
33. +3, 5
35. +3, 10
37. −28, 45

39. −14.08, 4.54
41. +2.5, 5
43. +16.17, 38.03
45. $+1\frac{1}{16}, 2\frac{15}{16}$
47. $+\frac{3}{16}, \frac{15}{32}$

49. −22, −18, −1, 0, +2, +4, +16
51. −15, −8, −3, 0, +3, +15, +17
53. −1.1, −1, −0.4, +1, +2.3, +17.8
55. $-6\frac{1}{4}, -6\frac{5}{32}, -1\frac{1}{2}, -1\frac{15}{32}, -1\frac{7}{16}, 0$

Exercise 6–4

1. 5   **3.** 4   **5.** 9   **7.** 1   **9.** 0   **11.** $2\frac{1}{2}$ **13.** 5.2   **15.** 0.023

Exercise 6–5

1. 15
3. 44
5. 25

7. −23
9. −29
11. 5

13. −3
15. −6
17. −22

19. −35
21. −18.8
23. −18.5

25. −13
27. $-20\frac{17}{32}$

29. −14.47
31. 1
33. 31.25

35. −14.06

Exercise 6–6

1. −2
3. 18
5. −8

7. 22
9. 0
11. −80

13. −7
15. −5
17. 39

19. 2.2
21. −2.2
23. 102.1

25. $3\frac{1}{4}$
27. $40\frac{51}{64}$

29. 15
31. −1
33. 11.15

35. −10

Exercise 6–7

1. −24
3. 24
5. −36
7. −35

9. 0
11. −13
13. −12.02

15. $-\frac{3}{4}$
17. 0
19. −8

21. 8
23. 10.81
25. −1

27. 0.38
29. $-2\frac{3}{8}$

Exercise 6–8

1. 2
3. −2
5. −7
7. 3

9. 8
11. 4
13. −3
15. −6

17. 120
19. −1.6
21. −0.68

23. −12
25. $-1\frac{5}{8}$

27. −5.6
29. −38.12

Exercise 6–9

1. 4
3. 8
5. −64
7. 16

9. −32
11. 103.82
13. 0.65
15. −0.61

17. −0.20
19. $+\frac{1}{4}$ or 0.25

21. $\frac{9}{16}$ or 0.56
23. 0.25
25. 0.06

27. 0.04
29. 0.04
31. 0.00
33. 0.00

Exercise 6–10

1. 3
3. −4
5. −3

7. 3
9. −5
11. −2

13. 1
15. −1
17. 1

19. 3
21. $\frac{1}{2}$

23. $-\frac{3}{8}$
25. 5.298

27. 3.354
29. −0.923

Exercise 6–11

1. −28
3. 33

5. 2.5
7. 3,257.72

9. −11.30
11. −35

13. 2,928
15. 1.97

17. 4.69
19. 1,136.34

Exercise 6–12A

1. 6.25×10^2
3. -3.789×10^3
5. 9.591×10^5
7. -6.3×10^{-4}
9. 9.5×10^{-6}
11. 2.03×10^6
13. -1.04×10^{-1}
15. 8.3×10^{-3}
17. 3.6×10^6 J
19. 3.77×10^{-7} kWh
21. 9.46055×10^{12} km

Exercise 6–12B

1. 3,000
3. -0.0473
5. 0.00005093
7. -0.0000002008
9. 4,005,200
11. 498,300
13. 0.00000007771
15. 61,070,000
17. 254,000,000 Å
19. 3,600 C
21. 0.000000505 hp

Exercise 6–12C

1. 1.28×10^9
3. -1.44×10^{10}
5. 1.35×10^{12}
7. 2.28×10^{-9}
9. 1.01×10^6
11. -4.77×10^{13}
13. 1.41×10^5
15. 4.61×10^7
17. -7.33×10^{10}
19. 3.28×10^9
21. 3.27×10^{10}
23. -7.10×10^{-3}
25. 6.40×10^5
27. 8.37×10^{-10}
29. 0.29 km/cycle
31. a. 0.0132 in
    b. 0.0199 ft
    c. 0.0129 in
33. 4.19×10^{-4} W

Exercise 6–13A

1. 625
3. -3.789×10^3
5. 959.1×10^3
7. -630×10^{-6}
9. 58×10^{-9}
11. 723×10^{12}
13. 930.0005×10^3
14. 35.7×10^6
17. 2.647768×10^6 J
19. 73.75616×10^{-9} lbf
21. 30.8374×10^{12} km

Exercise 6–13B

1. 5,000
3. -0.03124
5. 0.00013507
7. $-7,850,000$
9. 149 598 000 000 m
11. 0.000 000 000 000 000 000 160 21 J

Exercise 6–13C

1. 46.50×10^{15}
3. -108.12×10^{21}
5. 50×10^{15}
7. 42.5×10^{-18}
9. 30.77×10^{12}
11. 46.59×10^{-6}
13. 1.19×10^9
15. $-5.29 \times 10^0 = -5.29$
17. 8.01×10^{15}
19. -8.06×10^{-9}
21. 81.8807×10^{-15}

Unit Exercise and Problem Review

1. $-\$20,000$
3. $+18\%$
5. a. $+9$
    b. $+5.4$
    c. -9.2
    d. $+0.8$
    e. -4
    f. -2.8
    g. $+5.4$
    h. -5.8
7. -19
9. -7
11. 0
13. -4.7
15. 11
17. -6

19. 23
21. 11.554
23. -18
25. 8.68
27. 24
29. 5.47
31. -15
33. 450
35. -16.8
37. -0.15
39. 81
41. $7\frac{7}{8}$
43. 8.419488
45. 4
47. 8

49. -6
51. 0
53. -4.04317
55. 16
57. 64
59. -32
61. -0.008
63. $\frac{1}{8}$
65. $\frac{1}{100}$
67. $-194,406.41$
69. -3
71. 3.04
73. -4.37
75. $\frac{1}{2}$

77. -32
79. 145
81. 85
83. 4,896
85. -0.125
87. 66.83
89. $+\frac{1}{4}$
91. a. (1) $-\$87,000$
    (2) $-\$596,000$
    (3) $-\$164,000$
    (4) $+\$88,000$
    (5) $+\$93,000$
    (6) $+\$405,000$
  b. (1) $+\$105,000$
    (2) $-\$177,000$
    (3) $+\$108,000$

93. 9.76×10^5
95. 3.9×10^{-4}
97. 5,090,000
99. 2.13×10^{-10}
101. -1.20×10^{-6}
103. 3.28×10^{10}
105. 6.61×10^{-3}
107. 1.85×10^6
109. -618×10^{-9}
111. -0.000000571
113. -24.67×10^{-3}
115. 378.97×10^{-15}
117. 509.01×10^9
119. 9.06×10^6

SECTION II: Measurement

UNIT 7: PRECISION, ACCURACY, AND TOLERANCE

Exercise 7–4

1. **a.** 0.1″
   b. The range includes all numbers equal to or greater than 3.55″ and less than 3.65″.
3. **a.** 0.1 mm
   b. The range includes all numbers equal to or greater than 4.25 mm and less than 4.35 mm.
5. **a.** 0.001″
   b. The range includes all numbers equal to or greater than 15.8845″ and less than 15.8855″.
7. **a.** 0.001″
   b. The range includes all numbers equal to or greater than 12.0015″ and less than 12.0025″.
9. **a.** 0.01 mm
   b. The range includes all numbers equal to or greater than 7.005 mm and less than 7.015 mm.
11. **a.** 0.1 mm
    b. The range includes all numbers equal to or greater than 9.05 mm and less than 9.15 mm.

Exercise 7–5

1. 10.56 in   3. 87.3 ft   5. 1,472 mi   7. 8.001 in   9. 61 gal   11. 0.01 sq in

Exercise 7–6

1. 5   3. 5   5. 5   7. 5   9. 5   11. 2   13. 5   15. 3   17. 5   19. 4

Exercise 7–7

1. 5.05   3. 173   5. 123.0   7. 8.92   9. 70,108   11. 43.08   13. 0.0200   15. 818.0

Exercise 7–8

1. 44.8 mm   3. 170 ft   5. 7.6 mi   7. 0.006   9. 0.005   11. 10.378   13. 0.25   15. 755.0

Exercise 7–9

1. Absolute Error = 0.002 in
   Relative Error = 0.05%
3. Absolute Error = 0.2 lb
   Relative Error = 2%
5. Absolute Error = 0.14 cm
   Relative Error = 0.59%
7. Absolute Error = 3 ohms
   Relative Error = 3%
9. Absolute Error = 0.010 m^2
   Relative Error = 1.0%
11. Absolute Error = 2.6 m
    Relative Error = 0.14%

Exercise 7–10

1. Tolerance = $\dfrac{1}{32}$″
3. Minimum Limit = 16.74″
5. Tolerance = 0.0003″
7. Tolerance = 0.9 mm
9. 258.07 mm
11. 12.737 cm

Exercise 7–11

		Basic Dimension (inches)	Maximum Diameter (inches)	Minimum Diameter (inches)	Maximum Clearance (inches)	Minimum Clearance (inches)
1.	DIA A	1.4580	1.4580	1.4550	0.0090	0.0030
	DIA B	1.4610	1.4640	1.4610		
3.	DIA A	2.1053	2.1053	2.1023	0.0085	0.0025
	DIA B	2.1078	2.1108	2.1078		
5.	DIA A	0.9996	0.9996	0.9966	0.0071	0.0011
	DIA B	1.0007	1.0037	1.0007		

		Basic Dimension (millimetres)	Maximum Diameter (millimetres)	Minimum Diameter (millimetres)	Maximum Interference (millimetres)	Minimum Interference (millimetres)
7.	DIA A	32.07	32.09	32.05	0.10	0.02
	DIA B	32.01	32.03	31.99		
9.	DIA A	41.91	41.93	41.89	0.10	0.02
	DIA B	41.85	41.87	41.83		

Unit Exercise and Problem Review

1. a. 0.1 in
   b. The range includes all numbers equal to or greater than 5.25 in and less than 5.35 in.
3. a. 0.001 in
   b. The range includes all numbers equal to or greater than 1.8335 in and less than 1.8345 in.
5. a. 0.001 in
   b. The range includes all numbers equal to or greater than 19.0005 in and less than 19.0015 in.
7. a. 0.1 mm
   b. The range includes all numbers equal to or greater than 28.95 mm and less than 29.05 mm.

9. 20.917 mm
11. 2,449 mi
13. 13.997 in
15. 41 in
17. 5
19. 5
21. 5

23. 6
25. 6.07
27. 48,070
29. 0.870
31. 3.0006
33. 14
35. 0.005

37. 360,000
39. 0.24
41. Absolute Error = 0.003 in
    Relative Error = 0.05%
43. Absolute Error = 58 lb
    Relative Error = 1.1%

45. Absolute Error = 0.0005 cm
    Relative Error = 0.06%
47. Tolerance = $\frac{1}{16}''$
49. Maximum Limit = 2.781″
51. Tolerance = 0.6 mm
53. Minimum Limit = 78.66 mm

		Basic Dimension (inches)	Maximum Diameter (inches)	Minimum Diameter (inches)	Maximum Clearance (inches)	Minimum Clearance (inches)
55.	DIA A	1.7120	1.7120	1.7106	0.0044	0.0016
	DIA B	1.7136	1.7150	1.7136		
57.	DIA A	2.8064	2.8064	2.8050	0.0039	0.0011
	DIA B	2.8075	2.8089	2.8075		

		Basic Dimension (millimetres)	Maximum Diameter (millimetres)	Minimum Diameter (millimetres)	Maximum Interference (millimetres)	Minimum Interference (millimetres)
59.	DIA A	9.94	9.97	9.91	0.15	0.03
	DIA B	9.85	9.88	9.82		

61. Maximum value Length $A = 35\frac{23''}{32}$
    Minimum value Length $A = 35\frac{17''}{32}$

63. Maximum thickness = 2.82 mm
    Minimum thickness = 2.76 mm
65. Hole 5 is drilled out of tolerance.
    Hole 6 is drilled out of tolerance.

UNIT 8: CUSTOMARY MEASUREMENT UNITS

Exercise 8–2A

1. 4.25 ft
3. 7.083 yd
5. 1.2 mi
7. 3.70 ft
9. 18.9 yd
11. 0.675 mi

Exercise 8–2B

1. 6 ft 3 in
3. 1 mi 660 yd
5. 10 ft $7\frac{1}{2}$ in
7. 1 mi $165\frac{1}{3}$ yd

Exercise 8–2C

1. 72 in   **3.** 48.90 ft   **5.** 7,128 ft   **7.** 12.8 ft   **9.** 1,320 ft   **11.** 464 rd

Exercise 8–2D

1. 19 ft 6 in   **3.** 158 yd 1.2 ft   **5.** 697 rd 3.3 yd   **7.** 2 ft 8.4 in

Exercise 8–3A

1. 12 ft 9 in   **5.** 18 yd 2 ft   **11.** 3 mi 115 rd   **15.** 2 ft $4\frac{15}{16}$ in   **19.** 1 yd 2 ft 9.9 in
3. 19 ft $3\frac{1}{8}$ in   **7.** 23 yd 2 ft 3 in   **13.** 4 ft 3 in   **17.** 3 yd 0.5 ft   **21.** 1 rd 1 yd 1 ft
                  9. 5 rd 5 yd   **23.** 3 mi 250 rd

Exercise 8–3B

1. 14 ft 6 in   **7.** 19 yd 1 ft 6 in   **11.** 7 mi 294 rd   **17.** 14 yd$\frac{2}{3}$ ft   **21.** 13 yd 1 ft $6\frac{1}{2}$ in
3. 67 ft 4.5 in   **9.** 5 yd 1 ft $4\frac{1}{2}$ in   **13.** 3 ft 2 in   **19.** 7 yd 1 ft 3 in   **23.** 2 mi 100 rd
5. 133 yd 1 ft   **15.** 4 ft 6.975 in

Exercise 8–4

1. A = 32′-1″   **3.** 52 yd $2\frac{1}{4}$ ft   **5.** 1 ft $11\frac{1}{2}$ in   **7.** A = 30 ft   **9.** $12.15
    B = 38′11″   B = 6 ft 1 in
    C = 12′-9″   C = 4 ft 9 in
    D = 8′-4″

Exercise 8–5

1. 1.36 sq ft   **7.** 2.8 A   **13.** 338 sq in   **19.** 93,650 sq ft
3. 5.09 sq yd   **9.** 0.399 A   **15.** 38.7 sq ft   **21.** 165 sq yd
5. 2.5 sq mi   **11.** 0.0312 sq mi   **17.** 2,436.8 A   **23.** 697,000 sq ft

Exercise 8–6

1. 54 strips   **3.** 8 gal   **5.** 1,320 tiles   **7.** 59 panels

Exercise 8–7

1. 2.5 cu ft   **5.** 7.47 cu ft   **9.** 0.325 cu ft   **13.** 443 cu ft   **17.** 3,062 cu ft
3. 4.33 cu yd   **7.** 2.7 cu yd   **11.** 2,851 cu in   **15.** 472 cu in   **19.** 950 cu in

Exercise 8–8

1. 12.30 cu ft   **3.** 5.93 cu yd   **5.** 3,675 lb

Exercise 8–9

1. 3.26 qt   **7.** 709 cu in   **11.** 1.9 qt
3. 37.0 qt   **9.** 0.20 gal   **13.** 51 oz
5. 8.8 gal

Exercise 8–10

1. 10 qt   **3.** 30,700 gal   **5.** 33 min   **7.** 126   **9.** 25,100 gal

Exercise 8–11

1. 2.2 lb   **3.** 5,400 lb   **5.** 2.7 short tons   **7.** 2.72 lb   **9.** 240 lb

Exercise 8–12

1. 108 lb   **3.** 85 lb 3 oz   **5.** 9,500 lb   **7.** 0.24 oz

Exercise 8–13A

1. 1.02 mi/min   **3.** 35.8 rev/sec   **5.** 936 ft/hr   **7.** 320 lb/sq in

Exercise 8–13B

1. 5.32 short tons/sq ft   **3.** 29.0 cents/pt   **5.** 133 cu ft/hr

Exercise 8–14

1. a. 4.4 short tons
   b. 3.9 long tons

3. a. 304.3 ft^3
   b. 59,300 lb
   c. 25.62 hours or 3 days, 1.62 hours (assuming an 8 hour day)
   d. $4746.30

5. 434 cu ft

7. a. 1.95 in
   b. 0.374 in
   c. 0.491 in
   d. 3.87 in
   e. 0.855 in

Unit Exercise and Problem Review

1. 2.125 ft
3. 0.75 mi
5. 15 yd 2 ft
7. $24\frac{1}{2}$ ft
9. 81 in
11. 146 yd 2 ft
13. 12 ft 3 in
15. 10 ft $1\frac{3}{4}$ in

17. 2 yd $1\frac{1}{2}$ ft
19. 26 ft 3 in
21. 15 yd 1 ft 4 in
23. 3 yd $2\frac{2}{3}$ ft
25. 3.50 sq ft
27. 588 sq in
29. 5.000A
31. 48.0 sq ft
33. 104 sq in
35. 2.7 cu ft

37. 1,230 cu in
39. 1,026 cu in
41. 0.4 cu yd
43. 61.2 qt
45. 3.3 qt
47. 240 cu in
49. 1.6 pt
51. 2.1 lb
53. 21.6 long tons
55. 1,320 lb
57. 1,170 lb/sq ft
59. 2,220,000 cu ft/hr

61. 59.0 mi/hr
63. 119 cu ft/hr
65. A = 49' - 6"
    B = 4' - 0"
    C = 26' - 10"
    D = 12' - 10"
67. 24 pieces
69. 72 cu ft
71. 7.9 cu yd
73. 25,600 gal
75. 189 sheets

UNIT 9: METRIC MEASUREMENT UNITS

Exercise 9–1A

1. cm   **3.** m   **5.** cm   **7.** km

Exercise 9–1B

1. cm   **3.** cm   **5.** cm   **7.** m   **9.** mm

Exercise 9–2

1. 3.4 m
3. 50 m
5. 0.335 m
7. 84 m
9. 1,050 m
11. 148 m
13. 70 cm
15. 50 mm
17. 24 hm
19. 31.06 m
21. 7.35 dm
23. 0.616 km
25. 800 dm
27. 600 cm

Exercise 9–4

1. 12 cm
3. 148.95 m

5. Cost per piece:
   #105-AD: $0.1892
   #106-AD: $0.1428
   #107-AD: $0.2727

Cost per 2,500 pieces:
   #105-AD: $544
   #106-AD: $411
   #107-AD: $784

Exercise 9–5

1. 5 cm^2
3. 49 dm^2
5. 1 dm^2
7. 0.35 m^2
9. 800 dm^2
11. 4,800 cm^2
13. 208 cm^2
15. 4,400,000 dm^2

Exercise 9–7

1. 60 pieces   **3.** 120 t   **5.** 15 gaskets

Exercise 9–8

1. 2.7 cm^3
3. 0.94 m^3
5. 0.048 dm^3
7. 150 m^3
9. 0.07 dm^3
11. 5,000 cm^3
13. 800,000 cm^3
15. 5,230 mm^3
17. 1,030 cm^3
19. 106,000 mm^3

Exercise 9–10

1. 2.5 m^3
3. 8,400 kg

5. 91.40% Magnesium
   8.29% Aluminum
   0.13% Manganese
   0.18% Zinc

Exercise 9–11

1. 3.67 L
3. 23.6 cm^3
5. 5.3 L
7. 80 L
9. 83 L
11. 7.3 L
13. 29 L

Exercise 9–12

1. 18 L   3. 32 198 000 L   5. 76 min   7. 36 L   9. 90 000 L

Exercise 9–13

1. 1,720 mg   3. 2,600 kg   5. 2.7 metric tons   7. 40 mg   9. 23 g

Exercise 9–14

1. 61.2 kg   3. 55 g   5. 62.8 kg/cm^2 7. 0.96 g/°C

Exercise 9–15A

1. 1.29 g/mm^2 3. 0.000128 m^3/sec   5. 0.0870 hp/cm^2 7. $0.00903/g

Exercise 9–15B

1. 19 m/sec   3. 0.7596 km/min   5. 0.477 cents/mL

Exercise 9–16

1. 5.3 metric tons   5. 20 m^3 7. a. 49.2 mm   d. 126 mm
3. 1,100 kg   b. 13.6 mm   e. 26.4 mm
                                              c. 16.8 mm

Exercise 9–17A

1. 15.2 A   5. 8.2 Gb   9. 1.68 × 10^{12} b   13. 0.58 MHz
3. 0.75 kW   7. 0.38   11. 2.7 × 10^5 mW   15. 2.6 × 10^6 μA

Exercise 9–17B

1. 253 000 000 nm   3. 0.172 μg   5. 23 600 nL

Exercise 9–17C

1. 13.81 V   5. 3,000 KHz   9. 135 Gb   13. 0.368 Gs   17. 0.038 mF
3. 38.6 mA   7. 71.5 Ω   11. 0.7 A   15. 40 s

Exercise 9–18A

1. 30.5 cm   3. 991 mm   5. 3.94 ft   7. 59.14 km   9. 10 530 cm

Exercise 9–18B

1. 17 200 cm^2 3. 3.3 m^2 5. 160 acres   7. 201.8 ft^2

Exercise 9–18C

1. a. 4564 cm^3 3. 761.5 mL   5. 11.29 gal   7. 0.534 ft oz
   b. 4.564 L

Exercise 9–18D

1. 74.8 kg   3. 1241 g   5. 52.5 lb   6. 2.759 t

Unit Exercise and Problem Review

1. cm   19. 2401.9 m   37. 4600 dm^3 55. 0.97 km/min   73. 1.864 in
3. mm   21. 118.87 dm   39. 0.06 m^3 57. 320.4 Kg/dm^2 75. 19.7 m^2
5. cm   23. 5.32 cm^2 41. 1300 mL   59. 0.6348 km/min   77. 1.338 m^3
7. 80 mm   25. 1.466 m^2 43. 93.4 cm^3 61. 3.9 g/mm^2 79. 0.5096 qt
9. 2300 cm   27. 600 dm^2 45. 0.618 m^3 63. 0.094 kW   81. 282.0 kg
11. 800 m   29. 9000 m^2 47. 60 mL   65. 5 × 10^6 ns   83. 48 km
13. 122 mm   31. 0.028 m^2 49. 1.88 kg   67. 1.78 × 10^6 μA   85. 0.55 m^2
15. 76.6 mm   33. 2.4 cm^3 51. 2700 kg   69. 1.294 pm   89. 11.7 metric tons
17. 644.8 m   35. 7000 cm^3 53. 210 g   71. 3.744 m

UNIT 10: STEEL RULES AND VERNIER CALIPERS

Exercise 10–3

$a = \dfrac{5''}{32}$ $c = \dfrac{1''}{2}$ $e = \dfrac{13''}{16}$ $g = 1\dfrac{5''}{32}$ $i = \dfrac{5''}{64}$ $k = \dfrac{11''}{32}$ $m = \dfrac{41''}{64}$ $o = 1\dfrac{9''}{64}$

Exercise 10–4

$a = 0.08''$ $c = 0.4''$ $e = 0.8''$ $g = 1.28''$ $i = 0.09''$ $k = 0.38''$ $m = 0.84''$ $o = 1.27''$

Exercise 10–5

a = 5 mm   e = 37 mm   i = 6.5 mm   m = 44.5 mm
c = 20 mm   g = 57 mm   k = 17.5 mm   o = 63.5 mm

Exercise 10–7

1. 2.021″   **3.** 4.788″   **5.** 3.376″

Exercise 10–8

1. 30.82 mm   **3.** 60.52 mm   **5.** 52.42 mm

Unit Exercise and Problem Review

1. $a = \dfrac{3''}{32}$ $g = 1\dfrac{3''}{32}$ $m = \dfrac{47''}{64}$

$b = \dfrac{9''}{32}$ $h = 1\dfrac{11''}{32}$ $n = \dfrac{29''}{32}$

$c = \dfrac{7''}{16}$ $i = \dfrac{7''}{64}$ $o = 1\dfrac{5''}{32}$

$d = \dfrac{19''}{32}$ $j = \dfrac{17''}{64}$ $p = 1\dfrac{25''}{64}$

$e = \dfrac{23''}{32}$ $k = \dfrac{13''}{32}$

$f = \dfrac{7''}{8}$ $l = \dfrac{33''}{64}$

3. a = 4 mm   k = 19.5 mm
b = 15 mm   l = 31.5 mm
c = 23 mm   m = 43 mm
d = 34 mm   n = 49.5 mm
e = 45 mm   o = 62.5 mm
f = 56 mm   p = 72.5 mm
g = 62 mm   **5. a.** 77.82 mm
h = 71 mm   **b.** 33.76 mm
i = 2.5 mm   **c.** 50.16 mm
j = 11 mm   **d.** 9.42 mm

UNIT 11: MICROMETERS

Exercise 11–2

1. 0.598″   **3.** 0.736″   **5.** 0.157″   **7.** 0.589″   **9.** 0.738″   **11.** 0.022″

Exercise 11–4

1. 0.2749″   **3.** 0.4980″   **5.** 0.0982″   **7.** 0.3282″   **9.** 0.2336″   **11.** 0.5157″

Exercise 11–6

1. 7.09 mm   **3.** 5.69 mm   **5.** 9.78 mm   **7.** 0.34 mm   **9.** 3.12 mm   **11.** 24.93 mm

Exercise 11–8

1. 4.268 mm   **3.** 7.218 mm   **5.** 2.132 mm   **7.** 8.308 mm   **9.** 9.484 mm   **11.** 11.114 mm

Unit Exercise and Problem Review

1. a. 0.707″   **e.** 0.008″   **3. a.** 22.93 mm   **e.** 18.78 mm
b. 0.136″   **f.** 0.797″   **b.** 5.69 mm   **f.** 3.81 mm
c. 0.495″   **g.** 0.549″   **c.** 3.71 mm   **g.** 10.74 mm
d. 0.365″   **h.** 0.898″   **d.** 0.07 mm   **h.** 5.17 mm

SECTION III: Fundamentals of Algebra

UNIT 12:   INTRODUCTION TO ALGEBRA

Exercise 12–2

1. $a + 3$
3. $7 - d$
5. xy
7. $b \div 25$
9. $e + 12$

11. $102\,x$
13. $y - 75$
15. $a + b^2$
17. $3V - 12$
19. $\frac{1}{2}x - 4y$

21. $9m + 2n$
23. $x \div 25y$
25. a. $a + 6'' + b + 3a$ or $4a + b + 6''$
    b. $6'' + b + 3a + 5''$ or $3a + b + 11''$
    c. $3a + 5''$
    d. $a + 5'' + 3a + b + 6''$ or $4a + b + 11''$

27. $L \div N$
29. $B - C + D$
31. $0.785\,4D^2\,LN$
33. $\sqrt{X^2 + R^2}$

Exercise 12–3

1. a. 18
  b. 3
  c. 10
  d. 24
  e. 8

3. a. 105
  b. 52
  c. 48
  d. 7
  e. 9.6

5. a. 162.13
  b. 6.75
  c. 12.5
  d. 5.83
  e. 63.61

7. 0.5 min
9. 24°C
11. $9,485.85

13. 5,500 sq ft
15. 131.6 cm
17. 19.8 m

Unit Exercise and Problem Review

1. $6x + 12$
3. $\frac{1}{4}mR$
5. $d \div 14f$

7. $2M - \frac{1}{3}R$
9. $(F^2 + G) \div H$
11. $M \div C$

13. $3GH$
15. a. 26
  b. 135
  c. 3
  d. 48

17. a. 521.29
  b. 1.56
  c. 92
  d. 30.265

19. 0.5 A
21. 15.1 mm

UNIT 13:   BASIC ALGEBRAIC OPERATIONS

Exercise 13–2

1. $11a$
3. $-4x$
5. $21y$
7. $-34xy$
9. 0

11. $-8pt$
13. $15.2a^2b$
15. $1\frac{1}{4}xy$
17. $-2.91gh^3$

19. P
21. $-1\frac{7}{8}xy$
23. $6.666M$
25. $21x + 19y$

27. $10x + 7xy + 4y$
29. $4x + (-xy)$
31. $x^2y + (-xy^2) + (-15x^2y^2)$
33. $-0.4c + 3.6cd + 2.9d$
35. $6b^4 + (-6b^2c)$

37. $6.6x$
39. $4\frac{1}{2}B$

Exercise 13–3

1. $2x$
3. $-8a$
5. $-8y^2$
7. 0
9. $-11c^2$
11. $16M$
13. $-6.1P$

15. $0.1D$
17. c^2d
19. $1\frac{3}{8}H$
21. $-6.1xy$
23. g^2h
25. $-3P^2 + 5P$

27. 4
29. $-N + 11NS$
31. $-ab + a^2b - ab^2$
33. $9x^3 + 7x^2 - 9x$
35. $-\frac{1}{4}x + \frac{1}{4}x^2 - \frac{3}{8}x^3$
37. $10.09e + 15.76f + 10.03$

39. $A = 1\frac{3}{4}x$
    $B = \frac{7}{8}x$
    $C = \frac{1}{4}x$

$D = \frac{7}{8}x$
$E = 1\frac{1}{8}x$
$F = 2\frac{1}{4}x$

Exercise 13–4

1. $2x^3$
3. $-54c^7$
5. $27a^3b^3c^5$
7. $12c^5d^3$
9. $12P^8N^4$
11. $1.25x^4y^4$
13. 0

15. $\frac{15}{32}x^5y^3$
17. a^2b^2cd
19. $1.8F^3G^3$
21. $60x^3y^5n$
23. x^6y^3

25. $6x^2 + 2xy$
27. $3M^3 - 3M^2N$
29. $30c^3d^4 - 40c^2d^2$
31. $-r^4t^3s^3 + r^2t^3s^2 - r^5t^4s$
33. $-f^2g + 9fg^2 - 12fh$
35. $x^2 + 2xy + y^2$

37. $x^2 - 2xy + y^2$
39. $7a^5 + 21a + 12a^4 + 36$
41. $21a^3x^5 - 3ac^2x^2 + 7a^2cx^4 - c^3x$
43. $16x^4y^6 - 36x^2y^2$
45. a. $20N$
    b. $4.25\,(N - 35)$

47. a. $0.6xy$
    b. $1.25xy$
    c. $3xy$
    d. $5xy$
    e. $8xy$

Exercise 13–5

1. $2x$
3. 3
5. $5xy^4$
7. 1
9. $6a^3$
11. $-DM^2$
13. -8.4

15. $5cd$
17. $8g^2h$
19. xz^3
21. $4P^2V$
23. $8xyz$
25. $3x^2 + 5x$
27. $4x^2y - 2x$

29. $-14M + 12MN$
31. $4ab + 3a^2$
33. $0.3xy^3 + 0.1y^2$
35. $xy^3z - z^4 - xy$
37. $4a - 6a^2c - 8c^2$
39. $-3G - 7EG^2 + 9E^2H$

41. a. $2x$
 b. $0.5x$
 c. $1.2x$
 d. $9,600x^2$
 e. $1,500x^2$
 f. $4,000x^2$

43. $8x + \$300$

Exercise 13–6

1. a^2b^2
3. $9a^2b^2$
5. $8x^6y^3$
7. $-27c^9d^6e^{12}$
9. $49x^8y^{10}$
11. $a^6b^3c^9$
13. $-x^{12}y^{15}z^3$
15. $0.027x^{12}y^6$

17. $10.24M^6N^2P^4$
19. $\dfrac{27}{64}a^3b^3c^9$
21. $-27x^6y^{12}z^9$
23. $\dfrac{25}{64}a^8b^4c^{12}$
25. $a^2 + 2ab + b^2$
27. $x^4 + 2x^2y + y^2$

29. $9x^6 - 12x^3y^2 + 4y^4$
31. $x^4y^6 + 2x^3y^5 + x^2y^4$
33. $20.25M^4P^2 - 9M^2P^5 + P^8$
35. $a^{12} - 2a^6b^6 + b^{12}$
37. a. $125x^3$
 b. $0.008x^3$
 c. $1,157\dfrac{5}{8}x^3$
 d. $5.832x^3$

39. a. $4.16x^3$
 b. $0.065x^3; 0.07x^3$
 c. $0.69212x^3; 0.69x^3$
 d. $0.52x^3$

Exercise 13–7

1. $2abc$
3. $4cd^3$
5. $-4xy^3$
7. $5fg^4$
9. $7cd^3e^5$
11. $-5x^3y$

13. $0.8a^3c^4f$
15. $10\sqrt{xy}$
17. $12p^2\sqrt{ms}$
19. $\dfrac{1}{4}y^2\sqrt{x}$
21. $-4d^2t^3$

23. $3f^2\sqrt[3]{d^2e}$
25. $2h^2$
27. $c^2t^3d^{1/3}$
29. $\dfrac{2}{3}a^2c^3\sqrt{b}$

31. a. $4x$
 b. $0.3x$
 c. $\dfrac{2}{3}x$
 d. $\dfrac{3}{4}x$

33. a. $4.96x$
 b. $12.4x$
 c. $0.372x$
 d. $0.62x$

Exercise 13–8

1. $7a - 2a^2 + a^3$
3. $9b - 15b^2 + c - d$

5. $7y^2 - 12$
7. $-xy$

9. $-2c^3 + d - 12$
11. $-16 - xy$

13. 0
15. $34 - c^2d$

Exercise 13–9

1. $14 + 13a$
3. $20 - 49x^2y^2$

5. $7c^2 - 9d$
7. $6 + 5H^2 - 3H^4$

9. $y - xy + 5x$
11. $b^6 + 12$

13. $4D$
15. $-80a + 5a^4b^6$

17. $20c - c^3d^2 + 4c^2$
19. $-16x^2y$

Exercise 13–10

1. $2(10^2) + 6(10^1) + 5(10^0)$
 $200 + 60 + 5 = 265$
3. $9(10^4) + 0(10^3) + 5(10^2) + 0(10^1) + 0(10^0)$
 $90,000 + 0 + 500 + 0 + 0 = 90,500$
5. $2(10^1) + 3(10^0) + 0(10^{-1}) + 2(10^{-2}) + 3(10^{-3})$
 $20 + 3 + 0 + 0.02 + 0.003 = 23.023$
7. $4(10^3) + 7(10^2) + 5(10^1) + 1(10^0) + 1(10^{-1}) + 0(10^{-2}) + 7(10^{-3})$
 $4000 + 700 + 50 + 1 + 0.1 + 0 + 0.007 = 4751.107$
9. $1(10^2) + 6(10^1) + 3(10^0) + 0(10^{-1}) + 6(10^{-2}) + 4(10^{-3}) + 3(10^{-4})$
 $100 + 60 + 3 + 0 + 0.06 + 0.004 + 0.0003 = 163.0643$

11. 1_{10}
13. 5_{10}
15. 15_{10}
17. 11_{10}

19. 21_{10}
21. 53_{10}
23. 0.5_{10}
25. 3.75_{10}

27. 2.000_{10}
29. 9.3125_{10}
31. 71_{10}
33. 37_{10}

35. 35_{10}
37. 36_{10}
39. 1100100_2
41. 10111_2

43. 100_2
45. 1100010_2
47. 110_2
49. 100001110_2

51. 0.001_2
53. 1010.1_2
55. 10011.0001_2
57. 1.001_2

Unit Exercise and Problem Review

1. $-3x$
3. $8MP$
5. $14.7a^2c$
7. $-\dfrac{5}{8}x^2y^3$
9. $6x$
11. $1\dfrac{3}{8}ab$
13. $9a + 14b$
15. $6m + 3mn$
17. $-4.3F + (-4.9G)$
19. $-6P$
21. 0
23. $-0.7x^2$
25. $0.24H$
27. $\dfrac{5}{16}B$
29. $-cd^2$
31. $2R + 3R^2$
33. $7T - 5TW$
35. $-y^3 + 4y^2 - 4y$
37. $-\dfrac{3}{8}c - \dfrac{1}{4}d + 4\dfrac{1}{4} - cd$
39. $60x^3y^3$

41. $-20c^3d^3e^5$
43. $2.58x^4y^3$
45. $\dfrac{9}{32}a^4b^3c$
47. $c^3b^2d^3$
49. $-\dfrac{1}{10}P^4S^3$
51. $7D^5 - 7D^3H$
53. $-a^4b^3c^6 + a^2b^4c^4 - a^2b^5c^5$
55. $-20m^4 + 43m^2n^3 - 21n^6$
57. $0.64P^3S^3 + 0.4P^4S + 9.6S^2 + 6P$
59. $2y$
61. $-C$
63. $7m^3$
65. $5.2a^2b$
67. $12EF$
69. $\dfrac{1}{16}b^2$
71. $3x^2 - 2x$
73. $5CD^4 + 4C^2D^3$
75. $0.2FG^2 + 0.1F^2G$
77. $2x - 3x^2y - 8y^2$

79. $25a^2b^2$
81. $9M^8P^4$
83. $81M^{12}P^8T^{16}$
85. $37.21d^6f^2h^4$
87. $49x^8b^{12}c^2$
89. $0.16m^6n^6s^{12}$
91. $x^4 + 2x^2y + y^2$
93. $25a^4 + 40a^2b^2 + 16b^4$
95. $0.64P^4T^6 - 0.64P^2T^4 + 0.16T^2$
97. $F^{12} - 2F^6H^6 + H^{12}$
99. $3xy^2z$
101. $2MP^2T^3$
103. $0.4F^2H$
105. $11x\sqrt{y}$
107. $-4\sqrt[3]{d^2e}$
109. $-2a^2b\sqrt[5]{c^2}$
111. $-3a^2 - b + c^2$
113. $-xy$
115. $13 - P$
117. $25 - m^2r + 2r$
119. $18 + 31x$
121. $2M^2 - 3P$
123. $6 - 3D + 3D^2$
125. $6x - x^5y^2 + 2x^2 + 4y$

127. $-12m^2t$
129. 4_{10}
131. 27_{10}
133. 0.5625_{10}
135. 9.625_{10}
137. 10000_2
139. 1110101_2
141. 0.01_2
143. 1100.111_2
145. 114_{10}
147. 77_{10}
149. a. $0.75x$
  b. $0.15x$
  c. $0.25x$
151. $A = 2x + 3$ mm
  $B = 9x + 6$ mm
  $C = 1.5x + 1$ mm
  $D = 6x + 4$ mm
  $E = 0.7x + 3$ mm
  $F = 3.2x + 4$ mm
  $G = 4.2x + 4$ mm

UNIT 14: SIMPLE EQUATIONS

Exercise 14–3

1. $x + 20 = 35$
  $x = 15$
3. $4x = 40$
  $x = 10$
5. $20 \div x = 5$
  $x = 4$
7. $5x + x = 36$
  $x = 6$
9. $6x \div 3 = 12$
  $x = 6$
11. $10x - 4x = 42$
  $x = 7$
13. $5x + 4 - 2x = 34$
  $x = 10$
15. a. Let length of small piece $= x$
  $x + 3x = 12$ ft
  $x = 3$ ft
  b. $3x = 9$ ft

17. a. Let area of lot 1 $= x$
  $x + 2x + 2x + (x + 5{,}000$ sq ft$) = 65{,}000$ sq ft
  $x = 10{,}000$ sq ft
  b. 20,000 sq ft
  c. 20,000 sq ft
  d. 15,000 sq ft

19. a. Let x = number of calories for breakfast
  $x + 2x + 1\dfrac{1}{2}(2x) + 150$ cal $= 2{,}550$ cal
  $x = 400$ cal
  b. 800 cal
  c. 1,200 cal
21. $x + 3x + 5x = 27$ in
  $x = 3$ in
23. $x + 4x + x + 6x + x + 8x + x = 11$ in
  $x = 0.5$ in

25. a. $x + 0.8x + 1.3x + 2x + 0.9x = 120$ V
  $x = 20$ V
  b. 16 V
  c. 26 V
  d. 40 V
  e. 18 V
27. a. $\left(h + \dfrac{1}{4}\text{ in}\right) + h + \left(2h + \dfrac{3}{4}\text{ in}\right) = 9$ in
  $h = 2$ in
  $AB = 2\dfrac{1}{4}$ in
  b. $CD = 2$ in
  c. $EF = 4\dfrac{3}{4}$ in

Exercise 14–5

1. 10
3. 19
5. 4
7. 36
9. -22
11. -53
13. 34
15. -20.9
17. 18.8
19. 24.09
21. 0
23. $-\dfrac{5}{32}$
25. $-3\dfrac{1}{4}$
27. $-23\dfrac{1}{8}$
29. -17.101
31. 2.3 in
33. $\dfrac{7}{16}$ in
35. $61\dfrac{1}{2}°$
37. $\$194.75$
39. 7.22 cm
41. 115 W
43. 4.4286 in
45. 3.48 cm

Exercise 14–6

1. 40	**11.** 9	**21.** 109.861	**29.** $-\dfrac{1}{32}$	**37.** 53.3 mm
3. 13	**13.** 98.75	**23.** 1		**39.** $25.75
5. 3	**15.** 1.3	**25.** $18\dfrac{1}{2}$	**31.** 58.37 mm	**41.** 80.6 cm
7. −14	**17.** 8.00		**33.** 17′10″	
9. 20	**19.** 105.70	**27.** 0	**35.** 8.048 in	

Exercise 14–7

1. 8	**9.** −3.697	**17.** 0	**25.** 5	**33.** 8″	**41.** 3.5 yr
3. 4	**11.** 0	**19.** 7	**27.** 1	**35.** 15.8 ft	
5. −9	**13.** 6	**21.** −7.2	**29.** −2,469	**37.** 230 V	
7. 0.08	**15.** −8.8	**23.** 64	**31.** 25.7 mm	**39.** 10.5 A	

Exercise 14–8

1. 20	**11.** 18	**21.** 0.0832	**27.** 7.497	**35.** 44 mi
3. 63	**13.** 23.4	**23.** $3\dfrac{3}{4}$	**29.** 0.9	**37.** 12 V
5. 27	**15.** −6		**31.** 150°	**39.** 1,130 ft-lb
7. 0	**17.** 0	**25.** 2	**33.** 259 sq ft	**41.** 141.6 L
9. 36	**19.** 0.001			

Exercise 14–9

1. 4	**11.** 100	**17.** $-\dfrac{1}{2}$	**23.** 1.66	**31. a.** 5 cm
3. 9	**13.** $\dfrac{2}{3}$		**25.** 4.44	**b.** 0.4 yd
5. 3		**19.** $\dfrac{4}{5}$	**27.** 1.72	**c.** $\dfrac{2}{3}$ ft
7. 12	**15.** $\dfrac{3}{5}$	**21.** 0.3	**29.** −6.01	**d.** 0.3 m
9. 5				**e.** $\dfrac{1}{2}$ in

Exercise 14–10

1. 64	**11.** 0	**19.** $\dfrac{9}{64}$	**23.** $\dfrac{25}{64}$	**31. a.** 13 cm²
3. 1.44	**13.** −15.1		**25.** 23.4	**b.** 167 m²
5. 0.672	**15.** −0.216	**21.** $\dfrac{1}{16}$	**27.** 480	**c.** $30\dfrac{1}{4}$ sq in
7. 12.2	**17.** 0.001		**29.** 21,000	**d.** $\dfrac{9}{16}$ sq ft
9. −0.001				**e.** 0.0076 m²

Unit Exercise and Problem Review

1. 15	**21.** $8\dfrac{1}{4}$	**37.** −3.2	**53.** 225	**71.** $\dfrac{-2}{5}$
3. 27		**39.** −0.006	**55.** 216	**73.** 6
5. −33	**23.** $15\dfrac{1}{4}$	**41.** −377	**57.** −2	**75.** 3
7. 17		**43.** −8	**59.** 0.0001	**77.** 3.2 in
9. 18	**25.** 11	**45.** 22	**61.** 0.008	**79.** $350,000
11. 5.5	**27.** 294	**47.** $\dfrac{1}{8}$	**63.** −0.636	**81.** 60 V
13. 4.1	**29.** −5		**65.** 6.6	**83.** 0.3 Ω
15. 2.02	**31.** 81	**49.** 8	**67.** 16.3	**85.** 84.66 mm
17. −1.896	**33.** 4	**51.** −4	**69.** $\dfrac{25}{9}$ or $2\dfrac{7}{9}$	**85.** 4 ft
19. 0	**35.** 65			

UNIT 15: COMPLEX EQUATIONS

Exercise 15–1

1. 8	**7.** −2.59	**13.** 4.54	**19.** 3	**25.** −0.13
3. 7	**9.** 9	**15.** 7	**21.** 4	**27.** 0.77
5. 2	**11.** 10	**17.** 31.3	**23.** −5.2	**29.** 9.99

Exercise 15–3

1. 6 ft   **3.** 750 turns   **5.** 12.5 ft   **7.** 45   **9.** 42 amperes

Exercise 15–4

1. $\dfrac{A}{a}$

3. a. IR

   b. $\dfrac{E}{I}$

5. $180° - \angle A - \angle B$

7. $S + \dfrac{1.732}{N}$

9. $\dfrac{C_a + SF}{S}$

11. $\dfrac{1{,}000P}{E}$

13. a. $IR + E_c$

   b. $\dfrac{E_x - E_c}{I}$

15. a. $\sqrt{PR}$

   b. $\dfrac{E^2}{P}$

17. $\dfrac{D_o + d - 2a}{2}$

19. a. I^2R

   b. $\dfrac{P}{I^2}$

21. $M + 0.866P - 3W$

23. $\dfrac{L - 1.57D - 2x}{1.57}$

25. a. $\dfrac{nE - Inr}{I}$

   b. $\dfrac{nE - IR}{In}$

27. a. 3.8 m

   b. 18.6 ft

29. a. 17,700 Btu/min

   b. 6.4°F

Unit Exercise and Problem Review

1. 4
3. 7
5. −6.5
7. 7.5

9. −134
11. 8
13. −2
15. 0.25

17. 2.153
19. 248 Volts

21. a. $\dfrac{R}{1.155}$

   b. $\sqrt{\dfrac{A}{2.598}}$

23. a. $\sqrt{d^2 - b^2}$

   b. $\sqrt{d^2 - a^2}$

25. a. 4.30 in

   b. 21.21 cm

UNIT 16:   THE CARTESIAN COORDINATE SYSTEM AND GRAPHS OF LINEAR EQUATIONS

Exercise 16–2

1. (See Instructor's Guide)
3. (See Instructor's Guide)
5. (See Instructor's Guide)
7. (See Instructor's Guide)

9. (See Instructor's Guide)
11. (See Instructor's Guide)
13. (See Instructor's Guide)
15. (See Instructor's Guide)

17. (See Instructor's Guide)
19. (See Instructor's Guide)
21. (See Instructor's Guide)
23. (See Instructor's Guide)

Exercise 16–3

1. −3

3. $-\dfrac{1}{2}$

5. $-\dfrac{13}{3}$

7. $-\dfrac{7}{3}$

9. −4

11. $\dfrac{3}{7}$

13. −3

15. −11

Exercise 16–4

1. $y = \dfrac{1}{2}x + 5$

3. $y = \dfrac{3}{2}x + 10$

5. $y = -3x + 5$

7. $y = \dfrac{1}{5}x + 12$

9. $y = 2x + 4\dfrac{1}{4}$

11. $y = \dfrac{5}{8}x$

Exercise 16–5

1. $y - 4 = 3(x - 5)$ **3.** $y + 4 = \dfrac{1}{2}(x + 2)$ **5.** $y - 5.1 = -\dfrac{2}{5}(x + 3.25)$ **7.** $y = 3.4$

Exercise 16–6

1. $y = \dfrac{1}{2}x + \dfrac{1}{2}$

3. $y = -\dfrac{2}{3}x + 5$

5. $y = -\dfrac{2}{3}x + 3$

7. $y = -\dfrac{1}{2}x - 1\dfrac{1}{2}$

9. $y = -4x - 13$
11. $y = -x + 7$

13. $y = \dfrac{6}{7}x$

15. $y = -2x - 18$

Exercise 16–7A

1. $m = -\dfrac{1}{3}, b = -3\dfrac{1}{3}$

3. $m = \dfrac{1}{3}, b = 6\dfrac{2}{3}$

5. $m = 5, b = -21$

7. $m = -\dfrac{1}{2}, b = -4\dfrac{1}{2}$

9. $m = 2, b = 3$
11. $m = 8, b = 0$

13. $m = \dfrac{5}{3}, b = -\dfrac{2}{3}$

15. $m = -\dfrac{1}{3}, b = 2\dfrac{1}{3}$

17. $m = \dfrac{2}{5}, b = -3$

   (See Instructor's Guide)

19. $m = -\dfrac{6}{5}, b = -6$

   (See Instructor's Guide)

21. $m = \dfrac{-7}{4}, b = 5\dfrac{1}{2}$

   (See Instructor's Guide)

23. $m = 21, b = 0$

   (See Instructor's Guide)

Exercise 16–7B

1. a. (See Instructor's Guide)

   b. $y + 2 = \dfrac{1}{3}(x - 3)$

   c. $y = \dfrac{1}{3}x - 3$

3. a. (See Instructor's Guide)

   b. $y + 2 = -\dfrac{3}{4}(x + 5)$

   c. $y = -\dfrac{3}{4}x - 5.75$

5. a. (See Instructor's Guide)

   b. $y + 2 = 3(x + 5)$

   c. $y = 3x + 13$

7. a. (See Instructor's Guide)

   b. $y - 3 = 2.5(x - 1)$

   c. $y = 2.5x + 0.5$

Unit Exercise and Problem Review

1. (See Instructor's Guide)
3. (See Instructor's Guide)
5. (See Instructor's Guide)
7. $\dfrac{1}{2}$
9. $\dfrac{4}{9}$
11. 5
13. $y = \dfrac{1}{2}x + 6$
15. $y = -\dfrac{3}{4}x - 12$

17. $y = -3x + 12$
19. $y + 6.5 = -5(x - 7.1)$
21. $y + \dfrac{5}{8} = \dfrac{12}{5}\left(x - \dfrac{3}{4}\right)$
23. $y = -2x + 12$
25. $y = -\dfrac{3}{5}x - \dfrac{1}{5}$
27. $y = \dfrac{1}{5}x - 3\dfrac{2}{5}$
29. $m = -\dfrac{5}{7}, b = -3\dfrac{2}{7}$
31. $m = -9, b = -51$

33. $m = -\dfrac{1}{2}, b = 0$
35. $m = -\dfrac{1}{3}, b = 1\dfrac{2}{3}$
   (See Instructor's Guide)
37. $m = 6, b = 3$
   (See Instructor's Guide)
39. $m = -\dfrac{4}{5}, b = 0$
   (See Instructor's Guide)
41. (See Instructor's Guide)
42. (See Instructor's Guide)

UNIT 17: SYSTEMS OF EQUATIONS

Exercise 17–1

1. $(4, 2)$
3. $\left(-\dfrac{1}{2}, -1\right)$
5. $(3, -4)$
7. $(-4, 2)$
9. $(-2, 8)$
11. $(-1, 2)$
13. $(-2, -1)$
15. $(3, 6)$
17. $(-4, 7)$

Exercise 17–2

1. $x = 1, y = -3$
3. $x = 6, y = 3$
5. $x = 1, y = -2$
7. $x = -9, y = -1$
9. $x = 3, y = 5$
11. $x = 1, y = 3$
13. $x = 1, y = 2$
15. $x = 10, y = 4$
17. $x = 2\dfrac{4}{7}, y = 1\dfrac{2}{7}$
19. $x = 2, y = 1$
21. $x = -0.5, y = 0.25$
23. $x = -2, y = 2$

Exercise 17–3A

1. $x = 3, y = 2$
3. $x = 4, y = 3$
5. $x = 5, y = 1$
7. $x = 2, y = 3$
9. $x = -2, y = 5$
11. $x = -7, y = -8\dfrac{2}{3}$
13. $x = 4, y = 5$
15. $x = -3, y = -2$
17. $x = 5, y = 5$

Exercise 17–3B

1. $x = 1, y = 5$
3. $x = -1, y = -6$
5. $x = 3, y = 9$
7. $x = 20, y = -14$
9. $x = 6, y = 3$
11. $x = \dfrac{2}{3}, y = -2$
13. $x = 2, y = 1$
15. $x = -1.8, y = -2.2$
17. $x = 5, y = 4$
19. $x = 3, y = 2$
21. $x = 500, y = 200$

Exercise 17–4

1. Dependent (infinite solution sets)
3. Inconsistent (no solution set)
5. Dependent (infinite solution sets)
7. Consistent and independent (one solution set)
9. Inconsistent (no solution set)
11. Dependent (infinite solution sets)
13. Consistent and independent (one solution set)
15. Consistent and independent (one solution set)

Exercise 17–5

1. 8
3. -39
5. 24.03
7. -275.47
9. $74\dfrac{5}{8}$

Exercise 17–6

1. $x = 1, y = 5$
3. $x = -2, y = 4$
5. $x = \dfrac{37}{22} = 1\dfrac{15}{22}, y = \dfrac{3}{22}$
7. $x = x = -3.8, y = -9.8$
9. $\approx -0.7088, y \approx 0.3526$

Exercise 17–7

1. Numbers are 9 and 16
3. Cost is $7 and $12
5. Part A = 20.5 lb, Part B = 13 lb
7. Lengths are 16 in and 26 in

9. 6.25 liters of 18% acid solution
8.75 liters of 30% acid solution
11. 3,500 liters per hour by small pump
4,950 liters per hour by large pump
13. 80 bags of cement
18 bags of lime

15. 219.6 cm^3 of 12.80% acid solution
55.4 cm^3 of 26.20% acid solution
17. 26.4 gal of $8.60 per gal chemical
48.6 gal of $6.75 per gal chemical
19. Angle $A = 50°$
21. a. 6.25 gal
b. 16.25 gal

Unit Exercise and Problem Review

1. $(8, -2)$
3. $(-4, 0)$
5. $(-9, -1)$
7. $x = 0, y = 3$
9. $x = 5, y = 1$
11. $x = 5, y = 1$
13. $x = -35, y = -30$
15. $x = 2, y = 8$
17. $x = 3, y = 3$
19. $x = 2, y = 2$
21. $x = 3, y = 4$
23. $x = 4, y = 3$

25. $x = 2, y = 2$
27. $x = 2, y = -3$
29. Dependent (infinite solution sets)
31. Consistent and independent (one solution set)
33. Dependent (infinite solution sets)
35. Consistent and independent (one solution set)
37. $x = -6$
39. $\dfrac{1}{4}$

41. $x = 3.2, y = -5.1$
43. $D = 0$. The system is inconsistent. These two lines are parallel and there is no solution.
45. The two numbers are 9 and 14
47. 8.5 liters of 20.3% acid solution
17 liters of 35.6% acid solution
49. Length = 630 ft, Width = 335 ft
51. a. 4,600 items per day
b. 7,800 items per day

UNIT 18: QUADRATIC EQUATIONS

Exercise 18–2A

1. ± 12
3. ± 7.2
5. ± 2.24
7. ± 3.74
9. ± 8
11. ± 1.51
13. ± 1.69
15. ± 5.59
17. ± 1.13
19. ± 8
21. ± 8.06
23. ± 7.75
25. ± 8.57

Exercise 18–2B

1. 7.6 seconds
3. 73.2 mph
5. 492 ft/s
7. 10.6 cm
9. 20 cm/s
11. 51.0 ft

Exercise 18–3

1. 0, 10
3. $0, -\dfrac{2}{3} \approx -0.67$
5. 0, 5
7. 4, −0.75
9. 8, −3
11. 15, −5
13. 4, −0.75
15. 0.25, 0.75
17. 1, −0.75
19. 1.33, 1.5
21. $2.5, -\dfrac{1}{3} \approx -0.33$
23. 4, −5.5
25. 3, −0.25
27. 8, 0.75
29. 7, −3

Exercise 18–4

1. 0.54 mi
3. $l = 6.23$ m
5. $h = 39.65$ cm
7. $R = 5.500$ in
9. $w = 2.45$ cm
11. $h = 21.5$ cm
13. $d = 8.00$ in

Exercise 18–5A

1. 8 and 13, −8 and −13
3. 13 and 20, −13 and −20
5. 5 and 9, −5 and −9
7. 11 and 19, −11 and −19
9. 6.4 and 3.5, 3.5 and 6.4

Exercise 18–5B

1. Length ≈ 18.7 in
Width ≈ 11.5 in
3. 200 ft
5. 45 units at $265.63 or 83 units at $13.70
7. Length = 7.5 m
Width = 6.2 m
9. 181 ft

Unit Exercise and Problem Review

1. ± 8
3. ± 1.85
5. ± 5.12
7. ± 0.77
9. ± 5
11. ± 4.92
13. 0, 0.5
15. 2, −12
17. 24, 2
19. 1.5, −0.5
21. 3, 0.5
23. 0, 3
25. 1.34, −4.84
27. 5, −4
29. 0.8, −1
31. 7 and 11
33. 10 and 13
35. 5.6 and 11.8, −5.6 and −11.8
37. 8.39 in
39. Length ≈ 46.8 in
Width ≈ 36.3 in
41. $h ≈ 24.9$ cm

SECTION IV: Fundamentals of Plane Geometry

UNIT 19: INTRODUCTION TO PLANE GEOMETRY

Unit Exercise

1. Geometry is the branch of mathematics in which the properties of points, lines, surfaces, and solids are studied.
3. **a.** Quantities equal to the same quantities or to equal quantities are equal to each other.
   b. If equals are subtracted from equals, the remainders are equal.
   c. The whole is equal to the sum of all its parts.
   d. If equals are added to equals, the sums are equal.
   e. If equals are multiplied by equals, the products are equal.
   f. If equals are subtracted from equals, the remainders are equal.
   g. Quantities equal to the same quantities or to equal quantities are equal to each other.
5. (See Instructor's Guide)

UNIT 20: ANGULAR MEASURE

Exercise 20–4

1. 15.5°	7. 2.98°	13. 107.3067°	19. 7°45′	25. 44°37′	31. 0°32′18″
3. 59.2°	9. 256.32°	15. 1.0169°	21. 10°4′	27. 26°1′	33. 44°5′12″
5. 105.78°	11. 7.1417°	17. 312.9992°	23. 113°17′	29. 123°3′49″	35. 406°55′50″

Exercise 20–5A

1. 82°	7. 75°29′29″	13. 102°12′06″	19. 16°08′	25. 3°02′01″	31. 53°42′
3. 26°49′	9. 24°27′27″	15. 97°12′09″	21. 53°54′	27. 1°57′41″	33. 38°54′
5. 86°13′	11. 132°49′15″	17. 17°	23. 2°47′41″	29. 31°57′39″	35. 73°45′35″

Exercise 20–5B

1. 108°	7. 56°06′	13. 41°24′48″	19. 18°17′	25. 48°45′25″	31. 130°52′
3. 30°38′	9. 84°24′48″	15. 275°16′54″	21. 11°33′	27. 92°41′59″	33. 48°42′
5. 73°48′	11. 108°43′21″	17. 9°06′	23. 8°06′03″	29. 22°12′22″	

Exercise 20–6

1. $A = 25°$ $C = 54°$ $E = 93°$ $G = 23°$ $I = 81°$ 3. 45°   7. 70°   11. **a.** 90°
   $B = 42°$ $D = 77°$ $F = 11°$ $H = 46°$ $J = 87°$ 5. 65°   9. **c.** 60°   13. 30°

Exercise 20–7

1. 31°	5. 33°42′	9. 40°59′02″	13. 83°	17. 176°22′42″
3. 44°	7. 16°41′33″	11. 96°	15. 114°24′	19. 89°58′58″

Unit Exercise and Problem Review

1. 131°	13. 3°11′48″	23. 26°45′	33. 241°42′	43. 49.3967°	53. 27°
3. 60°29′	15. 30°01′03″	25. 10°41′36″	35. 18.5°	45. 207°45′0″	55. 3°40′12″
5. 19°23′44″	17. 108°42′	27. 30°34′12″	37. 109.8°	47. 80°55′23″	57. 45°56′56″
7. 81°29′35″	19. 159°57′36″	29. 5°15′23″	39. 67.9119°	49. 19°58′39″	59. 70°04′
9. 14°09′	21. 108°48′27″	31. 290°45′	41. 214.1239°	51. 196°30′37″	61. 1°50′39″
11. 36°19′11″					

UNIT 21: ANGULAR GEOMETRIC PRINCIPLES

Exercise 21–3

1. ∠A, ∠BAF, ∠FAB	11. ∠6, ∠DCB	21. Right	27. **a.** 72°	33. ∠GCB and ∠HBA;
3. ∠3, ∠CDE, ∠EDC	13. Acute	23. **a.** 330°	**b.** Acute	∠HBC and ∠DCG;
5. ∠5, ∠EFA, ∠AFE	15. Right	**b.** Reflex	29. **a.** 180°	∠FCB and ∠EBA;
7. ∠2, ∠FBC	17. Obtuse	25. **a.** 180°	**b.** Straight	∠EBC and ∠FCD
9. ∠4, ∠BCE	19. Acute	**b.** Straight	31. ∠1 and ∠8; ∠2 and ∠4;	
			∠3 and ∠9	

Exercise 21–4

1. $\angle 1 = 36°$
$\angle 2 = 60°40'$
$\angle 3 = 83°20'$
$\angle 4 = 104°$
$\angle 5 = 159°20'$
3. a. $\angle 2 = 121°$
$\angle 3 = 59°$
$\angle 4 = 121°$
$\angle 5 = 59°$
$\angle 6 = 121°$

$\angle 7 = 59°$
$\angle 8 = 121°$
b. $\angle 2 = 116°42'$
$\angle 3 = 63°18'$
$\angle 4 = 116°42'$
$\angle 5 = 63°18'$
$\angle 6 = 116°42'$
$\angle 7 = 63°18'$
$\angle 8 = 116°42'$

5. $\angle 1 = 95°$
$\angle 2 = 95°$
$\angle 3 = 85°$
$\angle 4 = 85°$
$\angle 5 = 85°$
$\angle 6 = 95°$
$\angle 7 = 85°$
$\angle 8 = 95°$
$\angle 9 = 85°$

$\angle 10 = 95°$
$\angle 11 = 85°$
$\angle 12 = 95°$
$\angle 13 = 95°$
$\angle 14 = 85°$
$\angle 15 = 53°$
$\angle 16 = 85°$
$\angle 17 = 53°$
$\angle 18 = 51°$

$\angle 19 = 76°$
$\angle 20 = 53°$
$\angle 21 = 76°$
$\angle 22 = 76°$
7. a. $\angle F = 94.95°$
$\angle G = 85.05°$
$\angle H = 85.05°$
b. $\angle F = 101°43'$
$\angle G = 78°17'$
$\angle H = 78°17'$

Unit Exercise and Problem Review

1. $\angle 6$, $\angle ABC$
3. $\angle 2$, $\angle DFH$
5. $\angle DFG$,
$\angle GFD$
7. Obtuse
9. Acute
11. Reflex
13. Acute

15. Straight
17. $\angle GAH$ and $\angle HAB$;
$\angle AGH$ and $\angle FGH$;
$\angle GHA$ and $\angle JHA$;
$\angle HJB$ and $\angle CJB$;
$\angle EJC$ and $\angle BJC$;
$\angle EJC$ and $\angle EJH$;
$\angle EJH$ and $\angle HJB$

19. $\angle 1$ and $\angle 5$;
$\angle 4$ and $\angle 8$;
$\angle 3$ and $\angle 7$;
$\angle 2$ and $\angle 6$
21. $\angle 1 = 97°$
$\angle 2 = 97°$
$\angle 3 = 83°$
$\angle 4 = 48°$

$\angle 5 = 35°$
$\angle 6 = 145°$
$\angle 7 = 82°$
$\angle 8 = 98°$
$\angle 9 = 117°$
$\angle 10 = 63°$
23. $\angle 1 = 92°20'$
$\angle 2 = 87°40'$

$\angle 3 = 92°20'$
$\angle 4 = 87°40'$
$\angle 5 = 87°40'$
$\angle 6 = 87°40'$
$\angle 7 = 37°$
$\angle 8 = 82°45'$
$\angle 9 = 60°15'$
$\angle 10 = 82°45'$

UNIT 22: TRIANGLES

Exercise 22–1

1. (See Instructor's Guide)
3. a. Isosceles

b. Scalene
c. Isosceles

d. Equilateral
e. Scalene

f. Isosceles
g. Equilateral

h. Isosceles

Exercise 22–2

1. $180°$
3. a. $26°21'$
b. $27°47'32''$

5. a. $83°$
b. $9°16'$

7. a. $45°$
b. $45°$

9. a. $25°39'$
b. $69°0'13''$

11. a. $47°$
b. $28°$

Exercise 22–5

1. a. 4.6 in
b. $38°46'$

3. a. 21.07 in
b. 22.72 in

5. a. $60°$
b. $60°$

c. 4.2 m

Exercise 22–7

1. 25 in
3. 24 mm

5. 60 mm
7. a. 8 cm

b. $51°$
9. 15.6 m

11. The wall and floor are not square.
13. 117 km

Unit Exercise and Problem Review

1. (See Instructor's Guide)
3. a. $140°46'$
b. $18°42'$
5. a. $16°$
b. $50°25'$
7. a. 30 mm
b. 12.2 mm

9. a. 2.5 m
b. 1.25 m
11. Since both the computed measurement and the actual measurement are 15 ft, the wall is square.
13. 35.0 ft
15. 46.28 mi/h

UNIT 23: CONGRUENT AND SIMILAR FIGURES

Exercise 23–1

1. a. *BC* and *EF*, *AC* and *DF*, *AB* and *DE*
b. *OM* and *OG*, *OK* and *OH*, *KM* and *GH*

c. *PT* and *PR*, *ST* and *SR*, *SP* and *SP*

Exercise 23–2

1. a. Polygons are similar. Corresponding sides are proportional, corresponding angles are equal, and the number of sides is the same.
  b. Polygons are not similar. Corresponding sides are not proportional.
  c. Polygons are not similar. Corresponding sides are not proportional.
  d. Polygons are similar. Corresponding sides are proportional, corresponding angles are equal, and the number of sides is the same.
  e. Polygons are similar. Corresponding sides are proportional, corresponding angles are equal, and the number of sides is the same.
  f. Polygons are similar. Corresponding sides are proportional, corresponding angles are equal, and the number of sides is the same.
3. a. 21.6 cm   **b.** 20 cm   **c.** 18.4 cm   **d.** 14.4 cm   **e.** 90.4 cm

Exercise 23–3

1. a. If two angles of a triangle are equal to two angles of another triangle, the triangles are similar.
  b. $AC = 36$ in
    $DE = 8$ in
3. a. If two sides of a triangle are proportional to two sides of another triangle and if the angles included between these sides are equal, the triangles are similar.
  b. 11.2 in
5. a. If two angles of a triangle are equal to two angles of another triangle, the triangles are similar.
  b. $HK = 1.12$ cm
    $LM = 1.85$ cm
7. $x = 2.77$ cm
   $y = 3.84$ cm
9. 19.2 mm

Unit Exercise and Problem Review

1. $\angle B$ and $\angle D$; $\angle BAC$ and $\angle DAC$; $\angle BCA$ and $\angle DCA$; AB and AD; BC and DC; AC and AC
3. HC and CD; FC and CE; FH and DE; $\angle F$ and $\angle E$; $\angle H$ and $\angle D$; $\angle FCH$ and $\angle ECD$
5. a. Polygons are not similar. Corresponding sides are not proportional.
  b. Polygons are similar. Corresponding sides are proportional, corresponding angles are equal, and the number of sides is the same.
7. a. CD is an altitude to the hypotenuse of right. $\triangle ACB$. If the altitude is drawn to the hypotenuse of a right triangle, the two triangles formed are similar to each other and to the given triangle.
  b. $\angle 1 = 59°$
    $\angle 2 = 31°$
    $\angle 3 = 31°$
9. 37.5 m

UNIT 24: POLYGONS

Exercise 24–1

1. a. Concave polygon   **d.** Convex polygon   **g.** Convex polygon   **i.** Not a polygon
  b. Convex polygon   **e.** Not a polygon   **h.** Not a polygon   **j.** Concave polygon
  c. Convex polygon   **f.** Convex polygon

Exercise 24–2

1. Yes. All parallelograms have four sides.
3. Yes. All rectangles have opposite sides parallel and equal.
5. Yes. All squares have equal and parallel opposite sides.
7. No. All the sides of a rectangle do not have to be equal.

9. a. 14 in   **11. a.** 3 ft
  b. 8 in   **b.** 115°
  c. 82°   **c.** 65°
  d. 98°

Exercise 24–4

1. a. 360°   **3. a.** 108°   **5. a.** 72°   **7. a.** 6   **9.** 56°01'
  b. 540°   **b.** 120°   **b.** 60°   **b.** 8
  c. 720°   **c.** 135°   **c.** 45°   **c.** 5
  d. 1,080°   **d.** 150°   **d.** 22.5° or 22°30'   **d.** 9

Exercise 24–5

1. Width $B = 2'7''$
   Width $A = 2'2''$
   Width $C = 3'0''$

Unit Exercise and Problem Review

1. Convex polygon
3. Not a polygon
5. Convex polygon
7. (1) regular, (2) quadrilateral
9. (1) equiangular, (2) hexagon

11. a. 11 cm
b. 38°
c. 52°
13. a. 17°
b. 73°
c. 73°

15. a. 90°
b. 144°
c. 156°
17. a. 90°
b. 40°
c. 14.4° or 14°24′

19. a. 10
b. 12
c. 20
21. a. 50°
b. 75°

23. a. $7\frac{5}{16}$ in
b. 15.94 m

UNIT 25: CIRCLES

Exercise 25–1

1. a. Chord
b. Diameter
c. Radius
d. Center
3. a. Segment
b. Sector
c. Arc

Exercise 25–4

1. a. 88.0 in
b. 103.0 cm
c. 23.6 ft
d. 19.5 m
e. 111.2 in
f. 223.7 mm
3. a. 92.8°
b. 119.2°
5. 6.0 in
c. 50.2°
d. 234.6°
7. 16.8 cm
9. 19.9 m
11. 750
13. 2,500 m
15. 145 ft

Exercise 25–7

1. a. 2.5 in
b. 2.5 in
3. a. 5.0 in
b. 3.3 in
5. a. 1.2 m
b. 1.35 m
7. a. 2.2 ft
b. 216.7° or 216°42′
9. 32.2 ft
11. 3.7 m

Exercise 25–10

1. a. ∠E = 45°42′
       ∠F = 24°52′
b. 43°40′
3. a. 32°
b. 54°29′
5. 1.913 in
7. 166 ft

Exercise 25–13

1. a. 35°
b. 145°
3. a. $\widehat{EF}$ = 98°
       ∠4 = 42°
b. $\widehat{GH}$ = 37°40′
       ∠3 = 52°30′
5. a. 61°12′
b. 90°
c. 57°36′
7. 95°18′
9. (See Instructor's Guide)
11. ∠1 = 25°
∠2 = 67°
∠3 = 48°
∠4 = 29°
∠5 = 48°
∠6 = 86°
∠7 = 94°
∠8 = 126°
∠9 = 54°
∠10 = 88°

Exercise 25–17

1. a. 13°
b. 17.3°
c. 32°
d. 104°
e. 13°33′
f. 105°
3. a. 25°
b. 24.57°
c. 108°
d. 11°
e. 13°5′13″
f. 35°
5. a. $\widehat{DH}$ = 13°
       $\widehat{EDH}$ = 26°
b. 154.16°
7. a. 15.40 cm
b. 32.26 cm
9. a. 19.10 m
b. 6.62 m
11. A = 0.400 in
B = 1.362 in

Unit Exercise and Problem Review

1. a. Chord
b. Diameter
c. Radius
d. Arc
e. Tangent
f. Tangent point
3. a. 100.5 cm
b. 16.3 in
c. 5.0 m
5. 2.4 m
7. 9.7 in

9. 224.1 revolutions
11. a. 1.7 radians
b. 0.3 radian
c. 3.4 radians
13. a. 2.9 m
b. 1.7 m
15. a. EH = 11.5 cm
       $\widehat{EGF}$ = 28 cm
b. $\widehat{EGF}$ = 41.0 cm
       HF = 17.9 cm

17. 8.0 cm
19. a. 39°
b. 42°
21. 5.7 in
23. a. 59°42′
b. 34°
25. a. 120°
b. 60°
c. 44°

27. a. 15°49′16″
b. 88°56′14″
c. 30°55′20″
29. a. 35.23°
b. 198.0°
c. 40.11°
31. ∠1 = 32°
∠2 = 58°
∠3 = 17°

∠4 = 15°
∠5 = 32°
∠6 = 20°
∠7 = 38°
∠8 = 21°
∠9 = 69°
∠10 = 122°
33. a. 15.30 m
b. 9.40 m

SECTION V: Geometric Figures: Areas and Volumes

UNIT 26:   AREAS OF COMMON POLYGONS

Answers may vary slightly due to rounding within the solution.

Exercise 26–1

1. 21 sq ft   **7.** 87.5 sq ft   **11.** 0.8 in   **15.** 0.9 mi   **19.** $675   **23.** 1,144 cm^2
3. 51 sq in   **9.** 14.0 mm   **13.** 205.2 sq ft   **17.** 4.88 sq ft   **21.** $180   **25.** 15.75 m
5. 8.7 cm

Exercise 26–2

1. 104 cm^2 **9.** 8.5 km   **15.** 0.1 sq yd   **21. a.** Wall area of square floor: 1,600 sq ft
3. 18.7 in   **11.** 311.1 sq mi   **17.** 1.72 sq in   Wall area of rectangular floor: 1,780 sq ft
5. 0.2 mi   **13.** 49.7 in   **19.** 546 cm^2 Wall area of parallelogram floor: 2,200 sq ft
7. 19.8 m^2 **b.** The square-shaped floor

Exercise 26–3

1. 104 sq in   **5.** 7.3 yd   **9.** 602.1 m^2 **13.** 0.5 yd   **17.** 9.98 acres   **21. a.** 2,490 bd ft
3. 2.7 m^2 **7.** 14.2 cm   **11.** 24.0 km   **15.** 8 ft   **19. a.** 24 sq in   **b.** $2,410
                                                                                          b. 1,200,000 lb

Exercise 26–5

1. 178.5 sq in   **7.** 0.2 km^2 **13.** 40 yd   **19.** 4.3 sq ft   **25.** 350 mm^2 **31. a.** 1,510 sq ft
3. 2.1 m^2 **9.** 28 ft   **15.** 0.3 mi   **21.** 1.2 sq yd   **27.** $7,710   **b.** 34.8 bundles
5. 0.2 mi   **11.** 28.4 cm^2 **17.** 11.6 m^2 **23.** 31.3 cm^2 **29.** 2744 cm^2 **c.** 21.1 ft

Unit Exercise and Problem Review

1. 157.5 sq in   **11.** 138.7 sq in   **21.** 8.3 ft   **31.** $b = 72$ ft   **41.** 370 sq ft   **c.** 23.0 L
3. 14 m   **13.** 1.9 km   **23.** 8.7 mm   **33.** 16.2 sq ft   **43. a.** 792 bd ft   **47.** 3.82 in
5. 0.6 mi   **15.** 24 ft   **25.** 472.5 sq in   **35.** 0.5 m^2 **b.** $586   **49.** 2,037 cm^2
7. 19.8 mm   **17.** 280 cm^2 **27.** 1.6 m   **37.** 2 sq ft   **45. a.** 275 m^2
9. 204 m^2 **19.** 2 yd   **29.** 0.5 km   **39.** 94.1 cm^2 **b.** 311 m^2

UNIT 27:   AREAS OF CIRCLES, SECTORS, SEGMENTS, AND ELLIPSES

Answers may vary slightly due to rounding within the solution.

Exercise 27–2

1. 153.9 sq in   **7.** 11.0 mm   **13.** 9 times greater   **19.** 631 sq yd   **25.** 5.0 kg
3. 188.7 sq ft   **9.** 0.4 sq mi   **15.** 2.13 times greater   **21.** 163 sq in   **27.** 103.4 sq in
5. 3.3 yd   **11.** 1.1 m   **17.** 39.69 times greater   **23.** $882   **29.** 50.6 cm

Exercise 27–3

1. 104.7 cm^2 **5.** 21.8 mm   **9.** 671.2 cm^2 **13.** 3,800 mm^2 **15. a.** 39 sq ft   **17.** 1.0 m^2
3. 706.9 sq in   **7.** 287.3°   **11.** 169.9°   **b.** $60   **19. a.** 1,050 sq ft
                                                                                                              b. 390 sq ft

Exercise 27–4

1. 8.3 sq ft   **3.** 57.5 mm^2 **5.** 0.13 km^2 **7.** $14,780   **9.** 133 sq in

Exercise 27–5

1. 55.0 sq in   **3.** 28.4 m^2 **5.** 5.4 yd   **7.** 718.6 sq in   **9.** 1.5 m   **11.** 487 sq ft   **13.** 52 lb

Unit Exercise and Problem Review

1. 615.75 sq in   **9.** 31.7 sq in   **17.** 402.9 sq yd   **25.** 1,050,000 lb   **31.** Minor axis = 50.0 ft
3. 3.78 yd   **11.** 5.2 in   **19.** 40.1 cm   **27.** 7.05 m^2 Major axis = 65.0 ft
5. 127.68 sq ft   **13.** 240.3 sq ft   **21.** 118.2 sq ft   **29.** 13 in
7. 1.76 times greater **15.** 72.5 sq yd   **23.** 16.8 m^2

UNIT 28:   PRISMS AND CYLINDERS: VOLUMES, SURFACE AREAS, AND WEIGHTS

Answers may vary slightly due to rounding within the solution.

Exercise 28–2

1. 1,000 cu in   **5.** 11.2 m^3 **7. a.** 409 cm^3 **9. a.** 800 cu in   **11. a.** 43.7 m^3
3. 66 lb   **b.** 3.3 kg   **b.** 230 lb   **b.** $4,220

Exercise 28–4

1. 653.6 cu in   **3.** 10.67 gal   **5.** 683.3 cm^3 **7. a.** 0.35 m^3 **9. a.** 848 m^3 **11.** 275.5 mL
                                                                                      b. 353 L   **b.** 283000 L

Exercise 28–5

1. 7 in   **3.** 8.5 ft   **5.** 4.94 in   **7.** 23.07 ft   **9.** 0.15 ft or 1.8 in

Exercise 28–6

1. a. 51 sq ft   **b.** 81 sq ft   **3.** 158 sq ft   **5.** 17.62 lb   **7.** 68 gal   **9. a.** 15.35 cm^2 **b.** $18,700

Unit Exercise and Problem Review

1. 1,672 cm^3 **5.** 183 sq in   **9.** 1.25 m   **13. a.** 385 cu in   **15.** 2.6 in
3. 1.8 ft   **7.** 65 lb   **11.** 970 sq ft   **b.** 117 lb

UNIT 29:   PYRAMIDS AND CONES: VOLUMES, SURFACE AREAS, AND WEIGHTS

Answers may vary slightly due to rounding within the solution.

Exercise 29–3

1. 920 cu ft   **3.** 352.8 cu in   **5.** 34,400 lb   **7.** 10,600 cu ft   **9.** 0.74 L

Exercise 29–4

1. 14.25 ft   **3.** 2.25 m   **5.** 2.69 m   **7.** 6.34 ft

Exercise 29–5

1. 920 sq in   **3. a.** 520 cm^2 **5. a.** 160 sq in   **7. a.** 1,690 sq ft   **9.** 12.5 m^2
                       b. 733 cm^2 **b.** 224 sq in   **b.** 78 bundles

Exercise 29–7

1. 2,137 cu ft   **3.** 18.8 L   **5. a.** 17.9 cu in   **b.** 9.93 oz   **7. a.** 11 cu yd   **b.** 29,000 lb

Exercise 29–8

1. 832 sq ft   **3. a.** 52.9 m^2 **b.** 73.2 m^2 **5.** 4.50 sq ft   **7.** 75.4 sq ft   **9.** 21.9 m^2

Unit Exercise and Problem Review

1. 0.331 m^3 **5.** 419 sq in   **9.** 7,700 cu ft   **13.** 35 L   **15. a.** 0.827 sq ft
3. 11.6 ft   **7.** 635 cm^3 **11.** 17.4 cm   **b.** 1.55 lb

UNIT 30:   SPHERES AND COMPOSITE FIGURES: VOLUMES, SURFACE AREAS, AND WEIGHTS

Answers may vary slightly due to rounding within the solution.

Exercise 30–2

1. 28.27 sq in   **5.** 784.3 sq ft   **9.** $4,020   **13. a.** The cylindrical tank requires more surface area material.
3. 24.63 m   **7.** 74.60 m^2 **11.** 3.2 oz   **b.** 20 m^2

Exercise 30–3

1. 33.5 m^3 **3.** 230 cu in   **5.** 457 cu in   **7.** 2,230 cm^3 **9.** 0.0369 lb   **11.** 6.4 h   **13.** 27 spheres

Exercise 30–4

1. 0.5 cu yd   **3.** 1,110 m^3 **5. a.** 619 sq in   **b.** 10.1 lb   **7.** 9.8 sq ft   **9.** 240 m^3

Unit Exercise and Problem Review

1. a. 129.7 sq in   **3. a.** 1.136 m^2 **5.** $3,210   **9. a.** 1.62 cu ft   **11.** 70 cu yd
    b. 138.9 cu in   **b.** 0.1138 m^3 **7.** 120 g   **b.** 7.61 sq ft

---

SECTION VI: Basic Statistics

UNIT 31: GRAPHS: BAR, CIRCLE, AND LINE

Exercise 31–2

1. a. Salaries: $720,000   **b.** $100,000
    Delivery: $160,000   **c.** 13%
    Depreciation: $120,000   **d.** $50,000
    Maintenance: $80,000   **3. a.** 28,600,000 vehicles
    Miscellaneous: $60,000   **b.** 6,200,000 trucks and
    Utilities: $60,000   buses
    Advertising: $20,000   **c.** 40%

Exercise 31–3

1. (See Instructor's Guide)   **3.** (See Instructor's Guide)

Exercise 31–4

1.

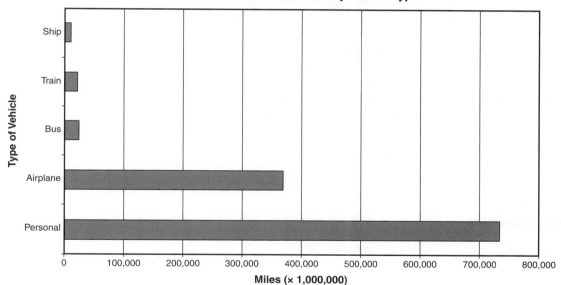

3.

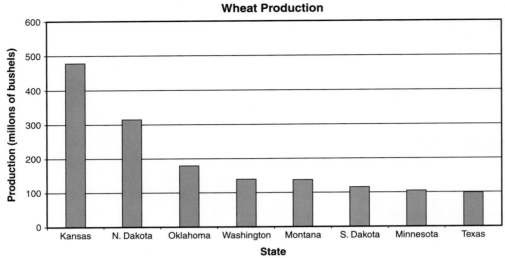

Wheat Production

5.

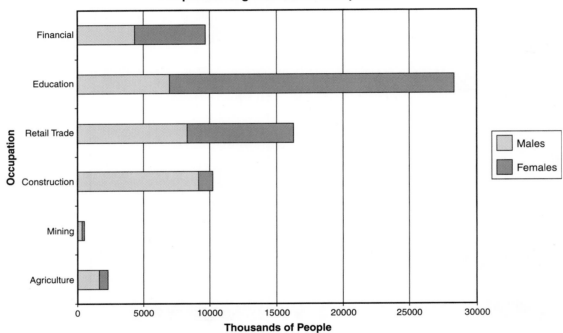

People Working in Selected Occupations

Exercise 31–5

1. (See Instructor's Guide)   3. (See Instructor's Guide)   7. **a.** 487,000,000
                                 5. (See Instructor's Guide)   **b.** 3,688,000,000
                                                                    c. 30,000,000

Exercise 31–6

1.

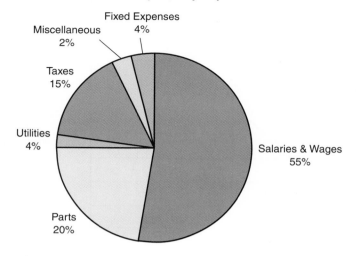

Junior's Auto Repair, July Expenses

3.

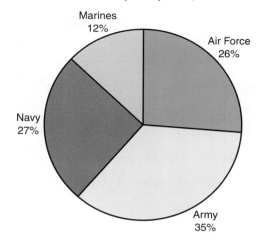

U.S. Military Manpower, 2003

Exercise 31–9

1. **a.** 2:00 PM: 17.0°C
   3:00 PM: 18.2°C
   4:00 PM: 20.6°C
   5:00 PM: 19.4°C
   6:00 PM: 18.6°C
   7:00 PM: 18.2°C
   8:00 PM: 16.0°C
   b. 18.3°C

c. **(1)** 3.6°C increase
   (2) 1.2°C decrease
   (3) 2.6°C decrease
3. **a.** 5 ohms: 5.4 amperes
   10 ohms: 3.8 amperes
   15 ohms: 3.0 amperes
   20 ohms: 2.6 amperes
   25 ohms: 2.4 amperes
   30 ohms: 2.2 amperes

b. 3 amperes: 15 ohms
   4 amperes: 9 ohms
   5 amperes: 6 ohms
   6 amperes: 4 ohms
c. 20 ohms
d. 7 ohms
e. 5 ohms

Exercise 31–11

1. (See Instructor's Guide)

Exercise 31–12

1.

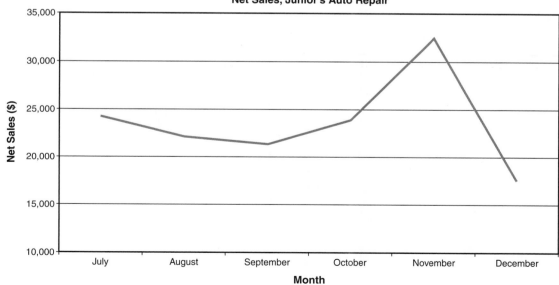

Net Sales, Junior's Auto Repair

3.

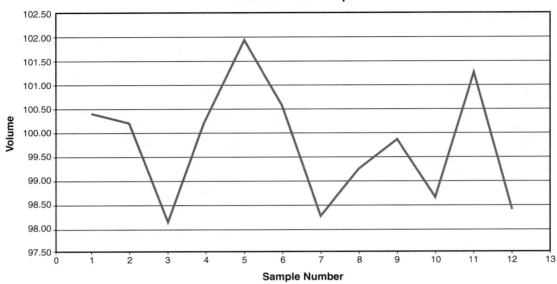

Volume of 100-uL Pipettes

Exercise 31–14

1. (See Instructor's Guide)   **3.** (See Instructor's Guide)

5.

Drill Diameters in Inches		0.250	0.500	0.750	1.000	1.250	1.500
Cutting Speeds in Feet per Minute		39.25	78.50	117.75	157.00	196.25	235.50
Change in Cutting Speeds in Feet per Minute	39.25	39.25	39.25	39.25	39.25		

Plotting data produces a straight-line graph.

Unit Exercise and Problem Review

1. a. 1st Quarter: 245,000 units
　　　2nd Quarter: 270,000 units
　　　3rd Quarter: 310,000 units
　　　4th Quarter: 305,000 units
b. 282,500 units
c. 46%

3. (See Instructor's Guide)
5. a. Jan: $18,200
　　　Feb: $21,000
　　　March: $20,800
　　　April: $22,000
　　　May: $21,200
　　　June: $23,200

b. $2,133 greater
c. 27%
7. (See Instructor's Guide)

9.

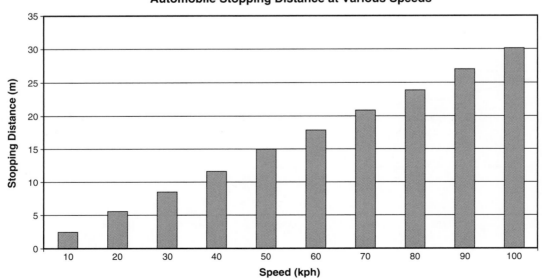

Automobile Stopping Distance at Various Speeds

11.

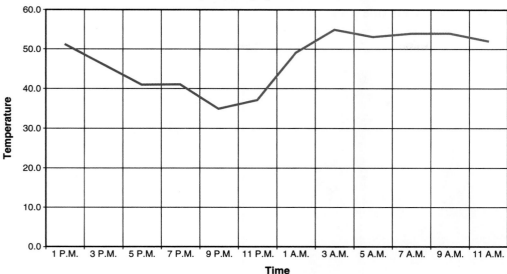

Temperature (°F) for 24-Hour Period

UNIT 32: STATISTICS

Exercise 32–1

1. $\{A\clubsuit, 2\clubsuit, 3\clubsuit, 4\clubsuit, 5\clubsuit, 6\clubsuit, 7\clubsuit,$
$8\clubsuit, 9\clubsuit, 10\clubsuit, J\clubsuit, Q\clubsuit, K\clubsuit\}$

3. $\{(1, 1), (1, 2), (1, 3), (1, 4), (1, 5), (1, 6), (2, 1),$
$(2, 2), (2, 3), (2, 4), (2, 5), (2, 6), (3, 1), (3, 2),$
$(3, 3), (3, 4), (3, 5), (3, 6), (4, 1), (4, 2), (4, 3),$
$(4, 4), (4, 5), (4, 6), (5, 1), (5, 2), (5, 3), (5, 4),$
$(5, 5), (5, 6), (6, 1), (6, 2), (6, 3), (6, 4), (6, 5),$
$(6, 6)\}$

5. $\{R_1R_2, R_1R_3, R_2R_3, R_1W_1, R_1W_2, R_2W_1, R_2W_2,$
$R_3W_1, R_3W_2, R_1G, R_2G, R_3G, W_1W_2, W_1G, W_2G\}$

7. $P(K\clubsuit) = \dfrac{1}{13}$

9. $P(4) = \dfrac{11}{36}$

11. $P(2\diamondsuit) = \dfrac{0}{6} = 0$

13. $P(5) = \dfrac{7}{28} = \dfrac{1}{4}$

15. $\dfrac{25}{1000} = 0.025$

Exercise 32–2

1. $\dfrac{6}{25}$

3. $\dfrac{1}{16}$

5. $\dfrac{3}{26}$

7. $\dfrac{1}{12}$

9. $\dfrac{1}{8}$

11. a. $\dfrac{9}{16} = 56.25\%$

b. $\dfrac{1}{16} = 6.25\%$

c. $\dfrac{3}{8} = 37.5\%$

Exercise 32–3

1. 4.9   **3.** 88.4   **5.** 5.36 μA   **7.** 61 mph   **9.** 463 "hits"

Exercise 32–5

1. a. median = 4.75
b. mode: none
c. $Q_1 = 2.7$,
Q_2 = median = 4.75,
$Q_3 = 6.9$

3. a. median = 90.5
b. mode: 96
c. $Q_1 = 82.5$,
Q_2 = median = 90.5,
$Q_3 = 96.0$

5. a. median = 5.345 μA,
mode = 5.34 μA
b. $Q_1 = 5.29$,
Q_2 = median = 5.345,
$Q_3 = 5.43$
c. $P_{35} = 5.31$
d. $P_{90} = 5.46$

7. a. median = 352 "hits",
mode: none
b. $Q_1 = 163.5$,
Q_2 = median = 352,
$Q_3 = 806$ "hits"

Exercise 32–6

1. a.

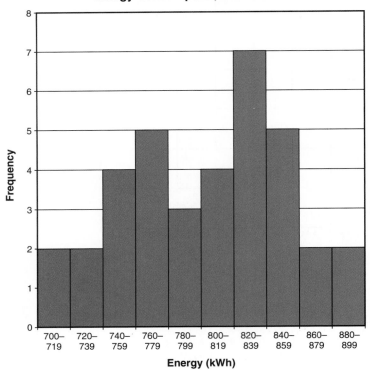

b. mean = 743.94 kWh

3. mean 45 years old

5. a.

Sulfur Oxide Emissions (tons)	Midpoint x	Tally	Frequency f	Product fx
5.0–9.9	7.45	LHT /	6	44.70
10.0–14.9	12.45	LHT LHT ////	14	436.25
15.0–19.9	17.45	LHT LHT LHT LHT LHT	25	538.80
20.0–24.9	22.45	LHT LHT LHT LHT ////	24	324.50
25.0–29.9	32.45	LHT LHT	10	32.45
30.0–34.9	32.45	/	1	61.90
			80	1501.00

b. mean = 18.7625

c.

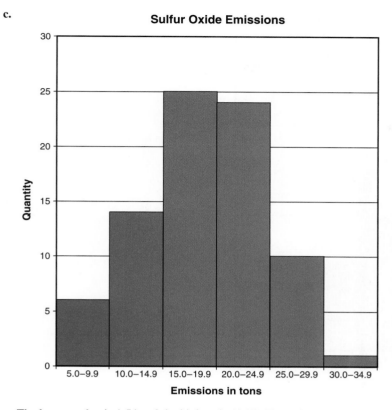

Sulfur Oxide Emissions

7. **a.** The lowest value is 1.54 and the highest is 11.78. Thus, the range is 11.78 − 1.54 = 10.24. There are several numbers of intervals that could be chosen, so your answers may vary. If 11 intervals are chosen, then the limits will be 1.00–1.99, 2.00–2.99, 3.00–3.99, . . . , 11.00–11.99.

Ignition Time (seconds)	Midpoint x	Tally	Frequency f	Product fx
1.00–1.99	1.495	LHT LHT	10	14.950
2.00–2.99	2.495	LHT ///	8	19.960
3.00–3.99	3.495	LHT LHT	10	34.950
4.00–4.99	4.495	LHT LHT /	11	49.445
5.00–5.99	5.495	LHT LHT ///	13	71.435
6.00–6.99	6.495	LHT ////	9	58.455
7.00–7.99	7.495	LHT ///	8	59.960
8.00–8.99	8.495	////	4	533.980
9.00–9.99	9.495	///	3	28.485
10.00–10.99	10.495	/	1	10.495
11.00–11.99	11.495	///	3	34.485
			80	416.600

b. mean = 5.2075

c.

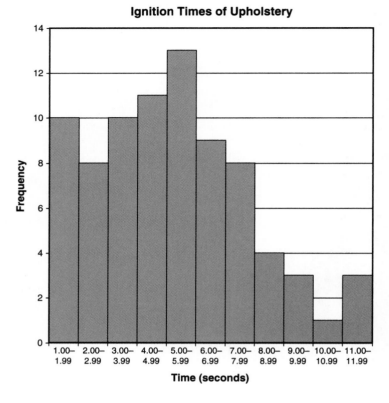

Ignition Times of Upholstery

Exercise 32–7

1. mean = 4.98; standard deviation = 2.23
3. mean = 55.36 μA; standard deviation = 0.07 μA

5. mean = 459 "hits"; standard deviation = 313.24 "hits"
7. mean = 45 years old; standard deviation = 11.2 years

9. mean = 5.21 sec; standard deviation = 2.60 sec

Exercise 32–8

1. a. $\bar{\bar{x}}$ = 2.53, UCL = 2.587, LCL = 2.473

b.

X-Bar Chart: Pin Diameters (mm)

c. The process is out of control at samples 8 and 9.

3. a. $\bar{\bar{x}} = 11.9889$, UCL $= 12.0180$, LCL $= 11.9598$

b.

X-Bar Chart: Shaft Lengths

c. The process is out of control twice: first because the first 10 samples are all above $\bar{\bar{x}}$ and again at samples 12, 13, and 14 because it is below the LCL.

5. a. $\bar{\bar{x}} = 41.3$, UCL $= 52.37$, LCL $= 30.22$

b.

X-Bar Chart: Motor Shaft Endplay

c. The process is never out of control.

Exercise 32–9

1. **a.** $\overline{R} = 01181$, UCL = 0.2367, LCL = 0

   b.

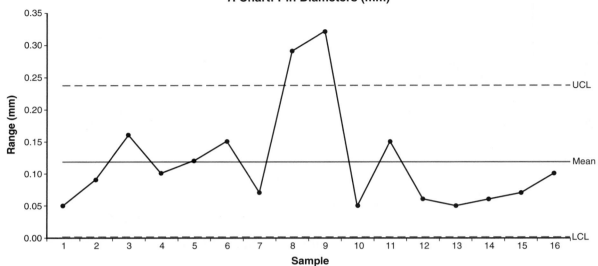

R Chart: Pin Diameters (mm)

c. The process is out of control at samples 8 and 9 because they are above the UCL.

3. **a.** $\overline{R} = 0.0505$, UCL = 0.1068, LCL = 0

   b.

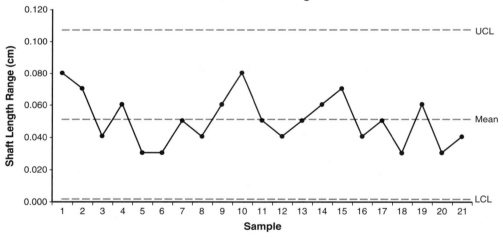

R Chart: Shaft Lengths

c. The process is never out of control.

5. a. $\overline{R} = 19.2$, UCL $= 40.59$, LCL $= 0$

b.

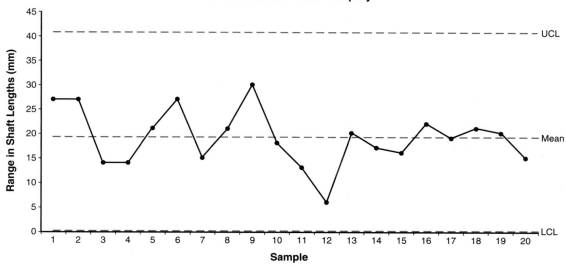

R Chart: Motor Shaft Endplay

c. The process is never out of control.

Unit Exercise and Problem Review

1. {1*H*, 2*H*, 3*H*, 4*H*, 5*H*, 6*H*, 1*T*, 2*T*, 3*T*, 4*T*, 5*T*, 6*T*}

3. $\dfrac{1}{2}$

5. $\dfrac{12}{13}$

7. $\dfrac{1}{6}$

9. $\dfrac{1}{6}$

11. $\dfrac{1}{25}$

13. 3,599.9

15. $Q_1 = 3{,}599.3$,
$Q_2 = \text{median} = 3{,}600.1$,
$Q_3 = 3{,}600.4$

17. a. $\overline{\overline{x}} = 3599.7$ rpm, UCL $= 3601.09$ rpm, LCL $= 3598.33$

b.

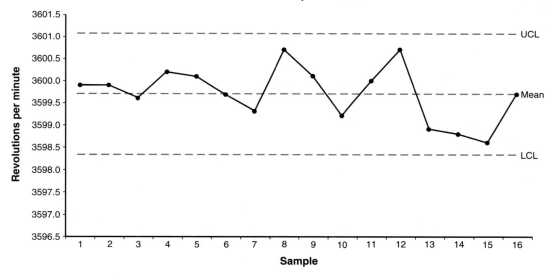

X-Bar Chart: Computer Disk Drives

c. The process is never out of control.

SECTION VII: Fundamentals of Trigonometry

UNIT 33:   INTRODUCTION TO TRIGONOMETRIC FUNCTIONS

Answers may vary slightly due to rounding within the solution.

Exercise 33–2

1. r is hypotenuse
x is adjacent
y is opposite
3. a is opposite
b is adjacent
c is hypotenuse

5. a is hypotenuse
b is opposite
c is adjacent
7. d is hypotenuse
m is opposite
p is adjacent

9. h is opposite
k is hypotenuse
l is adjacent
11. m is opposite
p is hypotenuse
s is adjacent

13. m is hypotenuse
r is adjacent
t is opposite
15. f is adjacent
g is hypotenuse
h is opposite

Exercise 33–3

1. $\sin \angle 1 = \dfrac{y}{r}$ $\cot \angle 1 = \dfrac{x}{y}$
$\cos \angle 1 = \dfrac{x}{r}$ $\sec \angle 1 = \dfrac{r}{x}$
$\tan \angle 1 = \dfrac{y}{x}$ $\csc \angle 1 = \dfrac{r}{y}$

3. $\sin \angle 1 = \dfrac{h}{g}$ $\cot \angle 1 = \dfrac{k}{h}$
$\cos \angle 1 = \dfrac{k}{g}$ $\sec \angle 1 = \dfrac{g}{k}$
$\tan \angle 1 = \dfrac{h}{k}$ $\csc \angle 1 = \dfrac{g}{h}$

5. $\sin \angle 1 = \dfrac{r}{s}$ $\cot \angle 1 = \dfrac{p}{r}$
$\cos \angle 1 = \dfrac{p}{s}$ $\sec \angle 1 = \dfrac{s}{p}$
$\tan \angle 1 = \dfrac{r}{p}$ $\csc \angle 1 = \dfrac{s}{r}$

7. a, b, d
9. a, c, d

Exercise 33–5A

1. 0.58779	**7.** 2.16222	**13.** 0.95747	**19.** 0.44594	**23.** 0.91955	**27.** 0.15983
3. 1.07237	**9.** 0.22940	**15.** 1.80926	**21.** 0.94147	**25.** 0.99402	**29.** 0.80667
5. 0.98163	**11.** 3.45553	**17.** 0.22250			

Exercise 33–5B

1. 2.20269	**5.** 9.31921	**9.** 18.61149	**13.** 1.06600	**17.** 1.00330	**21.** 2.98520
3. 1.53986	**7.** 1.32849	**11.** 8.31904	**15.** 3.20734	**19.** 19.18972	

Exercise 33–5C

1. 56.82°	**7.** 1.45°	**13.** 67.30°	**19.** 48.16°	**25.** 17.18°	**31.** 41.91°
3. 30.68°	**9.** 6.78°	**15.** 39.52°	**21.** 9.70°	**27.** 76.66°	**33.** 55.49°
5. 75.84°	**11.** 28.67°	**17.** 50.95°	**23.** 11.70°	**29.** 44.31°	**35.** 25.42°

Exercise 33–5D

1. 76°39′	**7.** 44°53′	**13.** 46°8′	**19.** 74°22′	**25.** 80°58′	**31.** 61°11′
3. 67°7′	**9.** 89°43′	**15.** 45°0′	**21.** 84°12′	**27.** 82°9′	**33.** 72°3′
5. 87°43′	**11.** 17°47′	**17.** 46°10′	**23.** 19°14′	**29.** 69°4′	**35.** 15°1′

Unit Exercise and Problem Review

1. a is opposite
b is adjacent
c is hypotenuse
3. e is adjacent
f is opposite
g is hypotenuse
5. h is hypotenuse
j is opposite
k is adjacent
7. a is opposite
d is hypotenuse
p is adjacent

9. $\sin \angle 1 = \dfrac{y}{r}$
$\cos \angle 1 = \dfrac{x}{r}$
$\tan \angle 1 = \dfrac{y}{x}$
$\cot \angle 1 = \dfrac{x}{y}$
$\sec \angle 1 = \dfrac{r}{x}$
$\csc \angle 1 = \dfrac{r}{y}$

11. $\sin \angle 1 = \dfrac{s}{m}$
$\cos \angle 1 = \dfrac{t}{m}$
$\tan \angle 1 = \dfrac{s}{t}$
$\cot \angle 1 = \dfrac{t}{s}$
$\sec \angle 1 = \dfrac{m}{t}$
$\csc \angle 1 = \dfrac{m}{s}$

13. 0.80902	**37.** 52.53°	**61.** 81°56′
15. 0.78823	**39.** 41.78°	**63.** 75°46′
17. 0.76046	**41.** 78.64°	**65.** 79°49′
19. 4.83590	**43.** 8.13°	**67.** 37°31′
21. 0.99314	**45.** 74.11°	**69.** 13°21′
23. 73.11306	**47.** 11.61°	**71.** 69°18′
25. 1.10338	**49.** 52.48°	
27. 1.58982	**51.** 13.34°	
29. 1.15249	**53.** 35.02°	
31. 1.01676	**55.** 67°55′	
33. 0.05824	**57.** 56°8′	
35. 1.08752	**59.** 84°31′	

UNIT 34: TRIGONOMETRIC FUNCTIONS WITH RIGHT TRIANGLES

Answers may vary slightly due to rounding within the solution.

Exercise 34–1

1. **a.** side y and side r are almost the same length
   b. side x is very small compared to side r
   c. side x is very small compared to side y
3. **a.** side y is very small compared to side r
   b. side x and side r are almost the same length
   c. side x is very large compared to side y

5. **a.** 45°
   b. 1.000 ...
   c. 1.000 ...
7. **a.** 1.000 ...
   b. 1.000 ...

c. 0
d. 0
9. tan 18°
11. cot 36°
13. csc 22°

15. cos 81°19′
17. csc 39.25°
19. sec 55°

Exercise 34–2

1. cot 73°
3. sin 64°
5. tan 55°
7. sin 0°
9. cot 23.5°
11. tan 82°50′
13. sec 89°22′
15. sin 84.11°
17. cot 0°
19. sin 89.99°
21. cos 55°
23. cot 56°
25. sec 43°
27. sin 12°
29. tan 1°20′
31. sec 0.2°

Exercise 34–3

1. 36°28′
3. 42°2′
5. 59.24°
7. 54.10°
9. $\angle 1 = 22°21'$
   $\angle 2 = 67°39'$
11. $\angle x = 29.50°$
    $\angle y = 60.50°$

Exercise 34–4

1. 5.313 in
3. 50.464 ft
5. 13.616 cm
7. 6.878 m
9. $d = 2.229$ in
   $e = 2.049$ in
11. $x = 7.285$ m
    $y = 2.480$ m

Unit Exercise and Problem Review

1. **a.** 0
   b. side x = side r
3. **a.** 45°
   b. 1.000 ...
   c. 1.000 ...
5. **a.** 0
   b. 1.000 ...
7. cot 43°
9. sin 19°30′
11. csc 79.12°
13. sin 9°
15. csc 65°
17. sec 90°
19. cos 38.12°
21. tan 40°
23. cos 44°
25. sec 45°
27. 38°41′
29. 6.773 cm
31. 38°24′
33. 55.70°
35. $\angle B = 17°30'$
    $x = 55.237$ in
    $y = 52.680$ in
37. $a = 9.349$ cm
    $b = 3.164$ cm
    $\angle 2 = 71.3°$

UNIT 35: PRACTICAL APPLICATIONS WITH RIGHT TRIANGLES

Answer may vary slightly due to rounding within the solution.

Exercise 35–1

1. 24°17′ and 65°43′
3. 7.08 m
5. 15.7 ft
7. 63°
9. 29.21 m
11. 48.1 ft
13. 67.38 m
15. 50 ft 8 in.

Exercise 35–2

1. 13°15′
3. 6.75°
5. 5.90 m
7. 28.98 mm
9. **a.** 46.16 m
   b. 31.20 m
11. **a.** 55°
    b. 58°
13. 30.4°
15. 5.82 m
17. **a.** 12′7″
    b. 22′0″

Exercise 35–3

1. 3.037 in
3. **a.** 4.08 m
b. 38.64°
5. 58°51′
7. 248 ft
9. 28.84 m
11. 0.928 in
13. **a.** 21.84°
b. 33.92°
c. 29.08 m
15. 57.88 mm

Unit Exercise and Problem Review

1. 10.64 in
3. 218.73 ft
5. 24.69 m
7. **a.** $\angle x = 76.36°$
   b. $AB = 3.86$ m
9. 47.70 mm
11. 6.631 in
13. 19.4°
15. 3.14 m
17. 56°4′
19. 81.72 ft

UNIT 36: OBLIQUE TRIANGLES: LAW OF SINES AND LAW OF COSINES

Answers may vary slightly due to rounding within the solution.

Exercise 36–3

1. Reference angle = 62°
   sin 118° = 0.88295
   cos 118° = −0.46947
   tan 118° = −1.8807
   cot 118° = −0.53171
   sec 118° = −2.1301
   csc 118° = 1.1326
3. Reference angle = 80°
   sin 260° = −0.98481
   cos 260° = −0.17365
   tan 260° = 5.6713
   cot 260° = 0.17633

sec 260° = −5.7588
csc 260° = −1.0154
5. Reference angle = 60°
   sin 300° = −0.86603
   cos 300° = 0.50000
   tan 300° = −1.7321
   cot 300° = −0.57735
   sec 300° = 2.0000
   csc 300° = −1.1547
7. Reference angle = 36°20′
   sin 216°20′ = −0.59248
   cos 216°20′ = −0.80558

tan 216°20′ = 0.73547
cot 216°20′ = 1.3597
sec 216°20′ = −1.2413
csc 216°20′ = −1.6878
9. Reference angle = 40°44′
   sin 139°16′ = 0.65254
   cos 139°16′ = −0.75775
   tan 139°16′ = −0.86115
   cot 139°16′ = −1.1612
   sec 139°16′ = −1.3197
   csc 139°16′ = 1.5325

11. Reference angle = 46.8°
    sin 313.2° = −0.72897
    cos 313.2° = 0.68455
    tan 313.2° = −1.0649
    cot 313.2° = −0.93906
    sec 313.2° = 1.4608
    csc 313.2° = −1.3718
13. 519.6 V
15. −120 V
17. −103.9 V
19. 300 V
21. −550 V

Exercise 36–5

1. Reference angle = 30°
   sin 510° = 0.50000
   cos 510° = −0.86603
   tan 510° = −0.57735
   cot 510° = −1.7321
   sec 510° = −1.1547
   csc 510° = 2.0000
3. Reference angle = 15°
   sin 555° = −0.25882
   cos 555° = −0.96593

tan 555° = 0.26795
cot 555° = 3.7321
sec 555° = −1.0353
csc 555° = −3.8637
5. Reference angle = 9°
   sin 531° = 0.15643
   cos 531° = −0.98769
   tan 531° = −0.15838
   cot 531° = −6.3138
   sec 531° = −1.0125

csc 531° = 6.3925
7. Reference angle = 37°
   sin 937° = −0.60182
   cos 937° = −0.79864
   tan 937° = 0.75355
   cot 937° = 1.3270
   sec 937° = −1.2521
   csc 937° = −1.6616
9. Reference angle = 11.6°
   sin 1,248.4° = 0.20108

cos 1,248.4° = −0.97958
tan 1,248.4° = −0.20527
cot 1,248.4° = −4.8716
sec 1,248.4° = −1.0209
csc 1,248.4° = 4.9732
11. 228.3 V
13. −304.3 V
15. 70.5 V
17. 723.9 V
19. 352.7 V

Exercise 36–8

1. 4.641 in
3. 37.153 cm
5. 3.941 in
7. a. 85.3°
   b. 50.532
9. a. 27°10′
   b. 15.059 in
11. a. 108°
    b. 12.265 in
    c. 8.347 in

Exercise 36–9

1. two solutions
3. one solution
5. one solution
7. two solutions
9. 51°30′
11. 61.73°
13. 57°19′
15. a. 19°23′
    b. 76°7′
17. a. 120°7′
    b. 32°43′
19. a. 18°59′
b. 102°11′
c. 6.898 in

Exercise 36–11

1. 3.283 in
3. 19.572 in
5. 361.556 ft
7. a. 6.833 in
b. 66°33′
c. 83°57′
9. a. 91.171 ft
   b. 47°37′
c. 34°23′
11. a. 9.760 in
b. 73°43′
c. 72°37′

Exercise 36–13

1. 22°59′
3. 69.51°
5. 132°53′
7. a. 79°15′
b. 72°18′
c. 28°27′
9. a. 102°1′
   b. 50°13′
c. 27°46′
11. a. 107.46°
b. 55.32°
c. 17.22°

Exercise 36–14

1. 50°51′
3. a. 60.00°
   b. 67.34°
c. 52.66°
5. 18.1 in
7. a. 3.9 in
b. 4.5 in
9. 11.981 cm
11. 125°44′
13. a. 45°34′
    b. 31°17′
    c. 12′0″
15. 68°24′
17. 352.21 ft
19. 18.03°
21. 100°11′

Unit Exercise and Problem Review

1. Reference $\angle = 65°$
sin 115° = 0.90631
cos 115° = −0.42262
tan 115° = −2.1445
cot 115° = −0.46631
sec 115° = −2.3662
csc 115° = 1.1034

3. Reference $\angle = 69°40'$
sin 290°20′ = −0.93769
cos 290°20′ = 0.34748
tan 290°20′ = −2.6985
cot 290°20′ = −0.37057
sec 290°20′ = 2.8778
csc 290°20′ = −1.0664

5. Reference $\angle = 40°$
sin 580° = −0.64279
cos 580° = −0.76604
tan 580° = 0.83910
cot 580° = 1.1918
sec 580° = −1.3054
csc 580° = −1.5557

7. Reference $\angle = 19.7°$
sin 739.7° = 0.33710
cos 739.7° = 0.94147
tan 739.7° = 0.35805
cot 739.7° = 2.7929
sec 739.7° = 1.0622
csc 739.7° = 2.9665

9. a. 154.3 V
b. −491.5 V
c. −91.9 V
d. 0 V
e. −102.6 V
11. 5.680 cm
13. 51.346 mm
15. 6.977 cm
17. a. 72°10′
b. 21.491 in
19. a. 121°50′
b. 28.344 in
21. a. 105.60°
b. 35.354 cm
c. 21.367

23. 1.850 in
25. 25.169 cm
27. 34.684 in
29. a. 8.966 cm
b. 68.33°
c. 76.87°
31. a. 6.441 in
b. 39°40′
c. 33°20′
33. a. 31.990 mm
b. 66.28°
c. 76.12°
35. 15.057 in
37. 97°54′
39. 49°49′, 58°50′, and 71°21′
41. 11.20 cm
43. 136.79 m

UNIT 37: VECTORS

Exercise 37–5

1. b, d, f   **3.** b, f   **5.** b, c   **7. a.** yes   **b.** no   **c.** no   **d.** yes

Exercise 37–6

1. 940 lb at 151°
3. 29 km at 52°
5. 160 km at 235°
7. 320 mi at 55°
9. 330 ft/s at 173°
11. 1000 m at 298°
13. 43,000 lb at 331°
15. 17 mi at 3°
17. 180 mi at 217°
19. 80 m/s at 91°
21. 450 m at 258°

Exercise 37–7

1. 34.1 mi at 208.7°
3. 373 km/h at 145.4°
5. 15,200 lb at 326.7°
7. 18.2 km at 216.3°
9. 8020 lb at 53.3°
11. 112 km at 140.3°
13. 1210 lb at 238.0°
15. 285 km at 290.0°
17. 2580 lb at 143.5°
19. 251 ft/s at 13.2°
21. Horizontal component: 361 lb at 0°
Vertical component: 384 lb at 90°
23. Horizontal component: 3.96 km at 180°
Vertical component: 17.1 km at 90°
25. Horizontal component: 621 mi at 180°
Vertical component: 330 mi at 270°
27. Horizontal component: 1040 lb at 0°
Vertical component: 10.9 lb at 90°
29. 3790 lb at 235.8°
31. 519 lb at 45.1°
33. 34.8 Ω at 36.3° or 34.8 Ω $\angle 36.3°$
35. 10,700 lb at 350°
37. a. 1540 lb at 238° (standard position)
b. 1.239 lb/sq ft

Exercise 37–8

1. 1140 lb at 83.3°
3. 62.3 mi at 164.4°
5. 762 km at 126.3°
7. 171 ft at 270.1°
9. 1450 ft at 116.6°
11. 4970 ft at 44.7°
13. 3360 lb at 251.8°
15. 1210 lb at 86.9°
17. 706 km at 132.2°
19. 1800 lb at 227.5°
21. 162 lb at 180°

Unit Exercise and Problem Review

1. c
3. a
5. 2600 lb at 214.0°
7. 100 ft/s at 136°
9. 5190 lb at 136.5°
11. 123 km at 146.5°
13. 2550 lb at 149.2°
15. 252 lb at 228.3°
17. 560 lb at 180°, 238 lb at 270°
19. 67.4 ft/s at 0°, 65.1 ft/s at 270°
21. 24.9 Ω at −48.1° or 24.9 Ω $\angle -48.1°$
23. 2800 lb at 131.3°
25. 2780 km at 192.1°
27. 269 ft at 150.1°

Index